Films That Work Harder

Films That Work Harder

The Circulation of Industrial Film

Edited by
Vinzenz Hediger,
Florian Hoof,
Yvonne Zimmermann,
with Scott Anthony

ROUTLEDGE
Routledge
Taylor & Francis Group
LONDON AND NEW YORK

This publication was made possible through a grant from the Open Access publication fund, Goethe Universität Frankfurt, with support from Philipps-Universität Marburg, by the Start-up Grant no. M4081571 from Nanyang Technological University, Singapore, and with support from DFG funded Institute for Advanced Studies Media Cultures of Computer Simulations (mecs), Leuphana Universität Lüneburg

Gefördert durch DFG Deutsche Forschungsgemeinschaft

mecs MEDIA CULTURES OF COMPUTER SIMULATION

ConTrust Vertrauen im Konflikt Ein Clusterprojekt des Landes Hessen

First published in 2024 by Amsterdam University Press Ltd.

Published 2025 by Routledge
4 Park Square, Milton Park, Abingdon, Oxon OX14 4RN
605 Third Avenue, New York, NY 10158

Routledge is an imprint of the Taylor & Francis Group, an informa business

ISBN: 9789462986534 (hbk)
ISBN: 9781041179436 (pbk)
ISBN: 9781003695653 (ebk)
NUR 670

Cover illustration: Film cans in Bay Village, Boston 1

Cover design: Kok Korpershoek

DOI 10.5117/9789462986534

Table of Contents

List of Illustrations

Introduction: A Sequel and a Shift

Vinzenz Hediger, Florian Hoof, Yvonne Zimmermann

This book is unique in that it is a sequel. Academic publications do not usually have sequels. Sequels in the film industry limit novelty as much as possible to minimize risk, whereas in science, like in art, risk-taking is a virtue and relevance flows from originality and novelty. This book, however, departs from its predecessor, *Films That Work*,[1] not through minimal variation, but through what the language of management calls diversification and expansion. In business, diversification and expansion result from a quest for profit. This book has a different motivation: It reflects the conviction that the work which the first book started was not finished, but only beginning. It also reflects the reality that a significant number of scholars, many of whom are now represented in this volume, shared this conviction and had already set out to consolidate and expand the field mapped by the first book.

The original *Films That Work* covered a wide variety of topics but was held together by what may be described as a shared research design. It combined a broad and bold working hypothesis with a manageable, relatively limited focus. The working hypothesis was that economic activity is inexorably tied to – i.e. dependent upon and shaped by – media and media infrastructures. Industrial films were, in other words, not merely an add on and, over time, a source of historical information but an indispensable part of industrial organization. This was a bold claim for the fields of social and economic history and, following the work of JoAnne Yates[2] and Susan Leigh Star,[3] less so for management and science and technology studies. Prior to *Films That Work*, Charles R. Acland and Haidee Wasson's *Useful Cinema*,[4] Lee

1 Hediger, Vinzenz, and Patrick Vonderau, editors. *Films That Work: Industrial Film and the Productivity of Media*. Amsterdam University Press, 2009.

2 Yates, JoAnne. *Control through Communication: The Rise of System in American Management*. Johns Hopkins University Press, 1989.

3 Star, Susan Leigh. "The Ethnography of Infrastructure." *American Behavioral Scientist*, vol. 43, no. 3, 1999, pp. 377–91.

4 Acland, Charles R., and Haidee Wasson, editors. *Useful Cinema*. Duke University Press, 2011.

Hediger, V., F. Hoof, Y. Zimmermann, with S. Anthony (eds), *Films That Work Harder: The Circulation of Industrial Film*. Taylor & Francis Group 2024
DOI: 10.5117/9789462986534_INTRO

Grieveson's *Cinema and the Wealth of Nations*[5] and Florian Hoof's *Angels of Efficiency*,[6] it was certainly a bold claim for film studies – a field which owed its existence to the demonstration that a canon of important art works exists in cinema and which, as a consequence, had barely paid attention to objects deemed artless like commissioned films, let alone to the nexus between business, industrial organization and film. The focus of *Films That Work* was manageable because it was limited to the firm as the privileged organizational entity; to the 1895–1970 time period with an emphasis on the post-war period of economic expansion; and to moving images as key elements in the "*Medienverbund*" of the firm, the media network which sustains the firm and makes its operation possible.

Films That Work Harder maintains the original working hypothesis and diversifies and expands this approach to include a broader array of organizations beyond the classical firm; to cover a broader geographical territory – in fact, all continents except Australia and Antarctica – and to expand the time frame from about 1970 to the contemporary transformations of the global economy. The most important modification, however, is the one indicated by the subtitle of the book: a shift from a focus on production to circulation or, if you will, a shift from the first volume of *Das Kapital* to the second.

Industrial films from the period of early cinema through what Anna Maria Falchero in this volume calls the "golden age" of industrial film, which coincides with *Les Trente Glorieuses* in France and the "economic miracles" in Germany and Italy, i.e. the post-war economic expansion between 1945 and 1973, tend to focus on processes of production. As Salomé Aguilera Skvirsky shows in her contribution to this volume and her recent book,[7] the process film is a genre unto itself, a cultural template of modernity which originates in visual representations of the eighteenth century and, by way of the industrial film, pervades other areas of cinema as well. In the spirit of the industrial film's focus on production, the subtitle of the first volume spoke of the "productivity of industrial cinema." But as Karl Marx writes in the *Grundrisse einer Kritik der politischen Ökonomie*, modern industry production is not an end in itself, but a means. Production serves to accumulate surplus value, which is the subject of Volume 1 of *Das Kapital*. However, that value is only realized in the sphere of circulation, which is the subject of the (unfinished and posthumously

5 Grieveson, Lee. *Cinema and the Wealth of Nations: Media, Capital, and the Liberal World System*. University of California Press, 2017.

6 Hoof, Florian. *Angels of Efficiency: A Media History of Consulting*. Oxford University Press, 2020.

7 Aguilera Skvirsky, Salomé. *The Process Genre: Cinema and the Aesthetic of Labor*. Duke University Press, 2020.

published) Volume 2. So, if indeed we are to fully understand how industrial organization relates to the moving image, we need to look not only at production, but at circulation, and at how circulation enables and informs production.

This claim seems even more valid from the vantage point of today's economy and the transformation brought about by the introduction of the standardized shipping container and the development of point-to-point supply chain logistics,[8] which runs parallel to and goes hand in hand with digitization. The classical factory, the focus of many, if not most, industrial films up to 1970, had a specific location close to raw materials such as iron ore and coal, proximity to markets and large inventories. From the eighteenth through the twentieth centuries, the classical factory – in exemplary fashion in the textile industry, which was at the forefront of industrialization in North-western Europe – greatly benefited from colonialism's exctractive supply chains for raw materials and human labor spanning Europe, Asia, Africa and the Americas, which also served to trade finished products back to the point of extraction. But in this system, the classical factory was the central node, a site of production organized around proprietary technologies and processes which were jealously guarded by their owners. It was also typically a national institution. "What is good for General Motors is good for America," Charles Erwin Wilson, president of General Motors, famously said during his Senate confirmation hearings to become Eisenhower's secretary of defense in 1953. Wilson said this to justify his decision not to divest himself of his important shareholding position in his company and was widely ridiculed for it, but he expressed, in the words of one commentator, an "undisputable fact of modern political life."[9] Container shipping has largely invalidated Wilson's equation. By dramatically lowering the cost and increasing the speed and reliability of transportation, containerization has replaced the classical factory model with global supply chains, which span the globe according to where labour and raw material inputs are cheapest and where demand is highest. General Motors took the lead in this process when they relocated most of their production away from Detroit and Flint, Michigan, to lower cost locations in the 1980s, leaving a trail of social and economic destruction which Michael Moore showcased in what was at the time the most successful theatrical documentary of all time, *Roger*

8 Levinson, Marc. *The Box: How the Shipping Container Made the World Smaller and the World Economy Bigger*. Princeton University Press, 2006; Klose, Alexander. *The Container Principle: How a Box Changes the Way We Think*. MIT Press, 2015.

9 Redish, Martin H., and Howard M. Wasserman. "What's Good for General Motors: Corporate Speech and the Theory of Free Expression." *George Washington Law Review*, vol. 66, no. 2, 1997–1998, pp. 235–96.

and Me from 1989. But this change from what Marc Levinson describes as the "massive industrial complexes in vogue since the turn of the twentieth century" to "networks of much smaller factories, linked by international supply chains and employing even smaller numbers of workers" affects not just the volume and location, but the types and forms of labour. Well-paid factory jobs disappeared, and in the new economy value comes "from innovation, design, and marketing, not from the physical process of tuning raw materials into finished goods."[10] What is more, as Anthony G. Hopkins writes, "the breakdown of complementarities between industrial metropoles and primary-producing peripheries, undermined vertical, hierarchic imperial systems," potentially furthering notions of racial equality and democracy.[11]

What is clear is that what economists and sociologists started calling globalization in the 1980s amounts to more than just an extension of trade and a proliferation of what William Cronon, in his discussion of the emergence of the futures market for grain in the nineteenth century, calls the "transmutation" of material objects and their displacement into "the symbolic world of capital."[12] From a media theory point of view one could argue that with containerization production, distribution and consumption finally fully catch up with the property of money as the primary medium of exchange, which according to Marx, still in the *Grundrisse*, is to split exchange into two acts, sale and purchase in different locations at different times. As early as 1990, at the very onset of the second wave of globalization, Arjun Appadurai described what the emerging economic and social order means in terms of disjuncture and difference in the new global cultural economy.[13] Considering the history of the industrial film and its long-standing focus on production and the firm, the demise of the classical factory alone is reason enough to shift the focus from production to circulation.

Film and media studies in the years since the publication of *Films That Work* has seen a turn towards the study of infrastructure in the work of Marta Braun, Charlie Keil, Lisa Parks, Nicole Starosielski, Gabriele Schabacher, Rahul Mukherjee and others.[14] A shift from production to circulation in

10 Levinson, Marc. *An Extraordinary Time: The End of the Postwar Boom and the Return of the Ordinary Economy.* Random House, 2016, p. 131.

11 Hopkins, A.G. "Globalisation and Decolonisation." *Journal of Imperial and Commonwealth History*, vol. 45, no. 5, 2017, pp. 241–42.

12 Cronon, William. *Nature's Metropolis: Chicago and the Great West.* W.W. Norton, 1991, p. 120.

13 Appadurai, Arjun. *Modernity at Last: Cultural Dimensions of Globalization.* University of Minnesota Press, 1996.

14 Braun, Marta, Charlie Keil, Rob King, Paul Moore, and Louis Pelletier, editors. *Beyond the Screen: Institutions, Networks and Publics of Early Cinema.* John Libbey, 2012; Parks, Lisa, and

the study of industrial film involves a similar focus on the material arrangements that sustain industrial organization and economic activity beyond the factory, both in the past and present. At the same time, the question of how cinematic forms act as part of such material arrangements remains central. Across the transformations sketched above, the moving image remains a constant factor of the "*Medienverbund*," which sustains industrial organization, and it remains what is perhaps the privileged format of corporate communication with so-called stakeholders, even as digital infrastructures and technologies supplant older media across organizations.

The circulation of industrial films has always been facilitated by small-gauge films and mobile projectors.[15] But while *Les Trente Glorieuses* and the "economic miracle" period may have been the golden age of the theatrical industrial film, the transition to digital platforms and the emergence of compression formats like MP4 – which is to sounds and images as the shipping container is to consumer goods, and which is a global standard approved and regulated by the same body, the International Organization for Standardization (ISO)[16] – has created more outlets for moving images which can be described as industrial films than at any other point in history. In particular, corporations use digital platforms like YouTube for global recruitment and to spread messages of sustainability. The latter have become de rigueur since the human and environmental costs of what J.R. McNeill and Peter Engelke call "the great acceleration" have become apparent, i.e. the phase of unprecedented growth since 1945, which has made humans the determining factor in the history of the environment and which is also known as the "Anthropocene."[17] In that sense, the study of the circulations of industrial film offers a perspective on media history more generally, as well as on environmental history and the study of the ecology and/of media.[18]

Nicole Starosielski, editors. *Signal Traffic: Critical Studies of Media Infrastructures*. University of Illinois Press, 2015; Starosielski, Nicole. *The Undersea Network*. Duke University Press, 2015; Schabacher, Gabriele. "Mobilizing Transport: Media, Actor-World, and Infrastructures." *Transfers*, vol. 3, no. 1, 2013; Mukherjee, Rahul. *Radiant Infrastructures: Media, Environment, and Cultures of Uncertainty*. Duke University Press, 2020.

15 Wasson, Haidee. *Everyday Movies: Portable Projectors and the Transformation of American Culture*. University of California Press, 2020.

16 Sterne, Jonathan. *MP3: The Meaning of a Format*. Duke University Press, 2012; Jancovic, Marek, Alexandra Schneider, and Axel Volmar, editors. *Format Matters: Standards, Practices, and Politics in Media*. Meson Press, 2020.

17 McNeill, J.R., and Peter Engelke. *The Great Acceleration: An Environmental History of the Anthropocene since 1945*. Belknap Press, 2014.

18 See, for example, Fuller, Matthew. *Media Ecologies: Materialist Energies in Art and Technoculture*. MIT Press, 2007.

What we propose to call the "circulations of industrial cinema" has three aspects: How industrial films represent circulation, how they operate as part of the infrastructures of circulation, and how they themselves circulate (that is, how distribution affects and potentially transforms their meaning and operation).

In sales, the role of film is usually limited to advertising rather than the visualization of process (unless the visualization of process serves advertising purposes, as a symbolic warranty of the quality of the product).[19] This makes the infrastructures and networks of circulation all the more significant if and when they come into view – for instance, in films about state-sponsored building projects, shipping and communication, but also in films about the primary medium of exchange, money. But it is equally as important to pay attention when they remain absent and their operation is simply assumed.

In the first volume, industrial cinema was discussed as a *parasite form*, i.e. a form of cinema that strategically adopts and inhabits other forms of cinema and adapts artistic forms and techniques to the requirements of specific pragmatic contexts or spaces of communication, to use Roger Odin's term,[20] with the ultimate goal of "disappearing in communication," as Michel Serres phrases it.[21] This is a property of infrastructures more generally speaking. Media infrastructures disappear into communication, and they only become perceptible when they break down, as Karen Ruhleter and Susan Leigh Star have argued,[22] or, following Fritz Heider, when established routines or formats of perception are thwarted.[23] In that sense, industrial films as part of the media infrastructure of circulation are particularly salient when they leave their designated space of communication. As they migrate across other spaces, in the new, decentralized and fluid forms of corporate communication, but also in new environments such as relational databases and digital archives, industrial films become *media boundary objects*,[24] i.e. objects that can mean different things to different people and

19 See Florin, Bo, Patrick Vonderau, and Yvonne Zimmermann. *Advertising and the Transformation of Screen Cultures*. Amsterdam University Press, 2021.

20 Odin, Roger. *Spaces of Communication*. Amsterdam University Press, 2022.

21 Serres, Michel. *Le parasite*. Grasset, 1980.

22 Star, Susan Leigh, and Karen Ruhleder. "Steps Toward an Ecology of Infrastructure: Design and Access for Large Information Spaces." *Information Systems Research*, vol. 7, no. 1, 1996, pp. 111–34.

23 Heider, Fritz. "Thing and Medium." *On Perception, Event, Structure, and Psychological Environment: Selected Papers*. International Universities Press, 1959.

24 Hoof, Florian. "The Media Boundary Objects Concept: Theorizing Film and Media." *MediaMatters: The Materiality of Media, Matter as Medium*, edited by Bernd Herzogenrath. Bloomsbury, 2015, pp. 180–200.

serve different purposes in different contexts, ranging from the original purpose of industrial organization to an afterlife in the sense of Warburg,[25] whether as historical markers or as raw material for artistic production.

The contributions to this book cover the three aspects of the circulations of industrial film – representation, operation, distribution – from a variety of angles and based on a broad range of case studies. The volume escapes the fate of what Hegel calls the "*schlechte Unendlichkeit*," the miserable infinity, of an endlessly proliferating series of case studies by largely sharing the modified heuristics which informs this sequel: that in industrial modernity, there is no production without circulation and neither without media, and that the moving image, as their enduring element, offers a privileged handle on both the analogue and digital media networks of industrial organization.

The book is organized in seven sections. Section 1, "Networks and Flows: Visualizing Value Chains," consists of four approaches to infrastructures and media networks beyond the factory. Furthermore, it tackles the afterlife of industrial film both as the circulation of meaning and by focusing on the material and political consequences that prevail over time. In Chapter 1, "The Aesthetics of the Global Value Chain: Container Shipping, Media Networks and the Problem of Visibility in the Global Sphere of Circulation," Vinzenz Hediger addresses the question of the form of infrastructures by analysing the corporate image strategy of Maersk, the world's largest container shipping provider. He develops the outline of an aesthetics of the global value chain by analysing three different, but interconnected types or styles of visibility: critical, operative and representational visibility. In Chapter 2, "Object Lessons and Infrastructural Imperialism," Lee Grieveson highlights the antecedents of the recent transformations of the world economy by showing how the US government in the 1910s and 1920s used industrial films to promote the construction of roads and the Pan-American highway system, with a view to facilitating the circulation of capital and new practices of economic imperialism during the period in which the US become the world's dominant industrial state. Thomas Turnbull adds an aspect of media theory in Chapter 3, "Energy and Industrial Film: Energo-Critical Registers," by proposing a thermodynamic reading of industrial film that addresses them as agents of form, content and transmission of energy. He draws on a wide range of industrial films from the 1920s through the 1960s to argue that industrial cinema can be analysed in ways that are congruent with the physicist's

25 Warburg, Aby. *Nachhall der Antike. Zwei Untersuchungen*. Edited by Pablo Schneider. Diaphanes, 2011; Didi-Huberman, Georges. *The Surviving Image: Phantoms of Time and Time of Phantoms: Aby Warburg's History of Art*. Penn State University Press, 2016.

concept of energy. In Chapter 4, "Digital Afterlife of Industrial Film: Weak Dispositives, Choice Architecture and the Distribution of Industrial Cinema," Florian Hoof develops a theoretical model to describe the digital afterlife of industrial film. Taking the German car industry as his example, he discusses the reappearance of industrial film as a cinematic experience structured by "weak dispositives" that are constantly being stabilized by relating them to "cinematic time" and "cinematic space."

Section 2, "Operative Iconographies, Industry and the Nation State," studies the implications, tensions and contradictions inherent in Charles E. Wilson's famous claim that corporate and national political interests align. Putting a focus on water and (media) infrastructures of hydropower, it illustrates how media and industrial film in particular turn the economic resource of water into a (national) cultural resource.[26] In Chapter 5, "Beautiful Luxembourg, Steel Works and a Swimming Pool: The Corporate Film *Columeta* and the Formation of a Corporate and a National Image," Ira Plein focuses on the steel industry in Luxembourg and shows how this industry, represented by the dominant company, ARBED, created and perpetuated an iconography that became operative also for the nation state of Luxembourg. In Chapter 6, "Hydropower for a Sealess Nation: Representation of Water Energy in Czech Visual Culture," Lucie Česálková explores the cultural imagination of rivers, waterways and generally water as a source of energy in Czech film and visual culture in the nineteenth and twentieth century. In particular, she traces landlocked Czechoslovakia's dream of ocean access and its water management projects across the economical paradigm shift from capitalist to nationalized socialist order. In Chapter 7, "Modern Water Sprites: History, People and the Landscape of Northern Sweden in Vattenfall's Film Production in the 1950s," Fabian Zimmer outlines a cultural history of hydroelectricity in Sweden by looking at films from the Swedish state-owned enterprise Vattenfall. Zimmer traces the narratives of hydroelectric modernity through the "golden age" of Vattenfall's film production to study their rhetoric and discursive contexts. In Chapter 8, "Taxonomy of Techniques: Visions of Industrial Cinema in Post-war Japan," Takuya Tsunoda discusses an infrastructure project in post-war Japan, which was the subject of *Sakuma Dam,* a series of highly popular industrial films for theatrical distribution made between 1954 and 1958. Tsunoda shows

26 Bartelheim, Martin, et al. "'ResourceCultures' – A Concept for Investigating the Use of Resources in Different Societies." *Persistent Economic Ways of Living: Production, Distribution, and Consumption in Late Prehistory and Early History,* edited by Alžběta Danielisová and Manuel Fernández-Götz. Archaeolingua Alapítvány, 2015, pp. 39–50.

how these films served to convey a new sense of industrial modernity both through the infrastructural changes they represented and through their form and technique as an epitome of state-ordained modernization and democratization. In Chapter 9, "The Power of Flows: The Spatiality of Industrial Films on Hydropower in Switzerland," the final chapter of this section, Yvonne Zimmermann draws on Manuel Castells' concept of "the space of flows" to explain how films on hydropower in Switzerland from the 1930s and 1950s mediate between the dynamic space of flows of technology and economy and the static places of society within a national framework that uses water – and its power – as a cultural resource.

Section 3, "Institutions and Distribution Frameworks: Archives, Festivals, Fairs," tackles the distribution aspect of the circulations of industrial film and how this relates to the conceptualization and the building of audiences, the creation of imagined communities, and technological frameworks of visual display. In Chapter 10, "Industry on Screen: The British Documentary in Distribution – British Transport Films: A Case Study," Steve Foxon provides an overview of the distribution of the British documentary film from the interwar period to the 1960s and its "aftermath" in the age of video and the digital. With a focus on the films made by British Transport Films, he describes the various target audiences that should be persuaded by these films and outlines a typology of distribution modalities and venues. In Chapter 11, "On the Red Carpet in Rouen: Industrial Film Festivals and a World Community of Filmmakers," Brian Jacobson discusses the history of industrial film festivals as key nodes in shifting global networks of cultural and economic exchange. Tracing the history of these events from early film festivals held in the 1950s in Europe, he argues that they were crucial to build and unite a community of capitalists with an international community of filmmakers. In Chapter 12, "Cinema and Industrial Design: Showmanship, Fairs and the Exhibition Film," Haidee Wasson discusses the history of form and circulation of industrial film in context of industrial fairs and exhibitions. By focusing on the 1939 World's Fair in New York City, she describes how cinema was an integral part of a wider array of presentational techniques but also how film stood out as a key element of such events, as one of the persistent and most voluble public interfaces for industrial and sponsored media.

Section 4, "Teaching Oneself and Others," focuses on pedagogy and governance in and through industrial films. It looks more closely on how these films were part of the co-creation and the circulation of knowledge and how they affected concepts of subjectivity. In Chapter 13, "Putting Films to Work: *System, the Magazine for Business*," Gregory A. Waller studies how this prominent American business periodical advised manufacturers and retailers on how to use moving pictures to boost sales and increase efficiency and managerial oversight in the workplace. Waller argues that *System* at this formative historical moment articulated the possibilities of film as an emerging and still largely untapped "force in the world of business." In Chapter 14, "New Media for the Schools of Tomorrow: The AV Instructional Films of Robert W. Wagner," Charles R. Acland looks at how new media and multimedia ensembles related to the educational realm in the post-WWII period. He parallels the work of Robert W. Wagner with ecological concepts from media theory that try to understand new environments of images and sounds. In Chapter 15, "We Must Know More Than We Can See: Images for Vocational Training and the Emergence of Cognitive Ergonomics," Guilherme Machado analyses techniques of video-based pedagogy and the emergence of a new type of ergonomics, which is focused on mental rather than bodily processes in a key moment in the broader history of industrial labour, the transition to the so-called "post-industrial" or service economy in the 1960s and 1970s. In Chapter 16, "Free Enterprise Film: Aims of Industry, Economic Propaganda and the Development of a Neoliberal Cinema," Scott Anthony discusses the filmmaking activities of Aims of Industry, an influential educational organization in Britain in the immediate post-war years. He analyses how Aims' library circulated a range of industrial and sponsored films and argues that in the consumerist, privatizing and post-industrial mode of cinema patronized by Aims some of the cultural origins of neoliberalism can be discerned.

Section 5, "Post/Colonial Industries and Third Industrial Cinemas," brings together approaches to industrial film and empire at the threshold of decolonization and shows how industrial film built and shaped colonial audiences. In Chapter 17, "Framing Local and International Sentiments and Sounds: Unilever and Royal Dutch Shell in a Changing Nigeria," Rudmer Canjels shows how the film department of Royal Dutch Shell, one of the oldest and most prolific units of any major company, used music to anchor their image and educational films in local and regional cultures and soundscapes to postulate a connection between enterprise and nation with a view to its continued operation in post-independence Nigeria. In Chapter 18, "Working through the End of Empire," Tom Rice discusses industrial film as a governmental approach to shape concepts of citizenship in the context of the colonial history of the British Empire and shows that film became increasingly important in the government's work to define and shape productive citizens and to formalize economic ties between colonizer and colonized. In Chapter 19, "Cinema-going on the Railway

Tracks: Transportation, Circulation and Exhibition of Information Film in Colonial India," Ravi Vasudevan addresses the role of railways as a mode of film distribution and a site of exhibition in colonial India. Discussing the logistics and programmes of the mobile exhibition by train, Vasudevan shows how the British administration used film to educate the rural population by intermingling educational films with travel films and other attractions to cultivate an audience. In Chapter 20, "The Latin American Process Film," Salomé Aguilera Skvirsky addresses a shared concern with process, i.e. with the material and bodily aspects of production, across a wide range of Latin American films and discusses the connection between post-war industrialization and the emergence of the New Latin American Cinema of the 1960s and 1970s.

Section 6, "Production Cultures and/of the Industrial Film: Amateurs and Professionals," demonstrates that, much as film has never been exclusively tied to the dispositive of cinema, industrial films are not exclusively tied to the factory. Rather, they need to be understood as cultural practices that exist beyond and constantly permeate organizational boundaries. In Chapter 21, "Soviet Industrial Film across Categories: Negotiating between Utility, Art and Science," Maria Vinogradova shows how despite the absence of an established category of "industrial film" within the Soviet film industry, such films were produced and spanned a broad variety of types, such as industrial process, educational, labour and advertising film. She explains how the focus on production processes reflects an economic system based on central planning. In Chapter 22, "'There Is No Life More Reckless and Adventuresome Than That of the Oil Prospector': ENI's Geologist-Filmmakers in Iran," Luca Peretti shows how the prospectors of Italy's state energy company ENI were also consummate amateur filmmakers who combined their geological exploration with the making of personal travelogues and family films, which complement the official presentation of the company's activities in Iran in films, such as Bernardo Bertolucci's three-part documentary *La via del petrolio*. In Chapter 23, "Industrial Film from the Home Studio: Amateur Cinema and Low-Budget Corporate Moving Image Culture in West Germany (1950 to 1977)," Alexander Stark takes amateur filmmaker Elisabeth Wilms, "the filming baker's wife" from the city of Dortmund in Western Germany, as his case study to show how amateur film practices intersected with the field of sponsored film in the 1950s to the late 1970s. In Chapter 24, "Movie and Industry in Italy: The 'Golden Age' of Italian Industrial Documentary (1950–1970)," Anna Maria Falchero offers a comprehensive account of industrial film production during the post-war boom era, with a focus on the main lines of film on output in terms of production arrangements, themes and distribution. And finally, in Chapter 25, "A Film That Doesn't Seem to Work: A Shot of Renault's Early Assembly Line (1920 to 1929) – A Case Study, Methodology and 3D Restitution for Film Analysis," Alain P. Michel traces the trajectory of *At the Renault Factory*, a film produced in the 1920s to showcase Renault's effort during WWI, through different storage and distribution formats. He discusses the implications of the loss of the film's original materiality as he reconstructs the film's original production context through a combination of new methodologies such as systematic cross-analysis and 3D technology.

Section 7, "Ephemeral Artistry: Ecologies of Authorship in Industrial Cinema," addresses a spectre which continues to haunt industrial film research despite the field's radically pragmatist orientation and focus on the nexus of cinematic form and industrial organization rather than expression: the spectre of authorship. In doing so, the section highlights one aspect of cinematic form in particular which has hitherto been largely neglected: sound. In Chapter 26, "Business and Art: Pharmaceutical Industries, Film Production and Circulation, and the French Film Production Company ScienceFilm, 1960–1980," Christian Bonah sheds light on the production and circulation of chemical-pharmaceutical films from the point of view of the film production company ScienceFilm, founded by French film director Éric Duvivier, a nephew of Julien Duvivier. Bonah shows how Duvivier connected his industrial filmmaking practice, which spans fifty years from 1950 to 2000, with contemporary film art while enlisting leading physicians as allies for contracting with industrial sponsors. In Chapter 27, "Transfer of Power: Films Officers in the British Coal Industry," Patrick Russell suggests that pragmatism and functionalism developed in the early phases of industrial film research by arguing for a concept of co-authorship. Based on a comprehensive survey of the British post-war film production in the energy field, Russell proposes an engagement with film style which remains attentive to moments of singularity rather than focusing exclusively on regular patterns. In Chapter 28, "Saudi Arabia's John Ford? Robert Yarnall Richie, *Desert Venture* and Ephemeral Authorship in Industrial Film," Martin Stollery offers an even more direct challenge to Rick Prelinger's suggestion that we should try to avoid the trap of auteurism this time around. He proposes the concept of "ephemeral authorship" to show how reviews and other paratexts treated Robert Yarnall Richie's *Desert Venture*, a film commissioned by Saudi-American oil giant ARAMCO, as a documentary work in the tradition of Robert J. Flaherty and John Ford, enlisting auteurist discourse to broaden the audience for a corporate film. In Chapter 29, "Sounds Industrial: Understanding the Contribution of Music and Sound in Industrial Films," Annette Davison focuses on a sample of promotional industrial films produced by the Shell Film Unit in the 1950s under the supervision of Arthur Elton to highlight the importance of sound and music in industrial films. She identifies and evaluates techniques of audiovisual rhetoric which are crucial to the work of industrial films but have so far been overlooked. In the final contribution to the volume, Chapter 30, "Creative Films for Creative Corporations: Music and Musicians in Experimental Italian Industrial Films," Alessandro Cecchi shows how Italian corporations in the 1950s and 1960s enlisted modernist composers like Angelo Francesco Lavagnino, Franco Potenza and Vittorio Gelmetti to produce musical scores which often featured electronic sounds and projected an aura of innovation and technological sophistication even as they resonated with some of the major trends of the contemporary musical avant-garde.

This book is partially based on a discussion and screening event held in Frankfurt in December 2015. Other contributions were then added to fully represent the current state of the field. Frankfurt is an apposite choice for a discussion on industrial film and circulation not just because of its role as the financial centre of Germany and, increasingly, Europe. Frankfurt was also the headquarters of the Thurn und Taxis postal service monopoly from 1810 to 1867, which was a key system of communication in the early phases of industrialization and became an agency of the Prussian state after the war of 1866, further boosting the industrial development of "*Gründerzeit*" Germany. Frankfurt is today the site of DE-CIX, the world's largest internet exchange in terms of average traffic throughput and the key node of the internet in Europe. It is, in other words, a good vantage point for further sequels.

The editors wish to thank the Deutsche Forschungsgemeinschaft (DFG) for funding. The book was made possible with the help of Nayang Technological University, ConTrust Frankfurt and MECS Lüneburg. And, last but not least, the editors wish to express their gratitude to our editorial assistants, Matthias Augsbach and Luca Schepers, without whose tireless effort this book would never have seen the light of day.

Works Cited

Acland, Charles R., and Haidee Wasson, editors. *Useful Cinema*. Duke University Press, 2011.

Aguilera Skvirsky, Salomé. *The Process Genre: Cinema and the Aesthetic of Labor*. Duke University Press, 2020.

Appadurai, Arjun. *Modernity at Last: Cultural Dimensions of Globalization*. University of Minnesota Press, 1996.

Bartelheim, Martin, et al. "'ResourceCultures' – A Concept for Investigating the Use of Resources in Different Societies." *Persistent Economic Ways of Living: Production, Distribution, and Consumption in Late Prehistory and Early History*, edited by Alžběta Danielisová and Manuel Fernández-Götz. Archaeolingua Alapítvány, 2015, pp. 39–50.

Braun, Marta, Charlie Keil, Rob King, Paul Moore, and Louis Pelletier, editors. *Beyond the Screen: Institutions, Networks and Publics of Early Cinema*. John Libbey, 2012.

Cronon, William. *Nature's Metropolis: Chicago and the Great West*. W.W. Norton, 1991.

Didi-Huberman, Georges. *The Surviving Image: Phantoms of Time and Time of Phantoms: Aby Warburg's History of Art*. Penn State University Press, 2016.

Florin, Bo, Patrick Vonderau, and Yvonne Zimmermann. *Advertising and the Transformation of Screen Cultures*. Amsterdam University Press, 2021.

Fuller, Matthew. *Media Ecologies: Materialist Energies in Art and Technoculture*. MIT Press, 2007.

Grieveson, Lee. *Cinema and the Wealth of Nations: Media, Capital, and the Liberal World System*. University of California Press, 2017.

Hediger, Vinzenz, and Patrick Vonderau, editors. *Films That Work: Industrial Film and the Productivity of Media*. Amsterdam University Press, 2009.

Heider, Fritz. "Thing and Medium." *On Perception, Event, Structure, and Psychological Environment: Selected Papers*. International Universities Press, 1959.

Hoof, Florian. *Angels of Efficiency: A Media History of Consulting*. Oxford University Press, 2020.

Hoof, Florian. "The Media Boundary Objects Concept: Theorizing Film and Media." *MediaMatters: The Materiality of Media, Matter as Medium*, edited by Bernd Herzogenrath. Bloomsbury, 2015, pp. 180–200.

Hopkins, A.G. "Globalisation and Decolonisation." *Journal of Imperial and Commonwealth History*, vol. 45, no. 5, 2017, pp. 729–45.

Jancovic, Marek, Alexandra Schneider, and Axel Volmar, editors. *Format Matters: Standards, Practices, and Politics in Media*. Meson Press, 2020.

Klose, Alexander. *The Container Principle: How a Box Changes the Way We Think*. MIT Press, 2015.

Levinson, Marc. *The Box: How the Shipping Container Made the World Smaller and the World Economy Bigger*. Princeton University Press, 2006.

Levinson, Marc. *An Extraordinary Time: The End of the Postwar Boom and the Return of the Ordinary Economy*. Random House, 2016.

McNeill, J.R., and Peter Engelke. *The Great Acceleration: An Environmental History of the Anthropocene since 1945*. Belknap Press, 2014.

Mukherjee, Rahul. *Radiant Infrastructures: Media, Environment, and Cultures of Uncertainty*. Duke University Press, 2020.

Odin, Roger. *Spaces of Communication*. Amsterdam University Press, 2020.

Parks, Lisa, and Nicole Starosielski, editors. *Signal Traffic: Critical Studies of Media Infrastructures*. University of Illinois Press, 2015.

Redish, Martin H., and Howard M. Wasserman. "What's Good for General Motors: Corporate Speech and the Theory of Free Expression." *George Washington Law Review*, vol. 66, no. 2, 1997–1998, pp. 235–96.

Schabacher, Gabriele. "Mobilizing Transport: Media, Actor-World, and Infrastructures." *Transfers*, vol. 3, no. 1, 2013.

Serres, Michel. *Le parasite*. Grasset, 1980.

Star, Susan Leigh. "The Ethnography of Infrastructure." *American Behavioral Scientist*, vol. 43, no. 3, 1999, pp. 377–91.

Star, Susan Leigh, and Karen Ruhleder. "Steps Toward an Ecology of Infrastructure: Design and Access for Large Information Spaces." *Information Systems Research*, vol. 7, no. 1, 1996, pp. 111–34.

Starosielski, Nicole. *The Undersea Network*. Duke University Press, 2015.

Sterne, Jonathan. *MP3: The Meaning of a Format*. Duke University Press, 2012.

Warburg, Aby. *Nachhall der Antike. Zwei Untersuchungen*. Edited by Pablo Schneider. Diaphanes, 2011.

Wasson, Haidee. *Everyday Movies: Portable Projectors and the Transformation of American Culture*. University of California Press, 2020.

Yates, JoAnne. *Control through Communication: The Rise of System in American Management*. Johns Hopkins University Press, 1989.

Section 1

Networks and Flows: Visualizing Value Chains

1 The Aesthetics of the Global Value Chain

Container Shipping, Media Networks and the Problem of Visibility in the Global Sphere of Circulation

Vinzenz Hediger

Abstract

Facilitated by containerization global value and supply chains have largely replaced the factory-in-one-place model of production prevalent from the industrial revolution up until the 1970s. But if there was an aesthetics of the factory, i.e. a generic way in which the visibility of the production process was organized in the interest of controlled transparency, operational efficiency and social legitimacy, is there a comparable aesthetics of the global value chain? Drawing on recent work in infrastructure and media studies, this contribution focuses on the social media, film and design work of Maersk, the world's largest container shipping company, to analyze three different, but interconnected regimes of visibility in and of global value chains: critical, operative and representational.

Keywords: infrastructure; visibility; containerization; globalization; social media; corporate communications

Factories are now like ships: they mutate strangely, masquerade, and sometimes sail away stealthily in the night in search of cheaper labor, leaving their former employees bewildered and jobless.
– Allan Sekula and Dave Sinclair, "Freeway to China (Version 2: For Liverpool)," 2000

Always act as if there were a crisis and the company was in a crisis.
– Maersk internal management guidelines, September 5, 1994

Hediger, V., F. Hoof, Y. Zimmermann, with S. Anthony (eds), *Films That Work Harder: The Circulation of Industrial Film*. Taylor & Francis Group 2024
DOI: 10.5117/9789462986534_CH01

Prologue: Sailing into Sight

On March 23, 2021, the *Ever Given*, a four-hundred-meter-long container ship carrying 20,000 TEU, twenty-foot equivalent unit containers, was hit by strong side winds and got stuck in the bank of the Suez Canal. The ship, operated by the Taiwanese Evergreen line and one of the largest of its kind, blocked the world's busiest waterway for six days. Close to four hundred ships lined up on either side of the canal, causing significant delays and an estimated nine billion dollars of losses. The Suez Canal Authority released photographs, and memes started circulating. Among them was one which identified the container ship as the main vehicle of global capitalism, and as something critical theory should study (albeit probably to little effect).

That suddenly everyone paid attention to container ships seemed to illustrate one of the properties of infrastructure identified by Susan Leigh Star, namely that it is usually transparent and only "becomes visible upon breakdown."[1] In the prologue to his 2010 film essay *The Forgotten Space*, which he co-directed with Noël Burch, Allan Sekula describes container shipping as "100,000 thousand invisible ships, 1.5 million invisible seafarers, binding the world together through trade." To the extent that these are indeed invisible, it is not so much because their owners hide them from view like the capitalists of the nineteenth century hid machines and working conditions behind factory walls. Rather, "the increasing reliability of [...] operation makes rapid diffusion feats unremarkable," as Vaclav Smil writes.[2] However, the infrastructures of global trade – which for the purposes of this contribution we take to include ships, ports, containers, traffic networks connecting the sea to the hinterland, media and knowledge – have in fact been highly visible for some time now, and have become increasingly so just around the time Sekula and Burch released their film.

Maersk Line, long the world's largest container shipping company with roughly eighty thousand employees, a turnover of €40 billion and a world market share of seventeen per cent in 2020, is a case in point. Unusually for a business-to-business company, Maersk has systematically built a social media presence in the years after the financial crisis of 2008–2009, which evolves around corporate videos and photographs providing inside

1 Star, Susan Leigh. "The Ethnography of Infrastructure." *American Behavioral Scientist*, vol. 43, no. 4, 1999, p. 382.

2 Smil, Vaclav. *Growth: From Microorganisms to Megacities*. MIT Press, 2019, p. xiv.

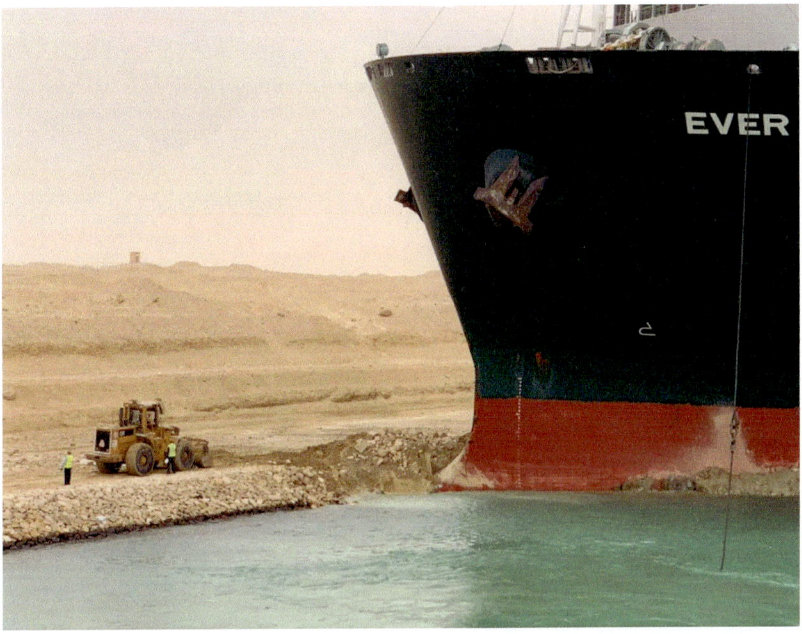

Fig. 1.1a. EVER GIVEN stuck in the bank. Image released by the Suez Canal Authority.

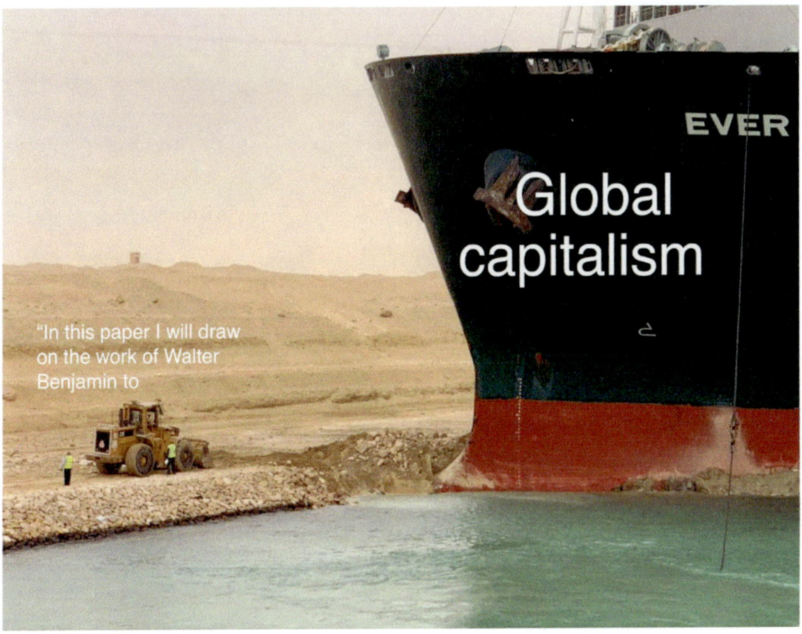

Fig. 1.1b. The memefication of the EVER GIVEN incident (author unknown).

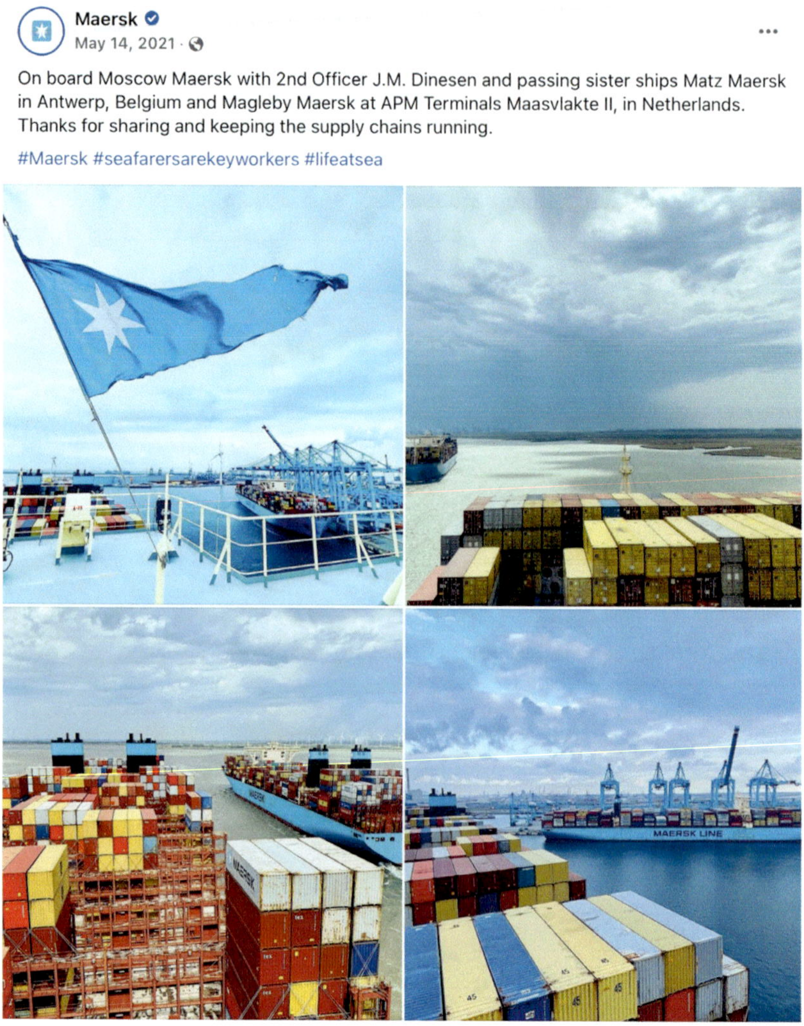

Fig. 1.2. Maersk crews share pictures via the company's Instagram account, May 15, 2021.

views of global supply chains with a documentary feel rather than the default glossiness of more conventional corporate image films.[3] Maersk was surprised, according to one executive, to find out that a business which it suspected people find "boring" was anything but in the eyes of a substantial

3 Agerdal-Hjermind, Annette. "The Enterprise Social Media Strategy: The Case of Maersk." *Communication & Language at Work*, no. 3, 2014, pp. 3–17; Katona, Zsolt, and Miklos Sarvary. "Maersk Line: B2B Social Media – 'It's Communication, Not Marketing'." *Berkeley Haas Case Series*, vol. 56, no. 3, 2014, pp. 142–56.

public.[4] Maersk's Facebook page had more than three million followers in May 2021, more than seven times the number of BASF, the world's largest chemical company, also a multinational and a company roughly twice the size in terms of turnover (but with a similar number of employees, roughly eighty thousand). Maersk's Instagram account had 234,000 followers, as opposed to BASF's combined 50,000 for its global and German accounts.

The combined impact of the 2008 financial crisis and climate change have led shipping companies to adopt new ship models with up to 25,000 TEU or twenty-foot equivalent unit containers, i.e. five times the capacity of an average container ship in the 1990s. These gigantic ships have become the object of a widespread fascination even as their carbon footprint is the focus of growing concerns. In 2009, the shipping industry's carbon footprint was slightly larger than that of Germany.[5] It has since decreased through a variety of measures including the lowering of travel speeds from twenty-four to as low as twelve knots,[6] and companies like Maersk and CMA-CGM currently work to fully decarbonize their fleets in the space of a few years rather than a decade or two. But in online forums, the ships themselves are the main concern. Ship-spotting videos of mega containerships are a popular YouTube genre. Discovery documentaries about the building of these ships and their operation often have an afterlife with millions of clicks on YouTube. One six-minute video made by a Maersk crew member offering a tour of his cabin in March 2019 had registered no fewer than 2.3 million views by May 2021.[7] Ship-spotting videos and photographs, whether sent in by employees or followers, are also an important part of the Facebook and Instagram feeds of Maersk.

Astride between the operative requirements of global supply chains and the widespread fascination with ships sits a website like MarineTraffic.com. Here volunteers provide data through an automated identification system (AIS) to track the movements of more than eighteen thousand ships worldwide on Google Maps in real time. MarineTraffic.com was started in 2007 as an academic project by Dimitri Lekkas, a professor of systems engineering at the University of the Aegean on the island of Syros in Greece,

4 Katona and Sarvary. "Maersk Line: B2B Social Media," p. 146.

5 Casdesus-Masanell, Ramon, Forrest L. Reinhard, and Frederik Nellemann. "Maersk Line and the Future of Container Shipping." Harvard Business School Case Study 9-712-449, 2012, p. 5.

6 Carriou, Pierre, Francesco Parola, and Theo Notteboom. "Towards Low Carbon Global Supply Chains: A Multi-trade Analysis of CO_2 Emission Reductions in Container Shipping." *International Journal of Production Economics*, 208, 2019, pp. 17–28.

7 https://youtu.be/JBH78Lt3Kbw.

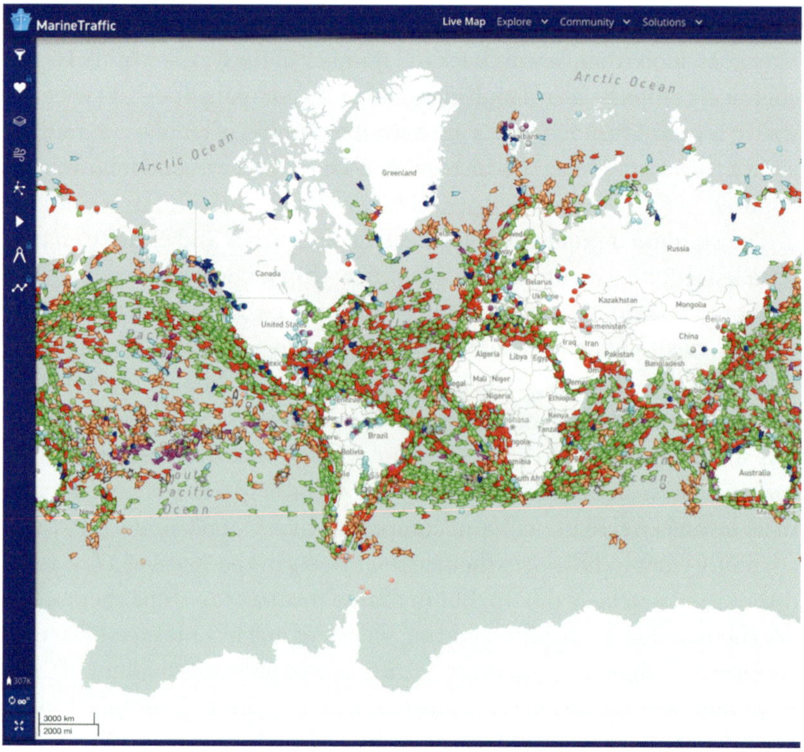

Fig. 1.3. The MarineTraffic.com landing page on May 10, 2021, 10:11 am CET (green: bulk cargo and container ships; red: tankers; orange: fishing ships; blue: passenger line ships; lilac: cruise ships and yachts).

the original hub of the Greek shipping industry. In itself an infrastructure in the sense that it is a boundary object which means different things to different people,[8] MarineTraffic.com caters to both industry professionals and shipping enthusiasts and combines tracking with options for uploading ship-spotting images and videos for a network of correspondents from around the world.

Among other things, this shows that visibility and invisibility, as Brian Larkin argues, "are not ontological properties of infrastructures." Rather, they "are made to happen as part of technical, political, and representational processes."[9] What is more, in the shipping industry "visibility" is also a

8 Star. "The Ethnography of Infrastructure."

9 Larkin, Brian. "Promising Forms: The Political Aesthetics of Infrastructures." *The Promise of Infrastructures*, edited by Nikhil Anand, Akhil Gupta, and Hanna Apel. Duke University Press, 2018, pp. 178–86.

technical term which refers to the ability to track assets in real time or, more specifically, "the data-based creation of transparency through tracking of assets or events and the generation of information and insights from that process."[10]

Visibility in this technical sense matters because what economists call "value creation," i.e. the input activities of a company to create value for customers, now no longer happens in self-contained factories located near vital resources and markets, but in dispersed locations across the globe chosen for cost considerations. The container made "global sourcing imaginable because it is cheap and logistically manageable," as historian Michael B. Miller writes.[11] Because of containers, the factory model of organization, which drove economic activity from the eighteenth to the late twentieth century, has largely been replaced with what has been known since the 1990s as global value chains.[12] These global value chains in turn rely on the operation of global supply chains, i.e. transportation and distribution networks which put resources and assets in the right place at the right time. Setting up and operating global supply chains is no longer just a matter of shipping and distribution. Swiss-born Napoleonic general and military theorist Antoine-Henri Jomini (1779–1869), a continuing influence on the US armed forces' doctrine of overwhelming force, coined the term "logistics" to describe "the practical art of moving armies" and supplying troops in the field.[13] Now it refers to the management of supply chains, the backbone of economic globalization, another term of relatively recent vintage.[14]

The *EverGiven* accident, then, reveals not so much that global value chains are hidden from sight. Instead, it points us to the ways in which these value chains and the infrastructures of global trade "as technical objects, take on form"[15] and become visible and readable. Brian Larkin has recently argued

10 Bauhaus, Henrik, Frederik Möller, Maleen Stachon, Christina Hoffmann, and Boris Otto. "Data-Driven Business Models in Logistics: A Taxonomy of Optimization and Visibility Services." *Proceedings of the 53rd Hawaii International Conference on System Sciences*, 2020, p. 5379, https://hdl.handle.net/10125/64403.

11 Miller, Michael B. *Europe and the Maritime World: A Twentieth Century History*. Cambridge University Press, 2012, p. 343.

12 Gereffi, Gary. *Global Value Chains and Development*. Cambridge University Press, 2018.

13 Van Creveld, Martin. *Supplying War: Logistics from Wallenstein to Patton*, 2nd ed. Cambridge University Press, 2004.

14 As Deborah Cowen's analysis suggests, as global supply chains and their networks have themselves increasingly become a security concern, the military origins of the term have come full circle in the twenty-first century. Cowen, Deborah. *The Deadly Life of Logistics: Mapping Violence in Global Trade*. University of Minnesota Press, 2014.

15 Larkin. "Promising Forms," p. 175.

that one "risks fundamentally misrecognizing the range of ways in which infrastructures address, order, and constitute political relations" if one fails to consider technics and aesthetics as mutually constitutive.[16] The study of industrial film has similarly assumed that technology and aesthetics are mutually constitutive. But it has also assumed that we risk missing out on the operational logics of both if we fail to address economic activity as the third term of the relation. Starting from the firm, the study of industrial film has assumed that every company is a media company, since all economic activity is made possible and shaped by media networks.[17] The global value chain expands the horizon of inquiry to an articulated sequence of companies interacting in and through media networks.

In what follows, I want to argue that global value chains are the object of, and shaped by, three different, but related types of visibility: *critical, operative and representational.* These share two commonalities: First, all are *aspirational.* In the world of global trade and globalized production no one ever sees enough – not the Marxist critic, not the supply chain manager, not the shipping customer, technology enthusiast or ecological activist. And, second, all compensate for their limitations by resorting to different types of what Susan Leigh Stars calls "*master narratives*" of infrastructures, stories told by a "voice [which] speaks unconsciously from the presumed center of things."[18] In modernity, which includes the compressed modernity of a country like South Korea, the default master narrative of infrastructure is the national epic: a heroic tale of achievement through bravery and technology. But the new technologies and practices of global communication and trade increasingly challenge that framework. Nicole Starosielski has shown how conflicting approaches to the operation of undersea cables can be sorted into narratives of connection, disruption, nodes and transmission.[19] Along similar lines, I want to argue that in critical visibility gives form to the infrastructures of global trade though *narratives of crisis*; operative visibility is tied to *narratives of efficiency*; and representational visibility produces

16 Ibid., p. 178.

17 Hediger, Vinzenz, and Patrick Vonderau, editors. *Films That Work: Industrial Film and the Productivity of Media.* Amsterdam University Press, 2009; Acland, Charles R., and Haidee Wasson, editors. *Useful Cinema.* Duke University Press, 2011. Lee Grieveson has offered a comprehensive view of cinema, government, and global trade networks in the first half of the twentieth century in *Cinema and the Wealth of Nations*, while Alexander Klose has discussed the "box" as a new paradigm of thinking in culture and the arts. Grieveson, Lee. *Cinema and the Wealth of Nations: Media, Capital, and the Liberal World System.* University of California Press, 2017; Close, Alexander. *The Container Principle: How a Box Changes the Way We Think.* MIT Press, 2015.

18 Star. "The Ethnography of Infrastructure," p. 384.

19 Starosielski, Nicole. *The Undersea Network.* Duke University Press, 2015.

contested, but similarly aspirational *narratives of growth, sustainability and inclusiveness,* which can also be read as globalization age reiterations of the modernist logic that "through infrastructures and circulation [...] development and modern subjectivity can be achieved" – albeit, as we will see, a reiteration in which global trade figures as a substitute for politics in the nation state.[20] With their adjacent narratives each of these three types of visibility constitutes what Jacques Rancière calls a "scene," a clearly delineated territory where we can study and verify the constituent elements of an organization, and which is not reducible to shared metaphysical principle or any other scene, let alone an underlying "ur-scene."[21] The task of an aesthetics of global value chains, then, is to provide a composite view of the form of their constituent elements as articulated through the three scenes of critical, operative and representational visibility. The task is further to account for how these three scenes exceed, challenge and transform the frames of the nation state and the firm, which provide the dominant form of infrastructure and the visibility of economic activity in the pre-1970s period.

Precisely to capture the transformation of these frames, and since shipping firms "have been singled out as key drivers of economic globalization," it seems legitimate to start with an individual firm.[22] Apart from its already mentioned proactive media work, Maersk, a family- and partnership-controlled multinational headquartered in Denmark, is an appropriate choice because of the size of the company and its historical role in establishing container shipping. The Maersk group also covers – or covered at one point – all elements of shipping infrastructure. As of May 2021, Maersk operated 708 ships and held a 16.9% world market share. Alongside the Maersk Line, the Maersk group includes APM, which operates container terminals, including the hub in Algeciras, Spain, which connects West Africa with the main routes into Europe and Asia, a terminal in Singapore and a string of terminals in West Africa, including Apapa, the container port of Lagos, Nigeria. Until it was absorbed under the man brand in 2020, Maersk also owned DAMCO, a freight-forwarding/supply chain logistics firm. Maersk also operated a shipyard until 2009, and from 1962 to 2017 it included Maersk Oil, a prospecting and drilling firm specializing in deep sea oil rigs.

20 Larkin. "Promising Forms," p. 190.

21 Rancière, Jacques. *La méthode de l'égalité. Entretien avec Laurent Jeanpierre et Dork Zabunyan.* Bayard, 2012, p. 99.

22 Peterson, Niels P., Stig Tenold, and Nicholas J. White, editors. *Shipping and Globalization in the Post-war Era: Context, Companies, Connections.* Palgrave Macmillan, 2019, p. 257.

If Brian Larkin is (partly) correct in suggesting that in "the study of infrastructure, form is both ubiquitously visible yet absent from analytic consideration," something similar can be said of shipping networks. The literature is growing fast now, but until recently systematic studies have been few and far between.[23] In the first section I want to review the emergence of container shipping from earlier networks with a focus on media and the organization of global trade. In particular I want to discuss the national epic as the default master narrative of infrastructure in its tension with the emergence of global value chains. In the second section I want to focus on what I propose to call the cinematic shape of trade and argue that the shift from factory to global value chain can be described as a shift from framing to tracking, with Maersk as the case study. And in my conclusion, I want to summarize my findings by discussing the notion of containerization as a revolution. Because of the antecedents of contemporary shipping networks and the current shape of the industry, the analysis is Euro- and US-centric. Further work will be required compensate for this defect.

1. From Empire to Logistics: Containerization and the Four Gs

Containerization can be seen as a part of the global history of the four Gs: The Great Transformation, the Great Divergence, the Great Acceleration and the Great Convergence. The Great Transformation is Karl Polanyi's term for the industrial revolution[24]; the Great Divergence describes the period of rapid growth in the nineteenth century which sees Europe and the United States pull ahead of other parts of the world, like India or China, which as

23 Marc Levinson, an economist by training, wrote his now classic book about container shipping as a history because the data were not sufficient for an economic study of the impact of containerization on the world economy (Levinson, Marc. *The Box: How the Shipping Container Made the World Smaller and the World Economy Bigger.* Princeton University Press, 2006). Similarly, in the introduction to their historical case study of Maersk, Jephson and Morgen note that "the theoretical and empirical literature on the role of transport is surprisingly weak. [...] [T]he role of transport remains ambiguous and subject to shallow interpretation" (Jephson, Chris, and Henning Morgen. *Creating Global Opportunities: Maersk Line in Containerization, 1973–2013.* Routledge, 2014, p. 4). Finally, Peterson et al. (*Shipping and Globalization in the Post-war Era*) consistently reference Michael B. Miller's 2012 cultural history of European shipping networks (*Europe and the Maritime World: A Twentieth Century History*) as a groundbreaking work to be emulated in further studies.

24 Polanyi, Karl. *The Great Transformation: The Political and Economic Origins of Our Time.* Beacon Press, 2001.

late as the mid-eighteenth century enjoyed comparable levels of economic welfare[25]; the Great Acceleration is the period of unprecedented economic and population growth after 1945 in which humans become the decisive factor in the ecology for the first time in history, a period now also known as the "Anthropocene"[26]; and the Great Convergence is the period of growth since 1990, which has seen the centre of the world economy shift to Asia as countries like South Korea, China or Indonesia progressively draw level with the wealthiest parts of the so-called West.[27] At the dawn of the Great Acceleration, containerization dissolves the factory, the main locus of the Great Transformation and the Great Divergence, into the global value chain, the driver of the Great Convergence. At the same time, containerization can be seen as part of the intertwined processes of decolonization and economic globalization, in which, as Anthony G. Hopkins argues, "'horizontal' concepts of racial equality and universal democracy, as well as the breakdown of complementarities between industrial metropoles and primary-producing peripheries, undermined vertical, hierarchic imperial systems."[28] As such, containerization substitutes the promise of a reliable, accessible and coherent global network for the circulation of goods for what John Darwin describes as the often haphazard and structurally violent practices of colonial domination of the "unfinished" European empires.[29] To situate this promise we have to first look more closely at the power differential containerization promises to overcome, and how that power differential, and its transformation, have been framed.

The Shipworm and the Steam Engine

Different from their landed predecessors, the European empires of the sixteenth to the twentieth centuries were built by sea.[30] Starting from an experience of cultural inferiority to the mature landed empires which they encountered on their voyages of "discovery" to the Americas, India and

25 Pomeranz, Kenneth. *The Great Divergence: China, Europe, and the Making of the Modern World Economy*. Princeton University Press, 2000.

26 Engelke, Peter, and J.R. McNeill. *The Great Acceleration: An Environmental History of the Anthropocene since 1945*. Harvard University Press, 2014.

27 Baldwin, Richard. *The Great Convergence: Information Technology and the New Globalization*. Belknap Press/Harvard University Press, 2016.

28 White, Nicholas J. "Thinking Outside 'The Box': Decolonization and Containerization." *Shipping and Globalization in the Post-war Era: Context, Companies, Connections*, edited by Niels P. Peterson, Stig Tenold, and Nicholas J. White. Palgrave Macmillan, 2019, pp. 68–69.

29 Darwin, John. *Unfinished Empire*. Bloomsbury, 2013.

30 Darwin, John. *After Tamerlane: The Global History of Empire since 1405*. Bloomsbury, 2008.

China,[31] Europeans progressively translated their technological superiority into a sense of cultural superiority from the second half of the eighteenth century and the Enlightenment onwards. Innovations in shipping and navigation, which in turn led to the invention of the steam engine, the railway and the steamship, laid the groundwork for their networks of global dominance. The connection between innovations in shipping and industrialization is complex, but in some ways quite straightforward.

Capitalism came about through the emergence of mass consumer markets in North-western Europe from the seventeenth century onward (with important input in terms of capital provided through the plunder of Bengal in 1757).[32] Land enclosures in seventeenth and eighteenth century Europe had created a landless underclass, which migrated to cities and provided the cheap labour required for industrialized mass production. But neither the emergence of mass markets nor the industrialization of production would have been possible without shipping. Inexorably intertwined with colonialism, capitalism as a system of industrialized innovation owed its emergence in large part to the growth of shipping networks. "Little of what Europeans made, sold, or consumed [...] was independent of overseas markets or sources of supply," and it was shipping networks and infrastructure which made mass industrial and consumer societies possible.[33] Ships brought raw materials, particularly cotton, from the various colonies to Europe and finished products back to the territories of extraction.

One of the most important innovation in the late eighteenth century was copper sheathing for ships, which prevented damage from tropical shipworms, a long-standing problem. Copper sheathing boosted the triangular trade of slaves to the Americas and goods to Europe by cutting "the death rates of slaves in the middle passage by half."[34] This turned the "planned disaster" of the middle passage – which incidentally, as Christina Sharpe points out, registers no mention in Allan Sekula and Noel Burch's film on shipping and capitalism[35] – into a more calculable risk, which in turn increased the supply of cotton, coffee and other trade goods to satisfy

31 Bitterli, Urs. *Die Wilden und die Zivilisierten. Grundzüge einer Geistes- und Kulturgeschichte der europäisch-überseeischen Begegnung*. C.H. Beck, 2004.

32 Plumpe, Werner. *Das kalte Herz. Kapitalismus: die Geschichte einer andauernden Revolution*. Rohwolt, 2019.

33 Miller. *Europe and the Maritime World*, p. 3.

34 Rönnbäck, Klas, and Peter M. Solar. "Copper Sheathing and the British Slave Trade." *Economic History Review*, vol. 68, no. 3, 2015, p. 806.

35 Sharpe, Christina. *In the Wake: On Blackness and Being*. Duke University Press, 2016, pp. 25–27.

demand in the emerging capitalist mass consumer markets in Europe.[36] The steam engine was originally developed to pump water out of copper mines in Cornwall.[37] From there it was adapted to power factory production lines, railway locomotives and ships.

The introduction of steamships in the mid-nineteenth century reduced travel time and made schedules more reliable, paving the way for what is generally referred to as the first era of globalization, a period of global free trade between 1870 and 1914, which also includes the Berlin Conference and the subsequent comprehensive colonial occupation of Africa. Barriers to trade for European companies were minimal, and the new trade networks powered by steamships and the telegraph – as well as, in the view of some economic historians, an abundance of coal[38] – consolidated the Great Divergence.

But while shipping has been an important factor of both the Great Transformation and the Great Divergence, it can be argued that the technological superiority of the West was initially a superiority of media technology. Ships as carriers of goods, people and ideas can, of course, be understood as media. Ships have been around for a long time, but the most important novelty in European expansion were improved instruments of navigation and what Bruno Latour calls "immutable mobiles," scalable and portable media like maps (even though Erhard Schüttpelz has rightly pointed out these technologies only reached reliable levels of standardization in the nineteenth century).[39] Infrastructures from the telegraph to channels and ports, which have been discussed in terms of a theory of media and communication by Karl Knies in the nineteenth and Harold Innis in the first half of the twentieth century,[40] further consolidated European and Western dominance. Starting in the mid-nineteenth century, shipping lines aligned with global telegraph networks. According to economic historian Stig Tenold "shipping lines can be considered – in tandem with the telegraph network – as the tentacles of empire." This is particularly true for the British

36 Sharpe. *In the Wake*, p. 27.

37 O'Sullivan, Mary. "Power & Profit: Copper Mines & Steam Engines in Late 18th century Cornwall." Working Paper, Paul Bairoch Institute of Economic History, 2021, https://www.unige.ch/sciences-societe/dehes/files/2516/1821/2863/Power_Profit_WP_2021.pdf.

38 Pomeranz. *The Great Divergence.*

39 Schüttpelz, Erhard. "Die medientechnische Überlegenheit des Westens. Zur Geschichte und Geographie der *immutable mobiles* bei Bruno Latour." *Mediengeographie. Theory–Analyse–Diskussion*, edited by Jörg Döring and Tristan Thielemann. transcript, 2009, pp. 67–110.

40 Knies, Karl. *Der Telegraph als Verkehrsmittel* [1857]. Baden-Baden, Nomos, 1996; Innis, Harold. *Empire and Communications*. Rowman & Littlefield, 2007.

Empire: "By 1914 the submarine cables that Britain controlled were almost twice as large as those of the next two powers (the United States and France) combined." A precursor of sorts to the Internet of Things, the "shipping lines were to cargo what the telegraph network was to information," and both were "a manifestation of British power, which contributed to preserving the very same dominance that it reflected."[41]

Until the onset of decolonization global shipping networks evolved around European companies and ports. The most important ports were Hamburg, Rotterdam, Antwerp, Liverpool and London, with London serving as the hub of finance, insurance, ship sales and charter. A limited number of large companies like HAPAG (Hamburg), Holland Amerika Line (Rotterdam), P&O Lines (London) or Holt (Liverpool) dominated the industry. They maintained transnational networks and cooperated in so-called "conferences," cartels with a regional focus designed to limit competition and reduce overcapacity. The conferences were suspended during the world wars but quickly resumed operation after both and disappeared only with containerization in the late 1990s. Britain also dominated the shipbuilding industry, with up to eighty per cent of new merchant and passenger ships in the early decades of the twentieth century coming from British yards.

As the alignment of shipping and telegraph networks suggests, shipping networks were knowledge networks. The infrastructures of these knowledge networks expanded beyond ships, paperwork and cable technology. Early access to information through oral communication could create a competitive advantage. Ports had always served as hubs of information, and pubs and clubs were important conduits of knowledge exchange. At least until the 1960s, letter writing was an important part of these networks. Agents sent out by shipping companies or forwarding agencies received detailed instructions in writing from their company heads.[42] The global networks of shipping were both cosmopolitan and Eurocentric, and colonial club culture facilitated the interaction between large shipping companies.[43] Clubbiness, writes Michael B. Miller, "was the transnationalism of a maritime community in which Europeans of a common formation predominated." Even in colonial outposts, "the cultural juxtaposition of territories and enclaves within a broader construct of

41 Tenold, Stig. "The Declining Role of Western Europe in Shipping and Shipbuilding 1900–2000." *Shipping and Globalization in the Post-war Era: Context, Companies, Connections*, edited by Niels P. Peterson, Stig Tenold, and Nicholas J. White. Palgrave Macmillan, 2019, p. 16.

42 Miller. *Europe and the Maritime World*, pp. 176ff.

43 Ibid., ch. 1.

cosmopolitanism, fit a globalism that remained fragmented, imperial, and Eurocentric."[44]

A master narrative which gives these networks shape can be found in the canon of early modern and modern European literature. Luís Vaz de Camões' *Os Lusíadas*, a sixteenth-century epic poem, turned Vasco da Gama's discovery of a sea route to India into Portugal's founding national epic and provided a narrative framework for future colonial enterprises,[45] while the novels of Joseph Conrad dramatize how the shipping networks of the colonial club culture era intersect with the violent modes of governance of the "unfinished" European empires. Conrad's first novel, *Almayer's Folly* (1895), tells the story of a failing Dutch trader married to a Malayan woman who considers himself superior to his extended family by virtue of his European origin. But he has never been to Europe and is treated with indifference and even contempt by the Dutch ship crews and trade agents on whom he depends for his survival. The novel, which was filmed by Chantal Akerman in 2011, can be read as a story about the limits and boundaries of infrastructure told from the perspective of someone who thinks he is part of a national epic but progressively finds out that he is not.

The First Innovation: Multinational Corporations meet Anti-colonial Imperialism at Sea

If the First World War led to a first shift of economic and cultural power towards the United States, the Second World War laid the groundwork for the erosion of European dominance in global shipping and trade in the second half of the twentieth century. Requisitioned merchant ships which supplied Allied forces on the Western Front played an important role in determining the outcome of the First World War. The Second World War brought with it innovations in logistics in Jomini's sense of the term – i.e. the deployment of armed forces in the field – which laid the groundwork for the transformation of the shipping industry after 1945 and the emergence of logistics in the now current sense of the term. As Miller writes: "All the pioneering ways of

44 Ibid., p. 193.

45 Camões and Da Gama now lie buried in symbolic graves next to each other at the entrance of the Cathedral of Belém, the historic point of departure of Portuguese ships where the Tejo flows into the Atlantic, while just outside of the Cathedral, on the former landing, is a monument to Portugal's colonial exploits, erected in the crucial year of decolonization, 1960, by Portuguese dictator Salazar with money from the South African Apartheid regime. On the meanings of this set-up, see Trouillot, Michel-Rolph. *Silencing the Past: Power and the Production of History*. Beacon Press, 2015, pp. 108–110.

moving cargo after the war had roots in logistical experiments during the conflict." These include in particular roll-on/roll-off (RORO) ferries modelled on allied landing crafts, palletization, which was widely used by US supply services, and containerization, "the great vitiator of traditions [which was] implied in the lashing of boxes to the decks of ships to maximize carrying capacity to global theaters."[46]

The first important innovation of the post-war period was the creation of the multinational shipping company. If the shipping networks of the pre-World War II period hat been informally cosmopolitan, if decidedly Eurocentric, the new shipping companies were legally and economically dispersed over several countries to take advantage of tax shelters and low-cost labour. As a result of wartime overproduction, the United States had more ships at its disposal than the government or American shipping companies knew how to operate. With funding from US banks, Greek shipowners Aristotle Onassis and Stavros Niarchos bought up the excess tonnage. Their companies became important conduits for the post-war oil boom, a key factor of the Great Acceleration, which was largely driven by consumer demand and the rapid spread of the private automobile as a means of transportation in Europe and the United States. While both were outsiders in the Greek shipping industry, Onassis and Niarchos had established international networks since the 1930s, continuing a centuries-old tradition of business in the Greek diaspora.[47] Onassis operated his shipping line from New York until he became the target of a pressure campaign orchestrated by US vice president Richard Nixon after he obtained a contract as the preferential carrier for Saudi oil.[48] At the behest of their American funders, Onassis and Niarchos had registered their ships in Liberia, Panama and Honduras, low-tax, low-regulation countries in the American sphere of influence. This gave the US government the option of requisitioning their fleets as reserve capacity for the US armed forces in the case of war.[49] Thus were born the so-called "flags of convenience," which shipping companies, including

46 Miller, *Europe and the Maritime World,* p. 288.

47 Harlaftis, Gelina. *Creating Global Shipping: Aristotle Onassis, the Vagliano Brothers, and the Business of Shipping, c. 1820–1970.* Cambridge University Press, 1970.

48 A.P. "A Plot on Onassis Charged in Article." *New York Times,* August 3, 1978, https://www.nytimes.com/1978/08/03/archives/a-plot-on-onassis-charged-in-article-nixon-burger-and-a-hughes-aide.html.

49 Harlaftis, Gelina, and Christos Tsakas. "The Role of Greek Shipowners in the Revival of Northern European Shipyards in the 1950s." *Shipping and Globalization in the Post-war Era: Context, Companies, Connections,* edited by Niels P. Peterson, Stig Tenold, and Nicholas J. White. Palgrave Macmillan, 2019, p. 187.

Maersk, continue to use today. The "flag state principle can be seen as an expression of national sovereignty and thus of the traditional so-called Westphalian system of international law,"[50] as Katharina Reiling writes. But the "slicing up of the value chain" in the post-war shipping industry and particularly in the wake of containerization has made it "difficult to determine the true nationality of a ship or a shipping company."[51]

Viewers of the 2013 Hollywood docudrama-thriller *Captain Phillips* (dir. Paul Greengrass), the first big-budget studio film set on a container ship, would gain little or no understanding of this background. The film stars Tom Hanks in the title role and tells the story of the *Maersk Alabama* hijacking off the coast of Somalia in 2009. Hanks reprises his role as the explorer, sufferer for and saviour of the American cause which he had also played in *Apollo 13* and *Saving Private Ryan*. At the time of the hijacking, the Taiwan-built ship, which had launched under a Danish flag, happened to be registered in the United States, which made the attack the first successful act of piracy against an American ship since 1839. In the film three US navy ships and a SEAL team come to the rescue of the *Alabama Maersk*. The US built up its ability to police the seas and wage war with aircraft carrier fleets at any point around the globe at (almost) any time in the course of the Korean War and has maintained it ever since.[52] The US Navy exemplifies the teachings of military strategist Alfred Thayer Mahan, who has been described as "the most influential American author of the nineteenth century" and who in his 1890 book *The Influence of Sea Power upon History, 1660–1783*, argued "that control of the seas is always crucial to countries seeking to achieve or maintain far-reaching power."[53] Mahan's doctrine of sea power aligns with a bipartisan trade policy which historian of diplomacy William Appleman Williams has memorably described in 1959 as "anti-colonial imperialism." Since the 1890s US policy has focused on maintaining access to foreign markets for American exports, if necessary at the price of "regime change," like the US-engineered coups in Guatemala and Iran in 1953 to the third Gulf

50 Reiling, Katharina. "The Emergence of Maritime Governance in the Post-war World." *Shipping and Globalization in the Post-war Era: Context, Companies, Connections*, edited by Niels P. Peterson, Stig Tenold, and Nicholas J. White. Palgrave Macmillan, 2019, p. 40.

51 Tenold. "The Declining Role," p. 33.

52 Cumings, Bruce. *The Korean War: A History*. Modern Library/Random House, 2010, ch. 8.

53 In books like *The Influence of Sea Power upon History, 1660–1783*, published in 1890 and "one of the most influential works of military history ever written in the United States," American admiral and military strategist Alfred Thayer Mahan developed the argument "that control of the seas is always crucial to countries seeking to achieve or maintain far-reaching power." See Kinzer, Stephen. *The True Flag: Theodore Roosevelt, Mark Twain, and the Birth of the American Empire*. Henry Holt, 2017, p. 23.

War in 2003.[54] In line with this policy, the US Navy became the guarantor of free passage for international commercial shipping after the Second World War, with a particular focus on the Gulf region and oil tankers since the 1970s.

The shipping industry has returned the favour where the occasion presented itself. In September 1990, for instance, following the Iraqi invasion of Kuwait, the CEO of Maersk, Mærsk Mc-Kinney Møller, "made the unilateral decision to offer garage space on two of Maersk Line's combined roll-on/roll-off (RORO) container ships to US authorities free of charge" in the build-up to Operation Desert Storm. Ostensibly, this was done in gratitude for the shelter the US had offered Maersk, then in charge of the New York office of his father's company, during the Second World War. The US Navy subsequently chartered Maersk ships to complete the build-up in the gulf.[55] Maersk also cooperated with the producers of *Captain Phillips* and documented the film shoot on its Facebook page. However, in the film no mention is made of the Danish origins of the ship's owner.

By remaining silent on the multinational character of the shipping industry *Captain Phillips* builds a patriotic narrative about the United States protecting free markets through military strength at sea, a national epic as master narrative of infrastructure. At the same time *Captain Phillips* points to another aspect of containerization: That its rise was related to the deregulation of global trade and finance, which started in the late 1970s and accelerated in the 1980s. Early in the film there is an altercation between the captain and a unionized crew member who insists that his union contract does not cover fighting hijackers. Organized labor as a form of cowardice in the face of the enemy: Hollywood's take on containerization in *Captain Phillips* is to frame it as the Reagan Revolution at sea.

The Second Innovation: Containerization, Decolonization and the Formalization of Knowledge Networks

The early stages of containerization coincide with the dissolution of the European colonial empires. The year in which the *Ideal-X*, a converted tanker, shipped the first containers from Newark, a manufacturing hub in New Jersey, to Houston – 1956 – is also the year in which Morocco and Tunisia gained independence from France. The year the first purpose-built container ship, the *Gateway City*, started operating – 1957 – is the year in

54 Williams, William Appleman. *The Tragedy of American Diplomacy*. W.W. Norton, 2009.

55 Jephson, Chris, and Henning Morgen. *Creating Global Opportunities: Maersk Line in Containerisation, 1973–2013*. Cambridge University Press, 2014, p. 192.

which Ghana became independent from Britain, the first country in West Africa to do so. But decolonization intersects with containerization beyond this historical coincidence.

Containerization was driven primarily by cost considerations. Loading and unloading ships used to require significant manpower, and in the post-war period labour costs grew significantly. In London loading costs doubled between 1960 and 1970s, for instance.[56] Loading and unloading ships also took a long time, to the point where ships often spent more time at or near harbour than at sea. Containers could be loaded at an inland factory and unloaded again at a warehouse in the port's hinterland, with significant time savings and at a fraction of the labour cost. Another cost factor driving containerization was insurance: Container shipping significantly reduced damage and theft, and thereby insurance rates. In the late 1960s and 1970s, New York harbour, once one of the busiest in the world, was replaced by the container terminals in Port Elizabeth, New Jersey. Taking advantage of point-to-point container shipping the textile industry left Lower Manhattan for cheaper locations inland or overseas. New York City lost seven hundred thousand jobs in the process.[57] The city's economy grew back when the financial service industry boomed as a result of the deregulation of global capital. At the same time artists, designers and music producers took over the vacant Manhattan factory lofts to create what Elizabeth Currid-Halkett calls "The Warhol Economy."[58] To frame it in terms of film history, containerization destroyed the world of *On the Waterfront* to make room for *Wall Street* and *Studio 54*.

Like the flag of convenience and the multinational shipping company, containerization originated from the initiative of newcomers and outsiders. American entrepreneur Norman McLean is considered the inventor of the container. McLean moved from trucking to shipping when he bought Pan-Atlantic in 1956, a shipping company which he renamed SeaLand in 1960, to highlight the company's distinctive service feature: point-to-point delivery on land without long layovers and loading time in port, made possible by container shipping. McLean was the owner and operator of the *Ideal-X* and the *Gateway City*, which was retrofitted from the *Sumter* in 1957,

56 Peterson, Niels P. "Shipping as a Knowledge Industry: Research and Strategic Planning at Ocean Group." *Shipping and Globalization in the Post-war Era: Context, Companies, Connections*, edited by Niels P. Peterson, Stig Tenold, and Nicholas J. White. Palgrave Macmillan, 2019, pp. 157–82.

57 Levinson, *The Box*.

58 Currid-Halkett, Elizabeth. *The Warhol Economy: How Fashion, Art, and Music Drive New York City*. Princeton University Press, 2006.

an American navy cargo ship built during the Second World War. The names are significant, whether by coincidence or whether the "imaginary force of history" wanted it so, as Foucault would say. Fort Sumter was the site of the opening salvo of the American Civil War, a war fought over the right to own human beings as property. The gateway city, as environmental historian William Cronon puts it, is "the principal colonizing agent of the western landscape", an urban hub of traffic and trade to and from the American West in the nineteenth century.[59] As long as people and goods moved mostly on rivers the gateway city was St. Louis, situated at the confluence of the Mississippi and Missouri rivers; with the advent of the railway, the role of gateway city shifted to Chicago. The West looms large in America culture in the 1950s and 1960s. In Hollywood the Western thrives, and in the political imaginary the West beckons as a libertarian utopia free of government interference (read: school integration, voting rights and other forms of civil rights legislation).[60] At the same time and in the words of John F. Kennedy, space becomes the "new frontier." Naming the first purpose-built containership *Gateway City* casts the sea as another new frontier and container shipping as the infrastructure of a new, de-territorialized form of American expansionism – precisely the alignment of global sourcing and American interventionism that the *Maersk Alabama* hijacking dramatizes.[61]

While the name of the first purpose-built container ship links containerization to the history of American settler colonialism, the United States is now no longer a player in the container shipping industry. SeaLand remained the only significant container line in the US and was absorbed by its long-time partner Maersk in 1999. Maersk, which was founded in 1904, had entered container shipping relatively late, in 1973. Through endogenous growth and mergers and acquisitions Maersk became the biggest shipping company in the world in 1996, a position which it held until January 2022, when MSC, a Swiss-Italian company overtook Maersk in terms of TEU capacity. In the former European-dominated shipping networks Maersk had been an outsider, operating ships in Europe and on a New York-Asia route through the Panama Canal since its opening. But a good part of Maersk's later growth came from the absorption of remnants of legacy lines from the colonial era. Apart from SeaLand in 1999, Maersk acquired P&O Nedlloyd in 2005,

59 Cronon, William. *Nature's Metropolis: Chicago and the Great West.* W.W. Norton, 2009, p. 307.
60 Richardson, Heather Cox. *How the South Won the Civil War: Oligarchy, Democracy, and the Continuing Fight for the Soul of America*. Oxford University Press, 2019.
61 Grandin, Greg. *The End of the Myth: From the Frontier to the Border Wall in the Mind of America*, Metropolitan Books, 2019.

which in turn had been a merger of the Dutch and British flag lines and their respective country's main colonial carriers. As Nicholas White writes, P&O, which was founded in 1837, had served as "the bloodline of the Raj" with its service to India in the nineteenth century.

Maersk's absorption and dissolution of the British and Dutch flag carriers into a multinational company points to another aspect of the colonial and decolonial legacy of container shipping. When the former colonies became independent, they tended to adopt the national epic narrative of infrastructure and worked to build their own national infrastructures to compete with other nations. Apart from dams, power stations, roads, factories and universities, these included flag carrier shipping lines. For instance, upon independence in 1957 Ghana launched the state-owned Black Star Line to compete with Europeans for cargo to Europe and the US. However, these lines were welded to an outgoing paradigm of shipping and were quickly overtaken by containerization. In former colonial ports like Lagos or Colombo, dock workers organized, causing regular disruptions and raising labour costs. As a consequence, multinational container shipping lines bypassed these unruly decolonized ports and directed trade to countries like Taiwan and South Korea instead, where authoritarian regimes held unions at bay to provide the low-cost, high-discipline labour which made global sourcing a profitable enterprise.[62] Taiwan and South Korea took advantage of containerization to move from import economy to import substitution and export economies in the 1960s and 1970, and, together with Singapore and Hong Kong, became important players in the global container shipping networks.[63] Perhaps somewhat ironically, the former European colonial powers also lost out. British shipping companies had taken an early strategic lead in container shipping in the 1960s. They were "in large part concerned with preserving Commonwealth links which were threatened by post-colonial globalization."[64] But the trade worked both ways and ultimately empowered decolonized ex-dominions like Singapore. This further contributed to the decline of the former colonial national flag carriers and prepared the ground for their merger with a multinational like Maersk. Among the ten largest container shipping companies today, the only legacy line which was already part of the global trading system in

62 White. "Thinking Outside 'The Box'."

63 On South Korea's industrialization, see Kuznets, Paul W. "Causes, Consequences, Relevance: South Korea's Industrialization." *Transformations in Twentieth Century Korea*, edited by Chang Yun-shik and Steven Hugh Lee. Routledge, 2006, pp. 89–105.

64 Ibid., p. 94.

the nineteenth century is Hapag Lloyd at number ten. Headquartered in Hamburg, Hapag-Lloyd operated 256 ships with a world market share of 7.3% in May 2021.

Decolonization thus not only coincided with but drove containerization, which in turn was facilitated by computerization. Specifically, computerization led to a standardization and formalization of knowledge and knowledge networks. As Miller writes, containerization "took away the quays and pubs, and replaced them with computers and logistics."[65] Company-specific, partly tacit knowledge about the efficient operation of cargo ships accumulated over decades became obsolete and was replaced with "knowledge that was explicit, codified and easily transmitted." By eliminating the competitive advantage of legacy lines, commodified knowledge lowered thresholds of entry and "transformed shipping from bottleneck to engine of global connectivity." By the 1980s, "with infrastructure in place and knowledge readily available, newcomers from Asia's developing countries found it relatively easy to enter the industry."[66] The consequence was a dramatic shift in the networks of global shipping. In the 1990s and early 2000s, the biggest shipping lines apart from Maersk were all Asian, notably COSCO (the People's Republic of China [PRC] state line), Evergreen (Taiwan), Ocean Network Express (Japan), the now defunct Hanjin shipping line (South Korea) and Pacific International Lines (Singapore). The market share of European lines shrank from close to one hundred per cent in 1880 to around twenty-five per cent in 2010.[67]

A similar shift happened in port operations and shipbuilding. While European ports continue to service European consumers Rotterdam, Europe's largest port, barely even makes the top ten now, with Shanghai topping the list, followed by Singapore and four other Chinese ports. Bespoke container ports have been built downstream from existing older ports, as in the case of Rotterdam and Shanghai, and in entirely new locations with cheaper land and good access to the hinterland like Algeciras or Felixstowe, which is now England's busiest port. London, the hub of the colonial shipping networks, is no longer a port town, and Liverpool, long Britain's second port, barely registers anymore. British shipyards had a quasi-monopoly from the mid-nineteenth century and the beginning of the steamship era until the Second World War. In the early 1960s Japan became the world's largest shipbuilder. Hyundai, South Korea's largest

65 Miller. *Europe and the Maritime World*, p. 352.
66 Petersson, Niels P. "Shipping as Knowledge Industry," p. 166.
67 Tenold. "The Declining Role," p. 21.

chaebol (a term for industrial conglomerates which translates as "wealth gang"), opened a shipyard in the south-east fishing village of Ulsan in 1974. Taking advantage of low labour costs, a highly educated workforce further disciplined by three years of mandatory military service and massive state subsidies, Hyundai Heavy Industries became the largest shipbuilder in the world by 1983, a position which it has retained ever since.[68] Maersk started ordering ships from Korea, Taiwan and Japan early on but maintained its own shipyard in Odense, Denmark, until 2009, when it closed down in the wake of the global financial crisis. The Triple-E Class ships with a capacity of 20,000 TEU, currently the biggest vessels operated by Maersk, were built by Daewoo, another *chaebol*, in Geoje, South Korea. The remaining European shipyards like Meyer in Papenburg, Germany, are now focused on war ships, cruise ships and megayachts for the world's super rich.

The life cycle of ships now also mostly ends in Asia. Around seven hundred container and bulk carriers are withdrawn from service every year. Ship breaking, i.e. the process of dismantling ships and recycling their components from scrap metal to furniture, is concentrated in Pakistan, India and Bangladesh, with Chittagong in Bangladesh alone processing forty per cent of the annual volume. If the emergence of Asian shipping lines and shipyards can be seen as both an indicator and a factor of the Great Convergence, the role of the Indian subcontinent as the ship-breaking hub of the world is better understood from a world-systems perspective. "Centrality in the world-system," writes sociologist R. Scott Frey, "allows countries to externalize their hazards or environmental harms on others."[69] These include extremely dangerous work conditions in the ship-breaking industry and environmental harm caused by toxic metals.[70] While ship-breaking proceeded out of sight for decades,[71] it has recently become more visible, like other parts of the infrastructure of global trade, through documentaries

68 "Militarized management is a major factor in accounting for the success of South Korea's export-oriented industrialization." Lie, John. "What Makes Us Great: *Chaebŏl* Development, Labor Practices, and Managerial Ideology." *Transformations in Twentieth Century Korea*, edited by Chang Yun-shik and Steven Hugh Lee. Routledge, 2006, p. 147.

69 Frey, R. Scott. (2015). "Breaking Ships in the World-System: An Analysis of Two Ship Breaking Capitals, Alang-Sosiya, India and Chittagong, Bangladesh." *Journal of World-Systems Research*, vol. 21, no. 1, 2015, p. 25.

70 Fakhruddin, A.N.M., M.I. Taluder, and M.A. Hossain. "Environmental Impacts of Ship Breaking and Recycling Industries of Sitakunda, Chittagong, Bangladesh." *Advances in Natural Sciences*, vol. 8, no. 1, 2015, pp. 51–58.

71 Cairns, George. "Postcard from Chittagong: Wish You Were Here?" *Critical Perspectives on International Business*, vol. 3, no. 2, 2007, pp. 266–79.

by filmmakers and artists such as Michael Glawogger, Yasmine Kabir and Hira Nabi and a dedicated genre of YouTube videos.[72]

The rise of Asian shipping was also the result of top-down economic policies initiated by authoritarian governments. The entry of Hyundai into the shipbuilding business is a case in point. As historian Bruce Cumings writes, in the early 1970s president Park Chun-hee "had heard that there was a big global demand for oceangoing tankers" because of the OPEC shake-up of the global oil market. Park was a former military officer who came to power in a coup in 1962 and pursued a relentless industrialization policy in close coordination with the country's large industrial conglomerates; he died in office at the hands of his own bodyguard in 1979. Acting on the information about the demand for tankers, Park summoned Chong Chu-yong, the founder of Hyundai, and basically ordered him to enter the shipbuilding business. Chung "flew off to Greece and landed two contracts for 260,000-ton oil tankers, by promising cheaper and quicker delivery than any other company." What he neglected to mention was that Hyundai, at this point mostly a car company, "had no shipyard to build them."[73] Nor did Chung have the capital or the staff. Barclay's bank lent Hyundai the money based on the contracts Chung brought back from Greece, and a team of Hyundai engineers were sent to Scotland to learn the craft. Hyundai then proceeded to open the shipyard in Ulsan and delivered the ships on time, under budget.

In a somewhat less adventurous fashion, the PRC entered container shipping with its state-run company COSCO after the turn to state capitalism in the 1980s. They did so with support from European companies, including Maersk.[74] In the 1990s and particularly after its ascension to the World Trade Organization in 2001, the PRC established itself as the "world's manufacturer" thanks to container shipping, while it also became the world's biggest operator of container ports. While South Korea under Park was focused on rapid industrialization to catch up with the West,

72 Michael Glawogger. *Working Man's Death*, Austria 2005 (episode 3 was filmed in Gadani, Pakistan; Yasmine Kabir, *The Last Rites*, Bangladesh 2016, 17 min (filmed at Chittagong), https://vimeo.com/144362610; Hira Nabi, *All That Perishes at the Edge of Land*, Pakistan 2018 (filmed at Gadani). I thank Bishnupryia Gosh and Pujita Guha for the references to Kabir and Nabi. For an example of a drone shot beaching video, see https://youtu.be/DLYIbmL09Lw.

73 At what cost to workers may be guessed from the fact that the Ulsan yard, a relatively high-paying employer, was the site of significant labour unrest in 1974. Cf. Cumings, Bruce. *Korea's Place in the Sun: A Modern History*. W.W. Norton, 2005, pp. 323–24.

74 Iversen, Martin Jes. "East Asiatic Company's Difficult Experiences with Containerization." *Shipping and Globalization in the Post-war Era: Context, Companies, Connections*, edited by Niels P. Peterson, Stig Tenold, and Nicholas J. White. Palgrave Macmillan, 2019, pp. 147ff.

containerization in China is part of a larger geopolitical project. China's current Belt and Road initiative aims to achieve through infrastructure projects across the world what the United States achieve though military dominance of the seas, namely secure stable access to world markets for domestic exports. The Chinese revolution had been a watershed moment in the move towards decolonization. Starting in the 1950s, Maoist China provided military assistance to anti-colonial insurgencies, but also invested in infrastructure, particularly in Africa.[75] Considering that PRC president Xi Jinping likes to style himself as the second coming of Mao, Belt and Road can also be understood as a reprisal and extension of Mao's infrastructure export policy with a new focus on Central Asia and Europe in addition to Africa.[76]

The New Ascendancy of the Multinational European Family Business and the Persistence of the National Epic

Even though the center of the world economy may be shifting to Asia and the Pacific, the most important development in the decade after the 2008 financial crisis was a shift of the centre of container shipping back to Europe. Container shipping is now dominated by three European carriers, which together with Hapag-Lloyd had a world market share of more than fifty per cent as of May 2021. Apart from Maersk, these include Mediterranean Shipping Company (MSC), a former Italian cruise operator which ran 589 container ships for a 16.1% world market share in May 2021 and is now headquartered for tax reasons in landlocked Geneva, Switzerland. They further include Compagnie Maritime Affrètment/Compagnie Générale Maritime (CMA-CGM), which operated 557 ships with a market share of 12.5% as of May 2021, about even with COSCO, the PRC's state line and the fourth-largest carrier. Incidentally, all three companies are family businesses. Maersk is still controlled by the Maersk family, MSC is owned by the Aponte family, and CMA-CGM was founded by Jacques Saadé, a Lebanese exile who fled the civil war in 1978. In 1996 CMA-CGM absorbed the former French state shipping company Compagnie Générale Maritime, but the Saadé family still controls the company. One of the narratives of the infrastructures of global trade in the twenty-first century, then, would be that of the ascendancy of the multinational family business. The Krupp family inspired Visconti's *La caduta degli dei*, tale of decadence, Nazism

75 Lovell, Julia. *Maoism: A Global History*. The Bodley Head, 2019, pp. 185–222.

76 Maçaes, Bruno. *Belt and Road: A Chinese World Order*. Oxford University Press, 2019.

and crime. The Murdoch family's internal power struggles were chronicled in *Succession*. Mærsk Mc-Kinney Møller the long-time company patriarch, is briefly alluded to in the figure of a power broker in *Borgen,* the Danish TV series about a female prime minister. The Aponte and Saadé families, by contrast, have so far managed to be boring enough to stay out of the entertainment industry limelight.

Meanwhile, the national epic persists elsewhere. South Korea, which moved from the periphery to the centre of the world system by inserting itself early and forcefully in the economy of global value chains, is a case in point. Over the last twenty-five years, TV dramas, pop music and films have become South Korea's most important export after cars, electronics and semi-conductors. In the transition to democracy in the 1990s the South Korean government relinquished its control and censorship of the film industry. Investing in new infrastructures for the media industries, the government recast film, television and music as a "central to the governments export-focused economic development strategies"[77] – both as exports themselves and carriers of what Joseph Nye has called "soft power," an accumulation of cultural capital which elevates the standing and reach of other Korean exports. South Korean domestic films now win major awards in Europe and the US while easily holding off the competition from Hollywood films at home. Two of most successful South Korean films to date domestically are both from 2014, *Ode to My Father* (No. 4), directed by Yoon Je-kyoon, which sold 14.2 million tickets in a country of 51 million, and Kim Han-min's *The Admiral* (No. 2), which sold 17 million tickets. Both films are historical dramas with a significant share of CGI effects, and both films have shipping storylines. *The Admiral* chronicles a historical sea battle with an unlikely Korean victory over imperial Japan. *Ode to My Father* is a Korean version of *Forrest Gump*: Yoon Deok-soo (Hwang Jung-min), the proverbial everyman, lives through all the major events of post-war Korean history. He steadfastly runs a small family business but never fulfils his lifelong dream, which is to sail the seven seas as the captain of a freighter ship.

Like the CGI effects for an earlier film by Yoon Je-kyoon, *Haeundae*, the effects in *Ode to My Father* were "an amalgamation of outsourced and regionalized visual effects," as Jihoon Kim writes. The outsourced parts were produced in Hollywood by Korean technicians under the guidance of American specialists (the effects for *Haeundae* cost $5 million, a fraction of

77 Kim, Joseph, and Seung-ho Kwon. "The Cultural-Industry Policies of the Korean Government and the Korean Wave." *International Journal of Cultural Policy*, vol. 20, no. 4, 2013, pp. 422–39.

the VFX budget of a Hollywood film).[78] Reversing the global value chain to their advantage, *Ode to My Father* and *The Admiral* assimilate Hollywood standards "into a perceptually seamless national spectacle."[79] *The Admiral* broadens the scope of a slate of recent historical dramas about the colonial period and travels back in history to celebrate an unlikely victory over Korea's colonial oppressor Japan as a historical predictor of sorts of current national strength. *Ode to My Father*, on the other hand, chronicles South Korea's compressed modernity in the aftermath of decolonization after the end of World War Two. Having barely escaped the advancing communists in North Korea by ship at the beginning of the film, the hero works as a shoeshine in Seoul early in the film. One day, Chong Chu-yong is his customer, the founder of Hyundai, who tells him about his dream of building ships – a dream which he makes come true. As Bruce Cumings reminds us, Chong's entry into shipbuilding was a matter of serendipity and top-down authoritarian economic policy. In *Ode to My Father* Chung appears as the man with a mission to make South Korea a shipping power from the outset. The scene establishes an important background narrative to the travails of Yoon Deok-soo, the everyman at the centre of the film. While he suffers though the various stages of South Korea's tortuous history of compressed modernity, the *chaebols* and the government, acting in concert in the interest of the country, secure the growth and welfare of the entire nation, which Yoon gets to enjoy with his extended family in old age, even as his longingly looks at the ships in the port of Seoul from the terrace of his apartment. Like *Os Lusíadas*, the ur-text of early modern European sea-faring literature, *Ode to My Father* casts the networks of global trade in the terms of a national heroic narrative, one which is not just perceptually, but ideologically seamless and in which the interests of the *chaebols* implicitly align with those of the country – an idea which still shapes economic policy in important (albeit not unproblematic) ways.[80] One challenge for an aesthetics of global value chains is to account for the persistence of this narrative even as it traces how globalization transforms the nexus of media networks and industrial organization in ways which transcend both nation and firm.

78 Kim, Jihoon. "Perceptual Realism and Digital VFXs in the Korean Blockbuster of the 2010s: *The Admiral: Roaring Currents* and *Ode to My Father*." *Journal of Popular Film and Television*, vol. 47, no. 1, 2019, pp. 9–20.

79 Ibid., p. 14.

80 Cf. Lim, Wonhyuk. "The *Chaebol* and Industrial Policy in Korea." *The Industrial Policy Revolution: The Role of Government Beyond Ideology*, edited by Joan Esteban, Justin Lin Yifu, and Joseph E. Stiglitz. Palgrave Macmillan, 2013, pp. 348–70.

2. From Framing to Tracking: The Cinematic Shape of Trade

Economists and economic historians associate containerization with the so-called second wave of globalization after 1990 in the wake of the GATT agreements and the creation of the WTO. The first wave of globalization, in which a liberal system of global trade coincides with the imperial phase of European colonialism, starts in the mid-nineteenth century and ends with the nationalist excesses of the First World War. However, as Michael B. Miller and others argue, the economic concept of the two phases of globalization is reductive. After the First World War, global trade contracted under a new tariff regime, but cultural globalization continued apace. Perhaps most notably, Hollywood emerged as the first truly global cultural industry capable of reaching audiences and setting trends across the world. The American film industry harnessed the advantage accruing from the size of its home market into global ascendancy through what Gerben Bakker has called an "escalation of quality." This was based on three factors: The new studios used downtown movie palaces as collateral to tap into Wall Street credit previously reserved for larger and more established industries; they built international networks of distribution to guarantee the widest possible exposure for their films; and they partially adapted the assembly line model of production from the automobile industry to rationalize production and standardize the quality of their output.[81] But Hollywood was not only an agent of cultural, but of economic globalization as well. While Hollywood looked to Detroit in the 1910s for their organizational models, it can be argued that the American film industry was at the forefront of the global movement to dissolve the self-contained factory model in the post-war period. Even as European economies embarked on their golden age and *Les Trente Glorieuses*, which also happened to be the golden age of industrial cinema,[82] Hollywood transitioned to flexible specialization, i.e. a system in which qualified specialists cooperate on individual projects rather than staying on a company's payroll on annual or longer-term contracts. The Paramount Decree of 1948, which outlawed the vertical integration of exhibition and distribution, and a tax raise for high salaries for employees made it attractive for stars, directors and producers to become self-employed

81 Bakker, Gerben. "The Quality Race: Feature Films and Market Dominance in the US and Europe in the 1910s." *The Classical Hollywood Reader*, edited by Steve Neale. Routledge, 2012, pp. 31–42.

82 See, for instance, Falchero, Anna Maria. "Movie and Industry in Italy: The 'Golden Age' of Italian Industrial Documentary (1950–1970)." This volume.

and work with studios on a case-by-case basis. Hollywood also anticipated global value chains by relocating production to Germany and Italy in the 1950s and 1960s, countries with excellent production infrastructures, high levels of technical proficiency and lower salary costs than the US.[83]

For entirely different reasons the automobile industry soon followed suit, but it was not Detroit, but Nagoya which took the lead. Toyota was the first company to introduce the *keiretsu* just-in-time delivery system in 1956, in which the factory served to assemble parts pre-produced by a network of suppliers, a system which cut inventory costs and distributed risk across the entire supply chain.[84] The *keiretsu* supply chains were organized mostly in regional clusters. Among the first industries to move from regional to global supply chains was the textile industry, which had driven much of the industrialization of North-western Europe from the late eighteenth century onwards. The pioneering companies included Li & Fung,[85] a Hong Kong-based firm, which started off as a trading house in 1906 and introduced transnational supply chain management in the late 1960s and early 1970s. As co-owner Victor Fung wrote in 2011, "We were there at the beginning of globalisation of labour-intensive manufactured goods. We were in the middle of it, but we didn't know what to call it until the academics told us it was globalisation!" Run by a group of Harvard MBAs, Li & Fung is now registered for tax reasons in the Bermudas. Their supply and production lines typically are structured as follows:

> the filling for a down jacket may come from China, the outer fabric from Korea, the zippers from Japan, the inner lining from Taiwan and the elastic, labels, Velcro and other trim from Hong Kong. The garment might be dyed in Southeast Asia, stitched in China, sent back to Hong Kong for quality control and finally packed for delivery in the United States or elsewhere.[86]

If global supply chains tend show a rhizome-like structure, global value chains can be visualized in a U-shaped graph which is somewhat euphemistically known as the "smiling curve." This curve indicates the level of value creation or profit margins along a given good's trajectory. Margins

83 Steinhart, Daniel. *Runaway Hollywood: Internationalizing Postwar Production and Location Shooting*. University of California Press, 2019.

84 Detroit had experimented with just-in-time inventories in the 1920s and again in the 1940s, but only moved to flexible specialization in the 1980s. Cf. Schwartz, Michael, and Andrew Fish. "Just-in Time Inventories in Old Detroit." *Business History*, vol. 40, no. 3, 1998, pp. 48–71.

85 https://www.lifung.com/.

86 Jephson. "Creating Global Opportunity," p. 157.

are highest at the design, development and branding stage (prefabrication), lowest in the manufacturing stage (fabrication), and rise again as the product reaches retailers and consumers (post-fabrication).[87] Geographically, the beginning and end of the smiling curve tend to be located in high-cost areas like Europe, the US or Japan, while the middle and low point is located in low-labour cost countries, particularly in Asia. The iPhone is the prime example: designed in California, manufactured and assembled mostly in the PRC, and sold again in brand stores across the more affluent parts of the world.

While the smiling curve was relatively flat in the early stages of containerization from the 1970s through the 1980s, it has become considerably more accentuated in the first decades of the twenty-first century. The deepening of the curve indicates that global trade may well make most people richer, but that it also exacerbates global inequality rather than abolishing it, at least for the time being. In that sense, the smiling curve is both an illustration of the global value chain as an operative reality and an aspirational concept: It speaks to a promise deferred for those who happen to be at the nadir of the curve, wherever that may be at any given moment. The economic policies of South Korea since the 1970s and the PRC since the 2000s, which have been focused on building global brands like Samsung or Huawei, can also be described as strategies to move from the manufacturing nadir to the prefab and post-fab ends of the smiling curve. Part of the problem of the visibility of global value chains is that the nadir of the smiling curve may be deepening, but it is also constantly on the move from one low-labour cost area to the next – in itself a constitutive element of what we may call their cinematic shape. But the core problem of visibility of global value chains is the dissolution of the boundaries of the form in the new order of production.

Images of/as Resources

In her 1959 book *The Theory of the Growth of the Firm*, economist Edith Penrose argues that the resources with which management is accustomed to work, as well as their experience and knowledge, shape the productive services a firm is capable of delivering. Knowledge and mental representations are, in other words, a key asset and element of the company's resources. Management tries to make the best of its resources in a dynamic interaction with its environment. In this process, "the environment is

87 Baldwin. *The Great Convergence*, pp. 154–60.

treated as an 'image' in the entrepreneur's mind of the possibilities and restrictions with which he is confronted, for it is, after all, such an 'image' which determines a man's behavior."[88] Images are interpretive frameworks "to make sense of the world," and an important part of management is "setting up new interpretive frameworks [...] for handling new types of problems."[89] These interpretative frameworks are not concentrated in the minds of top management, however. Rather, they are distributed across the firm in the shape of what contemporary management theory describes as "corporate culture": a "shared mental construct" which helps "coordinating distributed knowledge by infusing employees with firm-specific shared knowledge."[90]

Penrose's concept of the image has significant overlap with the concept of "social image" or image of society which Raymond Williams develops almost simultaneously in his 1961 book *The Long Revolution*. The social image is an interpretation, a way "of describing the organization and of conceiving relationships, necessary to establish the reality of social life but also under continual pressure from experience."[91] As Williams indicated by dedicating most of the rest of the book to a study of media and communication, such images and the knowledge are created through media. Similarly, the distributed knowledge of corporate knowledge cannot exist or be shared without the media networks of the firm. Photographs, films, factory plant maps, charts and diagrams are building blocks of mental frames and archives and carriers of firm-specific distributed knowledge.[92]

One operation of media in the production of corporate culture is to clearly mark the boundary between firm and market. Production can be broken down into different activities, which can be ranged according to similarity or dissimilarity. Dissimilar activities require different capabilities and imply different forms of organization, while similar activities can be subsumed under a coherent structure of governance. To put it differently, "capabilities are determinants of the boundaries of the firm." Process films can be understood as visual inventories of capabilities. Photographs and films of workers leaving the factory, incidentally, one

88 Penrose, Edith. *The Theory of the Growth of the Firm*. Oxford University Press, 1995, p. 5.

89 Foss, Nicolai J. "Edith Penrose, Economics and Strategic Management." *Contributions to Political Economy*, no. 18, 1999, p. 99.

90 Ibid., p. 100.

91 Williams, Raymond. *The Long Revolution*. Parthian, 2011, p. 127.

92 Hoof, Florian. *Angels of Efficiency: A Media History of Consulting*. Oxford University Press, 2020.

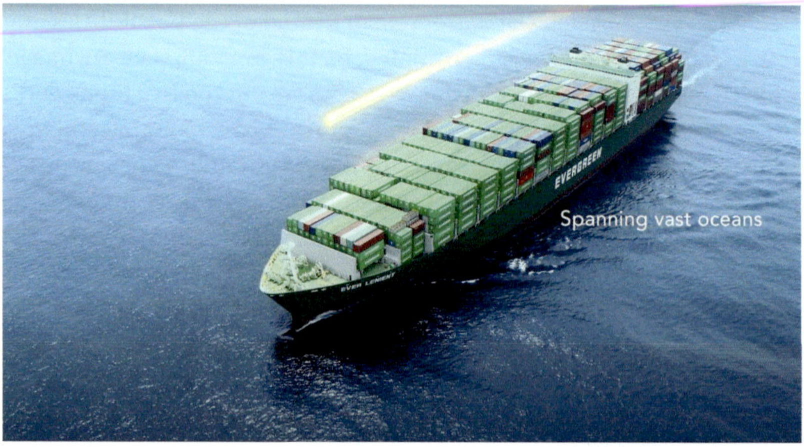

Fig. 1.4. Drone tracking shot of ship at sea. Evergreen line corporate video, 2019.[93]

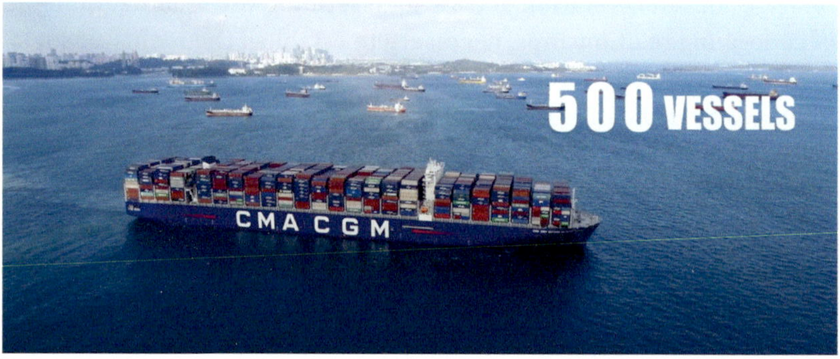

Fig. 1.5. Aerial tracking shot in a CMA-CGM video introducing the natural gas-powered line flagship *Jacques Saadé*, 2020.[94]

of the first motifs in the history of cinema, are both inventories of the company's resources – they show the workforce as a dynamic unit – and markers of the boundaries of the firm.[95] Another example are the panoramic

93 https://youtu.be/L6lVAMJzRGo.

94 https://youtu.be/jVVGfcTJDFo.

95 The Lumière film of workers leaving the Lumière factory in Lyon was part of the first public projection at the Paris Grand Café in December of 1895. Similar films were made by Mitchell & Kenyon, a pioneering film company from Blackburn in north-western England. Mitchell & Kenyon screened these films in their evening programmes to offer the workers in the pleasure of recognizing themselves on screen. On the occasion of the centennial in 1895, Harun Farocki subjected the Lumière films to a critical reading in a film essay. *Workers Leaving the Factory*, then, is an image which works in all three types of visibility: critical (Farocki), representational (Lumière, Mitchell & Kenyon), and operative (Krupp, Lumière).

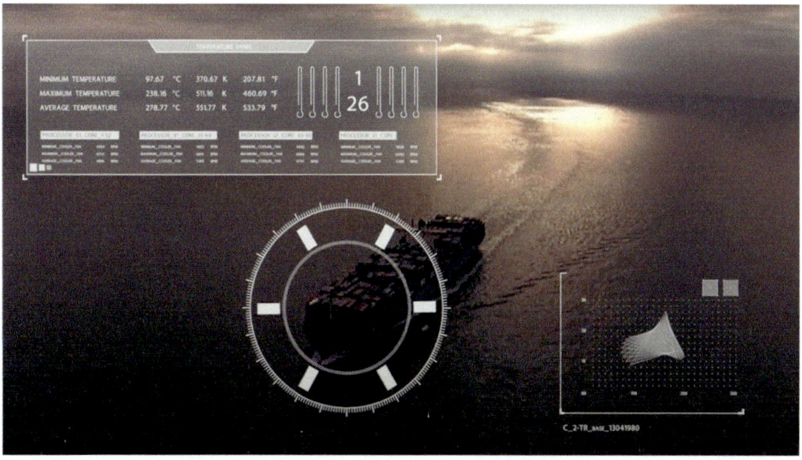

Fig. 1.6. Aerial tracking shot of a Maersk Triple E ship with tracking info graphics.[95]

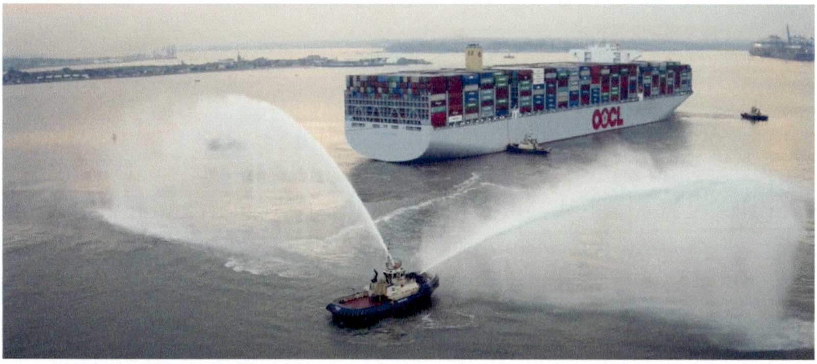

Fig. 1.7. Aerial tracking shot of *OOCL Hong Kong*'s maiden call at Felixstowe, 2017.[96]

360-degree photographs produced at the Krupp steel works in Essen at regular intervals starting 1862. For one thing, these serve to demonstrate the advanced media capabilities of Germany's largest steel producer, which created its own media department, the "Graphische Abteilung" in 1861.[98] In succession, the panoramic photographs document the growth of Krupp's main factory, as well as the growth of Essen as a company town, just beyond the boundaries of the factory. Benchmarks of growth and success, they

Cf. Toulmin, Vanessa. *Electric Edwardians: The Films of Mitchell & Kenyon*. British Film Institute, 2006, ch. 3; Harun Farocki, *Arbeiter verlassen die Fabrik* (D 1995, 36 min).

96 https://youtu.be/KEC5DQqCykI.

97 https://youtu.be/Moxpa7f1BqM.

98 Wolbring, Barbara. *Krupp und die Öffentlichkeit im 19. Jahrhundert*. C.H. Beck, 2000.

can also be read as a visualization at a glance of the company's assets and resources.[99]

The dissolution of the factory is a process in which production transcends the boundaries of the single firm and happens in sequences of successive, spatially distributed capabilities. As a consequence, devices such as panoramic shots of factory plants no longer map the knowledge base of production and are no longer sufficient to delineate an interpretive framework for management. In cinematic terms, this transformation can be described as a shift from framing to tracking. Perhaps the emblematic visual trope of the new order is the drone tracking shot of a container ship in movement, which appears repeatedly in corporate image films and across social network platforms: an epitome of the problem of tracking and an image of the – mental, social – image of the global value chain. To spell out its implications, I want to now turn to the three scenes of visibility of the infrastructures of global trade.

Scene 1 – Critical Visibility: Waking up to the Smiling Curve

The scene of critical visibility can be delimited as encompassing all narratives and argumentative templates which aim to reveal the hidden realities of the global value chain, including in particular ecological impacts of container shipping and the exploitation of labour. These narratives come in different genres and tonalities but share a focus on moments of crisis. Their temporality may be described as melodramatic: a wrong turn has been taken, and it may be too late to avert disaster, but we are certainly approaching the point of no return, or we are already at it. Their aspirational content is discovery, revelation and social change. In the combination of their melodramatic temporality and aspirational content narratives of crisis create what Raymond Williams calls a "structure of feeling" of urgency and looming crisis, a pattern which holds regardless of genre and tonality. It is the "culture of a period," in this case the period of global value chains, or at least an important part of it, and it "is firm and definite" yet "operates in the most delicate and least tangible parts of our activity."[100] By implication, critical visibility is shared visibility: it is not limited to individual decision-makers, but

99 Bosdorf, Ulrich, and Sigrid Schneider. "Ein gewaltiger Betrieb: Fabrik und Stadt auf den Kruppschen Fotografien." *Bilder von Krupp. Fotografie und Geschichte im Industriezeitalter*, edited by Klaus Tenfelde. C.H. Beck, 2000, pp. 123–50.
100 Williams. *The Long Revolution*, p. 69.

shared by groups, communities and that largest of political entities in modernity, the general public.

The narratives and templates of crisis that constitute the scene of critical visibility range from documentaries like Sekula and Burch's *The Forgotten Space* or Glawogger's *The Workingman's Death* to the work of a political economist like Benjamin Selwin, who in a recent article put the claim to the test that global value chains benefit everyone, even the low-paid workers at the nadir of the smiling curve, and in the light of his results proposed to re-label them as global poverty chains.[101] A case can be made that global supply chains have been the focus of critical visibility almost since they first emerged. Long before Western activists and artists started paying attention, performative forms of labour activism drew attention to life at the nadir of the smiling curve. The textile industry in South Korea in the early 1970s is a case in point. The self-immolation of textile worker Jeon Tae-il 1972, chronicled in Park Kwang-su's 1995 film *A Single Spark*,[102] was a signal event in South Korean labour history together with a preceding strike by female textile workers and served as an important point of reference in the country's long march towards democracy.[103]

In Europe and the United States, the scene of critical visibility for global trade took shape in the late 1980s and 1990s. As Allan Sekula and Dave Sinclair write, containerization and the expansion of production lines beyond the horizon of the classical factory left many in Europe and the United States, "jobless and bewildered." Michael Moore captured this bewilderment in a way which resonated with a broader public in his 1989 documentary *Roger & Me*. The film chronicled Moore's repeated attempts to meet the CEO of General Motors Roger Smith – one of whose predecessors had famously identified the interest of the company with those of the country – to hold him accountable for the devastation which the outsourcing of production into global supply chains had wrought on Detroit and Flint, Michigan. In terms of genre, Moore's documentary is also a comedy, with the filmmaker's repeated failure to secure a meeting with Smith as the main storyline.

101 Selwin, Benjamin. "Poverty Chains and Global Capitalism." *Competition & Change*, vol. 23, no. 1, 2018, pp. 71–97.

102 Available in a restored version on the Korean National Film Archive's YouTube channel: https://youtu.be/tTvVq_aeC6g.

103 Doucette, Jamie. "Minjung Tactics in a Post-Minjung Era? The Survival of Self-Immolation and Traumatic Forms of Labor Protest in South Korea." *New Forms and Expressions of Conflict at Work*, edited by Gregor Gall. Palgrave-Macmillan, 2013, pp. 212–32.

If Moore highlights the impact of outsourcing and the dissolution of the classical factory model on workers in the United States, the focus shifts to working conditions in sweatshops in South Asia and Southeast Asia in the 1990s. Here, documentary journalism, which has been part of the repertoire of media in liberal democracies since the 1920s, provides the narrative templates.[104] An article published in *Harper's* in 1992 showed the pay slip of an Indonesian worker employed by a Korean manufacturing firm that subcontracted for US-based company. The hourly wage was fourteen cents, below even the minimum wage in Indonesia. The US company, as it turned out, was Nike, incidentally one of Maersk's most important global brand customers.[105] The article marked the starting point of a pressure campaign in the United States which ended up becoming a threat to the brand value of Nike and expanded to other industries as well. Making the behaviour of consumers their focus, activists managed to force brand companies to ask subcontractors to move towards minimum safety standards and pay better wages, incidentally a move which gave the global brands and their affiliated subcontractors a competitive edge because their scale allowed them to better absorb the additional costs.[106] Ecological considerations were soon added to concerns about working conditions and became another focus of pressure group activists. Ecological consciousness as it emerged in the nineteenth century was already concerned with conservation and the protection of wildlife and "nature" in a broader sense from the fallout of industrial modernity. But it was only in the 1960s, as the Great Acceleration became a shared experience, that the temporality and aspirational content of critical visibility took shape in debates about ecology and conservation. Rachel Carson's book *Silent Spring* (1962), an indictment of the pernicious effects of pesticides, is a turning point for the United States. It shared the structure of feeling of the wrong turn, impending disaster and approaching point of no return; its impact was significant enough to actually change policy.[107]

104 Stott, William. *Documentary Expression and 1930s America*. University of Texas Press, 1973, chs. 10–13.

105 Greenberg, Josh, and Graham Knight. "Framing Sweatshops: Nike, Global Production, and the American News Media." *Communication and Critical/Cultural Studies*, vol. 1, no. 4, 2004, pp. 151–75; Bies, Robert J., and Jerald Greenberg. "Justice, Culture, and Corporate Image: The Swoosh, the Sweatshop, and the Sway of Public Opinion." *The Blackwell Handbook of Cross-Cultural Management*, edited by Martin J. Gannon and Karen L. Newman. Blackwell, 2002, pp. 320–34.

106 Frost, Stephen, and Richard Welford. "Corporate Social Responsibility in Asian Supply Chains." *Corporate Social Responsibility and Environmental Management*, vol. 13, no. 3, 2006, pp. 166–76.

107 Kroll, Gary. "The 'Silent Springs' of Rachel Carson: Mass Media and the Origins of Modern Environmentalism." *Public Understanding of Science*, vol. 10, no. 4, 2001, pp. 403–20.

From a film studies point of view, Western customers waking up to the depredations at the nadir of the smiling curve and the ecological implications of global value chains can be described in terms of the spectatorship of social documentary. As William Stott writes, social documentary "has an intellectual dimension to make clear, what the facts are, why they came about, and how they can be changed for the better." But it also has an emotional aspect: "feeling the fact may move the audience to wish to change it."[108] More specifically, "feeling the fact" involves what Judith N. Skhlar describes as "passive injustice": the experience of witnessing injustice without doing anything about it.[109] Social documentary creates a normative tension by addressing the viewers or readers as a citizen with a duty to not to remain complicit in passive injustice and act. Part of the reason why this mode of address has come to resonate so strongly and widely in the face of global value chains is that it casts the consumer as an activist who can effect change through choices in consumption, incidentally an attitude in which the enjoyment of one's own moral stance can become a powerful driving force.[110]

One of the outcomes of the broad resonance of critical visibility is that it has become part of the knowledge base and corporate culture of multinational corporations – not necessarily in the active mode of discovery through documentary form, but as a pre-emptive response to the possibility of such discovery. Large corporations have adopted sustainability as one of their publicly professed core values and made corporate social responsibility (CSR) an integral part of their governance and communication strategies, to the point where "sustainability" is now without doubt the most regularly used term in corporate social media campaigns, from chemistry giants like BASF to shipping companies like CMA-CGM and Maersk.[111] For instance, a Maersk video released on Instagram on June 5, 2021, on the occasion of the World Environment Day, outlines the main points of the company's sustainability policy, which includes responsible ship recycling – a testament to the fact that the ship breaking industries of South Asia have entered the realm of critical visibility – decarbonization and zero-dumping policies for ships at sea. A study in 2007 had shown that Maersk's communication about corporate social responsibility was relatively scarce in comparison with a retail giant like Wal-Mart, which

108 Stott. *Documentary Expression*, p. 20.

109 Skhlar, Judith N. *The Faces of Injustice*. Yale University Press, 1990, p. 41.

110 Chouliarki, Lilie. *The Ironic Spectator: Solidarity in the Age of Post-Humanitarianism*. Polity Press, 2012.

111 Hynan, Katherine A., and Anne H. Reilly. "Corporate Communication, Sustainability, and Social Media: It's Not Easy Being (Really) Green." *Business Horizons*, no. 57, 2014, pp. 747–58.

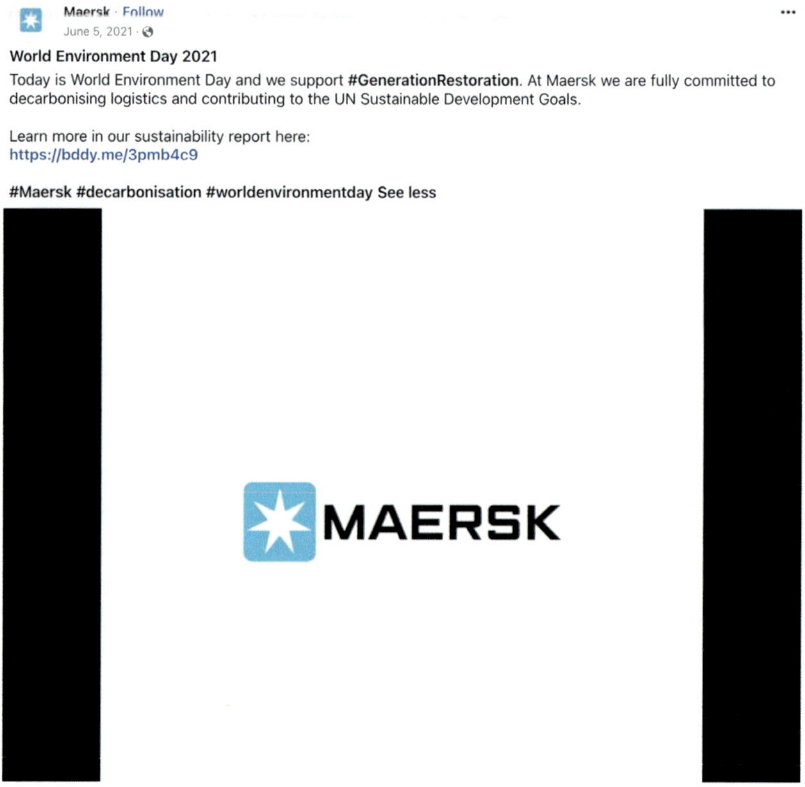

Fig. 1.8. Maersk Instagram video, released on World Environment Day, June 5, 2021.[114]

is understandable considering that shipping is a business-to-business industry; this has clearly changed.[112]

In a webinar from May 2021, Sanjay Vasudevan, who handles key accounts for Maersk, stated that the company responded to pressure from Nike and H&M, who are acutely conscious of their customers' preference for sustainable products and modes of shipping in accelerating the decarbonization of their fleet.[113] Critical visibility, in other words, is now part of the knowledge base of the operators of global value chains, and it can be said to directly shape the infrastructures of global trade. This also connects critical visibility to operative visibility.

112 Kampf, Constance. "Corporate Social Responsibility." *Corporate Communications*, vol. 12, no. 1, 2007, pp. 41–57.

113 https://youtu.be/uzXY_HpbTUM.

114 https://www.instagram.com/p/CPvNSODjTIQ/.

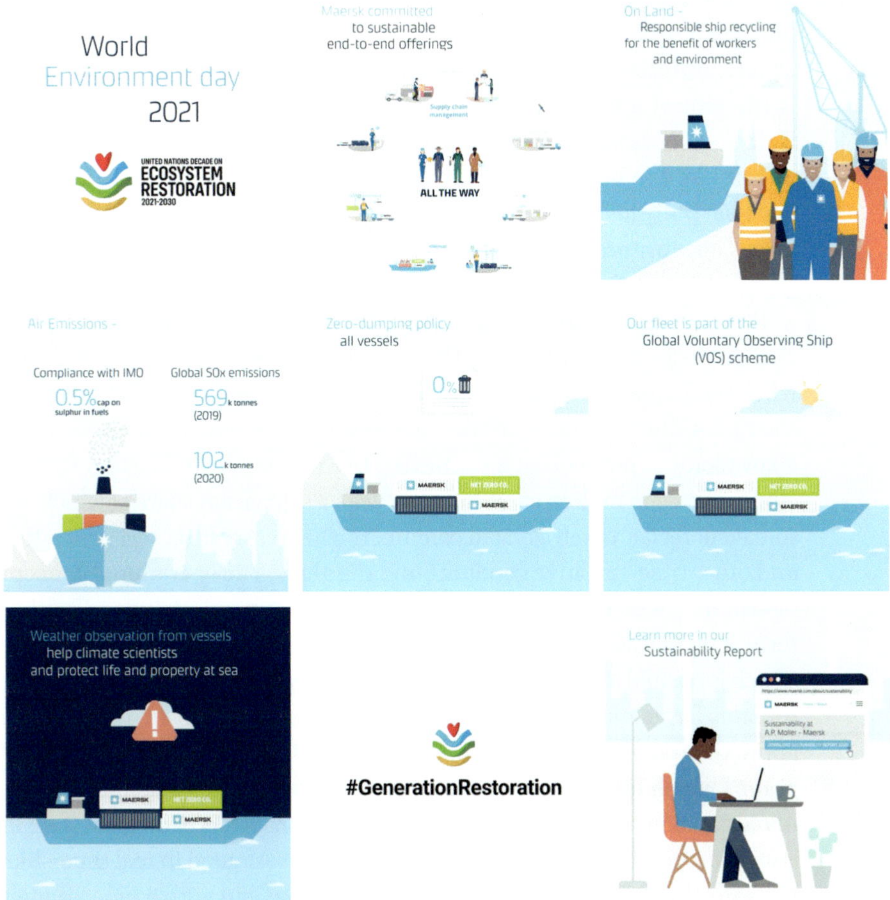

Figs. 1.9a-i. Screen shots from the Maersk Instagram video.

Scene 2 – Operative Visibility: Apprehending the Knowledge Base of the Global Value Chain

The scene of operative visibility is circumscribed by the extension of the global value and supply chains themselves. It is organized through narratives of efficiency, and their aspirational content is the optimization of performance. Like critical visibility, operative visibility is a form of distributed knowledge, but with a more limited scope: it is tied to decision-making along the supply chain and potentially involves all participants in the management of the chain, from cargo owners to port operators, freight forwarders, customs authorities and shipping firms. Getting lost at sea and finding your way home through cunning and skill, the plot of Homer's *Odyssey,* is one of the founding narratives of European subjectivity, understood as a technique of mastery of self and others. Building a civilization from a shipwreck using that technique is the plot of another enduring work of European modern literature, Defoe's *Robinson Crusoe* (even if in the hands of a twentieth-century writer like Michel Tournier the plot can become one of the subject at least partially losing itself to the forces it tries to master). Mastering the forces encountered at sea is also the drama that plays out in the scene of operative visibility.

From an operational perspective, the main problem of global supply chain management is knowing what happens where and when, and whether what was supposed to happen is actually happening in the right place at the right time. The ability to track assets and adjust process if necessary is a key factor of productivity. According to the company's website, Maersk alone moves 120 million containers per year, a significant share of which are so-called "reefer" or refrigerated containers which carry fruit, vegetables, fish and other time-sensitive goods around the globe. The operational concept of visibility, which refers to the ability to track assets, is, to use a Kantian term, a regulative idea. It invokes an ideal state of full information, but one which is posited as achievable. Decision-making in corporate management remains based on incomplete information. Penrose's "image" is always an approximation with a persistent element of uncertainty, as are tables and graphs which offer a view of the situation "at a glance" to help managers make decisions. "Visibility" in global supply chains refers to an ideal state in which it is possible to track every movement of every good in real time –like the ship-tracking function on MarineTraffic.com, but with a much higher resolution, in which ideally each individual container and its cargo can be located at any given moment.

Visibility in this operational sense is inextricably linked to media technologies. Historically, these have included letter writing, the telephone, telex machines, computers and floppy disks and the internet. One of the factors that allowed Maersk to become the world's largest container shipping company was an early investment in communications technology. Mærsk Mc-Kinney Møller was a personal friend of Thomas Watson Jr. and in 1970 became the first non-American to join the board of directors of IBM. Maersk used this connection as a sounding board for decisions concerning the development of the company's data operations, run through its own subsidiary, Maersk Data. The Maersk Telex System, later superseded by the Maersk Communication System, gave the company a matchless ability to

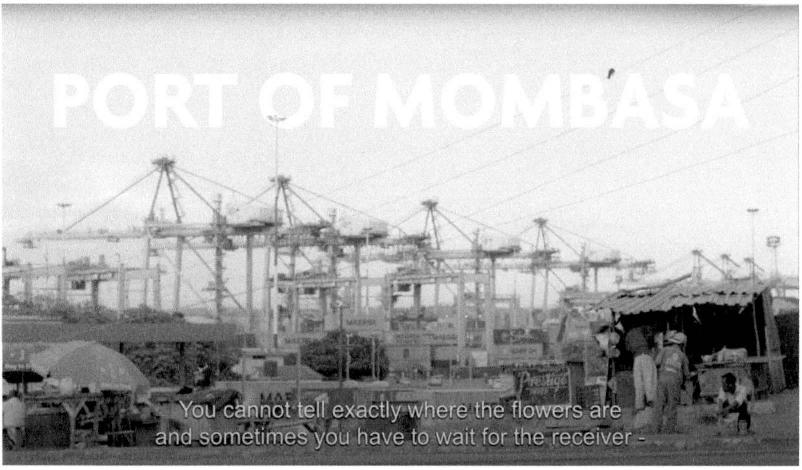

Fig. 1.10. The visibility problem in global supply chains: *The Paper Trail of a Container* (Maersk, 2015).[115]

communicate globally which became a "hidden differentiator" in the 1980s and 1990s.[116]

Even despite computerization and the internet, however, operative visibility, and with it the promise of efficiency, remains aspirational. Three main problems persist: (1) information about a given shipment continues to be stored in different company-specific systems along the supply chain which are difficult to coordinate, which creates delays and additional transaction costs; (2) end-to-end shipments still require about two hundred different documents and certificates to be completed; and (3) infrastructure can break down or operate according to different standards. In the early 2000s, for instance, a Nigerian-Liberian entrepreneur paid teenage boys to track down containers by their numbers with flashlights at night in Lagos' Apapa container port to speed up delivery.[117] As a consequence of these problems of coordination and operation, as late as 2010, seventy per cent reliability in terms of delivery of assets was enough to make Maersk the market leader. For most competitors, reliability was closer to fifty per cent. But the aspiration persists. Maersk's stated goal since the early 2010s was to get to ninety-five per cent, which "would allow our customers significantly to reduce their inventory buffer and increase our lead over our competitors."[118] From 2011 onwards, Maersk tried to achieve this goal by offering its customers

115 https://youtu.be/p8yH4e-Aafk.
116 Jephson. *Creating Global Opportunities*, p. 74.
117 https://youtu.be/mmi8crp7DM4.
118 Jephson. *Creating Global Opportunities*, p. 341.

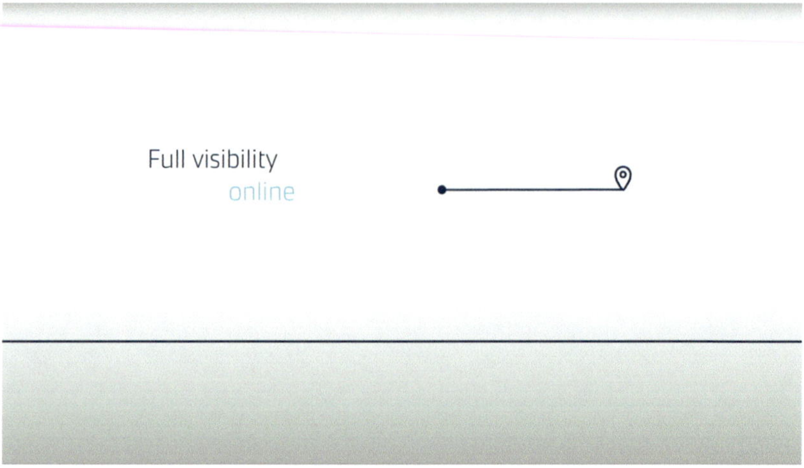

Fig. 1.11. Aspirational concepts: Advertisement for Maersk's Flex Hub, May 21, 2021.[119]

"Daily Maersk," a premium service with higher frequency of shipping and no storage time in ports, but demand remained low, and the service was abandoned in 2015.[120]

Incidentally, the 2008 global financial crisis may have contributed towards an improvement of operative visibility and reliability. In the wake of the crisis, libertarian-minded IT activists and denizens of the dark net moved to develop a new technology which would make financial transactions possible without the intermediary of banks or government institutions. The solution was blockchain, a distributed, immutable ledger designed to keep track of transactions and goods. Blockchains consist of blocks of transaction data linked by a cryptographic hash and marked by a time stamp. The time stamp authenticates the existence of the transaction data at the time of publication in order to obtain its hash. Each block contains the cryptographic hash of the previous block, which makes the chain unalterable and secure by design. Blockchains can be described as a technology for the industrial production of trust. They promise to lower transaction costs, including in particular the time and work invested in creating a bond of trust in conventional business relationships. Potentially, blockchains constitute what economist Jason Potts has described as a "fourth government technology" besides the state, the market and the firm.[121]

119 https://youtu.be/IrDXc-29gqA.

120 Porter, Janet. "Premium 'Daily Maersk' Service Abandoned." *Lloyd's Loading List*, March 5, 2015, https://www.lloydsloadinglist.com/freight-directory/news/Premium-Daily-Maersk-service-abandoned/61956.htm#.YL_YuG5CQXo.

121 Davidson, Sinclair, Primavera De Filippi, and Jason Potts. "Blockchains and the Economic Institutions of Capitalism." *Journal of Institutional Economics*, vol. 14, no. 4, 2017; Hediger, Vinzenz,

While blockchains have so far mostly been used for the mining and circulation of cryptocurrencies, public administrations have made attempts to use the technology to operate land registries, with varying success.[122] Since the mid-2010s, Maersk and IBM have built on their fifty-year alliance to develop a platform called "TradeLens," a label which suggests enhanced visibility and a close-up look at global supply chains. "TradeLens" aims to substitute the physical networks of shipping information with a single platform to which all actors provide data, a platform "populated by whoever is in control of the physical assets," according to TradeLens CEO Michael White. Ideally, on his platform, one set of data replaces the two hundred documents usually required to move a container from source to destination, and automated transaction protocols, so-called smart contracts, to move goods forward. To achieve this, the platform uses "chain code to drive cross-organizational workflow," i.e. to break open the "silos" of firm-specific information systems to produce "one consistent view of the supply chain."[123] TradeLens, in other words, ties together the firm-specific images in the sense of Penrose into an encompassing shared knowledge base which covers the entire supply chain.

Blockchains were originally designed to be "unpermissioned," i.e. accessible to everyone adhering to protocol. The Maersk/IBM platform uses blockchain as a permissioned ledger, which means that it is a proprietary technology operated by an ownership consortium, in this case Maersk and IBM, and accessible only to accredited users. However, with a view to establishing the platform as an industry standard, Maersk and IBM have made it available to their competitors. As of May 2021, CMA-CGM and MSC – which together with Maersk represent close to fifty per cent of world container traffic – had signed up, and TradeLens had been made available to Chinese users through China Unicom, the PRC's telecoms provider.[124] This created significant pressure on state actors, including customs authorities and other holdouts, to join. Whether blockchain technology will solve the problem of operative visibility in the long run remains open. However, a

and Jason Potts. "Die vierte Regierungstechnologie: Über Blockchain." *Zeitschrift für Medienwissenschaft*, no. 18, 2018, pp. 73–86.

122 McMurren, Juliet, Andrew Young, and Stefaan Verhulst. "Addressing Transaction Costs through Blockchain and Identity in Swedish Land Transfers." GovLab, 2018, https://blockchan.ge/blockchange-land-registry.pdf.

123 https://youtu.be/Xwqo_fwPEJo.

124 Maersk. "TradeLens Available in China through Collaboration with China Unicom Digital Tech." Press release, May 10, 2021, https://www.maersk.com/news/articles/2021/05/10/tradelens-available-in-china-through-collaboration-with-china-unicom-digital-tech.

platform like TradeLens is another step towards what Niels P. Peterson, Stig Tenold and Nicholas J. White propose to call "deterritorialization" through containerization: a development in which shipping companies "challenge the nation state in ways and to a degree that would be difficult for companies in more location-bound industries" by weakening and ultimately severing "the link between economic activity and the national jurisdiction."[125] Incidentally, a sense of loss of national jurisdiction over economic activity is one of the sources of populist movements which aim to "take back control". In that sense, the scene of operational visibility is no less a political site than that of critical visibility.

A much simpler platform like MarineTraffic.com, which is a low-cost bricolage of existing technologies (GPS, Google Maps), similarly speaks to the aspiration towards full operative visibility. But it is also a platform that connects to the third scene, representational visibility, by tracking ships and simultaneously hosting documents of another, more passionate form of tracking: ship spotting.

Scene 3 – Representational Visibility: The Politics of the Form of Global Trade

The scene of representational visibility is largely one of positive emotions: marvel at technology, confidence derived from hardships overcome, optimism derived from a sense of opportunity for all. Their narratives of growth, sustainability and inclusiveness address social and political concerns but are mostly driven by a seductive promise as "most people want to see more growth," to quote Vaclav Smil.[126] The audiovisual formats which make up the scene of representational visibility range from amateur films made by ship spotters, on-board footage shot by crew members with GoPro and other cameras to corporate films of container ships and automated ports and other elements of the global supply chain infrastructure. It is a graded field. Amateur footage is now often repurposed by corporations as part of their official communication, which is possible not least because the technical standards of amateur filmmaking, including the availability of drone cameras, are now so high that the difference in image quality between officially commissioned films and repurposed amateur footage has become negligible. Ship-spotting videos as a form of the enjoyment of what David Nye has called the "technological sublime" deserve their own longer

125 Peterson, Tenold, and White. *Shipping and Globalization in the Post-war Era,* p. 264.
126 Smil. *Growth,* p. x.

study.[127] In the following I want to focus on the corporate videos of Maersk instead, because of their volume and quality, but also because they offer a consistent narrative of growth, diversity and opportunity, which provides the overall structure of feeling for the scene of representational visibility. The primacy of trade over politics, which is implicity in operational visibility, becomes explicit in representational visibility: Ultimately, the promise of the infrastructure of container shipping is that access to frictionless trade will solve problems otherwise assigned to the realm of politics.

Maersk and the Aestheticization of the Space of Global Trade

In the wake of the spread of internet-capable personal devices, or smart phones, in the early 2010 Maersk became the first major shipping company to develop an intensive social media presence. As the company's director of social media, Samantha Adeluwoye, puts it, Maersk benefited from early mover advantage, Maersk's amibition was to to set a new standard in corporate communication for multiple audiences in an industry in which business-to-business relations prevail.[128] In the terms of this essay we can read this as an indicator that critical visibility had indeed become part of the knowledge base of the company: In a digital media environment which empowers individual users to make political statements and lowers the threshold for activism it becomes imperative even for a non-consumer brand like a logistics provider to cultivate a positive image. At the same time social media image work facilitates recruitment, an important concern for a global logistics provider. For instance, smart phone penetration in India, the Philippines, and Indonesia, three important source countries for low cost labor and services in the logistics industry, all were at or above 75% in 2022.

Reflecting the multiple target audiences the films, photographs, animation and texts Maersk publishes on their social media accounts (Facebook, YouTube, Instagram, Twitter) run the gamut of corporate communication formats from image films to historical materials and instruction videos for customers and employees, but also includes non-commissioned content submitted by crew members. Overall, Maersk's published content can be sorted according to the three Rs of corporate communication[129]: (1) *record*,

127 Nye, David. *American Technological Sublime*. MIT Press, 1996.

128 Interview with Maersk Global Social Media Content Manager Samantha Almon Adeluwoye, May 11, 2021.

129 Hediger, Vinzenz, and Patrick Vonderau. "Record, Rhetoric, Rationalization." *Films That Work: Industrial Film and the Productivity of Media*, edited by Vinzenz Hediger and Patrick Vonderau. Amsterdam University Press, 2009, pp. 35–50.

i.e. materials designed to document the operation of the company – these include films and segments about the history of company and container shipping, films about life at sea on Maersk ships, snapshots and films from company employees; (2) *rhetoric*, materials designed to induce cooperation from employees, shareholders and stakeholders – these include videos and animation addressing stakeholders like the Environment Day Instagram video, recruiting videos featuring successful employees telling their life stories or giving inside looks of Maersk's training programmes, videos featuring employees explaining the meaning of Maersk's core values to them personally, webinars featuring upper management explaining company policies like decarbonization; and (3) *rationalization*, i.e. materials designed to facilitate and optimize operations – these include instructional videos explaining the use of online tools addressed to customers and employees, e.g. videos introducing Filipino sailors to a new online registration tool which spares them long trips back to port which take time out of their family vacation, webinars about the introduction of TradeLens etc.

Maersk's audiovisual work stands out in comparison to the corporate films of its competitors such as Evergreen Lines in style and mode of address. The corporate image film of Evergreen from 2019 features multiple aerial tracking shots of ships in movements, containers on tractor trailers traversing landscapes, Asian crews at work and stock shots of a lily-white family of three generations celebrating a children's party at a table in a garden, all tied together with an animation of golden sparks moving through space, a visualization of the free flow of goods through container shipping. It is a competently made, smooth corporate image film remarkable not least for the suggestion that the primary beneficiaries of container shipping are upper-middle-class white people.[130] By contrast, Maersk films tend to tell stories of trade evolving around individual company employees and customers. They emulate the templates of documentary filmmaking rather than the tropes of more conventional corporate films. In terms of their tonality and visual finish they are clearly superior, for instance, to those of other multinational corporations like chemistry giant BASF.[131] And some of Maersk's films which can be found online are closer to documentary art films than corporate videos. For instance, one video entitled *On Board*

130 https://youtu.be/L6lVAMJzRGo.

131 See also Hediger, Vinzenz, and Florian Hoof. "The Agency of the Agency: Why Do All Corporate Image Films Look the Same?" *A History of Cinema without Names*, edited by Diego Cavallotti, Federico Giordano, and Leonardo Quaresima. Forum, 2016, pp. 209–18.

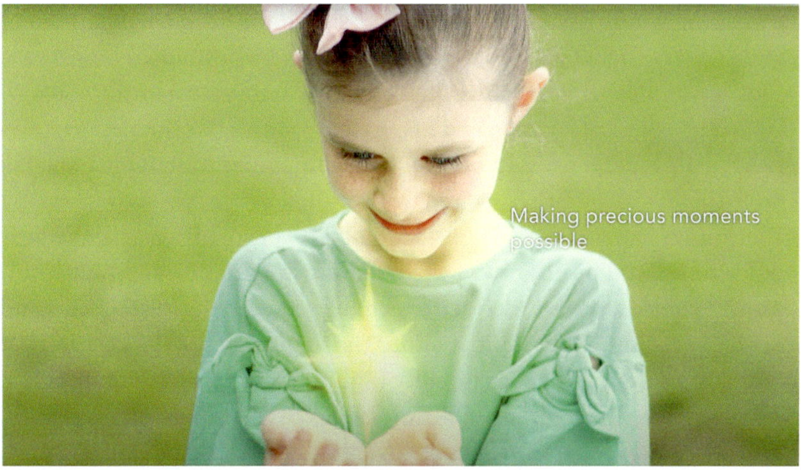

Fig. 1.12. Evergreen Line corporate video, 2019.

Edith Maersk consists of a fifteen-minute series of long takes of footage showing various aspects of life on board. Excerpts from these long takes show up as short clips in other corporate videos, but presented uncut, in sequence and without commentary and music, they are reminiscent of a film by James Benning.[132]

This difference in style can partly be explained by national origin. Among other things, Maersk is a Danish design project.[133] Organization has "always been an aesthetic phenomenon," writes Timon Beyes. But in the face of "the ubiquity of digital media and its corresponding devices that influence what and how we perceive and how we express or perform ourselves" organization is perhaps more than ever "tied to processes of aestheticization," including the aestheticization of space and built environments.[134] Maersk, however, has been aestheticizing the spaces of business since long before digital media became ubiquitous. It starts with the company's corporate identity. The logo springs from an episode in company lore when the company founder was lost at sea and saved by a seven-pointed star. The logo also refers to the seven seas and the seven

132 https://youtu.be/PHTgkVMoook.

133 Sornn-Friese, Henrik. "'Containerization in Globalization': A Case Study of How Maersk Line Became a Transnational Company." *Shipping and Globalization in the Post-war Era: Context, Companies, Connections*, edited by Niels P. Peterson, Stig Tenold, and Nicholas J. White. Palgrave Macmillan, 2019, p. 115.

134 Beyes, Timon. "Art, Aesthetics, and Organization." *A Research Agenda for Management and Organization Studies*, edited by Barbara Czarniawska. Edward Elgar, 2016, p. 115.

days of the week. Since 1955, all Maersk ship hulls have been painted in the distinctive light blue colour, which makes them instantly recognizable. Maersk was also the first shipping company to impose a uniform office design regardless of location and inspired by Danish interior design styles of the 1950s and 1960s.[135] Industrial film companies from which Maersk commissions corporate films receive detailed briefs specifying, among other things, the colour scheme with a palette of colours geared towards cooler tones and built around the company's signature light blue hull paint.[136] There is a clearly defined look to be achieved during shooting and in colour grading and post-production. The importance of the logo and the company's colour scheme is directly highlighted in the corporate films. The corporate films produced in 2015, which exist in continent-specific variations targeted at European, African and Asian audiences, opens with a montage of views of urban landscapes, travelling container ships and containers traversing landscape on tractor trailers cut together with a close-up view of a painter refreshing the paint on the company logo on one of the ships. Later, a similar montage involves shots of workers finishing the paint on a new container in Maersk's reefer container factory.[137] The company's motto and "core value" is "constant care," an exhortation reportedly handed down from company founder A.P. Møller to his son, Mærsk Mc-Kinney Møller, in a letter in 1946.[138] It is a motto that clearly also a applies to the company's aestheticization of space.

Time Lapse and Travelling Shots

Apart from the aerial drone tracking shots of moving ships and containers moving on trains and trucks, which can be read as an epitome of the image of the global value chain, one of the visual tropes recurring regularly in Maersk's corporate films – but also in many other contemporary corporate image films – is the time lapse shot of urban traffic nodes.

Globalization is often defined as the free circulation of people, goods, and ideas. In a world marked by the twin dynamics of global value chains and urban migration to fast-growing megacities, cities are the "factories of the future," as Richard Baldwin argues: urban space supersedes the factory

135 Jephson. "Creating Global Opportunity," p. 233.

136 Interview with Maersk Global Social Media Content Manager Samantha Almon Adeluwoye, May 11, 2021

137 https://youtu.be/k2IVNBCV4sc.

138 Jephson. "Creating Global Opportunity," p. 23.

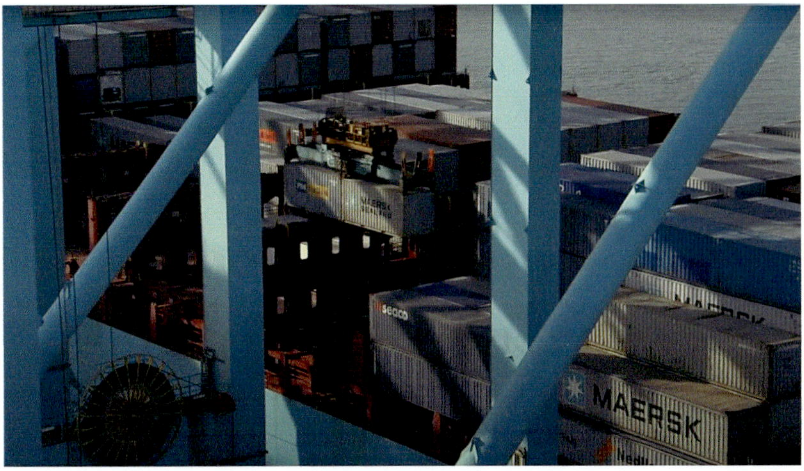

Fig. 1.13. Aerial tracking shot of container being unloaded from a Maersk ship with converging movements of drone and container. Maersk corporate video for Europe, 2011.[140]

as the site where the lines of economic activity converge and intersect.[139] If the image of workers leaving the factory can be read as a visualization of the workforce in the classical industrial era, then time lapse shots of urban traffic modes capture the productive energy of the global value chain economy in an image of unimpeded mobility. Often framed as a slow-moving tracking shot, these images of mobility are also images of layered temporalities. The time of the factory was ordered, hierarchical and homogeneous, organized in shifts and structured by the time clock. The factory's temporal regime was one of regularity and density, of a methodical life and constant activity.[140141] The shots of urban traffic nodes transfer these ideas of regularity and density to urban space, even as they suspend the spatial boundary of the factory. This is particularly true of tracking shots, which emphasize the space beyond the frame through camera movement. The same combination of suggestions of regularity and density in movement through space rather than in the closed site of the factory can be found tracking shots of ships, containers and other goods in movement. But the urban traffic shots usually involve people and vehicles moving at different speeds in the same space. Underscoring the different layers of temporality, Maersk corporate videos regularly include

139 Baldwin, *The Great Convergence.*

140 https://youtu.be/vD05HCJscio.

141 Snyder, Benjamin H. "From Vigilance to Busyness: A Neo-Weberian Approach to the Time Clock." *Sociological Theory*, vol. 31, no. 1, 2013, pp. 243–66.

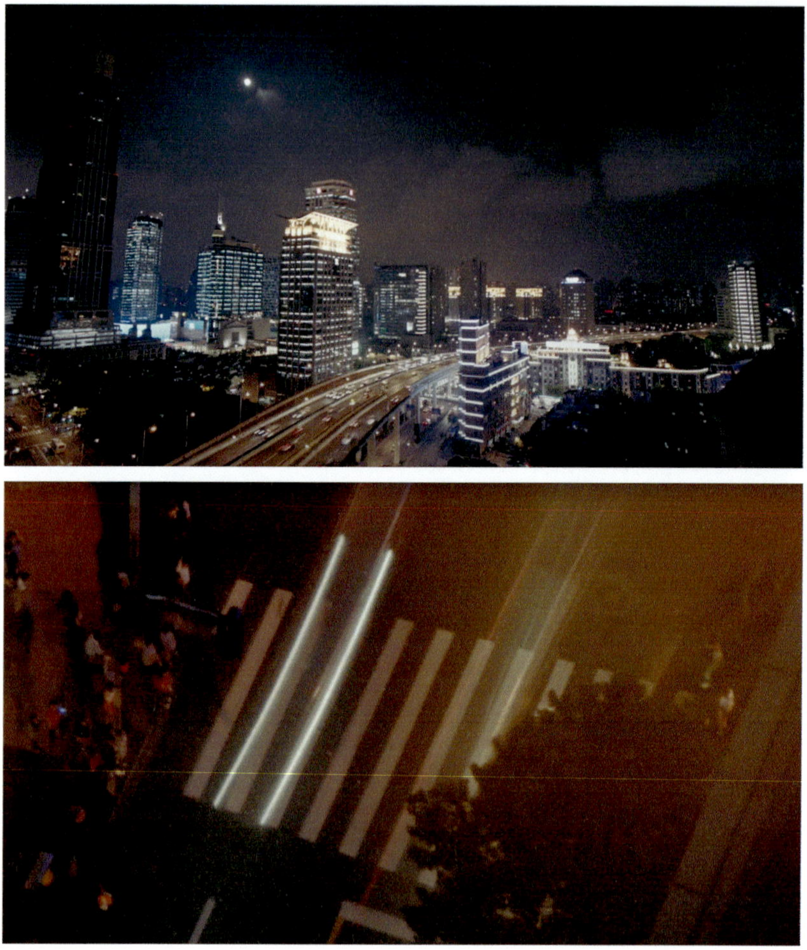

Figs. 1.14a and 1.14b. Time-lapse shots of urban traffic flows, from *We Are Maersk* (2015).

shots of vendors in local markets, usually women, with containers, which function as simultaneously concrete and abstract signifiers of constant activity and methodical life, moving past through urban space on tractor trailer in the background.

It is important to note that the factory space has not simply disappeared from corporate films. Quite to the contrary: Scenes of production, whether inside views of textile factories or harvest scenes on fruit plantations, feature regularly in these films. But they are framed by time lapse and tracking shots of people and goods in movement and figure as mere chain links in a broader order of limitless circulation.

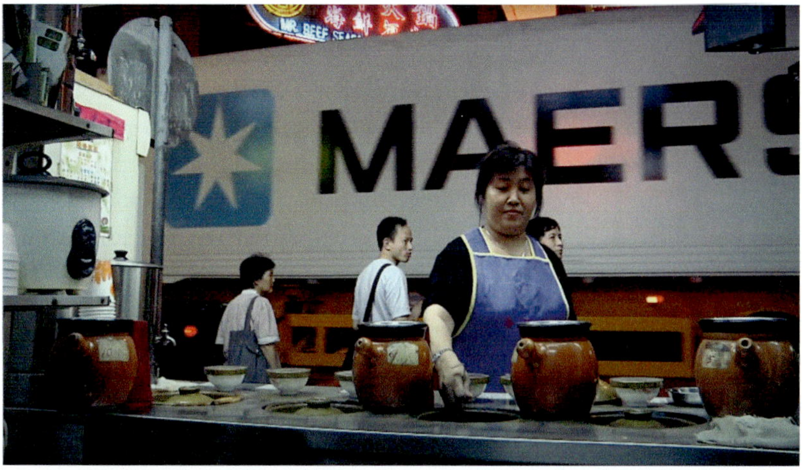

Fig. 1.15. Multi-layered temporalities: Market vendor and Maersk reefer container in Hong Kong, 2011.[142]

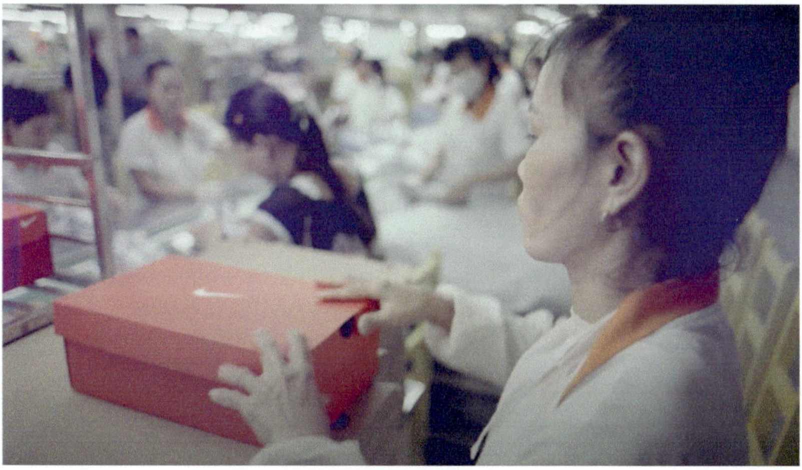

Fig. 1.16. Inside a Nike factory in China, from *We Are Maersk* (2015).[143]

Temporalities of Trade and the Politics of Infrastructure

Taken together, the time lapse shots of traffic nodes, the shots of vendors with traffic in the background and the factory shots in combination with tracking and time laps shots speak to "layers of temporal interdependency as a particular form of power," as Sarah Sharma writes, a power that also

142 https://youtu.be/-JOKwS64mRU.
143 https://youtu.be/k2IVNBCV4sc.

"plays out in private life." Rather than deploring the loss of a coherent public space, Sharma argues for a "power-chronography" which treats "all social spaces as transit spaces," an approach which "allows for recognition of the temporal contingency of political and public life."[144] It can be argued that what can be described as the documentary quality of Maersk corporate films lies in how they acknowledge such temporal contingency and articulate it by focusing on how the temporal interdependencies created by global trade play out in the private lives of the company's employees. Maersk's film work on ports and shipping in Africa are a case in point. Where the Evergreen corporate film shows ship crews at work and a white middle-class family connected by sprinkling animated stars, in one video Alphonse Bea, an executive for Maersk's freight forwarding service DAMCO working in Cameroon, talks about how he explains to his colleagues how geography and politics make logistics in Central Africa entirely different than in North America and Asia, fracturing the idea of one world held together by the infrastructures of global trade.[145] And when Maersk films discuss infrastructure, they argue in structural, but also in personal terms. Films about the container terminal which APM, Maersk's port-operating subsidiary, built in Monrovia, Liberia, after the end of the second civil war in 2003, focus on how the port transforms the life of employees, with Jacqueline Paye, a single mother who works as a crane operator, appearing at the centre of two films from 2014.[146] Numerous films – for instance, an entire series of short case studies entitled *The Heart of Trade* from 2017 – focus on case stories of small entrepreneurs and particularly women entrepreneurs from Africa succeeding in South-South trade thanks to access to Maersk's services.[147]

However, as much as power relations and differential temporalities are openly addressed in these films, it is ultimately coherent story of sustainable growth and inclusion through global trade and its infrastructures which brackets these films together and gives them a narrative coherence which matches their trademark visual style. Diversity and inclusiveness are a key element of that story. In May 2021, for instance, Maersk sent a number of containers with a rainbow finish on trips around the world as a statement in favour of equality and asked ship spotters and staffers to send in photographs

144 Sharma, Sarah. *In the Meantime: Temporality and Cultural Politics*. Duke University Press, 2014, p. 148.

145 https://youtu.be/iD7x5QOUk1k.

146 https://youtu.be/-MzEcvmmX9s; https://youtu.be/2jJaltdPO8k.

147 See for instance https://youtu.be/6ivPtgfKHeM; https://youtu.be/fHXrCRHGBJc; https://youtu.be/btcVnNCwnrM.

A.P. Moller - Maersk
1,347,863 followers
1yr • Edited •

+ Follow

Watch our rainbow container touring the beautiful tulip garden on Denkers BV truck of our customer Jan de Wit en Zonen B.V. in the Netherlands, "This week we gave our cooperation to Maersk's Rainbow Containers World Tour! 🌈 The containers were photographed at our bulb fields and at our tulip show garden. These containers represents Maersk's position on diversity and inclusion. The rainbow containers are also used for humanitarian work, such as the supply of aid freight to survivors of hurricanes in the United States. Jan de Wit en Zonen B.V. is very proud to be a part of Maersk's devotion to diversity and inclusion."

Thank you Jan de Wit en Zonen for your support and commitment to equality.

If you spot one of our rainbow containers on their voyage, take a picture and share on social media platforms using the #InclusionAllTheWay.

#Maersk #rainbowworldtour #diversityandinclusion #togetheralltheway #equalitymatters #rainbowcontainers

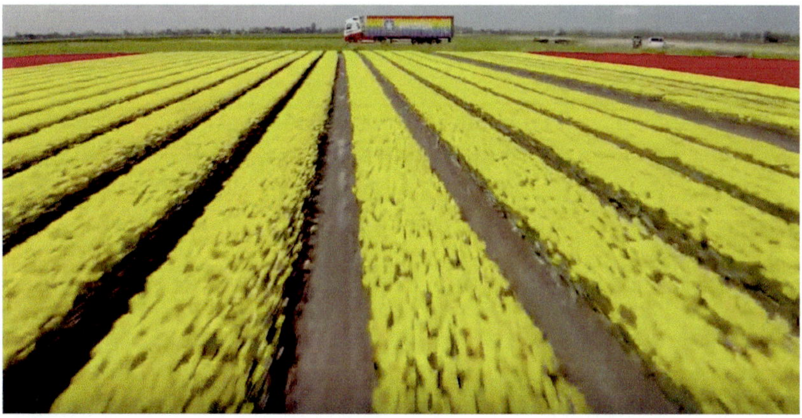

Fig. 1.17. A Maersk rainbow container touring the tulip fields outside of Amsterdam. Maersk Instagram account, May 19, 2021.

of the containers in different surroundings to the company's Instagram account. Corporate videos always take great care to stress the diversity and inclusiveness of Maersk's hiring policies by foregrounding local staff in Africa and Asia. Maersk's social media director, Samantha Almon Adeluwoye, is a London native and a person of colour, who appears as one of the faces of the company in their corporate communication

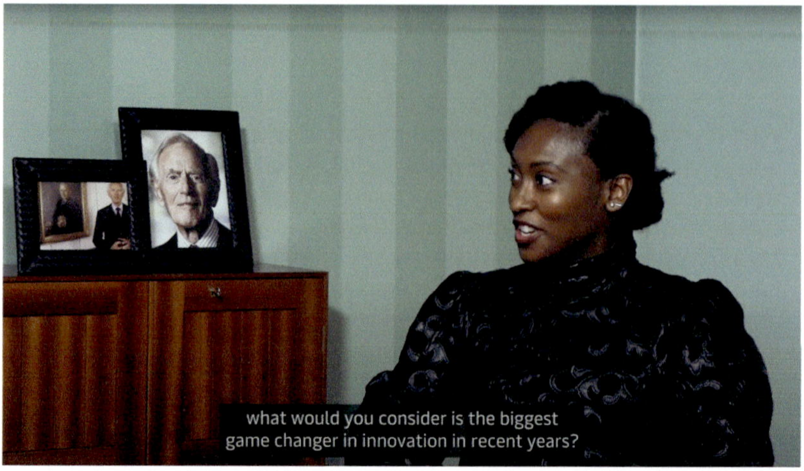

Fig. 1.18. Samantha Almon Adeluwoye interviewing Maersk CEO Søren Skou on innovation at Maersk under the watchful eyes of Mærsk Mc-Kinney Møller, 2020.[148]

On occasion, Eurocentric social and racial hierarchies still shine through, however. Two seemingly contradictory but equally inadequate perceptions about Africa – a continent in misery and in need of help, and "Africa rising", a continent full of promising start-ups which only lack proper logistics – remain part of the rhetorical arsenal of Maersk films. Another example is a Discovery channel documentary about the building of the Triple E 18,000 TEU ships at Daewoo shipyards in South Korea, narrated by a journalist with a Scottish accent, which only features Danish engineers as talking heads. It includes a montage in which a Danish engineer apparently threatens Korean staff with a shutdown of operations because of delays and, so it is implied, incompetence. Considering that this was filmed and edited at one of the world's most successful and productive shipyards shortly after Maersk had shuttered its own yard and shifted all its shipbuilding to East Asia, there is a whiff of imperial nostalgia in this montage. It is obviously an editorial choice of Discovery channel, but Maersk still put the segment on their YouTube channel.[149] Overall, however, a carefully curated narrative of diversity and inclusion prevails.

An important feature of the documentary aesthetic of Maersk corporate films is the use of experts as talking heads, particularly development economists from the region in question. They serve to provide scientific credentials to the narrative of growth and opportunity through infrastructure which

148 https://youtu.be/iThr7hI34-g.
149 https://youtu.be/RUfGCngF9DI.

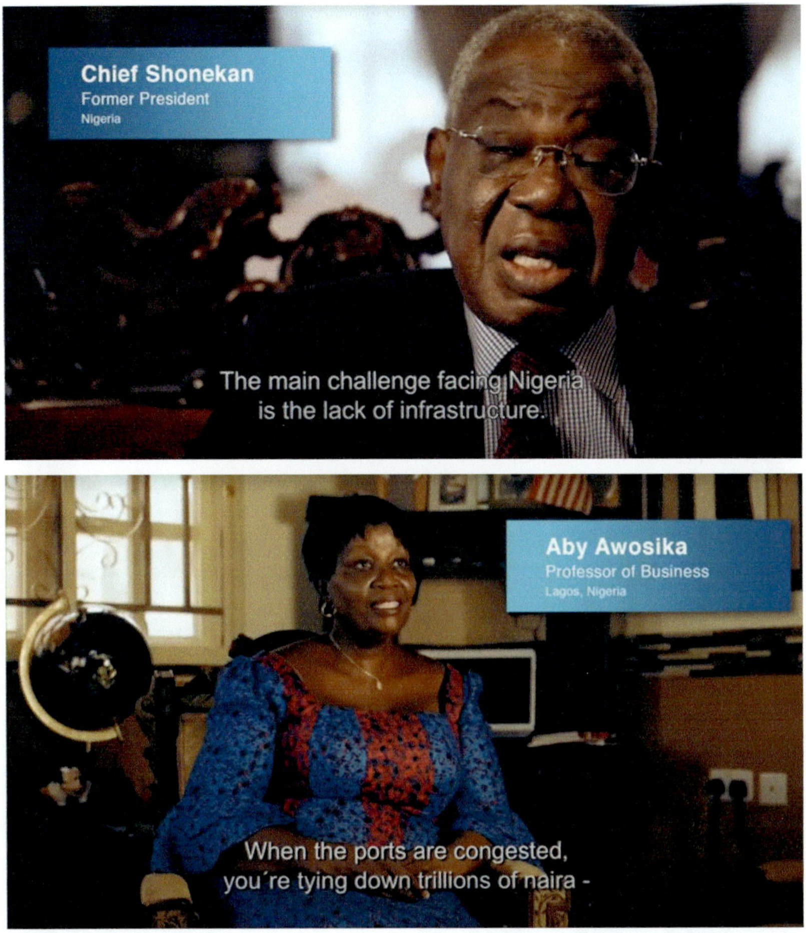

Figs. 1.19a and 1.19b. Economic experts: Former president of Nigeria Ernest Shonekan and Professor Aby Awosika explain Apapa container port.[150]

the films propound. A ten-minute documentary on Apapa container port in Lagos from 2011, which APM took over from the Nigerian government in 2006 and which is now the largest container terminal in West Africa, is entitled *Apapa Terminal's Impact on Nigerian Society* and features analysis from Ernest Shonekan, a Harvard MBA and former president of Nigeria, and Professor Aby Awosika. A 1966 film directed by Carlo Audisio, a veteran Italian industrial film director who also directed, among other things, a

150 For another example, see independent economist Eduardo Garcia explain the shift from Volkswagen to Kia in the Mexican automobile industry: https://youtu.be/n53_-N-e_1g.

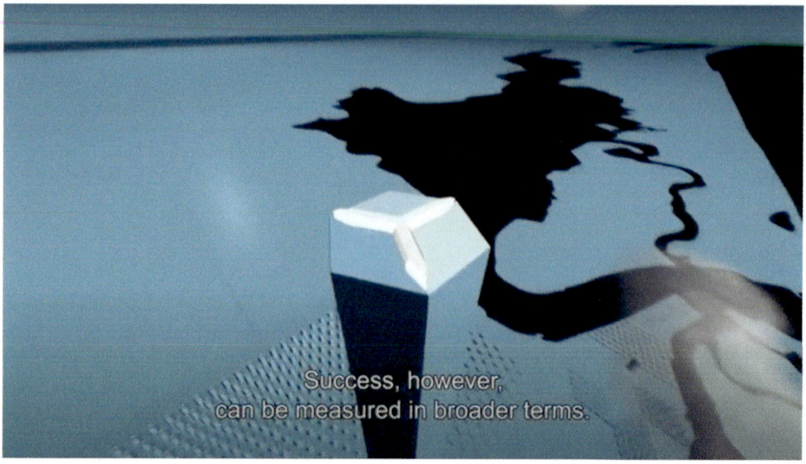

Fig. 1.20a. Computer graphics locate growth at the point of congestion: Map of Lagos with Apapa container port in the foreground right. APM corporate video, 2011.[151]

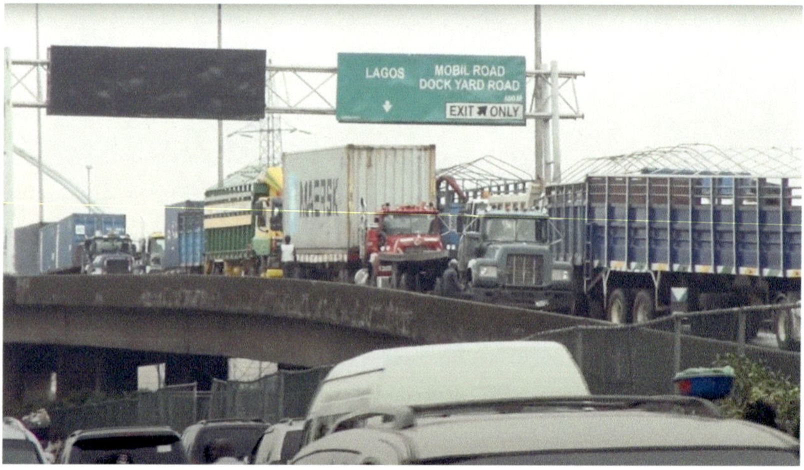

Fig. 1.20b. Congestion at Apapa port, 2018.[152]

theatrical documentary about the production process in the Italian film industry, chronicles the construction of Apapa port by an Italian company, culminating in a unabashedly nationalist celebration of Italian engineering prowess.[153] In the APM film, the nation state is, if anything, an obstacle to progress. The Nigerian government, Shonekan argues, needs to learn how to

151 https://youtu.be/mmi8crp7DM4.
152 https://youtu.be/sQ9HGJdanWk.
153 https://youtu.be/CTHrUxGdUmQ.

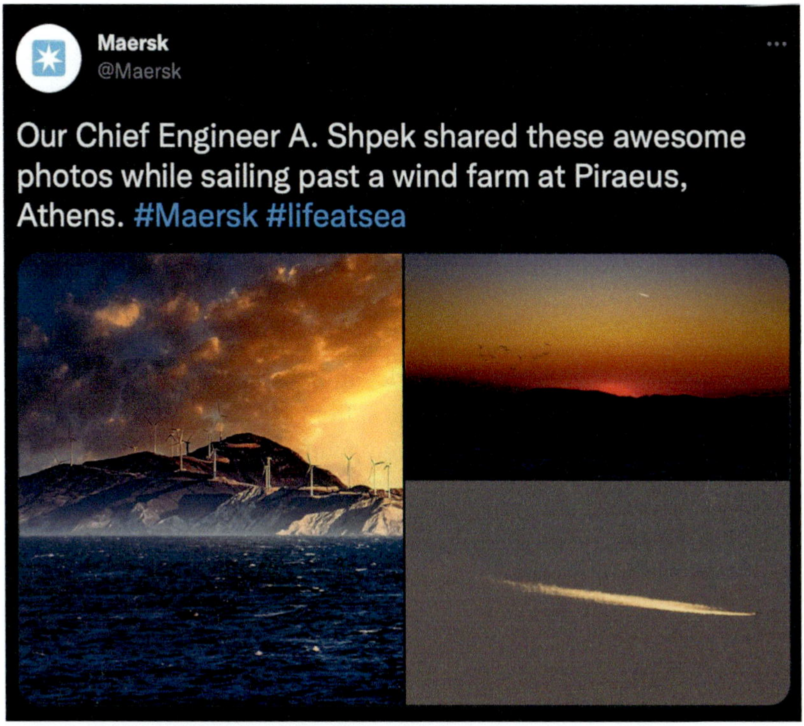

Figs. 1.21a and 1.21b. Sharing the thrill of life at sea via the Maersk twitter acount.

get out of the way of private business. The film also exemplified the tension between the promise and actual operation of infrastructure. An animated info graphic places a 3D growth column almost exactly at the hot spot of congestion on the freeways to and from the port. Congestion, as Brian Larkin notes, is almost a cliché of the current state of Lagos' urban infrastructure, but it is an enduring problem in Apapa, and one the film also talks about.

In exemplary fashion this film lays out how the contemporary narrative of infrastructure and global trade differs from the classical modernization narrative which, as Larkin summarizes it, posits that "through infrastructures and circulation [...] development and modern subjectivity can be achieved." First, the contemporary narrative assumes that subjectivities are fully formed and do not need to be developed; however, to realize their full potential they require an acknowledgment of diversity, a commitment to inclusiveness and proper infrastructure. Second, enterprise is superior to the nation state in providing, maintaining and improving infrastructure, particularly considering the transnational character of industrial organization in the age of global supply chains. And third, and this is the modified promise of infrastructure, once opportunity is guaranteed, i.e. once the infrastructure is in place and operative and entrepreneurs can fully realize their potential, growth, prosperity and equality will materialize as if automatically and on a global scale.

Raymond Williams argues that the dominant images of society in modernity tend to reduce society to two spheres of interest and versions of social relationship: "politics (the system of decision) and economics (the system of maintenance)."[154] Representational visibility of global supply chains, one could argue, presents an image of society in which politics is subsumed by economics: An image of society as a boundless firm, in which the proper operation of the global supply chain is the only political problem that remains to be addressed, because though it all other political problems will be solved. In that sense, the politics of the form of global trade can be summarized by a Maersk Instagram post from June 6, 2021: "Let trade take care of it and make a wish upon a seven-pointed star."

Conclusion: No Fragrance, Just Technology

In Michael Miller's history of shipping networks a freight forwarder reminisces about the challenge of loading and unloading ships and getting cargo to destination in the time before containerization:

> I belong to the very old guys who had really fun at the time you had general cargo because you had to work with your brains. Now "it's all container stupidity, it's just a box."[155]

Another of Miller's informants says that there was "nothing like something, an elephant or something, being swung on your ship," while yet another one points out "the irony that ports had once smelled of the world, but that in this moment of globalization, there was no fragrance, just technology."[156] Considering that containerization is routinely described as a revolution (to the point where the Wikipedia entry for "container revolution" redirects to "containerization"), these statements recall Talleyrand's famous dictum about the *ancient régime*: "He who did not live in the years before the Revolution cannot understand what the sweetness of living is."

But is containerization really a revolution? With an eye to the early investment of British legacy lines in containerization, which mainly aimed at maintaining their grip on the Commonwealth, Nicholas J. White argues

154 Williams. *The Long Revolution*, p. 138.
155 Miller. *Europe and the Maritime World*, p. 371.
156 Ibid., p. 372.

that for all "that talk of a 'container revolution' containerization was not that revolutionary."[157] But then revolutions may actually take longer than we assume and happen not where and how we expect to see them.

Apart from Hollywood, the second main factor of cultural globalization after the First World War was the Russian Revolution, which gave birth to the idea of the world revolution. Revolutions, dating back to the Glorious Revolution of 1688, which pitted an absolutist against a constitutional model of governance, can be described as – usually violent – confrontations between competing projects of modernization.[158] In that sense, the French Revolution can be seen as a contest between absolutist centralism, which had delegitimized landed nobility, and a democratic society based on a notion citizenship regardless of status or origin.[159] The Cold War, a decades-long contest between competing projects of modernization at a global scale, can similarly be seen as an extended revolutionary situation. Stalin gave up the Cold War was an extended revolutionary situation. Stalin gave up on the world revolution in 1943 to please his American allies and consolidate power at home, but the idea of world revolution found a second iteration in Maoism, a driving force of decolonization in the 1950s and a global political and cultural movement in the 1960s.[160] As the Cold War ended, the liberal democratic system of governance combined with the market economy seemed to emerge triumphant. The neoconservative project of "democracy promotion" was designed to take that victory to a truly global level, with the Middle East as its point of attack. Its historical novelty was that it administered revolutionary violence on a large scale through the United States military. But the neoconservative revolution foundered in Iraq, even as the universalist ambition of market-based liberal democracy ran up against what Dani Rodnik calls the "globalization paradox," i.e. the problem that of frictionless global trade, democracy and national sovereignty you can only have two at a time and not all three (a problem which representational visibility solves by taking national sovereignty out of the equation).[161]

The concept of "world revolution," then, is part and parcel of the history of globalization in the twentieth century, as is the failure of its successive

157 White. "Thinking Outside 'The Box'," p. 64.

158 Pincus, Steven. *1688: The First Modern Revolution*. Yale University Press, 2011.

159 Lefort, Claude. *L'invention democratique. Les limites de la domination totalitaire*. Fayard, 1994.

160 Lovell. *Maoism*.

161 Rodrik, Dani. *The Globalization Paradox: Democracy and the Future of the World Economy*. W.W. Norton, 2010.

iterations. As we have seen, at some crucial points containerization intersects with the neoconservative project, and it aligns with the neo-Maoism of current PRC president Xi Jinping. But there is a conceptual connection as well, i.e. one of form. For the proponents of what I propose to call critical visibility, containerization is a rather depressing story of enhanced exploitation and looming economic crisis, which may well lead up to a violent confrontation and transformation of the current social order, i.e. a revolution in the classical sense. The container industry itself tried out several other narratives, starting with the one which made the logistics of American settler colonialism its central metaphor in the 1950s and which is encapsulated in the name of the very first purpose-built container ship. Over the last decade in particular and led by Maersk, the industry has settled instead on the uplifting story about opportunity, welfare and growth in which trade supersedes politics. This, in fact, is a narrative which places containerization alongside the other grand twentieth-century projects for radical social and political change as their more palatable, non-violent alternative.

To quote Raymond Williams once again: "If it is pointed out, in traditional terms, that democracy, industry, and extended communication are all means rather than ends, I reply that this is precisely their revolutionary character, and that to realize and accept this requires new ways of thinking and feeling, new conceptions of relationships, which we must try to explore."[162] In that sense, the container industry's own "promise of infrastructure" can be described as as structure of feeling tied to an optimistic practical philosophy of history in which efficiency in logistics holds the promise of revolutionary outcomes without violence – or rather without violence other than the economic violence of what Schumpeter calls "creative destruction."

But at the same time, the primacy of trade over politics and the progressive removal of economic activity from national jurisdiction, which are the end state of operational visibility and supported by what this contribution has called representational visibility, help set the stage for another, populist and reactionary revolution, not the control revolution, but the "take back control" revolution. How much, and which kind of revolution is containerization really? To quote former PRC prime minister Zhou Enlai, who said in 1974 about the French Revolution (some say that he referred to the one of 1789, others that of May '68): "Too soon to tell."[163]

162 Williams. *The Long Revolution*, p. 14.
163 https://www.oxfordreference.com/view/10.1093/acref/9780191826719.001.0001/q-oro-ed4-00018657.

Works Cited

Acland, Charles R., and Haidee Wasson, editors. *Useful Cinema.* Duke University Press, 2011.

Agerdal-Hjermind, Annette. "The Enterprise Social Media Strategy: The Case of Maersk." *Communication & Language at Work*, no. 3, 2014, pp. 3–17.

A.P. "A Plot on Onassis Charged in Article." *New York Times*, August 3, 1978, https://www.nytimes.com/1978/08/03/archives/a-plot-on-onassis-charged-in-article-nixon-burger-and-a-hughes-aide.html.

Bakker, Gerben. "The Quality Race: Feature Films and Market Dominance in the US and Europe in the 1910s." *The Classical Hollywood Reader*, edited by Steve Neale. Routledge, 2012, pp. 31–42.

Baldwin, Richard. *The Great Convergence: Information Technology and the New Globalization.* Belknap Press/Harvard University Press, 2016.

Bauhaus, Henrik, Frederik Möller, Maleen Stachon, Christina Hoffmann, and Boris Otto. "Data-Driven Business Models in Logistics: A Taxonomy of Optimization and Visibility Services." *Proceedings of the 53rd Hawaii International Conference on System Scienc*es, 2020, p. 5379–88, https://hdl.handle.net/10125/64403.

Beyes, Timon. "Art, Aesthetics, and Organization." *A Research Agenda for Management and Organization Studies*, edited by Barbara Czarniawska. Edward Elgar, 2016.

Bies, Robert J., and Jerald Greenberg. "Justice, Culture, and Corporate Image: The Swoosh, the Sweatshop, and the Sway of Public Opinion." *The Blackwell Handbook of Cross-Cultural Management*, edited by Martin J. Gannon and Karen L. Newman. Blackwell, 2002, pp. 320–34.

Bitterli, Urs. *Die Wilden und die Zivilisierten. Grundzüge einer Geistes- und Kulturgeschichte der europäisch-überseeischen Begegnung.* C.H. Beck, 2004.

Bosdorf, Ulrich, and Sigrid Schneider, "Ein gewaltiger Betrieb: Fabrik und Stadt auf den Kruppschen Fotografien." *Bilder von Krupp. Fotografie und Geschichte im Industriezeitalter*, edited by Klaus Tenfelde. C.H. Beck, 2000, pp. 123–50.

Cairns, George. "Postcard from Chittagong: Wish You Were Here?" *Critical Perspectives on International Business*, vol. 3, no. 2, 2007, pp. 266–79.

Carriou, Pierre, Francesco Parola, and Theo Notteboom, "Towards Low Carbon Global Supply Chains: A Multi-trade Analysis of CO_2 Emission Reductions in Container Shipping." *International Journal of Production Economics*, no. 208, 2019, pp. 17–28.

Casdesus-Masanell, Ramon, Forrest L. Reinhard, and Frederik Nellemann, "Maersk Line and the Future of Container Shipping." Harvard Business School Case Study 9-712-449 (2012).

Césaire, Aimé. *Toussaint Louverture. La Révolution française et le problème colonial.* Présence Africaine, 1981.

Chouliarki, Lilie. *The Ironic Spectator: Solidarity in the Age of Post-humanitarianism*. Polity Press, 2012.

Close, Alexander. *The Container Principle: How a Box Changes the Way We Think*. MIT Press, 2015.

Cowen, Deborah. *The Deadly Life of Logistics: Mapping Violence in Global Trade*. University of Minnesota Press, 2014.

Cronon, William. *Nature's Metropolis: Chicago and the Great West*. W.W. Norton, 2009.

Cumings, Bruce. *The Korean War: A History*. Modern Library/Random House, 2010.

Cumings, Bruce. *Korea's Place in the Sun: A Modern History*. W.W. Norton, 2005.

Currid-Halkett, Elizabeth. *The Warhol Economy: How Fashion, Art, and Music Drive New York City*. Princeton University Press, 2006.

Darwin, John. *After Tamerlane: The Rise and Fall of Global Empires, 1400–2000*. Bloomsbury Publishing, 2009.

Darwin, John. *Unfinished Empire*. Bloomsbury, 2013.

Davidson, Sinclair, Primavera De Filippi, and Jason Potts. "Blockchains and the Economic Institutions of Capitalism." *Journal of Institutional Economics*, vol. 14, no. 4 (2017).

Doucette, Jamie. "Minjung Tactics in a Post-Minjung Era? The Survival of Self-Immolation and Traumatic Forms of Labor Protest in South Korea." *New Forms and Expressions of Conflict at Work*, edited by Gregor Gall. Palgrave-Macmillan, 2013, pp. 212–32.

Engelke, Peter, and J.R. McNeill. *The Great Acceleration: An Environmental History of the Anthropocene since 1945*. Harvard University Press, 2014.

Fakhruddin, A.N.M., M.I. Taluder, and M.A. Hossain. "Environmental Impacts of Ship Breaking and Recycling Industries of Sitakunda, Chittagong, Bangladesh." *Advances in Natural Sciences*, vol. 8, no. 1, 2015, pp. 51–58.

Florin, Bo, Patrick Vonderau, and Yvonne Zimmermann. *Advertising and the Transformation of Screen Cultures*. Amsterdam University Press, 2021.

Frey, R. Scott. (2015). "Breaking Ships in the World-System: An Analysis of Two Ship Breaking Capitals, Alang-Sosiya, India and Chittagong, Bangladesh." *Journal of World-Systems Research*, vol. 21, no. 1, 2015, pp. 25–49, https://doi.org/10.5195/jwsr.2015.529.

Foss, Nicolai J. "Edith Penrose, Economics and Strategic Management." *Contributions to Political Economy*, no. 18, 1999, pp. 87–104.

Frost, Stephen, and Richard Welford. "Corporate Social Responsibility in Asian Supply Chains." *Corporate Social Responsibility and Environmental Management*, vol. 13, no. 3, 2006, pp. 166–76.

Gereffi, Gary. *Global Value Chains and Development*. Cambridge University Press, 2018.

Grandin, Greg. *The End of the Myth: From the Frontier to the Border Wall in the Mind of America*. Metropolitan Books, 2019.

Greenberg, Josh, and Graham Knight. "Framing Sweatshops: Nike, Global Production, and the American News Media." *Communication and Critical/Cultural Studies*, vol. 1, no. 4, 2004, pp. 151–75.

Grieveson, Lee. *Cinema and the Wealth of Nations: Media, Capital, and the Liberal World System*. University of California Press, 2017.

Harlaftis, Gelina. *Creating Global Shipping: Aristotle Onassis, the Vagliano Brothers, and the Business of Shipping, c. 1820–1970*. Cambridge University Press, 1970.

Harlaftis, Gelina, and Christos Tsakas. "The Role of Greek Shipowners in the Revival of Northern European Shipyards in the 1950s." *Shipping and Globalization in the Post-war Era: Context, Companies, Connections*, edited by Niels P. Peterson, Stig Tenold, and Nicholas J. White. Palgrave Macmillan, 2019.

Hediger, Vinzenz, and Florian Hoof. "The Agency of the Agency: Why Do All Corporate Image Films Look the Same?" *A History of Cinema without Names*, edited by Diego Cavallotti, Federico Giordano, and Leonardo Quaresima. Forum, 2016, pp. 209–18.

Hediger, Vinzenz, and Jason Potts. "Die vierte Regierungstechnologie: Über Blockchain." *Zeitschrift für Medienwissenschaft*, no. 18, 2018, pp. 73–86.

Hediger, Vinzenz, and Patrick Vonderau. "Record, Rhetoric, Rationalization." *Films That Work: Industrial Film and the Productivity of Media*, edited by Vinzenz Hediger and Patrick Vonderau. Amsterdam University Press, 2009, pp. 35–50.

Hediger, Vinzenz, and Patrick Vonderau, editors. *Films That Work: Industrial Film and the Productivity of Media*. Amsterdam University Press, 2009.

Hoof, Florian. *Angels of Efficiency: A Media History of Consulting*. Oxford University Press, 2020.

Hynan, Katherine A., and Anne H. Reilly. "Corporate Communication, Sustainability, and Social Media: It's Not Easy Being (Really) Green." *Business Horizons*, no. 57, 2014, pp. 747–58.

Innis, Harold. *Empire and Communications*. Rowman & Littlefield, 2007.

Iversen, Martin Jes. "East Asiatic Company's Difficult Experiences with Containerization." *Shipping and Globalization in the Post-war Era: Context, Companies, Connections*, edited by Niels P. Peterson, Stig Tenold, and Nicholas J. White. Palgrave Macmillan, 2019.

Jephson, Chris, and Henning Morgen. *Creating Global Opportunities: Maersk Line in Containerisation, 1973–2013*. Cambridge University Press, 2014.

Kampf, Constance. "Corporate Social Responsibility." *Corporate Communications*, vol. 12, no. 1, 2007, pp. 41–57.

Katona, Zsolt, and Miklos Sarvary. "Maersk Line: B2B Social Media – 'It's Communication, Not Marketing'." *Berkeley Haas Case Series*, vol. 56, no. 3, 2014, pp. 142–56.

Kim, Jihoon. "Perceptual Realism and Digital VFXs in the Korean Blockbuster of the 2010s: *The Admiral: Roaring Currents* and *Ode to My Father*." *Journal of Popular Film and Television*, vol. 47, no. 1, 2019, pp. 9–20.

Kim, Joseph, and Seung-ho Kwon. "The Cultural-Industry Policies of the Korean Government and the Korean Wave." *International Journal of Cultural Policy*, vol. 20, no. 4, 2013, pp. 422–39.

Kinzer, Stephen. *The True Flag: Theodore Roosevelt, Mark Twain, and the Birth of the American Empire*. Henry Holt, 2017.

Knies, Karl. *Der Telegraph als Verkehrsmittel* [1857]. Baden-Baden, Nomos, 1996.

Kroll, Gary. "The 'Silent Springs' of Rachel Carson: Mass Media and the Origins of Modern Environmentalism." *Public Understanding of Science*, vol. 10, no. 4, 2001, pp. 403–20.

Kuznets, Paul W. "Causes, Consequences, Relevance: South Korea's Industrialization." *Transformations in Twentieth Century Korea*, edited by Chang Yun-shik and Steven Hugh Lee. Routledge, 2006, pp. 89–105.

Larkin, Brian. "Promising Forms: The Political Aesthetics of Infrastructures." *The Promise of Infrastructures*, edited by Nikhil Anand, Akhil Gupta, and Hanna Apel. Duke University Press, 2018, pp. 178–86.

Lefort, Claude. *L'invention democratique. Les limites de la domination totalitaire*. Fayard, 1994.

Levinson, Marc. *The Box: How the Shipping Container Made the World Smaller and the World Economy Bigger*. Princeton University Press, 2006.

Lie, John. "What Makes Us Great: *Chaebŏl* Development, Labor Practices, and Managerial Ideology." *Transformations in Twentieth Century Korea*, edited by Chang Yun-shik and Steven Hugh Lee. Routledge, 2006, pp. 138–52.

Lim, Wonhyuk. "The *Chaebol* and Industrial Policy in Korea." *The Industrial Policy Revolution: The Role of Government beyond Ideology*, edited by Joan Esteban, Justin Lin Yifu, and Joseph E. Stiglitz. Palgrave Macmillan, 2013, pp. 348–70.

Lovell, Julia. *Maoism: A Global History*. The Bodley Head, 2019.

Maçaes, Bruno. *Belt and Road: A Chinese World Order*. Oxford University Press, 2019.

Maersk. "TradeLens Available in China through Collaboration with China Unicom Digital Tech." Press release, May 10, 2021, https://www.maersk.com/news/articles/2021/05/10/tradelens-available-in-china-through-collaboration-with-china-unicom-digital-tech.

McMurren, Juliet, Andrew Young, and Stefaan Verhulst. "Addressing Transaction Costs through Blockchain and Identity in Swedish Land Transfers." GovLab (2018), https://blockchan.ge/blockchange-land-registry.pdf.

Miller, Michael B. *Europe and the Maritime World: A Twentieth-Century History*. Cambridge University Press, 2012.

Nye, David. *American Technological Sublime*. MIT Press, 1996.

O'Sullivan, Mary. "Power & Profit: Copper Mines & Steam Engines in Late 18th century Cornwall." Working Paper, Paul Bairoch Institute of Economic History, 2021, https://www.unige.ch/sciences-societe/dehes/files/2516/1821/2863/Power_Profit_WP_2021.pdf.

Peterson, Niels P. "Shipping as a Knowledge Industry: Research and Strategic Planning at Ocean Group." *Shipping and Globalization in the Post-war Era: Context, Companies, Connections*, edited by Niels P. Peterson, Stig Tenold, and Nicholas J. White. Palgrave Macmillan, 2019, pp. 157–82.

Peterson, Niels P., Stig Tenold, and Nicholas J. White, editors. *Shipping and Globalization in the Post-war Era: Context, Companies, Connections*. Palgrave Macmillan, 2019.

Penrose, Edith. *The Theory of the Growth of the Firm*. Oxford University Press, 1995.

Pincus, Steven. *1688: The First Modern Revolution*. Yale University Press, 2011.

Plumpe, Werner. *Das kalte Herz. Kapitalismus: die Geschichte einer andauernden Revolution*. Rohwolt, 2019.

Polanyi, Karl. *The Great Transformation: The Political and Economic Origins of Our Time*. Beacon Press, 2001.

Pomeranz, Kenneth. *The Great Divergence: China, Europe, and the Making of the Modern World Economy*. Princeton University Press, 2000.

Porter, Janet. "Premium 'Daily Maersk' Service Abandoned." *Lloyd's Loading List*, March 5, 2015, https://www.lloydsloadinglist.com/freight-directory/news/Premium-Daily-Maersk-service-abandoned/61956.htm#.YL_YuG5CQXo.

Rancière, Jacques. *La méthode de l'égalité. Entretien avec Laurent Jeanpierre et Dork Zabunyan*. Bayard, 2012.

Reiling, Katharina. "The Emergence of Maritime Governance in the Post-war World." *Shipping and Globalization in the Post-war Era: Context, Companies, Connections*, edited by Niels P. Peterson, Stig Tenold, and Nicholas J. White. Palgrave Macmillan, 2019.

Richardson, Heather Cox. *How the South Won the Civil War: Oligarchy, Democracy, and the Continuing Fight for the Soul of America*. Oxford University Press, 2020.

Rodrik, Dani. *The Globalization Paradox: Democracy and the Future of the World Economy*. W.W. Norton, 2010.

Rönnbäck, Klas, and Peter M. Solar. "Copper Sheathing and the British Slave Trade." *Economic History Review*, vol. 68, no. 3, 2015, pp. 806–29.

Schüttpelz, Erhard. "Die medientechnische Überlegenheit des Westens. Zur Geschichte und Geographie der *immutable mobiles* bei Bruno Latour." *Medienegeographie. Theory–Analyse–Diskussion*, edited by Jörg Döring and Tristan Thielemann. transcript, 2009, pp. 67–110.

Schwartz, Michael, and Andrew Fish. "Just-in-Time Inventories in Old Detroit." *Business History*, vol. 40, no. 3, 1998, pp. 48–71.

Selwin, Benjamin. "Poverty Chains and Global Capitalism." *Competition & Change*, vol. 23, no. 1, 2018, pp. 71–97.

Sharma, Sarah. *In the Meantime: Temporality and Cultural Politics*. Duke University Press, 2014.

Sharpe, Christina. *In the Wake: On Blackness and Being*. Duke University Press, 2016.

Skhlar, Judith N. *The Faces of Injustice*. Yale University Press, 1990.

Smil, Vaclav. *Growth: From Microorganisms to Megacities*. MIT Press, 2019.

Snyder, Benjamin H. "From Vigilance to Busyness: A Neo-Weberian Approach to the Time Clock." *Sociological Theory*, vol. 3, no. 1, 2013, pp. 243–66.

Sornn-Friese, Henrik. "'Containerization in Globalization': A Case Study of How Maersk Line Became a Transnational Company." *Shipping and Globalization in the Post-war Era: Context, Companies, Connections*, edited by Niels P. Peterson, Stig Tenold, and Nicholas J. White. Palgrave Macmillan, 2019.

Star, Susan Leigh. "The Ethnography of Infrastructure." *American Behavioral Scientist*, vol. 43, no. 4, 1999.

Starosielski, Nicole. *The Undersea Network*. Duke University Press, 2015.

Steinhart, Daniel. *Runaway Hollywood: Internationalizing Postwar Production and Location Shooting*. University of California Press, 2019.

Stott, William. *Documentary Expression and 1930s America*. University of Texas Press, 1973.

Tenold, Stig. "The Declining Role of Western Europe in Shipping and Shipbuilding 1900–2000." *Shipping and Globalization in the Post-war Era: Context, Companies, Connections*, edited by Niels P. Peterson, Stig Tenold, and Nicholas J. White. Palgrave Macmillan, 2019.

Toulmin, Vanessa. *Electric Edwardians: The Films of Mitchell & Kenyon*. British Film Institute, 2006.

Trouillot, Michel-Rolph. *Silencing the Past: Power and the Production of History*. Beacon Press, 2015.

Van Creveld, Martin. *Supplying War: Logistics from Wallenstein to Patton*, 2nd ed. Cambridge University Press, 2004.

White, Nicholas J. "Thinking Outside 'The Box': Decolonization and Containerization." *Shipping and Globalization in the Post-war Era: Context, Companies, Connections*, edited by Niels P. Peterson, Stig Tenold, and Nicholas J. White. Palgrave Macmillan, 2019, pp. 67–99.

Williams, Raymond. *The Long Revolution*. Parthian, 2011.

Williams, William Appleman. *The Tragedy of American Diplomacy*. W.W. Norton, 2009.

Wolbring, Barbara. *Krupp und die Öffentlichkeit im 19. Jahrhundert*. C.H. Beck, 2000.

About the Author

Vinzenz Hediger is professor of cinema studies at Goethe University Frankfurt, where he directs the Graduate Research Training Group "Configurations of Film" (www.konfigrationen-des-films.de). His publications include *Films That Work: Industrial Film and the Productivity of Media* (Amsterdam University Press 2009, with Patrick Vonderau) and *Essays zur Filmphilosophie* (Fink, 2015, with Lorenz Engell, Oliver Fahle, and Christiane Voss) and *Accidental Archivism. Shaping Cinema's Future with Remnants of the Past* (meson press, 2023, with Stefanie Schulte Strathaus). He is a co-founder of NECS – European Network for Cinema and Media Studies (www.necs.org) and the founding editor of *Zeitschrift für Medienwissenschaft* (www.zfmedienwissenschaft.de).

2 Object Lessons and Infrastructural Imperialism

Lee Grieveson

Abstract

This chapter uncovers the history of the US government's use of film in the interwar years to foster new domestic and transnational road networks. It explores the role and place of media in state-led efforts to generate circulation and foster forms of economic imperialism.

Keywords: infrastructure; governmentality; economic imperialism; liberal developmental media

Beginning in the 1910s, the US government made a series of films explicating precisely the methods by which roads could be constructed and the benefits this would bring to industry, government, and people. What follows explores this history in two parts. The first section examines the efforts of the Office of Public Roads in the US Department of Agriculture to use film to disseminate information on road-building practices in the US. In 1918 the office became the Bureau of Public Roads and thereafter made a number of films in collaboration with major automobile manufacturers that told stories about the varied benefits road building brought to people and communities. The second section explores the bureau's efforts in the mid-1920s to foster a proposed Pan-American road network. Moving pictures were significant to this too. Produced in collaboration with major industrial corporations, banking interests and some Central and South American governments, these films were made in particular for "distribution [...] in the countries of South, Central, and North America" to be used "in educating the people to the value of better highways."[1] Both are examples of the governmental use of media to

1 Pan American Confederation for Highway Education. *The Pan American Confederation for Highway Education*. 1927, p. 13; Highway Education Board. *Highways of Friendship*. 1924, p. 22.

Hediger, V., F. Hoof, Y. Zimmermann, with S. Anthony (eds), *Films That Work Harder: The Circulation of Industrial Film*. Taylor & Francis Group 2024
DOI: 10.5117/9789462986534_CH02

foster the circulation of capital and new practices of economic imperialism during the period the US become the world's dominant industrial state.

Roads, and the vehicles that travel across them, are best understood as political technologies that work together with economic logics to expand extraction and circulation. Domestically the 1916 Federal Aid Road Act supplied up to $75 million in federal monies to states to build new roads, principally farm-to-market roads. The 1921 Federal Aid Highway Act supported interstate road building, creating over 400,000 miles of highways in the 1920s in what has been described as the "golden age" of highway construction.[2] Quite clearly also the markets and materials of Central and South America came to be ever more important to the US in the context of its rapid industrialization in the latter parts of the nineteenth century. The completion of the Panama Canal in 1914 enabled expanded and accelerated movement of raw materials and manufactured goods. In 1915, the Panama-Pacific Exposition celebrating the completion of the canal saw the first demonstration of long-distance telephonic communication in a further illustration of the complex enmeshing of infrastructural, communicative and media forms integral to the rising political and economic significance of the US. In that same year an important Pan-American economic conference discussed economic ties, and thereafter a series of Pan-American conferences in the 1920s focused on travel infrastructures.[3] The central objective of the US in developing economic ties with Central and South America was in opening access to raw materials as well as new markets and investment opportunities.[4] Raw materials associated in particular with automobility were particularly prized, notably including oil in Colombia, Venezuela, Peru and Mexico and rubber in Brazil. The US government countered local efforts to nationalize mineral resources by organizing private loans from banks. Two examples can quickly illustrate this form of what has been called "dollar diplomacy."[5] In 1917 the Mexican government asserted sovereignty over mineral rights – including in particular oil but also the silver that

2 Seely, Bruce E. *Building the American Highway System: Engineers as Policy Makers*. Temple University Press, 1987, pp. 66–99.

3 Rosenberg, Emily S. *Spreading the American Dream: American Economic and Cultural Expansion, 1890–1945*. Hill and Wang, 1982, pp. 66–67.

4 I use "state" here rather than "government" to signal an interest in the persistence of state structures – such as the civil service, the military, the police, the legal system – that have greater continuity than governments placed in temporary control of the state. By doing so I follow a relatively common use of state as genus to government as species.

5 Nearing, Scott, and Joseph Freeman. *Dollar Diplomacy: A Study in American Imperialism*. Allen and Unwin, 1925.

perhaps made their way into celluloid – but the US government forced the country to grant legal guarantees of American property by organizing a banking consortium that promised much needed loans in return for such safeguards.[6] Colombia reformed its subsoil laws in 1919. The US government, working with Gulf Oil's attorney, Allen Dulles – the future director of the CIA and the secretive but influential Council on Foreign Relations – worked to safeguard US investment. The Department of Commerce cautioned American investors about loaning to Colombia until its government agreed to safeguard Gulf's properties – at which point Colombia received a loan from National City Bank.[7] In short: the state, large extractive corporations and finance capital colluded to safeguard the plundering of the material resources needed for economic expansion; indeed, perhaps also for the materials that enabled the visual record of that development.

Global expansion also in terms of markets was mandated by the contemporary idea that high levels of production needed new global outlets – the so-called "overproduction thesis" – and this shaped the practices of large corporations and trade associations like the United States Chamber of Commerce and the American Manufacturers' Export Association. To be sure, there was a history of US global economic expansion that predates the 1910s and 1920s, visible in the imperial agendas that sustained the Spanish-American War of 1898 and the annexation of Hawaii, Puerto Rico, Guam and the Philippines that were predicated both on access to raw materials and prized shipping connections to China.[8] Likewise, the Panama Canal can be seen in part as a material enactment of the Monroe Doctrine. But the developments in the 1920s in particular marked a (partial) shift away from a European-style colonialism and its territorial logics of power towards "economic imperialism" and what economic historian Emily Rosenberg calls the "liberal developmental paradigm" that sustained the US's rise to global hegemon in the interwar years.[9] It is a central tenet of my argument here

6 Grandin, Greg. "The Liberal Traditions in the Americas: Rights, Sovereignty, and the Origins of Liberal Multilateralism." *American Historical Review*, vol. 117, no. 1, 2012, pp. 87–88.

7 Rosenberg. *Spreading the American Dream*, pp. 130–31.

8 Schoonover, Thomas. *Uncle Sam's War of 1898 and the Origins of Globalization*. University Press of Kentucky, 2013.

9 Arrighi, Giovanni. *The Long Twentieth Century: Money, Power, and the Origins of Our Times*. Verso, 1994, pp. 269–99; Harvey, David. *The New Imperialism*. Oxford University Press, 2003; Rosenberg. *Spreading the American Dream*, pp. 7–15. The central features of this liberal developmental paradigm included faith in private free enterprise, support for fee trade, and growing acceptance of governmental activity to protect private enterprise and stimulate American participation in international economic exchange. The shift towards a different model of empire was only ever partial because US state elites regarded the protection of the Panama Canal and American

that film and media played varied roles in that complex and globally significant process.[10] What follows here concentrates on delineating the precise connections between the material and semiotic networks that underscored the US's centrality to an ever-expanding market-driven modernity.

***

In early 1914, the Office of Public Roads (OPR), located within the Department of Agriculture (USDA), produced a number of films that documented the testing and building of durable road surfaces. Amongst the films produced were *Road Building and Testing Scenes*, *Macadam Road Construction*, *Gravel Road Construction* and *Road Conditions and Maintenance* (all OPR/USDA c. 1914).[11] All of these were designated as "films designed for popular agricultural education," defined in internal USDA documents as those "designed (1) to illustrate progressive agricultural methods, (2) to arouse general interest and lead to community adoption of these methods, (3) to 'break ground' for the work of the personal demonstrator or county agent, and (4) to familiarize people with activities of the Department which directly touch their daily business."[12] Three "professional films" were also available, entitled *Testing Rock to Determine Its Value for Road Building*, *Rock Tests with Traction Dynamometer*, and *Cement and Concrete Tests* (all OPR/USDA c. 1914). "These films," the department's annual report stated, "are designed to show specialists the exact technical methods employed by the Department or to give those interested in an industry an accurate

economic interests abroad to be of paramount importance and acted accordingly, as visible in the examples of Mexico and Colombia. In 1922, Secretary of State Charles Evans Hughes sponsored a conference with the nations of Central America to discuss regional stability, guarantees for the protection of property, and methods for combating revolutions. Hughes supported the notion that governments should not recognize new governments formed from revolutions. In practice, this meant that the US frequently supported repressive standing governments, like it did in Nicaragua in 1926 when it went so far as to create a national guard to ensure stability without the further need for American intervention. Schmitz, David F. *The Triumph of Internationalism: Franklin D. Roosevelt and a World in Crisis, 1933–1941*. Potomac Books, 2007, pp. 11–12.

10 See also Grieveson, Lee. *Cinema and the Wealth of Nations: Media, Capital, and the Liberal World System*. University of California Press, 2018.

11 Department of Agriculture. "Report on Motion Picture Activities." October 6, 1915. National Archives and Records Administration II (NARA), College Park, RG16, General Correspondence of the Office of the Secretary, 1906–70, Box 214, p. 4. (Hereafter NARA materials from state bureaus and departments will be referenced by their record group (RG) number with a corresponding box number, date, and page number, where available.)

12 Ibid., p. 3.

idea of the best practice of their field."[13] Film was integrated in this way directly into the scientific testing procedures of the Office of Public Roads. "Educational" and technical "professional" road films from the OPR sought in this way to shape new road-building practices by specialists and to educate rural populations in the methods for constructing and maintaining farm-to-market roads.

Experimentation with surfaces and materials had become central to the scientific agenda of the Office of Public Roads since its creation in 1905.[14] OPR officials had close connections with the USDA's Bureau of Chemistry.[15] Government experts examined different materials, from asphalt to concrete to gravel to macadam; and different construction processes.[16] Expanding after 1905, partly to accommodate the arrival of the automobile, the OPR constructed a new laboratory to test materials and began to disseminate information about standards.[17] Its goals were consistent with those of the Department of Agriculture more broadly to orchestrate scientific knowledge in a robust bureaucratic form to foster a market economy in agriculture – here, specifically to enable the speedy distribution of crops to urban areas that could both sustain farm production and a very rapidly expanding urban society.

To accomplish these goals the Office of Public Roads constructed a number of roads as what they called "object lessons."[18] These were constructed for local rural communities who were encouraged to visit them to observe construction and maintenance practices and from this to move forward in constructing local roads themselves with assistance from the OPR. Celluloid's storage capacity, and indexicality, made film a useful supplement to the state use of extension models to facilitate new productive practices. *Concrete Road Construction*, for example, was described in a later USDA catalogue like this:

> *Illustrates* the essential requirements in the design of concrete pavements to provide the necessary stability to resist the destructive action caused by the volume, weight, and speed of modern motor traffic. Shows the

13 Ibid., p. 5.

14 The OPR grew out of an earlier Office of Road Inquiry. On the early history of both Offices, see Seely. *Building the American Highway System*, pp. 11–23.

15 Ibid., p. 21.

16 Holley, Irving Brinton. "Blacktop: How Asphalt Paving Came to the Urban United States." *Technology and Culture*, vol. 44, no. 4, 2003, pp. 703–33.

17 Seely. *Building the American Highway System*, p. 27.

18 Ibid., pp. 14–15, 27.

principal operations in the most advanced methods of concrete-pavement construction.[19]

Other films also "illustrated" methods of road construction advocated by the OPR to local communities and states.[20] Film was in this way a useful informational medium, capable of making clear – of concretely illustrating – the materials and methods needed for the construction of a new transportation infrastructure. Valued for its indexicality, its ontological ties to objects, and for its mobility (unlike those object lesson roads), film was put to work to help facilitate the new farm-to-market road construction that would connect rural and urban markets and so grease the wheels for the expansion of markets.

Later films by the Office of Public Roads, mostly after it became the Bureau of Public Roads in 1918, supplemented this agenda by extolling the virtues and benefits of road construction to rural communities and, in the later 1920s, urban workers. They did so in lockstep with the federal policies that began to establish new road networks beginning with the 1916 Federal Aid Road Act that supplied up to $75 million in federal monies to states to build new roads. The act proposed a model of federal/state collaboration whereby the federal government carried out fundamental research and provided matching funds, and newly formed state highway departments carried out the road construction.[21] Roads were mostly constructed at this point in order to enable farmers to move produce to markets, increasingly urban ones.

Rapid rises in individual car ownership, from 78,000 in 1905 to 3.3 million in 1916 and 8.1 million in 1920, also pushed federal and state governments to devote resources to road building. Automobile corporations and trade groups, like the American Automobile Association, lobbied for new road construction, including new intercity roads. Firestone, an important

19 *Motion Pictures of the United States Department of Agriculture.* 1935. RG33, Records of the Extension Service, Box 1, p. 33. My emphasis.

20 These films are listed in the 1935 catalogue. An internal 1917 memorandum listed the following films having been completed by 1917, all of which appear to follow the OPR's technical use of film to support road-building programmes: *Testing Rock to Determine Its Value for Road Building, Road Tests with Traction Dynamometer, Cement and Concrete Tests, Gravel Road Construction in Virginia, Macadam Road Construction in Maryland, Concrete Road Construction* (*Ohio Post Road*), *Bituminous Macadam Road Construction* (*Maine Road*). "List of Motion Pictures." RG16, General Correspondence of the Office of the Secretary, 1907–1970, Box 415, 1917.

21 The act and its consequence are discussed in Seeley. *Building the American Highway System*, pp. 46–65. Federal highway engineers came in practice to be the dominant partners. Ibid., pp. 47, 67.

early innovator in rubber tires with close connections to Ford, organized a transcontinental publicity tour in the summer of 1920 that included films, speakers and a handbook about the necessity of road construction.[22] In the wake of these developments the 1921 Federal Aid Highway Act "fulfilled some of the objectives of the automobile industry" in supporting limited interstate road building, as well as supplementing the extant focus on farm-to-market roads.[23] Following passage of the act, a series of state conferences with the Bureau of Public Roads and the American Association of State Highway Officials created new maps of how the road network interconnected, effectively creating a national road infrastructure.[24] State highway budgets increased from $430 million in 1921 to $1.3 billion in 1930.[25] Federal monies, state bonds and gasoline taxes supported the emergence of a vast automotive infrastructure thereafter, with deleterious consequences for the environment and global ecology but profitable ones for automobile manufacturers and related industries like steel, iron, rubber and the oil industry. Gasoline taxes beginning in 1919 were earmarked specifically for new road construction in a "self-replicating system" that was, Christopher Wells has argued, "designed explicitly to stimulate near-constant growth in American demand for gasoline."[26] By doing so it also mandated an increasingly aggressive foreign policy and military expansion, parts of which were explicitly designed to control the crucial resource of oil.

But for now I shall stick with a few of the films the bureau produced in the wake of the 1916 and 1921 federal legislation. I have written elsewhere about one of these films, called *The Road to Happiness*, and produced in collaboration with the Ford Motor Company.[27] It is worth briefly examining

22 Ibid., p. 53. Firestone sourced much of its rubber in Liberia. The company loaned large sums to the Liberian government, after receiving guarantees from the United States government that it would support the security of its investment. Writes Rosenberg: "Firestone Rubber, for all practical purposes, ran Liberia's economy." Rosenberg. *Spreading the American Dream*, p. 134.

23 Ibid., p. 63.

24 Ibid., pp. 71–88. These maps were supplemented by the so-called Pershing Map in 1922, created by the Bureau in consultation with the War Plans Division of the Army to plot military strategic routes. General of the Armies John Pershing presented this plan to Congress in 1922, and most of the 78,000 miles of roads requested were built with many of them becoming interstate highways. McNichol, Dan. *The Roads That Built America: The Incredible Story of the US Interstate System*. Sterling Publishing, 2006, p. 62.

25 Seely. *Building the American Highway System*, p. 73.

26 Wells, Christopher W. "Fuelling the Boom: Gasoline Taxes, Invisibility, and the Growth of the American Highway Infrastructure, 1919–1956." *Journal of American History*, vol. 99, no. 1, 2012, pp. 72–73. The federal government introduced the gasoline tax in 1919.

27 Grieveson, Lee. "The Work of Film in the Age of Fordist Mechanization." *Cinema Journal*, vol. 51, no. 3, 2012, pp. 35–37.

this though now from the perspective also of the state objectives that it embodied. *The Road to Happiness* (c. 1924) told the story of a young farmer who is instructed by his bad-tempered father to deliver eggs to market, but he misses the connecting train because the roads are so bad and his eggs are destroyed along the way. The father blames the boy for being lazy and dismisses his argument for the need for better roads. Later, at school, the boy is invited to enter a competition to write an essay about his local community and he writes about the need for good roads to enable the easy transportation of rural goods. The lesson learnt by the young farmer was about the commercial necessity of improving roads to enable the smoother flow of commerce and the integration of rural spaces with the national market. The young farmer's essay is the winning entry in the competition and he goes to accept his award from President Calvin Coolidge on the lawn of the White House.

Coolidge's appearance was possible because *The Road to Happiness* was produced with the cooperation of the Bureau of Public Roads.[28] In a memorandum in the National Archives in the US the director of the USDA's Extension Work, C.W. Warburton, recounts how the Bureau of Public Roads helped "in the location of suitable scenes for picturization and in the loaning of one or two members of the staff of that Bureau to act in the picture." Warburton revealed that the "Motion Picture Laboratory" of the USDA "was also used by the photographers of the Ford Motor Company in developing the film." Ford's resources mingled together with the state's to tell a story that carefully orchestrated allegiance to the cause for road building embodied in the young farmer. Policy was encoded in character and became narrative. Warburton concluded that the "film itself is entirely commendable and it is quite in keeping with those produced by the Department. As a matter of fact the Department might have produced the film in exactly its present form."[29] His remark in this respect makes clear that the film perfectly illustrated the mutuality of interests between state and corporation such that it was hard to tell which had produced it. Warburton's memorandum revealed that bureau officials acted in the film, in front of Ford's cameras, before the celluloid was developed with

28 "Latest Film Is Released." *Ford News*, March 15, 1924, p. 1. The sequence with President Coolidge is not in the extant version I have seen from the National Archives, but the *Ford News* reporting on it has a still from the scene. Officials of the Bureau of Public Roads were a little anxious that their support for the film compromised their ostensible neutrality in relation to industrial companies. Letters bat this back and forth, before concluding that the film was better advertised as a Ford film rather than as a collaboration between the company and the Bureau.
29 Warburton, C.W. Memorandum for Mr. W.A. Jump. April 2, 1924. RG16, Box 1070.

the hclp of USDA chemicals. *The Road to Happiness* embodied the goals of what Rosenberg calls "the cooperative state" that promoted industry and trade in the liberal belief in the mutuality of public and private interests.[30] Coolidge's appearance as the "good father" of the nation is fitting in this sense for it was he who famously pronounced, in a pithy articulation of the principles of liberal political economy, "the business of America is business."

Quite clearly, the films arguments about the necessity for a new material infrastructure for the transport of commerce were significant to the states interest in expanding markets that dovetailed also with the goals of commercial automobile manufacturers to facilitate the sale of expensive technology to new markets. What was at stake here was the innovation of a new technological space, a networked infrastructure shared out between corporate and state forces that would enable market expansion and the incorporation of rural America into an emergent consumer economy. Ford's happiness is akin to the states. The road to happiness is born from the union of the political and economic forces that prioritized the mobility of goods and capital.

Ford and the bureau disseminated the film through their own networks of distribution. The bureau circulated the film in particular through the Highway Education Board that had been set up by the chief of the bureau in 1923 to spread information about road building chiefly to schools. (The Highway Education Board is listed on the title card of the extant copy of the film, alongside the National Automobile Chamber of Commerce that functioned principally as a lobbying group for the automotive industry and that had close ties to government.[31]) In a press release to accompany the film on its way through these networks the bureau even imagined a future for the character of the young farmer. Escaping the grumpy backwards-looking father, anointed by the president as the good father of the nation, he "becomes a highway engineer," the bureau's PR had it, and

> is privileged to bring about such changes in the condition of the roads of his community that even the most sceptical opponents of road improvement are brought to a realization of the truth of the maxim which forms

30 Rosenberg. *Spreading the American Dream*, pp. 138–60.

31 See also "Latest Film is Released"; Warburton. Memorandum for Mr. W.A. Jump. Bureau of Public Roads chief Thomas Harris MacDonald set up the Highway Education Board as a public relations arm that produced material for schools, held nationwide contests, and had a speaker's bureau.

> the moral of the picture that "we pay for improved roads whether we have them or not, and we pay less if we have them than if we have not."[32]

In the context of the Highway Education Board's work with schools, the young farmer becomes an exemplar of the technocratic management of an emerging networked society.

Other films produced by the Bureau of Public Roads narrate a story of progress and happiness that culminates with the car and the paved road, as well as celebrating the advanced technology that enabled government engineers to build roads in treacherous conditions.[33] See, for example, *Highways of Service* (Bureau of Public Roads, c. 1927), *A Road Out of Rock* (Bureau of Public Roads, n.d.), *Building Forest Roads* (Bureau of Public Roads, n.d.) and *Highways and Skyroads* (Bureau of Public Roads, n.d.). Government and advanced construction technology intertwine, these films suggest, the one strengthening the other, establishing the networks that facilitate commercial and personal movement across the nation. "Education" through media sought to facilitate the kinds of mobility integral to this particular formation of liberal capitalism. The liberal state's efforts to establish new patterns of circulation for materials, objects, information and capital was supplemented, even partly facilitated, by film and/as education.

***

In these representations of technological advancement, as well as in their circulation, the state advertised and emphasized its power and modernity. This became particularly important with respect to the US government's efforts to establish a Pan-American road network that would enable the hemispheric circulation of raw materials and manufactured objects. In 1919 a Pan American Commercial Conference presented the need to disseminate information among Latin Americans about the superiority of US

32 Press Service. "'Road to Happiness,' New Motion Picture Showing Value of Highways to Community." Bureau of Public Roads, March 10, 1924. RG16, Box 1070. The press service distributed information about the film to "Farm Papers; Farm Editions of Daily Papers; Daily Newspapers (One in a city – all cities); Sunday Editions of Dailies; Washington Correspondents and freelance writers; County Seat papers; Trade papers as follows: Highway Publications; Motion Picture list; Auto journals." Press Service. Memorandum to Mr. Perkins, Motion Pictures. Bureau of Public Roads, March 11, 1924. RG16, Box 1070.

33 See *Highways of Service* (Bureau of Public Roads, c. 1927), *A Road Out of Rock* (Bureau of Public Roads, n.d.), *Building Forest Roads* (Bureau of Public Roads, n.d.), and *Highways and Skyroads* (Bureau of Public Roads, n.d.).

road-building methods as a prelude to exporting those methods, along with the machines that made roads and drove across them. Participants were told also of the effectiveness of "publicity films" to present US firms and products more effectively to South American markets.[34] In 1923 the efforts to foster road construction in South and Central America, and connecting roads across the continent, were addressed at the fifth Pan American Conference in Santiago, Chile. Delegates passed a resolution calling for a Pan American Highway Commission to be formed that would observe the US highway system and US means of financing, administering, constructing and controlling modern highways and to work to establish similar road networks.[35] Quickly thereafter the National Automobile Chamber of Commerce in Washington began arranging funding after the Bureau of Public Roads and the recently formed Department of Commerce endorsed the notion. It received funding from the bureau's Highway Education Board and pledges of $1,000 or more "from prominent bankers and leading automotive and road machinery manufacturers of the United States."[36]

Indeed, members of the Pan American Highway Commission's executive committee had strong ties to US automotive and banking interests. Roy D. Chaplin, chairman of the board of Hudson Motor Car Company, chaired the committee, which included also Fred I. Kent, vice president of Banker's Trust of New York, and W.T. Beaty, president of the Austin Manufacturing Company in Chicago, which made road maintenance equipment.[37] Amongst the advisory committee were representatives from the American Manufacturers Export Association, Atlas Portland Cement Company, Barber Asphalt Company, First National Bank of Boston, General Motors, Mack Trucks, Inc., Standard Oil, B.F. Goodrich Rubber Company, amongst others: all large corporations with clear economic interests in the expansion of road networks.[38] US-sanctioned encouragement of investment in transportation infrastructures filled the coffers of the companies that made the machines and materials for road building and automobiles as well as the banks that

34 Salvatore, Ricardo D. "Early American Visions of a Hemispheric Market in South America." *Transnational America: The Fading of Borders in the Western Hemisphere*, edited by Berndt Ostendorf. C. Winter, 2002, pp. 59–63.

35 Downes, Richard. "Autos over Rails: How US Business Supplanted the British in Brazil, 1910–28." *Journal of Latin American Studies*, vol. 24, no. 3, 1992, p. 565.

36 Pyke Johnson to Francis White, Latin American Div., US Department of State. December 8, 1923. Qtd. in Downes. "Autos over Rails," p. 565.

37 For a brief account of the Austin Manufacturing Company, see "F.C. Austin Manufacturing Company (1888–1934)." *Historical Construction Equipment Association*, http://archives.hcea.net/?p=creators/creator&id=229. Accessed July 16, 2012.

38 Highway Education Board. *Highways of Friendship*, p. 50.

invested in these construction projects. It is worth noting also that the US's support for Pan-American road networks supplanted previous British investment in expensive rail networks in Central and South America. The emergence of cheaper road-building projects, supported by the United States and US corporations, is a concrete example of the way US economic imperialism supplanted the extant economic exploitation orchestrated mostly by the British.[39]

Engineers and businessmen from Latin American countries were invited by the Pan American Highway Commission to consult with the Bureau of Public Roads and to travel on an expenses-paid excursion through the US focusing principally on highway construction and automobile manufacture. Coolidge gave an address of welcome to the delegates, when they arrived in June 1924: "We see today, more clearly than ever before, that the improved means of communication is not only one of the great forces making for cultural and economic advance, but it is also one of the basic factors in the development of Pan-American unity."[40] Likewise, Secretary of State Charles Evans Hughes told the delegates: "Highway construction enlarges the market for your products which in turn stimulates production and leads to general economic advance."[41] Liberal developmentalism was supplemented by a political argument that road systems – operating as what Coolidge described as communication networks – were "closely related" to what Hughes called "democratic development." The secretary of state accordingly proposed that "[i]mproved communication means the elimination of sectionalism and contributes in no small degree to the formation and construction of an intelligent and controlling public opinion."[42] Economic arguments dovetailed with political ones in a way that presented road networks as the magical solution that both kick-started economic development and stabilized political regimes. By doing so this would also ideally preclude the socialist reorganization of property rights that caused the United States and US corporations such problems in, for example, Mexico and Colombia, and that would later involve all sorts of expensive and troublesome coups needed to overthrow governments in Latin America that stretches at least from Guatemala in 1954 to oil-rich Venezuela in 2002, through Cuba, Chile, Nicaragua, Panama and so on.[43]

39 Downes. "Autos over Rails."

40 Coolidge, Calvin. Qtd. in Highway Education Board. *Highways of Friendship*, p. 1.

41 Hughes, Charles Evans. Qtd. in Highway Education Board. *Highways of Friendship*, p. 3.

42 Ibid.

43 All told, the US played a role in the overthrow of twenty-four governments in Latin America during the period of the Cold War alone. Coatsworth, John. "The Cold War in Central America,

Visits were arranged for the delegates to tour automobile manufacturers in Detroit, including meetings with General Motors and with Henry Ford himself, "for the serious study of the manufacture of the automotive vehicle designed to run over the highways they had seen in course of construction and maintenance."[44] Also included in the itinerary were visits to the rubber manufacturing companies located principally in Cleveland and Akron, Ohio. US automotive and rubber corporations had begun to invest in Latin America, notably in Brazil where Ford established an assembly plant in 1919 (and a disastrous rubber processing plant in 1928), and where the Goodyear Tire and Rubber Company, the United States Rubber Company, and the General Tire and Rubber Company all opened businesses in the 1920s.[45] The agents for US firms associated with road building began to criss-cross Brazil and other oil- and rubber-rich states, in particular, offering their wares, often with a letter of introduction from the US ambassador.[46] Corporations were helped by the "cooperative" United States to source the materials necessary for car and road construction from the mineral and plant resources of Latin America and then to sell the technologies of cars and roads back from whence much of the material came. It was in this way a textbook example of an economic imperialism that relied on the cheap extraction of primary materials and their transformation into expensive technologies that are exported back from the metropole to the periphery, even on the very road networks created from (and for) the endeavour.

Towards the end of the tour of Latin American engineers and businessmen, the USDA organized a film screening. Amongst the films screened in the summer of 1924 to help persuade these men to establish a network of roads connecting their countries to the US, and to buy US materials and machines to accomplish that, was a film of the tour itself. It was carefully planned. Fred Perkins of the USDA Motion Picture Laboratory had been appointed a member of the Pan American Highway Commission and had accompanied the delegates on their tour to film it. The resulting film was no doubt partly a diplomatically flattering use of government film technology, but it functioned also as a way of circulating the agenda of the commission across the hemisphere. "The motion picture record of the trip and a handbook on the underlying reasons for highway development," the official account

1975–1991." *Cambridge History of the Cold War, Vol. 3: Endings*, edited by Melvyn P. Leffler and Odd Arne Westad. Cambridge University Press, 2012, pp. 220–21.

44 Highway Education Board. *Highways of Friendship*, p. 10.

45 For a fascinating account of Ford's jungle rubber plant, see Greg Grandin. *Fordlandia: The Rise and Fall of Henry Ford's Forgotten Jungle City*. Metropolitan Books, 2009.

46 Downes. "Autos over Rails," pp. 570, 579–80.

of the tour observed, "are two outgrowths of the tour now in preparation, which will unquestionably find a wide circulation here as well as abroad."[47] It functioned, then, as a reminder of the "highways of friendship" built by the money invested by the US government and its industrial and financial collaborators, and as a mobile and visible way of circulating knowledge about the tour and the effort to orchestrate new road networks.

Also screened for the delegates was the Ford-financed *The Road to Happiness*, with the titles specially translated into Spanish. The film was to be "sent to their countries for use in educating the people to the value of better highways," a central part of the broader imperative to "arouse public interest" and provide "[e]ducation in the effects of highway development" in Latin America.[48] It is not clear if the film was produced with this purpose in mind, though it is possible given that the Highway Education Board began planning for the tour in late 1923, the film was made in March 1924, and the tour began in June 1924. The films' narrative of economic advancement to rural and peripheral areas, and of the progress and modernity brought by road networks could well have been seen by those delegates in allegorical and hemispheric terms. *The Road to Happiness* clearly connected well to the goals of the tour to convince the Latin American delegates of the advantages to be gained from building roads and using automobiles. The film's composite production, whereby it was financed from the money accruing from new mass assembly practices and made with equipment and personnel from the US government, makes abundantly clear the close ties between government and industry in orchestrating new transnational markets. Here, remarkably, was film as a "cooperative" endeavour between state and business operating as bridgehead to foreign economic policy and integrated directly with the economic imperialism that was repositioning the US as global hegemon. In this context, indeed, the movement of young progressive farmers out of the sphere of their old-fashioned, bad fathers towards that of the enlightened president of the US was invested with geopolitical resonance. Oedipal narrative was replayed as geopolitics and economic imperialism.

In the wake of the decisions made at the 1923 Pan American Conference, and following the forming of the Pan American Highway Commission in 1924, a Pan American Congress of Highways was organized in October 1925 in Buenos Aires. US interests dominated the congress. The Highway Education Board, an "associationalist" organization made up of members drawn from government, industry and finance capital, formed the executive committee

47 Ibid., p. 53.

48 Ibid., pp. 29, 28.

of the congress and shaped the agenda.[49] Accordingly, the central goal of the congress was to establish "a network of connecting highways for the American hemisphere," and its "primary purpose" was defined as "the dissemination of the fundamental principles of highway construction, finance, administration, and maintenance, and the advancement of the social and economic uses of the highway and the automotive vehicle."[50] Keeping with the film and extension practices of the Bureau of Public Roads, and those of the Highway Education Board in the summer of 1924, the congress proposed that these principles could be disseminated through "the distribution of a series of new films in the countries of South, Central, and North America."[51] Amongst these were Perkins' film of the 1924 tour, called *Wheels of Progress*, and a film called *Building Roads in North America* that appears to have been one of the Bureau of Public Roads' "popular agricultural education" films demonstrating "progressive" methods of road construction "to arouse general interest and lead to community adoption of these methods."[52] *Roads for All America*, a film of the conference itself, was subsequently used, the director of the United States Bureau of Foreign and Domestic Commerce observed, "to inform American firms of road conditions in Latin-America."[53]

By the early 1920s, then, various films produced by the Bureau of Public Roads, and interested corporations, were put into circulation principally by the US to help expand material networks of circulation across the continent. Here was cinema being used to elaborate infrastructural pathways of circulation for American capital and machines. The films were a form of governmental modernism literally in the service of the construction of the material means to compress space and time to enable the expansion of markets.

***

49 I refer here to historian Ellis Hawley's account of the emergence of the "associational state" in the US in the 1920s that was predicated on the establishment of a cooperative system of self-regulating business associations and an expanded state bureaucracy in order to foster modern industrial and commercial practices and reconcile the emergent corporate form with the traditions of non-statist liberal government. See Hawley, Ellis. "Herbert Hoover, the Commerce Secretariat, and the Vision of an 'Associative State,' 1921–1928." *Journal of American History*, vol. 61, no. 1, 1974, pp. 116–40; and idem. *The Great War and the Search for a Modern Order: A History of the American People and Their Institutions, 1917–1933*. St. Martin's Press, 1979, pp. 36–37.

50 Pan American Confederation for Highway Education. *The Pan American Confederation for Highway Education: Its Aims and Purposes, Constitution and By-laws*. 1925, pp. 12, 5.

51 Ibid., p. 13.

52 Department of Agriculture. "Report on Motion Picture Activities," p. 3.

53 Klein, Julius. "What Are Motion Pictures Doing for Industry?" *Annals of the American Academy of Political and Social Science*, vol. 128, November 1926, p. 83.

Was this project successful? The question is not easy to answer. The success of the road-building efforts was only ever partial, hampered always by the parlous economic situation in many states in Latin America and by imperatives for those states to build domestic roads.[54] It would be overtaken in the later 1920s by the emergence of air networks, and the creation of the airline – with support from the US Air Force and capital from J.P. Morgan – that would come to be called, fittingly, Pan American Airlines.[55] But it is quite clear, as economic historians emphasize, that US economic hegemony over Latin America increased during the 1920s, and this supplanted British interests.[56] Road networks replaced British-financed and British-built rail networks. Visible in these films is the establishment of the road networks and infrastructure of automobility that has profoundly damaged our collective environments. Economic imperialism formed through a complex "cooperative" effort between state, corporations, finance capital and associated lobbying groups would be a crucial precondition for US global economic hegemony and a liberalized world order from circa mid-century onwards. The films and institutions examined here were made to facilitate US economic expansion, playing a small role, then, in the coming of the American/automobile century and the catastrophic consequences that has had for the world we now precariously live in.

My thanks to Vinzenz Hediger. For my friend Tom.

Works Cited

Arrighi, Giovanni. *The Long Twentieth Century: Money, Power, and the Origins of Our Times*. Verso, 1994.

Coatsworth, John. "The Cold War in Central America, 1975–1991." *Cambridge History of the Cold War, Vol. 3: Endings*, edited by Melvyn P. Leffler and Odd Arne Westad. Cambridge University Press, 2012, pp. 201–21.

Department of Agriculture. "Report on Motion Picture Activities." October 6, 1915. National Archives and Records Administration II, College Park, RG16, General Correspondence of the Office of the Secretary, 1906–70, Box 214.

54 Salvatore, Ricardo D. "Imperial Mechanics: South America's Hemispheric Integration in the Machine Age." *American Quarterly*, vol. 58, no. 3, 2006, p. 677.

55 Josephson, Matthew. *Empire of the Air: Juan Trippe and the Struggle for World Airways*. Harcourt Brace, 1944.

56 Rosenberg. *Spreading the American Dream*, in particular pp. 122–37.

Downes, Richard. "Autos over Rails: How US Business Supplanted the British in Brazil, 1910–28." *Journal of Latin American Studies*, vol. 24, no. 3, 1992, pp. 551–83.

"F.C. Austin Manufacturing Company (1888–1934)." *Historical Construction Equipment Association*, http://archives.hcea.net/?p=creators/creator&id=229. Accessed July 16, 2012.

Grandin, Greg. *Fordlandia: The Rise and Fall of Henry Ford's Forgotten Jungle City*. Metropolitan Books, 2009.

Grandin, Greg. "The Liberal Traditions in the Americas: Rights, Sovereignty, and the Origins of Liberal Multilateralism." *American Historical Review*, vol. 117, no. 1, 2012, pp. 68–91.

Grieveson, Lee. *Cinema and the Wealth of Nations: Media, Capital, and the Liberal World System*. University of California Press, 2018.

Grieveson, Lee. "The Work of Film in the Age of Fordist Mechanization." *Cinema Journal*, vol. 51, no. 3, 2012, pp. 25–51.

Harvey, David. *The New Imperialism*. Oxford University Press, 2003.

Hawley, Ellis. *The Great War and the Search for a Modern Order: A History of the American People and Their Institutions, 1917–1933*. St. Martin's Press, 1979.

Hawley, Ellis. "Herbert Hoover, the Commerce Secretariat, and the Vision of an 'Associative State,' 1921–1928." *Journal of American History*, vol. 61, no. 1, 1974, pp. 116–40.

Highway Education Board. *Highways of Friendship*. 1924.

Holley, Irving Brinton. "Blacktop: How Asphalt Paving Came to the Urban United States." *Technology and Culture*, vol. 44, no. 4, 2003, pp. 703–33.

Josephson, Matthew. *Empire of the Air: Juan Trippe and the Struggle for World Airways*. Harcourt Brace, 1944.

Klein, Julius. "What Are Motion Pictures Doing for Industry?" *Annals of the American Academy of Political and Social Science*, vol. 128, November 1926, pp. 79–83.

"Latest Film Is Released." *Ford News*. March 15, 1924, p. 1.

"List of Motion Pictures." National Archives and Records Administration II, College Park, RG16, General Correspondence of the Office of the Secretary, 1907–70, Box 415, 1917.

McNichol, Dan. *The Roads That Built America: The Incredible Story of the US Interstate System*. Sterling Publishing, 2006.

Motion Pictures of the United States Department of Agriculture. 1935. National Archives and Records Administration II, College Park, RG33, Records of the Extension Service, Box 1.

Nearing, Scott, and Joseph Freeman. *Dollar Diplomacy: A Study in American Imperialism*. Allen and Unwin, 1925.

Pan American Confederation for Highway Education. *The Pan American Confederation for Highway Education*. 1927.

Pan American Confederation for Highway Education. *The Pan American Confederation for Highway Education: Its Aims and Purposes, Constitution and By-laws.* 1925.

Press Service. Memorandum to Mr. Perkins, Motion Pictures. Bureau of Public Roads, March 11, 1924. National Archives and Records Administration II, College Park, RG16, Box 1070.

Press Service. "'Road to Happiness,' New Motion Picture Showing Value of Highways to Community." Bureau of Public Roads, March 10, 1924. National Archives and Records Administration II, College Park, RG16, Box 1070.

Rosenberg, Emily S. *Spreading the American Dream: American Economic and Cultural Expansion, 1890–1945.* Hill and Wang, 1982.

Salvatore, Ricardo D. "Early American Visions of a Hemispheric Market in South America." *Transnational America: The Fading of Borders in the Western Hemisphere,* edited by Berndt Ostendorf. C. Winter, 2002, pp. 45–64.

Salvatore, Ricardo D. "Imperial Mechanics: South America's Hemispheric Integration in the Machine Age." *American Quarterly,* vol. 58, no. 3, 2006, pp. 662–91.

Schmitz, David F. *The Triumph of Internationalism: Franklin D. Roosevelt and a World in Crisis, 1933–1941.* Potomac Books, 2007.

Schoonover, Thomas. *Uncle Sam's War of 1898 and the Origins of Globalization.* University Press of Kentucky, 2013.

Seely, Bruce E. *Building the American Highway System: Engineers as Policy Makers.* Temple University Press, 1987.

Warburton, C.W. Memorandum for Mr. W.A. Jump. April 2, 1924. National Archives and Records Administration II, College Park, RG16, Box 1070.

Wells, Christopher W. "Fuelling the Boom: Gasoline Taxes, Invisibility, and the Growth of the American Highway Infrastructure, 1919–1956." *Journal of American History,* vol. 99, no. 1, 2012, pp. 72–81.

About the Author

Lee Grieveson is professor of media history at University College London. Most recently, he is the author of *Cinema and the Wealth of Nations: Media, Capital, and the Liberal World System* (University of California Press, 2018), and co-editor (with Haidee Wasson) of *Cinema's Military Industrial Complex* (University of California Press, 2018).

3 Energy and Industrial Film

Energo-Critical Registers

Thomas Turnbull

Abstract

Given growing interest in environmental aspects of film, this chapter proposes an energo-critical approach to industrial film, notable for depicting combustive processes, by suggesting four registers for reading such thermodynamic imagery: First the sensorial, which quantifies aspects of energy conversion. Second, energy is recognized in a physicist's sense, as a capacity to do work, pointing to the augmentation and denigration of human labour. Third, the imperative to save energy via increasing efficiency, a central characteristic of industrial film, manifest in portrayals of power transfers and transformations. Fourth the motif of friction is affirmed as various off-screen resistances impose themselves on the extractive processes of industry. Far from comprehensive, these registers offer starting points for reading energy's increasing pertinence back into film.

Keywords: thermodynamics; sensorium; efficiency; labour; friction

This chapter reflects upon the role of energy in industrial film. These reflections were prompted by my attendance at "Films That Work: The Circulations of Industrial Cinema," a conference held in Frankfurt in December 2015. Many of the films screened documented the productive activities of private and publicly owned fuel companies, including Royal Dutch Shell, British Petroleum (BP), the Arabian-American Oil Company (ARAMCO), Italy's state-owned Ente Nazionale Idrocarburi (ENI), the Czech Pražská železářská společnost (PŽS), and Britain's National Coal Board (NCB), and there were two panels on "The Politics of Performance: Energy and Work." Alongside industrial films from these fuel companies,

Hediger, V., F. Hoof, Y. Zimmermann, with S. Anthony (eds), *Films That Work Harder: The Circulation of Industrial Film*. Taylor & Francis Group 2024
DOI: 10.5117/9789462986534_CH03

there were also a number of films relevant to the topic of energy in so far as the products they depicted were prominent consumers of energy both in their production and use, from American automobile firm Chevrolet to Italy's Fiat, and from the heavy industry of South Yorkshire's United Steel to the brewing processes of Dublin's Guinness. In the following discussion, drawing on films screened at this conference and others, my intention is to argue that industrial cinema can be analysed in at least four ways that are congruent with the physicist's concept of energy, understood as a capacity for work, in what I term a "thermodynamic reading" of industrial film.

In proposing these films can be addressed from the perspective of thermodynamics, I suggest the study of industrial film might benefit from considering the significance of energy as an aspect of form, content, and transmission.[1] What significance does the concept of energy, an assessment of a system's potential to do work, dictated by the laws of thermodynamics – laws of energy conservation and dissipation – hold for industrial film? Geographer Andrew Barry has recently argued that the concept of energy has been curiously under interrogated by social theorists, despite a growing interest in new kinds of materialist analysis and the reaffirmation of the obdurate physicality of the world as a locus of its agency. His proposed solution to this oversight is to suggest that researchers should document how energy, as it is understood by physicists and engineers, becomes folded into social life, primarily via measurements of conversion and efficiency, and manipulations in its form.[2] However, energy also acts upon society in more diffuse, indirect ways. Laws of energy have both shaped and been shaped by culture in significant ways, as historians of science and more recent eco-critics have demonstrated. These fields, offer a number of observations appropriate to the study of industrial film. Drawing on this scholarship, I argue here that a thermodynamic reading of industrial film can be pursued on at least four registers: those relating to the senses, the derivation of work, efficient use, and as a source of friction.

Can seemingly inviolable physical laws be considered as cultural artefacts? In their reading of Paul Theroux's novel *The Mosquito Coast* (1981), historians of science Ian Higginson and Crosbie Smith discussed the influence of the culture of New England Puritanism on the central protagonist, Allie Fox, who

1 A debt is owed to Thomas Pringle, who offered advice. His work in this same direction helped inspire and clarify my own thinking. Pringle, Thomas. "Photographed by the Earth: War and Media in Light of Nuclear Events." *NECSUS European Journal of Media Studies*, vol. 3, no. 2., 2014, pp. 131–54.

2 Barry, Andrew. "Thermodynamics, Matter, Politics." *Distinktion: Journal of Social Theory*, vol. 16, no. 1, 2015, pp. 110–25.

has committed his working life to the creation of a monstrous ice-making machine called "Fat Boy." Rather than doomsday, this machine is a means by which Fox is able to pit himself against the universe's tendency toward entropy, disorder and chaos. In their reading, Higginson and Smith suggest the novel rejects "an 'essentialist' history of thermodynamics (in which the science has some sort of independent, monolithic and timeless existence) in favour of a cultural history in which the science itself is shaped by the cultures in which it is practiced."[3] As a Calvinist, Fox considers entropy as imperfection, and ice making as means of personal salvation. Their thermodynamic reading of Theroux's book built on Smith's contextual understanding of the science of energy in nineteenth-century Glasgow, in which a local culture of entrepreneurial engineers, Presbyterian enthusiasm for personal thrift and an eschatology of humankind's sinfulness helped establish a particular understanding of energy and its imperative to pursue perfectibility in an imperfect world. In Smith's account, culture lent credibility to and borrowed from the science of energy.[4]

A few years before Smith, historian Gillian Beer noted the dissonant cultural implications prompted by the discovery of a "progressive" evolutionary theory around the same time as the more pessimistic science of thermodynamics. In 1852 Glaswegian engineer William Thomson had written of the "universal tendency in nature to the dissipation of mechanical energy," which led him to predict the eventual death of the Sun and, consequently, of life on Earth. Beer noted how these competing eschatologies generated a vast amount of imaginative literary work that reflected the anxieties prompted by a universe of seemingly depleting power, from George Eliot's *Middlemarch* (1871) to H.G. Wells' *The Time Machine* (1895).[5] Beer's work emerged independently around the same time as a form of cultural studies referred to as "ecocriticism." Prompted by environmental concerns, since the 1990s ecocriticism has undertaken the interdisciplinary task of "ecologizing" cultural artefacts, reaffirming the environmental composition and significance of literature, art, film, *inter alia*, whilst also identifying the presence of "the environment" in areas where it is less obviously present.[6]

3 Higginson, Ian, and Crosbie Smith. "'A Magnified Piece of Thermodynamics': The Promethean Iconography of the Refrigerator in Paul Theroux's *The Mosquito Coast*." *British Journal for the History of Science*, vol. 32, no. 3, 1999, pp. 325–42.

4 Smith, Crosbie. *The Science of Energy: A Cultural History of Energy Physics in Victorian Britain*. Athlone Press, 1998.

5 Beer, Gillian. "'The Death of the Sun': Victorian Solar Physics and Solar Theory." *Open Fields: Science in Cultural Encounter*. Oxford University Press, 1996.

6 Garrard, Greg, editor. *The Oxford Handbook of Ecocriticism*. Oxford University Press, 2014.

A sub-field of ecocriticism soon emerged that could be termed energo-criticism, in that its proponents sought to discern how the science of energy both expressed and informed the cultural milieu from which it emerged. For example, literary critic Barri Gold suggested Alfred Tennyson's *In Memoriam* (1851), a romantic elegy for his recently deceased friend, Arthur Hallam, could be interpreted as a "brilliant work of thermodynamics," which drew on the science of energy to mourn his loss whilst also celebrating the conservation of his friend's spirit in the form of a "diffusive power."[7] Gold's work joined a growing body of thermodynamic interpretations of literature, which sought to discern the cultural significance of energy in the work of a range of authors, from Joseph Conrad to Thomas Pynchon.[8]

Film studies also entered a period of ecological reflection, with the editor of a recent collection of essays calling for the adoption of an "ecocritically informed standpoint toward all forms of cinematic productions, not just those that overtly focus on environmental issues."[9] Adrian J. Ivakhiv's *Ecologies of the Moving Image* (2013) has called for an "ecophilosophy of the cinema," an engagement with the ecological implications of film, and a reconsideration of it as media that does not just document environments, but is an agent of environmental change in and of itself. An ecology of film, in Ivakhiv's approach, should not only involve the "material ecologies" of film production and broadcasting; it should also consider "perceptual ecologies," moving images that have the capacity to alter our perception of, or relation to, the world.[10] Amid this ecocritical turn, there has been an increasing emphasis on the role of energy in film, as both a subject and means of understanding the physicality of film production, its materiality, projection, and the reception of filmic images.

Of direct relevance to industrial film, Nadia Bozak has argued that "[c]inema has always demonstrated an awareness of its industrial self and therefore a connection to the environment, the realm from which it derives its power, raw materials and, often enough, subject matter." What is often considered an ephemeral and largely immaterial medium is, Bozak argues, grounded in a material and energetic substrate. She

7 Gold, Barri. *Thermopoetics: Energy in Victorian Literature and Science*. MIT Press, 2010, p. 56.

8 MacDuffie, Allen. "Joseph Conrad's Geographies of Energy." *English Literary History*, vol. 76, no. 1, 2009, pp. 75–98; Seed, David. "Order in Thomas Pynchon's 'Entropy'." *Journal of Narrative Technique*, vol. 11, no. 1, 1981, pp. 135–53.

9 Willoquet-Maricondi, Paula, editor. *Framing the World: Explorations in Ecocriticism and Film*. University of Virginia Press, 2010.

10 Ivakhiv, Adrian. *Ecologies of the Moving Image: Cinema, Affect, Nature*. Wilfrid Laurier University Press, 2013.

points out that we can rightly consider the cinematic image as "fossilized light, thus practically and metaphorically equating cinema with the geological dimensions of the naturally derived fuels (fossilized sunlight) that continue to enable industrial society and culture."[11] At least one reviewer of Bozak's work has suggested she marks the prelude to a coming "energy era in film studies."[12]

Bozak suggests the material and energetic embeddedness of film often remains overlooked in its popular forms, aside from overtly "environmental" films such as Al Gore's *Inconvenient Truth* (2006), owing to presumed disinterest of consumers and the diktats of commercialism. In industrial film, in which often the extraction and transformation of raw materials is the primary subject, we find many exceptions to this oversight. In support of this argument, we can also point to Bozak's notion of the "resource image," the idea that cinematic depiction does not merely turn material things into ephemeral moving forms – film becomes a resource in and of itself. Resource images render "visible the subordination of nature as the root of industrial culture."[13] This subordination occurs in the resource making of cinema: the manipulation of the photochemical process, light and the mechanisms of recording and projection allow the storage and conveyance of moving images over space and time.

Nature's subordination to industry is, of course, readily apparent in industrial film, as raw materials – through manual and mechanical exertion – are rendered into useable products. In industrial film we directly encounter what Bozak terms the "image of a resource," the documentation of the material and energetic basis of productive processes that "displays rather than disguises the energy economy's less than sublime origins."[14] If post-war industrial film was intended to celebrate industry's artful manipulation of matter and pursuit of material progress, the eschatological implications of contemporary environmental concerns has considerably altered the reception of such images. Industrial film can now be considered as testament to society's misplaced faith in energy-driven perpetual growth.[15]

11 Bozak, Nadia. *The Cinematic Footprint: Lights, Camera, Natural Resources*. Rutgers University Press, 2012.

12 Banita, Georgiana. "Prelude to the Energy Era in Film Studies" [review of *The Cinematic Footprint: Lights, Camera, Natural Resources* by Nadia Bozak]. *Imaginations*, vol. 3, no. 2, 2012, pp. 208–09.

13 Bozak. *The Cinematic Footprint*, p. 54.

14 Ibid., p. 127.

15 Mitchell, Timothy. "Economentality: How the Future Entered Government." *Critical Inquiry*, vol. 40, no. 4, 2014, pp. 479–507.

Energy as Sensorium

In its close documentation of the transformation of raw materials into secondary goods, industrial film provides accounts of physical changes in state achieved via the exertions of human and mechanical means for directing energy. As sociologist and business historian James Beniger has noted, manufacturing, at its most abstract, is simply "the synthesis of matter and energy into more organized forms."[16] For film and media scholar Florian Hoof, beside their promotional role, industrial films can be seen as repositories of evolving managerial knowledge regarding these organizational processes. Management consultants adopted film as a means of governing "processes of labour" and, we can suggest, flows of energy.[17] In the film *Master Hands* (1936) we see the stages of manufacture by which a Chevrolet was assembled in the Flint, Michigan, automobile plant.[18] We see a worker, off-screen, apply a matchstick to ignite a taper, and then a blowtorch bursts into flame and is inserted into a furnace. We see torrents of molten metal, and the eventual casting of the automobile's engine. All this productive power results from the thermic energy of a single matchstick. As Beniger explains, in manufacturing a "system can only sustain work if its internal energy is purposively organized in a heat gradient."[19] Production requires the transfer of energy from one system to another, according to differences in chemical potential or temperature. We see energy in the consequences of these transfers, the result of its successful conversion or wasteful diffusion.[20] In a volume documenting the first "Films That Work" conference, Vinzenz Hediger suggested a crisis in visibility had occurred in industrial film at the cusp of the information age, as manufacturing became directed at the manipulation of symbols rather than forms. In discussing computerized control in German industry, specifically Krupp, he pointed

16 Beniger, James. *The Control Revolution: Technological and Economic Origins of the Information Society*. Harvard University Press, 1986.

17 Hoof, Florian. *Angels of Efficiency: A Media History of Consulting*. Oxford University Press, 2020, p. 15.

18 The film was the work of the Detroit-based Jam Handy Organization, which specialized in making "soft sell" educational films on behalf of industry that contained an underlying commercial imperative. Prelinger, Rick. "Smoothing the Contours of Didacticism: Jam Handy and His Organization." *Learning with the Lights Off: Educational Film in the United States*, edited by Devin Orgeron et al. Oxford University Press, 2011.

19 Beniger. *The Control Revolution*, p. 46.

20 Shove, Elizabeth. "Revealing the Invisible: Sociology, Energy, and the Environment." *International Handbook of Environmental Sociology*, edited by Michael Redclift and Graham Woodgate. Edward Elgar, 1997.

out that the physics of production were no longer readily apparent, so film required animated annotation to explain how computerized industry worked. Computers were animated as if they were heat engines, in an attempt at instructive communication he terms "thermodynamic kitsch." By contrast, he notes, in "the thermodynamic age work, and the control of work, were essentially tied to visibility."[21] Production required acts of combustion or mechanical manipulation. Whilst the central protagonist, energy, remains invisible, via acts of transduction, there is no crisis of visibility in more physical productive processes.

In industrial film, the sensorial aspects of energy conversion come to the fore. In *Substance and Function* (1910), a systematic account of the role of concepts in scientific reasoning, philosopher of science Ernst Cassirer suggested the emergence of the modern concept of energy meant energy now superseded matter as the fundament of objective reality. In fact, it could even explain human perception. Quoting chemist Wilhelm Ostwald, Cassirer noted that: "What we see is nothing but radiating energy, which effects chemical changes in the retina of our eye that are felt as light."[22] Cassirer's use of Ostwald was intended to remind the reader that alongside conversion efficiencies energy is a bearer of "sensuous properties" which are perceived in qualitative terms. Sound, movement, heat and light are the primary ways that energy is experienced sensorially. Beginning from this observation, it can be argued that certain examples of industrial film can be thought of as the resensualisation of things normally relegated to quantitative processes of conversion in industrial production.

Energy is not a single definable entity, as geographer Vaclav Smil has sought to affirm, it is "rather an abstract collective concept, adopted by nineteenth-century physicists to cover a variety of natural and anthropogenic phenomena."[23] In industrial film we see these diverse phenomena in action. In *Master Hands* we are shown the searing heat of molten metal,

21 Hediger, Vinzenz. "Thermodynamic Kitsch: Computing in German Industrial Films, 1928/1963." *Films That Work: Industrial Film and the Productivity of Media*, edited by Vinzenz Hediger and Patrick Vonderau. Amsterdam University Press, 2009.

22 I thank Daniela Russ for drawing my attention to this aspect of Cassirer's work, and for her ongoing friendship and collaboration. Ostwald, Wilhelm. *Vorlesungen über Naturphilosophie*. 1901, pp. 159–60; Cassirer, Ernst. *Substance and Function and Einstein's Theory of Relativity*. Translated by William and Marie Swabey. Open Court, 1923. [Translation of *Substanzbegriff und Funktionsbegriff: Untersuchungen über die Grundfragen der Erkenntniskritik*. Verlag Bruno Cassirer, 1910.]

23 Smil, Vaclav. *Energy: A Beginners Guide*. Oneworld Publications, 2006.

the relentless motion of assembly lines, the clanging of mechanical casts, the glow of diffusing sparks as air is blown through molten metal. In British Petroleum's playful *Divertimento* (1968) we see the microscopic chemical refinement of hydrocarbons, the delicate geometrical and fractal like forms they assume at this scale. In Michelangelo Antonioni's *Sette canne, un vestito* (1949) we witness the transubstantiations of industrial chemistry, as reeds are first heated in vast steaming crucibles, then cleaned in bubbling tanks of chemically infused water. This solution is then manipulated in a continual process of refinement to extract fibres to create rayon, a cutting-edge fabric of post-war modernity.

In industrial film we see energy being converted into qualitatively different states with an arrestingly sensorial aspect. To paraphrase historian Anson Rabinbach, in documenting processes of transduction, we are reminded that modernity is primarily an achievement powered by the combustion of carbon.[24] Jean-François Lyotard, in one of his few pronouncements on film, compared the medium of film to the energetic but extravagant combustion of fireworks. Under capitalism such pyrotechnics could assume a function, which sprang "from the effort to eliminate aberrant movements, useless expenditures, differences of pure consumption. This film is composed like a unified and propagating body, a fecund and assembled whole transmitting instead of losing what it carries."[25] Following Lyotard, we might consider industrial film as a form of utilitarian pyrotechnics, a series of edited images intended to reveal productive conversions, via motion and light, in a sequence intended to both instruct and enchant with regard to the miracles of production.

Energy as Work

As stated, energy is an assessment of a system's ability to "do work." This curiously anthropocentric definition is why American physicist Percy Bridgman suggested of all nature's laws, thermodynamics "smells more of its human origin."[26] This conflation of meaning began in the nineteenth century, as engineers and natural philosophers sought to quantify the

24 Rabinbach, Anson. *The Human Motor: Energy, Fatigue, and the Origins of Modernity*. University of California Press, 1990.

25 With thanks to Dawid Kasprowicz for directing me to this reference. Lyotard, Jean. "Acinema." *Narrative, Apparatus, Ideology: A Film Theory Reader*, edited by Philip Rosen. Columbia University Press, 1986, pp. 349–59.

26 Bridgman, Percy. *The Nature of Thermodynamics*. Harper and Row, 1941.

power of steam engines in terms of labour. As Norton Wise stated, "steam engines, like all other engines, produced work by energy conversion. Engines therefore embedded both the natural philosopher's 'energy' and the political economist's 'labour value' as work."[27] In a strange conflation of engineering and cosmology, the growing credibility of thermodynamics meant the human principles of "work" and "waste" assumed universal import, becoming laws of "conservation" and "dissipation."[28] It was this realization, as Rabinbach documented, that led to the emergence of a "science of work" in late nineteenth-century society, a discipline aimed at optimizing the "human motor."[29]

Drawing on energy's conceptual origins, we can address the idea that energy is not only a prosthesis to human labour but its replacement. Rabinbach believes Marx most clearly affirmed a belief in an equivalence between labour and machines. By the time he wrote *Capital* (1867), Rabinbach argues, Marx considered efficient machinery, if in the right hands, as the means by which labour would be emancipated from perpetual servitude.[30] The expenditure of energy on film must surely be considered in its relation to labour. Chevrolet's *Master Hands* (1936), like many other industrial films, shows human labour augmented by combustion; at one point a boiler's dial shows fluctuations in the physiologically grounded units of "horsepower" (fig. 3.1). Each unit of expended energy is in some way equivalent to, and therefore possibly a replacement for, animal or human labour. Whether such machinic energy leads to labour's emancipation is, of course, a question of politics.

What *Master Hands* ignored, for obvious reasons, was an ongoing conflict between labour and industry at the plant it depicted. That same year, workers at Chevrolet joined others in a violently opposed general automotive workers' strike. Its Flint Michigan factory was considered a "slave plant" by trade unionists, while General Motors, Chevrolet's holding company, was known for its poor working conditions. Smoke, gasses, bubbling pots of cyanide solution, and static electricity assailed workers. On the assembly line, they carried out repetitious tasks at an unforgiving rate, driven by a bonus system considered a "racket" and "predicated on getting every bit

27 Wise, Norton. "Mediating Machines." *Science in Context*, vol. 2, no. 1, 1988, pp. 77–113.

28 Smith, Crosbie, and Norton Wise. "Work and Waste: Political Economy and Natural Philosophy in Nineteenth Century Britain." *History of Science*, vol. 27, no. 3, 1989, pp. 263–301.

29 Rabinbach. *The Human Motor*; see also Curtis, Scott. *The Shape of Spectatorship: Art, Science, and Early Cinema in Germany*. Columbia University Press, 2015.

30 Rabinbach. *The Human Motor*, p. 73.

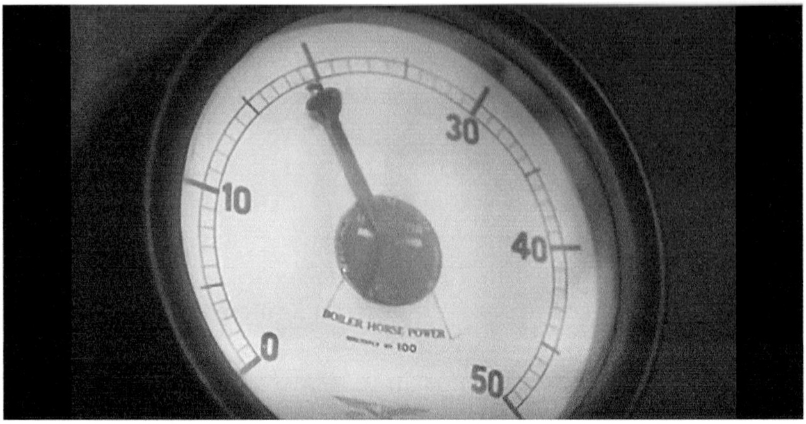

Fig. 3.1. Boiler horse power gauge. Still frame from *Master Hands* (1936). © Handy (Jam) Organization, archival footage supplied by Internet Archive (archive.org) in association with Prelinger Archives.

of work out of a person [...] that he could stand."[31] As Marx had predicted, without political resistance, industrial work "confiscates every atom of freedom, both in bodily and intellectual activity."[32] Human labour would be subsumed by the pursuit of capital. As energy and mechanization allowed the increasing automation of industrial processes, humans' role on the factory floor was increasingly limited to those activities that could not be automated.[33] In pursuit of efficiency and profit, workers found themselves "deskilled" by machines, which in turn, justified their lower wages.[34] Far from the harmonious energetic augmentation of human labour depicted in *Master Hands*, increasingly efficient industrial conversions were almost invariably at the expense of the worker.

Energy as Efficiency

The nineteenth-century discovery that energy dissipated as it was converted made its efficient use an imperative. Hence, as the science of work evolved,

31 West, Kenneth. "'On the Line': Rank and File Reminisces of Working Conditions and the General Motors Sit-down Strike of 1936–1937." *Michigan Historical Review*, vol. 12, no. 1, 1986, pp. 57–82.

32 Rabinbach. *The Human Motor*, p. 74.

33 Ford engineering executive Del Harder first applied the term "automation" to manufacturing in 1947. He did not mean automated servo-mechanical or computational control, but rather an increase in electro-mechanical, hydraulic, and pneumatic means of production. See Noble, David. *America by Design: Science, Technology and the Rise of Corporate Capitalism*. Oxford University Press, 1979.

34 Idem. *Forces of Production: A Social History of Industrial Automation*. Transaction, 2011.

its adherents emphasized the need for conservation. First proposed by American mechanical engineer Frederick W. Taylor in 1911, the principles of scientific management were encapsulated by his dictum:

> Science, not rule of thumb.
> Harmony, not discord.
> Cooperation, not individualism.
> Maximum output, in place of restricted output.
> The development of each man to his greatest efficiency and prosperity.

Taylorism was a science of work, intended not only to optimize worker efficiency through measurement, but also to dampen potential labour disputes in a period of growing industrial unrest. Harmony, Taylor claimed, could be achieved by increasing productivity for the manager and increasing prosperity for the worker, primarily through the analysis of the worker's output over time, and the development of new working practices or systems of incentive that could increase individual effort and overall productivity.[35] For Taylor, the measurement of energy provided a uniform unit to analyse the operations of a given worker's movements in the course of a work shift. Both mechanical and human action could be quantified in "foot-pounds of energy exerted for each man each day," allowing clear cut comparison between the efficacy of different workers and different configurations of labour.[36]

As Hoof details, the principles of Taylorism were integrated with filmic technologies by industrial engineer and consultant Frank B. Gilbreth, who used film to capture worker's movements as a sequence of images. Melding graphical depictions of labourers in motion with Taylor's promise of increasing labour's output via its measurement, Gilbreth sold a purportedly scientific approach to optimizing industrial productivity based on analysing film.[37] Again, energy was at the fore, as Gilbreth and his wife Lillian would pitch their motion film method as a means of achieving something called a "fatigue study" in which pictorial analysis of movement over time could be used to minimize the "unnecessary waste of human energy" that dissipated

35 Maier, Charles. "Between Taylorism and Technocracy: European Ideologies and the Vision of Industrial Productivity in the 1920s." *Journal of Contemporary History*, vol. 5, 1970, pp. 27–61.
36 Taylor, Frederick W. *The Principles of Scientific Management*. Harper and Brothers, 1911, pp. 55–57.
37 Hoof. *Angels of Efficiency*. As Hoof notes, Gilbreth's children would gently satirize their parent's work in the novel *Cheaper by the Dozen* (1948), which was made into a feature film in 1950.

in every misdirected effort, exaggerated movement, and even improperly spent period of leisure time.[38]

Whilst such forms of management had been intended to quell class tensions, disruptions in industrial flow often resulted from breakdowns in labour relations. In such situations, as we saw at Chevrolet's Michigan plant, rather than a salve, the physiological and psychological implications of the finely tuned rationalization of workers' embodied dispositions were often resented and bitterly opposed on occasion. Such scientism could be seen as part of the dehumanizing processes of alienation that capital inflicted on labour.[39] If efficient energy use was central to this managerial ethos, one way of manifesting opposition, proposed institutionalist economist Thorstein Veblen, was the "conscientious withdrawal of efficiency."[40] For Veblen, withdrawal did not have to involve full-scale strikes, of the kind that later occurred at Chevrolet, but could mean an array of sabotaging micro-aggressions, from "surreptitious manoeuvres of delay, obstruction, friction and defeat."[41] But he also argued managerial interventions could be considered acts of sabotage if wrongly motivated by the pursuit of pecuniary advantage rather than the "conservation of economic energy."[42]

Returning to the manifestation of such concerns in film, in Fiat's *Sotto i tuoi occhi* (1931) we see not only the rationalization of automobile manufacturing, but how the factory itself, a monumental hippodrome designed by Mattè Trucco, was a concrete manifestation of scientific management. Trucco had studied Taylor's work on the advice of Giovanni Agnelli, Fiat's founder. The Lingotti plant realized Taylorist principles of flow: raw materials entered at the bottom of the building, and assembly continued up each floor of the building, until the cars reached the roof, where the climax of the film shows a number of recently made Fiat 522s being dramatically test-run in formation on a steeply banked track overlooking Turin, their six-cylinder internal combustion engines producing the equivalent motive energy of

38 Ibid., ch. 3; see also Gilbreth, Frank, and Lillian Gilbreth. *Fatigue Study: The Elimination of Humanity's Greatest Unnecessary Waste: A First Step in Motion Study.* Sturgis & Walton Company, 1916.

39 Karns Alexander, Jennifer. *The Mantra of Efficiency: From Waterwheel to Social Control.* Johns Hopkins University Press, 2008.

40 Knoedler, Janet. "Veblen and Technical Efficiency." *Journal of Economic Issues*, vol. 31, no. 4, 1997, pp. 1011–26.

41 Veblen, Thorstein. *The Engineers and the Price System.* Viking, 1921.

42 White, Ron D. "Growth versus Conservation: A Vebelenian Perspective." *Journal of Economic Issues*, vol. 12, no. 2, pp. 427–33; Veblen. 1900. Qtd. in Mirowski, Philip. *Against Mechanism: Protecting Economics from Science.* Rowman & Littlefield, 1988, p. 1.

fifty or sixty horses.[43] Clearly the pursuit of efficiency documented in this industrial film not only employed the visual rhetoric of productive transductions in energy, but also showed how such processes could be housed in elegant architectural forms that were themselves intended to maximize the efficiency with which energy was converted in processes of production.

In his study of the depiction of electrical appliances in the promotional films of Électricité de France (EDF), historian Yves Bouvier suggests the ways in which energy suppliers chose to present themselves offers a means to "deconstruct the cultural representations associated with energy and to analyse how these representations emerged within the scope of business strategies."[44] In many industrial films industry represented itself via elegant, smooth productive flows and the unbroken conveyance of a companies' products. In so doing, industrial film conveyed the seamless logic of managerial rationality, and the corporation could present itself as a benevolent guardian against inefficiency. The resolution of disruption was the theme of *La pattuglia del Passo San Giacomo* (1954), in which Ermanno Olmi, filming for Edison-Volta, Italy's national power company, presented a romanticized depiction of power line repair in the Italian Alps. The film documents a patrol team of engineers, whose work and good humour are in harmony with a pristine snow-covered landscape and its rural people, grateful for but unchanged by electrification.

The reinstatement of energy supply resonated with Italy's post-war reconstruction. Film historian Paulo Bonifazio explains that Edison-Volta had its own film unit, Sezione Cinema Edison-Volta (SCE), and that the majority of Olmi's films for SCE portrayed a beneficent managerial paternalism that it was hoped would diffuse through the nation. Such paternal benevolence was evident to Olmi, whose father and mother had worked for Edison-Volta, and who joined the company at fifteen – they even gave him his first camera. Bonifazio suggests Edison-Volta's aim was not to sell their product, as they were Italy's sole electricity supplier at the time. Rather, their aim was to sell the benefits of hydropower technology to rural Italians, hence the film depicts technology in harmony with nature, against a quiet, unobtrusive soundtrack filled with running water and crisp snow.[45] Olmi's film reiterated an older

43 Kirk, Terry. *The Architecture of Modern Italy, Volume 2: Visions of Utopia, 1900–Present*. Princeton Architectural Press, 2005.

44 Bouvier, Yves. "Filming Electrical Consumption: EDF's Promotional Films (1946–2004)." *Past and Present Energy Societies: How Energy Connects Politics, Technologies and Cultures*, edited by Nina Möllers and Karin Zachmann. Transcript Verlag, 2012.

45 Bonifazio, Paulo. *Schooling in Modernity: The Politics of Sponsored Films in Post-war Italy*. University of Toronto Press, 2014.

idea that electricity held a utopian potential in contrast to the dirty and labour-intensive direct use of coal. During the New Deal Era American critic and philosopher Lewis Mumford (1895–1990) argued that the use of electricity could lead to a "neotechnic" society in which electricity-driven prosperity would engender social and natural order. In his *Technics and Civilization* (1934), Mumford described an electrified world in which "[t]he same qualities prevail in the power station or the factory as in the kitchen or the bathroom of the individual dwelling. In any one of these places one could eat off the floor."[46] This image of a pristine productive environment, in which processes flow without hindrance from an incompliant nature or labour force, in which interruptions in productive flows are skilfully and harmoniously corrected, seems a persistent narrative in the rhetoric of industrial film.

Energy as Friction

But just as new energy technologies displaced or degraded labour rather than replacing it, the same could be said of new forms of energy supply. In post-war Italy, and for Western Europe in general, the American-led recovery, the so-called "Marshall Plan," was largely predicated upon a shift from coal to oil as the primary feedstock for generating electrical power. Around ten per cent of the funding Europe received in Marshall Plan aid was recycled back into the coffers of American owned oil companies funnelling oil from the Middle East.[47] Italy had become the "refinery of Europe," the unloading point for oil tankers arriving in the Mediterranean via the Suez Canal.[48] Historian Timothy Mitchell argues this shift in fuel dependency was partly intended to break the control of increasingly nationalized and often socialist coal mining industries had across much of Europe: coal depended upon miners' labour power, whereas, once infrastructures were in place, oil could be pumped and piped from wells with relative ease.[49] By 1956, around 65 million tons of oil came by tanker from the Persian Gulf each year, another 35 million came via the new Trans-Arabian pipeline and the older British-owned "Iraqi Petroleum Company" pipeline.[50] Such

46 Carey, James, and John Quirk. "The Mythos of the Electronic Revolution." *American Scholar*, vol. 39, no. 3, 1970, pp. 395–424.

47 Painter, David. "The Marshall Plan and Oil." *Cold War History*, vol. 9, no. 2, 2009, pp. 159–75.

48 King, Russell. *The Industrial Geography of Italy*. Routledge, 2015.

49 Mitchell. *Carbon Democracy: Political Power in the Age of Oil*. Verso, 2013.

50 OEEC. *Europe's Need for Oil: Implications and Lessons for the Suez Crisis*. Organisation for European Economic Co-operation, 1958.

globe-spanning infrastructure, such as pipelines, often become sites of "friction," echoing Veblen's notion of the withdrawal of efficiency. Borrowed from physics, anthropologist Anna Tsing suggests friction encapsulates the "awkward, unequal, unstable and creative qualities of interconnection across difference" that globalization entails.[51] Here it is suggested the concept of friction, often absent from the smooth processes shown in industrial film, particularly with regard to oil, provides one means of addressing energy as an aspect of industrial film.

In *Desert Venture* (1958 [1947]) the consequences of the new global geography of energy were made clear. An orientalist image of Saudi Arabia, the site of the Arabian-American Oil Company's (ARAMCO) newly discovered reserves is shown to the viewer via a sequence of shots contrasting the physical labour of Arabs with the highly technologized work of the American oil company. Cultural frictions are absent, the two nations work harmoniously, exchanging smiles and each in awe at the peculiarities of the other. Such beneficent cooperation could be seen as a stage-managed attempt to soften ARAMCO's control over this premodern nation. America's first Saudi Arabian ambassador had described ARAMCO as an "octopus," whose tentacles extended in every domain. He warned Washington that this private American company would dominate the Middle East. Alert to such accusations, ARAMCO spent millions of dollars in portraying an image of partnership rather than parasitism. *Desert Venture* failed in this respect, as Crown Prince Ibn Saud considered the film an insult, due to its presentation of ARAMCO as the Saudis' saviour.[52]

Off-screen frictions are readily apparent elsewhere in "petrofilms," as film scholar Mona Damluji terms them.[53] She argues petrofilms are a distinct genre of industrial film, unconcerned with selling the company's product – oil – so much as selling the image of a civilizing, modernizing, and altruistic company that can smoothly extract a nation's oil and leave progress in its wake. Much like *Desert Venture,* she describes how the Anglo-Iranian Oil Company (AIOC) put out *Persian Story*[54] (1951), a rosy take on Iran's response to the extraction of its oil, which ignored the fact that Iranian oil workers had just begun to protest against the exploitation and their

51 Tsing, Anna. *Friction: An Ethnography of Global Connection.* Princeton University Press, 2005.

52 Vitalis, Robert. *America's Kingdom: Mythmaking on the Saudi Oil Frontier.* Stanford University Press, 2006.

53 Damluji, Mona. "The Image World of Middle Eastern Oil." *Subterranean Estates: Life Worlds of Oil and Gas,* edited by Hannah Appel et al. Cornell University Press, 2015.

54 Note: not screened at "Films That Work."

unsanitary living arrangements. Their strike, a conscious withdrawal of efficiency, eventually encouraged Prime Minister Mohammad Mossadegh to nationalize Iranian oil, so beginning a move toward oil nationalism in the Middle East that precipitated the crisis of the 1970s and continues to provoke geopolitical tensions. As Damluji notes, a work of fiction, the film provided British audiences with an upbeat account of the Iranian oil industry that wholly ignored the ongoing Iranian oil crisis.

Similarly ameliorant, scholar Georgiana Banita argues that Bernardo Bertolucci's *La via del petrolio* (1967), in its artfully detached portrayal of oil's conveyance from Iran to Europe, demonstrates "blissful immunity from the political critique that has shaped the late twentieth century's energy conscience."[55] Funded by Italy's Ente Nazionale Idrocarburi (ENI), *La via del petrolio* depicts the monolithic infrastructures of oil transport, its extraction, pumping, and shipping through the Suez Canal to the refineries of Genoa, and on to Switzerland and Bavaria. Though frictions are alluded to, particularly the Suez Crisis of 1956, during which the Suez Canal was blockaded by Egyptian president Abdel Nasser, which forced tankers to take a 9,600-mile detour while delivering Europe's oil, they were presented as assailable. In Bertolucci's film, once more, the rhetoric of petrofilms suggest a harmonious geopolitics. This is the legitimating function of such films, attempts to render extractive relations free from political critique. However, from our contemporary perspective, in which the oil-rich Middle East has been riven with conflict, and into which human energy use has transformed the climate and other aspects of the Earth system, these past attempts to evade critique appear naïve if not malevolent. To return to Ivakhiv, in light of our current energetic condition, we can reaffirm his statement that films "make a cut in time (and in space), but this cut is artificial and ultimately unsustainable; the real will always return in one form or another."[56]

In the novelist W.G. Sebald's *Die Ringe des Saturn*, during a hallucinatory journey through Suffolk, England, the unnamed protagonist observes that "[c]ombustion is the hidden principle behind every artefact we create."[57] This chapter has argued that the combustive origins of our contemporary condition were graphically affirmed in industrial film. Combustion, the energetic feedstock of industrial production, imposed obligations on the

55 Banita, Georgiana. "From Isfahan to Ingolstadt: Bertolucci's *La via del petrolio* and the Global Culture of Neorealism." *Oil Culture*, edited by Ross Barrett and Daniel Worden. University of Minnesota Press, 2014.

56 Ivakhiv. *Ecologies of the Moving Image*, p. 305.

57 Sebald, W.G. *The Rings of Saturn*. Translated by Michael Hulse. Harvill, 1998. [Translation of *Die Ringe des Saturn. Eine englische Wallfahrt*. 1995.]

managerial class that found expression in industrial film. In proposing a tentative thermodynamic reading of industrial film, it is suggested the laws of energy can be read in at least four registers. First, industrial film re-evokes the sensorial qualities of productive transductions in energy. Combustion is not hidden, as Sebald observed, but affirmed, reminding the viewer of the miraculous plasticity that humankind's control over heat transfer is able to achieve, and the visual spectacle such productive pyrotechnics entail. Second, the augmentation of human labour with power derived from energy resources are readily apparent. The controlled use of energy extends the human capacity for labour to a superhuman extent. These films emphasize the harmonious nature of this co-production. No more so than in the third register, in which the visual rhetoric of efficient energy transfer is foregrounded, with imagery of industry engaging in unbroken transductions, with energy applied purposively through the heat gradient. But fourth, and finally, a thermodynamic reading of such demonstrative efficiency ignores the second law of thermodynamics. Entropy, or friction, is apparent in what lies off camera, be it conscious withdrawal of efficiency, organized labour, waste, finitude or the transformation of the atmosphere.

Works Cited

Banita, Georgiana. "From Isfahan to Ingolstadt: Bertolucci's *La via del petrolio* and the Global Culture of Neorealism." *Oil Culture*, edited by Ross Barrett and Daniel Worden, University of Minnesota Press, 2014.

Banita, Georgiana. "Prelude to the Energy Era in Film Studies" [review of *The Cinematic Footprint: Lights, Camera, Natural Resources* by Nadia Bozak]. *Imaginations*, vol. 3, no. 2, 2012, pp. 208–09.

Barry, Andrew. "Thermodynamics, Matter, Politics." *Distinktion: Journal of Social Theory*, vol. 16, no. 1, 2015, pp. 110–25.

Beer, Gillian. "'The Death of the Sun': Victorian Solar Physics and Solar Theory." *Open Fields: Science in Cultural Encounter*. Oxford University Press, 1996.

Beniger, James. *The Control Revolution: Technological and Economic Origins of the Information Society*. Harvard University Press, 1986.

Bonifazio, Paulo. *Schooling in Modernity: The Politics of Sponsored Films in Post-war Italy*. University of Toronto Press, 2014.

Bouvier, Yves. "Filming Electrical Consumption: EDF's Promotional Films (1946–2004)." *Past and Present Energy Societies: How Energy Connects Politics, Technologies and Cultures*, edited by Nina Möllers and Karin Zachmann. Transcript Verlag, 2012.

Bozak, Nadia. *The Cinematic Footprint: Lights, Camera, Natural Resources*. Rutgers University Press, 2012.

Bridgman, Percy. *The Nature of Thermodynamics*. Harper and Row, 1941.

Carey, James, and John Quirk. "The Mythos of the Electronic Revolution." *American Scholar*, vol. 39, no. 3, 1970, pp. 395–424.

Cassirer, Ernst. *Substance and Function and Einstein's Theory of Relativity*. Translated by William and Marie Swabey, Open Court, 1923. [Translation of *Substanzbegriff und Funktionsbegriff: Untersuchungen über die Grundfragen der Erkenntniskritik*. Verlag Bruno Cassirer, 1910.]

Curtis, Scott. *The Shape of Spectatorship: Art, Science, and Early Cinema in Germany*. Columbia University Press, 2015.

Damluji, Mona. "The Image World of Middle Eastern Oil." *Subterranean Estates: Life Worlds of Oil and Gas*, edited by Hannah Appel, Arthur Mason, and Michael J. Watts, Cornell University Press, 2015.

Garrard, Greg, editor. *The Oxford Handbook of Ecocriticism*. Oxford University Press, 2014.

Gilbreth, Frank, and Lillian Gilbreth. *Fatigue Study: The Elimination of Humanity's Greatest Unnecessary Waste: A First Step in Motion Study*. Sturgis & Walton Company, 1916.

Gold, Barri. *Thermopoetics: Energy in Victorian Literature and Science*. MIT Press, 2010.

Hediger, Vinzenz. "Thermodynamic Kitsch: Computing in German Industrial Films, 1928/1963." *Films That Work: Industrial Film and the Productivity of Media*, edited by Vinzenz Hediger and Patrick Vonderau. Amsterdam University Press, 2009.

Higginson, Ian, and Crosbie Smith. "'A Magnified Piece of Thermodynamics': The Promethean Iconography of the Refrigerator in Paul Theroux's *The Mosquito Coast*." *British Journal for the History of Science*, vol. 32, no. 3, 1999, pp. 325–42.

Hoof, Florian. *Angels of Efficiency: A Media History of Consulting*. Oxford University Press, 2020.

Ivakhiv, Adrian. *Ecologies of the Moving Image: Cinema, Affect, Nature*. Wilfrid Laurier University Press, 2013.

Karns Alexander, Jennifer. *The Mantra of Efficiency: From Waterwheel to Social Control*. Johns Hopkins University Press, 2008.

King, Russell. *The Industrial Geography of Italy*. Routledge, 2015.

Kirk, Terry. *The Architecture of Modern Italy, Volume 2: Visions of Utopia. 1900–Present*. Princeton Architectural Press, 2005.

Knoedler, Janet. "Veblen and Technical Efficiency." *Journal of Economic Issues*, vol. 31, no. 4, 1997, pp. 1011–26.

Lyotard, Jean. "Acinema." *Narrative, Apparatus, Ideology: A Film Theory Reader*, edited by Philip Rosen. Columbia University Press, 1986, pp. 349–59.

MacDuffie, Allen. "Joseph Conrad's Geographies of Energy." *English Literary History*, vol. 76, no. 1, 2009, pp. 75–98.

Maier, Charles. "Between Taylorism and Technocracy: European Ideologies and the Vision of Industrial Productivity in the 1920s." *Journal of Contemporary History*, vol. 5, 1970, pp. 27–61.

Mirowski, Philip. *Against Mechanism: Protecting Economics from Science*. Rowman & Littlefield, 1988.

Mitchell, Timothy. *Carbon Democracy: Political Power in the Age of Oil*. Verso, 2013.

Mitchell, Timothy. "Economentality: How the Future Entered Government." *Critical Inquiry*, vol. 40, no. 4, 2014, pp. 479–507.

Noble, David. *America by Design: Science, Technology and the Rise of Corporate Capitalism*. Oxford University Press, 1979.

Noble, David. *Forces of Production: A Social History of Industrial Automation*. Transaction, 2011.

OEEC. *Europe's Need for Oil: Implications and Lessons for the Suez Crisis*. Organisation for European Economic Co-operation, 1958.

Painter, David. "The Marshall Plan and Oil." *Cold War History*, vol. 9, no. 2, 2009, pp. 159–75.

Prelinger, Rick. "Smoothing the Contours of Didacticism: Jam Handy and His Organization." *Learning with the Lights Off: Educational Film in the United States*, edited by Devin Orgeron, Marsha Orgeron, and Dan Streible. Oxford University Press, 2011, pp. 338–55.

Pringle, Thomas. "Photographed by the Earth: War and Media in Light of Nuclear Events." *NECSUS European Journal of Media Studies*, vol. 3, no. 2., 2014, pp. 131–54.

Rabinbach, Anson. *The Human Motor: Energy, Fatigue, and the Origins of Modernity*. University of California Press, 1990.

Sebald, W.G. *The Rings of Saturn*. Translated by Michael Hulse. Harvill, 1998. [Translation of *Die Ringe des Saturn. Eine englische Wallfahrt*. 1995.]

Seed, David. "Order in Thomas Pynchon's 'Entropy'." *Journal of Narrative Technique*, vol. 11, no. 1, 1981, pp. 135–53.

Shove, Elizabeth. "Revealing the Invisible: Sociology, Energy, and the Environment." *International Handbook of Environmental Sociology*, edited by Michael Redclift and Graham Woodgate. Edward Elgar, 1997, pp. 261–73.

Smil, Vaclav. *Energy: A Beginners Guide*. Oneworld Publications, 2006.

Smith, Crosbie. *The Science of Energy: A Cultural History of Energy Physics in Victorian Britain*. Athlone Press, 1998.

Smith, Crosbie, and Norton Wise. "Work and Waste: Political Economy and Natural Philosophy in Nineteenth Century Britain." *History of Science*, vol. 27, no. 3, 1989, pp. 263–301.

Taylor, Frederick W. *The Principles of Scientific Management*. Harper and Brothers, 1911.
Tsing, Anna. *Friction: An Ethnography of Global Connection*. Princeton University Press, 2005.
Veblen, Thorstein. *The Engineers and the Price System*. Viking, 1921.
Vitalis, Robert. *America's Kingdom: Mythmaking on the Saudi Oil Frontier.* Stanford University Press, 2006.
West, Kenneth. "'On the Line': Rank and File Reminisces of Working Conditions and the General Motors Sit-down Strike of 1936–1937." *Michigan Historical Review*, vol. 12, no. 1, 1986, pp. 57–82.
White, Ron D. "Growth versus Conservation: A Vebelenian Perspective." *Journal of Economic Issues*, vol. 12, no. 2, pp. 427–33.
Willoquet-Maricondi, Paula, editor. *Framing the World: Explorations in Ecocriticism and Film*. University of Virginia Press, 2010.
Wise, Norton. "Mediating Machines." *Science in Context*, vol. 2, no. 1, 1988, pp. 77–113.

About the Author

Thomas Turnbull is a researcher at the Max Planck Institute for the History of Science in Berlin, where his focus is the history of energy in all its forms, and he works as part of a wider research group called "Anthropocene Formations." Recent publications include "Energy, History, and the Humanities: Against a New Determinism" (*History and Technology*, 2021) and "Toward Histories of Saving Energy: Erich Walter Zimmermann and the Struggle against 'One-Sided Materialistic Determinism'" (*Journal of Energy History/Revue d'histoire de l'énergie*, 2020).

4 Digital Afterlife of Industrial Film

Weak Dispositives, Choice Architecture and the Distribution of Industrial Cinema

Florian Hoof

Abstract
This chapter looks at films of the German car industry and their reappearance in post-cinematic exhibition spaces as an exemplary case to theorize the "afterlife" of non-theatrical film. It is argued that the circulation of such films within digital cultures can be understood as a relation between the persistence of visual forms and changing modes of perception. This results in a mode of cinematic experience that is structured by "weak dispositives," as part of the exhibition space as choice architecture. The chapter extends existing mostly historical scholarly work on non-theatrical film to concepts of cinematic experience, digital "nudging" and the politics of digital media distribution.

Keywords: digital media distribution; non-theatrical film; afterlife of film; nudging; cinematic experience; museum

The "Reappearance" of Industrial Film

The last century saw extensive changes in the way industrial societies and media are operating. This is being described as the rise of the "post-industrial society,"[1] the "knowledge society"[2] or the "networked information

1 Bell, Daniel. *The Coming of Post-industrial Society*. Harper Colophon Books, 1974.
2 Drucker, Peter. *The Age of Discontinuity: Guidelines to Our Changing Society*. New York, Harper and Row, 1969; Machlup, Fritz. *The Production and Distribution of Knowledge in the United States*. Princeton University Press, 1962.

Hediger, V., F. Hoof, Y. Zimmermann, with S. Anthony (eds), *Films That Work Harder: The Circulation of Industrial Film*. Taylor & Francis Group 2024
DOI: 10.5117/9789462986534_CH04

society."[3] Communication devices, strategies and aesthetics used to organize, create markets and value,[4] or to provide for self-assurance within industrial societies shifted as well.[5] This in particular affected industrial film a once important medium of industrial organizations that went through significant transformations since its heyday in the twentieth century. Industrial films are no longer produced in huge quantities, they are no longer screened in cinemas and national industrial film distribution bodies have vanished. Indeed, as a phenomenon of the Second Industrial Revolution and tied to early organizational forms of mass production, they still matter. In fact, the emerging digital media distribution infrastructure has made industrial films more widely available than ever before.

While over time industrial film ceased to be solely a medium of organization and gradually reappeared as a palimpsest of the past, used as an index for historical and aesthetic meaning, it became a vibrant field of research for film and media studies. Within this field approaches mostly situate industrial film in a historical perspective, for example, as "useful cinema"[6] commissioned for a specific organizational purpose, as a means of rhetoric or rationalization or as part of a wider media network or visual culture.[7] This enabled us to better understand how such films mattered in stabilizing organizational structures, social concepts and discourses in industrial societies.

Building on this rich body of research and taking this diagnosis as a point of departure, I pivot the existing historical approaches and extend the perspective on the "afterlife"[8] of industrial film: on films that are no longer

3 Castells, Manuel. *The Information Age: Economy, Society and Culture, Vol. I: The Rise of the Network Society*. Blackwell, 1996.

4 Hutter, Michael. *The Rise of the Joyful Economy: Artistic Invention and Economic Growth from Brunelleschi to Murakami*. Routledge, 2015.

5 Beniger, James. *The Control Revolution: Technological and Economic Origins of the Information Society*. Harvard University Press, 1986.

6 Acland, Charles R., and Haidee Wasson, editors. *Useful Cinema*. Duke University Press, 2011.

7 Hoof, Florian. *Angels of Efficiency: A Media History of Consulting*. Oxford University Press, 2020; Hediger, Vinzenz, and Patrick Vonderau. "Record, Rhetoric, Rationalization: Industrial Organization and Film." *Films That Work: Industrial Film and the Productivity of Media*, edited by Vinzenz Hediger and Patrick Vonderau. Amsterdam University Press, 2009, pp. 35–49; Elsaesser, Thomas. "Die Stadt von Morgen. Filme zum Bauen und Wohnen in der Weimarer Republik." *Geschichte des dokumentarischen Films in Deutschland, Band 2: Weimarer Republik 1918–1933*, edited by Klaus Kreimeier, Antje Ehmann, and Jeanpaul Goergen. Reclam, 2005, pp. 381–409.

8 Warburg, Aby. *Nachhall der Antike. Zwei Untersuchungen*. Edited by Pablo Schneider. Diaphanes, 2011; Warburg, Aby. *Der Bilderatlas Mnemosyne*. Edited by Martin Warnke and Claudia Brink. Akademie Verlag, 2000. Didi-Huberman, Georges. *The Surviving Image: Phantoms of Time and Time of Phantoms: Aby Warburg's History of Art*. Penn State University Press, 2016.

part of the functionalist setting in which they were once commissioned, produced and screened and that started to reappear in a decisively different economic, cultural and social context. I specifically focus on industrial films produced by German automakers throughout the twentieth century: Moving images that originated in the context of production regimes such as Fordism in the 1920s, the car racing Nazi propaganda films of the 1930s or the waves of automation in the 1970s. Taking the Mercedes-Benz Museum, the alleged "Louvre of car museums,"[9] as an exemplary example, I develop a theoretical model to describe the afterlife of industrial films. It specifies the reappearance of industrial film as a cinematic experience structured by "weak dispositives" that are constantly being stabilized by relating them to cinematic time and space. Finally, I have a look at the politics of decentralized post-cinematic dispositives and digital distribution intermediaries.

Persistence of Visual Forms and Changing Modes of Perception

The afterlife of industrial films touches upon two related constitutive aspects: the persistence of visual forms and the changing modes of perception. Both dimensions relate to two strands in film and media theory that to date intersect only partially.

First, the *persistence of visual forms*: How do aesthetic values, forms and formats that were established throughout the long history of industrial film change when industrial film reappears or is reanimated at present time? This strand relates loosely to a body of work that has looked into the history of screen technologies and aesthetics in context of industrial film, exhibitions, world fairs and (automobile) industry pavilions.[10]

Second, the *changing modes of perception*: How does industrial film originally produced for quite different film dispositives and occasions including cinemas, exhibitions, small-gauge projection devices for salesmen, crash test research or tools for industrial rationalization, relate to contemporary modes of (post-)cinematic experience that have been said to be going through a shift from a centralized to a "decentralized gaze"[11]? This

9 Williams, Stephen. "Touring the Temples of German Automaking." *New York Times*, December 31, 2009, www.nytimes.com/2010/01/03/automobiles/03museums.html.

10 Wasson, Haidee. "Cinema and Industrial Design: Showmanship, Fairs and the Exhibition Film." This volume; Huhtamo, Erkki. "Screen Tests: Why Do We Need an Archaeology of the Screen?" *Cinema Journal*, vol. 51, no. 2, 2012, pp. 144–48.

11 Casetti, Francesco. "Back to the Motherland. The Film Theatre in the Postmedia Age." *Screen*, vol. 52, no. 1, 2011, pp. 1–12.

strand refers to a body of work that focuses on "digital media distribution"[12] and on the "relocation"[13] of the cinematic experience from the movie theatre to "digital media intermediaries."[14]

One way to describe what happens to industrial films when they have left their original context of production and projection is to focus on processes of relocation of film and how this alters the way those films are appreciated. The concept of relocation differentiates between a period of time when the cinematic experience was part of the movie theatre as the dominant and centralized film dispositives and the contemporary situation of mobile and ubiquitous screens. From time to time, industrial films were screened in movie theatres, but probably more significant is its history connected to a wide range of different film dispositives and formats, including exhibition and fair films[15] or vocational training films.[16] Films produced by the car industry were screened in the futuristic twentieth-century world fair pavilions,[17] but also include scientific crash test footage or small-gauge product presentation films for sales agents. The latter quite literally were produced to be relocated by using the compact setting of a film reel and a mobile film projector to present bulky products throughout the country. Here, the relocation of film is not so much an exception but the only reason why such industrial films exist.[18] This is only one case out of the variety of dispositives that ever since were used for quite diverging purposes. But it shows that the afterlife of industrial film is hard to grasp by arguments

12 McDonald, Paul, Courtney Brannon Donoghue, and Timothy Havens, editors. *Digital Media Distribution. Portals, Platforms, Pipelines,* NY University Press, 2021.

13 Casetti. "Back to the Motherland."; Casetti, Francesco. *The Lumière Galaxy: Seven Keywords for the Cinema to Come.* Columbia University Press, 2015.

14 Cunningham, Stuart and David Craig, *Social Media Entertainment. The New Intersection of Hollywood and Silicon Valley.* New York University Press, 2019, pp. 115–147.

15 Griffiths, Alison. *Wondrous Difference: Cinema, Anthropology, & Turn-of-the-Century Visual Culture.* Columbia University Press, 2002; Acland, Charles R., and Haidee Wasson. "Introduction: Utility and Cinema." *Useful Cinema*, edited by Charles R. Acland and Haidee Wasson. Duke University Press, 2011.

16 Hoof, Florian. "Between Recognition and Abstraction: Early Vocational Training Films." *The Image in Early Cinema: Form and Material*, edited by Scott Curtis, Philippe Gauthier, Tom Gunning, and Joshua Yumibe. Indiana University Press, 2018, pp. 111–19; Hoof, Florian. "Between the Frontlines: Vocational Training Films, Machine Guns and the Great War." *Cinema's Military Industrial Complex*, edited by Lee Grieveson and Haidee Wasson. University of California Press, 2018, pp. 177–91.

17 Wasson, Haidee. "The Other Small Screen: Moving Images at the New York's World Fair, 1939." *Canadian Journal of Film Studies*, vol. 21, no. 1, 2012, pp. 81–103.

18 Wasson, Haidee. *Everyday Movies: Portability and the Transformation of American Culture.* University of California Press, 2021.

that focus on the place and the dispositives of moving images to describe a recent shift towards a post-cinema age. Thus, to better understand the contemporary circulation of industrial film, such dispositive-driven concepts need to be supplemented by approaches that also consider the aesthetics and visual forms that derive from industrial film. Here, a perspective on the afterlife of forms offers a lens on the contemporary circulation of industrial film as a process of "reappearance" of visual forms that derive from a specific production context as "useful films" or "utility films" in an industrial organization. These visual forms have their own inner logics and history that persist, at least in part, when they reappear in today's digital distribution and media networks. To grasp the afterlife of industrial film as the reappearance of visual forms that stem from a completely different time, we need an understanding of time that "disorientates history"[19] and refrains from a chronological understanding of time. From this angle, the reappearance of industrial film is an arbitrary, contingent, unstable and multidimensional process.

An exemplary case is the reappearance of industrial films in the Mercedes-Benz Museum. As the value of Mercedes-Benz cars does not lie in the future but in the past, in the tradition of the company as *the* inventor of the automobile, industrial films are repurposed as part and parcel of a so-called "history marketing" strategy. Films that once were produced to grip the imagination of the viewers and to show them futuristic scenarios of car modernity reappear in fragments, as "mumm[ies] of time,"[20] to stabilize a specific history of car modernity. As this is only one aspect amongst many, the reappearance of industrial film is not restricted to the efforts of a history-marketing strategy. The afterlife of industrial film has to be understood as an "ensemble of processes characterized by tensions,"[21] as a site of unpredictable and suppressed conflicts between persisting visual forms and the dynamics of changing modes of visual perception that collide within industrial films. Industrial films once used to show futuristic scenarios are now experienced differently. Its visual forms reappear, but in a different, anachronistic way, not least because of the decisive shifts that occurred to the way moving images are perceived. To account for the productivity of

19 Didi-Huberman, Georges. *The Surviving Image: Phantoms of Time and Time of Phantoms: Aby Warburg's History of Art.* Penn State University Press, 2016 (95 dt. version).

20 Bazin, André. "Ontologie des photographischen Bildes" [1945]. *Was ist Kino? Bausteine zur Theorie des Films,* edited by Hartmut Bitomsky, Harun Farocki, and Ekkehard Kaemmerling. DuMont, 1975, p. 25.

21 Didi-Huberman, Georges. *The Surviving Image: Phantoms of Time and Time of Phantoms: Aby Warburg's History of Art.* Penn State University Press, 2016, p. 12.

these tensions is key to understand the afterlife of industrial cinema that is characterized by an anachronism between image, time and perception of moving image.

Childs of Mechanics: Moving Images – Moving Cars

The history of the Mercedes-Benz Group AG is closely linked to the history of the moving image. Film was used to sell their products as early as the 1910s and 1920s when industrial films show trucks driving through muddy meadows at the banks of the Neckar River just across the company headquarters in Stuttgart-Untertürkheim; or when people dressed in modernist, Bauhaus-style clothing were filmed during a factory tour.[22] The close ties between film and Mercedes-Benz are reflected as well on the organizational level of the company. Today, its large corporate film archive with its inventory of more than ten thousand industrial and non-fiction films is a subdivision of the public relations department of Mercedes-Benz Classic. Not surprisingly, this tight coupling was also reflected in the construction of the Mercedes-Benz Museum, a €150 million architectural extravaganza that opened its doors in 2006 (see fig. 4.1). The architects and designers conceptualized it as a "time machine,"[23] according to a carefully planned "screenplay."[24] Film, architecture and objects on display were used to create a cinematic experience that would let the "film start rolling in the head of the visitor."[25] According to the museum's designers, the aim of the museum is, consequently, to create an "aura"[26] not despite but through the means of mechanical reproduction.[27] And this concept seems to work quite seamlessly. Visitors frequently describe their experiences in the museum as a "mash up of scenes from *Blade Runner*, *Brazil* and *Star Wars*."[28] In the museum, moving images and moving cars

22 "Produktion im Werk UT," 1925, 7025–01, Corporate Archiv Mercedes-Benz Group AG.

23 Van Berkel, Ben, and Caroline Bos. "Das Museum des Einundzwanzigsten Jahrhunderts." *Buy Me a Mercedes Benz. Das Buch zum Museum*, edited by UNSTUDIO and HG Merz. Actar Publishers, 2006, p. 8.

24 Van Berkel, and Bos. "Das Museum des Einundzwanzigsten Jahrhunderts," p. 29. Translated by the author.

25 HG Merz. "Mehr Aura, weniger Inszenierung." *Bauwelt*, no. 16–17, 2005, p. 36. Translated by the author.

26 HG Merz. "Mehr Aura, weniger Inszenierung," pp. 34–37.

27 Benjamin, Walter. "Das Kunstwerk im Zeitalter seiner technischen Reproduzierbarkeit." *Walter Benjamin: Schriften. Band I*, edited by Theodor W. Adorno. Suhrkamp, 1974, pp. 366–405.

28 Martin, Lester. "Porsche & Mercedes-Benz Museums in Stuttgard [sic]" [blogpost]. *Jazzy Earl*, September 18, 2017, martin.atlassian.net/wiki/spaces/jazzyearl/blog/2017/09/10/122978305/

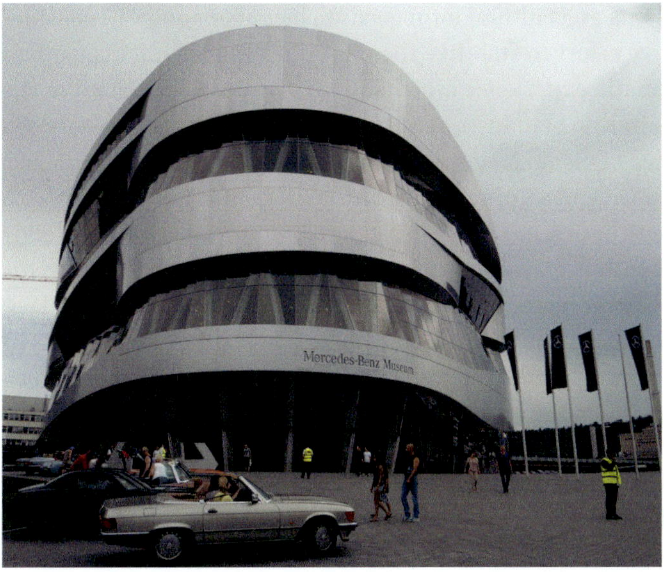

Fig. 4.1. The Mercedes-Benz Museum.

appear as a natural modernist fit. Or, to paraphrase Éric Rohmer, if film is a "child of mechanics,"[29] then cars seem to be not only in a close relationship with moving images but its identical twin.

While the museum is constructed and perceived as a deeply cinematic experience, cinematic features and infrastructures are scarce. The alleged cinematic experience does not correspond to cinematic apparatuses and moving images that occupy a central or hegemonic position in the museum's space comparable to the situation of the cinema dispositive in a movie theatre: neither are there black-boxed projection rooms, nor big cinema screens. Rather, moving images including racing films and industrial production footage are screened by what I will define in the following paragraphs as "weak dispositives." Weak dispositives offer the opportunity to engage with moving images, but they also adhere to structural conditions that are not comparable to the cinema dispositive or to extensive screen installations and audiovisual projections in the context of exhibitions or world fairs.

How then can such a museum feel so overwhelmingly cinematic, although almost all of the moving images are displayed by "weak dispositives"? My argument is that this mixture of different aesthetics, and the explicit

Porsche+Mercedes+Benz+Museums+in+Stuttgard. Accessed September 25, 2018.

29 Rohmer, Eric. "Die 'Summe' von Bazin." *André Bazin: Was ist Kino? Bausteine zur Theorie des Films*, edited by Hartmut Bitomsky, Harun Farocki, and Ekkehard Kaemmerling. DuMont, 1975, p. 11.

display of cinema as a cultural form needs to be understood as a constant and productive effort to stabilize the museum's weak film dispositives and the cinematic experience attached to it. Or to reformulate it in the terms of the afterlife of industrial films: it is a constant struggle between the persistence of visual forms and an altered mode of film experience. Consequently, the museum's architecture combines industrial film with blockbuster aesthetics, movie set design architecture, aesthetics of early cinema, optical illusions of moving images at the brink of the cinematic age, and aesthetics and technology of contemporary screens and displays. This effort to prevent the cinematic machine from falling apart points us to the core of the afterlife of industrial film, a constant tension between modes of post-cinematic experience and the persistence of visual forms.

Industrial Film Dispositives: Installations, Elevator Cars and Micro Cinemas

My analysis starts with three exemplary cases of weak dispositives that are used throughout the museum to project industrial film: elevator cars, film installations and so-called "micro cinemas." After entering the museum, the visitor takes one of three elevators, an "architectural time capsular"[30] "that looks like it came out of *Star Wars*"[31] (see fig. 4.2). It brings the visitors directly to the top floor where the entrance of the museum exhibition is located. Each elevator car is equipped with a window that resembles the format and ratio of a cinema screen.

This widescreen window frames the gaze of the visitor and guides them to a concrete wall on the opposite site of the building. The wall is forty-four metres tall and functions as a vertical screen for moving images. A large beamer mounted on top of the elevator car projects them onto this wall as soon as the elevator car is set into motion. The images are moving and at the same time moved by the elevator going up. Holes in the concrete wall once necessary for the construction workers to transport and assemble the concrete modules even seem to mimic the perforation of a celluloid film strip. A sound system inside the sound-dampened elevator car accompanies the moving images. There are different film loops in existence that alternate on

30 Williams. "Touring the Temples of German Automaking."

31 Sunder, Kalpana. "The Ultimate Shrine – The Mercedes Benz Museum." *Huffington Post*, October 20, 2014, http://www.huffingtonpost.com/kalpana-sunder/mercedes-benz-museum_b_6009530.html.

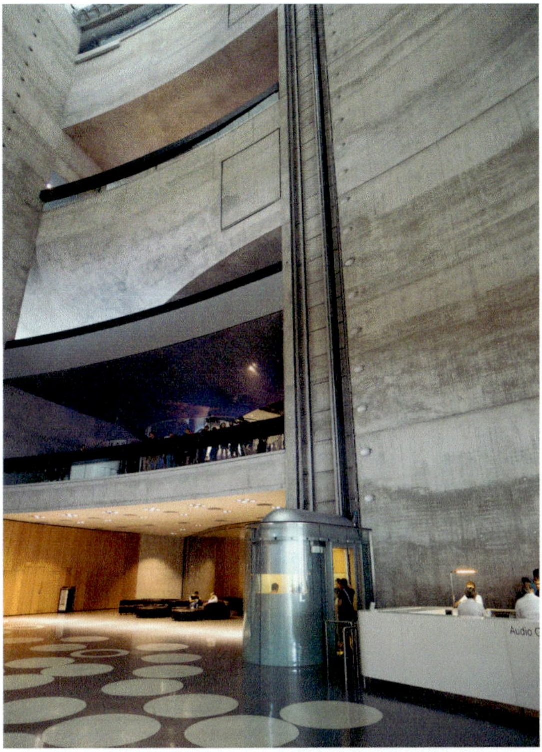

Fig. 4.2. Elevator cinema.

a random base. While focusing on topics such as car racing or commercial vehicles, they all share a similar temporal structure as they are meant to function as audiovisual time machines. The film loops all start with contemporary images and then go literally back in time to the advent of the automobile age.

One exemplary case displays the history of car racing. Starting with "screaming engine"[32] sounds, images and moving images of contemporary Formula One racing, the projected images and sounds go back into the history of car racing of the 1930s and 1920s (see fig. 4.3). The film footage used stems from newsreels, racing films and industrial films that are partly kept in the Mercedes-Benz corporate film archive. The thirty-second ride back in time ends with images of horse carriages, and finally with horses accompanied by the galloping sound of horse hooves.

After exiting the elevator, the visitor passes another weak dispositive, a narrow passage that is equipped with a film installation named *Descending from the Horse* (see fig. 4.4).

32 HG Merz. "Exhibition Design." *Buy Me a Mercedes Benz*, p. 332. Translated by the author.

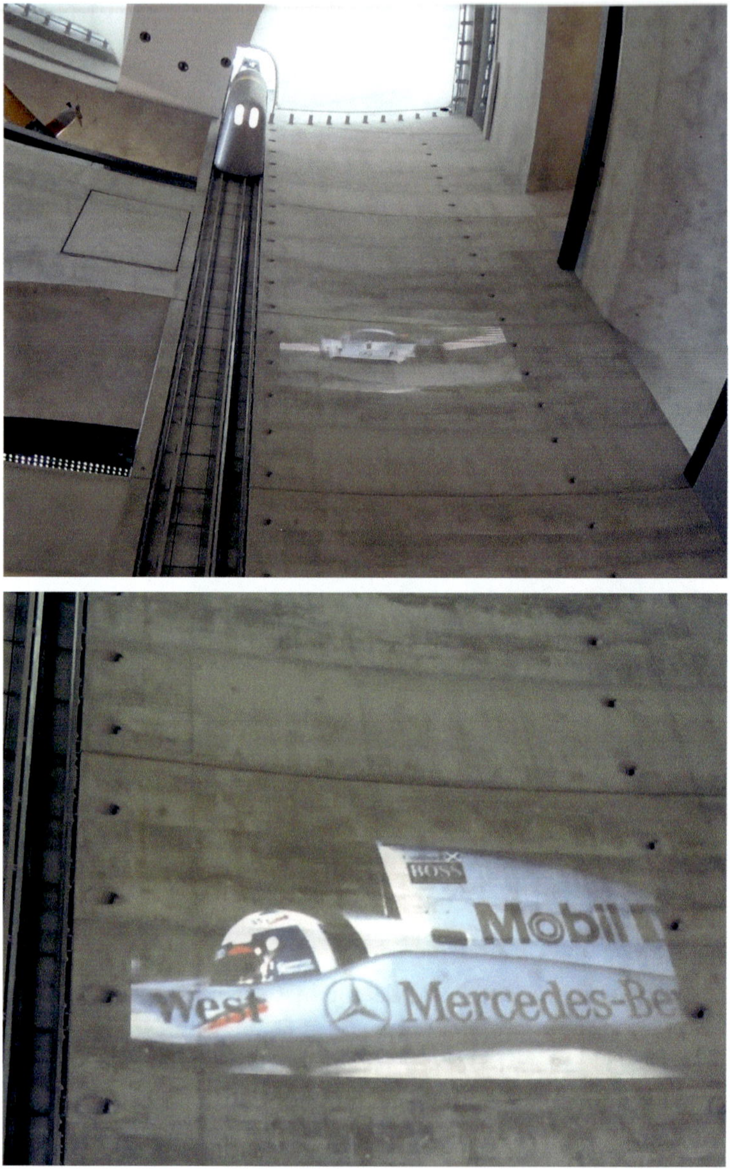

Fig. 4.3. Elevator cinema: Car racing.

It encompasses a life-size horse model and a video installation consisting of ten flat screens mounted onto the wall. The installation shows a rider on a galloping horse who, at the end, stops and descends from his horse. The style of the moving images is not so much inspired by Eadweard Muybridge's iconic motion studies, it rather is an almost exact replica of his aesthetics.

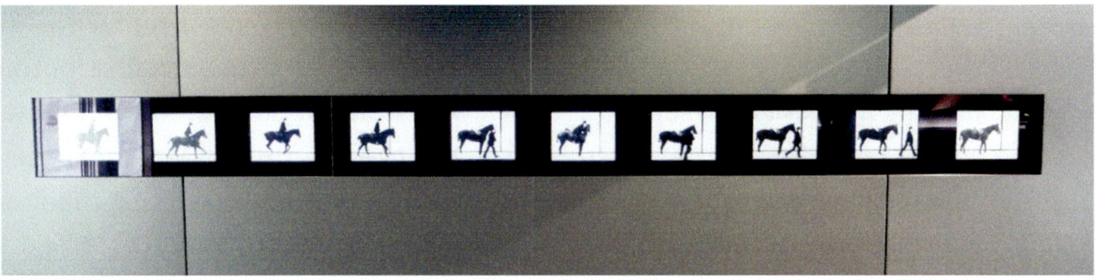

Fig. 4.4. Film Installation: *Descending from the Horse.*

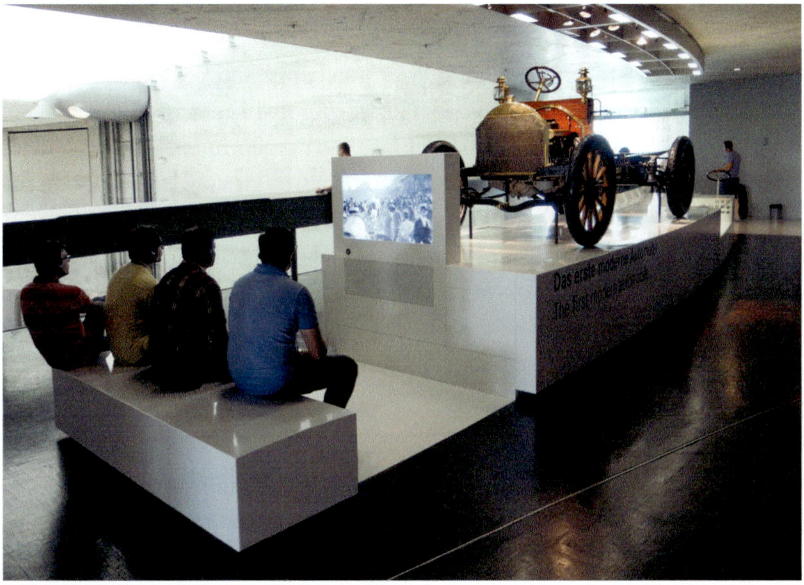

Fig. 4.5. Micro cinema.

Only after passing through the elevator car and the installation on the top floor of the museum does the actual exhibition space begin. From here, the exhibition path spirals down through the building. In seven so-called "legend rooms" the history of the company and its products, cars and combustion engines is told in a linear way, starting in 1886 and ending in the present.

In these legend rooms further weak dispositives, so-called "micro cinemas," display industrial and other non-fiction films. The micro cinemas are situated at the fringes of each of the mythos rooms. According to the museum's designers, micro cinemas are part of a "hands-on" approach that could be distinguished from a "don't touch" approach.[33] The visitor's audio guide

33 See HG Merz and Patrick Wais. "Ausstellungen konzipieren und gestalten." *Schauplätze dreidimensionaler Markeninszenierung. Innovative Strategien und Erfolgsmodelle erlebnisorientierter*

(an intrinsic part of the museum visit) can be actively synchronized with the films that are on display. Besides offering a voice-over narration for the moving images it also separates the viewers from the surrounding museum space. According to the museum designers, the ten-minute documentaries "shed light on key topics of the museum" and provide for "surprising changes of perspective between politics and corporate history, the history of transport and brand history."[34] Such topics include the collaboration of Mercedes-Benz with the Nazi regime and its entanglement with slave labour during the third Reich amongst others. The short films consist of historical footage that stems from industrial films such as early production footage from the 1920s, from newsreels and other non-fiction films. Sources are the German Federal Archives and the Mercedes-Benz corporate film archive. The concept of micro cinemas as a means to integrate the perception of moving images into the museum space is complemented by a similar viewing station in the last legend room that displays the history of car racing, the most famous and prestigious part of the Mercedes-Benz history. Here, moving images of the *Silberpfeil* (Silver Arrow) and the *Rekordwagen* (a high-speed racing car) are displayed. The micro cinemas in this room show historical footage of car racing, including high-speed racing attempts of the 1930s.

Weak Dispositives, Cinematic Space and Cinematic Time

Weak Dispositives

Micro cinemas, video installations and elevator cars exemplify the generic principle of how moving images are used throughout the museum. The weak dispositives are carefully integrated into the overall architecture and the interior design of the museum. They allow the visitors to float freely through the museum, but at certain instances, by offering specific choices, they take control of the visitor for a certain amount of time. The elevator car restricts the mobility of the visitor and the window ratio offers an obvious option to guide their gaze towards a screen. A noise damping system isolates the visitors from external noise and creates a feeling of cosiness comparable to the situation when one sits next to a damped wall inside a cinema auditorium. The build-in surround sound system provides a concise

Begegnungskommunikation, edited by Nicolai O. Herbrand. Edition neues Fachwissen, 2008, pp. 250–54.

34 HG Merz. "Exhibition Design," p. 409. Translated by the author.

soundscape – quite similar to the situation in a movie theatre. But at the same time, the behaviour codes of movie theatres do not apply, there is no darkness, the length of the stay inside the elevator car only lasts for thirty seconds. But still, there are functional elements of a cinematic dispositive: The position of the viewer is restricted, a screen, a sound system and the architecture create a specific atmosphere and enable specific sightlines. A situation quite alike can be found within the installation *Descending from a Horse*, that is situated at the narrow entrance to the exhibition every visitor has to pass. While it might increase the likelihood that visitors will choose to engage with the moving images it has no overall determining force. The same applies for the micro cinemas as optional dispositives located at the fringes of the exhibition rooms.

Weak dispositives do not provide for a situation of strict determination that is attributed to the cinema dispositive by concepts such as the apparatus theory. They lack the "privileged conditions of effectiveness – no exchange, no circulation, no communication with any outside"[35] necessary for an isolated and captivating perception of moving images. Weak dispositives are no rigid technological apparatuses nor sturdy architectural structures comparable to cave walls or movie theatres. Rather than that, they are examples of cinematic "choice architecture" altering "people's behaviour in a predictable way without forbidding any options."[36] They are capable of "nudging" the visitors with more or less determination into a cinema dispositive-like setting by creating an ensemble of possible choices through the means of architecture and design. They offer the option to watch a certain sequence of moving images but at the same time it is also "easy [...] to avoid"[37] it. Weak dispositives can become cinematic and captivate the viewer, but unlike Plato's cave, people are not strictly chained to the wall.[38] Instead, they function as "obligatory passage points"[39] or "media boundary objects"[40] the visitor needs to pass but that allow for a certain flexibility of

35 Baudry, Jean-Louis, and Alan Williams. "Ideological Effects of the Basic Cinematographic Apparatus." *Film Quarterly*, vol. 28, no. 2, 1974, p. 44.

36 Thaler, H. Richard, and Cass A. Sunstein. *Nudge: Improving Decisions about Health, Wealth, and Happiness*. Yale University Press, 2008, p. 6.

37 Ibid.

38 Baudry and Williams. "Ideological Effects," p. 44.

39 Callon, Michel. "Elements of a Sociology of Translation: Domestication of the Scallops and the Fishermen of St. Brieuc Bay." *Power, Action and Belief: A New Sociology of Knowledge?*, edited by John Law. Routledge, 1986, pp. 196–233.

40 Hoof, Florian. "The Media Boundary Objects Concept: Theorizing Film and Media." *Media Matter: The Materiality of Media, Matter as Medium*, edited by Bernd Herzogenrath. Bloomsbury, 2015, pp. 180–200.

action. It is obligatory to eventually enter an elevator car and to walk past an installation that is mounted into a narrow passage or a micro cinema viewing station. But the situation comes with weak determining forces concerning time, space, mode of perception and individual action. Nevertheless, as choice architecture they incorporate institutional "strategies"[41] that make certain actions more likely than others. In a way, weak dispositives organize moving images according to the logics of relational databases as one object amongst others. They are discreetly stored away on a hard drive, and their visibility and accessibility on screens is managed by the "gentle power"[42] of weak dispositives. Consequently, watching one of these decentralized moving images becomes the result of an individual choice, but a choice that is restricted by the alternatives that are integrated into the fabric of the museum's architecture.

Moving images are widely available within the museum and the visitors regularly describe their experience in the museum as cinematic, but the cinematic experience of the visitor is not determined by the film dispositives. Rather, as the visitors are walking freely through the building, it is an experience that is constantly stabilized by the architecture that gives overall cohesion to the "heterogenous ensemble"[43] of weak dispositives. To better understand the fragile character of such a cinematic experience the perspective needs to be extended on the overall architecture and interior design.[44] This lens then not only encompasses the perception of moving images but also the selection and fragmenting of the cinematic experience by the performative act of walking through the museum's space.[45] The visitor wanders through a cluster of weak cinematic dispositives that provide for an overall cinematic impression because they at the same time relate to the concept of time and space of the building's architecture. This results in a performative act of perception that has been described, for example, when discussing the nineteenth-century arcades in Paris.[46] Here, quite similar to the "choice architecture" in the Mercedes-Benz Museum, Walter Benjamin

41 De Certeau, Michel. *The Practice of Everyday Life*. University of California Press, 1984, p. 94.

42 Thaler, H. Richard, and Cass A. Sunstein. *Nudge: Improving Decisions about Health, Wealth, and Happiness*. Yale University Press, 2008, p. 8.

43 See Hoof, Florian. *Engel der Effizienz. Eine Mediengeschichte der Unternehmensberatung*. Konstanz University Press, 2015, p. 31; Foucault, Michel. *Power/Knowledge: Selected Interviews & Other Writings 1972–1977*. Edited by Colin Gordon. Pantheon Books, 1980, pp. 194–210.

44 For a similar argument, see Casetti, Francesco. "Mediascapes: A Decalogue." *Perspecta 51: Medium*, edited by Shayari de Silva, Dante Furioso, and Samantha Jaff. MIT Press, 2019, pp. 21–43.

45 De Certeau, Michel. *The Practice of Everyday Life*. University of California Press, 1984, p. 101.

46 Benjamin, Walter. *Das Passagen-Werk, Vol. 1/2*, edited by Rolf Tiedemann and Hermann Schweppenhäuser. Suhrkamp, 1982.

described the architecture and design of arcades as a form of exercising power where the constructions take over the role of the unconsciousness.[47]

For the nineteenth-century arcades, the flaneurs' experience of modernity was based on architecture enabled through cutting edge metallurgy and engineering. The Mercedes-Benz Museum is based on digital simulation modelling and is made of armoured concrete. The museum's designers explicitly refer to this space as an example of "retroactive smoothing,"[48] a concept developed by Gilles Deleuze and Félix Guattari to describe spaces that create uncertainty and uneasiness because of their unclear and ever shifting position between openness and closeness.[49]

Within this museum space that on purpose offers an irritating amount of choices, cinema and film culture function as means of orientation by providing for cinematic experience. The spacial design of the exhibition rooms is modelled according to existing movie set designs from film history. The concept of time within the exhibition is related to markers and a periodization of time that derive from film and media history. These two elements, "cinematic space" and "cinematic time," create an aesthetic environment that stabilizes the cinematic experience and thus enable the weak dispositives to flourish and to function. To focus on the interplay between these elements is key to better understand how the cinematic experience within the museum is created and to shed light on the afterlife of industrial film. In the following I will describe in more detail how weak dispositives are stabilized by placing them within "cinematic space" and contextualize them within "cinematic time."

Cinematic Space

The entire design and architecture of the museum seems like a walkable movie, a mash up from different movie set designs, such as those from the *James Bond 007* film series, *2001: A Space Odyssey* (1968), *Star Wars* (1977), *Blade Runner* (1982) or *Brazil* (1985). Here, cinematic space does not so much relate to a "fictional space" that is based on diegetic strategies to create space experiences on flat cinema screen. Rather, architectural concepts used for film set design are reconstructed as parts of the museum's structure and

47 Benjamin, Walter. "Paris, die Hauptstadt des XIX. Jahrhunderts." *Das Passagen-Werk, Vol. 1*, edited by Rolf Tiedemann and Hermann Schweppenhäuser. Suhrkamp, 1982, p. 46.

48 Betsky, Aaron. "Automobilität. Das Mercedes-Benz Museum." *Buy Me a Mercedes Benz*. Actar Publishers, 2006, p. 19.

49 Deleuze, Gilles, and Félix Guattari. *A Thousand Tableaus: Capitalism and Schizophrenia*. Continuum, 2004, p. 532.

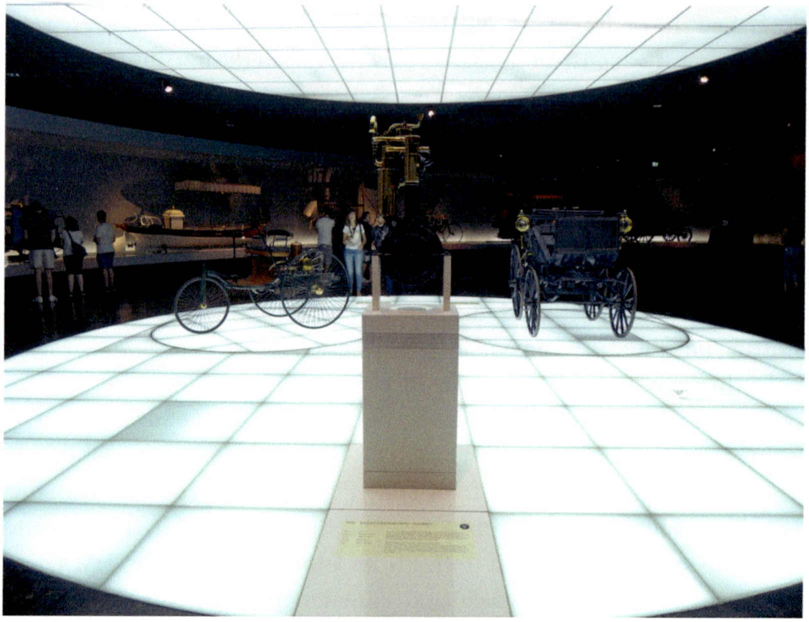

Fig. 4.6. Legend Room One: The Dawn of the Automobile Age.

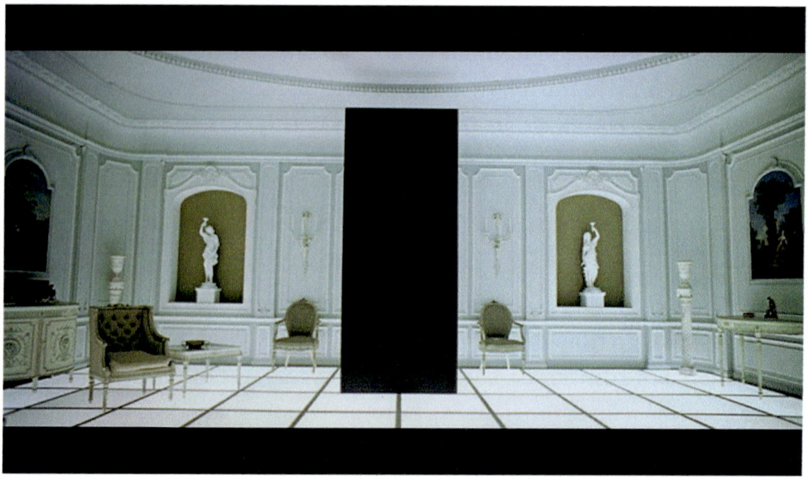

Fig. 4.7. Cinematic space becomes museum space: *2001: A Space Odyssey*.

thus made walkable. This concept somehow pivots artistic interventions and concepts, such as in Janet Cardiff and George Bures Miller's documentary project *Alter Bahnhof Videowalk* (2012) for Documenta 13. The visitors are not guided by screens through an existing architectural ensemble or space; rather, the whole architecture is modelled as a cinema space in the first place.

Screens simply become redundant. The difference between the imaginary and the real collapses or, as André Bazin puts it, "Every image is to be seen as an object and every object as an image."[50] It is a museum walk guided by movie set design where cinematic spaces created for the flat cinema screen are turned into the museum's architecture. This aesthetic strategy is used throughout the building in every single exhibition room.

An exemplary case is the first legend room that celebrates Mercedes-Benz as the inventor of the combustion engine (see fig. 4.6). The exhibition space does not only look like the white-cube aesthetics from Stanley Kubrick's *2001: A Space Odyssey* (1968), it is actually modelled according to Ken Adam's set design.[51] To be more precise, according to the last sequences of the movie when a black monolith appears in a white-cube setting (see fig. 4.7).

Inside the museum, there is no arbitrary black monolith on display, but instead we encounter the so-called "Standuhr," the world's first single-cylinder combustion engine. It is situated in a transparent monolithic block that is placed in the centre of a circular room. The museum borrows the same aesthetics and white-cube light system Adam used to visualize the dawn of something new to proclaim the birth of the automobile age. Cinema provides not only for concepts of space but is also used as a metaphor for modernity. The museum extensively relies on architectural and spatial concepts that refer to iconic cinematic scenes and spaces from film history. The exhibition rooms appear as three-dimensional versions of movie sets that have been made walkable.

The curved architecture of Legend Room Six further exemplifies this architectural principle. It derives from the Pentagon War Room set design Ken Adam constructed for *Dr. Strangelove* (1964) (fig. 4.8) and refers to other set designs, such as the SPECTRE base Adam created for the James Bond movie *You Only Live Twice* (1967). These film sets rely on aesthetics of abstraction that were inspired by Art Nouveau, American Style and Bauhaus design and not so much by ideas of realism. Adam created cinematic environments that would "look better than the real thing."[52] They are in line with the Hollywood tradition of production design he aimed towards "enhancing or stylizing reality. Fooling the audience, but in a nice way."[53] Ultimately, the purpose of his set design was not about showing real life, which is, as Adam puts it,

50 Bazin, André. *What Is Cinema? Vol. 1.* Edited and translated by Hugh Gray. University of California Press, 2005, pp. 15–16.

51 HG Merz. "Exhibition Design," pp. 346–53.

52 Adam, Ken. Qtd. in Christie, Ian. "Designing Dreams and Nightmares: Ken Adam on His Finest Sets." *Sight & Sound*, March 14, 2016, www.bfi.org.uk/news-opinion/sight-sound-magazine/comment/obituaries/designing-dreams-nightmares-ken-adam-his-finest.

53 Ibid.

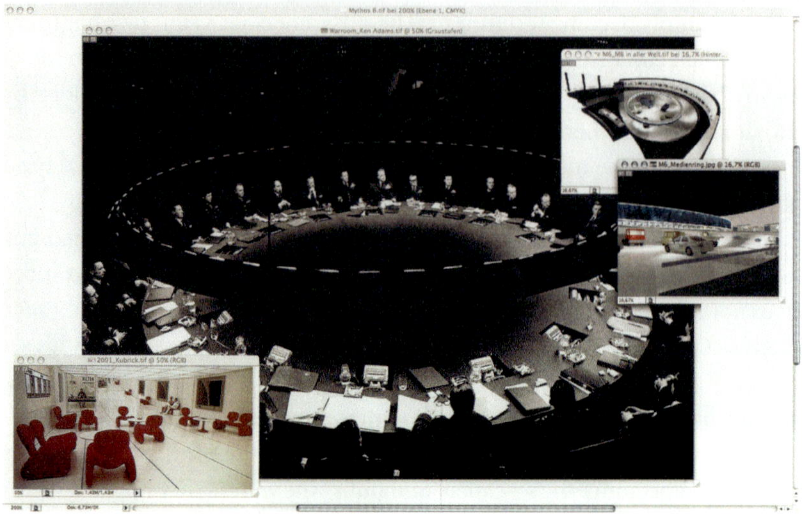

Fig. 4.8. The War Room set design turned into museum space: *Dr. Strangelove.*

"frustrating," rather it was about "entertainment."[54] He created spaces of "stylized reality" that would ultimately be constructed and planned according to the needs of the film camera.[55] Consequently, such spaces allow for a wide range of possible sightlines and turn each gaze of the visitor into a multitude of possible camera angles. The implications are multifold. For the visitors, the stylized reality of movie set design works as mean of orientation and pleasure. The museum space feels familiar because aesthetics and architecture connect to the "memories as spectators"[56] in the visitor's mind and ultimately forms a collective cinematic experience. Visitors feel cinematic because the architecture allows personal gazes that mimic aesthetic experiences that stem from film culture. At the same time, walking through a movie set allows the visitors to become cameramen and photographers themselves. It makes it easy to shoot cinematic photos of a stylized reality. They not only engage with the dimension of perception but also with the dimension of a performative entanglement with the cinematic aesthetics of the museum space. This specific structure of the museum space aims towards a spreading of the cinematic aesthetics via "social media entertainment intermediaries,"[57] such

54 Kissling-Koch, Petra. *Macht(t)räume. Der Production Designer Ken Adam und die James Bond Filme*. Bertz+Fischer, 2012, p. 202.

55 Ibid., pp. 167–75.

56 Casetti. "The Relocation of Cinema." *European Journal of Media Studies*, vol. 1, no. 2, 2012, p. 19.

57 Cunningham and Craig, *Social Media Entertainment*. pp. 115–147.

as Instagram, YouTube, Twitter, TicTok or Facebook. The museum becomes a hub of circulation of moving images and of cinematic aesthetics within digital film culture. It turns the visitors into multipliers and opens a plain playing field to "expand the potential meanings that get attached to a brand by inserting it into a range of unpredicted contexts of use."[58] And it creates a cinematic atmosphere and performative action that stabilizes and interconnects the weak film dispositives that are spread throughout the museum.

Cinematic Time

The concept of "cinematic space" described so far connects to "cinematic time" as the second trope that stabilizes the cinematic experience. Here, the history of film, media and cinema is used as symbolic markers in time to situate the history of the automobile age in a broader chronological temporality. As described above, the visitors step into an elevator car that functions as an audiovisual time machine first. It transfers them back to the dawn of the automobile age that coincides with the beginning of the visual culture of moving images. This is marked by the installation that mimics Eadweard Muybridge's visual aesthetics at the brink of the cinematic age. The dawn of the automobile age is then symbolized by the movie set design that derive from the intro and the final sequence of *2001: A Space Odyssey* (1968), one of the most iconic cinematic scenes in recent film history. Moving images and moving cars as centrepieces of modernity are then further paralleled throughout the museum. The history of film, cinema and popular culture is used as a chronological timeline to contextualizes the history of the Mercedes-Benz company in a symbolic macro-temporality. Landmark film productions and the Hollywood star system, including actors such as Charlie Chaplin or Buster Keaton but also the fate of the German film industry during the third Reich, are utilized to situate certain historical developments into the context of a symbolic, almost abstract cinematic time. Besides the macro-temporality of the museum as a walkable chronological timeline on the micro level, cinematic time is used to contextualize specific historical developments of relevance for the company.

The legend room "Elegance at the Abyss" is an exemplary case for the importance of cinematic time to create and stabilize such micro temporalities. It showcases the invention of the diesel engine and at the same time

58 Jenkins, Henry, Xiaochang Li, and Ana Domb Krauskopf, with Joshua Green. *If It Doesn't Spread It's Dead: Creating Value in a Spreadable Marketplace*. Convergence Culture Consortium, 2008, p. 22.

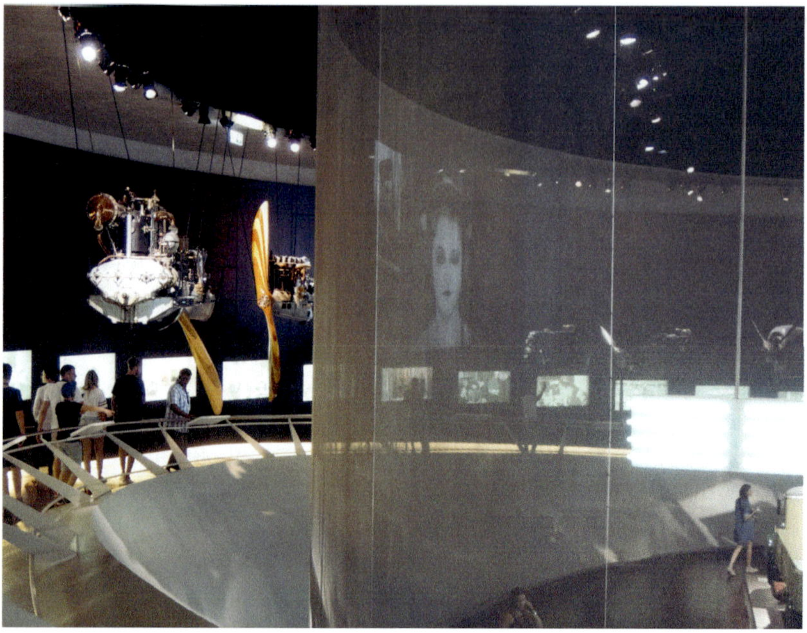

Fig. 4.9. Film installation "Fragments."

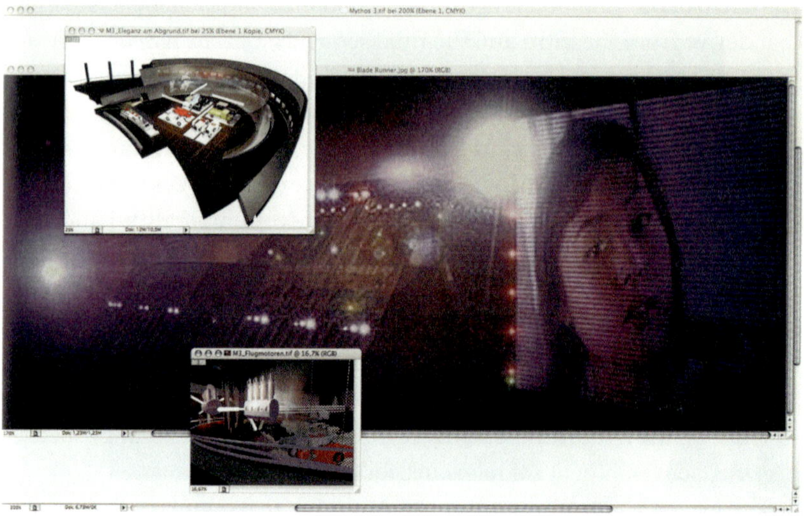

Fig. 4.10. Cinematic screens become museum screens: *Blade Runner.*

the company's history of Nazi collaboration and use of slave labour during the Third Reich. The main part of the room is devoted to the history of the diesel engine, while the political dimension of the time between 1933 and 1945 is integrated into weak film dispositives. The more controversial (or, as the museum's designers call them, the "more complex") aspects are both integrated into micro cinemas situated at the fringe of the room and displayed in an installation named "Fragments." The installation, albeit situated quite prominent in the centre of the exhibition space, is not clearly visible as it is projected onto a curved, semi-transparent screen (see fig. 4.9 and 4.10). Here, the exhibition designers relied on the design of public screens that derive from the science fiction movie *Blade Runner* (1982), a movie set design influenced by Italian futurism and architects such as Antonio Sant'Elia.

Alike the overall concept of the museum, the film montage of the installation follows a chronological time structure. It starts with footage from a German newsreel that shows Berlin in the roaring 1920s and then turns to scenes depicting the fire that burned down the German parliament as well as book burnings. This rather conventional approach to depict historical developments then turns into abstract "cinematic time." With the beginning of the Nazi terror regime, moving images no longer include "realistic" images from newsreels or documentaries. Rather, the installation then parallels the history of the Mercedes-Benz company with the history of the German film industry by showing iconic scenes from movies by renowned directors. This includes Nazi collaborators like Leni Riefenstahl and Walter Ruttmann as well as émigrés such as Wilhelm Thiele and Fritz Lang, who had to flee Germany during the Third Reich. Films on display are amongst others include *Metropolis* (1927), *Berlin Symphonie einer Großstadt* (1927), *Die weisse Hölle vom Piz Palü* (1929), *Die Drei von der Tankstelle* (1930), *Der blaue Engel* (1930), *Die Olympischen Spiele. Fest der Nationen* (1938), and *Triumph des Willens* (1935). Images from fiction and documentary films are used to stabilize the cinematic time as a framework to make sense of historical incidences. The audiovisual museum's designers *teamstratenwerth* responsible for the installation describe it as a "*Sittengemälde*," as a moral and social portrayal of life. According to them it shows the impact of Nazi Germany on society, and how individual choices were made. Some of the film directors collaborated with the Nazi regime, others had to flee Germany. Instead of directly relating the exhibition to the time of the third Reich, the history of the German film industry and its fate during the fascist period is used to establish an abstract framework of cinematic time.

The exhibition is based on micro and macro temporalities of cinematic time that provide for a stylized concept of modernity as exemplified by the media technology and aesthetics of the moving image and parallelizes it with the development of the modern concept of mobility based on cars that are run by the combustion engine. This concept of cinematic time strengthens a linear conception of time and functions as a medium of abstraction.

Cinematic Experience

The legend room "Elegance at the Abyss" discussed above is also an exemplary case that shows how cinematic experience is a result of the interrelatedness of cinematic space, cinematic time, and weak dispositives. The room is dominated by an abstract, dark and futuristic set design that derives from the movie *Blade Runner*. This cinematic space and its aesthetics of stylized reality is combined with the concept of a symbolic cinematic time when the installation "fragments" parallels the history of the German film industry with the history of the Mercedes-Benz company during the Third Reich. Controversial topics including the history of slave labour and the notorious collaboration of Mercedes-Benz with the national socialists are reintegrated into micro cinemas that are placed at the fringes of the exhibition room. As part of those weak dispositive the controversial history of the Mercedes-Benz company becomes an optional section that exists but that is almost bypassed by the carefully arranged screenplay that guides the visitor through the museum's choice architecture. The cinematic experience isolates the controversial role of the Mercedes-Benz company and relocate it into a symbolic and at the same time fuzzy realm of cinematic time and space. Film as technology, as a mode of representation and experience, and film and media history as a metaphor and marker in time are bound together to establish a physical environment to surprise and to impress the visitors but also to guide their focus of attention towards certain objects and to obscure the gaze towards sensitive issues by offering specific choices enabled and determined by architecture.

The only museum that can claim to map the *whole* history of the automobile age uses the most iconic cinematic scenes in recent film history to declare and to symbolize the dawn of the age and to describe how automobility as a centrepiece of modernity unfolded throughout the twentieth century. To build the museum experience on aesthetics that derive from film and cinema is a well-informed decision made by the experts who planned and constructed the museum, people with a wide and deep knowledge about film history and post-structuralist theory. They rely on film and cinema culture to construct the museum as a cinematic machine. The reason for using the vast number of cinematic references, as the architects argue, lies in the fact that the aesthetics of film are not restricted to its movies but have already found their way into the collective memory of society. Thinking about film aesthetics in this way shares some similarities with Francesco Casetti's argument that the cinematic experience is not restricted to its material conditions but lies in "our memory as spectators."[59] This

59 Casetti. "The Relocation of Cinema," p. 19.

"cinematographicity"[60] is what the museum's designer, HG Merz, refers to, when he argues that it is specifically the collective memory of cinematic forms that enables the museum space to activate them.[61]

The theoretical question at stake here inquires what it actually means when cinematic experience that stems from the collective memories of film history reappears or is reanimated in a place such as the Mercedes-Benz Museum. What happens to cinema and film history when used as symbolic historical context for the display of the history of the automobile age? How does the agency of the cinematic experience shift when it reappears outside the 'classical' cinema dispositive?

Cinema-as-Mythos

To be able to understand the reappearance of the cinematic experience in places such as the Mercedes-Benz Museum we need to look how "cinema" is addressed within this context. Cinema is not referred to as separated entities such as a technological apparatus, a cultural form of experiencing or a mode of perception consisting of screens and moving images. Cinema no longer represents a specific mode of perception or a set of media technologies. Rather, as cinematic time and space it appears as a "secondary semiotic system,"[62] as a mythos that provides for orientation but at the same time represents a fuzzy and ubiquitous experience of modernity. Cinema-as-mythos turns into a symbolic form of modernity, a cypher like the automobile is a cypher for mobility, prosperity and freedom. According to Roland Barth's concept of mythos, this alters the epistemological status of cinematic time and space. He argues that the semiotic chains and maps of meaning a mythos consists of are turned into material to stabilize the mythos as a secondary semiotic system.[63] In the case of cinema-as-mythos, it builds upon the knowledge and experience that evolved over time in film and cinema history. Such material that stabilizes the mythos of cinema as a significant building block of modernity loses flexibility and mobility and diminishes their power as maps-of-meaning. Consequently, turning cinema into a mythos comes with specific epistemic shifts. Barth's claim that maps-of-meaning are turned into static forms of material is one way to account for the shifts that occur when the cinematic experience reappears as part of the afterlife of industrial film.

60 Ibid.

61 HG Merz. "Exhibition Design."

62 Barthes, Roland. *Mythen des Alltags*. Suhrkamp, 1964, p. 92.

63 Ibid., pp. 92–93.

It provides a lens to distinguish between different forms of determination that structure the politics of weak dispositives.

The Politics of Weak Dispositives

How cinematic time, cinematic space and weak dispositives are utilized to structure the narrative of the museum towards the company's history is an exemplary case that highlights the politics of weak dispositives. Mercedes-Benz is here a specific interesting case, due to its nimbus as the inventor of the automobile and as the world's most successful and innovative participant in the history of Grand Prix motor racing. But the "golden era" of the company's motorsports division in the 1930s was also directly funded by the fascist regime in Germany.[64] Furthermore, Mercedes-Benz was an early and active supporter of the fascist movement before they came to power. And finally, Adolf Hitler himself was not only "a car craze but much more a Mercedes craze"[65] person. This rather controversial historical constellation led to the nowadays iconic racing films of the *Silberpfeil* and the *Rekordwagen* that are one of the main reasons why the Mercedes-Benz car company has such a unique reputation. Consequently, the Mercedes-Benz Museum at the same time has to facilitate an international audience with a historical narrative to maximize potential customers in the international market for luxury cars first and associate itself with the German concept of *Vergangenheitsbewältigung* (the struggle to deal with its past) in relation to the Holocaust. How those different aspects are approached results in strategies that allows us to better grasp the politics of weak dispositives.

Of course, from a historical perspective industrial films have always been a site of power struggle and politics. Early industrial films presented to the public spectacular attractions, such as previously unseen scenes of machinery and modern industrial production to create trust in industrial mass production. In the course of history, they were frequently used for corporate or government propaganda.[66] Those types of industrial films that were aiming towards a broader audience were typically shown in movie theatres or in temporarily erected film pavilions that would privilege the

64 Brauchitsch, Manfred von. *Ohne Kampf kein Sieg*. Verlag der Nation, 1974, p. 28.

65 Pohl, Hans, Stephanie Habeth, and Beate Brüninghaus, editors. *Die Daimler-Benz AG in den Jahren 1933 bis 1945. Eine Dokumentation*. Steiner, 1986, p. 25. Translated by the author.

66 Grieveson, Lee. *Cinema and the Wealth of Nations: Media, Capital, and the Liberal World System*. University of California Press, 2018.

moving images on display. Films about automobility produced for and screened at industrial fairs and exhibitions, such as the Futurama exhibit at the 1939 World's Fair in New York City,[67] would grip the imagination of the visitors and show them a "a new world."[68] Here, moving images appear as powerful representations that can even overwhelm the subject.

But unlike the politics of the cinematic dispositive that concentrates the attention of the spectator towards moving images projected on a privileged screen, the politics of weak dispositives are about a cinematic experience of dispersion. In the Mercedes-Benz Museum, moving images are discreet entities, scattered around the museum space, while the spectators are able to move between them. The politics of the weak dispositives are no longer about propaganda and censorship, about showing or not showing; rather, it is about managing and installing a refined hierarchy of possible access, a rigged index that points towards moving images in different ways. It is not about the overwhelming effect of film as a representational map of meaning but about turning the film dispositive into choice architecture as means of orientation and navigation. Cinema is understood and utilized as an entity, as a machine of the imaginary that consists of "weak dispositives," and is held together by the dimensions of "cinematic space" and "cinematic time." As an overarching and at the same time elusive concept it stabilizes the cinematic experience inside the museum by giving orientation and reducing contingency and uncertainty in an environment that is characterized by "retroactive smoothing."[69] Along it goes a shift of agency from the moving image to the surrounding structures that continuously stabilizes the cinematic experience by offering discrete choices.

Afterlives: Reappearance, Industrial Films and Anachronistic Times

This chapter has focused on the digital afterlife of industrial film as an unstable twofold process that consists of the reappearance of visual forms and changing modes of perception. At first glance, the industrial films that reappear in the Mercedes-Benz Museum seem to contradict this very idea

67 Wasson, Haidee. "The Other Small Screen: Moving Images at the New York's World Fair, 1939." *Canadian Journal of Film Studies*, vol. 21, no. 1, 2012, pp. 81–103.

68 Betsky. "Automobilität. Das Mercedes-Benz Museum," pp. 20–21.

69 Deleuze, Gilles, and Félix Guattari. *A Thousand Tableaus: Capitalism and Schizophrenia.* Continuum, 2004.

of unpredictability. Here, industrial film is situated in a carefully crafted cinematic machine, a combination of weak dispositives, cinematic space and cinematic time. As a part of the museum's choice architecture they attempt to control the afterlife of the moving image by turning their semiotic chains and maps of meaning into static forms of material that stabilizes the mythos of modernity. But to which degree can the politics of weak dispositives turn industrial film into material for an overarching choice architecture? How does the perception change if we revisit the museum by focusing on the contingencies that arise from the tensions between the reappearance of forms that, according to Aby Warburg and Georges Didi-Huberman, are bound to anachronistic times and the shifting ways how aesthetic forms are appreciated, approached and perceived?

The decision to design the Mercedes-Benz Museum according to an overarching screenplay that parallels cinema and cars as symbols of modernity was made almost twenty years ago – at a time when car sales were thriving and the German car industry was successfully entering the growing Asian markets for luxury cars. Since then, the image of the German car industry has undergone some tectonic shifts. Besides ecological issues inevitably tied to the combustion engine, Dieselgate (the Volkswagen emissions scandal), the largest corporate scandal in post-war Germany, involving all major German automakers, including Mercedes-Benz, fundamentally changed the public perception of the industry.

The museum's screenplay of German engineering as masters of efficiency and precision is now complemented with the public reputation of fraud, white-collar criminal activity, software manipulation and cheating to defeat regulatory devices. The concept of "car modernity" and the abstract aesthetics within the museum space are no longer perceived as flawless and uncontested, but eventually as an economic and ecological dead end. This series of events has altered the way the Mercedes-Benz Museum and its aesthetics derived from cinematic time and space are perceived and interpreted.

As a paradigmatic case of how perception might change, let's return once more to the mythos room "Elegance at the Abyss." For some the combined exhibit of the history of the diesel combustion engine with the unspeakable history of the company during the Third Reich seems to be from the start a quite elusive decision. But under the impression of the Dieselgate revelations, it appears as an almost visionary decision. This single room now seamlessly combines the early success of the Mercedes-Benz company that started in 1933 and was the result of a ruthless collaboration between a fascist terror regime and the largest corporate crime scheme in Germany after 1945.

Perhaps even the overarching design concept of the museum – the decision of the museum's designers to make extensive use of cinematic set designs of operations and control rooms – now seems to reveal a deeper meaning. The cinematic space and its aesthetics that derive from Ken Adam's "stylized reality" might no longer connect the history of the Mercedes-Benz company to the flawless concept of modernity. Rather, one perhaps will be reminded of the purposes these operation and control rooms had in the original context as part of a movie set design. Almost all the cinematic spaces used as blueprints for the museum's spaces were originally populated and used by cinematic villains as their nerve centre of attempts to conspire against the world on a big scale. So finally, the anachronistic and uncontrollable times bound to the reappearance of forms puts the history of Mercedes-Benz back in close company with the cinematic imaginaries of the pure evil represented by villains such as Dr. No, Ernst Stavro Blofeld, Auric Goldfinger, or the organization SPECTRE. Since Mercedes-Benz was, at least temporarily, a partially criminal organization, it has every right to encapsulate the history of its approach to marketing with the architecture of operation rooms of pure evil as provided by film history.

Works Cited

Acland, Charles R., and Haidee Wasson, editors. *Useful Cinema*. Duke University Press, 2011.

Barthes, Roland. *Mythen des Alltags*. Suhrkamp, 1964.

Baudry, Jean-Louis, and Alan Williams. "Ideological Effects of the Basic Cinematographic Apparatus." *Film Quarterly*, vol. 28, no. 2, 1974, pp. 39–47.

Bazin, André. "Ontologie des photographischen Bildes" [1945]. *André Bazin: Was ist Kino? Bausteine zur Theorie des Films*, edited by Hartmut Bitomsky, Harun Farocki, and Ekkehard Kaemmerling. Dumont, 1975, pp. 21–27.

Bazin, André. *What Is Cinema? Vol. 1*. Edited and translated by Hugh Gray. University of California Press, 2005.

Bell, Daniel. *The Coming of Post-industrial Society*. Harper Colophon Books, 1974.

Beniger, James. *The Control Revolution: Technological and Economic Origins of the Information Society*. Harvard University Press, 1986.

Benjamin, Walter. "Das Kunstwerk im Zeitalter seiner technischen Reproduzierbarkeit." *Walter Benjamin: Schriften. Band I*, edited by Theodor W. Adorno. Suhrkamp, 1974, pp. 366–405.

Benjamin, Walter. *Das Passagen-Werk, Vol. 1/2*, edited by Rolf Tiedemann and Hermann Schweppenhäuser. Suhrkamp, 1982.

Benjamin, Walter. "Paris, die Hauptstadt des XIX. Jahrhunderts." *Das Passagen-Werk, Vol. 1*, edited by Rolf Tiedemann and Hermann Schweppenhäuser. Suhrkamp, 1982, pp. 45–59.

Betsky, Aaron. "Automobilität. Das Mercedes-Benz Museum." *Buy Me a Mercedes-Benz*. Actar Publishers, 2006, pp. 10–23.

Brauchitsch, Manfred von. *Ohne Kampf kein Sieg*. Verlag der Nation, 1974.

Callon, Michel. "Elements of a Sociology of Translation: Domestication of the Scallops and the Fishermen of St. Brieuc Bay." *Power, Action and Belief: A New Sociology of Knowledge?*, edited by John Law. Routledge, 1986, pp. 196–233.

Casetti, Francesco. "Back to the Motherland: The Film Theatre in the Postmedia Age." *Screen*, vol. 52, no. 1, 2011, pp. 1–12.

Casetti, Francesco. *The Lumière Galaxy: Seven Keywords for the Cinema to Come*. Columbia University Press, 2015.

Casetti, Francesco. "Mediascapes: A Decalogue." *Perspecta 51: Medium*, edited by Shayari de Silva, Dante Furioso, and Samantha Jaff. MIT Press, 2019, pp. 21–43.

Casetti, Francesco. "The Relocation of Cinema." *European Journal of Media Studies*, vol. 1, no. 2, 2012, pp. 5–34.

Castells, Manuel. *The Information Age: Economy, Society and Culture, Vol. I: The Rise of the Network Society*. Blackwell, 1996.

Christie, Ian. "Designing Dreams and Nightmares: Ken Adam on His Finest Sets." *Sight & Sound*, March 14, 2016, www.bfi.org.uk/news-opinion/sight-sound-magazine/comment/obituaries/designing-dreams-nightmares-ken-adam-his-finest.

De Certeau, Michel. *The Practice of Everyday Life*. University of California Press, 1984.

Deleuze, Gilles. *Das Zeit-Bild. Kino 2*. Suhrkamp, 1997.

Deleuze, Gilles, and Félix Guattari. *A Thousand Tableau: Capitalism and Schizophrenia*. Continuum, 2004.

Didi-Huberman, Georges. *The Surviving Image: Phantoms of Time and Time of Phantoms: Aby Warburg's History of Art*. Penn State University Press, 2016.

Drucker, Peter. *The Age of Discontinuity: Guidelines to Our Changing Society*. Harper and Row, 1969.

Elsaesser, Thomas. "Die Stadt von Morgen. Filme zum Bauen und Wohnen in der Weimarer Republik." *Geschichte des dokumentarischen Films in Deutschland, Band 2: Weimarer Republik 1918–1933*, edited by Klaus Kreimeier, Antje Ehmann, and Jeanpaul Goergen. Reclam, 2005, pp. 381–409.

Grieveson, Lee. *Cinema and the Wealth of Nations: Media, Capital, and the Liberal World System*. University of California Press, 2018.

Griffiths, Alison. *Wondrous Difference: Cinema, Anthropology, & Turn-of-the-Century Visual Culture*. Columbia University Press, 2002.

Hediger, Vinzenz, and Patrick Vonderau. "Record, Rhetoric, Rationalization: Industrial Organization and Film." *Films That Work: Industrial Film and the*

Productivity of Media, edited by Vinzenz Hediger and Patrick Vonderau. Amsterdam University Press, 2009, pp. 35–49.

HG Merz. "Exhibition Design." *Buy Me a Mercedes-Benz*. Actar Publishers, 2006, pp. 309–571.

HG Merz. "Mehr Aura, weniger Inszenierung." *Bauwelt*, no. 16–17, 2005, pp. 34–37.

HG Merz, and Patrick Wais. "Ausstellungen konzipieren und gestalten." *Schauplätze dreidimensionaler Markeninszenierung. Innovative Strategien und Erfolgsmodelle erlebnisorientierter Begegnungskommunikation*, edited by Nicolai O. Herbrand. Edition neues Fachwissen, 2008, pp. 250–54.

Hoof, Florian. *Angels of Efficiency: A Media History of Consulting*. Oxford University Press, 2020.

Hoof, Florian. "Between Recognition and Abstraction: Early Vocational Training Films." *The Image in Early Cinema: Form and Material*, edited by Scott Curtis, Philippe Gauthier, Tom Gunning, and Joshua Yumibe. Indiana University Press, 2018, pp. 111–19.

Hoof, Florian. "Between the Frontlines: Vocational Training Films, Machine Guns and the Great War." *Cinema's Military Industrial Complex*, edited by Lee Grieveson and Haidee Wasson. University of California Press, 2018, pp. 177–91.

Hoof, Florian. *Engel der Effizienz. Eine Mediengeschichte der Unternehmensberatung*. Konstanz University Press, 2015.

Hoof, Florian. "The Media Boundary Objects Concept: Theorizing Film and Media." *Media Matter: The Materiality of Media, Matter as Medium*, edited by Bernd Herzogenrath. Bloomsbury, 2015, pp. 180–200.

Huhtamo, Erkki. "Screen Tests: Why Do We Need an Archaeology of the Screen?" *Cinema Journal*, vol. 51, no. 2, 2012, pp. 144–48.

Hutter, Michael. *The Rise of the Joyful Economy: Artistic Invention and Economic Growth from Brunelleschi to Murakami*. Routledge, 2015.

Jenkins, Henry, Xiaochang Li, and Ana Domb Krauskopf, with Joshua Green. *If It Doesn't Spread It's Dead: Creating Value in a Spreadable Marketplace*. Convergence Culture Consortium, 2008.

Kissling-Koch, Petra. *Macht(t)räume. Der Production Designer Ken Adam und die James Bond Filme*. Bertz+Fischer, 2012.

Machlup, Fritz. *The Production and Distribution of Knowledge in the United States*. Princeton University Press, 1962.

Martin, Lester. "Porsche & Mercedes-Benz Museums in Stuttgart [sic]" [blogpost]. *Jazzy Earl*, September 18, 2017, martin.atlassian.net/wiki/spaces/jazzyearl/blog/2017/09/10/122978305/Porsche+Mercedes+Benz+Museums+in+Stuttgart. Accessed September 25, 2018.

Pohl, Hans, Stephanie Habeth, and Beate Brüninghaus, editors. *Die Daimler-Benz AG in den Jahren 1933 bis 1945. Eine Dokumentation*. Steiner, 1986.

Rohmer, Éric. "Die 'Summe' von Bazin." *André Bazin: Was ist Kino? Bausteine zur Theorie des Films*, edited by Hartmut Bitomsky, Harun Farocki, and Ekkehard Kaemmerling. DuMont, 1975, pp. 7–18.

Steinle, Matthias. "Das Archivbild. Archivbilder als Palimpsete zwischen Monument und Dokument im audiovisuellen Gemischtwarenladen." *MEDIENwissenschaft*, vol. 22, no. 3, 2005, 295–309, https://doi.org/10.25969/mediarep/6957.

Sunder, Kalpana. "The Ultimate Shrine – The Mercedes-Benz Museum." *Huffington Post*, October 20, 2014, http://www.huffingtonpost.com/kalpana-sunder/mercedes-benz-museum_b_6009530.html.

Thaler, H. Richard, and Cass A. Sunstein. *Nudge: Improving Decisions about Health, Wealth, and Happiness*. Yale University Press, 2008.

UNSTUDIO, and HG Merz, editors. *Buy Me a Mercedes-Benz. Das Buch zum Museum*. Actar Publishers, 2006.

Van Berkel, Ben, and Caroline Bos. "Das Museum des Einundzwanzigsten Jahrhunderts." *Buy Me a Mercedes-Benz*. Actar Publishers, 2006, pp. 1–9.

Warburg, Aby. *Der Bilderatlas Mnemosyne*. Edited by Martin Warnke and Claudia Brink. Akademie Verlag, 2000.

Warburg, Aby. *Nachhall der Antike. Zwei Untersuchungen*. Edited by Pablo Schneider. Diaphanes, 2011.

Wasson, Haidee. "Cinema and Industrial Design: Showmanship, Fairs and the Exhibition Film." This volume.

Wasson, Haidee. *Everyday Movies: Portability and the Transformation of American Culture*. University of California Press, 2021.

Wasson, Haidee. "The Other Small Screen: Moving Images at the New York's World Fair, 1939." *Canadian Journal of Film Studies*, vol. 21, no. 1, 2012, pp. 81–103.

Williams, Stephen. "Touring the Temples of German Automaking." *New York Times*, December 31, 2009, www.nytimes.com/2010/01/03/automobiles/03museums.html.

About the Author

Florian Hoof is Associate Lecturer in Film and Media Studies and Research Associate in the research initiative ConTrust, part of the Cluster of Excellence Normative Orders, both at the Goethe University Frankfurt. He is the author of *Angels of Efficiency. A Media History of Consulting* (Oxford University Press 2020), *Alternative Sports and Media Distribution. Infrastructures, Logistics, and the Circulation of Moving Images* (Palgrave Macmillian 2024), and co-editor of *Films that Work Harder. The Circulation of Industrial Cinema* (Amsterdam University Press 2023).

Section 2

Operative Iconographies, Industry and the Nation State

5 Beautiful Luxembourg, Steel Works and a Swimming Pool

The Corporate Film *Columeta* and the Formation of a Corporate and a National Image

Ira Plein

Abstract
Before the sovereignty of the financial sector, the steel company ARBED (founded 1911) was considered the genuine backbone of the Luxembourg economy and was perceived as a company of "national identity." This chapter argues that ARBED's corporate image was strongly shaped through the 1921 corporate film *Columeta*. Carefully designed for a modular use to be presented to different audiences, the film compiled the right motifs at the right time in order to communicate the company's claim of national and societal importance, making *Columeta* a stepping stone for integrating industrial motifs into the national iconography. The chapter will outline the circulation and interaction of corporate, national and social motifs in different media, and their subsequent inclusion into the national Luxembourgian self-presentation.

Keywords: Luxembourg; steel industry; corporate communications; industrial film; national iconography

Luxembourg's last blast furnace was shut down at the site of Esch-Belval in 1997. The industrial wasteland was added to the list of Luxembourg's national monuments and subsequently converted into a signboard of the country's self-presentation, where the remains of the steel industry form the backdrop for Luxembourg's present and anticipated future of service sectors and education. Today, the site houses Luxembourg's university, banks, a concert hall and a shopping centre, and visitors can book a guided tour

Hediger, V., F. Hoof, Y. Zimmermann, with S. Anthony (eds), *Films That Work Harder: The Circulation of Industrial Film*. Taylor & Francis Group 2024
DOI: 10.5117/9789462986534_CH05

at "Belval – City of Science & Blast Furnace."[1] It was the mining and steel industry that initially brought wealth to the Grand Duchy of Luxembourg.[2] Steel making and industrialization rose late in Luxembourg, but quickly dominated the country's economy, until the international steel crisis in the 1970s made room for the upcoming financial sector.

Looking back at a long tradition of iron making in preindustrial times, Luxembourg had been a predominantly agricultural country until the late nineteenth century. Over the course of the century, it was the (re)discovery of vast iron ore reserves, the affiliation to the German Customs Union, the (rather late) installation of railroads, and, not least, the adaption of the Gilchrist-Thomas process, which enabled the exploitation of the highly phosphoric local iron ore for steel making, and thus lead to the rise of the steel industry.[3] At the turn of the twentieth century, the industry, with its new industrial sites and its high demand for workers, had transformed agricultural landscapes and communities in the country's South into an industrial region.

The foundation of the Luxembourgian corporation ARBED (Aciéries réunis de Burbach-Eich-Dudelange) in 1911 counteracted the dominance of German capital in Luxembourg's steel sector, which only dissolved after World War I. In 1919, ARBED became a partner in the newly founded Société Métallurgique Terres Rouges, in order to take over the steel works and ore mines of the German Gelsenkirchener Bergwerks-Aktiengesellschaft in Esch-sur-Alzette, and thus dominated two-thirds of the Luxembourgian steel production.[4] Subsequently, ARBED became the "synonym and the symbol of the Luxembourgian steel industry."[5] For decades, ARBED was perceived as company of "national identity" and considered the genuine

1 For touristic endeavours of industrial heritage, see, for instance, Edwards, J. Arwel, and Joan Carles Llurdés I Coit. "Mines and Quarries: Industrial Heritage Tourism." *Annals of Tourism Research*, vol. 23, no. 2, 1996, pp. 341–63.

2 In peak times Luxembourg's metallurgical sector had a share of over 31.8% in the gross domestic product (1960), employed half of all industrial employees (1966) and was the country's largest taxpayer (1967). See Trausch, Gilbert. *L'ARBED dans la société luxembourgeoise*. ARBED/ Victor Buck, 2000, p. 35.

3 For an overview on the evolution of the metallurgical sector in Luxembourg from the Iron Age until the early twentieth century, see Knebeler, Christoph, and Denis Scuto. *Belval: Passé, présent et avenir d'un site Luxembourgeois exceptionell (1911–2011)*. Éditions Le Phare, 2010, pp. 22–45.

4 The perception of ARBED as the Luxembourgian leader continued despite the predominance of Franco-Belgian capital in the steel industry, which replaced the German capital after the war. Terres Rouges was fully incorporated by ARBED in 1937. Its competitor HADIR (Société des Haut Fourneaux et Aciéries de Differdange, St. Ingbert et Rumelange) was incorporated in 1967; see Trausch. *L'ARBED dans la société luxembourgeoise*, pp. 31–35.

5 Ibid., p. 35.

backbone of the Luxembourg economy, until the financial sector took the economic lead and ARBED eventually was incorporated into the world's leading multinational steel corporation, Arcelor Mittal, in a series of mergers from 2002 to 2007.[6]

How did the image of ARBED as a national corporation evolve? This chapter investigates the creation and consistency of ARBED's corporate image, which was strongly shaped through the 1921 corporate film *Columeta* and that was maintained for decades.[7] Moreover, I will outline the circulation and interaction of corporate, national and social motifs, and their subsequent inclusion into the national Luxembourgian self-presentation.

Drawing on the concept of *ResourceCultures*, the employed motifs can be mapped as the *communication of resources* that ARBED placed as significant components for its corporate identity as a productive, efficient and responsible Luxembourgian company. Going beyond an understanding of resources as tangible economic means, such as raw materials, technologies or capital, and including intangible social or cultural resources, such as networks, groups, institutions or education, Bartelheim et al. understand "resources as a base of or means to create, maintain or transform social relations, units and identities on the background of culturally shaped beliefs and practices."[8] In this sense, I will look at examples of visual communications that established and maintained a value-creating iconography around the industrial sphere in Luxembourg. The employed motifs underwent shifting valuations and were alternately appropriated over time, until they propagated a corporate, and to a large extent, a national identity. To substantiate my argument, I will take a look at media presentations of the steel industry in films, photography, print brochures and exhibitions from 1889 until the 1960s.[9]

6 Lorang, Antoinette. "D'Heichiewen. Les hauts-fourneaux." *Lieux de mémoire au Luxembourg: usages du passé et construction nationale*, edited by Sonja Kmec et al., 2nd ed. Éditions Saint Paul, 2008, pp. 215–20; see also *La sidérurgie luxembourgeoise. Un siècle d'histoire et d'innovation/ Steelmaking in Luxembourg: A Century of History and Innovation*. Arcelor Mittal, 2011, pp. 81–83.

7 I thank Charles Barthel and Gilles Regener from the National Archives of Luxembourg and Alessandra Luciano from the Centre national de l'audiovisuel (CNA), Luxembourg, for their support in accessing the respective files. In addition, I thank the members of the FAMOSO team and André Linden for their discussions on the topic.

8 Bartelheim, Martin, et al. "'ResourceCultures' – A Concept for Investigating the Use of Resources in Different Societies." *Persistent Economic Ways of Living: Production, Distribution, and Consumption in Late Prehistory and Early History*, edited by Alžběta Danielisová and Manuel Fernández-Götz. Archaeolingua Alapítvány, 2015, pp. 39–50, quotation on p. 39. Unless otherwise stated, all translations are my own.

9 As could be expected, the corporate image was not always accepted without opposition, e.g. by the labour movement, see: Plein, Ira. "'Der tägliche Weg zur Schicht'. Aspekte zur proletarischen Kunst Albert Kaisers in der Zwischenkriegszeit." *100 Joer fräi Gewerkschaften 1916–2016*, edited by

Columeta (1921–1922)

After Luxembourg's exit from the German Customs Union, ARBED and Terres Rouges, in 1920, founded the joint distribution company Comptoir Métallurgique Luxembourgeois (Columeta) to develop new markets. Soon after, Columeta produced the homonymous corporate film *Columeta* (1921–1922). During this period of political and economic rearrangement, the film *Columeta* served as a cornerstone for the successful establishment of a comprehensive corporate image. The film shaped a corporate image, consisting of (1) beautiful views of Luxembourg, (2) the Luxembourgian steel industry, and (3) the industry's social works, which determined the industry's self-presentation for decades. At the same time, the association of industry with picturesque views of Luxembourg was a stepping stone for the successful integration of industrial motifs into the national iconography.

It should be stated at the outset that there is no such thing as a definitive version of *Columeta*. The film's production included several steps, involved different parties and we do not have any information on the actual making of subsequent versions.[10] ARBED first commissioned the film from the Paris-based operator Gustave Labruyère (L'art cinématographique) in 1921. As some sequences shot by Labruyère were defective, ARBED commissioned Edmund Epkens from the Rheinische Filmgesellschaft in Cologne to complete the film.[11] This version was 1,400 metres long, tinted and mounted with French title cards. The correspondence files about the production of the corporate film date back to January 1921, while the first public screenings took place in Luxembourg in October of the same year. This initial French version served as a master copy for another eight copies in six languages, which were sent to the company's international branch offices, with the aim of supporting the corporate propaganda and Columeta's global distribution of steel products. Additional short films and sequences were produced in 1922

Frédéric Krier et al. OGB-L/Le Phare, 2016, pp. 83–97; a detailed analysis of visual communications and media as a part of a societal discourse on industrialization is provided in my dissertation: *Propaganda for Steel and Nation: Images and Counter-images to Economic and Social Progress in Luxembourg Media of the Interwar Period.*

10 Based on the existing archival documents, historian Charles Barthel largely reconstructed the genesis of *Columeta*. See Barthel, Charles. "*Columeta*: L'aventure du premier film publicitaire de l'Arbed." *Hemecht*, vol. 50, no. 2, 1998, pp. 177–206.

11 Before Epkens, the Luxembourgian Alfred Heinen was commissioned to execute the additional shots, but the result was not satisfactory. Heinen later assisted Edmund Epkens. See ibid.

by Pathé.[12] The following outline of *Columeta* refers to a preserved script of the initial version from 1921 and to a reconstructed version of *Columeta*, made in 1997 by the Centre national de l'audiovisuel (CNA) in Luxembourg and titled *Vu Feier an Eisen* (*Of Fire and Iron*).[13]

After the initial completion of *Columeta* in November 1921, the Luxembourg headquarters of ARBED sent a circular to the international branch offices, announcing the delivery of a five-reel film from the Rheinische Filmgesellschaft Cologne. The letter roughly sketches the content of the five reels:

> I.- The first one presents views of Luxembourg City, of our new headquarters and our ARBED and Terres Rouges plants, situated in the Grand Duchy;
> II.- The second one treats the extraction of the minerals, the arrival of the coke & the blast furnaces;
> III.- The third, the Thomas steel work;
> IV.- The fourth, the different mill trains;
> V.- The fifth, the social works.[14]

In addition, the letter clearly expresses the film's objective:

> As we do not have the intention of making a film of purely technical qualities, but rather a propaganda film, with the objective to give an idea of the company's importance and to popularize the name "Columeta," it will be of interest first and foremost for the general public and then only for the technicians.[15]

After introducing Columeta as the exclusive sales company of all products of the Luxembourgian steel companies ARBED and Terres Rouges, a second title

12 Another undated script foresees a version of *Columeta* in four parts, where the Luxembourg sequences and the social works frame the Pathé films of ingot production and mill trains. See undated script. Archives nationales de Luxembourg (ANL), Arbed 03-00003. An actual realization of another version is neither documented nor reported.

13 In some parts, the 1997 reconstruction differs from the script. A precise reconstruction of the 1921 version is not possible at the moment, since the preserved film reels, which were found at ARBED in 1997, are not yet fully accessible. The untinted parts in the reconstructed version of 1997 might consist of the 1922 short films by Pathé. The nitrate positive films are preserved at the Centre national de l'audiovisuel, Luxembourg. The reconstructed version, *Vu Feier an Eisen*, is available on the DVD *Le Luxembourg sidérurgique. 8 films autour de l'acier* (CNA, 2004).

14 Projet de lettre. November 12, 1921. ANL, Arbed 03-00001.

15 Ibid.

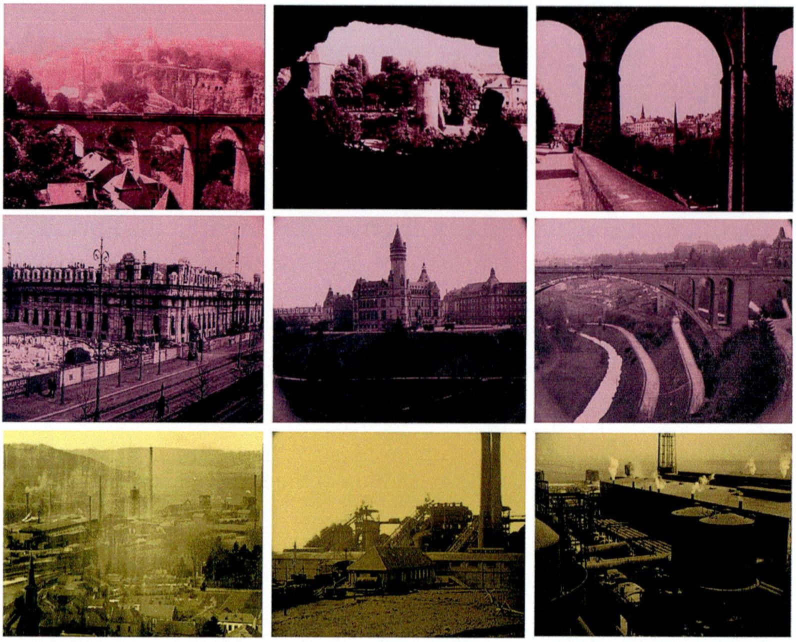

Fig. 5.1. Frames from *Vu Feier an Eisen*, a reconstructed version of *Columeta* (1921–1922), reel I. Courtesy of Centre national de l'audiovisuel (Luxembourg)/Arcelor Mittal.

card announces the city of Luxembourg as the capital of the Grand Duchy. The following sequences present traditional touristic motifs of Luxembourg City as a travelogue of "living picture postcards" (e.g. views of the old bridge, the remains of the former fortress city, scenes of children playing and of people enjoying the panoramic view of the old town).[16] These scenes are followed by a long panning shot over the Vallée de la Pétrusse – beginning with the old bridge and ending with the new Pont Adolphe – linking the old town with Luxembourg's new "carte de visite," the Plateau Bourbon.[17] An iris wipe interrupts the panning shot and affords the viewer a closer look at the future headquarters of

16 See Jung, Uli. "Städtebilder und Lokalaufnahmen." *Geschichte des dokumentarischen Films in Deutschland, Band 1: Kaiserreich 1895–1918*, edited by Uli Jung and Martin Loiperdinger. Reclam, 2005, pp. 274–97; and Deeken, Annette. "Geschichte und Ästhetik des Reisefilms." *Geschichte des dokumentarischen Films in Deutschland, Band 1: Kaiserreich 1895–1918*, edited by Uli Jung and Martin Loiperdinger. Reclam, 2005, pp. 298–323, esp. p. 302. The well-received touristic film *Das malerische Luxemburg* (*Picturesque Luxembourg*), shot by Felix Medinger in 1912, also combines views of old and new of Luxembourg City; see Lesch, Paul. "Film et tourisme au Luxembourg à la Belle Epoque." *Ons Stad*, vol. 88, 2008, pp. 58–61.

17 Thewes, Guy. "La silhouette de la ville." *Lieux de mémoire au Luxembourg: usages du passé et construction nationale*, pp. 253–58, quotation on p. 256.

ARBED, then still under construction.[18] After focusing in on the commissioners' headquarters, the panning shot continues to the Spuerkess (savings bank) and the national railways company, before it ends with Pont Adolphe. Title cards that give basic information about the company, including the annual sales and range of products, identify the company's headquarters, and list Columeta's international branch offices and production sites complement the scenes. Further panning shots link the city views to views of ARBED and the Terres Rouges steel works in Luxembourg: thus, the only Luxembourgian "landscapes" to be perceived in the film are the industrial sites.[19]

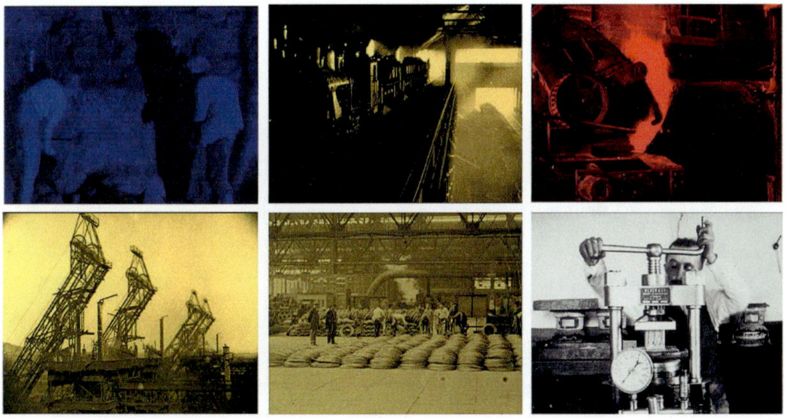

Fig. 5.2. Frames from *Vu Feier an Eisen*, a reconstructed version of *Columeta* (1921–1922), reel II, III and IV. Courtesy of Centre national de l'audiovisuel (Luxembourg)/Arcelor Mittal.

With reels II, III and IV, the largest part of the film is dedicated to the process of steel production, shown in the common practice of temporal succession from the ore mine to the final product at the different ARBED and Terres Rouges sites (fig. 5.2). Title cards explain the production sequences. Reel II shows the production of pig iron – from underground and daylight ore mining, the transport to the steel plant by train, to impressive shots of the charging of the blast furnaces that illustrate the plant's size and grandeur. Reel III presents the production of steel using the Gilchrist-Thomas process, including the spectacle of sparking Bessemer converters and of glowing-hot iron and steel. The film is tinted accordingly: blue for underground scenes, yellow for daylight settings,

18 The iris wipe can only be seen completely in one of the accessible original sequences: CNA, FCDARBED000104.

19 Loiperdinger, Martin. "Industriebilder." *Geschichte des dokumentarischen Films in Deutschland, Band 1: Kaiserreich 1895–1918*, edited by Uli Jung and Martin Loiperdinger. Reclam, 2005, pp. 324–32, esp. p. 331.

and red for interior shots of the plants. The reel ends with scenes that show techniques of mechanical and chemical quality control.[20] Reel IV shows the mill trains, where steel beams, rails, plates, and wire are shaped according to the rhythm of the machines. It concludes with the forging of an ingot by a huge steam hammer, showing several workers busy moving the ingot. This sequence was presented with a simple skip frame, so that the ingot was shaped into a thinner bloom within seconds. This scene of labour is immediately (without a title card) followed by an imposing scene of a swimming pool, with men and boys enjoying themselves in front of the steel works in Dudelange (fig. 5.3).[21]

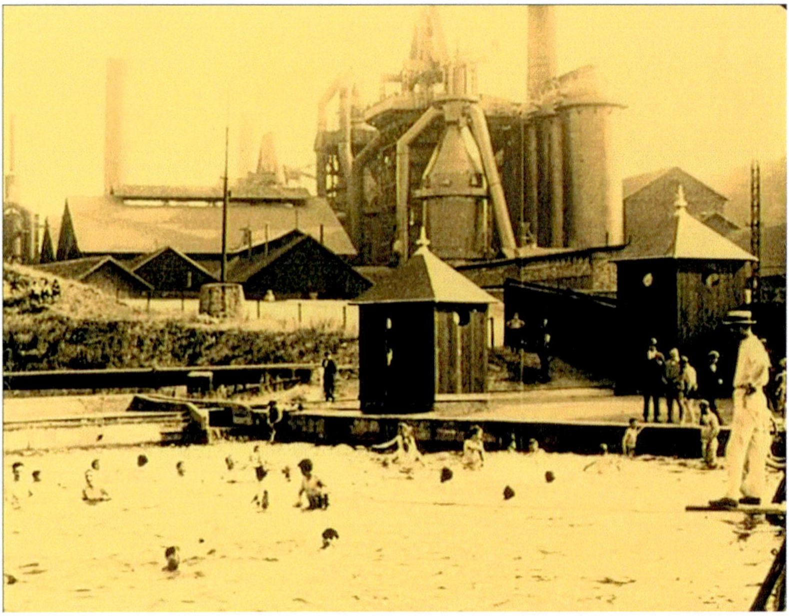

Fig. 5.3. Frames from *Vu Feier an Eisen*, a reconstructed version of *Columeta* (1921–1922), end of reel IV. Courtesy of Centre national de l'audiovisuel (Luxembourg)/Arcelor Mittal.

The fifth reel opens with a title card, introducing the social works as follows:

> To improve the situation of their workers and staff, Arbed and Terres Rouges created numerous social institutions, of which the following

20 Scenes of quality control have been listed in the original script, but were striked out. The (untinted) footage in the reconstructed version of 1997 might derive from the 1922 Pathé films.
21 This order is documented in the script (ANL, Arbed 03-00003) and a preserved clip (CNA, FCB 300, 4D). In another preserved copy (CNA, FCA 636, 2E) the outdoor swimming pool sequence follows on the indoor swimming sequence at the Institut Emil Metz in Dommeldange, without indicating the actual different locations.

are to be emphasized: the workers' housing estate Emil Mayrisch; the "*maison des enfants*" [sanatorium] in Dudelange; the open air school in Dudelange; the Institut Emile Metz [vocational school] in Dommeldange.

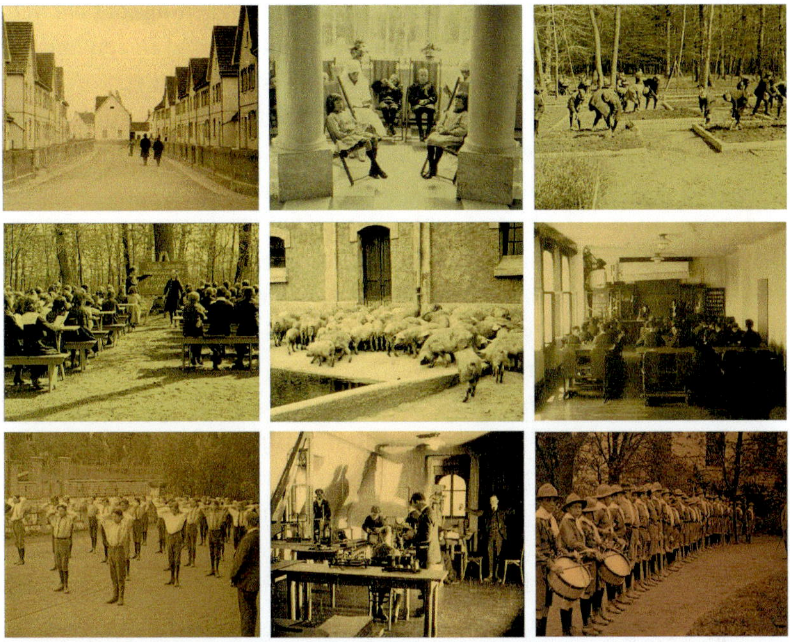

Fig. 5.4. Frames from *Vu Feier an Eisen*, a reconstructed version of *Columeta* (1921–1922), reel V. Courtesy of Centre national de l'audiovisuel (Luxembourg)/Arcelor Mittal.

The sequences indeed present scenes of caring and education: newly built houses and nurses who take care of children's meals, hygiene and recreation time, as well as schooling in the classroom and physical training at the sanatorium. In the presentation of the open-air school, the viewer cannot miss the children's gratitude.[22] It is written on the chalkboard and shown in a full-screen shot: "Nous sommes reconnaissantes à nos bienfaiteurs" ("We are grateful to our benefactors"). Further scenes show the provision of milk and a company-owned farming project. The detailed sequences on the vocational school Institut Emile Metz present the school building,

22 For detailed research on open-air schools in Luxembourg, see Thyssen, Geert. "Engineered Communities? Industry, Open-air Schools, and Imaginaries of Belonging (c. 1913–1963)." *History of Education & Children's Literature*, vol. 10, no. 2, 2005, pp. 297–320; and Hadžalić, Irma. "Sick and Weak but Made of Steel: Luxembourgian Open-air Schools and Other Responses to the Spread of Tuberculosis at the Beginning of the 20th Century." *Revista da História e Historiografia da Educação*, vol. 1, no. 1, 2017, pp. 44–64.

theoretical schooling and physical education, the psycho-technological laboratory and the indoor swimming pool in the basement.[23] The film ends with the students' boy scout activities, followed by a final title card with information on the respective Columeta branch office.

To date, it has been possible to trace only a few screenings of *Columeta*. The film was shown to selected audiences in October and December 1921, each time with orchestral accompaniment: first, to the directors of the steel works, engineers and employees of the headquarters in Luxembourg City, second, to employees and representatives of popular education associations (*Volksbildungsvereine*) in Dudelange, and, finally, to workers in Dudelange, Esch-sur-Alzette and Esch-Belval.[24] The long version and the additional technical short films by Pathé were sent to Rio de Janeiro to be presented at the World Fair in 1922–1923, where the company contributed to the Luxembourgian fair stand with a private screening room – the "Cinema Columeta."[25] In Basel, the film was presented as *Die Eisenindustrie in Luxemburg* (*Columetafilm*) (*The Iron Industry in Luxembourg*) to Gottlieb Imhof's newly founded Studienkommission für Schulkinematographie (Study Commission for School Cinematography) in 1922.[26] Other than a letter in which Pathé is offering to distribute the film in their affiliated cinemas, there is no documentation about screenings in public cinemas.[27]

Copies were sent to the branch offices of Columeta in Paris, London, Madrid, Rotterdam and Basel. However, the already mentioned letter to the branch offices explained that the total length of one hour and fifteen minutes was too long for the general public. The executives were advised to show the complete five-reel version "only to an audience that is especially interested in this genre of representation" such as, for example, school children or groups of engineers.[28] Screenings for the general public in commercial cinemas should feature only one reel of *Columeta* in the programme of shorts. If reels II, III, IV or V were presented, it was emphasized

23 See Herman, Frederik. "Forging Harmony in the Social Organism: Industry and the Power of Psychometric Techniques." *History of Education*, vol. 43, no. 5, 2014, pp. 592–614; Priem, Karin, and Frederik Herman. "Hautnah: Materialität der Moderne und sensomotorische Ansätze der Berufsbildung im 'Zeitalter des Stahles'." *Die Sache(n) der Bildung*, edited by Christiane Thompson, Rita Casale, and Norbert Ricken. Ferdinand Schöningh, 2017, pp. 213–39.

24 See Barthel. "*Columeta*," pp. 195–196.

25 See Lorang. *L'image sociale de l'Arbed à travers les collections du Fonds du logement*. Le fonds pour le développement du logement, 2009, p. 18.

26 See Gertiser, Anita. "Domestizierung des bewegten Bildes." *montage/av*, vol. 15, no. 1, 2006, pp. 58–73, esp. p. 63.

27 See Barthel. "*Columeta*," p. 200.

28 Projet de lettre.

that they each *must* be combined with and preceded by the beginning of reel I, including title card six.[29] These are precisely the Luxembourg City scenes, ending with the panning shot over the Vallée de la Pétrusse. Thus, the company prescribed a modular screening and perception, where it was the Luxembourg City views, not the industrial scenes, that were seen as indispensable elements – an emphasis that warrants closer analysis.

The Luxembourg scenes were not part of the very first draft concept in early 1921, but soon included, possibly at the recommendation of Gustave Labruyère, in order to "create a picturesque atmosphere around the new headquarters."[30] However, one reason for commissioning shots by Edmund Epkens and the Rheinische Filmgesellschaft was the fact that footage like the panning shot of the Plateau Bourbon and the panoramic views of the steel works by Gustave Labruyère had turned out badly.[31] Shortly before Epkens was to shoot the additional scenes in July 1921, Willy Heymers, head of the Municipal Cinematography Department in Cologne and in charge of assessing and implementing educational films, sent a note of advice, concerning the added value of industrial films.[32] Heymers advises to "make the shots of industry as long as possible, so that the eye has enough time for contemplation – and, furthermore, to give the widest possible latitude to pictures of the city [of Luxembourg]; then the industrial film becomes in a certain sense an educational film. It thereby also becomes more valuable for your propaganda purposes."[33] In the letter, which might have been intended to support ARBED's decision for commissioning the Rheinische Filmgesellschaft, Heymers further recommended Epkens for having the best experience to arrange the shots accordingly.

The Luxembourgian motifs can be understood as a visual claim and manifestation of Columeta's importance for the country, which it already expressed in its company name Comptoir Métallurgique Luxembourgeois. This may also have been the reason for the visual absence of any production sites outside the country. While a title card in reel I listed all production sites of ARBED and Terres Rouges in Luxembourg, France, the Saar region, and

29 Ibid.

30 See Barthel. "*Columeta*," p. 182.

31 Ibid., pp. 185–87.

32 Due to restoration work, the original files are currently not accessible, but Heymers' function can be inferred from the online listing and description of the material at the Historical Archive of the City of Cologne. Holding 608 (Culture Department) A 149, call number: Best. 608 (Kulturdezernat) A 149.

33 Handwritten letter. Addressee unknown. July 7, 1921. ANL, Arbed 03–00001. Signed by "Willy Heymers Leiter der Städt, Lichtspiele, Köln."

the Rhineland, the film itself featured only views of the Luxembourgian sites (Dommeldange, Dudelange, Esch-Terres Rouges, Esch-Arbed (Schifflange), Belval-Terres Rouges).[34] The presentation of the Luxembourgian sites of Terres Rouges, which had previously been in German hands, surely also bolstered the company's claim of being *the* new steel force of Luxembourg. The emphasis on the ARBED headquarters' location in the panning shot might have served the same objective: The Luxembourg government had reserved the prestigious lots at the new quarter of Plateau Bourbon for companies of national economic importance, and in the case of the ARBED headquarters had even abandoned its plans to build a museum on the site.[35]

Still, the benefits of the Luxembourg views are manifold: The picturesque framing worked for the entire film. With Luxembourg City views in the beginning and the company's social works at the end, the technical imagery of production is framed by pleasant views and lively scenes, which may well have enhanced the film's attractiveness for the audience. Moreover, by linking the cultural heritage of the old town to the prestigious new quarter of Plateau Bourbon, the sequence of the city scenes functions as a display of historical progress from tradition to modernity. This ongoing process – and progress – of modernization is continued in modern industrial landscapes, state-of-the-art industrial production and the industry's social reform initiatives, with an emphasis on the health and education of future generations. The motifs are subtly interwoven in the progression of the film. National and industrial motifs merge in the succession from the city views to the panoramic views of the Luxembourgian industrial sites, which lead over to the industrial sequences that picture the whole range of production. The swimming pool scene then visually associates images of leisure and hygiene with the steel plant, links social life to the industrial technosphere, and logically segues into a depiction of the industry's social works.[36]

Ultimately, *Columeta* constituted a visual synthesis of different themes, addressing a wide range of audiences, helped shape ARBED's comprehensive corporate image as a company of national importance that ensured economic prosperity and social reform initiatives for Luxembourg, and offered an optimistic vision of the company and the country's future. For a corporate audience, the film features the new distribution company Columeta as

34 The shots of the works in Burbach (Saar) had been executed by Edmund Epkens. The reason for their abundance in the film is not documented; see Barthel. "*Columeta*," pp. 190–93.

35 Philippart, Robert L. *Luxembourg. Historicisme et identité visuelle d'une capitale*. Éditions Saint Paul, 2007, pp 42–44; see also Lorang. *L'image sociale*, pp. 22–25.

36 For an analysis of the links between human bodies and the industrial technosphere, see Priem and Herman. "Hautnah."

the dominant Luxembourgian steel trader. The multiple industrial sites, production processes and quality controls suggested huge capacity and state-of-the-art technologies. The presentation of the social works displayed the companies as responsible and modern employers, who – in times of rationalization – care for the well-being of their current and future workforce, which is presented as healthy, skilled and disciplined – and thus able to master modern technologies, guarantee the quality of the products and ensure the continuity of production. *Columeta*, presumably the first feature-length film with a Luxembourgian subject matter that was distributed globally, also communicated a certain image of Luxembourg. To an audience abroad, *Columeta* provided an insight into Luxembourg's cultural heritage and its modern steel industry, and showed a prospering society with good living conditions and well-equipped healthcare and educational institutions – which, it suggests, are linked to the steel industry. To a Luxembourgian audience, *Columeta* delivered a message of national pride in the economic power and prosperity of the Luxembourgian steel industry and thus presented a political statement of stability and independence after World War I.[37] The images of the company's social works, presenting modern initiatives of reform and the well-being of the workers, could also counter the demands of the labour movement after the strike of 1921.[38] Until today, ARBED is commonly acknowledged as a pioneer when it comes to the establishment of social relief initiatives in Luxembourg.[39] Thus, *Columeta* perfectly expressed the company's self-image as a benefactor and a forerunner in providing urgently needed social services. Moreover, the emphasis on schooling in the presentation of the social works conveys the promise of social advancement and better living conditions through vocational education, which at the time was not yet mandated by law.[40] The detailed presentation of the vocational school Institut Emile Metz propagated

37 Following the recent threat of annexation, this message was also addressed at Luxembourg's neighbouring countries; see Barthel. "*Columeta*," p. 206.

38 See ibid., p. 205

39 See, for example: Lorang. *L'image sociale*; Kieffer, Simone. "Les oeuvres sociales de l'ARBED (1900–1930) (Aciéries Réunies de Burbach-Eich-Dudelange)." MA thesis, Marc Bloch University, 2003; Schmitz, Nadine. "La politique sociale de l'ARBED/ARBED's Social Policy." *La sidérurgie luxembourgeoise. Un siècle d'histoire et d'innovation/Steelmaking in Luxembourg: A Century of History and Innovation*. Arcelor Mittal, 2011, pp. 149–51; Novella, Enric J. "Tuberculosis and Political Economy: Industrial Wealth and National Health in the Grand Duchy of Luxembourg, c. 1900–1940." *Social History of Medicine*, vol. 31, no. 2, 2018, pp. 308–27, https://doi.org/10.1093/shm/hkx004; Herman. "Forging Harmony."

40 See Herman. "Forging Harmony."

the theoretical, physical and moral formation of a new "workers' elite," which had an immediate impact on Luxembourg's societal stratification.[41]

Prescribing a modular use of the five individual reels in commercial cinemas abroad, the company might even have envisioned tailor-made screenings to specific audiences. The indispensable Luxembourg City scenes (which included the title cards with the most important information on the company) in each version linked the company to the national context and could be complemented either with scenes of industrial production or with those of the social initiatives. In the case of reel IV, where the production sequence ended with scenes of leisure and hygiene in the swimming pool, the trinity of national, industrial and social motifs was even communicated in a short version.

The corporate image, which was established in *Columeta* in the early 1920s, was extensively used for corporate communication in various media during the interwar period and remained dominant for the company's self-presentation for decades.[42] Even under German occupation during World War II, when ARBED published the illustrated book *Stahl aus Luxemburg* (Steel from Luxembourg) with the Heinrich Hoffmann publishing house in Munich in 1942, it upheld the well-established corporate image as Luxembourgian industry.[43]

The Steel Industry as a Motif of National Cultural Heritage

Linking the industrial views to the picturesque views and the cultural heritage of Luxembourg in *Columeta* can be seen as a coherent strategy to

41 See Priem, Karin, and Frederik Herman. "'Sensuous Geographies' in the 'Age of Steel': Educating Future Workers' Bodies in Time and Space (1900–1940)." *Fabricating Modern Societies: Education, Body and Mind in the "Age of Steel"*, edited by Karin Priem and Frederik Herman. Brill, 2019; Herman. "Forging Harmony"; Priem and Herman. "Hautnah"; Herman, Frederik, Karin Priem, and Geert Thyssen. "Körper_Maschinen? Die Verschmelzung von Mensch und Technik in Pädagogik, Industrie und Wissenschaft." *Jahrbuch für Historische Bildungsforschung*, vol. 20, 2015, p. 61.

42 Columeta and ARBED published a number of corporate books and postcards that communicated the corporate image. In 1922, an illustrated brochure was exclusively dedicated to the presentation of the history of ARBED's social works. In 1929, the animated film *Une Grande Industrie dans un Petit Pays* (*A Great Industry in a Small Country*) demonstrated ARBED's – and Luxembourg's – global economic power through numerous animated statistical graphics. For further examples on ARBED's publicity package with special regards to its social works, see Lorang. *L'image sociale.*

43 The motifs in the book comprise a Luxembourg City view, a Luxembourgian industrial landscape, steel production "from the ore mine to the final product," and the industry's social works. Luxembourg's annexation to the newly established German administrative division "Gau Moselland" is only mentioned in the introduction; see Muth, F., et al. *Stahl aus Luxemburg.* Verlag Heinrich Hoffmann, 1942, p. 9.

promote ARBED as a Luxembourgian company. The national iconography had been formed during the nineteenth century and was strongly connected to the emerging tourism. Landscape painters had established a canon of *vues pittoresques* (picturesque views) of the capital's and the country's cultural heritage, which strongly reflected the romantic desires of foreigners for what "had to be visited and seen" in Luxembourg.[44] Following this painterly tradition, the photographer Charles Bernhoeft published the album *Le Grand-Duché de Luxembourg* (The Grand Duchy of Luxembourg) between 1889 and 1891.[45] The album was dedicated to the future Grand Duke Adolphe and was intended to comprise images of "all important aspects" in Luxembourg.[46] When Bernhoeft, in 1889, included the plate *Hauts fourneaux de Dudelange* (Blast furnaces of Dudelange) (fig. 5.5) in the album, his attempt to broaden the national imagery was criticized by an anonymous critic in the Luxembourgian press.[47]

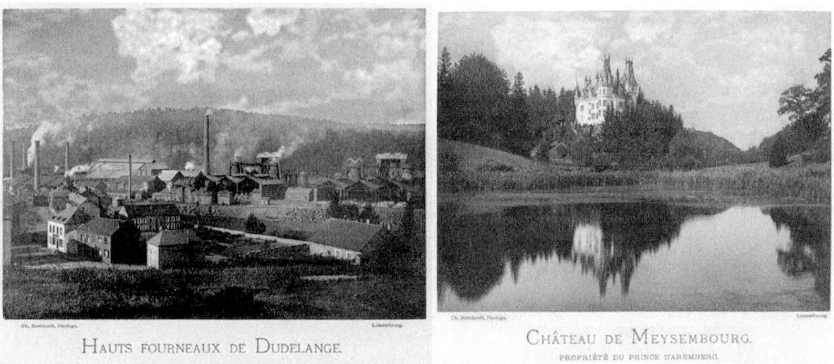

Fig. 5.5. Charles Bernhoeft, *Hauts fourneaux de Dudelange* (Blast furnaces of Dudelange), collotype, from the album series *Le Grand-Duché de Luxembourg* (1889–1891). © Musée National d'Histoire et d'Art - Luxembourg.

Despite the careful assimilation of the industrial site to the pictorialist style of the series, the critic pointed out that this very modern plant, with its smoking chimneys, had nothing in common with the ancient monuments

44 An example is the *Album Pittoresque du Grand Duché de Luxembourg* (Picturesque album of the Grand Duchy of Luxembourg), published in 1857 by the Luxembourgian painter Jean-Baptiste Fresez; see Linden, André, and Guy Thewes. *Greetings from Luxembourg: A Journey into the World of Tourism*. Musée d'histoire de la ville, 2008, p. 46.

45 See Thill, Edmond. "Charles Bernhoeft: Photographe de la Belle Époque." *Charles Bernhoeft: Photographe de la Belle Epoque*, edited by idem. Imprimerie Centrale S.A., 2014, pp. 93–412, esp. p. 124.

46 Ibid., pp. 138, 155.

47 See Herman, Frederik, and Ira Plein. "Envisioning the Industrial Present: Pathways of Cultural Learning in Luxembourg (1880s–1920s)." *Paedagocica Historica*, vol. 53, no. 3, 2017, pp. 268–84.

and cultural heritage. Bernhoeft did not include the plate in his later albums of Luxembourgian views and thus may have adapted to the dominant desire for traditional picturesque national motifs.[48] More than a decade later, however, Bernhoeft entered into an apparently profitable business, when he published some motifs of the recently built Differdange plant as postcards.[49] At the turn of the century, the steel industry became an increasingly popular motif for the emerging mass medium of the illustrated postcard.[50] Though still not accepted as "official" national motifs, industrial sites thus became a part of the popular range of Luxembourgian motifs, which surely made it possible for the general public to familiarize itself with, and aesthetically appreciate, the new industries and technologies.[51]

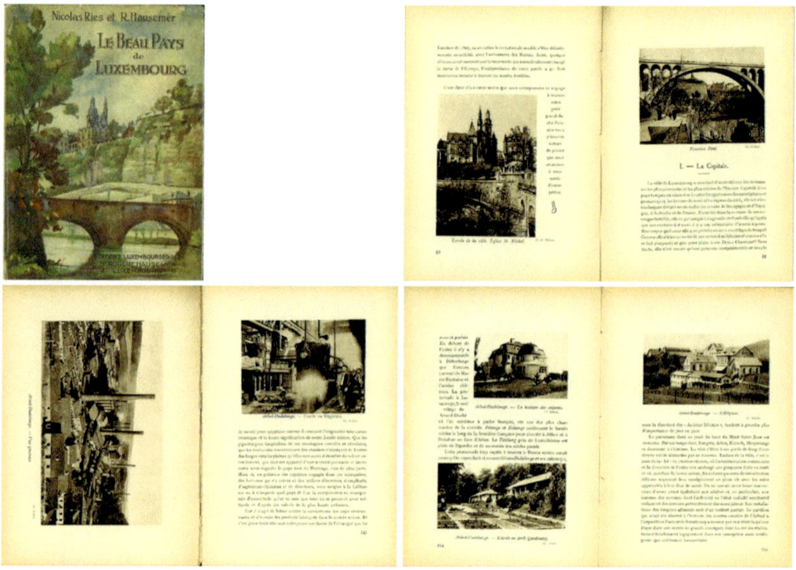

Fig. 5.6. Nicolas Ries and Robert Hausemer, *Le beau pays de Luxembourg* (The beautiful country of Luxembourg, 1932).

It was only after World War I, when ARBED came to dominate Luxembourgian steel production, and after the *Columeta* film that the steel industry

48 Thill. "Charles Bernhoeft," p. 144.

49 See Herman, and Plein. "Envisioning the Industrial Present," pp. 276–80.

50 Views of Differdange rank as the fourth most common motifs among the more than 1,300 postcard motifs published by Luxembourgian photographer Jacques Marie Bellwald around 1900, see ibid.

51 The steel industry was also aesthetically elevated in contemporary Luxembourgian literature. See Millim, Anne-Marie. "Schooling the Gaze: Industry and Nation-Building in Luxembourgish Landscape-Writing, 1900–1940." *Journal of European Studies*, vol. 44, no. 2, 2014, pp. 151–69.

was irrevocably presented and accepted as part of the national iconography. A cornerstone for the elevation of industrial motifs to national value was laid with the richly illustrated travel guide *Le beau pays de Luxembourg* (The beautiful country of Luxembourg) in 1928 (fig. 5.6). Photographs of industrial sites and even of ARBED's social works are incorporated into the visual canon of national sites and cultural heritage in an explicit context of "beauty."

Possibly planned as a film adaption of the travel guide, a film of the same title, *Le beau pays de Luxembourg*, was commissioned by the government for the 1935 World Fair in Brussels, where the steel industry featured prominently in the presentation of the country's economy.[52] After a first screening in Luxembourg, the response of the press was unanimous: the film by a Belgian operator, which was supposed to serve as touristic propaganda for the country, was seen as technically bad.[53] Moreover it had failed to represent the country's beauty, which was considered as an impossible undertaking in fifteen minutes.[54] It was further criticized that "our metallurgy" was featured in a manner that did not represent Luxembourg's leading position in Europe's steel production.[55] Describing his vision of "*the* film about Luxembourg," the film critic Evy Friedrich said that it would need to be technically accurate and require a detailed knowledge of the country, so that one could choose "only the best" and the "most characteristic" topics of the country.[56] Not least, a film about Luxembourg would require more time to capture the landscape and to leave the audience time "to await the highlights."[57]

As it turned out, Evy Friedrich became the assistant of the Luxembourgian cineaste René Leclère, who directed a forty-seven-minute film for the World Fair in Paris 1937, entitled *Il est un petit pays* (*It Is a Small Country*).[58] In this film, a couple takes the viewer on a touristic car ride through different regions of the Grand Duchy, highlighting its natural beauty and cultural heritage, but also pointing out its modern features, for example, the new radio station or life in the capital city. When the journey goes south, to

52 The film is not preserved. For the 1935 World Fair, see Mousset, Jean-Luc, and Ulrike Degen, editors. *Un petit parmi les grands: Le Luxembourg aux Expositions universelles de Londre à Shanghai (1851–2010)*. Musée national d'histoire et d'art, 2010, pp. 189–205.

53 M., Dr. "Der Propagandafilm über Luxemburg." *Luxemburger Wort*, February 15, 1935, p. 8.

54 Ibid.

55 M., N. "'Le beau pays de Luxembourg': Ein touristischer Propagandafilm." *Tageblatt*, February 15, 1935, p. 7.

56 F., E. [Friedrich, Evy]. "Ein Film über Luxemburg." *Tageblatt*, February 15, 1935, p. 7.

57 Ibid.

58 Lesch. *René Leclère, pionier du cinéma luxembourgeois*. CNA, 1999, pp. 46–56, esp. p. 49.

the industrial region of Luxembourg, the film's commentary announces the steel industry with lyrics. After citing a poem on the ambivalence of Luxembourgian landscapes, accompanied by a long shot of a barren landscape, the voice-over introduces the industrial scenes as follows: "And here a poetry of another sort, that of work and of effort." The sequence on the steel industry features views of steel works and production scenes inside the plants, which are accompanied by a mixed audio background of machine sounds and music, before the journey moves on through rural landscapes to Luxembourg City. The film was screened on the occasion of the World Fair in Paris, where the steel industry was an essential part of the fair's media system consisting of the film, an official brochure and the pavilion's exterior and interior design (fig. 5.7).[59] Hence, the steel industry was aesthetically elevated and finally fully established as part of Luxembourg's canonical cultural heritage – in the entire exhibition and in a film commissioned by the government, which became a big success and was praised in the Luxembourg press as a "real Luxembourgian film."[60]

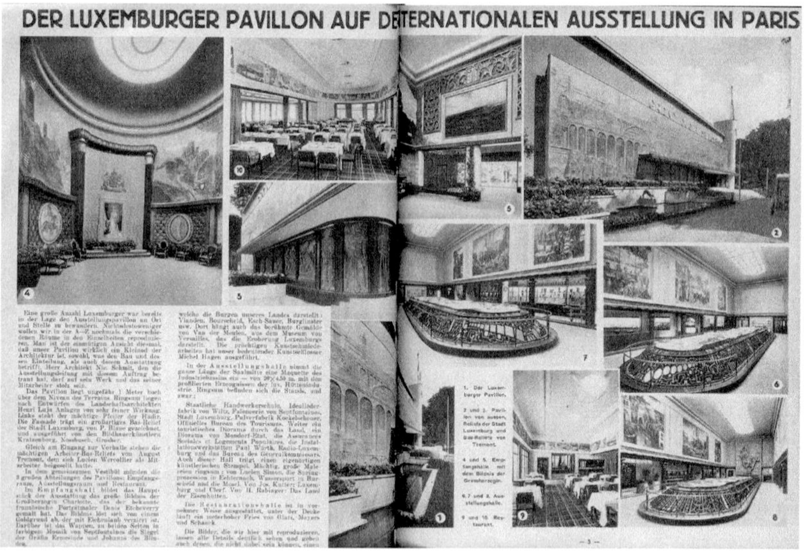

Fig. 5.7. Article on the Luxembourgian pavilion at the 1937 World Fair in Paris in *A-Z – Luxemburger illustrierte Wochenschrift,* no. 23 (1937), pp. 4–5.

59 The main exhibition room featured samples of steel products and a huge model of the southern industrial region underneath a wall with large-scale paintings of Luxembourgian motifs, which included an industrial landscape; see also Mousset, and Degen. *Un petit parmi les grands*, pp. 206–29.
60 Lesch. *René Leclère*, p. 49.

A *Columeta* "Remake" in the 1960s?

In the 1960s, midway through *Les Trente Glorieuses*, the glorious thirty years of economic prosperity between 1945 and 1974, ARBED produced another corporate film, which appears like a slightly extended, modernized colour remake of *Columeta* and demonstrates the consistency and continuity of the company's self-presentation over time. The corporate image with its three main motifs (Luxembourg, industry, social works) remained unchanged. What differed from 1921 is the extent to which ARBED explicitly associated itself with Luxembourg and the well-being of the entire country. While *Columeta* had connected the industrial landscapes and production scenes to the picturesque views of Luxembourg City, the striking Eastmancolor film with the telling title *Aciéries dans un parc* (*Steel Works in a Park*) takes the spectator on a journey through the Grand Duchy, pointing to touristic regions and local traditions. In keeping with the title, the film unfolds an image of a prospering and state-of-the-art steel industry, which is situated in a landscape of flourishing beauty and peaceful living – all of it presented in a colourful and aesthetically pleasing way.

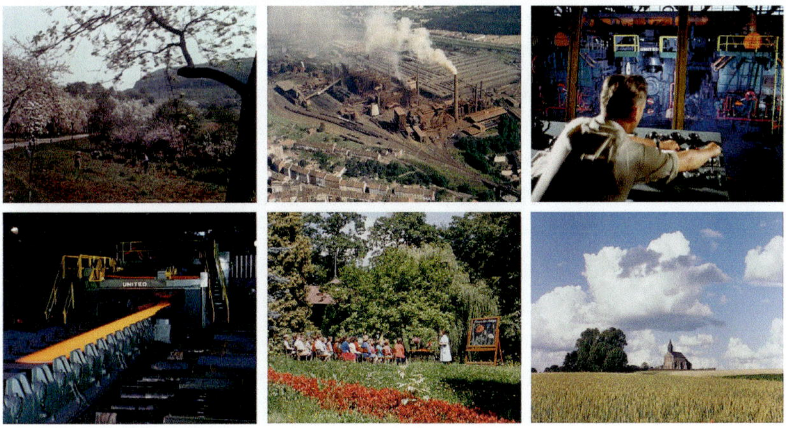

Fig. 5.8. Frames from *Aciéries dans un parc* (dir. Georges Lust, 1960s, 45 min). Courtesy of Centre national de l'audiovisuel (Luxembourg)/Arcelor Mittal.

The introduction presents blossoming landscapes of the north, where farmers cultivate the "good land," views of cities and the dancing procession of Echternach (which honours the missionary and the city's founder Willibrord), castles, "charming villages" and "wild valleys." The commentary introduces the viewer to the country's history, including its folkloristic and Christian traditions and the origins of its dynasty. Several scenes present the remains of ancient ironworks and point to the age-old tradition of iron production in preindustrial times, also featuring iron tomb slabs and the sculpture of a saint holding hammer and pliers. The steel industry is thus presented as a native part of Luxembourg's cultural heritage, which is also deeply rooted in Catholic tradition. The commentary tells the history of iron and steel production in Luxembourg until the present day, with scenes of blacksmiths in a village linking this historical tradition to contemporary craftsmanship. A train, loaded with charcoal, takes the viewer on a ride through the countryside, traversing Luxembourg City via the old bridge before it arrives in the industrial south of Luxembourg. After several views of the plants, the production sequences at ARBED show modern production lines and workers who control the machines from the safe distance of control rooms. The film's commentary stresses ARBED's global expansion as well as its importance for Luxembourg by emphasizing the economic prosperity, which the industry had brought to the entire country and posits Luxembourg's importance for Europe, as hosting one of the headquarters of the European Community. The well-being of the working class is featured in scenes of leisure-time activities such as dancing or playing ninepins in a city café or canoeing in the wild. The section on the social works presents the still existing vocational school Institut Emil Metz and the open-air school. The commentary explicitly asserts ARBED's societal authority in the past – "Long before the state legislature intervened, ARBED took the initiative to improve the workers' fate through social works and human solidarity" – and further stresses that ARBED still cares for its former employees, its adult workers and their children, and that this engagement goes "a lot further than the law stipulates." The end of the film refers to the transformation of the country, which is depicted as going hand in hand with the rise of the steel industry. While showing scenes of rural meadows and grain fields, the commentary explains that Luxembourg's agriculture has not suffered from the steel works. Quite the contrary, the Thomas slag, a by-product of steel production, had fertilized the country's soil – including those lots that could not be cultivated before. The film's last comment points out that Luxembourg has supplemented agriculture with rolling mills, ensuring Luxembourg's prosperity, before a long panning shot shows the city of Esch-sur-Alzette in a valley and ends with a view of a steel plant.

Aciéries dans un parc gives proof of the striking consistency of ARBED's corporate image, roughly forty years after it was formed through *Columeta*. Was it not about time for something new? We do not have any records about ARBED's intentions, the precise context or even precise dating of the film's production, or its actual screenings.[61] Hence, looking at Luxembourg's national film production after World War II until the early 1960s turns out to be a helpful approach to shed light on some aspects of the film's reaffirmation of national, industrial and social motifs at the time. During the post-war period, the steel industry was featured in numerous touristic and/or economic propaganda films.[62] Yet, it had lost its exclusive position of being the only industry featured in films of Luxembourg's self-presentation. Luxembourgian author André Linden has underlined that the "strong nationalist thrust of the post-war era at first favoured images of national harmony" in Luxembourgian films.[63] This brought along the uplift of agriculture as a topic of national economy and, for some time, the separation of economy and tourism in the national film production of the 1950s.[64]

However, at the 1958 World Fair in Brussels, the steel industry clearly dominated the Luxembourgian pavilion, a modernist construction of steel and glass.[65] This very modern architecture was criticized by a prominent voice: Joseph Petit, the author of the Luxembourg's official fair brochure, deplored the lack of Luxembourg's (natural) beauty in the Luxembourg pavilion, and stated: "We could have said it with more flowers."[66] For the World Fair, the government commissioned the film *Luxembourg 1958. Pays des châteaux et des forêts* (*Luxembourg, 1958: Country of Castles and Forests*). While the steel industry was featured among other industries, it was pictured as the driving force for "the wealth and the fortune of all Luxembourg"

61 It is likely that *Aciéries dans un parc* was produced before 1967, since in that year, ARBED incorporated its smaller competitor, HADIR, including the steel plant of Differdange, which is not mentioned in the film. The film's commentary gives a total steel production figure of 3.3 million tons. According to Chomé, Arbed produced 3.565 million tons in 1964; see Chomé, Felix. *ARBED. Un demi-siècle d'histoire industrielle 1911–1964*. ARBED, 1972, p. 150.

62 Many of these films were created under the auspices of the government or "supervised by the Department of Economic and Tourism Expansion" and directed by the Luxembourgian Philippe Schneider; see Linden, André. "Lux et vox (... 1929–1979 ...) – 'Sons et Lumières' of Luxembourg Cinema." *Lëtzebuerger Kino: Aspects of Luxembourg Cinema*, edited by Jean Back and Viviane Thill. CNA, 2005, pp. 13–57, esp. p. 18.

63 For an extensive analysis of the Luxembourg film production from the 1920s to the 1970s, see Linden. "Lux et vox," pp. 13–57, quotation on p. 37. An analytical inventory of a hundred Luxembourgian films from the 1920s to the 1970s is provided in the book's appendix, pp. 234–39.

64 See ibid., p. 18.

65 See Mousset, and Degen. *Un petit parmi les grands*, pp. 249–63.

66 See Linden, André. "'Nous aurions pu dû le dire avec plus de fleurs'. Mises au point à l'ombre de l'Atomium." *Un petit parmi les grands: Le Luxembourg aux Expositions universelles de Londre à Shanghai (1851–2010)*. Musée national d'histoire et d'art, 2010, p. 342.

and being "one of the master builders and major partners in the European Coal and Steel Community, the seed of a united Europe." Steel production sequences follow the presentation of the Echternach dancing procession and of the Codex Aureus of Echternach, an illuminated manuscript created in the city's Benedictine monastery in the eleventh century.

The film presents an incisive turning point by linking the well-being of the population to the state rather than to the steel industry: it contains a long scene of a public swimming pool and scenes of urban housing, a fun fair, and the Maison des Enfants (which was founded by ARBED and featured in *Columeta*), but does not explicitly mention ARBED as a pioneer in social welfare and reforms.

The state's "claim" to the topic (and the visuals) of social well-being is then plainly expressed in the seventeen-minute film *Le droit au soleil* (*The Right to the Sun*), commissioned by the government in 1959. Concentrating exclusively on social affairs, it contrasts the "bad old days," when the poor and the disabled had depended on charity, to the present-day achievements of the state's social legislation, which ensures social protection for "all" its citizens.[67] The 1960 school film *Visage du Luxembourg* (*The Face of Luxembourg*) equally highlights workers' housing estates, retirement homes, and schools as the outcome of the state's social legislation. Again, the steel industry is featured as the country's driving economic force, but the commentary also expresses concerns about the economic changes to which the steel industry is able to adapt to, while the agricultural sector has difficulties with keeping up.

Thus, at a time when the steel industry was economically dominant, the national films promoted preindustrial traditions and heritage, such as agriculture and Christianity, and a prominent critic was asking for "more flowers" rather than technological modernity in the Luxembourgian self-presentation, *Acièries dans un parc* appears like an adjustment to this atmosphere, through its emphasis on ARBED's rootedness in Luxembourgian heritage such as craftsmanship, preindustrial iron making and Christian traditions, ARBED's benefits for agriculture, and the use of colourful beauty – be it blossoming meadows or the colourful presentation of steel production. At the same time, the film goes beyond national boundaries and highlights Luxembourg's and the industry's share in the European process. Furthermore, *Acièries dans un parc* subtly counters the state's recent claim to the social motifs by explicitly reminding the audience of ARBED's social

67 See Linden. "Lux et vox," p. 19.

welfare and reform achievements in the past and its ongoing commitment to corporate responsibility.[68]

Conclusion

By linking national imagery to the modernity of technological and social progress, *Columeta* in 1921 created a synthesis of motifs that established ARBED's corporate image, which was maintained into the 1960s. The inclusion of touristic views of "beautiful Luxembourg" adapted to the country's traditional iconography and further displayed a progression from Luxembourg's past to the present of industrial technology and social reforms – a link that emphasized the industry's importance for Luxembourg and suggested an optimistic outlook to the country's future. The swimming pool scene in *Columeta* and the extensive display of the industry's social works linked societal well-being with the industrial sphere. The promise of economic progress, which was communicated by the industrial motifs, was subsequently included into the Luxembourgian self-presentation of the interwar period.

In the late 1950s, when the state asserted that the social well-being of its citizens was due to state legislation, thereby laying claim to the topic and the visuals that had previously been closely linked to the steel industry, the corporate representation of ARBED and the national representation of Luxembourg converge to the extent that both communicate national, technological and social motifs as significant components of a corporate, or a national identity, respectively. Thus, *Columeta* not only established the constituent motifs for the corporate image of Luxembourg's major steel company, but also to a large extent of Luxembourg's national self-image.[69] The film thus was a precursor of a modernized self-presentation of Luxembourg that presented the country as a harmonious synthesis of tradition

68 This claim to historic importance and social responsibility was also manifested in films that paid homage to the life of ARBED's former director, Emil Mayrisch, and were produced by ARBED (*Le Temps de Chateau*, 1962; *Emil Mayrisch*, 1963) and Telé Luxembourg (*Emil Mayrisch, l'homme et son œuvre*, 1963); see ibid., p. 25; and the appendix of *Lëtzebuerger Kino: Aspects of Luxembourg Cinema*, p. 238.

69 As Yvonne Zimmermann has shown, corporate films have also had a strong impact on the nation's branding in Switzerland; see Zimmermann, Yvonne. "'What Hollywood Is to America, the Corporate Film Is to Switzerland': Remarks on Industrial Film as Utility Film." *Films That Work: Industrial Film and the Productivity of Media*, edited by Vinzenz Hediger and Patrick Vonderau. Amsterdam University Press, 2009, pp. 101–17.

and modernity. With *Acièries dans un parc* ARBED once again adapted to the national imagery and also reacted to the current atmosphere of the time. The extensive display of iron and steel production as a native part of Luxembourg's cultural heritage inscribed the industry even more into the national iconography and historiography. The film explicitly reconfirmed ARBED's economic and social achievements of the past and the present, which once again enabled the communication of its corporate image – and also an image of Luxembourg – to the broadest possible audience(s).

Works Cited

Bartelheim, Martin, Roland Hardenberg, Thomas Knopf, Anke K. Scholz, and Jörn Staecker. "'ResourceCultures' – A Concept for Investigating the Use of Resources in Different Societies." *Persistent Economic Ways of Living: Production, Distribution, and Consumption in Late Prehistory and Early History*, edited by Alžběta Danielisová and Manuel Fernández-Götz. Archaeolingua Alapítvány, 2015, pp. 39–50.

Barthel, Charles. "*Columeta*: L'aventure du premier film publicitaire de l'Arbed." *Hemecht*, vol. 50, no. 2, 1998, pp. 177–206.

Chomé, Felix. *ARBED. Un demi-siècle d'histoire industrielle 1911–1964*. ARBED, 1972.

Deeken, Annette. "Geschichte und Ästhetik des Reisefilms." *Geschichte des dokumentarischen Films in Deutschland, Band 1: Kaiserreich 1895–1918*, edited by Uli Jung and Martin Loiperdinger. Reclam, 2005, pp. 298–323.

Edwards, J. Arwel, and Joan Carles Llurdés I Coit. "Mines and Quarries: Industrial Heritage Tourism." *Annals of Tourism Research*, vol. 23, no. 2, 1996, pp. 341–63.

F., E. [Friedrich, Evy]. "Ein Film über Luxemburg." *Tageblatt*, February 15, 1935, p. 7.

Gertiser, Anita. "Domestizierung des bewegten Bildes." *montage/av*, vol. 15, no. 1, 2006, pp. 58–73.

Hadžalić, Irma. "Sick and Weak but Made of Steel: Luxembourgian Open-air Schools and Other Responses to the Spread of Tuberculosis at the Beginning of the 20th Century." *Revista da História e Historiografia da Educação*, vol. 1, no. 1, 2017, pp. 44–64.

Handwritten letter. Addressee unknown. Signed by "Willy Heymers Leiter der Städt, Lichtspiele, Köln," July 7, 1921. Archives nationales de Luxembourg, Arbed 03–00001.

Herman, Frederik. "Forging Harmony in the Social Organism: Industry and the Power of Psychometric Techniques." *History of Education*, vol. 43, no. 5, 2014, pp. 592–614.

Herman, Frederik, and Ira Plein. "Envisioning the Industrial Present: Pathways of Cultural Learning in Luxembourg (1880s–1920s)." *Paedagocica Historica*, vol. 53, no. 3, 2017, pp. 268–84.

Herman, Frederik, Karin Priem, and Geert Thyssen. "Körper_Maschinen? Die Verschmelzung von Mensch und Technik in Pädagogik, Industrie und Wissenschaft." *Jahrbuch für Historische Bildungsforschung*, vol. 20, 2015, pp. 47–75.

Historical Archive of the City of Cologne. Holding 608 (Culture Department) A 149, call number: Best. 608 (Kulturdezernat) A 149.

Jung, Uli. "Städtebilder und Lokalaufnahmen." *Geschichte des dokumentarischen Films in Deutschland, Band 1: Kaiserreich 1895–1918*, edited by Uli Jung and Martin Loiperdinger. Reclam, 2005. Reclam, 2005, pp. 274–97.

Jung, Uli, and Martin Loiperdinger, editors. *Geschichte des dokumentarischen Films in Deutschland, Band 1: Kaiserreich 1895–1918*. Reclam, 2005.

Kieffer, Simone. "Les oeuvres sociales de l'ARBED (1900–1930) (Aciéries Réunies de Burbach-Eich-Dudelange)." MA thesis, Marc Bloch University, 2003.

Knebeler, Christoph, and Denis Scuto. *Belval: Passé, présent et avenir d'un site Luxembourgeois exceptionell (1911–2011)*. Éditions Le Phare, 2010.

La sidérurgie luxembourgeoise. Un siècle d'histoire et d'innovation/Steelmaking in Luxembourg: A Century of History and Innovation. Arcelor Mittal, 2011

Lesch, Paul. "Film et tourisme au Luxembourg à la Belle Epoque." *Ons Stad*, vol. 88, 2008, pp. 58–61.

Lesch, Paul. *René Leclère, pionier du cinéma luxembourgeois*. Centre national de l'audiovisuel, 1999.

Linden, André. "Lux et vox (... 1929–1979 ...) – 'Sons et Lumières' of Luxembourg Cinema." *Lëtzebuerger Kino: Aspects of Luxembourg Cinema*, edited by Jean Back and Viviane Thill. Centre national de l'audiovisuel, 2005, pp. 13–57.

Linden, André. "'Nous aurions pu dû le dire avec plus de fleurs'. Mises au point à l'ombre de l'Atomium." *Un petit parmi les grands: Le Luxembourg aux Expositions universelles de Londre à Shanghai (1851–2010)*. Musée national d'histoire et d'art, 2010, pp. 336–43.

Linden, André, and Guy Thewes. *Greetings from Luxembourg: A Journey into the World of Tourism*. Musée d'histoire de la ville, 2008.

Loiperdinger, Martin. "Industriebilder." *Geschichte des dokumentarischen Films in Deutschland, Band 1: Kaiserreich 1895–1918*, edited by Uli Jung and Martin Loiperdinger. Reclam, 2005, pp. 324–32.

Lorang, Antoinette. "D'Heichiewen. Les hauts-fourneaux." *Lieux de mémoire au Luxembourg: usages du passé et construction nationale*, edited by Sonja Kmec, Benoît Majerus, Pit Péporté, and Michel Margue, 2nd ed. Éditions Saint Paul, 2008, pp. 215–20.

Lorang, Antoinette. *L'image sociale de l'Arbed à travers les collections du Fonds du logement.* Le fonds pour le développement du logement, 2009.

M., Dr. "Der Propagandafilm über Luxemburg." *Luxemburger Wort*, February 15, 1935, p. 8.

Millim, Anne-Marie. "Schooling the Gaze: Industry and Nation-Building in Luxembourgish Landscape-Writing, 1900–1940." *Journal of European Studies*, vol. 44, no. 2, 2014, pp. 151–69.

M., N. "'Le beau pays de Luxembourg': Ein touristischer Propagandafilm." *Tageblatt*, February 15, 1935, p. 7.

Mousset, Jean-Luc, and Ulrike Degen, editors. *Un petit parmi les grands: Le Luxembourg aux Expositions universelles de Londre à Shanghai (1851–2010)*. Musée national d'histoire et d'art, 2010.

Muth, F., Muth, L., and Kurt Schraudenbach. *Stahl aus Luxemburg*. Verlag Heinrich Hoffmann, 1942.

Novella, Enric J. "Tuberculosis and Political Economy: Industrial Wealth and National Health in the Grand Duchy of Luxembourg, c. 1900–1940." *Social History of Medicine*, vol. 31, no. 2, 2018, pp. 308–27, https://doi.org/10.1093/shm/hkx004.

Philippart, Robert L. *Luxembourg. Historicisme et identité visuelle d'une capital.* Éditions Saint Paul, 2007.

Plein, Ira. "'Der tägliche Weg zur Schicht'. Aspekte zur proletarischen Kunst Albert Kaisers in der Zwischenkriegszeit." *100 Joer fräi Gewerkschaften 1916–2016*, edited by Frédéric Krier, Jacques Maas, Arnaud Sauer, and Denis Scuto. OGB-L/Le Phare, 2016, pp. 83–97.

Priem, Karin, and Frederik Herman. "Hautnah: Materialität der Moderne und sensomotorische Ansätze der Berufsbildung im 'Zeitalter des Stahles'." *Die Sache(n) der Bildung*, edited by Christiane Thompson, Rita Casale, and Norbert Ricken. Ferdinand Schöningh, 2017, pp. 213–39.

Priem, Karin, and Frederik Herman. "'Sensuous Geographies' in the 'Age of Steel': Educating Future Workers' Bodies in Time and Space (1900–1940)." *Fabricating Modern Societies: Education, Body and Mind in the "Age of Steel"*, edited by Karin Priem and Frederik Herman. Brill, 2019.

Projet de lettre. November 12, 1921. Archives nationales de Luxembourg, Arbed 03-00001.

Schmitz, Nadine. "La politique sociale de l'ARBED/ARBED's Social Policy." *La sidérurgie luxembourgeoise. Un siècle d'histoire et d'innovation/Steelmaking in Luxembourg: A Century of History and Innovation.* Arcelor Mittal, 2011, pp. 149–51.

Thewes, Guy. "La silhouette de la ville." *Lieux de mémoire au Luxembourg: usages du passé et construction nationale*, edited by Sonja Kmec, Benoît Majerus, Pit Péporté, and Michel Margue, 2nd ed. Éditions Saint Paul, 2008, pp. 253–58.

Thyssen, Geert. "Engineered Communities? Industry, Open-air Schools, and Imaginaries of Belonging (c. 1913–1963)." *History of Education & Children's Literature*, vol. 10, no. 2, 2015, pp. 297–320.

Trausch, Gilbert. *L'ARBED dans la société luxembourgeoise.* ARBED/Victor Buck, 2000.

Undated film script. Archives nationales de Luxembourg, Arbed 03-00003.

Zimmermann, Yvonne. "'What Hollywood Is to America, the Corporate Film Is to Switzerland': Remarks on Industrial Film as Utility Film." *Films That Work: Industrial Film and the Productivity of Media*, edited by Vinzenz Hediger and Patrick Vonderau. Amsterdam University Press, 2009, pp. 101–17.

About the Author

Ira Plein is a lecturer in media studies at the University of Trier. She studied art history and media studies and, in 2019, completed her dissertation project "Propaganda for Steel and Nation: Images and Counter-images to Economic and Social Progress in Luxembourg Media of the Interwar Period" (publication forthcoming) under joint supervision at the University of Luxembourg and the University of Trier. Her further publications cover various types of visual communications in the industrial realm in interwar Luxembourg, both in the industrial sphere and the labour movement, and the politically informed montage art works of Alice Lex-Nerlinger in interwar Berlin.

6 Hydropower for a Sealess Nation

Representation of Water Energy in Czech Visual Culture

Lucie Česálková

Abstract

Post-war Czechoslovakia preferred to rely on water resources in its energy policy. This study analyses the role of art and media in the country's communication of its turn to hydropower. Rivers are approached as environtechnical landscapes that can (re)define modern national identities since they served as lines of communication, economic commodities, and a representational medium, helping to conceptualize ideas of the nation. Accordingly, media and art relating to the construction of dams supported the circulation of knowledge and exploited the symbolic potential of dams as energy sources and as a solution to the national trauma associated with the lack of access to the sea. Industrial films about hydropower contributed to understanding the profound economic paradigm shift from capitalist to socialist order and significantly shaped ideas on the possibilities of turning nature into a source of energy.

Keywords: resource cultures; environtechnical landscapes; water energy; post-WWII reconstruction; Czechoslovakia

A poetic documentary about the Vltava River, *Příběh staré řeky* (*The Story of an Old River*, 1957) by director Jiří Lehovec, was created by chance. Lehovec, one of the avant-garde-minded Czech filmmakers of the interwar period, participated artistically for several years in other cinematic pieces on a similar topic before completing this film – Svatopluk Studený's *Povážské stupně* (*Váh Degrees*, 1955) and *Vodní díla v Československu* (*Waterworks in Czechoslovakia*, 1956), commissioned by the Technoexport company as an audiovisual promotion of Czech hydropower and especially water turbines abroad. In parallel with these informative promotional films with a technical

Hediger, V., F. Hoof, Y. Zimmermann, with S. Anthony (eds), *Films That Work Harder: The Circulation of Industrial Film*. Taylor & Francis Group 2024
DOI: 10.5117/9789462986534_CH06

focus, Lehovec thought about a film that would portray the phenomenon of the Vltava River more comprehensively, not just in terms of the energy potential acquired by modern water management technology. As the author of the photographic screenplay for *Botič, potok chudých* (*Botič, the Stream of the Poor*, 1935) and the director of a movie about the progress in regulating rivers and dam building, *Zkrocený živel* (*Tamed Element*, 1938), Lehovec viewed the river phenomenon for a long time, primarily as a vehicle of social life and as a part of the landscape. *The Story of an Old River* links these two interests with other dimensions of the river's meaning, namely poetic, industrial, technological and economic.

It is symptomatic of Lehovec's approach to emphasize the river in its ambivalent unity. He reflects on the eradication of old villages and the creation of a new hydropower plant (a new energy source) with the same calmness and humility, aware of the naturalness of the change in the historical cycle of the river. He is fascinated by the poverty of former villages, by their churches and the traditional peculiarity of some houses, by the romantic countryside, as well as by the dangerous work of the rafters, the design of the giant dam, the hydropower plant's technological equipment and the vicissitudes of the water element. Considering all partial aspects surrounding the river at that stage of its preparation for the exploitation of the energy potential in overall harmony, Lehovec understands the Vltava as an "organic machine."

Environmental historian Richard White uses the same metaphor in 1995 to explain the meanings of the Columbia River in the United States. White notes the growing industrial mechanization of the river, however, he does not interpret it as evidence of the triumph of technology and the human element of nature, but rather in the sense of new trends in engaging nature and converting its energy to other forms. Humans, according to White, in this case did not destroy, but rather changed nature, and their impacts on the river are undistinguishable from those of environmental forces. "To say that there should be thousands of chinook and sockeye salmon passing upriver on a given day is perhaps to miss the point. If this were the old Columbia River system there should be salmon, but this is a different river. It is not the river the salmon evolved in. This new river produces carp and shad,"[1] claims White, freed from the need to evaluate certain aspects of the river's development either as negative or positive. That is to say, the river is both natural and cultural; it is an organic machine. Sara B.

1 White, Richard. *Organic Machine: The Remaking of the Columbia River*. Hill and Wang, 1995, p. 90.

Pritchard offers a profound deepening of White's concept by accessing the French Rhône River as an envirotechnical landscape, defined by specific envirotechnical systems and regimes and adding culturally constructed beliefs to the interplay of factors influencing its function.

According to Pritchard, rivers and their management are interwoven with political questions, economic debates, cultural ideas, and social struggles. More importantly, in her approach the history of the river reveals that "ideas about nature, technology, and their relationship infused conceptions of the nation."[2] Developments of certain envirotechnical landscapes thus, in her opinion, can be analysed as redefinitions of the identity of modern national states. The image of the river, or any other envirotechnical landscape, changing throughout history, however, is constantly constructed and reconstructed by external agents – politicians, scientific and technical experts, environmentalists, journalists, artists, as well as ordinary people.

In the Czech national imagination and consciousness, the Vltava River plays a similarly crucial role. But not only the Vltava alone. As a part of Central Europe, Czechoslovakia strongly identified itself with the whole Central European waterway system, and its regulation was one of the most prominent political and technological priorities. Despite the evidence that the construction of dams and the regulation of river flow generally reduces the intake of sediment and thus increases the likelihood of erosion of the river bed, including gravel, which is thought to contribute to creating an ideal environment for spawning, and the undisputable fact that dam and hydropower plant construction projects meant the flooding of former villages and often violent destruction of indigenous communities, in the public discourse, this social and environmental damage was balanced by the advantages that the reduction of flooding or conversely dramatic drought has for farmers. In the set of influences rectifying the image of rivers and other envirotechnical landscapes like ambivalent phenomena, media and the arts play a vital role. Therefore, the explanation of why, despite these negative aspects, waterpower plants and dams are still perceived as an unproblematic part of everyday Czech life is hiding in the cultural imagination that surrounds rivers, waterways and, generally, water as a source of energy.

In order to highlight the role of the social and cultural background in the economic processes from a historical perspective, contemporary archaeology

2 Pritchard, Sara B. *Confluence: The Nature of Technology and the Remaking of the Rhône*. Harvard University Press, 2011, p. 9.

and historical anthropology introduced the concept of *ResourceCultures*.[3] This concept enables historians and anthropologists to point out that different things were considered as significant (re)sources by different cultures and societies at different times, and go beyond traditional resource categorization such as renewable/non-renewable, limited/unlimited, recyclable/non-recyclable etc. From this perspective, resources are understood as "parts of networks composed of people, objects, technologies, and knowledge,"[4] and can be explained within the complex of political-economic relations, scientific and technological developments, geographic conditions, social and cultural dynamics, as well as practices of usage.

Although historical anthropology by its nature deals with very old cultures and societies, the concept of *ResourceCultures* is relevant to recent history as well. Above all, it is bridging political-economic and sociocultural perspectives on resources, and as such it provokes explanations of resource use as being embedded not only in political and economical decisions, but also in technological developments, knowledge formation, as well as in the cultural beliefs and social practices of specific communities. To understand how the meanings are assigned to resources through cultural beliefs and social practices, and to explore the dynamics of this symbolic construction in its complexity is no less useful in the modern history of Western societies, where the construction and validation of symbolic ideas and beliefs was importantly catalysed by culture, arts, and media. Cultural artefacts and media as vehicles of the national imagination and knowledge formation can, if conceived as an insight into complex political, economic, social and technological relations, reveal how the concept of a particular resource in a given cultural context transformed and what factors contributed to dampen the controversy related to it.

Water Culture

Besides coal, twentieth-century Czechoslovak energy politics was most closely identified with water, the reason being the importance of water resources (rivers, ponds and lakes) in the national culture and imagination.

3 Bartelheim, Martin, et al. "'ResourceCultures' – A Concept for Investigating the Use of Resources in Different Societies." *Persistent Economic Ways of Living, Production, Distribution, and Consumption in Late Prehistory and Early History*, edited by Alžběta Danielisová and Manuel Fernández-Götz. Archaeolingua Alapítvány, 2015, pp. 39–50.

4 Ibid.

As a state whose borders do not touch the sea and from which all the rivers only flow out and none enter it, Czechoslovakia always struggled not only with specific natural conditions, but also with many national frustrations. As a small, densely populated country often plagued by floods, but also severe droughts, a country with no colonial ambitions, Czechoslovakia dreamed about its own sea. The idea of the Czech sea, evolving as a specific concept since at least the mid-nineteenth century, cumulated in a series of sometimes conflicting small Czech national ambitions – from seafaring to political projects, economic recovery solutions, and business plans, to visions of tourism development.

Visual culture historian Petra Hanáková explains how culture has become a projection screen for unfulfilled longing for a Czech sea with an example from a popular Czech mock-biopic film, *Jára Cimrman, ležící spící* (*Jára Cimrman, Lying, Sleeping*, 1983). Hanáková recalls the scene where the main character, a Czech genius and in a way a caricature of a national hero, decides to lay the sea at the feet of the Czech capital – Prague. In this scene, set in 1897, an enormous panorama is extended between two Renaissance houses in one corner of the Old Town Square in Prague, and a patriotic speech at the opening of the event reminds the crowd that "this sea was not given to us by nature as with other nations, but by the genius of the Czech artist." Cimrman's sea – depicted as a panorama – Hanáková argues, functions here as both a "tangible screen, and elusive ideal, a screen infested with dreams, ambitions and desires, a screen which offers a condensed metaphor for visual culture working as a compensation for lack, inadequacy, and trauma." In the imagery of a Czech sea, Hanáková reveals the distress of a small nation, a small country at the mercy of neighbouring political and cultural superpowers throughout its history, as well as the effort to compensate for this smallness and oppression through an investment in culture.[5]

This scene from the film emphasizes the compensation role of a culture which here actually literally replaces what for various external reasons does not exist. Culture, here in the form of a panorama's canvas, was at the time an entirely new medium of an enhanced and immersive vision. The playful imagination of its creator (an artist) represents a conflict between progress (visual novelty) and tradition (its performance and theme), but it also soothes and alleviates the internal discord and the social trauma felt by a nation whose borders do not touch the sea. Moreover, Cimrman typically

5 Hanáková, Petra. "The Seas and Skies of Bohemia: Projections and Inventions of the National Space." *EUtROPEs: The Paradox of European Empire*, edited by John W. Boyer and Berthold Molden, University of Chicago Press, 2015, pp. 335–58.

does not stimulate the imagination of its viewers through the return to the past, to history, but wants to bring them to the amazement of utopia by displaying the improbable.

In conceptualizing Central Europe, to which Czechoslovakia (or the Czech lands) belong, the idea of a water(way) has wound like a red thread since at least the late eighteenth century, proving that culturally defined envirotechnical landscapes importantly contribute to the national imagination. The polarity of the Rhine and the Danube, more recently elaborated, for example, by Claudio Magris, was for the first time introduced by historian and publicist Constantin Frantz as crucial to his model of spacious Germanness.[6] German economist Friedrich List then proposed his concept of Central Europe as based on the idea of a customs union interconnected through a complex transportation system,[7] and as Henry Cord Meyer reminds us in his *Mitteleuropa*, such thoughts about the importance of ports and canal building were present in the Central European discourse throughout the entire nineteenth century.[8] Although considered more or less as hopeless visions, such projects as Rhine-Main-Danube, Oder-Danube or Elbe-Moldau-Danube canals were enthusiastically discussed in contemporary newspapers, and even the intellectual history of the Czech National Revival shows us how deeply the Revivalist concept of Czech national identity was rooted in Johann Georg Herder's idea of the *Mitte/mittel*, the centre/middle, considered as Slavic, and spread from Balaton to Balt, along the Danube, from Sava to Volga,[9] and how it in various text formats shaped popular Czech imagery in the nineteenth and early twentieth century.

A waterway network was also crucial for the very first visions of an integrated Central Europe, which in the early 1920s was enjoyed by Czech diplomatic patronage, business, as well as the intellectual elite.[10] "To define our relationship to the sea" was a topic of the first public debates in the independent Czechoslovakia after 1918 and led to the founding of the Czechoslovak Shipping Company in 1920. Stating that "it would be a mistake to believe that the sea is of no importance for us just because it does

6 Ehmer, Manfred. *Mitteleuropa: Die Vision des politischen Romantikers Constantin Frantz*. Tredition, 2012.

7 List, Friedrich. *Das nationale System der politischen Ökonomie*. Cotta, 1841; Brechtefeld, Jörg. *Mitteleuropa and German Politics, 1848 to the Present*. Macmillan, 1996.

8 Meyer, Henry Cord. *Mitteleuropa in German Thought and Action, 1815–1945*. Springer, 1955, p. 75.

9 Kollár, Ján. Qtd. in Macura, Vladimír. *Znamení zrodu a české sny*. Academia, 2015, p. 159.

10 Janáč, Jiří. *European Coasts of Bohemia: Negotiating the Danube-Oder-Elbe Canal in a Troubled Twentieth Century*. Amsterdam University Press, 2012.

not neighbour us,"[11] the company brought together domestic elites across disciplines, from entrepreneurs to tourists, university professors and artists,[12] to join them in a nationwide societal movement, whose leaders openly reflected on the longing for the sea as a part of the national imagination, and through its activities wanted this idea to escape from the realm of fantasy.[13]

What is important, however, is that since the very beginning of the company's activities, close links to culture were declared to be an irreplaceable part of its identity. Primarily they looked for visual culture, mainly cinema. The company celebrated its establishment in one of the most luxurious cinemas in Prague's city centre, Světozor, with slides and films accompanying the opening speech and a lecture explaining the plan of creating a Central-European waterway network and the possibilities of hydropower. The ceremony was attended by the president, Tomáš Garrigue Masaryk, who also gave a donation to the company and affirmed the prestigious status of the event. Topical films depicting the Vltava River, such as *Inspekční cesta ministra Staňka z Prahy do Ústí nad Labem* (*Minister Staněk's Inspection Trip from Prague to Ústí nad Labem*) and the well-known *Svatojánské proudy* (*St. Johann's Rapids*, Antonín Pech, 1912) were screened, together with a film about serial shipbuilding in the United States.[14] Since this event, the paths of culture, namely film and hydropower in Czechoslovakia, have remained admirably intertwined.

There were two major proponents of the Central European waterway regulation project: Hugo Vavrečka and Antonín Smrček. Vavrečka, the Czech ambassador in Vienna, joined Dominik Čipera and the well-known businessman Jan Antonín Baťa to become directors of the Baťa company in 1932, which was involved in commerce (especially shoe manufacture) and media production. Together, they developed a vision of a dramatic increase in the use of water transport in business and industry, and they promoted this vision at the national and international levels.[15]

11 Müldner, Josef. *Námořní společnost československá, její vznik a úkoly*. Námořní společnost československá, 1920, p. 13.

12 Ibid.

13 "It is not only the suppressed, subdued desires of a small, landlocked nation, distant from the sea, that leads swimmers, experts, and other interested persons to the constitution of the Czechoslovak Shipping Company, but also especially severe economic considerations. The programme of the company [...] therefore does not consist of dreams, fantasies, or romanticism, neither does the company work for profit. Its intentions are purely cultural. The company wants to help educate ordinary people and spread its noble idea," ibid., p. 15.

14 The ceremony took place on April 29, 1920. Ibid., p. 17.

15 Similarly, trade interests intersect with waterway development worldwide throughout history. See, for example, Innis, Harold. *The Fur Trade in Canada*. University of Toronto Press, 2001 (originally published in 1930).

Antonín Smrček, the second important voice in the water regulation and water energy debates of the 1920s and 1930s, was a professor of hydraulic engineering at the Brno University of Technology. In addition to being a member of a number of national and international water management commissions, he himself implemented many waterworks, whose construction he filmed. Emphasizing technical perfection, the beauty of the materials and the utility of the buildings, Smrček represented constructivism in Czech film. His works, such as *Stavba přehrady u Vranova nad Dyjí* (*Construction of the Dam at Vranov upon Thaya*), *Stavba přehrady u Kníniček* (*Construction of Kníničky Dam*), *Laboratoř Střekovský jez* (*Laboratory at Střekov Weir*) etc. appeared in the Czech collection at the International Photographic Exhibition in The Hague in 1928, as well as in several avant-garde Czech film collections screened both in Czechoslovakia and abroad. Smrček captured his waterworks from an unusual angle with a sophisticated sense of the materiality of bodies of water, and he introduced several tropes that would reappear in Czech imagery surrounding water energy later on, namely calling it "white coal" to highlight the fact that Czech coal is rare and of poor quality, while hydropower creates energy using otherwise "unproductive" land. Most crucial was the way in which Smrček's films defined the relationship between water management and nature – which was understood as "taming." Taming – via the construction of locks and reservoirs – meant new work for unemployed people in the 1930s and the tamed, docile countryside thus became fulfilled by new values and meanings.

Vavrečka and Smrček were key proponents of the construction of the Danube-Oder-Elbe canal in the interwar period, which is often described as a "laboratory of Europe"[16] and whose planning developed throughout the twentieth century, in each period with different ambitions. The canal was developed "at the junction of transnational system network building and national development," and thus its potential building carried different meanings for the state that built it and for those whom it would enable to connect to new territories. Negotiations about balancing costs and advantages were therefore always accompanied by conflicts of interest at the international political level. Discussions about canal construction projects and visions surrounding it were closely associated with the current international political direction of the country. In the dramatic Czechoslovak history in the twentieth century, frameworks legitimizing this project transformed in relation to changes in state policies. As the country transformed from a part of Austro-Hungarian territory to independent state, to Nazi occupied

16 Janáč. *European Coasts of Bohemia*, p. 20.

land and, after the war, to a member of the Soviet Bloc, the idea of the canal and its position within Central Europe changed as well.

The idea of Danube-Oder-Elbe water management was viewed in the international dimension as a promise of connections to foreign markets and symbolic approximated seas. In the 1920s and 1930s, the "Czech sea" was most often linked with the Adriatic Sea, as the seaside destination that was the closest and most affordable in terms of traffic and culture.[17] Besides bilateral cultural and trade exchanges, the idea of the canal, then, was associated with the conquest of the sea and with the development of transport and tourism, Czech integration into European structures and open, international communication. The media followed this tendency in various ways, one of them being the utopian vision of a water empire. The short educational film *Přístav v srdci Evropy* (*The Harbour in the Heart of Europe*) from 1946 represented such a water empire agenda. Recycling footage from 1930s and 1940s travelogues and educational films, *The Harbour in the Heart of Europe* significantly contributed not only to a debate on the national identity's relationship with industrial development, but also to the circulation of the imagery and meanings in industrial films.

There were four main sources for *The Harbour in the Heart of Europe*, four groups of films whose footage the film recycled. These were classroom films by Jaroslav Novotný, travelogues and documentaries from Ceylon and India by Alexander Hackenschmied (aka Hammid), the above-mentioned dam construction films by Antonín Smrček, and Baťa's unpublished instructional films about the benefits of water transportation. These all led to the project's background, namely to Zlín. The very idea of the film as well as the project it promoted has its roots in early 1940s Zlín, a city that was the home and base of the famous Czech shoe company Baťa. The film calls for the construction of a canal between three major rivers, thanks to which, together with other improvements, the whole country would be directly connected with key European seas and with all the countries throughout Europe. After World War II, however, the idea fostered earlier by a single capitalist firm was adopted by the state as a national project aimed to support the recovery of the post-war economy, despite the fact that in the new left-wing and rather

17 "The Czech sea is a fairy-tale dream of the Czech people. This tale has already been made into a reality by the engine, the airplane, the radio, and a distant vision. But it would also be nice if we had our own piece of a small sea and, thanks to it, a way opened to the endless Czech waters, so that our children could play on the sunny sea beach while still being at home." Nesnídalová, Marie. "Iluse českého moře – Koupání v Jugoslávii." *Český lid*, vol. 17, no. 251, September 9, 1928, p. 1; Kvapil, Jaroslav. "Na nové parádní lodi Jihoslovanské po 'našem' moři." *Pestrý týden*, vol. 6, no. 28, July 11, 1931, p. 8.

socialist conditions the Baťa company was criticized for, among other things, being a symbol of the exploitation of workers by capitalists.

The Harbour in the Heart of Europe confirms its role in the debate about hydropower's significance for the post-war economy, and its rootedness in Central European water imagery, on two interrelated levels. First, by articulating the "lack of access to the sea" motive, and second, by illustrating the transnational advantages of the canal project. The commentary clearly links the necessity of building the Danube-Oder-Elbe canal with the historical injustice of Czechoslovakia's geographical position in the very heart of Europe and lacking a coastline. Followed by an illustrative animation, it states:

> Because of Father Czech's[18] decision, however, we have become a purely inland nation. We are the heart of Europe, but we are gripped like in an iron ring, we lack the sea. Let's make sure that worldwide there are only six separate civilized countries whose borders do not touch the sea. These are in South America: Bolivia and Paraguay, in Europe: Switzerland, Austria, Hungary and Czechoslovakia. Fate did not favour us when it placed us among such unfortunate world peculiarities. I wish we were an island...

The solution to how to resuscitate the lonely heart of Europe was the canal project. It would be a major artery in the centre of the continent, thanks to which Czechoslovakia would connect with the south and the north, with the east and the west of Europe. This moving image of Czechoslovakia as being the beating heart of Europe, whose main arteries and veins are rivers through which knowledge and commerce easily flow in all directions, clearly emphasized the importance of river transportation and water energy for the post-war national economy. Similarly, it corresponded to the Czechoslovak political straddling between East and West during the Third Republic (the era of Czechoslovak history between 1945 and 1948), where at least part of the Czechoslovak political elites still believed that the country could be a strategic bridge between the Eastern and Western spheres of influence.[19]

18 "Grandfather Czech" is a father of Czech nation, who according to a legend, together with his brother, Lech, wandered in an area of today's Central Bohemia, the area around Prague, and seeing the mediocre landscape and climate there, he decided to settle there and establish a village. Lech, conversely, wasn't so satisfied, and headed north, to find his dreamland in the area of today's Poland.

19 Brenner, Christiane. *"Zwischen Ost und West": Tschechische politische Diskurse 1945–1948*. R. Oldenbourg Verlag, 2009.

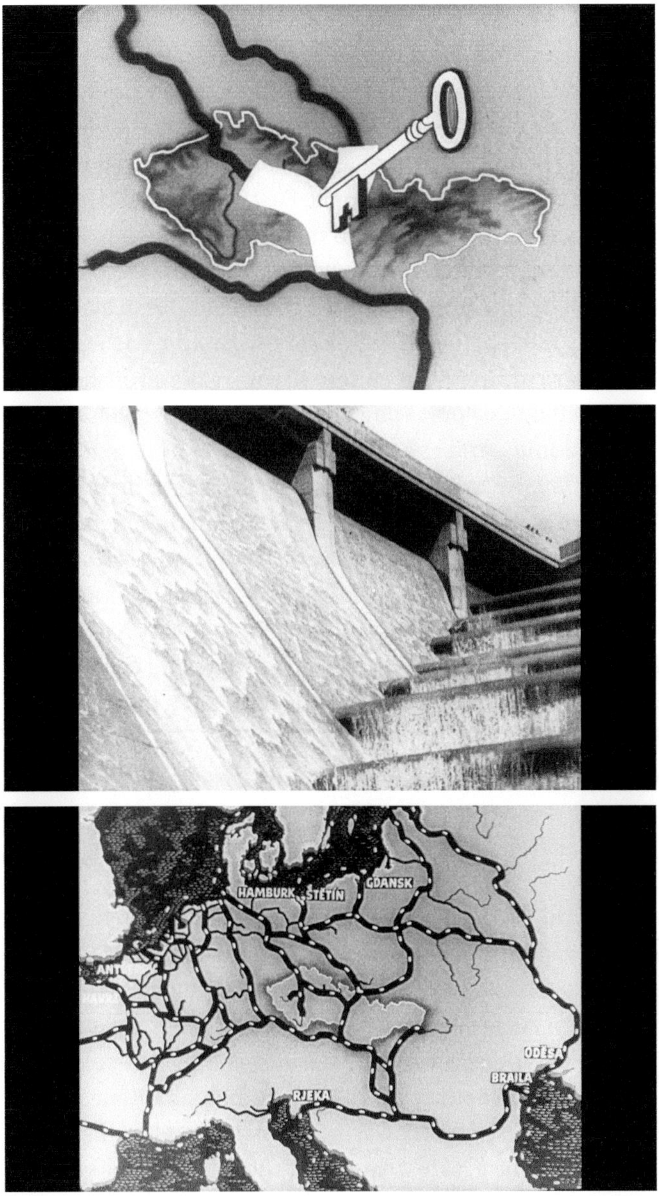

Figs. 6.1a–6.1c. Film stills from *Přístav v srdci Evropy* (*The Harbour in the Heart of Europe*, Drahoslav Holub, 1946).

The early post-war political situation in Czechoslovakia, however, despite the global peace ethos, which called for the values with which the canal has been associated in earlier periods, was not in favour of the project. The interests of national protectionism outweighed the interests of international understanding and cultural trade and exchange. A new water management policy accentuated the energy dimension of water flow regulation, which for a few years completely choked debates on the implementation of the canal. By the time of the restoration of the idea in the 1950s, it no longer had the same spirit as before the war, when the canal was called the laboratory of Europe and the way to the (Czech) sea, or just after the war, when it symbolized a "European crossroads." New canal projects aimed to be the "gateway to the East" instead.[20] As histories of major energy eras show us, turning to new sources of energy was a global trend: "The post-war economic upswing was associated with the global substitution of hydrocarbons for coal, the global rise of electricity generation (including nuclear fission), mass car ownership, and extensive energy subsidies in agriculture."[21] The shift to natural gas and oil, however, was importantly related to colonial power networks. In Czechoslovakia, in a land without colonies, and even without a sea, a similar shift took place, but instead of hydrocarbons, the country turned to hydropower. Rethinking the role of water management in the national economy, from the regulation of water flows to the emphasis on energy, was a clear signal of a new incoming socialist economic order, which had fundamentally transformed the dynamics of industrial and social relations in the state. Under this new order the interests in energy, transport and communications interweaved, and the interpretation of the national relation to water resources changed.

Dam Culture

After the Second World War, thanks to new industrial developments and new demands on transportation, communication and energy sources, the hidden imagery of the Central European water empire reappeared in new energy politics. Post-war energy debates were shaped by two complementary theses. Czechoslovakia as a country which is not energy self-sufficient, must firstly educate its people to use coal sparingly and, simultaneously, to explore the uses of new energy sources, especially water and, in the future, probably also atomic.[22] The greatest failure of the Czechoslovak energy system was considered to be the wasteful use of energy and the inefficient use of raw resources, both crucial problems for a country that lacked raw

20 Janáč. *European Coasts of Bohemia*, p. 149.

21 Smil, Václav. *Energy in World History*. Westview Press, 1994, p. 241.

22 According to research in contemporary literature on energy politics and the magazine *Energetika*, published by Průmyslové nakladatelství (Industrial Publishing).

materials.[23] According to comparative research, Czechoslovakia gained less energy than other states from a comparable amount of resources. Latent awareness of the need to embrace the energy revolution has resonated through the debate. Such a revolution, however, should be guided by the knowledge that "economic life cannot be stopped, or immediately rebuilt for new sources of energy."[24] The new situation was thought of as a new challenge and focused its research mainly on water energy during the early post-war period.

The late 1940s and early 1950s thus represent a transitional period in Czech history in which the media in general and film in particular collaborated on a major economical paradigm shift. As historians of the economy and industrial revolutions remind us, when we look at all great economic paradigm shifts in history, they share one common denominator – that in a particular moment in time, three defining technologies emerge and converge to create a new infrastructure and change the way economic life is organized.[25] Those three defining technologies are new communication technologies to more efficiently manage economic activity, new sources of energy to more efficiently power economic activity, and new transportation logistics to more efficiently move economic activity.

Around the end of the 1940s Czechoslovakia tended towards such a significant shift. An economic boost was seen as a prerequisite for the post-war recovery across the political spectrum. In parallel to the construction of new industrial centres, projects to transform agriculture and build new transportation networks, the government promoted research into new energy sources. The establishment of the Ostrava-Karviná industrial centre together with the building of the entirely new city of Havířov was followed by improvements in the transportation infrastructure, represented by the Railway Line of Friendship and the Railway Line of Youth projects. The first connected the capital city of Prague with Košice in eastern Slovakia and the second was important for Slovak inner transport infrastructure. Both projects were realized with the help of people's brigades, consisting mostly of young volunteers. All these efforts were legitimized by profound transmedia campaigns, in which the press, radio and film were used alongside literature,

23 Ibler, Jaroslav. "Úkoly energetiky v socialistickém hospodářství I." *Elektrotechnický obzor*, vol. 38, no. 4–5, 1949, pp. 2–4. Presented also as a part of lecture series called "Energy Course" at Czech Technical University.

24 Michalec, Jaroslav. *Rozvoj energetiky v ČSR: Příspěvek k sestavení dlouhodobého plánu energetiky*. Průmyslové vydavatelství, 1956, p. 21.

25 Rifkin, Jeremy. *The Third Industrial Revolution: How Lateral Power Is Transforming Energy, the Economy, and the World*. Palgrave Macmillan, 2011, p. 35.

drama and painting to educate audiences about the advantages of these transformative projects. The communist government was very well aware of the power of media and arts to educate new socialist citizens, and a centralized system of nationalized media and art production enabled it to regulate and control themes, genres and the focus of all output across art and media.

In 1951, the Czech government Act on New Economic Developments was issued to speed up the current slow pace of hydropower plant construction. This was followed by the 1952 Act on United Energy Balance, which also recognized the important position of hydropower. The shift to hydropower was aimed at conserving exhaustible coal mine reserves.[26] These developments were clearly inspired by the Soviet state plan for the electrification of Russia, the GOELRO, which established hydropower as a key source for the electrification of the country. The GOELRO plan was implemented during a ten- to fifteen-year period. According to the plan, Soviet territory was divided into eight regions, with distinct development strategies due to the specific features of each region and included construction of a network of thirty regional power plants, including ten large hydroelectric power plants and numerous electric-powered large industrial enterprises. In reality, only three out of ten hydroelectric stations were built by 1930 (the Volkhov, the Svir and the Dnieper hydroelectric stations) and the country's main energy source turned out to be oil and natural gas.[27] GOELRO was closely related to the Great Plan for the Transformation of Nature. This plan, proposed by Joseph Stalin in the second half of the 1940s, covered land development, agricultural practices and water projects to improve agriculture in the nation.[28] Czechoslovak engineers were directly inspired by the Soviet plan and compared its possibilities with the approach of industry in Switzerland, similarly lacking any original sources of energy and depending mostly on hydropower.

Established in January 1953, the Czech Committee for the Transformation of Nature settled on the construction of water reservoirs as one of its priorities. Although proclaiming its inspiration in the Stalinist plan, the

26 Michalec, Jaroslav. *Energetická bilance ČSR*. Elektrotechnický svaz československý, 1948.

27 Among other films devoted to the Soviet achievements in electrification, Dziga Vertov's *The Eleventh Year* (1928), according to John MacKay, elaborates the author's conviction that movement can be conceptualized in terms of energy flow and proves that his conception of cinema was rooted in the idea of productivism and transcendental materialism. See MacKay, John. "Film Energy: Process and Metanarrative in Dziga Vertov's *The Eleventh Year* (1928)." *October*, vol. 121, Summer 2007, pp. 41–78.

28 Michalec. *Rozvoj energetiky v ČSR*.

committee soon realized that in a small country with totally different geographical and natural conditions, the Great Plan for the Transformation of Nature would not be possible to implement with the same grandiosity as in the Soviet Union.[29] A thorough analysis of the state of the water resources and problems with water pollution in Czechoslovakia ultimately provided the basis of the state's water management plan.[30] The parameters of this plan were consistent with the 1952 Act on United Energy. The goal of water managers to build reservoirs and the intention of energy engineers to build hydropower plants thus converged into one effort, which was a radical transformation – that is, a taming or subjugation – of nature.[31]

In the 1950s, Czechoslovakia thus engaged in the rapid construction of hydropower plants, dams, and water reservoirs. All the big hydropower plants on the Vltava River and on the Váh River were built in this period, followed by many other smaller water management projects. Thanks to the overall control of media and arts production, the government was able to cover these construction endeavours in various formats distributed through various channels. Similar to the case of building new industrial centres and new transportation networks, construction sites for waterworks became the settings of novels, dramas and films, creating a new genre, the so-called "production drama."

The production dramas produced at the time (plays, literature and films set in factories and construction sites) engaged in reportage and displayed similar motives and tropes. The journalistic character of the documentary features and production dramas reflected contemporary "ideas about what is and is not reality and realism."[32] The idea was that art should "shape the present" and therefore it naturally embraced stories of about current building projects. This approach was rooted in the works of Czech social novels from the 1930s, such as Marie Majerová's utopian novel *Přehrada* (Dam, 1932).[33] Majerová not only denied the epic storyline and intensified simultaneity of journalistically tuned episodes from reality, but also portrayed the romance of transforming nature as coinciding with technical

29 For detailed description of the whole process of the implementation of the Transformation of Nature plan in Czechoslovakia, see: Olšáková, Doubravka, and Arnošt Štanzel. "Kafkaesque Paradigms: The Stalinist Plan for the Transformation of Nature in Czechoslovakia." *In the Name of the Great Work: Stalin's Plan for the Transformation of Nature and Its Impact in Eastern Europe*, edited by Doubravka Olšáková. Berghahn Books, 2016, pp. 43–125.

30 Rosík, January. "Státní vodohospodářský plán a zásady pro další plánovité řízení vodního hospodářství." *Vodní hospodářství*, vol. 4, no. 4, 1954, pp. 99–103.

31 A clear example of this general approach is given in a popular science book: Hoch, Alois Adalbert. *Živly pracují za nás*. Mladá fronta, 1955.

32 Janoušek, Pavel. *Studie o dramatu*. Ústav pro českou a světovou literaturu, 1993, p. 112.

33 Several re-editions of Majerová's novel were published in the 1950s.

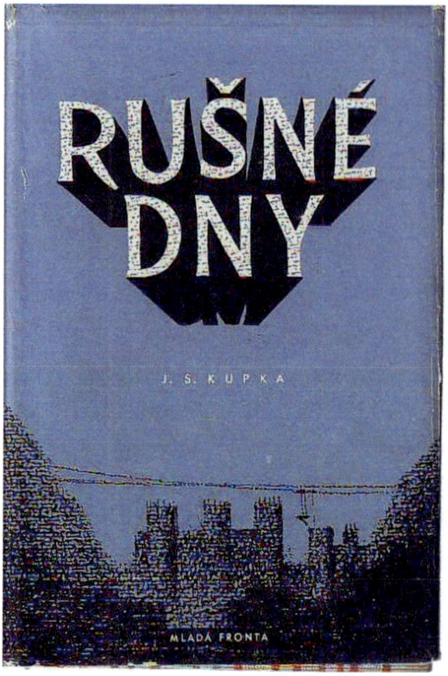

Fig. 6.2a. Cover of J.S. Kupka's novel *Rušné dny* (Busy days, 1955).

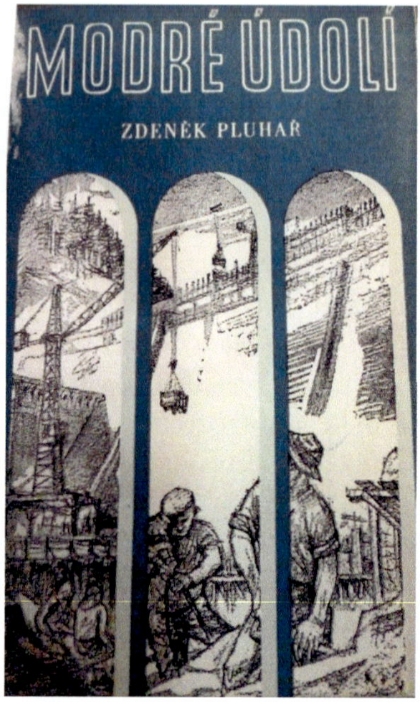

Fig. 6.2b. Cover of Zdeněk Pluhař's novel *Modré údolí* (Blue valley, 1954).

progress. In works of the 1950s, especially those created around 1952 and 1953, however, a constructivist ethos and a commitment to and enthusiasm for involvement in the construction of the new society were complemented by concerns about the struggle with obstacles and with internal and external enemies. The feature film *Priehrada* (*Dam*, Paĺo Bielik, 1952), for example, introduces both types of characters: those who believe in the importance of dam construction and those who criticize it.

Specific hybrid genres were developed within the transmedia campaign to inform the people about hydropower construction. Contemporary press covered the big construction projects, mostly in narrative reports, focusing on technical details, enumerating quantitative data and reminding people of the benefits of the transition to hydropower. Production novels, dramas and fiction films told stories about the construction of hydropower plants and dams. Poetic documentaries and popular scientific and educational films were made on the same topic. Painters created large canvases of the structures that reveal the hydropower plants' inner mechanisms. The human element in the stories was mostly suppressed, as if this aspect was a part of the goal to tame nature. With a basis in real stories and by clinging to the facts, fictional genres resembled non-fictional ones. At the same time, these works were intentionally created with both aesthetic and non-aesthetic goals.

The absorption of topics, themes and plots from everyday industrial life, along with an aesthetic approach to technology and mechanical processes, led to the emergence of various hybrid art forms, which in the centralized socialist media ecology worked in synergy to support the introduction of a new energy politics within the new economic shift.

In 1953, writer and director Dušan Kodaj wrote a production novel, *Oravska priehrada* (Oravska Dam), followed by a documentary film of the same name about the construction of the Orava dam and reservoir in Slovakia. In addition to reports in the press, this construction project was the subject of a painting by Eduard Světlík (*Stavba Oravské přehrady od betonárny* [*Construction of Orava Dam as Viewed from the Concrete Plant*, 1952, oil on canvas]), a triptych drawing by Dezider Milly (*Triptych Oravská priehrada* [*Orava Dam Triptych*, 1949, ink on paper]) and the above-mentioned feature film by Pal'o Bielik. Similarly, press, film, literature and painting co-created the image of the Váh Cascade Project[34] and the construction of dams and reservoirs on the Vltava River. The documentary film *Příběh staré řeky* (*The Story of an Old River*, 1956) by Jiří Lehovec, together with the feature film by prominent director Martin Frič, *Povodeň* (*Flood*, 1958) and the educational film *Kronika Slapské přehrady* (*Slapy Dam Chronicle*, Olga Růžičková, 1960) represented the construction of the Slapy dam, reservoir and hydropower plant, together with the novel *Rušné dny* (Busy days, 1955) by Jiří Svetozar Kupka and František Bílek's painting *Stavba Slapské přehrady* (*Slapy Dam Construction*, 1958, oil on canvas). The same set of outcomes surrounded the construction of the Lipno dam, reservoir and hydropower plant: useful films like *Vodní díla v Československu* (*Waterworks in Czechoslovakia*, Svatopluk Studený, 1956) and *Vodní energie* (*Water Energy*, Josef Pinkava, 1951) recombined shots of films previously made. Stories from dam construction also shared similar motives, such as the fight with nature, the destruction of the old to make way for the new etc.[35]

A combination of facts with technological optimism and the mythologization of the whole construction process resulted in a factual symbolism

34 Jozef Šturdík's canvas of Váh Dam was followed by documentary films *Považské stupně* (*Váh Degrees*, Svatopluk Studený, 1955) and *První vážský stupeň* (*First Váh Degree*, Štefan Ondrkal, 1957).

35 In the 1950s a specific sub-genre emerged within the socialist production novel, namely the "dam novel" (*přehradní román*), which established these key characteristics. This massive trend can be represented by the translation of Leonid Leonov's *Dravá řeka* (*Wild River*, Соть, 1928–1929), and novels such as *Modré údolí* (*Blue Valley*, Pluhař, Zdeněk, 1954), *Závod ve stínu* (*The Enterprise in Shade*, Sedláček, Květoslav František, 1954), *Rušné dny* (*Busy Days*, Kupka, Jiří Svetozar, 1955) etc. For more, see Hodrová, Daniela. "Žánrový půdorys budovatelského románu." *Vztahy a cíle socialistických literature*, edited by Hana Hrzalová and Radko Pytlík. ÚČSL, 1979, pp. 121–41.

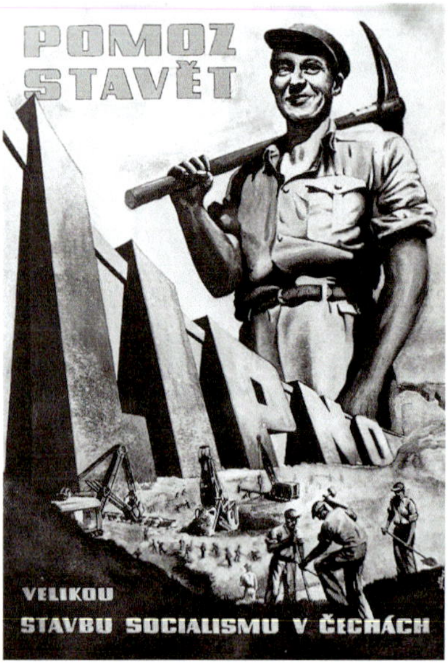

Fig. 6.3a. Recruiting poster for the construction of Lipno hydropower plant, 1952.

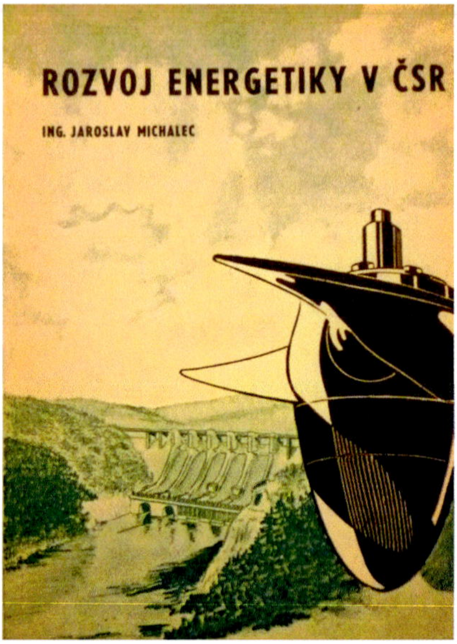

Fig. 6.3b. Cover of the economic research report *Rozvoj energetiky v ČSR* (Energy development in the ČSR, 1950).

as an approach which, based on current topics or stories, celebrated construction and technological progress and developed across factual as well as fictional arts and media. Visual culture emphasized the arch of the dam together with the water turbines and electric generators that became symbols of the most efficient method of energy generation (these images appeared on the covers of books and in brochures about new energy developments and plans, for example), while the previous era was associated with the image of the "coal barons," opportunists who did not belong to the new socialist order. While trying to establish new energy trends, however, the arts and media did not reject the old methods and procedures. A country focusing on heavy industry did not close its mines and steel mills, but instead celebrated its miners as "heroes of labour." Just as with the coverage of mines in the media emphasized the aspect of workers' labour, the new waterworks were viewed as Czech engineers' technological endeavour. Turbines designed by Viktor Kaplan were especially commemorated as a Czech contribution to hydropower engineering, which had both a technical as well as an artistic value. The ten-paddle wheel of Kaplan's turbine, later installed into the Orlík hydropower plant, was a key element in the Czechoslovak pavilion at EXPO '58 in Brussels and was awarded the gold medal.

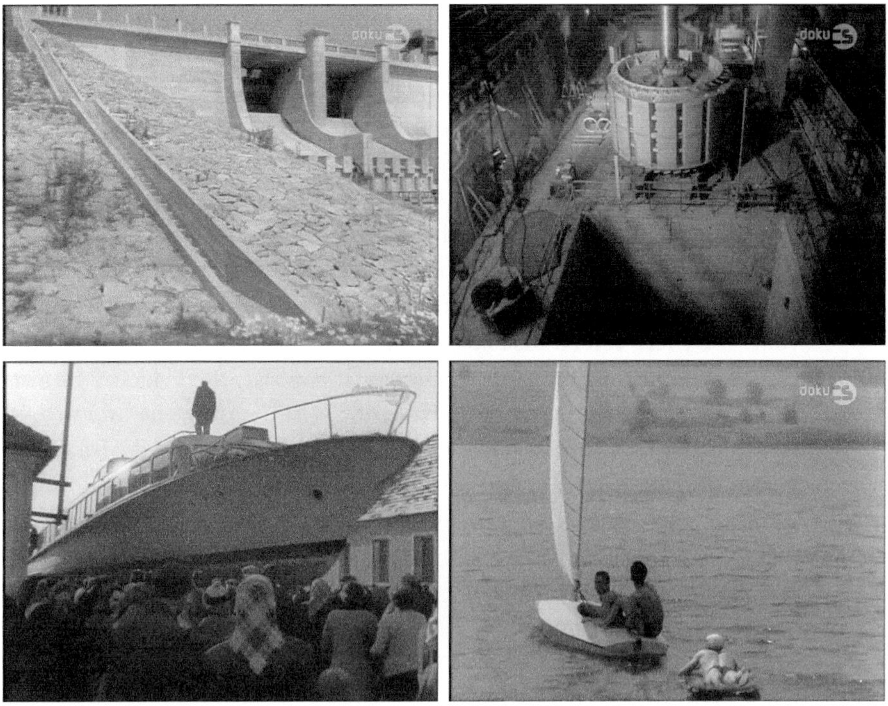

Figs. 6.4a–6.4d. Film stills from *Lidé nad Čertovou stěnou* (*People over the Devil's Wall*, Emanuel Kaněra, 1962), showing the construction of the dam, a look at the machinery inside and its recreational benefits.

Representations of hydropower in Czech culture carried contradictory meanings. Emphasizing the size of dam construction and technical maturity, linking artistic and useful goals, and facts with symbols, it depicted technology as an art and the water element as an organic machine.

The imagery and rhetoric associated with the symbol of the dam proves how multi-layered and complex the phenomenon of hydropower is. The dam is a big fortress, evidence or, better, a symbol of gigantic construction and socialist technological and scientific endeavour. The dam is a symbol of how man's rationality and scientific development can transform nature. The dam is a wall behind which, however, hides a beautiful holiday resort and, as such, the dam gives the sea to the Czech nation. The dam is a symbol so big that it hides all negative environmental and social consequences these constructions had (from fish extinctions to the eradication of villages and the relocation of entire communities of people).

Dams and reservoirs, around which recreational resorts developed, brought, along with the help of the media, the Czech people closer to the sea. Just like the articles that appeared in the 1920s and 1930s claiming that the sea could be reached through technical innovations such as "an engine, an airplane, a radio and a distant vision," the media of the 1950s called all water surfaces close to dams and hydroelectric power stations "seas" (Czech sea, Šumava sea). Of these two ways of getting closer to the sea via the media, the second had the effect of redirecting attention away from the problems that were linked to water management projects, in favour of a positive interpretation of the process as a natural transformation. One article about the origins of the Czech (Šumava) sea, the Lipno dam, claimed: "Where the birds nest this summer, fish will swim in the coming year."[36] Others voiced pleasure about the transformation in which the romantic landscape disappears beneath the water's surface.[37] Joining the poetic approach were plenty of tourist films from the late 1950s[38] that presented artificial lakes created by dams as tourist destinations, places of recreation and relaxation, and that used the metaphor of the sea to increase the attractiveness of particular resorts and, above all, divert attention away from the social and environmental context of their construction.

Conclusion

The nationalized economy of post-war Czechoslovakia established a synergy of arts and media and built an efficient communication tool to promote new endeavours in energy supply, namely by hydropower plants, and supported new shifts in transportation and mobility associated with tourism. Between 1945 and 1948, the Czechoslovak president Edvard Beneš had compromised with the Communist Party to avoid a post-war coup, hoping that the democratic process would restore a more equitable distribution of power. Beneš had negotiated the Soviet alliance, but at the same time he strived to establish Czechoslovakia as a "bridge" between East and West, capable of maintaining contacts with both sides. After the communist coup d'état in 1948, however, the country's geopolitical situation profoundly changed, while the shift in water energy imagery from canal to dam illustratively follows these geopolitical processes. Exemplified in the clear move from projects of (centrifugal) canal to (centripetal) dam, to a fortress in a way, the trend of the country's geopolitical isolation prevailed over efforts towards

36 "Les ustupuje šumavskému moři." *Rudé právo*, vol. 36, no. 276, October 3, 1956, p. 1.

37 "U nejvyššího 'Gottwaldova stupně míru'." *Mladá fronta*, vol. 10, no. 187, August 9, 1952, p. 1.

38 Miškuv, Pavel, *Oravské more*, 1959; Šulc, František, *Za Oravou za Váhom*. 1959 etc.

international circulation of knowledge and capital. Besides mountains and spas, both former highlights among domestic tourist attractions, artificial lakes beside dams became new popular destinations, easily accessible by automobiles, and were associated with new trends in the socialist lifestyle.

Water energy both as a complex ideological concept and energy solution for the future in the Czech case positioned art and media into the role of servant to the state. Yet even such works offered variable explanations of the importance of rivers and bodies of water. Traditionally considered as borders, rivers and bodies of water were represented in Central European thought as lines of communication, economic commodities and a representational medium, and had an irreplaceable role within the process of conceptualizing ideas related to the nation and national identity. As such, all these media and artistic forms supported the circulation of knowledge about scientific background and the energy benefits of hydropower, and exploited the symbolic potential of dam construction as an important energy source, a specific work environment and, last but not least, an important solution to the national trauma associated with the lack of access to the sea. Industrial films on hydropower, however, not only contribute to our understanding of the profound economical paradigm shift from capitalist to a nationalized socialist order, but also tell us more about a cinematic and broader visual imagery connected with water in general. Through establishing a new topoi of the dam, these films heavily profit from the pre-war avant-garde constructivist movement (represented in the Czech environment by the technical films of Antonín Smrček), poetic documentaries about water sports and other leisure activities, social documentaries about work and labour, and Czech feature films in which rivers and water were associated with lyrics, melancholy and eroticism, such as Josef Rovenský's *Řeka* (*River*, 1933) and Gustav Machatý's *Extáse* (*Ecstasy*, 1932). As mentioned in the introduction, the poetic documentary that can be considered emblematic in this sense was named by its director, Jiří Lehovec, *The Story of the Old River*. By that name, however, Lehovec did not aim to evoke nostalgia, but rather to emphasize the aspect of time, crucial for the medium of film itself. The story that is being told, and, therefore, lasts, and deals with the replacement of the old by the new, represented the river as a vehicle of historical flow and change.

This article was written with institutional support for the long-term conceptual development of a research organisation provided by the Ministry of Culture, Czech Republic.

Works Cited

Bartelheim, Martin, Roland Hardenberg, Thomas Knopf, Anke K. Scholz, and Jörn Staecker. "'ResourceCultures' – A Concept for Investigating the Use of Resources in Different Societies." *Persistent Economic Ways of Living: Production, Distribution, and Consumption in Late Prehistory and Early History*, edited by Alžběta Danielisová and Manuel Fernández-Götz. Archaeolingua Alapítvány, 2015, pp. 39–50.

Brechtefeld, Jörg. *Mitteleuropa and German Politics, 1848 to the Present*. Macmillan, 1996.

Brenner, Christiane. *"Zwischen Ost und West": Tschechische politische Diskurse 1945–1948*. R. Oldenbourg Verlag, 2009.

Ehmer, Manfred. *Mitteleuropa: Die Vision des politischen Romantikers Constantin Frantz*. Tredition, 2012.

Hanáková, Petra. "The Seas and Skies of Bohemia: Projections and Inventions of the National Space." *EUtROPEs: The Paradox of European Empire*, edited by John W. Boyer and Berthold Molden. University of Chicago Press, 2015, pp. 335–58.

Hoch, Alois Adalbert. *Živly pracují za nás*. Mladá fronta, 1955.

Hodrová, Daniela. "Žánrový půdorys budovatelského románu." *Vztahy a cíle socialistických literature*, edited by Hana Hrzalová and Radko Pytlík. ÚČSL, 1979, pp. 121–41.

Ibler, Jaroslav. "Úkoly energetiky v socialistickém hospodářství I." *Elektrotechnický obzor*, vol. 38, no. 4–5, 1949, pp. 2–4.

Innis, Harold. *The Fur Trade in Canada*. University of Toronto Press, 2001.

Janáč, Jiří. *European Coasts of Bohemia: Negotiating the Danube-Oder-Elbe Canal in a Troubled Twentieth Century*. Amsterdam University Press, 2012.

Janoušek, Pavel. *Studie o dramatu*. Ústav pro českou a světovou literaturu, 1993.

Kvapil, Jaroslav. "Na nové parádní lodi Jihoslovanské po 'našem' moři." *Pestrý týden*, vol. 6, no. 28, July 11, 1931, p. 8.

"Les ustupuje šumavskému moři." *Rudé právo*, vol. 36, no. 276, October 3, 1956, p. 1.

List, Friedrich. *Das nationale System der politischen Ökonomie*. Cotta, 1841.

MacKay, John. "Film Energy: Process and Metanarrative in Dziga Vertov's *The Eleventh Year* (1928)." *October*, vol. 121, Summer 2007, pp. 41–78.

Macura, Vladimír. *Znamení zrodu a české sny*. Academia, 2015.

Meyer, Henry Cord. *Mitteleuropa in German Thought and Action, 1815–1945*. Springer, 1955.

Michalec, Jaroslav. *Energetická bilance ČSR*. Elektrotechnický svaz československý, 1948.

Michalec, Jaroslav. *Rozvoj energetiky v ČSR: Příspěvek k sestavení dlouhodobého plánu energetiky*. Průmyslové vydavatelství, 1956.

Müldner, Josef. *Námořní společnost československá, její vznik a úkoly*. Námořní společnost československá, 1920.
Nesnídalová, Marie. "Iluse českého moře – Koupání v Jugoslávii." *Český lid*, vol. 17, no. 251. September 9, 1928, p. 1.
Olšáková, Doubravka, and Arnošt Štanzel. "Kafkaesque Paradigms: The Stalinist Plan for the Transformation of Nature in Czechoslovakia." *In the Name of the Great Work: Stalin's Plan for the Transformation of Nature and Its Impact in Eastern Europe*, edited by Doubravka Olšáková. Berghahn Books, 2016, pp. 43–125.
Pritchard, Sara B. *Confluence: The Nature of Technology and the Remaking of the Rhône*. Harvard University Press, 2011.
Rifkin, Jeremy. *The Third Industrial Revolution: How Lateral Power is Transforming Energy, the Economy, and the World*. Palgrave Macmillan, 2011.
Rosík, January. "Státní vodohospodářský plán a zásady pro další plánovité řízení vodního hospodářství." *Vodní hospodářství*, vol. 4, no. 4, 1954, pp. 99–103.
Smil, Václav. *Energy in World History*. Westview Press, 1994.
"U nejvyššího 'Gottwaldova stupně míru'." *Mladá fronta*, vol. 10, no. 187, August 9, 1952, p. 1.
White, Richard. *Organic Machine: The Remaking of the Columbia River*. Hill and Wang, 1995.

About the Author

Lucie Česálková is an associate professor at the Department of Film Studies, Charles University, and a researcher and editor at the National Film Archive, Prague. Currently, she participates in the research of the consumer imagination of the Czechoslovak communist dictatorship, and leads a research group which is part of the international project "Visual Culture of Trauma, Obliteration and Reconstruction in Post-World War II Europe" (supported by HERA). She focuses on the history of non-fiction and documentary film, and the history of film distribution and movie-going. In 2014, she published a monograph in Czech, *Atoms of Eternity: Czech Short Film of the 1930s–1950s*, and in 2019, together with Kateřina Svatoňová, *The Dictator of Time: (De) contextualizing the Phenomenon of Laterna Magika*.

7 Modern Water Sprites

History, People and the Landscape of Northern Sweden in Vattenfall's Film Production in the 1950s

Fabian Zimmer

Abstract
In the 1950s, the Swedish State Power Board Vattenfall commissioned a large number of short films documenting the construction of water power plants in northern Sweden. Focusing on a selection of these films, I argue in this chapter that Vattenfall's film production must be seen in the context of a larger shift in Vattenfall's PR strategies. Starting in the 1950s, Vattenfall used film as a central medium for mitigating conflicts surrounding the construction of water power plants. Mobilizing new and ambivalent narratives of modernity, the films were designed to depict Vattenfall as an environmentally and socially friendly company – an image that was perfectly embodied by the allegory of the modern water sprite.

Keywords: hydropower; Sweden; technological sublime; anti-dam protest; landscape; industrial film

Introduction: The Water Sprite

In the old times it was said that he who borrowed the fiddle from the näck got a strange power into his playing.... The *näck*, or *strömkarl*, was a water sprite in Scandinavian folk tales, a creature who could usually be seen at a rapid or a waterfall in the boreal forests, playing his fiddle in the waning light of a long northern summer evening. He who borrowed the fiddle or learned how to play it from the water sprite could make people dance to his playing like no other – but he would also be unable to stop unless someone wrested the instrument from his cursed hands. In other variations of the tale, the

Hediger, V., F. Hoof, Y. Zimmermann, with S. Anthony (eds), *Films That Work Harder: The Circulation of Industrial Film*. Taylor & Francis Group 2024
DOI: 10.5117/9789462986534_CH07

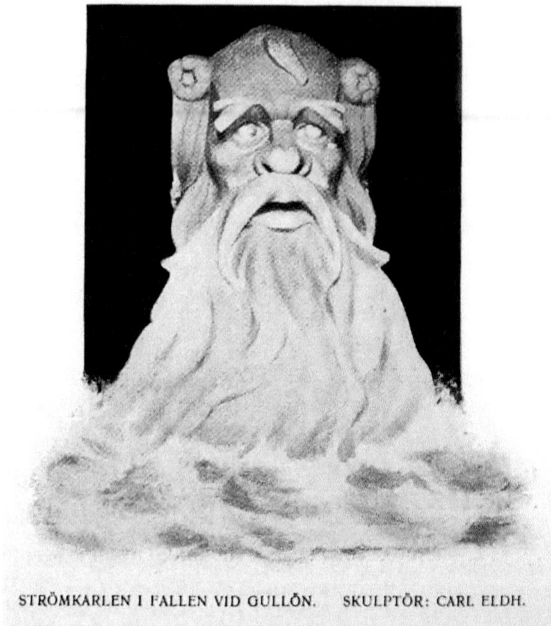

Fig. 7.1. The water sprite at Trollhättan (Erik Josephsson/Carl Eldh, 1910).

water sprite lured the incautious into the rapids with his playing or caused trouble for the millers who used the rapids to grind their flour.[1] So the old tales went that captivated the fantasies of nineteenth-century Scandinavian romantics. And although the water sprite embodied a grave sense of danger, it is no wonder that he featured once more when the Scandinavian waterfalls began to be exploited for the production of hydroelectricity.

His most prominent reappearance was at the Trollhättan hydropower plant, where a "giant water sprite surfacing out of the water masses," a four-metre-high granite sculpture, was erected at the weirs.[2] Trollhättan was the first power plant built by the State Power Board (Kungliga Vattenfallsstyrelsen, later Vattenfall AB), from 1906 onwards, and state officials spared no expense to make the constructions as representative as possible. The sculpture was most probably intended as an "admonition about the complexity of nature's exploitation" by its creators.[3] Popular imagination,

1 Cf. Stattin, Jochum. *Näcken: Spelman eller gränsvakt?*, 2nd ed. Carlsson, 1992.

2 Josephsson, Erik. "Statens kraftverk vid Trollhättan." *Teknisk Tidskrift: Arkitektur*, vol. 41, 1911, p. 62.

3 Brunnström, Lasse. *Estetik och ingenjörskonst: Den svenska vattenkraftens arkitekturhistoria*. Riksantikvarieämbetet, 2001, p. 35.

however, interpreted it as a "victory monument" – a monument to the victory of mankind over wild nature.[4]

Fifty years later, in the mid-1950s, Vattenfall's PR efforts had shifted away from ornamental architecture towards the more easily distributable media of film and photography. Nevertheless, the water sprite continued to be a popular figure in these PR materials. He was charged with new allegorical meanings as the company faced new technological, cultural and social challenges.

Among Vattenfall's film archive, which encompasses over a thousand titles produced between 1915 and today, the 1950s stand out as a "golden age."[5] Almost seventy films were produced during this decade, encompassing unedited material, a newsreel series and polished documentaries and educational films for both Vattenfall's staff and the general public.[6] Taking the figure of the water sprite as a vantage point, I aim here to demonstrate the great potential of this film corpus as a source for a cultural history of hydroelectricity. I focus on a selection of four documentary films from the 1950s, all about thirteen minutes in length: *Strömkarl* (*Water Sprite*, 1956), *Öden bortom horisonten* (*Destinies beyond the Horizon*, 1956), *Den nya sjön* (*The New Lake*, 1957), and *Det nya ansiktet* (*The New Face*, 1958).[7] Surprisingly, very little scholarly or other work has yet been published on Vattenfall's film archive.[8] Yet, the following chapter discerns the 1950s as a crucial turning

4 Svensson, Ove. *Carl Eldhs Strömkarlen: Symbolen, myten och motivet: En berättelse om Strömkarlens tillblivelse och gestaltning.* Kraftverkshistoria. Vattenfall Västsveriges Pensionärsförening, 1999, p. 75.

5 Jacobsson, Lars. *Vattenfalls filmproduktion genom tiderna.* December 17, 2015. Unpublished manuscript. I thank the author for sending me a copy of his manuscript.

6 About sixty of these films have recently been digitalized and made available by Vattenfall's cultural heritage bureau at: historia.vattenfall.se/sv/media-archive/video (accessed May 29, 2017). I am indebted to Per Israelsson at Vattenfall's archives for providing me with an inventory list, from which I take the numbers.

7 *Öden bortom horisonten.* 1956. Dir.: Gösta Werner, prod.: Ivan Christoferson for Svensk Filmindustri, camera: Sten Dahlgren, premiered on November 12, 1956, in Stockholm; *Strömkarl.* 1956. Dir.: Hasse Skogsberg & Ivan Christoferson, prod.: Charlie Cederholm for Europa Film, manus.: Carl Olov Mellander, camera: Ingvar Borild, premiered on December 26, 1956, in Stockholm; *Den nya sjön.* 1957. Dir.: Ivan Christoferson, manus.: Charlie Cederholm & Carl Olov Mellander, camera: Ingvar Borild, produced for Europa Film, premiered on February 9, 1959, in Stockholm; *Det nya ansiktet.* 1958. Dir. & manus.: Alex. Jute, prod.: Charlie Cederholm & Ivan Christoferson, camera: Karl-Erik Alberts. See the respective entries in *Svensk Filmdatabas*, sfi.se/sv/svensk-filmdatabas. Accessed May 19, 2017.

8 Besides an older brochure by Vattenfall's historian Nils Forsgren. *Vattenfall visar: Glimtar ur filmarkivet.* Vällingby, 1984, there has been some interest in these films more recently. See my detailed reading of a large number on hydropower films by Vattenfall and other agencies in Zimmer, Fabian. *Hydroelektrische Projektionen: Eine Emotionsgeschichte der Wasserkraft im*

point in Vattenfall's PR policy, making this decade a very interesting period to study. The bulk of this chapter then focuses on the rhetoric of the films and their discursive contexts. I argue that all four films can be understood as an attempt by Vattenfall to "engineer consent" in a cultural environment that grew increasingly critical to water power constructions.[9] To counter public discontent, the company fashioned new narratives of hydroelectric modernity in which history, people and landscapes figured in ways that were unprecedented in earlier discourses on hydroelectricity.

"Vattenfall Visar": Public Relations at the Swedish State Power Board

Vattenfall was founded in 1909 as the "State Power Board" with the task to "manage" the state's waterfalls. The Power Board remained a government office until 1991, but from its very beginning it enjoyed a considerable freedom of action and operated rather like a commercial enterprise.[10] Nevertheless, its propagandistic activities up until the 1950s clearly bore the mark of a state management agency. As Eva Jakobsson has pointed out, the State Power Board was formed by a techno-scientific elite of engineers who put planning and rationality at the centre of their ideology.[11] Correspondingly, most of Vattenfall's publications had a rather scientific tinge to them. So, still in 1948, for the celebration of "[f]our decades of the State Power Board," a 350-page book was published that collected forty rather descriptive articles on Vattenfall's history, facilities and organization – following a pattern very

Industriefilm. Deutsches Museum. Abhandlungen und Berichte, n.s., vol. 36. Wallstein, 2022. *Det nya ansiktet*, which has had something of a reputation among Swedish historians for a while, now figures as an allegory of Swedish industrial media culture in the post-war years in Björkin, Mats. *Post-war Industrial Media Culture in Sweden, 1945-1960: New Faces, New Values*. Amsterdam University Press, 2022. An earlier analysis of this film is found in Krohn Andersson, Fredrik, *Kärnkraftverkets poetik: Begreppsliggöranden av svenska kärnkraftverk 1965–1973*, Eidos, vol. 24, Stockholms universitet, 2012.

9 On the metaphor "engineering consent," see Zimmermann, Yvonne. "The Power of Flows: The Spatiality of Industrial Films on Hydropower in Switzerland." This volume.

10 For a detailed analysis of Vattenfall's foundation and early history, see Jakobsson, Eva. *Industrialisering av älvar: Studier kring svensk vattenkraftutbyggnad 1900–1918*. Avhandlingar från Historiska institutionen i Göteborg 13. Gothenburg, 1996, pp. 74–90; for the "management" quote see ibid., p. 88 and passim. The Swedish Parliament, that decided on Vattenfall's budget, uniformly supported Vattenfall's projects up until the 1960s; cf. Vedung, Evert, and Magnus Brandel. *Vattenkraften, staten och de politiska partierna*. Nya Doxa, 2001.

11 Cf. Jakobsson. *Industrialisering av älvar*, pp. 68–90.

similar to previous publications. The few photographs in the book had for the most part been in use since the Power Board's inception.[12]

What a difference then, when eleven years later, for the fiftieth anniversary, Vattenfall's press officer, Charlie Cederholm, hired Lennart Nilsson, the photographer who would become world famous a few years later for his photographs of human embryos, to produce a photo book. This book, entitled *Strömkarl*, rather resembles the printed version of a film – it has no pagination; photographs in a variety of formats are stringed together associatively, connected by short text commentaries that can be read out like a voice-over. The book's foreword explicitly rejects the aesthetics and structure of its predecessors:

> [This book does not] attempt to outline the course of development in our history. With the help of the photographs we have merely tried to capture something of the atmosphere and activity which are the Board's own. Technical explanations are therefore seldom to be found in this book – it is people and their environment which are predominant in these pages.[13]

Unsurprisingly, Cederholm was also the driving force for Vattenfall's film production that expanded massively during the 1950s. Together with Ivan Christoferson, he led the production of most of these films. Both Cederholm and Christoferson were employed at Vattenfall's press office since 1949. They belonged to a generation that had not experienced the beginnings of water power exploitation in Sweden and they both had a journalistic, not a technical background. Therefore, they approached technology with fresh eyes. They saw water power not as a technical matter for select electrical professionals but as something that needed to be popularized among the broad public. Moreover, a strong need to create and coordinate a collective identity among the workforce had arisen within the company, as the number of employees amounted to over 14,000 during the 1950s. The press office answered this need with the employee magazine *Vi i Vattenfall* (*We in Vattenfall*) that was published from 1948 onwards. The printed propaganda was accompanied by a large number of films that were announced under the label "Vattenfall visar" ("Vattenfall shows"). So much did the PR efforts

12 Kungl. Vattenfallsstyrelsen, editor. *Statens Vattenfallsverk under fyra decennier*, 2nd ed. Stockholm, 1948. For the inauguration of the Trollhättan plant, the Power Board even produced a folio publication in three heavy volumes that appeared between 1911 and 1916.

13 Cederholm, Charlie, and Lennart Nilsson. *Strömkarl: En krönika om vattenfall, dess miljöer och människor jubileumsåret 1959*. Stockholm, 1959. The quote is from the preface to the English translation (the Swedish version of this paragraph reads the same). Idem. *The Power Team: A Chronicle of the Swedish State Power Board, Its Background and People: Jubilee Year 1959*. Stockholm, 1960.

rely on film that at the beginning of the 1960s state auditors urged the press department to cut down the expenses for film production.[14] I will now take a closer look at four of the films produced during the 1950s.

Narratives of Modernity and the Crisis of the Sublime

> "In the old times it was said that he who borrowed the fiddle from the *näck* got a strange power into his playing. In the site of this power, in the rapids, now lives the water sprite of our times. The modern water sprite also plays an instrument." With these words, spoken in voice-over, the film *Strömkarl* begins. It opens with a motif that is central to many of the 1950s films by Vattenfall: *Den nya tiden* (*The New Times*). Two of the other films, *Den nya sjön* (*The New Lake*) and *Det nya ansiktet* (*The New Face*), already announce the distinction between the "old" and the "new" in their titles. Thus, from the start, these films frame their plots as narratives on modernity.

Strömkarl is essentially a "process film," connecting the natural resource, the production and distribution of electricity and its consumption in one (almost) linear narrative.[15] Much of the imagery that the film presents highlights the modernity of this process. The production scenes are mainly shot in Kilforsen, a very recent plant that opened in 1953.[16] Here, according to the commentary, its operation happens "almost automatically" and the camera takes delight in depicting all the details of the control panels. The effortless mastery of nature is then shown, as one of the employees opens the dam weirs with the simple turning of a small toggle switch – "his light finger movement is transformed into a giant gesture." Just like electricity production, distribution is also depicted as both a frictionless and a monumental endeavour. The camera follows the undulating panoramas created by the power lines, which the commentary calls "electricity's national highways"

14 Very little is known about Vattenfall's press office so far. The above account is based on Jacobsson, *Vattenfalls filmproduktion genom tiderna*; on Cederholm's short biography in Paul Harnesk, editor. *Vem är vem: Stor-Stockholm*, 2nd ed. Bokförlaget Vem är Vem AB, 1962, p. 247; and Christoferson's in Helge Jonsson, et al. "Ivan Christoferson, till mine." November 9, 2010. *Allehanda*, www.allehanda.se/slakt-o-vanner/dodsfall/ivan-christoferson-till-minne. Accessed May 11, 2017. A more detailed account can now be found in Zimmer. *Hydroelektrische Projektionen*, pp. 120–24.

15 Cf. Gunning, Tom. "Before Documentary: Early Nonfiction Films and the 'View' Aesthetic." *Uncharted Territory: Essays on Early Nonfiction Film*, edited by Daan Hertogs and Nico de Klerk. Stichting Nederlands Filmmuseum, 1997, pp. 17–18.

16 Brunnström. *Estetik och ingenjörskonst*, pp. 241–42.

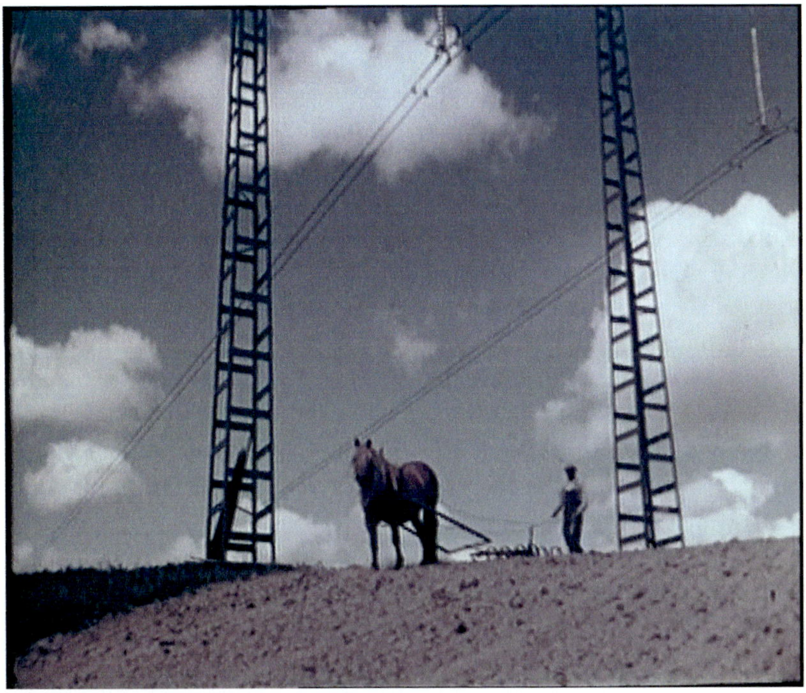

Fig. 7.2. Old times and new times in *Strömkarl* (1956).

and the film repeatedly plays with the geometries of electrical towers shot from up close in high contrast against the blue sky. Their gigantic dimensions are shown in a striking shot, that again embodies the contrast between the "old" and the "new": A farmer with a horse-drawn plough in the foreground is dwarfed by the pylons of the modern electrical towers in the background.

This imagery must be seen in the tradition of what David Nye has described as the "technological sublime": A mode of perception that is integral to the culture of modernity and that highlights the monumental and dramatic in the confrontation with new technologies. Shared "experiences of awe and wonder" are at the heart of the technological sublime, behind which human labour frequently becomes invisible or appears outdated.[17] Indeed, from its very inception, the electrotechnical modernization of the Swedish landscape was perceived and described as the sublime and dramatic subjugation of nature to human will.[18] Thus,

17 Nye, David E. *American Technological Sublime*. MIT Press, 1994. The quote is on p. xvi; on rendering labour invisible, see p. 123.

18 Cf. Brunnström. *Estetik och ingenjörskonst*, pp. 21–28; Sörlin, Sverker, and Christer Nordlund. "Modernizing the National Landscape." *Kenyon Review*, n.s., vol. 25, no. 3–4, 2004, pp. 301–15.

when in 1911 the first state power plant at Trollhättan was inaugurated, the State Power Board celebrated this as the beginning of "a new period."[19] Spectacular opening ceremonies accompanied the inaugurations and the plants were ennobled by monumental architectural design – as I have noted in the introduction.

As much as the sublime tradition is visible in *Strömkarl*, what is most remarkable about this film is how much it departs from that tradition. In fact, the sublime only plays a subordinate role in this film and the others that I analyse, as I will demonstrate in the course of this chapter. What is equally remarkable is that electricity consumption is notably absent from almost all films in the 1950s. While *Strömkarl*'s narrative indeed ends with consumption, only one minute out of eleven is dedicated to this topic. Also the elaborate praise and demonstration of lighting, electric motors and household gadgets that was prominent in Vattenfall's films in the 1930s and 1940s is missing.[20] Rather, the main focus of the films is on production and distribution, and, first and foremost, on construction.

This apparent shift might seem surprising, as it falls into a period in which consumer society was rapidly growing in Sweden just as in other industrialized countries. Yet, an essential reason for the shift can be found exactly in the expansion of consumerism: By the 1950s Sweden was one of "the best electrified countries in the world," as *Strömkarl* proudly explains; it was among the countries with the highest per capita consumption of electricity.[21] Vattenfall thus was not facing problems in selling the electricity it produced but rather in meeting the increasing demand for electricity. Management had expected demand to decrease after the Second World War, just like it had after the First. Aggravated by the difficulty to procure building materials in the post-war period, construction therefore lagged behind the development of demand. Vattenfall thus deemed it necessary to "keep up an intensive construction work," especially in the Swedish north, where many rivers were still unharnessed.[22] Therefore, it became the essential task of Vattenfall's films in the 1950s to promote the further construction of power stations and distribution lines. But the construction of power plants, an uncontroversial topic at the beginning of the twentieth century, faced a

19 Hansen, Fredrik Vilhelm. "Foreword." *Trollhättan: Dess kanal- och kraftverk. Historik och beskrifning*, vol. 1, edited by Kungl. Vattenfallsstyrelsen, 1911.

20 Several films were produced by Vattenfall in collaboration with FERA (Föreningen för elektricitetens rationella användning, Association for the Rational Utilization of Electricity) in the 1930s and 1940s.

21 According to the foreword in Cederholm, and Nilsson. *The Power Team*.

22 Kungl. Vattenfallsstyrelsen. *Statens Vattenfallsverk under fyra decennier*, p. 35.

growing opposition from the 1940s on, and made promotional action even more urgent for Vattenfall.

At the beginning of the war in September 1939, Vattenfall and private lobby organizations had pressed for an adaption of the Swedish water laws that would facilitate the regulation and exploitation of the Swedish rivers in order to secure the national electricity supply. Shortly thereafter, parliament passed a corresponding amendment on "provisional water regulation." The constructions built on the basis of this law, by contrast, were "hardly provisional measures," as Margareta Biörnstad has pointed out,[23] and they quickly aroused protest. It was a protest that previously had been remarkably absent: for the first time in Swedish water power history, companies had to refrain from exploiting certain stretches of rivers due to public opposition. From 1940 onwards, local protests originating in northern Sweden, where most power plants were projected, succeeded to gain attention on a national level.[24] By the early 1950s the protest gradually took the shape of a protest movement that managed to gain a foothold, not only in public opinion, but also on a legal level. A new law on environmental legislation passed in 1952 made two important organizations in the protest against water power, the Swedish Society for Nature Conservation (Svenska Naturskyddsföreningen) and the Association for Cultural Heritage (Samfundet för Hembygdsvård), advisory bodies in state committees concerning major changes in the Swedish landscape. In 1956, after six years of local protest supported by these two and several other organizations, a planned regulation of Lake Ottsjön was toppled by the water court. For the first time the court rejected the planned regulation of a lake over environmental concerns.[25]

23 Biörnstad, Margareta. *Kulturminnesvård och vattenkraft 1942–1980: En studie med utgångspunkt från Riksantikvarieämbetets sjöregleringsundersökningar.* Kungl. Vitterhetsakademien Historie och Antikvitets Akademien, 2006, p. 48.

24 The increasing protest against hydropower stations in Sweden from the 1940s on has lately been well researched. See Sundin, Bosse, and Richard Pettersson. "Sigurd Curman, Nämforsen och det 'kultiverade kraftverket'." *När dalen var röd och kusten blev hög*, edited by Erland Mårald, Skrifter från forskningsprogrammet Landskapet som arena 10, Umeå, 2008; an English summary is found in Sundin. "From the Technological to the Antiquarian's Landscape: Hydroelectric Power and the Design and Discovery of the Landscape." *Icon*, vol. 10, 2004; Biörnstad. *Kulturminnesvård och vattenkraft 1942–1980*, pp. 48–51 and 87–104; Vedung, and Brandel. *Vattenkraften, staten och de politiska partierna*, p. 53.

25 For the developments in the 1950s, see Anshelm, Jonas. *Vattenkraft och naturskydd: En analys av opinionen mot vattenkraftutbyggnaden i Sverige 1950–1990.* Tema teknik och social förändring 18, Linköping University Press, 1992, pp. 13–35; Vedung, and Brandel. *Vattenkraften, staten och de politiska partierna*, pp. 60–97; for a brief survey in English, see Nilsson, Kjell. "Where Industry Meets Nature: How Public Concern Has Influenced the Design of Swedish Industrial Landscapes during the 20th Century." *Landscape and Urban Planning*, vol. 23, no. 1, 1992, pp. 38–40.

Rewriting Hydroelectric Modernity: History, People and Landscapes

The strategy chosen by Vattenfall to address these legitimization issues in their films was, instead of overriding the protest with a straightforward rhetoric of progress and modernity, to propose a revised narrative of modernity. The narratives of the films depart from the mere celebration of water power's sublimity and its benefits to the consumer. Instead, they incorporate the problematic aspects of modernity, while still retaining a clear pro-exploitation message.

> "All development is painful and costly. Everyone has to sacrifice something for it. [...] Today this development is necessary and inevitable, even if we know that it often leads to grave human problems. [...] We have created them ourselves and we ourselves have made them – unsolvable." *Öden bortom horisonten* (*Destinies beyond the Horizon*) concludes with these words. The tone of the speaker is slow and calm throughout the film, as if to be demonstratively conciliatory. And the film was screened as "a short film by Gösta Werner," a renowned director and film theorist of the 1940s and 1950s but was actually produced under the auspices of Vattenfall's Christoferson.[26] However, contemporary spectators had no difficulty seeing behind the curtain and grasping the actual message of the film. One journalist described the film as nothing but "pure and simple genuine defence propaganda for the vandalism and the destruction of invaluable aesthetic values that is conducted by the power plant building stakeholders."[27]

Indeed, in several of Vattenfall's films from the 1950s a rhetoric resembling the one in *Öden bortom horisonten* can be discerned: They argue for continued exploitation from a perspective that Jonas Anshelm has identified as "industrial fatalism."[28] The term has been coined by sociologist Ulrich Beck and designates a perspective on society and modernization that is perfectly mirrored in the above quote: Conflicts, problems, and dangers created by the progress of civilization, according to this view, cannot be solved; further industrialization is nevertheless necessary – it is seen as a "destiny beyond the horizon" that cannot be influenced by humans.[29]

The most outspokenly pedagogical among these films is *Det nya ansiktet*, which focuses on the aesthetic implications of landscape change caused by

26 Cf. Jacobsson. *Vattenfalls filmproduktion genom tiderna*.

27 Ibid., quoted from *Bildjournalen*, a magazine on film, music, and culture, aimed at younger audiences.

28 Anshelm. *Vattenkraft och naturskydd*, p. 15.

29 Cf. Beck, Ulrich. *Gegengifte: Die organisierte Unverantwortlichkeit*. Suhrkamp, 1988, pp. 103–09.

the construction of hydropower plants. It approaches the issue in a rather daring manner: comparing the construction of dams and power plants with a car accident. In order to do so, the main narrative of the film depicts a loving couple at home, quite fittingly watching a television programme on natural disasters and landscape change. While the commentary explains that also the beauty of a human face can be impaired by forces beyond human control, the plot thickens: Wife and husband say goodbye, she steps into a car and drives off at breakneck speed. The car speeds out of sight and a crashing sound can be heard, immediately followed by images from a blasting at a dam construction site. The film then explains in lengthy sequences how landscaping measures during and after dam construction are undertaken. It regularly cuts back to the couple's story, showing a plastic surgeon planning the operation of the woman's injured face and showing the distressed husband in the hospital. However, the bare and sterile rock landscapes, just like the injury of the woman's face (which is not shown in contrast to the construction sites) form "only a transitional stage" according to the commentary. At the end of the film the couple is happily reunited. The woman's face is different, but no less beautiful than before, so the commentary holds. In the final frame of the film the woman's new face is cross-faded with the new face of the landscape – now a hydroelectric dam – and the title of the film becomes evident.

Many conservationists perceived the conflict between the exploitation and the protection of nature as "truly tragic in complete conformity with the classical definition, that says that tragic is a conflict which cannot be solved without the demise of one party."[30] *Det nya ansiktet* addresses the gravity of this conflict. Through the metaphor of the car accident, it gives expression to the emotionally and even bodily painful impact that the encroachments on the landscape could have. At the same time however, it uses a rhetoric that continuously pulls the tragedy back into normality. This rhetoric relies, on the one hand, on the comparison with naturally occurring landscape changes shown in the TV programme at the beginning of the film – naturalizing the clearly human-inflicted encroachments on the landscape. On the other hand, it relies on the comparison with the inherent risks of modernity, in this instance: driving a car – thereby absorbing the man-made encroachments into a logic of risk. By conceptualizing it as a risk, the tragedy of the encroachment on the landscape is rationalized and takes on commensurability with the positive effects of exploitation. "If

30 Curman, Sigrud. "Landskapsvård vid kraftverksbyggen." *Svensk bygd och natur*, edited by John Nihlén, 2nd ed. Kulturhistoriska förlaget, 1955, p. 413.

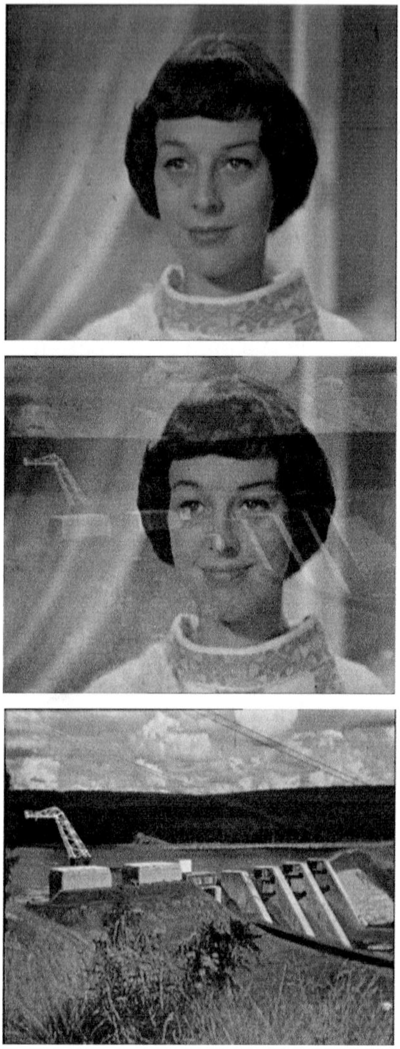

Fig. 7.3a–7.3c. The "new face" as demonstrated in *Det nya ansiktet* (1958).

every intervention is balanced by a positive value," the commentator thus asks, "have we suffered a loss?" Seen in this perspective, landscape change is no longer a "sacrifice"; rather, "we have to sacrifice our one-sided way of seeing," as the commentator demands in an overbearing first person plural. "We must accept interventions in nature in order to obtain what we need from it; just like we have to come to terms with changes that nobody can be blamed for."

The comparisons are obviously flawed; yet, they perfectly demonstrate the "organized irresponsibility" that Ulrich Beck has identified as a central characteristic of industrial fatalism: there is purportedly nobody to be blamed for the exploitation of rivers; it is the individual's task to cope with the changes. In this context, the figures of the plastic surgeon and the landscape architect are central: They embody the rational handling of risks and natural hazards and are thus presented as trustworthy and professional, yet non-responsible figures to help the individual.

While *Det nya ansiktet* focuses on the aesthetics of landscape only, *Öden bortom horisonten* and *Den nya sjön* phrase their call for acceptance in much more general terms. They focus on the economic, social and emotional impacts of the regulation of lakes on those affected. Or, as the commentary in *Öden bortom horisonten* puts it, on the "human factor in the struggle for the kilowatts." Unlike the loving couple in *Det nya ansiktet*, the characters in these two films are not clumsily allegorical figures but believable characters in realistic settings. Both films are explicitly set in northern Sweden, where, as mentioned before, most lake regulations were undertaken during the 1950s, and *Öden bortom horisonten* refers to actual lakes and settlements where the regulation of lakes had taken place, while the characters in *Den nya sjön* have proper names.

Looking at *Öden bortom horisonten,* it is striking that almost the first half of the film is used to embed the call for acceptance in an elaborate account of the history of the Swedish north, spanning from the first settlements after the last ice age to industrialization and the present day. This was unprecedented in Sweden's water power history. The construction of the first state-built water power plant in northern Sweden, at Porjus in the 1910s, took place without any consideration of local concerns or the history of the place. Porjus was simply considered "wilderness," an empty, unhistorical place that would acquire meaning and relevance only through its industrialization.[31] Just as northern Sweden had to be acknowledged as a place with a living *present* and a substantial population after the protests arising in the 1940s, it also started to be appreciated as a place with a living *past* due to new archaeological findings. In 1942, a new law for the care of ancient monuments and relics of the past obliged construction enterprises to finance field surveys if there was reason to believe that the constructions would destroy relevant traces of the national past. This law, together with

31 Cf. Össbo, Åsa, and Patrik Lantto. "Colonial Tutelage and Industrial Colonialism: Reindeer Husbandry and Early 20th-century Hydroelectric Development in Sweden." *Scandinavian Journal of History*, vol. 36, no. 3, 2011, pp. 325–48; this is not to say that colonial policies in northern Sweden ended in the 1950s. Cf. Öhman, May-Britt. *Taming Exotic Beauties: Swedish Hydropower Constructions in Tanzania in the Era of Development Assistance, 1960s–1990s.* KTH, 2007, pp. 52–80.

the intense water power construction activity, sparked archaeological and ethnographical surveys of unprecedented dimensions, whose findings brought a fundamental rewriting of the history of northern Sweden: Apparently many areas, especially along the rivers and lakes, had been populated since thousands of years. The present landscape turned out to be a cultural landscape shaped by Sámi reindeer herding as well as more recent settlements.[32]

Both *Öden bortom horisonten* and *Den nya sjön* use this deep history of northern Sweden to highlight continuity and coherence: Livelihoods have changed before, people have come and gone. Seen from a historical perspective, the latest changes are only another small step in the history of northern Sweden. *Den nya sjön* includes this historical perspective in just one brief sequence in which it shows an archaeological excavation, but it is here that the rhetoric of continuity appears in the most compact and striking manner: The sequence is introduced with the image of a huge construction machine digging up the ground. A match cut then links the machine with the hand of a female archaeologist carefully digging with a small trowel. The sequence ends with a shot of the archaeologists having tea near a rediscovered ancient fireplace. Cutting to the fires in which the cleared vegetation from the lake shores is burned, the film then returns to the construction works, evoking the seamless continuity of past and present landscape uses.

Öden bortom horisonten and *Den nya sjön* have a similar message, but they are designed for different audiences. *Öden bortom horisonten* aims to make the challenges of regulating lakes in northern Sweden understandable to a Stockholm audience unfamiliar with how people live in the north. To do so, it compares this type of encroachment with the redevelopment of Norrmalm, Stockholm's central city, that was started in 1955 and entailed the complete rerouting of inner-city traffic, the extension of the metro network and the replacement of approximately seven hundred historical buildings with a modern skyscraper quarter. *Den nya sjön*, on the other hand, apparently aims at audiences in northern Sweden potentially affected by the regulation of lakes. In this film, people are at the centre of attention.

The first half of the film depicts how the news of a planned lake regulation come to an affected village. The villagers react with scepticism: "One started to guess hither and thither: 'Here will be the water.' 'No, that high!' 'No, even higher! It can be as high as the roof of your house there!' It was

32 Cf. Sundin. "From the Technological to the Antiquarian's Landscape"; Biörnstad. *Kulturminnesvård och vattenkraft 1942–1980*.

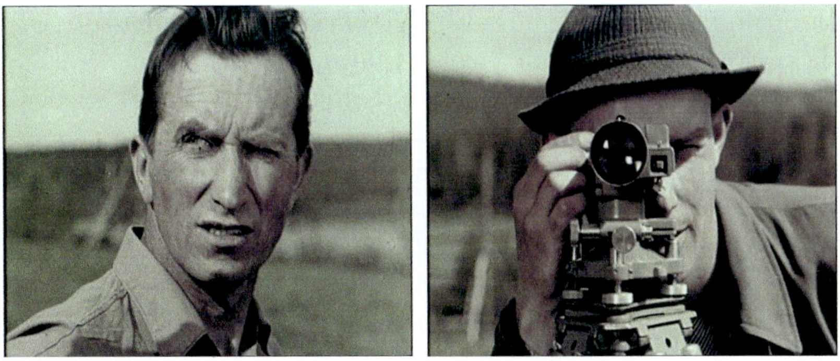

Fig. 7.4. An anxious gaze is met with a measuring eye in *Den nya sjön* (1957).

no wonder that almost everyone was a little anxious." Of course, the film does not stop there: The anxious gaze of one of the villagers shot in close-up that accompanies the above quote is juxtaposed in a match cut (once more) with the measuring eye of one of Vattenfall's land surveyors. The frame introduces a long sequence portraying the detailed process of measuring and describing in which a living landscape is objectified into words, numbers and maps, and eventually filed conveniently in dozens of folders. Yet another match cut concludes the sequence: One of the folders is carefully studied by one of Vattenfall's executives in the planning office, who then puts it back on its shelf, covering up the camera's lens. The folder is then taken out of the shelf again in the affected village, where another copy is held. The folders and the scientific knowledge of the place thus become a token of reliability and mutual trust in a situation of insecurity. Yet the rhetoric of industrial fatalism is close at hand here as well: The decision whether or not the lake is to be regulated is not made by Vattenfall but by the water law court – "impartially," as the film highlights. The water court's proceeding and final decision, however, are deliberately left out of the frame. The court appears only as a fatal power standing outside the human sphere in which Vattenfall and the affected riparian people operate *together*: "What do we know about the future? Not much," the commentary suggests in reference to the imminent court hearing, thereby once more demonstrating Vattenfall's organized irresponsibility.

However, this bleak outlook is immediately absorbed by a positive vision of how the regulation of lakes and the "new times" that have come with them have changed living conditions elsewhere. This vision shows the construction works and the machinery, all accompanied by cheerful music, while the commentary highlights the general advantages of the regulation of lakes (namely "good job opportunities" and "bridges and roads"). However, the main aim of this second half of the film is to praise Vattenfall's local office – "where one can both complain and get help with one's problems." In elaborate before-and-after comparisons, the film presents various "solutions" that have been found for problems with water supply, timber floating, fishing, washing facilities and rehousing – and thereby underscores Vattenfall's capability to adapt to local and individual problems. Comparable to the plastic surgeon or the landscape architect in *Det nya ansiktet,* the local office in *Den nya sjön* is thus depicted as another professional, yet non-responsible help, supporting the individual in better accepting changes in the landscape that are deemed inevitable.

Modern Water Sprites

Let us now turn back to the water sprite and the homonymous film *Strömkarl.* In comparison to the films I have just examined, *Strömkarl* is less openly rhetorical as it deals with the less obviously political topics of production and distribution, leaving construction out of the frame for the most part. Yet, seen against the background of the discourses I have outlined above, it becomes evident that this film too aims to engineer consent by presenting a new narrative of hydroelectric modernity. It is in this context that the figure of the water sprite takes on new allegorical meanings.

Strömkarl presents the company Vattenfall by showing "people and their environment," just like Cederholm puts it in the almost programmatic preface to the eponymous book quoted above.[33] Unlike *Det nya ansiktet, Den nya sjön* or *Öden bortom horisonten, Strömkarl* doesn't focus on the affected people, but on Vattenfall's employees, whom it continuously designates as "modern water sprites." It depicts them in rather idealized ways, as their representation perfectly embodies the "operating culture" (*driftskultur*) that Vattenfall demanded from its employees: According to an instruction handbook first published in 1941, this included "a sense of responsibility, discipline, vigilance and deliberateness," along with "calm" and "sense of order," among others.[34] Accordingly, the film's commentary highlights that a modern power station "is operated almost automatically – but under constant vigilance." And indeed, whenever possible, the machines are shown conscientiously watched over by one or several "water sprites." Among the

33 Cederholm, and Nilsson. *The Power Team.*

34 Kungl. Vattenfallsstyrelsen, editor. *Handbok för driftpersonal vid Statens kraftverk: Del IV: Mätning och skydd, Kraftverksdrift, Säkerhet,* 2nd ed. Stockholm, 1943, p. 381.

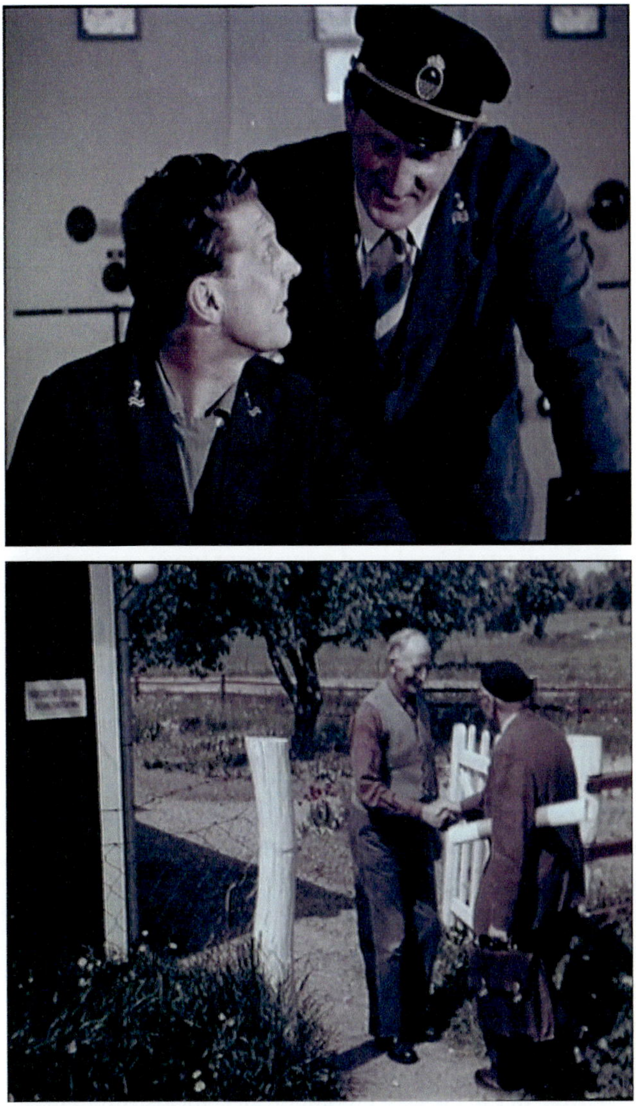

Figs. 7.5 and 7.6. Modern water sprites in *Strömkarl* (1956).

work force, the film hereby aims to foster identification with this working ethos. To a lay audience, it demonstrates that the "water sprite" is a reliable and professional figure – much like the local office or the landscape architect.

However, the visual rhetoric of the film shows much more than serious and professional water sprites. Rather, the film makes an effort to show a friendly, modest, and approachable work force. The employees in the control rooms are depicted as ordinary people (only dressed in nifty blue-yellow uniforms), who chat and laugh – quite contrary to the commentary, which simultaneously claims that the power stations are "operated in silence." While *Strömkarl*'s depiction of electricity distribution almost exclusively focuses on machinery, humans get involved again at the very end of the distribution network: The film shows two elderly men shaking hands in a countryside setting – one of them, the livelier of the two, is an electrical professional from Vattenfall coming to assist a local distribution cooperative, "where they don't possess all necessary expertise"; the other one represents the cooperative and thereby electricity's consumers. In their handshake, electricity and the consumer meet at eye level. This is the helpful water sprite; electricity with a human face.

As we have already seen, *Strömkarl* introduces the water sprite as a musical creature. The opening commentary claims that "the modern water sprite also plays an instrument. He plays on strings that singingly span the bridges of the mountains and the sounding bodies of the lakes." This "instrument" is visualized first by a shot of a generator tower inspected by a Vattenfall staff member and then a shot of power lines spanning over hills in the background and a lake in the foreground. In its closing sequence, the film returns to this musical metaphor. Showing a young Vattenfall employee taking a stroll over a weir, the commentary becomes lyrical: "Whenever we ask for light and power, the water sprite is in position. For him, it might be a lonely routine, but it does happen sometimes that he falls under the spell of this union of natural power and technological magic." The camera then pans to the left, showing the white water below the weir, while the commentary passes into a poem praising electricity as "a servant, quiet and discreet." As it proclaims: "he comes to your house with speech and song," a female hand is shown that turns on the radio and the music from the film's opening scene comes in. The image then cuts to a calm river scene in the evening light and the poem concludes: "And maybe, who knows,/even now as well,/when we've crept into our pleasant downy sheets,/he sings for us,/ the modern water sprite." In the musical metaphor, all comes together in the allegory of the water sprite: Electricity and Vattenfall's staff – both servants, "quiet and discreet"; the singing of the waterfall and the music from the electric radio; the water sprite's fiddle and the technical landscapes of the power lines. Technology, humans and nature are harmoniously united as the company Vattenfall, its employees, electricity, landscape and music all converge in the allegory of the water sprite.

The water sprite sculpture at Trollhättan had been a sublime, extraordinary figure. It can be read as an allegory on the first half-century of the industrialization of Swedish waterfalls, during which the country's modernization was often perceived and fashioned in the dramatic forms of the sublime. Several historians have noted that the post-war years, in contrast, saw a drastic change in the aesthetics of hydroelectricity. Lasse Brunnström has stated that an "anonymization" of hydropower stations took place from the 1950s on, both in terms of design and in terms of their cultural significance. He traces this back to economic reasons, an increasing mundanity of electricity, as well as to the closing of the national grid that made the individual station lose its singular importance.[35] Sverker Sörlin and Christer Nordlund have made the complementary argument that during the 1950s the planners' and the public interest turned from the individual station to landscaping and to the social context of electricity production, as a manifestation of Swedish welfare ideals.[36] The revised narrative of modernity that I have traced in Vattenfall's films resonates well with these findings: The modern water sprite is no longer a sublime figure to be gazed at by overwhelmed humans. Rather, he is embodied by humans themselves, engaging with one another in changing environments. The modern water sprite announces the departure to a more mundane modernity that is no longer captivated by a dramatic conquest of nature but caught by the problems it has created itself.

Works Cited

Anshelm, Jonas. *Vattenkraft och naturskydd: En analys av opinionen mot vattenkraftutbyggnaden i Sverige 1950–1990*. Tema teknik och social förändring 18, Linköping University Press, 1992.

Beck, Ulrich. *Gegengifte: Die organisierte Unverantwortlichkeit*. Suhrkamp, 1988.

Biörnstad, Margareta. *Kulturminnesvård och vattenkraft 1942–1980: En studie med utgångspunkt från Riksantikvarieämbetets sjöregleringsundersökningar*. Kungliga Vitterhetsakademien Historie och Antikvitets Akademien, 2006.

Björkin, Mats. *Post-war Industrial Media Culture in Sweden, 1945-1960: New Faces, New Values*. Amsterdam University Press, 2022.

Brunnström, Lasse. *Estetik och ingenjörskonst: Den svenska vattenkraftens arkitekturhistoria*. Riksantikvarieämbetet, 2001.

Cederholm, Charlie, and Lennart Nilsson. *The Power Team: A Chronicle of the Swedish State Power Board, Its Background and People: Jubilee Year 1959*. Stockholm, 1960.

Cederholm, Charlie, and Lennart Nilsson. *Strömkarl: En krönika om vattenfall, dess miljöer och människor jubileumsåret 1959*. Stockholm, 1959.

35 Brunnström. *Estetik och ingenjörskonst*, p. 238.

36 Sörlin, and Nordlund. "Modernizing the National Landscape," pp. 306–07.

Curman, Sigrud. "Landskapsvård vid kraftverksbyggen." *Svensk bygd och natur*, edited by John Nihlén, 2nd ed. Kulturhistoriska förlaget, 1955, pp. 413–39.

Forsgren, Nils. *Vattenfall visar: Glimtar ur filmarkivet*. Vattenfall AB, 1984.

Gunning, Tom. "Before Documentary: Early Nonfiction Films and the 'View' Aesthetic." *Uncharted Territory: Essays on Early Nonfiction Film*, edited by Daan Hertogs and Nico de Klerk. Stichting Nederlands Filmmuseum, 1997, pp. 9–24.

Hansen, Fredrik Vilhelm. "Foreword." *Trollhättan: Dess kanal- och kraftverk. Historik och beskrifning*, vol. 1, edited by Kungl. Vattenfallsstyrelsen, 1911.

Harnesk, Paul, editor. *Vem är vem: Stor-Stockholm*, 2nd ed. Bokförlaget Vem är Vem AB, 1962.

Jacobsson, Lars. *Vattenfalls filmproduktion genom tiderna*. December 17, 2015. Unpublished manuscript.

Jakobsson, Eva. *Industrialisering av älvar: Studier kring svensk vattenkraftutbyggnad 1900–1918*. Avhandlingar från Historiska institutionen i Göteborg 13. Historiska institutionen, 1996.

Jonsson, Helge, Tor Eklind, and Agneta Nilsson. "Ivan Christoferson, till mine." November 9, 2010. *Allehanda*, www.allehanda.se/slakt-o-vanner/dodsfall/ivan-christoferson-till-minne. Accessed May 11, 2017.

Josephsson, Erik. "Statens kraftverk vid Trollhättan." *Teknisk Tidskrift: Arkitektur*, vol. 41, 1911, pp. 60–66.

Krohn Andersson, Fredrik. *Kärnkraftverkets poetik: Begreppsliggöranden av svenska kärnkraftverk 1965–1973*. Eidos 24, Stockholm University Press, 2012.

Kungl. Vattenfallsstyrelsen. *Statens Vattenfallsverk under fyra decennier*, 2nd ed. Stockholm, 1948.

Kungl. Vattenfallsstyrelsen, editor. *Handbok för driftpersonal vid Statens kraftverk: Del IV: Mätning och skydd, Kraftverksdrift, Säkerhet*, 2nd ed. Stockholm, 1943.

Nilsson, Kjell. "Where Industry Meets Nature: How Public Concern Has Influenced the Design of Swedish Industrial Landscapes during the 20th Century." *Landscape and Urban Planning*, vol. 23, no. 1, 1992, pp. 33–45.

Nye, David E. *American Technological Sublime*. MIT Press, 1994.

Öhman, May-Britt. *Taming Exotic Beauties: Swedish Hydropower Constructions in Tanzania in the Era of Development Assistance, 1960s–1990s*. KTH, 2007.

Össbo, Åsa, and Patrik Lantto. "Colonial Tutelage and Industrial Colonialism: Reindeer Husbandry and Early 20th-century Hydroelectric Development in Sweden." *Scandinavian Journal of History*, vol. 36, no. 3, 2011, pp. 325–48.

Sörlin, Sverker, and Christer Nordlund. "Modernizing the National Landscape." *Kenyon Review*, n.s., vol. 25, no. 3–4, 2004, pp. 301–15.

Stattin, Jochum. *Näcken: Spelman eller gränsvakt?*, 2nd ed. Carlsson, 1992.

Sundin, Bosse. "From the Technological to the Antiquarian's Landscape: Hydroelectric Power and the Design and Discovery of the Landscape." *Icon*, vol. 10, 2004, pp. 94–109.

Sundin, Bosse, and Richard Pettersson. "Sigurd Curman, Nämforsen och det 'kultiverade kraftverket'." *När dalen var röd och kusten blev hög*, edited by Erland Mårald, Skrifter från forskningsprogrammet Landskapet som arena 10, Umeå, 2008, pp. 103–37.

Svensk Filmdatabas, sfi.se/sv/svensk-filmdatabas. Accessed 19 May 2017.

Svensson, Ove. *Carl Eldhs Strömkarlen: Symbolen, myten och motivet: En berättelse om Strömkarlens tillblivelse och gestaltning*. Kraftverkshistoria. Vattenfall Västsveriges Pensionärsförening, 1999.

Vedung, Evert, and Magnus Brandel. *Vattenkraften, staten och de politiska partierna*. Nya Doxa, 2001.

Zimmer, Fabian. *Hydroelektrische Projektionen: Eine Emotionsgeschichte der Wasserkraft im Industriefilm*. Deutsches Museum. Abhandlungen und Berichte, n.s., vol. 36. Wallstein, 2022.

Zimmermann, Yvonne. "The Power of Flows: The Spatiality of Industrial Films on Hydropower in Switzerland." This volume.

About the Author

Fabian Zimmer is a postdoctoral researcher at Technische Universität Berlin. He works at the intersection of cultural history, environmental history, and the history of technology. After studying at the Universities of Heidelberg and Lund, he was a doctoral student at Deutsches Museum, the Rachel Carson Center and LMU Munich, where he earned his PhD in 2020. His first book on the PR and film work of hydropower companies in 1950s Europe was published with Wallstein in 2022.

8 Taxonomy of Techniques

Visions of Industrial Cinema in Post-war Japan

Takuya Tsunoda

Abstract

This chapter, a case study of the *Sakuma Dam* series, presents three interlocking points: its geopolitical imbrication with state-endorsed modernization and democratization; a genealogy of science film as motion photography of life; and a seemingly incongruous connection between industrial film and the legacy of post-war leftist filmmaking. The central argument revolves around investigating industrial film as a point of convergence between the increasingly technical conditions of existence and a material understanding of cinema. The dual functions of technologies as infrastructural innovations and cinematic mediation techniques point toward epistemological inquiries about cinema as technological phenomenon. A taxonomy of techniques, an epistemic and reflexive process that transformed this dual function into an object of analysis, was entwined with the post-war resurgence of science film production in Japan.

Keywords: Japan; Japanese cinema; documentary; industrial film; tacit knowledge; useful cinema

"[A]s it is also a film about men of our time – contemporaries whom a short plane ride would enable us to meet face-to-face – Kurosawa is entitled to draw upon world cinematic language," writes André Bazin in his review of Akira Kurosawa's *Ikiru* (1952) that appeared in *Cahiers du cinéma* in March 1957, immediately following the Kurosawa retrospective at the Cinémathèque.[1] Bazin's literal reference to an aerial voyage typifies the

1 Bazin, André. "Petit journal du cinéma: *Vivre*." *Cahiers du cinéma*, vol. 69, March 1957, p. 2; idem, *The Cinema of Cruelty: From Buñuel to Hitchcock*, edited by François Truffaut, translated

Hediger, V., F. Hoof, Y. Zimmermann, with S. Anthony (eds), *Films That Work Harder: The Circulation of Industrial Film*. Taylor & Francis Group 2024
DOI: 10.5117/9789462986534_CH08

critic's enthusiasm for experiencing and discovering the world through cinema, perhaps with a little bit of "jetlag." Japanese films indeed reached Europe "like that of light from distant stars,"[2] when Kurosawa's *Rashomon* (1950) took the stage at the 1951 Venice International Film Festival, followed by Kenji Mizoguchi's *The Life of Oharu* (*Saikaku ichidai onna*, 1952), *Ugetsu* (*Ugetsu monogatari*, 1953), and *Sansho the Bailiff* (*Sanshō dayū*, 1954), which split the top prize with Kurosawa's *Seven Samurai* (*Shichinin no samurai*, 1954) again at Venice.

The year 1954 was also when Ishirō Honda's *Godzilla* (*Gojira*), a science fiction giant monster spectacle, was produced and released by Tōhō Studios – the same studio where Kurosawa built his career. The "nuclear dinosaur" soon gained iconic status internationally when the re-edited Hollywood version, *Godzilla, King of the Monsters* (1956), hit the American and global markets in the mid-1950s. While the future Japanese canon started to comprise an atlas of world cinema, domestic audiences in Japan were infatuated with cinematic "remapping" of the nation and its geographical transformation. *Sakuma Dam* (*Sakuma damu*, 1954–1958), an unprecedentedly massive-scale industrial film series that documented the construction of Japan's largest dam on the Tenryū River Valley in Aichi Prefecture, was a record-breaking hit as a feature theatrical documentary and drew nearly six million viewers across the country. The series was the first Japanese feature documentary shot with the Eastmancolor and it laid the foundation for what can be described as the non-fiction construction spectacle, a commercially viable industrial feature aimed at theatrical release.

Takeji Takamura, a young neophyte documentarian with in-depth knowledge of jurisprudence, was talent-spotted to direct the series, which was fully commissioned by the Electric Power Development Company (Dengen kaihatsu, or Denpatsu in short), the nation's largest public utility established in 1952 as Japan was about to resume its sovereignty at the end of the Allied occupation.[3] The crew members of the film also included Noriaki Tsuchimoto and Kazuo Kuroki, who would become the central figures of the post-war new cinema movement in Japan. Crucially, the documentary was produced by Iwanami Productions (Iwanami Eiga Seisakusho, est. 1950), which evolved from a major provider of sponsored educational and

by Jane Pease and Rose Kaplan. Seaver Books, 1982, pp. 200–204. Quote from p. 202.

2 Bazin, *The Cinema of Cruelty*, pp. 184–85.

3 The foundation of Denpatsu was the pre-war government monopoly Japan Electric Generation and Transmission Company (Nippon Hassōden), which had been dissolved under the order of GHQ as part of democratic reform of infrastructure.

PR films into a key player in the new cinemas of the 1960s in Japan. To put it differently, *Sakuma Dam* points toward pivotal socio-historical nodes between cinema and its local/global markets, moving image medium and its role in stirring the popular imagination, and the politics of industrial enterprise and the new cinemas' political radicalism.

This chapter examines several nexuses of concern about industrial film related to the ongoing transformation occurring across a broad spectrum of media technologies as well as social and infrastructural developments in post-war Japan. Taking up *Sakuma Dam* as a case study that both typifies and diverges from traits and functions of sponsored films in Japan, the chapter presents four interlocking points as a way to position industrial film in relation to its popularity with contemporaneous audience, geopolitical imbrication with state-endorsed modernization and democratization, a genealogy of science film as motion photography of life, and seemingly incongruous connection between industrial film and the legacy of post-war leftist filmmaking. Against an overly schematized concept of sponsored films as regulatory and manipulative media works for top-down governance and advertisement, I highlight larger questions concerning cinema, (filmic) technique and systems of knowledge formation. The central argument revolves around an investigation of industrial film serving as a point of convergence between increasingly technical conditions of existence in post-occupation Japan and a material understanding of cinema, which called into being (or "engineered") techniques that enabled viewers to interface with cinema as a technical image medium. Films on infrastructural developments, often reinforced by various film techniques, rendered and showcased the nation in the process of technologizing itself. The dual function of technologies as both infrastructural innovations in content and techniques of cinematic mediation points toward epistemological inquiries about cinema as technological phenomenon. What I describe as a taxonomy of techniques, an epistemic and reflexive process that transformed this dual function into an object of analysis, was entwined with the post-war resurgence of science film production in Japan.

"The 'Post-war' Is Already Over"

In 1956, the introductory chapter of the Japanese government's official *Economic White Paper* (*Keizai Hakusho*) famously included the statement: "The 'post-war' is already over" (*"mohaya 'sengo' de wa nai"*). It had been widely circulated as a trendy phrase that marked the symbolic end of the post-surrender economic recovery in Japan. Indeed, 1955 was the year the annual gross national product (GNP) surpassed the pre-war peak for the first time.[4] The strong economic ties between the US and Japan around this period led to the establishment of the Japan Productivity Center (Nihon seisansei honbu), a crucial long-range industrial planning organization that soon became the major sponsor of global-scale technical missions. The year 1955 also saw Japan's admittance to the General Agreement on Tariffs and Trade (GATT). Japan was no longer recovering from the war's devastation, but growing towards the future.

The era's upbeat spirit, which was commonly shared by the populace in Japan, was also defined by industrial investment in costly new technologies – literally becoming the engine of growth. The term *kaihatsu*, or "[technical, industrial and infrastructural] development," symbolized the improvement of people's lives, and was closely tied to the advent of post-war consumer culture. Accordingly, the mid-1950s saw popular home appliances dominating the markets; electric washing machines, refrigerators and televisions were especially revered as the "three divine appliances" (*sanshu no jingi*) by every middle-class family who dreamed of owning them.[5] The accelerated technical innovations of the nation's high-growth period prompted the formation of mass consumption culture. Referred to as the "age of household electrification" (*katei denka no jidai*), the period was nourished by the rapid development of hydroelectric and nuclear power stations. The industrial efforts by Japanese governmental bodies and corporations were thus directed toward dam and nuclear power plant construction, further fuelling the consumer revolution of the time. From a macro perspective, the public-private partnership that endorsed electrification of the nation motivated Japan's post-war (technological) modernization. From a micro standpoint, the increasingly technologically *wired* household was becoming an individually configured infrastructure whose connecting network was ensured by electricity.

The rapid electrification of urban life in Japan instigated the emerging boom of industrial films. In 1953 the National PR Film Festival (Zen Nihon PR Eiga Konkūru) was inaugurated. This newly established venue accepted submissions not from production companies but from the sponsoring institutions and corporations. The juries mostly comprised financially powerful names in and around the industries, including governmental officials, ad

4 For insightful historical accounts on Japan's post-war economic recovery, please see essays compiled in Gordon, Andrew, editor. *Postwar Japan as History*. University of California Press, 1993.

5 Dower, John W. "Peace and Democracy in Two Systems." Gordon, *Postwar Japan as History*, edited by Andrew Gordon. University of California Press, 1993, p. 16.

agency or film studio executives, and university professors of economics.[6] While the adjectival use of "PR" here was inclusive, counting in industrial documentaries as well as educational/training/instructional films, the concept of "public relations" in post-war Japan was introduced and evolved in tandem with the occupation policies administered by the US-led Allied Forces (1945–1952) in order to nurture Japan as a Cold War ally. The English acronym "PR" was thus imbued with a newly instituted notion of the "public" that formed the basis of democratization, the security industry, educational reform and information management both at local and national levels. The number of industrial and educational films produced in post-occupation Japan continuously rose toward a peak with an annual total of 970 titles in 1960.[7] And the unprecedented popularity of industrial documentaries was further intensified by the concurrent dam developments, which had been one of the favoured subject matters of PR films produced by various documentary studios of the era.[8] Most of these films were disseminated widely through commercial and non-commercial channels of distribution, including theatres, trade fairs and school auditoriums. The "age of household electrification" was also the "age of massive dam construction" ("*daidamu kensetsu no jidai*"),[9] whose images and narratives – photographs, postal stamps, brochures and other cultural productions and practices – constituted crucial mediascapes occurring in the period.

Making Democracy Visible

Located in the central region of Japan's main island, Sakuma Dam was then the largest post-war public works project in Japan and was a symbolic

6 Yoshihara, Junpei. *Nihon tanpen eizōshi: Bunka eiga, kyōiku eiga, sangyō eiga* [History of short film in Japan: Culture film, educational film, industrial film]. Iwanami Shoten, 2011, p. 195.
7 Kusakabe, Kyūshirō. *Eizō o tsukuru hito to kigyō: Iwanami Eiga no sanjūnen* [The man and the studio that produced the image: Thirty Years of Iwanami Productions]. Mizuumi Shobo, 1980, p. 110.
8 Kiroku Productions (Kiroku Eigasha) made *Ikari Dam* (*Ikari damu*) in 1952 and the film was re-edited in 1956. In 1954, Tokyo Cinema (Tokyo Shinema) also produced *Awano Village* (*Awanomura*), a colour short that documented the electrification of a village in Fukushima. The film was sponsored by Tohoku Electric Power (Tōhoku Denryoku), a newly founded electric utility owned by the government. Hanabusa Productions (Hanabusa Eiga) put together *Ikawa Goro Dam* (*Ikawa Gorō damu*) in 1957.
9 Machimura, Takashi. "'Kaihatsu' o egakanai kaihatsu eiga" ["The construction film that does not represent construction"]. *Iwanami eiga no 1-oku furēmu* [Images of post-war Japan: The documentary films of Iwanami Productions], edited by Yoshiyuki Niwa and Shun'ya Yoshimi. Tokyo, 2012, pp. 123–43.

hydroelectric power development whose potential had been explored since the beginning of the twentieth century. What had previously been an impossible blueprint began to move forward when the aforementioned Electric Power Development Company (or Denpatsu) was established as a governmental agency in 1952 under the supervision of GHQ (General Headquarters, lead by Douglas MacArthur), and through foreign aid loans from the United Nations.[10] The five-year construction involved the relocation of nearly three hundred households located along the river valley. The ceremony that celebrated the completion of the dam in October 1957 was attended by a number of government officials and members of the imperial family, including Emperor Hirohito himself. The Sakuma Dam project has been often compared to the Tennessee Valley Authority (TVA) in the 1930s United States, a federally owned corporation (est. 1933) known as America's effort to facilitate the idea of public ownership of utilities and to boost economic development during the Great Depression.[11]

The unparalleled commercial success of the documentary *Sakuma Dam* in part derived from the unprecedented scale of the sponsored film project, whose shooting schedule alone had spanned nearly five years. *Sakuma Dam* consists of a series of feature length films, each of which was theatrically released separately "along with" the progress of construction itself: *Chapter One* (*Dai ichibu*) in 1954, *Chapter Two* (*Dai nibu*) in 1955 and *Chapter Three* (*Dai sanbu*) in 1957. A ninety-six-minute re-edited version was released as *Sakuma Dam: Highlights* (*Sakuma damu: sōshū-hen*) in the following year and was also distributed internationally with English narration.[12] The serialized releases situated contemporaneous audiences as witnesses, about which I will return later in this chapter, of the gradual and ongoing building process of the monumental dam rather than the retroactive – and thus narrativized – recounting of an existing architecture. For this reason, the *Sakuma Dam* saga, especially *Chapter One* and *Chapter Two*, in which the construction had not yet been completed, was structurally distinctive from the great majority of other industrial documentaries, in which a building record would conclude typically with the completion of a project.

10 The foundation of Denpatsu was the pre-war government monopoly Japan Electric Generation and Transmission Company (Nippon Hassōden), which had been dissolved under the order of GHQ as part of democratic reform of infrastructure.

11 Toba, Koji. "Damu kaihatsu to kiroku eiga" ["Dam development and documentary"]. *Iwanami eiga no l-oku furēmu* [Images of post-war Japan: The documentary films of Iwanami Productions], edited by Yoshimi Shun'ya. Niwa Yoshiyuki pp. 145–62.

12 The shot analyses in the following sections are from the 1958 version, unless otherwise noted.

The absence of narrative enclosure also characterizes the cinematic rendition of human actants, especially the image of the relocated villagers. As was indicated earlier, the dam affected three hundred households located in the vicinity of the construction site.[13] And *Sakuma Dam* points toward their existence; it does so, however, by soberly filming and "studying" the villages and their customs, keeping the villagers within their disappearing ecological networks of livelihood. As if conducting data visualization for an environmental survey, the camera records the "indigenous" culture of Tomiyama village, which is about to be submerged permanently. Such a paucity of human elements in *Sakuma Dam* can be deciphered as an indication that the documentary epic was an embodiment of capitalism that concealed the condition of the suffering. However, sociologist Takashi Machimura points out that the face of the victimized was, paradoxically, one of the common tropes typically used in numerous industrial films of the period in Japan. Especially for dam constructions, the sacrifice made by dislocated villagers was often an integral part of narrativizing the industrial procedure. The votive decision of the weak – for the greater good and to electrify Japan – was glorified as the "icon of industrial development," through which the importance of exploitation would be further dramatized and reinforced.[14]

To make the victimized visible was in fact inherent to the many sponsoring organizations legitimizing industrialization. The image of the dislocated also easily fit into the newly introduced post-war notion of the "public" as well as democratic ethos, wherein the citizen's consensus would preside over private interests.[15] In other words, visualizing the villagers rendered democracy visible. As noted above, the term PR entailed geopolitical implications of post-war Japan, in which an emerging notion of the public was highly imbricated with (America-led) governmental techniques in administering and managing civic and corporate bodies.[16] And various images of the sacrificial were often ingrained in the organizational logic

13 For the detailed account of the Denpatsu's negotiation with Toyone Village, Tenryū Village, and Tomiyama Village (as well as governmental interventions), please see Machimura, Takashi, editor. *Kaihatsu no jikan, kaihatsu no kūkan – Sakuma damu to chiiki shakai no hanseiki* [Time of industrial growth, space of industrial growth – Fifty years of Sakuma Dam and its region]. Tokyo Daigaku Shuppankai, 2006.

14 Machimura. "'Kaihatsu' o egakanai kaihatsu eiga," p. 129.

15 Please see Upham, Frank K. "Unplaced Persons and Movements for Place." *Postwar Japan as History*, edited by Andrew Gordon. University of California Press, 1993, pp. 325–46.

16 Please see Ha, Kyung Jin. *Paburikku rirēshonzu no rekishi shakaigaku* [Historical sociology of public relations]. Iwanami Shoten, 2017.

of industrial films. In a broader context, the logic and rhetoric of a victimhood that reaffirmed the position of post-surrender Japan as war victim also prevailed in cultural and political fields. So-called post-war "victim consciousness" (*higaisha ishiki*) foregrounded the suffering to obscure or erase the question of responsibility for the wartime imperialist aggression toward other Asian nations. Such collective amnesia was characteristic not only of the political discourse endorsed by the conservative leadership but of various cultural works and activities by progressives, including a group of leftist filmmakers in the 1950s as the Japanese Communist Party was itself going through factional struggles.

The post-war Japanese mainstream film industry in the 1950s was not immune to a red purge carried out under McCarthyism, which led to an independent production boom formed and sustained by the personnel expelled from the major studios.[17] Exemplified by socialist realist aesthetics and a teleological narrative, these independently produced fiction films, many of which garnered critical acclaim from various media, often centred around exposing the perceived "reality" and its contradictions in post-war society, such as social injustice and the plight of the impoverished. Underlined was, however, the legitimacy of self-victimization fuelled with the dramatic rendition of anti-authoritarian sentiment, which was then rigorously critiqued by an emerging generation of the Japanese New Wave players, such as Nagisa Ōshima. The gradual dissolution of post-war socialist realism in part corresponded to a global shift from the old left to the new left. Curiously, a rhetorical pattern that displayed social-public benefits in industrial film grew in tandem with the political ambitions of the immediate post-war left in Japan. And it should be stressed that *Sakuma Dam*, which bore neither the self-enclosing narrative nor the iconic image of the self-sacrificial heroism of the dislocated, deviated from both the dominant tradition of the industrial film and the old left, and instead possessed a peculiar rapport with the uncompromising politics of the radicalized new left.

Destruction, Excavation, Cartography and Deixis

Rather than the victimized, *Sakuma Dam* is characteristically dominated by the recurring images of technologies and their arrangements/utilization: a variety of monstrous machines –dump trucks, excavators, power shovels

17 For a concise history of Japanese cinema, please see Yomota, Inuhiko. *What Is Japanese Cinema? A History*. Translated by Philip Kaffen. Columbia University Press, 2019.

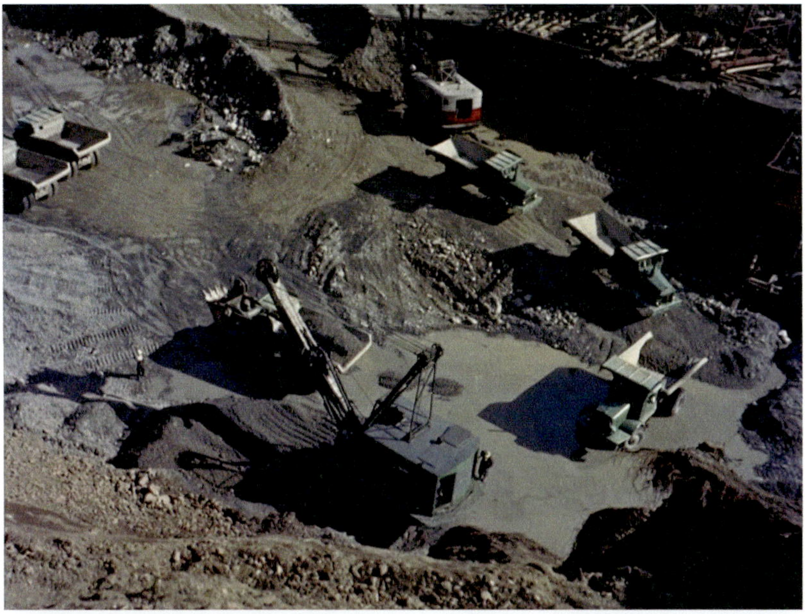

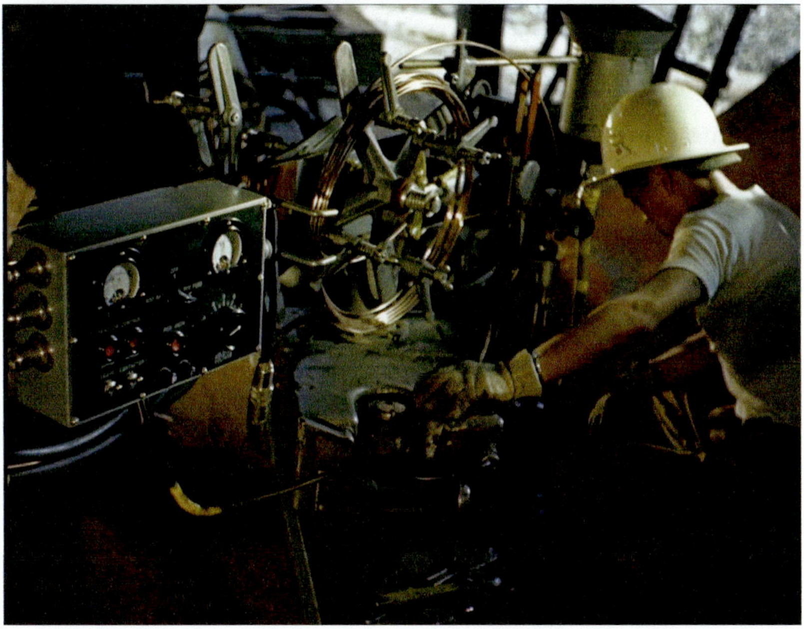

Figs. 8.1 and 8.2. From *Sakuma Dam*. Picture courtesy Documentary Film Preservation Center, Japan.

as well as various control apparatuses mostly imported from the United States – and tunnel blasting (figs. 8.1 and 8.2). One of the climactic spectacles of the documentary series revolves around the blocking of the Tenryū River and a series of explosions due to blasting to create drain tunnels to divert its flow, that is to say, destruction and excavation. The watercourse, which once had the nickname "Unruly Tenryū" (*abare Tenryū*), meandered through deep forests in the Japanese Central Alps, and blocking it off required exhausting and repetitive reclamation work. The documentary underscores the assiduous and destructive procedure of blasting and scraping nearby fields and transporting sediment. The film does not erase the repetitive nature of a series of steps that demanded mechanical precision and technical sophistication.

The governing organization principle of the sequences that highlight the destruction sites is temporal, a characteristic Tom Gunning and Salomé Aguilera Skvirsky point out as a major representational strategy in "process films," which typically detail "the successive stages of industrial or handicraft manufacturing."[18] The gradual progress, which alone consumed nearly two years, comes to a climax with the very moment of the sealing off of the river. This instant of geographic transformation is also underlined with narration – "March 28, 1954, the Tenryū River halted its flow" – that (re)addresses its permanent change to the "witnessing" of viewers with a bird's-eye perspective, which also enframes the diegetic workers as onlookers (fig. 8.3). Throughout the film, the voice-over narrates the passing of time in a manner reminiscent of dating a legal document; accordingly, the spectatorial engagement with the film as both witness and record endows this cathartic moment of the shutting down of the river with a juridical effect. To put it differently, the indexical reference to the geographical transformation "in motion" is open to both the past and the present. The two temporalities inscribed in the moving image are further indexed with the film's expository narration that frequently uses singular proximal demonstratives – such as "this" and "that" – to direct the viewer to the image. The indexical status of the cinematic image (in the semiotics of Charles Sanders Peirce) operates both as trace and deixis, which signal both the corroboration of an existence from the past (i.e. trace) and a deictic pointer to its referent (whose meaning is contingent upon each contextual situation).[19] *Sakuma Dam*'s juridical form of address enables the viewer to witness the moment "here and now."

18 Gunning, Tom. "Before Documentary: Early Nonfiction Films and the 'View' Aesthetic." Uncharted Territory: Essays on Early Nonfiction Film, edited by Daan Hertogs and Nico de Klerk. Stichting Nederlands Filmmuseum, 1997, pp. 9–24. Quote from p. 13. Please also see Aguilera Skvirsky, Salomé. "The Latin American Process Film." This volume.
19 Doane, Mary Ann. "The Indexical and the Concept of Medium Specificity." *Differences*, vol. 18, no. 1, 2007, pp. 128–52, esp. pp. 136–41.

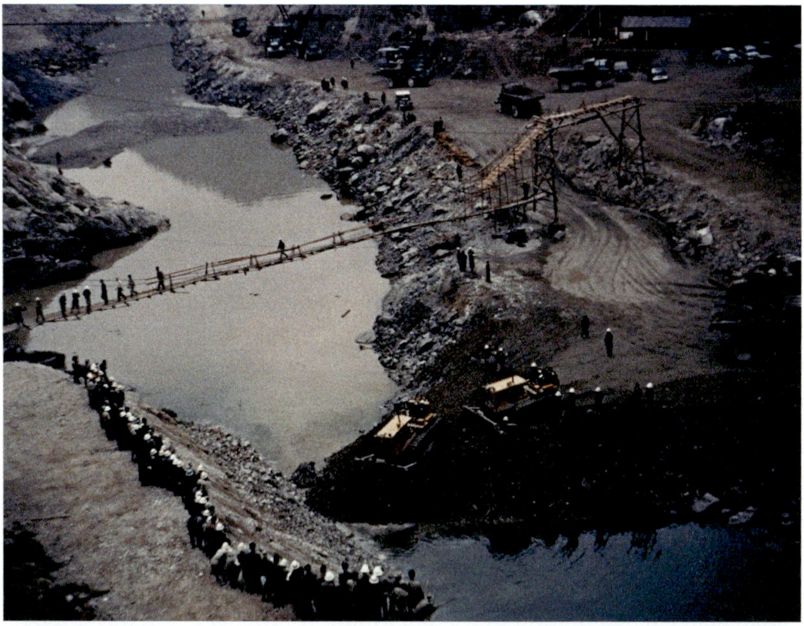

Fig. 8.3. Blocking and diverting the Tenryū River in *Sakuma Dam*. Picture courtesy Documentary Film Preservation Center, Japan.

Similarly, the cathartic closure of the river does not reveal itself as a narrative closure, which would be anchored firmly by an absorptive experience of cinema; it rather shares its affinity with what Jane Gaines calls as "the pathos of fact" surrounding an inherently indexical documentary image, whose evidential power of material condition animates the spectator's affective circuit of the film viewing experience.[20] Developing after Eisenstein's notion of "political pathos," Gaines explicates how "[s]ites of suffering" can be "rendered as *sights* of suffering" in indexed image, which in return *moves* and galvanizes the spectator's political consciousness and commitment "to transform the world."[21] Unlike the formulaic political yearning of the socialist realist views, for Gaines this oscillation between *evidence* and *aspiration* serves as a pivotal point of contact between melodrama(tization of documentary) and (photographic) realism, or a pathos-triggered affective

20 Gaines, Jane. "Documentary Radicality." *Revue canadienne d'études cinématographiques/ Canadian Journal of Film Studies*, vol. 16, no. 1, 2007, pp. 5–24.

21 Ibid., pp. 6, 9. Also, please see Zhen, Zhang. "Toward a Digital Political Mimesis: Aesthetics of Affect and Activist Video." *DV-Made China: Digital Subjects and Social Transformations after Independent Film*, edited by Zhang Zhen and Angela Zito. University of Hawai'i Press, 2015, pp. 316–46.

dimension of viewing bodies often sidelined or obscured under the banner of the "critique of realism."[22]

Gaines' analysis further foregrounds a politically charged experiential connection between the viewer and the moving image, in which any given "moment" is constantly turned "into its progression" and which allows one to trace "that moment as it is made to move."[23] Her primary intent here is to reposition the rhetoric of radical politics within the long-standing leftist documentary filmmaking practice and discourse, and the moving image would "come back out again" as *movement* on the bodies of energized spectators, *mobilizing* actions in and on the real world.[24] *Sakuma Dam*, a government-sponsored PR feature for theatrical release, does not fall into the category of radical documentary, deliberately avoiding presenting the site as "suffering" and taking little notice of the dislocated. Yet, it is crucial that the film that promotes hydroelectric power generation makes the viewer engage with waves of semiosis, signifying the material conditions of the infrastructure in the making. Despite its ostensive political and institutional goal to show off the grandeur of power and technology, the viewer's participation in the documentary image – and its cathartic "enhancements" – operates as sense testimony.

The non-human scale of the infrastructural manoeuvres also makes the documentary close to Ishirō Honda's *Godzilla* (*Gojira*, 1954), the aforementioned monster spectacle released the same year. Importantly, *Sakuma Dam* was scored by Akira Ifukube, an acclaimed composer whose best-known works include none other than the soundtrack of *Godzilla*. Chon A. Noriega, in his psychoanalytic study of the Godzilla franchise, illustrates how the Japanese has tend to sympathize with the destructive dinosaur "as a tragic hero" beyond a "comical icon" widely circulated and reappropriated in popular culture in a global scale.[25] Stuck in the Cold War geopolitical staging of military-economic governance, Japan in the 1950s was well under the United States' nuclear presence, epitomized by the *Lucky Dragon* incident, in which a Japanese tuna boat (named *Daigo Fukuryū Maru* [Lucky Dragon 5]) was hit by the fallout of US H-bomb testing at Bikini Atoll in March 1954. Godzilla's emergence, which corresponded to this shocking "return of the repressed" in post-occupation Japan, can be psychoanalytically situated as a "therapeutic" re-enactment of

22 Gaines. "Documentary Radicality," p. 6.

23 Ibid., p. 12.

24 Ibid., p. 12.

25 Noriega, Chon A. "Godzilla and the Japanese Nightmare: When *Them!* Is US." *Hibakusha Cinema: Hiroshima, Nagasaki and the Nuclear Image in Japanese Film*, edited by Mick Broderick. Kegan Paul International, 1996, pp. 54–74.

trauma. The radioactive dinosaur is not merely the unfathomable Other to be annihilated but a symbolic object that exhibits the very process of dealing with the repressed anxieties. Noriega contends that Japan in 1954 is itself "a transitional monster caught between the imperial past and the post-war industrial future, aroused by United States H-bomb tests."[26]

If *Godzilla* integrates its genre trope – the monster spectacle of mass destruction – into the therapeutic re-enactment of the problematic in an attempt to cope with trauma, the cathartic witness of the river closing in *Sakuma Dam* is also participation of different sort, which allows one to engage with the duality of infrastructures, that is to say, magnificent infrastructural materials that comprise the substrata of the phenomenal world and a larger infrastructural system in the making. Brian Larkin describes this duality as the "peculiar ontology" of infrastructures, which are fundamentally "things and also the relation between things."[27] The film about electrification of the nation makes the materials and foundations of the infrastructure and its facility (iron, soils, cable wires and so on) visible and sensible, simultaneously situating the viewer in the midst of ongoing infrastructural constructions. The revealed process of "system building" here was also entangled with one's spatio-temporal experience of cinema, another institutionalized structure or *dispositif* that encompasses a heterogeneous constellation of film, built space, social, political and discursive practices whose configuration is historically specific. Japan in 1954, as a transitional monster, was inseparable from this disposition, in which cinema as an aggregation of various cluster of mechanisms, infrastructural making in transition and the viewers (as actants) were themselves comprising elements.

The Method of Science Film, Taxonomy of Techniques and the Politics of Scale

Noriaki Tsuchimoto, one of the most prominent documentary filmmakers in post-war Japan, laid stress on *Sakuma Dam* in his 1988 essay on documentary history, characterizing the film as a magnum opus, whose "method" would align itself with that of "science film" (*kagaku eigateki shuhō*).[28] Tsuchimoto,

26 Ibid., p. 62.
27 Larkin, Brian. "The Politics and Poetics of Infrastructure." *Annual Review of Anthropology*, vol. 42. 2013, pp. 327–43. Quote from p. 329.
28 Tsuchimoto, Noriaki. "Shiron dokyumentarī eiga no sanjūnen" ["My perspective: Thirty years of documentary film"]. *Koza Nihon eiga 7: Nihon eiga no genzai* [Japanese film lecture

now mostly known for his documentaries in the 1970s on the mercury poisoning incident in Minamata, remained deeply involved in the Japanese Communist Party and its activism throughout the early 1950s and joined Iwanami Productions in 1956 at the start of his filmmaking career. Despite his persistent empathy toward the radical left, Tsuchimoto paradoxically recognized the governmental infrastructure film as an admirable variation of scientific inquiry detached from directorial intervention to the rapport between the camera and the subjects. For him, the film documented each stage of the "work schedule" (*kōtei*) not as a "follow-up" (*atooi*) story but as a record of progress, and what was "enhanced" was sober yet minute precision of the operation flow, itself grounded upon a rigorously calculated schedule.[29]

Tsuchimoto's observation of operational precision highlighted several key film techniques devised in *Sakuma Dam*. As was indicated above, *Sakuma Dam* sequentially presents the temporarily determined stages of the infrastructure project, sharing its stylistic traits with process films, which typically demonstrates "a series of chronological, sequentially ordered steps."[30] Aguilera Skvirsky stresses that such traits can be located extensively as a "genre-determining feature," from early sponsored film to art fiction cinema from the 2000s.[31] As the film rhythmically and almost mathematically renders blasting for drain tunnels, sequences are, for instance, typically cued with various shots of dynamite installations, followed by a close-up of a detonator. An operator – rather than a labourer – pulls the lever. There is a pause and then a blast. The explosion itself occasionally functions as formal punctuation, a release from a prolonged suspense. Historian Jinshi Fujii also points out that the white rising smoke that covers up the frame is stylistically repurposed as a fade-out, an optical effect to create a transition between scenes.[32]

Dust and debris indeed fly off toward the camera in several shots, which directly assault the spectatorial gaze as immediate and disruptive spectacles; explosions break both the mountain surfaces and the fourth wall. Shizuo Komura, the chief cinematographer of the documentary, recalls

series 7: Japanese cinema now], edited by Tadao Satō et al. Iwanami Shoten, 1988, pp. 248–69. Quote from p. 251.

29 Ibid., p. 250.

30 Aguilera Skvirsky, Salomé. "Realism, Documentary, and the Process Genre in Early New Latin American Cinema." *The Routledge Companion to Latin American Cinema*, edited by Marvin D'Lugo et al., Routledge, 2017, pp. 119–32. Quote from p. 126.

31 Ibid., pp. 126–29.

32 Fujii, Jinshi. "'Damu o tsukuru' noka 'damu ga dekiru' noka" ["'Dam is constructed' or 'dam comes into being'"]. *Iwanami eiga no l-oku furēmu* [Images of post-war Japan: The documentary films of Iwanami Productions], edited by Yoshiyuki Niwa and Shun'ya Yoshimi. Tokyo, 2012, p. 111.

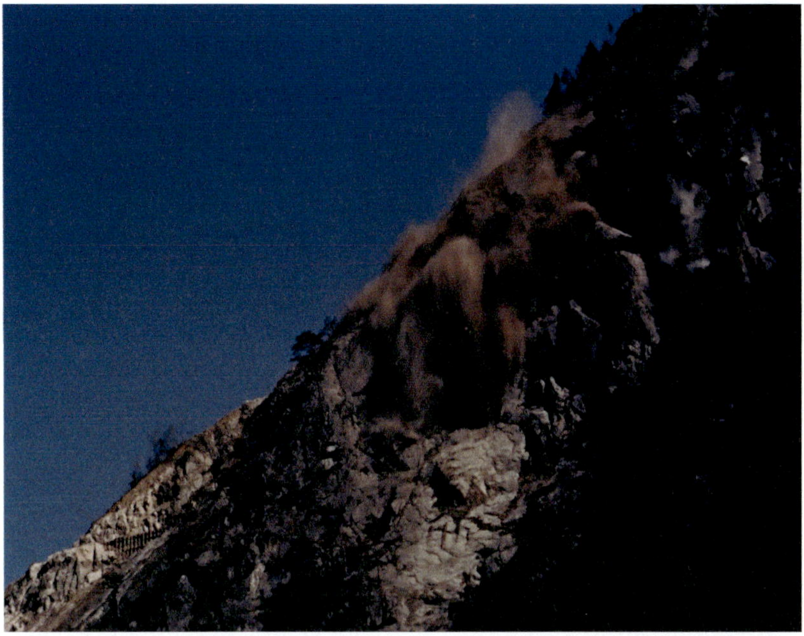

Fig. 8.4. Blasting in *Sakuma Dam*. Picture courtesy Documentary Film Preservation Center, Japan.

that the filming of blasting involved "deadly" endeavours, attempting to get as close to explosions as possible and wiring up the lighting equipment (combinations of 3K and 5K in dark tunnels) to secure enough lighting exposure for Kodachrome film.[33] However, the moments of explosions, or the mise en scène that potentially jeopardized the lives of those filming, are efficiently and intelligibly edited. Violent interference with nature is tamed, turning them into tuneful spectacle under control. In Tsuchimoto's essay, the construction "work schedule" (*kōtei*) is rhetorically correlated with the formal "structure" (*kōsei*) of the documentary, and this implicit parallel is more than a paronomastic play. The "processual" filmic syntax is here doubled with the arithmetical construction work flow. To put it differently, the organizational logic of editing, which parallels the flow of construction process, serves as an epistemic template that itself configures film techniques.[34] The viewer learns both the infrastructural project made visible in content and film techniques made recognizable in the film.

33 Komura, Shizuo. "Sakuma damu to watashi" ["Sakuma dam and me"]. *Iwanami eiga no 1-oku furēmu* [Images of post-war Japan: The documentary films of Iwanami Productions], edited by Yoshiyuki Niwa and Shun'ya Yoshimi. Tokyo, 2012, pp. 95–97.

34 Tsuchimoto. "Shiron dokyumentarī eiga no sanjūnen," pp. 250–51.

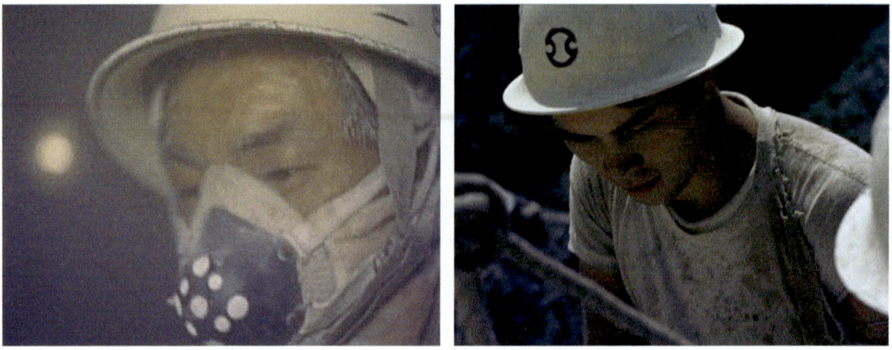

Figs. 8.5–8.6. Workers' faces in *Sakuma Dam*. Pictures courtesy Documentary Film Preservation Center, Japan.

Tsuchimoto's reference to the formal structure of *Sakuma Dam* also pointed to a use of the close-up, another film technique that he drew attention to in his observation of a worker covered with (toxic) white pulverized dusts in the documentary.

American cutting-edge methods, gigantic cranes, and powerful bulldozers were under full operation, reminding us of the change from the era of pickaxes and rope basket [*tsuruhashi to mokko no jidai*]. Yet, the film [*Sakuma Dam*] also shows white dust covering a head and face of a worker drilling holes in a rocky tunnel, including the close-up of his eyelashes that have now turned gray.[35]

Tsuchimoto added that the close-up images would now be the overt "*photographic evidence* of silicosis" (*keihaibyō*), a lung disease common to construction workers caused by the inhalation of toxic dust containing silica (figs. 8.5–8.6). The remark, on the one hand, connotes the filmmaker's political-empathic reading of the representations of labourers as ambivalent images of critique, and pinpoints the government-led massive enterprise behind the dam project. On the other hand, the contrast he draws between American technologies and Japanese on-site labourers (covered with industrial dust) is also intertwined with the binary oppositions in terms of "scale," namely, between the perceived totality of the grandeur and the anatomical details of the image. The close-up on the eyelashes does not abstract itself merely as a metonymical figure that indicates America-Japan geopolitical tension or a capital-labour exploitive relation; it also holds the revelatory quality that enabled Tsuchimoto to *trace* the evidential power inherent to the specific film technique: the

35 Ibid., p. 251.

close-up. To rephrase, the face of the Japanese worker is not rendered as a miniature diminished in awe of largeness but is discerned as an enlarged microcosm that solicits minute and analytical observation. What his remark suggests is a taxonomical approach to the moving-image medium and its technique, reflexively contrasting aerial view/long shot and close-up. The close-up here does more than signifying *what* is shown; it is also pedagogical and processual, teaching one a technique of *how* to close in to the microscopic attributes of what is shown. This discovery of the technique – as an epistemic mediator – is what Tsuchimoto sees as the implicit connection between the industrial PR documentary and science film, whose "arithmetical" perfection in *tracing* the microscopic "living cell organism in motion" (*seimei no kotai no undō no kurōzu appu*) had once made him want to become a cinematographer rather than a director.[36] The film that records the infrastructure in the making here is also "engineering" the arithmetical approach to film techniques.

The former reading of the close-up as an image of critique is premised upon a semiosis of significance, which strives for the symbolic inscribed on the worker's face. The latter presents the possibility of cinematic mediation operating in tandem with categorical study of film techniques, through which the viewer explores the terrain between the technical images of infrastructural making (i.e. dam construction) and the techniques of revealing (film form). If the former largely centres around a politics of representation, the latter points toward a politics of cinematic scale, drawing from a work on the close-up by Mary Ann Doane. The close-up, which has claimed a unique status throughout the histories and theories of cinema, tends to elaborate a politics of the "face," a fragment of the body that simultaneously serves as an undecipherable/unreadable material surface as well as a window into interiority. However, it is also the term that designates "enlargement," foregrounding its pivotal roles in relation to the spatial configurations, scale and (dis)proportion. Doane's study reveals that a politics of the close-up is entwined with these seemingly quantitative and apolitical operations of scales, which in fact encompass deeply political questions of perspective/point of view, spatiotemporal rationalization, narrative construction and capitalist desire for commodification and possession.[37] Doane argues that the politics of scale, as "potential semiotic threat," undermines a schematized logic of continuity and identification and instead "empower[s] the spectator

36 Ibid., p. 250.

37 Doane, Mary Ann. "The Close-Up: Scale and Detail in the Cinema." *Differences*, vol. 14, no. 5, 2005, pp. 89–111.

as analyst of, rather than vessel for, meaning."[38] I stress that such empowerment effectively destabilizes assumed autonomies of various spaces among the diegetic, representational and spectatorial, once again evoking the question of *dispositif.* Beyond the politics of representation and its textual specificity, the politics of scale illustrates reflexive interrelations among the infrastructure in the making, technological condition of film production, and the film's own techniques of mediation.

Crucially, it is also here that Tsuchimoto makes brief yet critical comments on post-war Japan and the film, in which the magnificent infrastructural construction, for him, is "mirroring" (*eijiru*) the "resurrection of a defeated Japan."[39] If one persists in labelling the filmmaker as belonging to the radical left, his seemingly celebratory remark on the politico-economical centralization occurring in immediate post-occupation Japan has to be questioned as problematic or dismissed as frivolous. However, the cinematic inquiry of science once again operates as the linchpin of the relation Tsuchimoto suggests between war and Japan. The ambivalent ethical, existential and military implications of aerial cartographic views in the post-World War II period were addressed by various figures outside the Japanese context. It is Paul Virilio, for instance, who mapped out how scientific and mechanized modes of modern warfare converged on the aerial vision from military aircraft.[40] Virilio's homology between the operation of human eyes and that of aerial reconnaissance as a military ocular machine is often associated with the murderous vision of modern warfare. Erwin A. Gutkind, a German-born architect and urban planner, also spelled out in his 1956 essay that aerial vision, which had once made humans completely invisible, would simultaneously make collective human responsibility for the fragile planet visible. As a visual critique of humanity's destructive exploitation of nature, aerial vision ultimately reflects "the moral conscience of mankind."[41] I argue that the camera's miniaturizing, macroscopic and distanced vision of the wartime "killing machine" (in Virilio's term) came to be rearticulated into a form of visual

38 Ibid., p. 107.

39 Tsuchimoto. "Shiron dokyumentarī eiga no sanjūnen," p. 251.

40 Virilio, Paul. *War and Cinema: The Logistics of Perception.* Translated by Patrick Camiller, Verso, 1989, esp. ch. 2: "Cinema Isn't I See, It's I Fly."

41 Gutkind, Erwin A. "Our World from the Air: Conflict and Adaptation." *Man's Role in Changing the Face of the Earth,* edited by William L. Thomas. University of Chicago Press, 1956, pp. 1–44. See p. 11. *Also quoted by Amad, Paula. "From God's-Eye to Camera-Eye: Aerial Photography's Post-humanist and Neo-humanist Visions of the World." *History of Photography,* vol. 36, no. 1, 2012, pp. 66–86. See p. 75.

and perceptual expansion with microscopic views of the living organism in motion in post-war science documentaries in Japan.

In the specific case of post-war Japanese documentary, the series of cartographic bird's-eye shots, exemplified by the above-mentioned river closing in *Sakuma Dam*, was prefigured by the aerial vision in *The Effects of the Atomic Bomb on Hiroshima and Nagasaki*, the 1946 documentary made by a group of filmmakers at the Japan Film Company (Nippon Eigasha, or Nichiei for short), a former government-run monopoly where large-scale war documentaries and newsreels had been mass produced. The production of the film, which recorded the epicentres of the nuclear bomb attacks and their impacts on inhabitants in Hiroshima and Nagasaki in the months immediately following the events, was itself haunted, deeply intertwined with the Japanese government and US occupational tension and confusion. Furthermore, the production process was stopped, subsumed and resumed repeatedly. The finished film – 165 minutes in length – was also confiscated and suppressed by multiple ruling bodies for decades.[42] This doomed film, as Abé Mark Nornes argues, literally had a *maboroshi* or "phantom-like" status surrounding its existence.[43]

The production context and textual presence, at the same time, demystifies the spectral status of this documentary on atomic warfare.[44] The film was, as its blunt title implies, conceived as a scientific investigation of the *effects* of the atomic bombs on different sites and *hibakusha*, or A-bomb victims, conducting a meticulous (and often inhuman) survey about various impacts on nature and living tissues caused by radioactive particles. Along with extremely horrific representation of "pain" inflicted by the bomb, the film contained ample cartographic images of the evaporated cities recorded from an aerial perspective, which, as Nornes suggests, would correspond to the A-bomb's "point of view."[45] If macroscopic shots of the

42 One of the earliest fruitful attempts to release the film publicly was pursued by a group of civilian activists, and in April of 1968, NHK and NET broadcasted the shorter version of the film. Please see Yoshihara. *Nihon tanpen eizōshi*, p. 101.

43 Nornes conducts insightful study on *The Effects of the Atomic Bomb on Hiroshima and Nagasaki*, pairing it with Fumio Kamei's another monumental post-war documentary *A Japanese Tragedy* (1946). Please see Nornes, Abé Mark. *Japanese Documentary Film: The Meiji Era through Hiroshima*. University of Minnesota Press, 2003, pp. 183–219.

44 The filming crew was largely comprised of Nichiei directors and cinematographers, including Sueo Itō, Dairokurō Okuyama, Shigeru Miki, and Kiyoji Suzuki. Please see Yoshihara. *Nihon tanpen eizōshi*, pp. 98–101.

45 Nornes. *Japanese Documentary Film*, p. 211. Nornes also used this expression at a lecture held in conjunction with the conference "Legacies of Leftism in Film and Media Theory: East Asia and Beyond" at Columbia University on February 29, 2019.

scientifically motivated documentary rendered the ultimate apocalyptic vision of death, annihilation and destruction, the way Tsuchimoto relates the science film method to *Sakuma Dam* and his reference to the "resurrection" of post-war Japan also take on a new meaning. Kiyoji Suzuki, one of the camera crewmembers for *The Effects of the Atomic Bomb on Hiroshima and Nagasaki*, noted in 1949 that the wartime air raids had literally burnt out all the photo-microscopic equipment.[46] The sense of heavy losses and industrial unrest in the early post-war years certainly underscored the break from the preceding wartime works and practices.

History of microcinematography goes back to the early twentieth century, when Julius Ries, a Swiss biologist, produced one of the first time-lapse micro-cinematographic records of the cellular development of the sea urchin in 1907.[47] Capturing organic cells "in motion" was also a major agenda for physiologists such as Louise Chevreton and Frederic Vlès, who made a short film of the sea urchin's cellular transformation from egg to larva, rigorously adopting Étienne-Jules Marey's chromophotographic method.[48] Such European precedents were soon imported to Japan by M. Pathé and other ruling organizations in both public and private sectors.[49] Domestic studios investing in production and distribution of science film, such as the Jūjiya Culture Film Division (Jūjiya Bunka Eigabu) and Riken Science Film (Riken Kagaku Eiga), also started to appear in the late 1920s and the 1930s partly in response to an influx of UFA *kulturfilms* from Germany.[50] Just as UFA underwent Nazification in the 1930s under the government's protectionist measures, the budding science film culture in Japan became largely integrated into the wartime mobilization of the entire film industry, especially after the 1939 implementation of the Film Law (Eigahō), which required all theatres nationwide to include short documentaries in every programme.[51]

46 Suzuki, Kiyoji. "Bisokudo satsuei no keiken to hansei" ["Experience and reflection upon time-lapse photography"]. *Eiga gijutsu* [Film technique], vol. 5, August 1949. Also quoted in Yoshihara, *Nihon tanpen eizōshi*, p. 324.

47 For an insightful analysis of a live-cell imaging technology and its development, see Landecker, Hannah. "The Life of Movement: From Microcinematography to Live-Cell Imaging." *Journal of Visual Culture*, vol. 11, no. 3, 2012, pp. 378–99.

48 Ibid., pp. 381–82. Also see Kelty, Christopher, and Hannah Landecker. "A Theory of Animation: Cells, L-Systems, and Film." *Grey Room*, no. 17, 2004, pp. 30–63.

49 For M. Pathé, please see High, Peter B. "Umeya Shokichi: The Revolutionist as Impresario." Nagoya University, n.d., pp. 124–25.

50 Tanigawa, Yoshio. *Nihon no kagaku eigashi* [History of science film in Japan]. Yuni Tsūshinsha, 1978, pp. 26–33.

51 For brief overview of the Film Law and its impact on the documentary industry, please see Nornes, Abé Mark. *Japanese Documentary Film: The Meiji Era through Hiroshima*. University

Tsuchimoto, however, prompts us to recognize that the very attempt to make the living cells – or "life" – visible in motion fully blossomed in Japan after surrender, corresponding to the technological rigor of the early post-war (life) science films that marked progress and the "resurrection" from the devastation of the war. The early major post-war science shorts, Dairokurō Okuyama's *Living Bread* (*Ikiteiru pan*, 1948),[52] *Ecosystem of Tuberculosis* (*Kekkaku no seitai*, 1952) and Nikichi Ōta's *The Life of Rice* (*Ine no isshō*, 1950) all employed the phase-contrast imaging technology and time-lapse techniques that enabled the camera to record a living microorganism in motion instead of a stained (therefore dead) specimen fixed to prepared slides.[53] The aforementioned Nichiei was corporatized in 1946 and came to stand at the forefront of the field of science film production along with a few other studios, most notably Tokyo Cinema (est. 1954) and Iwanami Productions (est. 1950), which, as I noted, produced *Sakuma Dam*. Tsuchimoto's account on *Sakuma Dam* should be intertwined with this redemptive perspective of life and birth in the close-up of the microscopic details. To put it differently, the filmmaker's admiration for *Sakuma Dam* does not derive entirely from its subversive quality that captured the perceived evidence of hard labour; rather the documentary is for him a moving-image catalogue of the post-war infrastructural innovations, in which the evidence of material conditions is simultaneously presented as the indexical connection of motion photography to life.

Conclusion

Sakuma Dam Chapter One alone attracted more than three million viewers when theatrically released in 1954. This commercial success as a feature documentary became the cornerstone of a non-fiction infrastructure construction genre for theatrical release. For instance, the development of the Kurobe Dam (Kurobe damu), the tallest arch dam in Japan to date, was also made into a series of films by Nichiei. Sponsored by the third-sector Kansai Electric Power Company (Kansai Denryoku or KEPCO, for short), the films recororded eight years of construction process between 1956 and

of Minnesota Press, 2003, especially pp. 61–69. Also, Kasza, Gregory J. *The State and the Mass Media in Japan, 1918–1945*. University of California Press, 1993.

52 As noted above, Okuyama was one of the crew members of *The Effects of the Atomic Bomb on Hiroshima and Nagasaki*.

53 Yoshihara. *Nihon tanpen eizōshi*, pp. 324–27.

1963 and were released in sequential instalments in 1957, 1958, 1961 and 1963. Kurobe Dam was also known as "Kuroyon Dam" (Kuroyon damu) for housing the then largest hydropower plant, Kurobe No. 4 (dai-yon). Accordingly, Nichiei's documentary saga has often been referred to as the *Kuroyon* series and was widely popularized in Kinema Junpo and other major film journals in Japan.[54]

One crucial point to be made is the way in which a rhetoric of infrastructural innovation in content also brought formal, technical and experiential features to light that reflected a transforming condition of highly mediated societies. In 1961 the third chapter of the *Kuroyon* series was shot and released in CinemaScope, followed by an enormous commercial success of *The Sands of Kurobe* (*Kurobe no taiyō*), a fictional dramatization of the Kurobe Dam project as a serial novel in 1964 and its film adaptation in 1968. Casting the then peerless stars Toshirō Mifune and Yūjirō Ishihara, the film, with a 196-minute running time, dominated the box office and became the top-grossing film of the year.[55] In 1969, Nippon Television – Japan's first commercial TV station – also put together a televised drama of the same title, recasting Ishihara as well as other major names from the film version. A newly available widescreen format, or a technological innovation in part as a countermeasure against the emergence of television, was implemented in the documentary series to display infrastructural innovation in content, which then migrated to mainstream media enterprises, including television. This type of transmedia strategy is not entirely new, given that histories of cinema have always been inseparable from those of other media. However, post-war industrial films in Japan, beyond the localized boundaries and spaces of cinema, became deeply entangled with a range of epistemic transformations and changing mediascapes occurring in the period. To put it differently, the question at stake is not necessarily about corporate structures (determined by a market-driven economy) or ideology inter-textually woven into representations. Rather, the industrial films comprised the industrial, infrastructural, organizational and technological relationalities, from which various forms or systems of knowledge emerge. Such systems had often been configured with and preconditioned by film (and other

54 For a brief account of the *Kuroyon* series, please see Yoshihara. *Nihon tanpen eizōshi*, pp. 214–16.

55 A popular writer, Shōji Kimoto, penned the novel, which was first serialized in the *Mainichi* newspaper in 1964. Both Ishihara and Mifune founded their own film production studios in the 1960s: Ishinara Promotions in 1963 and Mifune Productions in 1963. The two studios co-produced *The Sands of Kurobe*.

media) as technologies and techniques of mediation as well as of medial participation, which constituted particular *dispositif* in post-war Japan. This historically and geopolitically specific *dispositif* points toward open-ended media environment or media-infused world building, where social actants remains analytical, processual and reflexive.

Works Cited

Aguilera Skvirsky, Salomé. "Realism, Documentary, and the Process Genre in Early New Latin American Cinema." *The Routledge Companion to Latin American Cinema*, edited by Marvin D'Lugo et al. Routledge, 2017, pp. 119–32.

Amad, Paula. "From God's-Eye to Camera-Eye: Aerial Photography's Post-humanist and Neo-humanist Visions of the World." *History of Photography*, vol. 36, no. 1, 2012, pp. 66–86.

Bazin, André. *The Cinema of Cruelty: From Buñuel to Hitchcock*, edited by François Truffaut, translated by Jane Pease and Rose Kaplan. Seaver Books, 1982, pp. 200–04.

Bazin, André. "Petit journal du cinéma: *Vivre*." *Cahiers du cinéma*, vol. 69, March 1957, p. 2.

Doane, Mary Ann. "The Close-Up: Scale and Detail in the Cinema." *Differences*, vol. 14, no. 3, 2003, pp. 89–111.

Doane, Mary Ann. "The Indexical and the Concept of Medium Specificity." *Differences*, vol. 18, no. 1, 2007, pp. 128–52.

Dower, John W. "Peace and Democracy in Two Systems." *Postwar Japan as History*, edited by Andrew Gordon. University of California Press, 1993.

Fujii, Jinshi. "'Damu o tsukuru' noka 'damu ga dekiru' noka" ["'Dam is constructed' or 'dam comes into being'"]. *Iwanami eiga no l-oku furēmu* [Images of post-war Japan: The documentary films of Iwanami Productions], edited by Yoshiyuki Niwa and Shun'ya Yoshimi. Tokyo, 2012.

Gaines, Jane. "Documentary Radicality." *Revue canadienne d'études cinématographiques/Canadian Journal of Film Studies*, vol. 16, no. 1, 2007, pp. 5–24.

Gordon, Andrew, editor. *Postwar Japan as History*. University of California Press, 1993.

Gunning, Tom. "Before Documentary: Early Nonfiction Films and the 'View' Aesthetic." *Uncharted Territory: Essays on Early Nonfiction Film*, edited by Daan Hertogs and Nico de Klerk. Stichting Nederlands Filmmuseum, 1997, pp. 9–24.

Gutkind, Erwin A. "Our World from the Air: Conflict and Adaptation." *Man's Role in Changing the Face of the Earth*, edited by William L. Thomas. University of Chicago Press, 1956, pp. 1–44.

Ha, Kyung Jin. *Paburikku rirēshonzu no rekishi shakaigaku* [Historical sociology of public relations]. Iwanami Shoten, 2017.

High, Peter B. "Umeya Shokichi: The Revolutionist as Impresario." Nagoya University, n.d., pp. 124–25.

Kasza, Gregory J. *The State and the Mass Media in Japan, 1918–1945*. University of California Press, 1993.

Kelty, Christopher, and Hannah Landecker. "A Theory of Animation: Cells, L-Systems, and Film." *Grey Room*, no. 17, 2004, pp. 30–63.

Toba, Koji. "Damu kaihatsu to kiroku eiga" ["Dam development and documentary"]. *Iwanami eiga no l-oku furēmu* [Images of post-war Japan: The documentary films of Iwanami Productions], edited by Yoshiyuki Niwa and Shun'ya Yoshimi. Tokyo, 2012, pp. 145–62.

Komura, Shizuo. "Sakuma damu to watashi" ["Sakuma dam and me"]. *Iwanami eiga no l-oku furēmu* [Images of post-war Japan: The documentary films of Iwanami Productions], edited by Yoshiyuki Niwa and Shun'ya Yoshimi. Tokyo, 2012, pp. 95–97.

Kusakabe, Kyūshirō. *Eizō o tsukuru hito to kigyō: Iwanami Eiga no sanjūnen* [Film-makers and Their Company: Thirty Years of Iwanami Productions]. Mizuumi Shobo, 1980.

Landecker, Hannah. "The Life of Movement: From Microcinematography to Live-Cell Imaging." *Journal of Visual Culture*, vol. 11, no. 3, 2012, pp. 378–99.

Larkin, Brian. "The Politics and Poetics of Infrastructure." *Annual Review of Anthropology*, vol. 42, 2013, pp. 327–43.

Machimura, Takashi. "'Kaihatsu' o egakanai kaihatsu eiga" ["The 'construction' film that does not represent construction"]. *Iwanami eiga no 1-oku furēmu* [Images of post-war Japan: The documentary films of Iwanami Productions], edited by Yoshiyuki Niwa and Shun'ya Yoshimi. Tokyo, 2012, pp. 123–43.

Machimura, Takashi, editor. *Kaihatsu no jikan, kaihatsu no kūkan – Sakuma damu to chiiki shakai no hanseiki* [Time of industrial growth, space of industrial growth – Fifty years of Sakuma Dam and its region]. Tokyo Daigaku Shuppankai, 2006.

Noriega, Chon A. "Godzilla and the Japanese Nightmare: When *Them!* Is US." *Hibakusha Cinema: Hiroshima, Nagasaki and the Nuclear Image in Japanese Film*, edited by Mick Broderick. Kegan Paul International, 1996, pp. 54–74.

Nornes, Abé Mark. *Japanese Documentary Film: The Meiji Era through Hiroshima*. University of Minnesota Press, 2003.

Suzuki, Kiyoji. "Bisokudo satsuei no keiken to hansei" ["Experience and reflection upon time-lapse photography"], *Eiga gijutsu* [Film technique], vol. 5, August 1949.

Tanigawa, Yoshio. *Nihon no kagaku eigashi* [History of science film in Japan]. Yuni Tsūshinsha, 1978, pp. 26–33.

Tsuchimoto, Noriaki. "Shiron dokyumentarī eiga no sanjūnen" ["My perspective: Thirty years of documentary film"]. *Koza nihon eiga 7: Nihon eiga no genzai* [Japanese film lecture series 7: Japanese cinema now], edited by Tadao Satō et al. Iwanami Shoten, 1988, pp. 248–69.

Upham, Frank K. "Unplaced Persons and Movements for Place." *Postwar Japan as History*, edited by Andrew Gordon. University of California Press, 1993, pp. 325–46.

Virilio, Paul. *War and Cinema: The Logistics of Perception*. Translated by Patrick Camiller. Verso, 1989.

Yomota, Inuhiko. *What Is Japanese Cinema? A History*, translated by Philip Kaffen. Columbia University Press, 2019.

Yoshihara, Junpei. *Nihon tanpen eizōshi: Bunka eiga, kyōiku eiga, sangyō eiga* [History of short film in Japan: Culture film, educational film, industrial film]. Iwanami Shoten, 2011.

Zhen, Zhang. "Toward a Digital Political Mimesis: Aesthetics of Affect and Activist Video." *DV-Made China: Digital Subjects and Social Transformations after Independent Film*, edited by Zhang Zhen and Angela Zito. University of Hawai'i Press, 2015, pp. 316–46.

About the Author

Takuya Tsunoda is assistant professor of Japanese cinema and media at Columbia University. His primary research centres on the interplay between institutions and media, technologies and sociocultural practices, various modes of reflexivity, science and material culture, and representation and knowledge formations. He is currently working on a book that examines the history and theory of audiovisual education and its relation to the new cinemas of the 1960s in Japan.

9 The Power of Flows

The Spatiality of Industrial Films on Hydropower in Switzerland[1]

Yvonne Zimmermann

Abstract
This chapter looks at two exemplary Swiss films from the 1930s and 1950s to study the circulation of narratives, rhetoric strategies and motifs in industrial films on hydropower. Drawing on Manuel Castells' concept of "the space of flows," the chapter focuses on the spatiality of these films and shows how they mediate between the dynamic space of flows of technology and economy and the static places of society. The films negotiate space within a national framework and use water power and landscape as cultural resources to shape ideas about national identity. Arguably, these films are not as much representations and records of spatial transformations as they are actors in the process of negotiating the transformation of places into spaces of flow.

Keywords: space; energy; flow; hydropower; landscape; national identity

Until the mid-1960s, when with the introduction of nuclear power a new competitor among the energy resources emerged, hydropower was by far the most important resource in Switzerland. Hydropower was also a significant topical resource in regard to film: No other matter fuelled industrial film production as much as water power. For example, the 1956 film catalogue of the Schweizerische Zentrale für Handelsförderung, a private association founded in 1927 to promote Swiss industry in foreign markets (nowadays

1 An earlier, shorter version of this essay was published under the title "Negotiating Landscape: Engineering Consent on the Exploitation of Water Power in Swiss Corporate Films." *Cultural Heritage and Landscapes in Europe/Landschaften: Kulturelles Erbe in Europa*, edited by Christoph Bartels and Claudia Küpper-Eichas. Deutsches Bergbau-Museum, 2008, pp. 180–91.

Hediger, V., F. Hoof, Y. Zimmermann, with S. Anthony (eds), *Films That Work Harder: The Circulation of Industrial Film*. Taylor & Francis Group 2024
DOI: 10.5117/9789462986534_CH09

operating under the name OSEC Business Network Switzerland), lists eighteen titles from the electric industry, followed by thirteen titles from the machine and metal industry, eleven from the food industry and five titles each from the textile industry and the chemical industry.[2] The number of films reflects the economic and political importance of hydropower in Switzerland: The country is poor in mineral resources; it lacks coal, gas, and oil. The only natural resource to supply the country with energy in the first half of the twentieth century is – water.

However, the large number of films dedicated to hydropower is more than evidence of the economic importance of water and the electric industry in Switzerland. It is first and foremost a sign for and symptom of the branch's major need in communication. The use of water power requires technological implementations and infrastructure that bring about major spatial changes and affect the living conditions of animals and humans alike. The instalment of water reservoirs went along with the flooding of villages and valleys and the resettling of local inhabitants. Huge dams, power stations and power supply systems left significant and permanent traces in the landscape. The rising controversies about such massive interventions in traditional living environments presented a vital threat to the development of the electric industry. According to French science historian Bruno Latour, scientific and technological innovations are only going to be successful if scientists and engineers manage to shift, translate and connect existing social interests with their own interests – a process that Latour calls "alignment of interests" and "shaping up allies."[3] Also, economic prosperity is only possible in alliance with consumers. When conflicts threatened economic expansion, industrial films would be used as "trouble shooters" in moments of crisis to negotiate between producer and consumer, between industry and society. Films about electricity can therefore be understood as tools for "the engineering of consent" on economic expansion, to use a term that is generally ascribed to US advertising pioneer Edward L. Bernays.[4] In this sense, the films worked on the "sociocultural adaptation" of consumers to the needs of the industry, which, according to Friedrich Mörtzsch, head of the German press and PR department of AEG in the 1950s, is the objective of industrial

2 OSEC. *Catalogue des films suisses industriels et économiques/Katalog schweizerischer Filme industriellen und wirtschaftlichen Inhalts.* Office suisse d'Expansion commerciale, 1956.

3 Latour, Bruno. *Science in Action: How to Follow Scientists and Engineers through Society.* Harvard University Press, 1987, pp. 157, 86–87; see also idem. *The Pasteurization of France.* Harvard University Press, 1988.

4 See Bernays, Edward L. "The Engineering of Consent." *Annals of the American Academy of Political and Social Science*, vol. 250, no. 1, March 1947, pp. 113–20.

films.[5] Vinzenz Hediger and Patrick Vonderau refer to this purpose as rhetoric or governance and count it among the three main functions of industrial films within the organizational context of corporations, the other two being record and rationalization.[6] However, social consent was not to be engineered by overtly displaying the main interest of the producers, which was – and still is – economic profit. Instead, corporate films produce *surplus* profit to induce cooperation from society. They portray economic development as social benefit, arguing that what is good for the electric industry is good for society.

To show how industrial films align interests and "sell" hydropower as social and cultural benefit and to better understand how resources are used in the process of "engineering consent" in industrial films is the main goal of this chapter. For this purpose, the analytical concept of "ResourceCultures" proposed by Martin Bartelheim et al. is helpful, for it expands the economic notion of resources to include not only raw materials of economic value – water, in our case – but also a large variety of other means, material and immaterial, tangible and intangible, human and non-human, that are used in different social contexts as social and cultural resources.[7] Resources in this perspective are understood as "a base of or means to create, maintain or transform social relations, units and identities on the background of culturally shaped beliefs and practices."[8] This approach expands the initially economically defined category of resources and opens it up for research in the humanities. While the concept of "ResourceCultures" was developed in the field of anthropology to bridge the gap between anthropology and historical sciences, it can also be productive in cinema and media studies to analyse (from a comparative perspective) the place and role of (audio) visual media in shaping and negotiating "ResourceCultures" in specific historical and social contexts.

In the following, I focus on films about hydropower that were sponsored by Swiss corporations and associations in the first half of the twentieth century. Starting from a micro-level analysis of formal and narrative

5 Mörtzsch, Friedrich. *Die Industrie auf Zelluloid. Filme für die Wirtschaft.* Econ, 1959, p. 22.

6 Hediger, Vinzenz, and Patrick Vonderau. "Record, Rhetoric, Rationalization: Industrial Organization and Film." *Films That Work: Industrial Film and the Productivity of Media*, edited by Vinzenz Hediger and Patrick Vonderau. Amsterdam University Press, 2009, pp. 35–49.

7 Bartelheim, Martin, et al. "'ResourceCultures' – A Concept for Investigating the Use of Resources in Different Societies." *Persistent Economic Ways of Living: Production, Distribution, and Consumption in Late Prehistory and Early History*, edited by Alžběta Danielisová and Manuel Fernández-Götz. Archaeolingua Alapítvány, 2015, pp. 39–50.

8 Ibid., p. 39.

patterns, recurring rhetoric strategies and visual motives, I wish to examine how the films mobilize natural resources as well as cultural resources to regulate social relations and identities. Thereby, spatial dimensions are of particular importance, not only because of the massive spatial impact of the hydroelectric industry. As Bartelheim et al. underline, spaces – understood as landscapes – are both shaped by specific "ResourceCultures" and in turn influence the emergence and structuring of these "ResourceCultures."[9] In my analysis, I put a particular emphasis on *energy flows*; that is, on the circulation of energy represented in the films and on the transformation that energy undergoes in this process. Manuel Castells' concept of "the space of flows" and his observations on the relations between technology, space and society will thereby serve as an analytical framework to examine how industrial technology, spearheaded and promoted by informational technology – film in our case – reshapes spatial dynamics and how it works to influence social and political processes.

I will mainly draw on two prototypical films that serve as exemplary cases to study the formal and rhetoric strategies within the filmic spaces of flows: The first example is *Dienstbare Kraft/Forces domptées* (*Useful Power*), a joint production of the Swiss machine and electric industry sponsored in 1938 by the Schweizerische Zentrale für Handelsförderung (Office Suisse d'Expansion Commerciale, OSEC). The second example, *Drei in einem Boot: Technik und Natur* (*Three in a Boat: Technology and Nature*), was sponsored by Elektrizitätswerke des Kantons Zürich (Electric Power Company of the Canton of Zurich, not mentioned in the credits) in the early 1950s. *Dienstbare Kraft* mainly targeted foreign audiences, albeit it also circulated in Switzerland, whereas *Drei in einem Boot* exclusively addressed a vernacular audience in the German-speaking part of Switzerland. Both films negotiate space within a national framework and use water power and landscape in concert as cultural resources to shape ideas about national identity.

The Source of National Energy: Alpine Landscape as Cultural Resource

Switzerland's factual poverty in mineral resources is at the origin of a figure of speech that could be called "the rhetoric of lack." The rhetoric of lack in mineral resources (coal, oil and gas) declares lack a typical Swiss feature and addresses it as a national specificity. This lack, so the argument goes, is in

9 Ibid., p. 41.

need of compensation, which can only be achieved by optimizing the use of all available resources. The rationalization of the use of available resources, natural and human, is thus presented as a necessity and national duty. This rhetoric of lack is considerably nurtured by films on water power and prevails in industrial films that promote national economic development at least into the 1960s. It emerged during World War One, when Switzerland ran short of coal supply from abroad. This was the moment when the nation's dependency on electricity became sorely evident and when water, the so-called "white coal," was "discovered" as *the* raw material to secure the country's energy independence. As a consequence, the exploitation of hydropower was heavily pushed during and after the war. And so was the electrification of the railway network. Not surprisingly, the connection between electricity and transportation also became an important topic in corporate films. Silent films from the early 1920s, such as *Electrification de la ligne du Gothard/Der elektrische Betrieb auf der Gotthardbahn* (*Electrification of the Gotthard Railway Line*, 1922, Eos Film) sponsored by Brown Boveri, and *L'électrification des chemins de fer suisses/Die Elekrifikation der Schweiz. Eisenbahnen* (*The Electrification of the Swiss Federal Railway Network*, attributed to Charles-Georges Duvanel, 1921–1926) sponsored by the Swiss Federal Railway Network, both end on lavish sequences of trains travelling through picture-book alpine landscape, crossing bridges in dizzy heights and passing by well-known tourist attractions such as the Castle of Chillon, Lucerne's Chapel Bridge and the memorial of the national hero Wilhelm Tell in Altdorf. These moving pictures are industrial films, travelogues and tourist films at the same time and were often co-sponsored by multiple interest groups from the machine, electric and tourist industries. I will come back to the relations between networks of electricity and railway networks in regard to Castells' concept of the space of flows and the circulation of energy.

Since the turn of the twentieth century, private electric industry came more and more under the supervision of the cantons and, in 1908, under state supervision. Hand in hand with the growing governmental control went a nationalization of discourses on electricity. From 1914 onwards, water power had been promoted as a weapon to fight for national independence and as a tool to enable economic progress.[10] What coal meant to Germany or Great Britain, "white coal" meant to Switzerland. Electricity was at the same time mobilized as economic and cultural resource and became a major factor for the national economy and identity.

10 See Gugerli, David. *Redeströme: Zur Elektrifizierung der Schweiz 1880–1914*. Chronos, 1996.

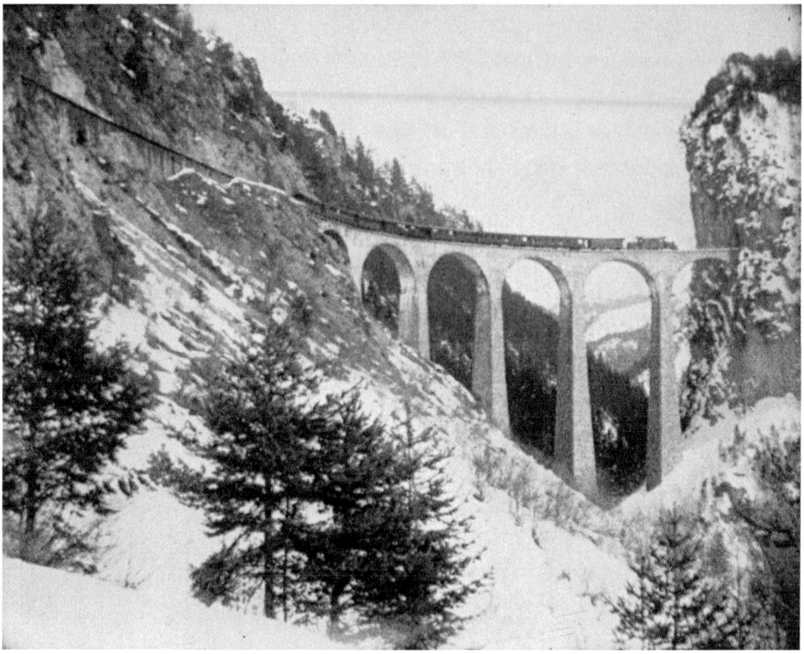

Fig. 9.1. Screenshot from *L'électrification des chemins de fer suisses/Die Elektrifikation der Schweiz. Eisenbahnen* (*The Electrification of the Swiss Federal Railway Network*, 1921–1926) (Cinémathèque suisse Lausanne).

The integration of electricity into contemporary national discourses was facilitated by the close connection between hydropower and another pillar of Swiss national identity, the mountains, for it is in the mountains that the country's energy supply originates. Together with the country's (actually not so) "glorious" medieval past, the alpine landscape is the very foundation national identity is built on. Such a combination of time and space, of history and topography, was a construct of the mid-nineteenth century, when the confederation founded in 1848 sought political stabilization through the ideological fabrication of a national myth. It was the moment, according to historian Eric Hobsbawm, when tradition was invented.[11] Or, to put it differently, it was the moment when the mountains turned into a cultural resource for national identity in Switzerland.

In the 1930s, the threatening geographical proximity of totalitarian regimes in Germany and Italy induced a phenomenon called *Geistige*

11 Hobsbawm, Eric, and Terence Ranger, editors. *The Invention of Tradition*. Cambridge University Press, 1983; see also Marchal, Guy P., and Aram Mattioli, editors. *Erfundene Schweiz. Konstruktionen nationaler Identität*. Chronos, 1992; and Marchal, Guy P. *Schweizer Gebrauchsgeschichte: Geschichtsbilder, Mythenbildung und nationale Identität*. Schwabe, 2006.

Landesverteidigung (spiritual defence of the country). In an attempt to distinguish the country from other nation states, Switzerland "armed" itself with whatever was thought and above all declared to be "typically" Swiss. On a political level, the *Geistige Landesverteidigung* was a compromise between divergent political wings to unify forces, marked by the integration of the socialist party into governmental responsibility. In the field of culture, the *Geistige Landesverteidigung* focused on the cultivation of traditions deemed "typically" Swiss: the medieval past, rural and alpine agrarian origins (in contrast to modern urbanism), folklore, humanitarian commitment, democracy and neutrality.[12] In this context, the mountains as *the* emblem of the country gained further significance. If with industrialization and mass production, on the one hand, and the introduction of the nation state, on the other, alpine landscape turned into a functional category, it is the 1930s that best illustrate how mountains worked as "useful" symbol and cultural resource with multiple and dynamic layers of values that could – and would – be adapted to the specific needs of various interest groups. In the context of the late 1930s, alpine landscape served as a cultural resource for its political meaning (as a warrant for liberty, democracy, neutrality and the will to self-defence), for its social significance (as a stronghold of tradition and sanity), and for its economic value (as a quality mark for trade).

The Circulation of Energy and the Space of Flows

Since mountains are at the same time the natural water reservoir and the national power source, Switzerland's "national" energy has its source in the "national" landscape. It is therefore no coincidence that almost all films about electricity begin where the water comes from, in the symbolic mountain space. This is also the case with *Dienstbare Kraft*, a film by Charles-Georges Duvanel about the production and consumption of electricity in Switzerland. The film opens with a spectacular pan across a majestic chain of mountain peaks covered with snow that captures the majesty and beauty of a landscape in a pure condition, untouched by civilization. This aesthetic contemplation of landscape in the iconographic tradition of nineteenth-century landscape painting and photography is

12 See Mooser, Josef. "Die 'Geistige Landesverteidigung' in den 1930er Jahren. Profile und Kontexte eines vielschichtigen Phänomens der schweizerischen politischen Kultur in der Zwischenkriegszeit." *Schweizerische Zeitschrift für Geschichte*, vol. 47, no. 4, 1997, pp. 685–708.

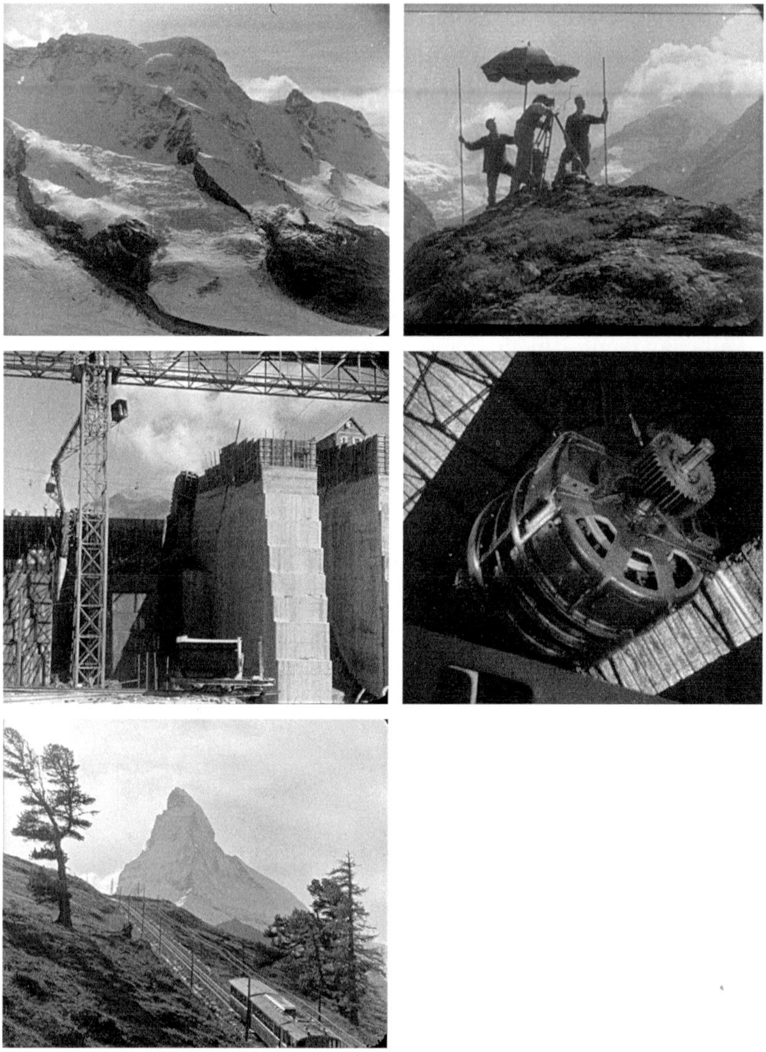

Figs. 9.2.-9.6. Screenshots from *Dienstbare Kraft/Forces domptés (Useful Power)*, sponsored by OSEC in 1938 (Cinémathèque suisse Lausanne).

followed by views of melting ice and snow, which initiate the filmic space as a space of flows. Trickles of water gather into small brooks that finally become torrential rivers and waterfalls. After the alpine "birth" of the natural resource and its growing strength as it flows down into the valley, the human hand begins to interfere. A river is dammed, creating a storage reservoir in order to domesticate, so the voice-over commentary holds, the natural, often devastating power of water by transforming it into useful electric energy.

Dienstbare Kraft is a classical process film, as Tom Gunning has termed this genre of industrial films, for it follows the narrative pattern of the industrial transformation of raw material into consumable goods step by step.[13] The narrative moves from the opening scenes of raw material – in this case, water – through various stages of production to culminate in a scene of delightful consumption. This trajectory from raw material to consumable good enacts, according to Gunning, a basic narrative of industrial capitalism.[14] *Dienstbare Kraft* presents this trajectory as a flow that moves across space in a very particular way. In a literal sense it *circulates*, that is, it moves in a circle or circuit. Scenes from the electric and machine industries show the construction and workings of a hydroelectric power station, the fabrication of Pelton turbines and current cables as well as the electric equipment of a locomotive are followed by a train journey alongside beautiful lakes, across daring bridges and uphill through tunnels. The railway trip takes the audience back to where the film started: to the mountains and into the symbolic "national" landscape.

Dienstbare Kraft, itself a shortened, re-edited and sound recorded version of the silent film *Unsichtbare Kraft/Puissance invisible* (*Invisible Force*), directed by Duvanel in 1934,[15] refined previous narratives of water power and established a narrative pattern that would dominate Swiss films about electricity at least until the end of the 1960s: the pattern of energy flow as a *cycle*. This cycle is presented as a closed circuit in which electricity starts from the mountains to undergo a transformative and shape-shifting process and to return to its origin, to the mountains. Its initial and final points are short-circuited in an alpine landscape.

Manuel Castells' concept of "the space of flows" introduced in 1983 and most prominently promoted in his 1989 book *The Informational City* is helpful to better understand the narrative and rhetoric strategies that

13 Process films were the first highly standardized genre of corporate films. They were widespread in early cinema under the name of *scène d'art et d'industrie* or *Industriebild*. The early standardization is owed to a pre-figuration of the films' narrative pattern in preceding media such as booklets, lantern slides, and photographs. See Zimmermann, Yvonne. "Vom Lichtbild zum Film. Anmerkungen zur Entstehung des Industriefilms." *Montage AV*, vol. 15, no. 1, 2006, pp. 74–88. See also Skvirsky, Salomé Aguilera. *The Process Genre. Cinema and the Aesthetic of Labor*. Duke University Press, 2020.

14 Gunning, Tom. "Before Documentary: Early Nonfiction Films and the 'View' Aesthetic." *Uncharted Territory: Essays on Early Nonfiction Film*, edited by Daan Hertogs and Nico de Klerk. Stichting Nederlands Filmmuseum, 1997, pp. 17–18.

15 *Unsichtbare Kraft* was the third film produced by OSEC. It circulated in a long silent version and in a short sound version in 35 mm and 16 mm prints.

are at work in the film.[16] In *The Informational City* Castells studies the intimate relations between new technologies, industrial innovations, spatial transformations and social outcomes operating within the logic of capital. Castells identifies three modes of development, determined by technical relations between labour and matter, namely the agrarian, the industrial, and the informational. While Castells focuses on the informational mode of development that emerged in the context of economic restructuring in the United States during the late 1970s and 1980s, I would like to use Castells' concept of "the space of flows" for the analysis of the industrial mode of development that characterizes the period under study here, for films on hydropower such as *Dienstbare Kraft* do not just illustrate or reproduce the industrial mode; rather, they construct and thus *produce* it.

Castells holds that each industry has a unique spatial logic, and that since the emergence of the informational mode of development, spatial logic obeys an increasingly "placeless logic of industrial activity [...] made up of *networks* and *flows*."[17] As a consequence, the space of places is superseded by the space of flows:

> The flows of power generate the power of flows, whose material reality imposes itself as a natural phenomenon that cannot be controlled or predicted, only accepted and managed. This is the real significance of the current restructuring process, implemented on the basis of new information technologies, and materially expressed in the separation between functional flows and historically determined places as two disjointed spheres of human experience.[18]

Claiming that "[p]eople live in places, power rules through flows,"[19] Castells identifies a conflict between social relations rooted in the static character of place versus those that emerge from the dynamics within the power-holding space of flows. Castells thus points to a tension between place as a static category and the dynamic space of networks and flows – and to the power relations bound to it. My argument here is that the unique spatial logic of the electric industry has always been placeless. Electricity is more an infrastructure than a product, and its production, distribution, management

16 Castells, Manuel. *The Informational City: Information Technology, Economic Restructuring, and the Urban Regional Process.* Blackwell, 1989.
17 Ibid., p. 71. Original emphasis.
18 Ibid., p. 349.
19 Ibid.

and consumption is organized by a placeless logic of networks and flows. For this reason, the concept of the space of flows that for Castells is typical for the informational mode of development can also be used as an analytical framework to study the electric industry.

A film like *Dienstbare Kraft* negotiates between static place, understood as the social living space of people, and the dynamic space of flows of technology and economy envisioned in the circulation of electricity in its metamorphosing material shapes. Places represent a social category, whereas the space of flows is the domain of technology and business practice. In regard to electricity, Castells' claim that the flows of power generate the power of flows seems particularly pertinent. It may (or may not) be a coincidence that the term "power" is polyvalent, signifying power in the sense of force, strength, vigour or agency, but also power in the sense of energy. Energy, in all cases, *is* power – and strength, force and agency. Castells also addresses the material reality of power. The subject matter of energy is invisible. One of the tasks of a film like *Dienstbare Kraft* is to make hydropower *visible*. The classical filmic techniques of slowing down, accelerating or magnifying images to make visible phenomenon that are invisible to the human eye fail to work in regard to energy. Power can only be made visible through *materialization*. Power inhabits materials while flowing through them. This is to say that while power is physically present in different materials, it is also visually represented through different materials. This explains its ever shape-shifting, metamorphosing character. In the case of hydropower, it is stored in ice, snow and water, it runs through transmission lines, and it is at work in pumps, light bulbs, household appliances and railways. Power is materialized in and visualized through all these objects. Like energy, railways operate as networks and flows. When electrified, railway networks become part of the networks and flows of energy. Railways, in their placeless logic of traversing space, are therefore emblematic materializations of energy – hence the entanglement between electricity and travel in industrial films.

Castells' claim that the material reality in the space of flows imposes itself as a natural phenomenon can also be productively related to *Dienstbare Kraft*. First of all, the film's stereotypical beginning and end in the mountains and the return of water to its origin in form of electricity can be understood as the opening and closing of a cinematic space of flows. The circular narrative pattern of energy flows suggests that water in the space of flows undergoes a natural transformation into electricity. The production, circulation and consumption of hydropower are presented as an organic process. This seemingly organic process *naturalizes* the technical effort as well as the economic interest involved in the use of water power. Water power is presented as a natural phenomenon, a law of nature that, as Castells holds, cannot be controlled, but can only be accepted and managed.

The emergence of the dynamic space of flows affects the static places in it. *Dienstbare Kraft* uses iconographic images to argue that the spatial transformation of places into a space of flows does not damage, let alone destroy places. Instead, the transformation of places renders places like the mountains *accessible*. The travel on the Gornergrat mountain railway (the first electrified rack railway in Switzerland which opened in 1898) into the midst of the four-thousand-metre-high mountain peaks of the Monte Rosa region at the end of the film is a scene of delightful consumption typical for process films. A modern transportation system fuelled by electricity develops the alpine landscape for consumption. Thanks to mountain railways, the physical appropriation of the symbolic alpine space, once the privilege of an exclusive group of trained climbers, is turned into a commodity for tourist mass consumption. The cultural resource is made available to everybody as first-hand experience. The surplus value and social benefit from spatial transformations brought about by the industrial use of hydropower then is, so the visual argument of *Dienstbare Kraft* goes, the *democratization* and *commodification* of specific places. These places are at the same time cultural resources. Democratic access to the alpine landscape in this context also means democratic appropriation of a symbolic space. In this sense, electricity offers a trip to the heart of Switzerland, to the cradle of national identity. It is therefore no coincidence that the film closes on shots of the Matterhorn, *the* emblem of the country known all over the world. In this way, the film also partakes in the branding of a certain place – the mountains – as a unique selling point of Switzerland and Swiss industry abroad. Interestingly enough, it is with a specific *place* that the film legitimizes and promotes the space of (energy) flows.

Explicit evidence for a strategic use of the mountains as cultural resource and key visual in films on hydropower can be found in the production documents of the film *Gebändigte Kraft/Conquête de l'énergie* (*Domesticated Power*) that replaced *Dienstbare Kraft* in 1948. The main sponsor was again OSEC, seconded by private corporations from the machine, metal and electric industries. The declared purpose of the film was twofold. To a general audience, it wished to illustrate the worldwide need for energy. At the same time, it served as a business-to-business communication tool to promote the Swiss electric industry abroad.[20] To achieve these goals, the

20 Minutes. June 13, 1946. Cinémathèque suisse, OSEC, Gebändigte Kraft.

sponsors unanimously agreed that the film must end in the mountains and that it is essential "to accord ample room to the landscape, since this typical element in Switzerland is an element of success abroad."[21] Very clearly, films on hydropower partake in the economization of a specific place, the mountains, and its symbolic and aesthetic value as a cultural resource. Within the space of flows, we can learn from the film, it is *places* that are negotiated with the aim to align interests between industry and society.

However, *Dienstbare Kraft*, like all OSEC-sponsored films, does not only negotiate between economic and social interests; it also represents state interests, even if this is not always obvious. OSEC was granted a special credit of 25,000 Swiss francs by the government in 1929 to start producing and sponsoring films.[22] Despite its private legal status, OSEC received substantial financial support from the state for marketing measures, which included industrial films that promoted particular industry sectors primarily abroad. *Dienstbare Kraft* circulated internationally in German, French and English versions and was screened in 1938 and 1939 in movie theatres in England as well as at international exhibitions, trade fairs and special screenings in Lyon, Stockholm, Kabul, Teheran, New York, Buenos Aires and Bombay.[23] But OSEC films were also exhibited at home, where they would be screened in the programme of shorts shown in commercial movie theatres. In addition, *Dienstbare Kraft* was exhibited in the Zentralkino, the main cinema of the Schweizerische Landesausstellung (Swiss National Exhibition)

21 My translation. The original quote reads as follows: "[D]'accorder une bonne place au paysage, étant donné que cet element typique en Suisse est un facteur de succès pour l'étranger." Minutes. April 25, 1947; and letter from OSEC to Gamma Film. June 19, 1946. Cinemathèque suisse, OSEC, Gebändigte Kraft. For more information on *Gebändigte Kraft*, see Zimmermann, Yvonne. "Industriefilme." *Schaufenster Schweiz: Dokumentarische Gebrauchsfilme 1896–1964*, edited by idem. Limmat, 2011, pp. 368–70.

22 OSEC. *Rapport sur l'action de l'OSEC dans le domaine de la propagande filmée*, pp. 2, 6–7. Typo script. Cinémathèque suisse, Dossier OSEC; on OSEC's involvement in film production see Jaques, Pierre-Emmanuel. "La propaganda nationale par le film: Albert Masnata et l'Office suisse d'expansion commerciale." *Revue historique vaudoise*, no. 115, 2007, pp. 65–78; and idem and Zimmermann. "Dokumentarischer Film in der Schweiz im historischen Überblick (1896–1914)." *Schaufenster Schweiz*, pp. 84–127; on OSEC's role in exhibition culture, see Debluë, Claire-Lise. *Exposer pour exporter: Culture visuelle et expansion commerciale en Suisse (1908–1939)*. Éditions Alphil-Presses universitaires suisses, 2015.

23 Information on the distribution in OSEC. *Rapports de l'Office suisse d'expansion commerciale, Zurich et Lausanne.* Office suisse d'expansion commerciale, 1938, p. 17; and OSEC. *Rapports de l'Office suisse d'expansion commerciale, Zurich et Lausanne.* Office suisse d'expansion commerciale, 1939, p. 20; OSEC. *Rapport sur l'action de l'OSEC dans le domaine de la propagande filmée*. Office suisse d'expansion commerciale, December 1942, p. 7. Typo script. Cinémathèque suisse, Dossier OSEC.

in Zurich in 1939. This national exhibition of Swiss achievements in culture and economics took place in a delicate moment when the country faced the outbreak of World War Two. The *Geistige Landesverteidigung* was at its peak of popular resonance. The display of alpine landscape in combination with the achievements of native industry propelled national pride and aimed at reinforcing national identity. In a similar vein, Schweizer Woche (Swiss Week), an association founded in 1917 with the goal to promote Swiss goods in domestic markets, used *Dienstbare Kraft* for screenings in schools. From October 1938 to May 1939, Schweizer Woche organized seventy non-theatrical screenings of the film and thereby reached 11,500 spectators.[24]

The circulation and exhibition of *Dienstbare Kraft* lends itself as an example to illustrate that OSEC films had a quasi-official status. They represented, if more or less unacknowledged, the interests of the state. Hence, one can suspect, the films' concurrent emphasis on the national as a political and social factor at home and as an economic factor abroad. In fact, OSEC films were among the very first films in Switzerland to be produced with state money. Unlike in many other countries, the Swiss government was very reluctant to overtly invest in film and other "propagandistic" measures and preferred indirect intervention and hidden control instead. Official state funding of film production was not introduced in Switzerland until 1963. It is therefore noteworthy that the state sponsoring of film production has a long, yet somewhat obscured tradition, and that this tradition, as in most other countries, is rooted in the field of useful cinema and not in film as art or entertainment.

Multiplying Places in the Space of Flows

While *Dienstbare Kraft* was first and foremost a marketing tool to boost the export of Swiss products from the machine and electric industry in foreign markets, *Drei in einem Boot* (directed by Victor Borel) served to develop the native industrial production of electricity at home. And while *Dienstbare Kraft* negotiates places within a space of flows in an implicit, if not subliminal way, *Drei in einem Boot* uses explicit rhetoric to promote the use of water power. It even spells out the usually covered intent to "align interests" in Latour's sense. The film was shot during the construction of the hydroelectric power station in Rheinau near Schaffhausen (1952–1957) – a

24 Schweizerische Zentrale für Handelsförderung/Office suisse d'expansion commerciale, editor. *Wirtschaftliche Mitteilungen/Informations économiques*, vol. 18, no. 27, July 5, 1939, p. 136.

project that at the same time marked the peak of contemporary social opposition against new projects of the electric industry.

After World War Two, the use of water power had been heavily intensified in Switzerland. Between 1950 and 1960, current production increased over eighty per cent. The maximal output was doubled and the storage capacity of reservoirs almost tripled.[25] Such growth was possible only through a rise in capacity. However, the developments in the electric industry met with opposition from traditional organizations such as Schweizerischer Bund für Naturschutz (Swiss Association for the Protection of Nature) and Schweizerische Vereinigung für Heimatschutz (Swiss Association for Homeland Protection), both founded at the beginning of the twentieth century. For conservative and patriotic reasons, these groups feared the destruction of the country's landscape heritage and fought against the "spoiling" of the homeland's beauty.[26] Additional opposition came from organizations formed on specific occasions, such as the Rheinau Committee, which tried to prevent that project from taking shape.

Facing such fierce opposition, the Elektrizitätswerke des Kantons Zürich (Electric Power Company of the Canton of Zurich), operator of the future Rheinau power station, felt forced to react. *Drei in einem Boot* was one of the measures undertaken. According to the OSEC film catalogue, the film wished to outline the possibilities to develop the production of electricity "while still providing for the natural beauties that are to be preserved."[27] Being one among many elements in a massive propaganda campaign, the film's main function was to defeat a federal popular initiative "to protect the electricity landscape Rheinfall-Rheinau" that demanded the withdrawal of the concession for the power station.[28]

The film presents a fictional dispute between an environmentalist and an engineer. In the process it grants the side opposed to the project a good deal of screen time to make its argument. Examples of particular places being consumed by excavators and destroyed by industrial exploitation

25 See Skenderovic, Damir. "Die Umweltschutzbewegung im Spannungsfeld der 50er Jahre." *Achtung: die 50er Jahre! Annäherungen an eine widersprüchliche Zeit*, edited by Jean-Daniel Blanc and Christine Luchsinger. Chronos, 1994, pp. 119–46.

26 Although pollution of the environment increased exponentially with the economic boom and technological progress after World War Two, ecological concerns were not predominant until the 1970s when "modern" environmental protection movement set in.

27 My translation. The original quote reads as follows: "[E]n tenant compte des beautés naturelles qu'il convient de sauvegarder." OSEC. *Catalogue des films suisses industriels et économiques/ Katalog schweizerischer Filme industriellen und wirtschaftlichen Inhalts*, p. 7.

28 My translation. The original title of the initiative was "Zum Schutz der Stromlandschaft Rheinfall-Rheinau."

– represented by views of coal mines from abroad – are displayed and the technological progress is critiqued as a pure expression of cold calculation and disrespect for Mother Nature. Being the mouthpiece of the industry, the engineer admits that "sins" had been committed in the past, but refutes this argument for the present. Firstly, he recalls the impact of technology on modern life and its being taken as a matter of course, indicating the "naturalization" of technology in everyday live, in a sequence that outlines what would happen if the wheel of time was turned back. We see trains standing still, workers of the watch industry being dismissed, a man and a woman freezing and starving in a dim, candlelit living room, and a tramway being dragged by a horse. Secondly, the environmentalist's major objection against the Rheinau power station – the draining of the Rhine Falls – is refuted. On visiting the spectacular site together, the engineer reassures the viewers that despite the upriver use of water, the largest falls in Europe will remain intact. The conservation of the Rhine Falls, a tourist attraction and an element of patriotic pride, is used here to provide evidence of the respect that the industry pays to the space of places and its national landscape heritage. After inspecting the building site of the Rheinau power station and a boat trip on the Rhine (steered by a local inhabitant whose support of the plans marks the final step in convincing the environmentalist), a voice-over narrator sets in and praises water power as clean energy far more compatible with environmental concerns than any other energy resource. To underline this point, the film contrasts the use of oil and coal abroad with the use of "white coal" at home. Whereas foreign countries have to sacrifice their landscape, so the argument goes, the beauty of Switzerland remains intact. This claim is visually evidenced by a shot of travellers on a motorbike passing by an alpine reservoir hemmed by a chain of snowy mountains.

The development of the space of flows for the use of hydropower is further promoted as the only means to "win the war against hunger, cold, and deprivation of freedom in our way." The "we" that the narrator addresses refers to a national community "united" in the fight against communism. After a short period in World War Two, the concept of the *Geistige Landesverteidigung* was reactivated in the Cold War by conservative and right-wing politicians to call again for the cultivation of traditions and values that were proclaimed "typically" Swiss.[29] In contemporary Cold War rhetoric, *Dienstbare Kraft* declared the development of the space of flows a

29 See Imhof, Kurt. "Das kurze Leben der geistigen Landesverteidigung: Von der 'Volksgemeinschaft' vor dem Krieg zum Streit über die 'Nachkriegszeit' im Krieg." *Krise und sozialer Wandel, 2: Konkordanz und Kalter Krieg: Analysen von Medienereignissen in der Schweiz der Zwischen- und*

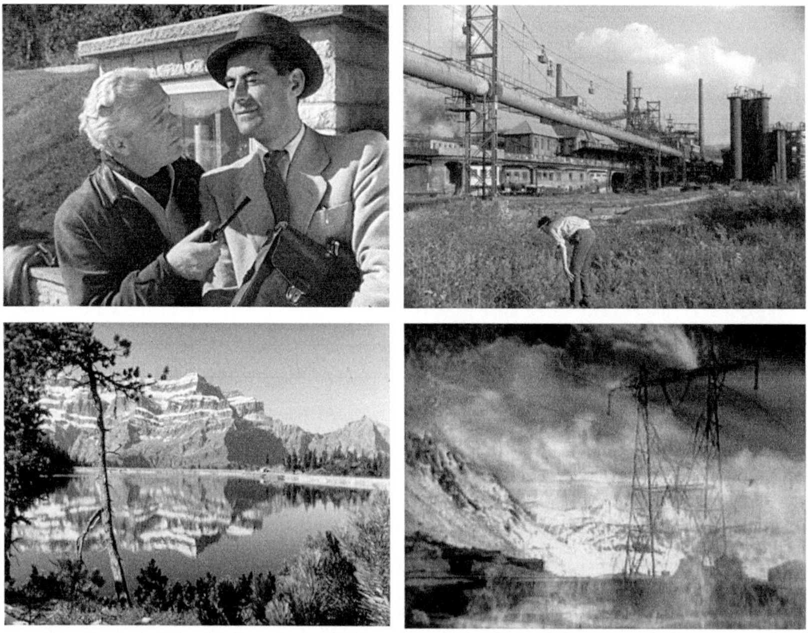

Figs. 9.7.-9.10. Screenshots from *Drei in einem Boot (Three in a Boat)* from the early 1950s (Cinémathèque suisse Lausanne).

national duty. Thus, negotiation between the space of flows and the space of places is shifted and transferred from the economic interests of the electric industry to national politics and patriotic sentiment. In order to maintain national independence, so the commentary concludes, support is needed from "everybody who really loves nature, the homeland, and its inhabitants."

Another line of argument that is used to align interests between industry and society is closely linked to the one discussed above, but is also comparable to the rhetoric strategy we saw at work in *Dienstbare Kraft*, for the development of the space of flows is again negotiated with arguments about *places*. But while *Dienstbare Kraft* argues with the democratization of places, *Drei in einem Boot* claims that the industrial transformation of places *ameliorates* those places. For this, *Drei in einem Boot* takes up a subject that has become a visual motif in both photographic and filmic representations of the use of water power, the water reservoir that mirrors surrounding mountains. Ever since the first films on hydropower that can be found in archives and that date back to the early 1920s, this visually attractive and highly symbolic motif of the water reservoir, surrounded by mountains that

Nachkriegszeit, edited by idem. Seismo, 1996, pp. 19–83; and idem. "Wiedergeburt der geistigen Landesverteidigung. Kalter Krieg in der Schweiz." *Krise und sozialer Wandel*, 2, pp. 173–247.

are reflected on the surface of the water, has become the very emblem for the industrially developed use of water power in Switzerland. With this motif, particular places have become *the* icons of the space of flows of the electric industry. These places are the result of spatial transformation and are used as new – or rather developed – cultural resource to visually prove that the space of flows does not destroy, but instead enhances and empowers, the space of places. With the instalment of reservoirs, places develop into attractive environments for plants, animals and humans alike.

The artificial lakes are not just technological reproductions of existing places. Instead, the creation of artificial lakes transforms places into reflecting mirrors of places. By reflecting the place's beauty on the surface of the water, the transformed place becomes a layer for landscape (mountains, in this case) to represent and reproduce itself. In *Landscape and Power*, W.J.T. Mitchell explains the special value that is attributed to landscape with lakes or reflecting pools as follows: "The reflection exhibits Nature representing itself to itself, displaying an identity of the Real and the Imaginary that certifies the reality of our own images."[30] The correspondence between real and imagined landscape thanks to reflection validates the "image" of landscape and certifies its value as cultural resource. Thus, so the visual rationale has it, the transformation of places into the space of flows neither destroys nor synthetizes places. Instead, it turns places into a medium that represents – and at the same time reproduces – the values of the places. Artificial lakes are exhibited as both a self-reflecting and a self-reproducing cultural resource. Transformed places, so the message of the film goes, allow natural places to unfold, or, to put it another way, spatial transformation develops places in that it multiplies the aesthetic and symbolic value of places. In this way, the transformation of places creates a surplus value – a value that is once again certified by its being mediated through filmic representation and reproduction.

In film and photography, it is much easier to visualize static places than dynamic flows. This instance may be another reason why films on hydropower negotiate places in order to engineer consent on the development of the space of flows. Interestingly enough, in its closing shot, *Drei in einem Boot* tries to visualize the space of flows within a single image. What the film comes up with is a superimposition of three shots: of mountains, of a pylon with transmission lines, and of streams of water. It is an attempt to condensate the flows of power and the power of flows in one single

30 Mitchell, W.J.T. "Imperial Landscape." *Landscape and Power*, edited by idem, University of Chicago Press, 1994, pp. 15.

emblematic image. However, this image has not made it into the repertoire of iconographic images on industrial development in Switzerland. But still, on October 5, 1954, the Swiss populace dismissed a federal popular initiative to withdraw the concession for the Rheinau power station, together with an initiative that demanded the extension of the people's rights regarding the granting of concessions to water rights.[31]

Conclusion

As a means to secure economic development and growth for its sponsors, films about electricity negotiate spatial transformation from a space of places into a space of flows in order to engineer social consent on economic exploitation. The films mediate between the dynamic space of flows of technology and economy and the static places of society and thereby "sell" economic exploitation as social benefit. In their argumentation, they heavily draw on cultural resources to legitimize the use of economic resources. The two examples discussed are typical in their producing surplus values ascribed to spatial transformation. In *Gebändigte Kraft*, the surplus value is *access* to the cultural resource of the mountains, which goes hand in hand with the democratization and commodification of alpine landscape, while *Drei in einem Boot* promises the *development* of alpine landscape through the multiplication of the beauty and the symbolic value of specific places. In both cases, the transformation of the space of places into a space of flows renders places available for consumption. The space of flows, in other words, turns places into commodities. The surplus values that the films offer to audiences in exchange for their consent obey the capitalist logic of rationalization: Democratization minimizes the expense to access symbolic places, whereas multiplication maximizes the profit from places. The space of flows established for economic development thus produces an increase (both in quantity and in quality) of consumable aesthetic and symbolic places. In this way, corporate films about the exploitation of hydropower are not so much representations and records of spatial transformations; rather, they are actors in the process of negotiating the transformation of places into spaces of flow. Thereby, it is striking to what extent private corporations and business associations cultivated cultural resources and a "national" iconography.

31 Schweizerische Bundeskanzlei. "Abgestimmte Volksinitiativen." Bundeskanzlei BK, www.admin.ch/ch/d/pore/vi/vis_2_2_5_7.html. Accessed December 1, 2016.

Works Cited

Bartelheim, Martin, Roland Hardenberg, Thomas Knopf, Anke K. Scholz, and Jörn Staecker. "'ResourceCultures' – A Concept for Investigating the Use of Resources in Different Societies." *Persistent Economic Ways of Living: Production, Distribution, and Consumption in Late Prehistory and Early History*, edited by Alžběta Danielisová and Manuel Fernández-Götz. Archaeolingua Alapítvány, 2015, pp. 39–50.

Bernays, Edward L. "The Engineering of Consent." *Annals of the American Academy of Political and Social Science*, vol. 250, no. 1, March 1947, pp. 113–120.

Castells, Manuel. *The Informational City: Information Technology, Economic Restructuring, and the Urban Regional Process.* Blackwell, 1989.

Debluë, Claire-Lise Debluë. *Exposer pour exporter: Culture visuelle et expansion commerciale en Suisse (1908–1939).* Éditions Alphil-Presses universitaires suisses, 2015.

Gugerli, David. *Redeströme. Zur Elektrifizierung der Schweiz 1880–1914.* Chronos, 1996.

Gunning, Tom. "Before Documentary: Early Nonfiction Films and the 'View' Aesthetic." *Uncharted Territory: Essays on Early Nonfiction Film*, edited by Daan Hertogs and Nico de Klerk. Stichting Nederlands Filmmuseum, 1997, pp. 9–24.

Hediger, Vinzenz, and Patrick Vonderau. "Record, Rhetoric, Rationalization: Industrial Organization and Film." *Films That Work: Industrial Film and the Productivity of Media*, edited by Vinzenz Hediger and Patrick Vonderau. Amsterdam University Press, 2009, pp. 35–49.

Hobsbawm, Eric, and Terence Ranger, editors. *The Invention of Tradition.* Cambridge University Press, 1983.

Imhof, Kurt. "Das kurze Leben der geistigen Landesverteidigung. Von der 'Volksgemeinschaft' vor dem Krieg zum Streit über die 'Nachkriegszeit' im Krieg." *Krise und sozialer Wandel, 2: Konkordanz und Kalter Krieg. Analysen von Medienereignissen in der Schweiz der Zwischen- und Nachkriegszeit*, edited by Kurt Imhof. Seismo, 1996, pp. 19–83.

Imhof, Kurt. "Wiedergeburt der geistigen Landesverteidigung. Kalter Krieg in der Schweiz." *Krise und sozialer Wandel, 2: Konkordanz und Kalter Krieg. Analysen von Medienereignissen in der Schweiz der Zwischen- und Nachkriegszeit*, edited by Kurt Imhof. Seismo, 1996, pp. 173–247.

Jaques, Pierre-Emmanuel. "La propaganda nationale par le film: Albert Masnata et l'Office suisse d'expansion commerciale." *Revue historique vaudoise*, no. 115, 2007, pp. 65–78.

Jaques, Pierre-Emmanuel, and Yvonne Zimmermann. "Dokumentarischer Film in der Schweiz im historischen Überblick (1896–1914)." *Schaufenster Schweiz:*

Dokumentarische Gebrauchsfilme 1896–1964, edited by Yvonne Zimmermann. Limmat, 2011, pp. 84–127.

Latour, Bruno. *The Pasteurization of France.* Harvard University Press, 1988.

Latour, Bruno. *Science in Action:. How to Follow Scientists and Engineers through Society.* Harvard University Press, 1987.

Marchal, Guy P. *Schweizer Gebrauchsgeschichte. Geschichtsbilder, Mythenbildung und nationale Identität.* Schwabe, 2006.

Marchal, Guy P., and Aram Mattioli, editors. *Erfundene Schweiz. Konstruktionen nationaler Identität.* Chronos, 1992.

Mitchell, W.J.T. "Imperial Landscape." *Landscape and Power*, edited by W.J.T. Mitchell University of Chicago Press, 1994, pp. 5–34.

Mooser, Josef. "Die 'Geistige Landesverteidigung' in den 1930er Jahren. Profile und Kontexte eines vielschichtigen Phänomens der schweizerischen politischen Kultur in der Zwischenkriegszeit." *Schweizerische Zeitschrift für Geschichte*, vol. 47, no. 4, 1997, pp. 685–708.

Mörtzsch, Friedrich. *Die Industrie auf Zelluloid. Filme für die Wirtschaft.* Econ, 1959.

OSEC. *Catalogue des films suisses industriels et économiques/Katalog schweizerischer Filme industriellen und wirtschaftlichen Inhalts.* Office suisse d'Expansion commerciale, 1956.

OSEC. *Rapports de l'Office suisse d'expansion commerciale, Zurich et Lausanne.* Office suisse d'expansion commerciale, 1938.

OSEC. *Rapports de l'Office suisse d'expansion commerciale, Zurich et Lausanne.* Office suisse d'expansion commerciale, 1939.

OSEC. *Rapport sur l'action de l'OSEC dans le domaine de la propagande filmée.* Office suisse d'expansion commerciale, 1942.

Schweizerische Bundeskanzlei. "Abgestimmte Volksinitiativen." Bundeskanzlei BK, www.admin.ch/ch/d/pore/vi/vis_2_2_5_7.html. Accessed December 1, 2016.

Schweizerische Zentrale für Handelsförderung/Office suisse d'expansion commerciale, editor. *Wirtschaftliche Mitteilungen/Informations économiques*, vol. 18, no. 27, July 5, 1939.

Skenderovic, Damir. "Die Umweltschutzbewegung im Spannungsfeld der 50er Jahre." *Achtung: die 50er Jahre! Annäherungen an eine widersprüchliche Zeit*, edited by Jean-Daniel Blanc and Christine Luchsinger. Chronos, 1994, pp. 119–46.

Skvirsky, Salomé Aguilera. *The Process Genre. Cinema and the Aesthetic of Labor.* Duke University Press, 2020.

Zimmermann, Yvonne. "Industriefilme." *Schaufenster Schweiz: Dokumentarische Gebrauchsfilme 1896–1964*, edited by Yvonne Zimmermann. Limmat, 2011, pp. 242–381.

Zimmermann, Yvonne. "Negotiating Landscape: Engineering Consent on the Exploitation of Water Power in Swiss Corporate Films." *Cultural Heritage and*

Landscapes in Europe/Landschaften: Kulturelles Erbe in Europa, edited by Christoph Bartels and Claudia Küpper-Eichas. Deutsches Bergbau-Museum, 2008, pp. 180–191.

Zimmermann, Yvonne. "Vom Lichtbild zum Film. Anmerkungen zur Entstehung des Industriefilms." *Montage AV*, vol. 15, no. 1, 2006, pp. 74–88.

About the Author

Yvonne Zimmermann is professor of media studies at Philipps University Marburg, Germany. She is the editor and co-author of a volume on "useful cinema" in Switzerland (*Schaufenster Schweiz: Dokumentarische Gebrauchsfilme 1896–1964*, 2011), the co-author of *Advertising and the Transformation of Screen Cultures* (with Bo Florin and Patrick Vonderau, 2021) and the co-editor of *How Film Histories Were Made: Materials, Methods, Discourses* (with Malte Hagener, 2024). She has published widely on industrial film, "useful cinema," and non-theatrical film.

Section 3

Institutions and Distribution Frameworks:

Archives, Festivals, Fairs

10 Industry on Screen

The British Documentary in Distribution – British Transport Films: A Case Study

Steve Foxon

Abstract

This chapter explores the struggle that British documentary film units faced in getting their films seen. It examines the essential relationship between sponsored filmmaking and its distribution. It analyses the distribution methods of the pioneer documentary units and focuses on the methods adopted by one industrial film unit, British Transport Films, dedicated to the needs of nationalized transport and a cousin to the industrial film unit of nationalized coal. This historical analysis highlights the essential relationship, where distribution is key to conveying the message and ensuring the success of a production. The demise of the industrial film unit is then shown to go hand in hand with the demise of the distribution networks as corporate communications change direction.

Keywords: industrial film; British documentary; transport; British Transport Films; Anstey, Edgar (1907–1987); distribution

Whatever the purpose, we are all wasting our time unless the recipients of the message, the chosen audience, can be reached.

– Edgar Anstey[1]

1 Anstey, Edgar. "Skill, Time and Money Needed If Audience Is to Be Impressed." *The Times*, June 27, 1974, p. 16.

Hediger, V., F. Hoof, Y. Zimmermann, with S. Anthony (eds), *Films That Work Harder: The Circulation of Industrial Film.* Taylor & Francis Group 2024

DOI: 10.5117/9789462986534_CH10

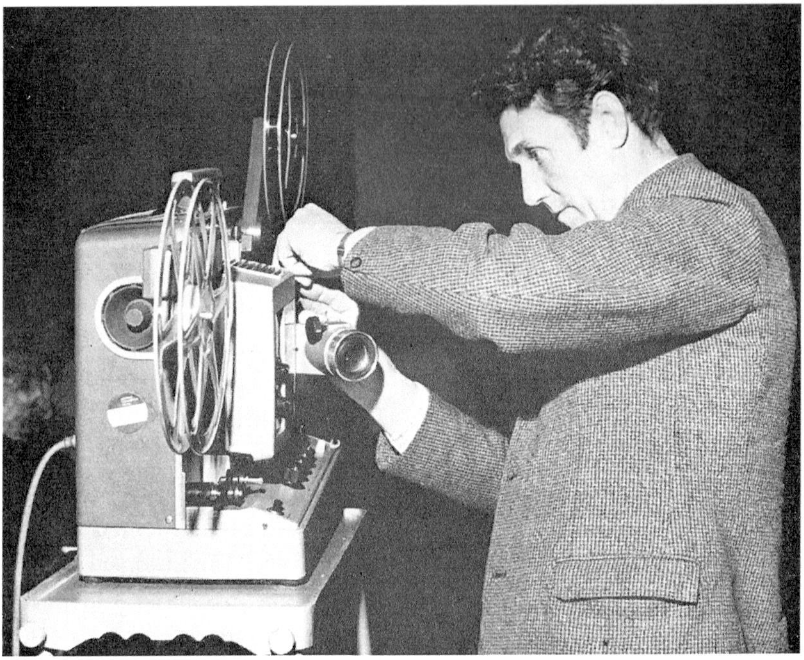

Fig. 10.1. Lights dimmed and projector ready for a BTF non-theatrical film show, c. 1970.

It is hard to imagine the significance and scale of audiovisual needs within large organizations during the pre-email and internet age. Marketing films to generate revenue, staff training films and internal newsreels were all essential communication tools for large companies, yet the films produced could only be considered successful if they could reach the audience for which they were made. The challenges and opportunities facing nationalized transport within the UK during the 1950s and subsequent decades represent a significant UK example of how the documentary film could be put to work in large industry. British Transport Films (BTF), the film-producing department of the British Transport Commission, successfully implemented a structured method of film distribution from the very start of its existence in 1949, ensuring that all films produced by the unit reached their target audience. This study examines the development of documentary film distribution in the UK and examines the methodology behind the successful distribution of documentary within the UK transport industry. It demonstrates that organization and structure were a result of a long learning curve in how to unify films with their chosen audience and that distribution was a key consideration even before the production of the film itself.

The Development of Distribution for the Documentary Film

The pre-war documentary practitioners never had a successful or harmonious relationship with the British film industry. Distribution of their films was a key sticking point. Before World War II, documentary distribution in the UK had been arranged on a film-by-film basis with chosen venues. The non-theatrical field was very new and in the earliest throes of development. The predominant General Post Office Film Library, which was to influence many of the post-war industrial libraries, was not established with a view to generating significant income or with a view to repaying production costs; indeed much of its distribution was free of charge. Theatrical cinema distribution in this pre-war period involved having to convince the cinema trade that a film was worth showing. Certain industrial documentary films, notably *Night Mail* (1936) and *North Sea* (1938) from the GPO Film Unit, and *The Future's in the Air* (1937) from Imperial Airways, had enjoyed wide commercial distribution upon release, finding a place within cinema programmes amongst the newsreel and the supporting feature, although it was claimed that these films did not recover their production costs owing to the low prices paid by the exhibitors for this sort of film.[2] The cinema programme of the 1930s was deemed to be very formulaic and strictly commercial, with little room for the emerging documentary save for the entrepreneurial cinema manager willing to experiment, providing it was on favourable terms for them. Documentaries such as *Song of Ceylon* (1934), *To-day We Live* (1937) and *The Londoners* (1939) also received significant showings, slowly garnering their reputation largely in press reviews and at film society and specialist screenings. There are no known examples in the UK of a pre-war documentary film ever standing with a commercial footing, returning its costs or generating a profit.

During the war, however, with very few exceptions, the distribution of documentary film was taken outside the concerns of the documentary production units. With all significant documentary production being turned over to the war effort the Ministry of Information (MOI) became a central point of governmental film commissioning and distribution. The MOI arranged its own theatrical distribution contracts with the cinema trade through the Kinematograph Renter's Society, and built up its own free non-theatrical distribution through the Central Film Library.[3]

2 Rotha, Paul. *Documentary Film Focus*. Albyn Press, 1947, p. 11.

3 A model based entirely on Grierson's non-theatrical activities of the GPO Film Library but with significant momentum built up in the ensuing years before 1939.

By 1946, the Central Office of Information (COI) had replaced the MOI, becoming the UK government's marketing and communications agency. With much of the UK documentary production originating with, or for, government departments the COI found itself playing a key role in the representation of British documentary film. The COI Films Division structured itself into two forms of distribution to ensure its official films met with their desired audiences: commercial and non-theatrical. The COI had contracted a significant number of its films to major distributors in 1947, which were reported as securing for the Exchequer, between seventy and eighty per cent of the gross rentals paid by the cinemas. These films included: *Children on Trial*, *Now It Can Be Told*, *English Criminal Justice*, *Defeated People*, *Fenlands*, *Man*, *One Family*, *Cyprus Is an Island*, *Diary for Timothy*, and the *This Is Britain* cinemagazines (all 1947).[4] However, in the 1947 year's review by its director, Robert Fraser noted that whilst relations were considered to be good with major distributors it was highlighted that documentary films were not popular with exhibitors and their commercial distribution was very difficult.[5] The most commercially successful theatrical arrangements of the government's films had been the feature documentaries showing efforts on the front line: *Target for Tonight* (1941), *Coastal Command* (1942), *Desert Victory* (1943), *Tunisian Victory* (1943), *Western Approaches* (1944), *Burma Victory* (1945), *Journey Together* (1945), and *The True Glory* (1945). The public had been happy to pay to see these films because they were dramatized news, with all the physical excitement and dramatic action of battle, raid and shipwreck but without the danger. Technically they were very well-made with high production values; they were not, however, in the traditional line of peace-time documentary, and with a few exceptions, they were not made by technicians from the pre-war Grierson-influenced documentary group. Documentary practitioners of industrial, social and informational documentaries were largely engaged during the war in producing for non-theatrical exhibition via the Central Film Library.[6] Their theatrical distribution was free by arrangement with the cinema trade and not on the commercial terms of the MOI's feature documentaries.

On the non-theatrical side, many COI productions were aimed at specialist audiences such as those found in factories, schools, women's

4 Fraser, Robert. *Report on the Work of the Central Office of Information, London*. 1947. The National Archives (TNA), London, INF12/1584.

5 Ibid.

6 Most notably Arthur Elton and Edgar Anstey.

organizations and youth groups etc. In the first six months of the COI it was claimed that 25,000 non-theatrical film shows were given in the UK. This represented a key audience for the COI who reported 10,000 shows per year for the various government departments such as agriculture, food, health and labour; 12,000 shows a year in factories, and 15,000 in towns and villages to general audiences.[7] The closure of the Crown Film Unit (the government's own internal documentary unit, the successor to John Grierson's pioneering GPO Film Unit and often regarded as the flagship of post-war British documentary) in 1952 symbolized for many contemporary writers of British documentary an end to a tradition of government film sponsorship. On the contrary, government sponsorship through the COI and through the post-war nationalized industries actually increased to greater levels. Non-theatrical audiences had established a demand for documentary, though one that that was unfortunately based on low cost or free distribution, paving a way for audience-specific and targeted sponsored films.

In the late 1940s a popular misunderstanding had manifested itself amongst the industry and public alike that the documentary film should be considered as a short film. Paul Rotha, a champion of more lengthy productions, had claimed that this "short film" standpoint had done much to damage the development of the documentary film and its desirability with audiences, he argued vigorously to separate the documentary medium from that of other short films, claiming that the motive underlying the production of advertising films, for which screen space is hired by the advertiser, could not be remotely linked with the founding principles of documentary production.[8] Though that may have been true for Rotha, the industrial films that were to come in the ensuing years from the Shell Film Unit, and the nationalized industries of coal and transport clearly directed an element of their production output towards a more commercial standpoint – not necessarily with a view to recovering costs but certainly to instil a message and create an appetite for the theme. Arthur Elton and Edgar Anstey (representing the film units of Shell and British Transport, respectively) had both gained significant experience producing films for private sponsors prior to the war, and found it far easier to accept more commercial ambitions for their post-war documentary ideology.

7 Fraser. *Report on the Work of the Central Office of Information*.

8 Rotha. *Informational Film Yearbook*. Albyn Press, 1946, p. 14.

Reaching the Audience: British Transport Films – A Case Study

Just one year after the nationalization of transport by the Labour government, the British Transport Commission founded British Transport Films in 1949 to produce sponsored films generally falling into one of three categories; informational, traffic-promoting and staff instructional. The Commission stated that it would be its policy to use film as a major arm in the field of publicity and information.[9]

At the helm of the film unit as chief officer of films was Edgar Anstey, one of Grierson's pre-war acolytes who, along with Arthur Elton, had been assigned by John Grierson to produce films for industrial sponsors outside of the government-restricted GPO Film Unit. The Transport Act of 1947 brought under one unified government umbrella all transport within the UK except for airways, creating a number of executives, London Transport, British Railways, British Road Services, British Waterways, Ports, Hotels and a number of bus companies, collectively referred to as the British Transport Commission.

With railways as the dominant executive within the Commission, much of the work for BTF was focused on railways with the railway information film designed to keep the general public abreast of development, to report railway problems and the steps taken to solve them. The over-riding intention was to win sympathy of the public and to obtain cooperation in their desire to improve the efficiency of the railway service.[10] The main purpose behind the railway traffic-promoting films, on the other hand, was to directly increase revenue sales by persuading the public to use the railway for both business and pleasure. Frequently the two ambitions merged and examples such as *Service for Southend* (1957) and *Blue Pullman* (1959) fall across these two key objectives. Edgar Anstey, having witnessed the struggles of securing distribution in the pre-war Grierson led years and who had experienced the development of the non-theatrical markets first-hand during the war, was well versed in producing films that could lend themselves to more than one key target audience.

Clearly it is wrong to make a film, however precisely it may express the message of the sponsor, if the subject-matter or the nature of the treatment precludes it from adequate distribution. The production of a film of good quality represents a considerable expenditure (ranging from £1,000 per

9 British Transport Commission. *British Transport Films: Annual Report and Accounts for 1954*. 1955, p. 119.

10 Anstey, Edgar. "Bringing Railways to the Public." *The Film User*, vol. 9, no. 105, July 1955.

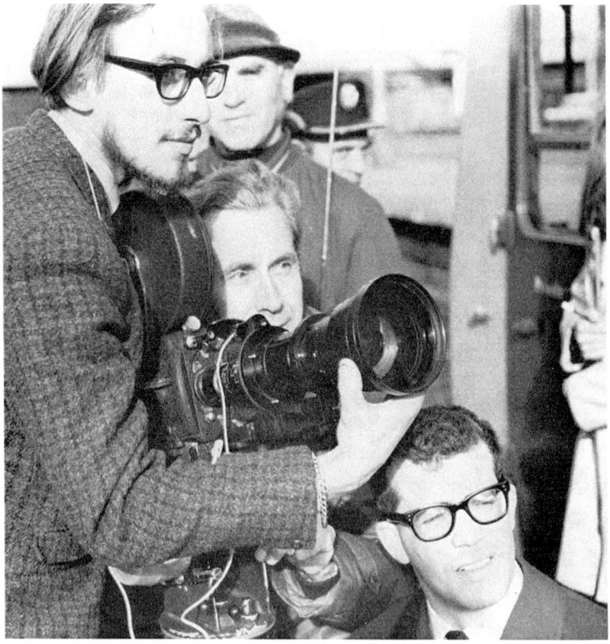

Fig. 10.2. The BTF unit on location at Marylebone railway station, London, during 1969.

ten-minute reel to £5,000 per ten-minute reel in exceptional circumstances). We regard the average cost of reaching each audience member as a major factor in planning our film programme.[11]

For transport, the sponsored film proved essential to a wider purpose than a simple targeting of theatrical and non-theatrical audiences, and the distribution of a film was actively planned for and managed as a key stage in the life and usefulness of a film even before its production had commenced.

The successful distribution of a film made use of several of the five established routes available to BTF in reaching its desired audience:

1) **Theatrical:** BTF continued a tradition of aiming prestige sponsored films at the theatrical cinema trade. Many of the libraries films gained a commercial cinema screening and in this way returned a varying proportion of their production cost. As seen throughout the war years and before, it is unlikely that any production returned its true production costs but a measure of return was adequate to appease Anstey's superiors; financial return being considered as secondary, in the opinion of the Commission, to the magnitude of audience size being reached.

BTF had a most notable success in gaining theatrical screen time, between the cinema-owning giants of the Rank Organization and the Associated

11 Idem. "How We Use Films for Transport." May 1955. Speech delivered to UIC Congress.

British Picture Corporation (ABC) with their respective "Look at Life" and "Pathé Pictorial" supporting programmes,[12] there was little room at the cinema for the aspirations of many industrial film units who felt their documentaries were equally worthy of the big screen.[13] BTF had learnt an important key lesson very early on in its theatrical screenings and one that stayed with it throughout its productive years, serving it well when it came to attracting bookings from the cinema trade: Not all cinema-going audiences wished to see its films! In 1952, BTF's thirty-minute survey of planning and efficiency upon the railway network, *Train Time*, had been touring theatrically with the feature film *The Gift Horse* (UK, 1952), starring Trevor Howard and Richard Attenborough. One of BTF's editing team, John Legard, who later became chief editor at BTF, decided to see how the film was fairing on the big screen at the Odeon, Kensington. To his disappointment the audience, who had been there to be entertained by seeing officialdom as played out by the actors of the day, had not anticipated on actually seeing thirty minutes of real government officialdom in all its sponsored glory extolling the virtues of the nation's railway system. The full house responded with an indignant slow-hand clap, which was reported back to Anstey the following morning. Aside from the realization of the need for entertainment, a small change in the company's logo inserted on theatrical release films took place. No more were the words "British Transport Films" displayed; instead, a lozenge-shaped logo with the acronym "BTF" was inserted, a less subtle nod to its official government stature. Upon reflecting over twenty-one years at the helm of BTF, Edgar Anstey wrote that publicity, public relations and instruction are, however, not entertainment – although a measure of entertainment must be present if audience attention is to be held.[14]

2) Television: The BBC television service frequently broadcast internally produced documentaries in their post-war scheduling and happily transmitted sponsored films from BTF under the auspices of public information.[15]

12 "Look at Life" and "Pathé Pictorial" supported short films, thereby thwarting the aspirations of many for the cinema distribution of their documentaries and short films. Cheaply made by the UK cinema-owning giants they occupied programme space alongside a newsreel and supporting short feature film prior to the main exhibited feature, thus ensuring that the cinema giants did not have to share their earnings by taking on additional exhibition agreements bar that of the main and supporting feature.

13 Russell, Patrick, and James Piers Taylor, editors. *Shadows of Progress: Documentary Film in Post-war Britain.* Palgrave Macmillan, 2010, p. 20.

14 Anstey, Edgar. "British Transport Films Rolls on for 21 Years." *The Film User*, vol. 24, June 1970.

15 It is not recorded to what extent, if any, discourse emerged from broadcasting-sponsored documentaries on the BBC. Under the banner of public information it was considered acceptable and within the BBC's remit to broadcast on such topics. Though it is clear that films such as those

Revenue for television transmissions was far less than that gleaned from theatrical distribution but the significantly large nationwide audience that could be reached simultaneously far outweighed any shortcomings in revenue.

Bookings received via the theatrical and television outlets were to be considered mutually exclusive. Cinema renters were not keen to hire films which had appeared on television and a film which had completed a theatrical cinema tour would equally be of low interest to be considered acceptable to the BBC.

The year 1955 was a significant one for British Railways as it marked the introduction of the first "Modernization Plan" – a plan to introduce radical efficiencies to the railway network, introduce modern diesel traction and eliminate steam powered locomotives from the network by 1970. The modernization proposals brought an audience distribution coup for the film unit, and cemented the start of an excellent working relationship with the BBC in the shape of a short five-minute film, *Signpost* (1955). The film demonstrated that factual information to the advantage of British Railways was acceptable by the BBC for television broadcast. A few days prior to the debate in Parliament on the plan for the modernization of British Railways, the unit had prepared a brief record of Britain's first electrified main line railway which could indicate, to the television watching public, the shape of things to come. A discussion with the BBC indicated that the film should be ready for broadcast as soon as the final speeches in the debate had been made and, in this way, the rather difficult technical and financial terms of the Parliamentary debate were brought alive to those following the debate on radio and television. The film was, of course, directly propagandist in style but considered acceptably so by the BBC and those at BTF in view of the national character of British Railways and the national importance of the subject.

In later years when colour transmissions were being aired specifically for television engineers and retail outlets selling new colour television sets, BTF colour films as well as many other industry-sponsored films were given a reprise airing. These transmissions, known as the BBC2 Trade Test Transmissions, were televised around the scheduled daily programmes and aired from 1956 to 1973 offering many titles that could be repeated, often over seventy-five times.

from British Transport and Central Office of Information benefitted from televised airtime, illustrious documentary units from the private sector (such as Shell Oil) found less success in television broadcast prior to the advent of Independent Television in 1955.

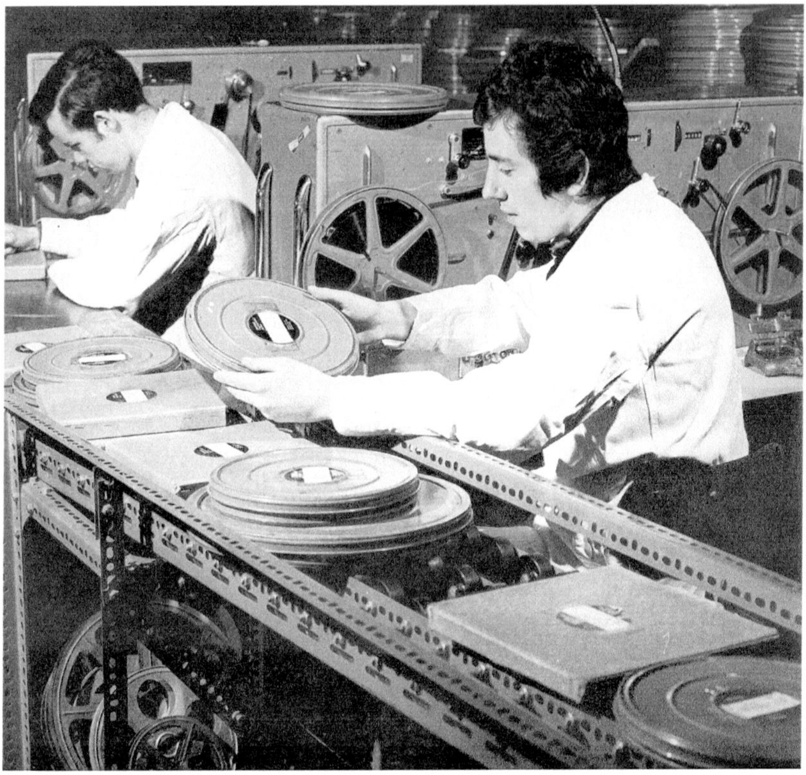

Fig. 10.3. Checking 16 mm prints on completion of a screening. As many as two hundred copies of each title were held in the BTF Film Library.

3) BTF Film Library: 16 mm copies of BTF films were made available for hire non-theatrically to suitable groups, clubs and societies owning their own projectors, from the BTF Film Library in much the same vein and along the same lines as those established at the GPO and Central Film Libraries.

If a film was destined to feature on a theatrical tour, its status in the library was held until completion of the theatrical circuit. The library was helped in its ambition to keep film rental charges to a minimum (often free of charge) by their ability to ship films throughout the country via the railway's own parcels service.

4) Film Shows: In addition to the hiring of 16 mm film prints, film shows were given free to invited public audiences by means of British Railways' fleet of over one hundred 16 mm projectors distributed at regional offices throughout Britain. The films service also controlled three cinema coaches which were reported to be in constant use throughout the railway regions. Two of the coaches were intended for screening films to staff only at stations and depots. The third cinema coach, however, was available for screening

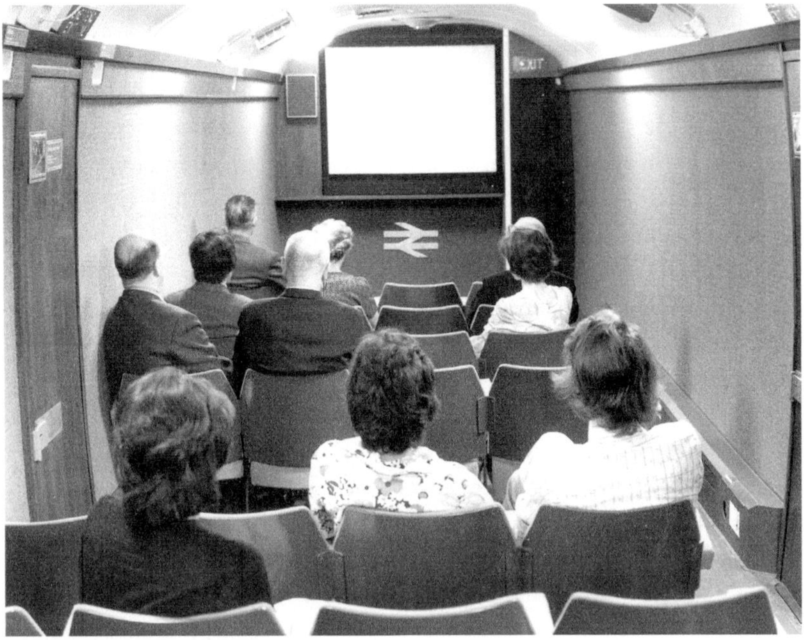

Fig. 10.4. British Rail's prestige cinema coach, which was launched in 1975, could be used on special trains to show films at 100 mph!

films on the move as part of a train (i.e. special trains hired by organizations) or used by the railway for prestige events.[16]

5) **Overseas distribution:** Films were also distributed to overseas audiences through the cinemas, by television and through non-theatrical channels. Such distribution had been arranged through the British Railways overseas offices, the International Union of Railways, through the overseas agencies of the government and through commercial distributors. Special overseas versions of many prestige films were prepared with soundtracks and commentary specially recorded in a multitude of different languages.

The Audience

Some staggering audience statistics were reported. A snapshot of BTF distribution at the end of 1954 reveals that the twenty films in British theatrical distribution were seen by slightly more than a million people. The twenty-nine films available for the same year in non-theatrical distribution were

16 Willmott, Alan J. *British Railways Report on Cinema Coaches*. BRB, 1969.

seen during one month by 750,000 people.[17] This figure for non-theatrical coverage was reported as growing steadily year by year and theatrical showings (always variable and unpredictable in nature and extent) were forecast in time to be somewhat diminished in relative importance compared to the value of the non-theatrical market. The high but infrequent audience figures for television transmissions certainly swell the total audience numbers and they do not sit very comfortably alongside the numbers from the traditional film-screening distribution methods. The Commission's film *Signpost* received an evening transmission by the BBC and on that single occasion it was reported that there had been over four million viewers.[18]

The subject of railways and the wider transport industry had proved itself to be an acceptable topic as marketable entertainment, and Anstey, given the unit's experience with *Train Time* at the Odeon, Kensington, was very aware that the amount of direct advertising they contain must be carefully restricted. This consideration, however, did not apply, to the same extent, to films destined to be shown through the non-theatrical channels. That said, an element of care and control was given in respect to the large and increasing numbers of non-theatrical audiences regularly accepting Commission films – if groups were to continue to find them attractive, advertising would have to remain unobtrusive. This is most evident in the case of the films aimed at schools. The classroom had been a staple user of industrial documentary film libraries and BTF were all too aware that the classroom audience represented customers and potential recruitment opportunities. In general, however, the Commission appeared to have made a conscious decision to make no sharp distinction between theatrical and non-theatrical styles.

The public relations officers within the British Transport Commission were quick to capitalize on the use of film to sell the transport services. The impact of non-theatrical shows was boosted in a variety of ways. Short trailers advertising particular railway services could be included in the programmes; leaflets were available to hand out to the audiences; at larger events a commercial representative could be present and on hand to answer questions or put forward specific sales arguments. In this way the screening of the films in the programme were used not so much for direct selling but to invoke an "appetite" indirectly generating a desire to travel or to formulate a climate of opinion favourable to railways. The short trailers, leaflets and the commercial representatives could then take advantage of the mood created by the film and do the selling without the need for the

17 British Transport Commission. *British Transport Films*, p. 124.

18 Anstey. "Bringing Railways to the Public."

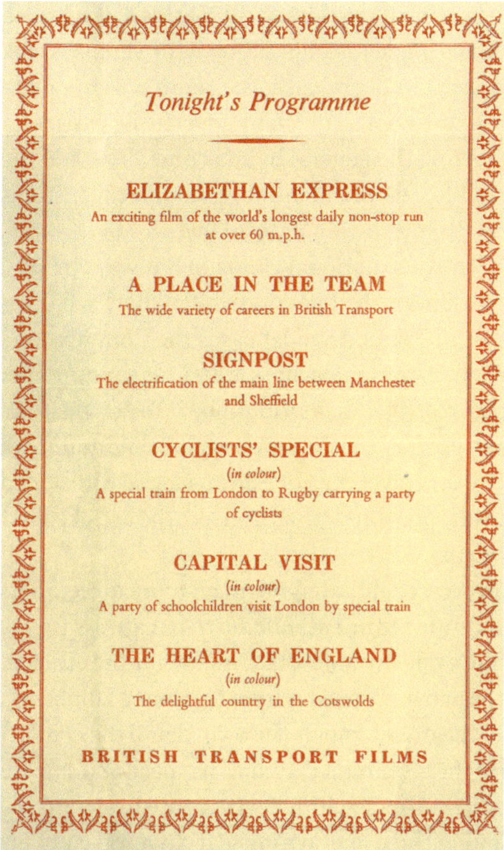
Tonight's Programme

ELIZABETHAN EXPRESS
An exciting film of the world's longest daily non-stop run at over 60 m.p.h.

A PLACE IN THE TEAM
The wide variety of careers in British Transport

SIGNPOST
The electrification of the main line between Manchester and Sheffield

CYCLISTS' SPECIAL
(in colour)
A special train from London to Rugby carrying a party of cyclists

CAPITAL VISIT
(in colour)
A party of schoolchildren visit London by special train

THE HEART OF ENGLAND
(in colour)
The delightful country in the Cotswolds

BRITISH TRANSPORT FILMS

Fig. 10.5. A BTF non-theatrical programme for 1955 creates an appetite for travel.

film to resort to direct advertising. This method of marketing could often be associated with the traffic-promoting type film – films that offered beautiful landscapes, seaside panoramas and party travel conveniences. Films like *This Year London* (1951) and *Capital Visit* (1955) were made to promote travel in terms of group excursions – in the former a factory staff outing, in the latter a party of school children. These films were shown first in the cinemas and then cascaded to the non-theatrical library to help sell the kind of travel they depict. Addressing issues around railway usage, Anstey reported that travel-promoting films are helping to solve one of the most serious problems on the railways – how to spread and minimize the economic burden of facilities brought into existence to meet short periods of peak demand.[19]

19 Anstey, Edgar. "How Films Serve the World's Biggest Employer." *Industrial Screen*, vol. 3, no. 2, May–June 1959, p. 92.

The travel films deal for the most part with particular areas of the British Isles, although a few productions made in conjunction with Thomas Cook & Son covered subjects overseas. In every case, the remit was to create interest in the attractions of the area not so much by advertising particular resorts, hotels etc., but by showing the general holiday amenities and the attractions of the local way of life. Historical buildings and local customs often find a place, together with interesting natural features of the area. In addressing the congress of the International Union of Railways in 1955, Anstey said: "To gain good distribution, it appears essential that such films should nowadays be produced in colour. In most cases the Commission's films have been photographed in 16 mm Kodachrome and 'blown-up' either to Technicolor or Eastman Color for 35 mm distribution."[20] To demonstrate the modernity of filmmaking, portray the truest possible scene and instil the strongest possible appetite for travel the colour travelogue film added a prestige not yet available to all sponsored films. Indeed, the age of colour did not reach television until 1967.

The information film, however, could often be much more direct in its propaganda and still rank as entertainment. *Elizabethan Express* (1954), technically an information film exclusively devoted to showing the efficient work behind the running of the non-stop express from London to Edinburgh, enjoyed a long and successful theatrical tour. It shows in detail the comfort and up-to-date meals served in the restaurant cars and the speed and smooth running of the train. All of these direct-selling features are drawn to the viewer's attention by employing the use of whimsical rhyming couplets and light popular verse.

The unit's traffic-promoting films for the most part deal with passenger travel though films were made to show efficiencies in the handling of freight traffic, produced with a view to increasing revenue in particular fields. For example, *Farmer Moving South* (1952), which proved very popular on the BBC, receiving a number of television transmissions, shows a whole farm together with its livestock being moved by rail from Yorkshire to Sussex. A prime example of the unit's output proving effective with audiences outside of the specific original target market.

In its instructional filmmaking the Commission made some considerable headway towards solving the basic problem of reaching the employee in a large and dispersed industry. Some 900,000 employees made British Transport the largest employer in the world, presenting some integration challenges and opportunities for the sponsored film. Employees were

20 Idem. "How We Use Films for Transport."

shown training films of new techniques and modern methods that would be winging their way to the regional areas from mobile projectors, via the cinema coaches, instructional trains or from screenings in the Commission's training schools and colleges. The cost of producing instructional films was significantly less than the prestige films and reported to rarely exceeding £1,000 per ten-minute reel.[21] Whenever possible, to save on resources, the unit happily produced educational filmstrips as alternatives to film if it was felt this medium was better suited.[22] Many of the filmstrips could also be hired from the BTF Film Library and some proved very popular within schools and railway clubs. Colour, music and sound effects could all be omitted unless considered necessary to the instruction being given. A successful example of the training films is the driver training series *Diesel Train Driver* (1959). Produced as a series of five films, it forms part of a complete visual package which included other films and filmstrips and was designed to interlock precisely as part of a course of instruction. A series of internal newsreel-style films were also produced to keep staff abreast of developments and to boost morale. Regular updates continued almost until the close of the unit in the early 1980s.

Decline, Closure and the Digital Age

From the late 1960s there is a notable shift in BTF's target audience and subject matter. The decline in production of the prestige sponsored film is a response to a national decline in the theatrical supporting programme at the cinemas, no doubt heavily influenced by the widespread popularity of television. British Transport's prestige films slowly give way to corporate communications as the new filmmaking order of the day. By the 1970s external trends began to affect the rail industry and these factors are obviously reflected in the change of subject matter for the film unit. The motor car had become commonplace within UK families and any planned holiday would more often than not involve the family car rather than rail travel. The taste for continental holidays was also beginning to take hold. Though there are notable exceptions which still attracted a prestige

21 Idem. "How Films Serve the World's Biggest Employer," p. 93.

22 Educational filmstrips were long considered by school pupils as the poor substitute to a moving image film. Filmstrips were the precursor to slide shows and were used in classrooms to demonstrate process or relay history, often with supporting notes and scripts to be read out by the school teacher or a visiting professional.

cinema audience such as *Wires over the Border* (1974) and *Overture One-Two-Five* (1979), the vast majority of film production for the 1970s becomes geared to communicating to the staff of the railway and towards internal railway modernization and safety subjects. Despite this notable change in production, the film library remained in healthy active use and new film prints appear to be ordered for many of the popular titles right in to the 1980s in a library that by this time contained over five hundred titles. Even after British Railways closed its own film library in the mid-1980s many of the films transferred to the Government Central Film Library for continued availability to non-theatrical audiences, clearly boosted by the many hundreds of railway enthusiasts keen to fill their clubs' regular programme.

Rebranded in 1983 to CFL Vision, the Central Film Library remained as the last bastion of active film and video distribution in the non-theatrical field, actively distributing BTF (and COI) films until its own closure in March 2000, surprisingly outliving British Railways itself, which had been privatized in stages between 1994 and 1997. The BTF Film Library and the films produced under the nationalized industries of coal and rail survive and flourish today in the care of the British Film Institute's National Archive – many of the films can be seen theatrically in retrospectives of the sponsored film, available to own on DVD for home use and now garnering an enthusiastic following across digital platforms.

Conclusion

The challenges and opportunities that faced nationalized transport within the UK during the 1950s and beyond lent themselves well to the application of industrial documentary both for commercial entertainment, information and for staff training and communication. Film proved to be an exceptionally useful tool to convey a unified corporate message in an industry with over 900,000 employees. Edgar Anstey embraced the lessons of those pioneering documentary days which had allowed him the opportunity to implement a structured method of film distribution from the establishment of the film unit in 1949, ensuring that all films produced by the unit reached their target audience. The concept of a documentary film covering its cost, however, was not completely addressed in the post-war field. Grierson had established a wider distribution method outside of the cinemas in his forming of the non-theatrical GPO film library but at a price – that audience consumption did not generate income and audiences expected non-theatrical screenings at extremely low cost or, for the most part, free of charge. Like Grierson before him, Anstey had created a unit within a government department with a remit to capture the widest possible audience. As the medium had developed, greater opportunities to reach new audiences had emerged and Anstey's unit was adaptive with the foresight to consider distribution and audience at the very outset, even before the film had entered its production phase. The unit capitalized and exploited secondary audience possibilities in the pre-production planning stage and post-production opportunities as they arose, such as television transmissions beyond first release. The successes of BTF speak for themselves, numerous awards tell of a prestigious unit consistently pairing high quality industrial filmmaking with key target audiences for the purposes of training, marketing and entertainment. Perhaps the failure for British documentary was that like Grierson before him, Anstey could only make "some" of his unit's work financially solvent enough to cover all costs and return a profit. Perhaps it is unfair to saddle a successful (in every other respect) film unit with this criticism – after all, BTF was well and truly lodged within government sponsorship, just like the GPO had been in Grierson's day. Covering the unit's costs was not a requirement of the remit – reaching the widest possible audience, however, was!

Works Cited

Anstey, Edgar. "Bringing Railways to the Public." *The Film User*, vol. 9, no. 105, July 1955, p. 334.

Anstey, Edgar. "British Transport Films Rolls on for 21 Years." *The Film User*, vol. 24, June 1970, p. 38.

Anstey, Edgar. "How Films Serve the World's Biggest Employer." *Industrial Screen*, vol. 3, no. 2, May–June 1959, p. 92.

Anstey, Edgar. "How We Use Films for Transport." May 1955. Speech delivered to UIC Congress.

Anstey, Edgar. "Skill, Time and Money Needed If Audience Is to Be Impressed." *The Times*, June 27, 1974, p. 16.

British Transport Commission. *British Transport Films: Annual Report and Accounts for 1954*. 1955.

Fraser, Robert. *Report on the Work of the Central Office of Information, London*. 1947. The National Archives (TNA), London, INF12/1584.

Rotha, Paul. *Documentary Film Focus*. Albyn Press, 1947.

Rotha, Paul. *Informational Film Yearbook*. Albyn Press, 1946.

Russell, Patrick, and James Piers Taylor, editors. *Shadows of Progress: Documentary Film in Post-war Britain.* Palgrave Macmillan, 2010.
Willmott, Alan J. *British Railways Report on Cinema Coaches.* BRB, 1969.

About the Author

Steven Foxon is a curator (non-fiction) at the BFI National Archive. He is the lead on the conservation and reappraisal of major British documentary works, such as the GPO Film Unit and the British Transport Films collections as well as works of renowned individual documentary practitioners, such as John Grierson, Humphrey Jennings, and Geoffrey Jones.

11 On the Red Carpet in Rouen

Industrial Film Festivals and a World Community of Filmmakers

Brian Jacobson

Abstract

This chapter examines the cycle of international industrial film festivals that began in Rouen, France, in 1960 and continued in Turin, West Berlin, Madrid, London, Venice, Lisbon, Vienna and a host of other European cities in the 1970s. It focuses on the 1960 festival in Rouen, which screened 160 films from eighteen nations to more than five hundred visitors. It argues, first, that such festivals helped stimulate interest in a critical but neglected movement – an alternative, industrial "new wave" – that many observers hoped would help revive European film industries and, second, that they fostered a form of cultural cosmopolitanism with ramifications that went well beyond the film business itself to both reflect and animate shifting configurations of global capital.

Keywords: film festivals; French cinema; French history; industrial film; corporate film; transnational cinema

In January 1978, the French film journal *L'avant-scène cinéma* detoured from its usual focus on international art cinema and classics of film history to turn its readers' attention to what it termed *un autre cinéma*, the industrial film or *film d'entreprise*.[1] After tracing a tidy auteur history of this "other cinema" from the Lumières to Jean Epstein, Jean Painlevé, Alain Resnais, Georges Franju and Jean-Luc Godard, the issue turned to a lesser-known group of "industrial artisans," including Robert Enrico, Sydney Jézéquel, Jean Guglielmi, Robert Ménégoz and Carlos Vilardebo. In the work of these artisans, the editors argued, "art does not assert itself with pride but rather is put humbly in the service of craft." But they also encouraged readers to find the art in an industrial form in which "the industry, if we look carefully, has never managed entirely to stifle the art."[2]

1 "Un autre cinéma." *L'avant-scène cinéma*, no. 199–200, January 1978, p. 3.

2 Ibid.

Hediger, V., F. Hoof, Y. Zimmermann & S. Anthony (eds), *Films That Work Harder: The Circulation of Industrial Film*. Taylor & Francis Group 2024
DOI: 10.5117/9789462986534_CH11

This tension between art and industry – what Paul Rotha described as cinema's "great unresolved equation" – was, of course, not new.[3] For the editors of *L'avant-scène cinéma*, working in the long wake of the Nouvelle Vague and midcentury art cinema, cinema's status as art could be taken for granted; what needed to be remembered was the other half of the equation. On the cover (fig. 11.1), that half was incongruously rendered in nineteenth-century form with images of trains, mills and traditional forms of agriculture. Inside, the iconic image of Chaplin caught in the cogs from *Modern Times* (1936) appeared opposite the introductory text, which included a reminder, citing André Malraux, that "cinema is [also] an industry" and that its industrial character went beyond the Hollywood dream factory system or its studio-based equivalents (fig. 11.2).[4] The films highlighted in the pages that followed best illustrated the point: in boardrooms and on factory floors, in mine shafts and on oil rigs, in train cars and on cargo ships, filmmakers had founded a new kind of film industry by forging a new kind of industrial art.

While even the most ardent of cinephiles might have needed some convincing about these films' importance, the sentiment must have rung true to many film industry veterans. These films and their makers were hardly unknown to the film business or to business at large. For the previous two decades their work had been celebrated in France and across Europe at yearly national and international industrial film festivals. The product of growing interest in both the French film industry and among the purveyors of the industries shown on their screens, these festivals embodied their films' synthesis of art and industry. They showcased the cultural work of the government agencies and major corporations that made an "other cinema" possible, from Shell, BP and VW, to Siemens, Bayer and Heineken, to Fiat, Rolls-Royce and Renault. They highlighted fierce national rivalries and anxieties about the statuses of both national economies and film economies in reconfigured post-World War II hierarchies. And they became critical sites of exchange, not only among film professionals but also the industrialists who employed them.

Like the more often celebrated festivals in Cannes, Venice and Berlin, industrial film festivals created key nodes in shifting global networks of cultural and economic exchange. They did so, however, with a unique focus on uniting a community of capitalists and what one journalist – writing for

3 The quote appears without attribution in Nowell-Smith, Geoffrey. "General Introduction." *The Oxford History of World Cinema*, edited by idem, Oxford University Press, 1996, p. xix.
4 "Un autre cinéma."

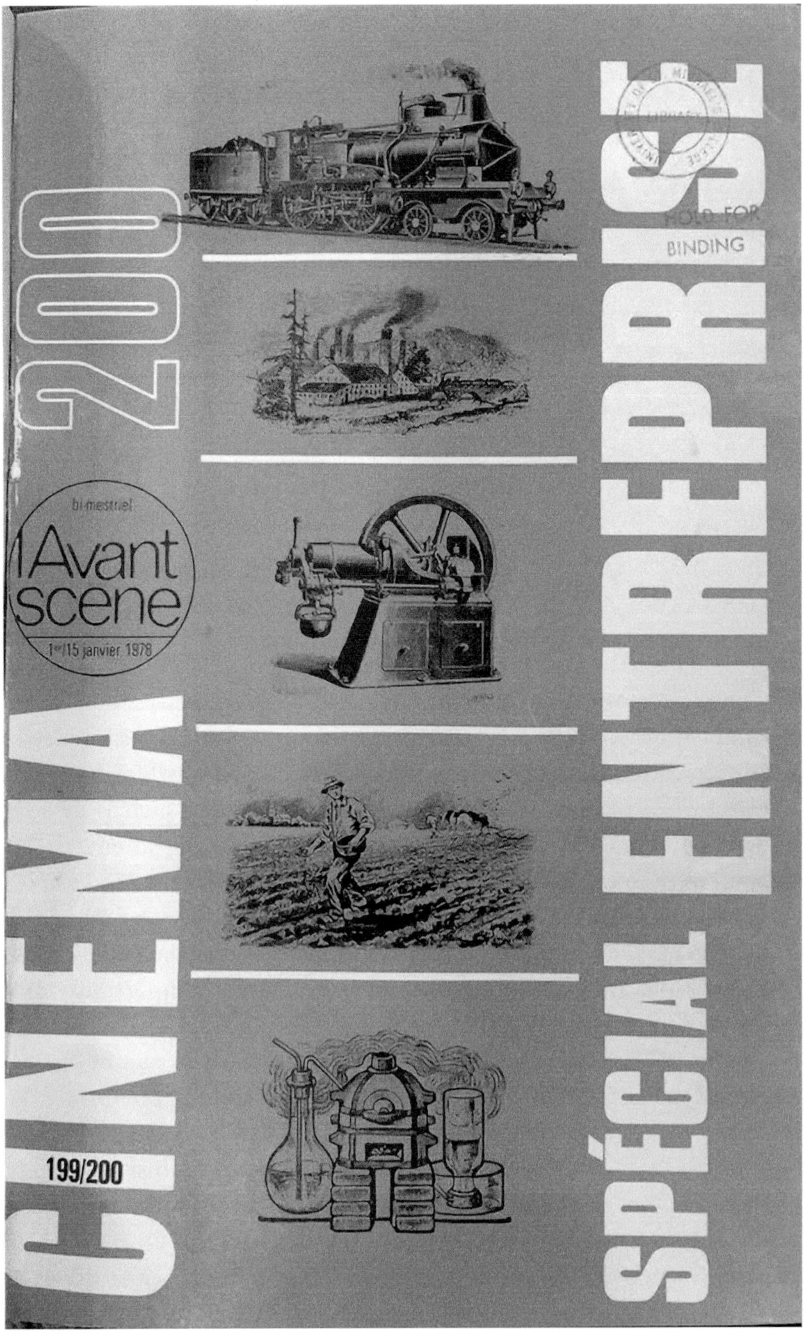

Fig. 11.1. Cover of *L'avant-scène cinéma*.

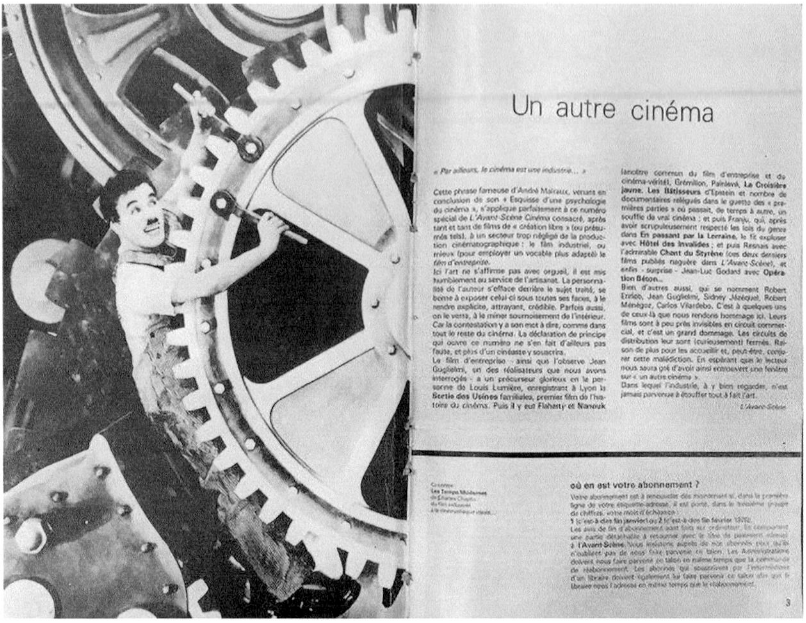

Un autre cinéma

« Par ailleurs, le cinéma est une industrie... »

Cette phrase fameuse d'André Malraux, venant en conclusion de son « Esquisse d'une psychologie du cinéma », s'applique parfaitement à ce numéro spécial de *L'Avant-Scène Cinéma* consacré, après tant et tant de films de « création libre » (ou présumés tels), à un secteur trop négligé de la production cinématographique : le film industriel, ou mieux (pour employer un vocable plus adapté) le *film d'entreprise*.

Ici l'art ne s'affirme pas avec orgueil, il est mis humblement au service de l'artisanat. La personnalité de l'auteur s'efface derrière le sujet traité, se borne à exposer celui-ci sous toutes ses faces, à le rendre explicite, attrayant, crédible. Parfois aussi, on le verra, à le miner sournoisement de l'intérieur. Car la contestation y a son mot à dire, comme dans tout le reste du cinéma. La déclaration de principe qui ouvre ce numéro ne s'en fait d'ailleurs pas faute, et plus d'un cinéaste y souscrira.

Le film d'entreprise – ainsi que l'observe Jean Guglielmi, un des réalisateurs que nous avons interrogés – a un précurseur glorieux en la personne de Louis Lumière, enregistrant à Lyon la **Sortie des Usines familiales**, premier film de l'histoire du cinéma. Puis il y eut Flaherty et **Nanouk** (ancêtre commun du film d'entreprise et du cinéma-vérité), Grémillon, Painlevé, **La Croisière jaune**, **Les Bâtisseurs** d'Epstein et nombre de documentaires relégués dans le ghetto des « premières parties » où passait, de temps à autre, un souffle de vrai cinéma ; et puis Franju, qui, après avoir scrupuleusement respecté les lois du genre dans **En passant par la Lorraine**, le fit exploser avec **Hôtel des Invalides** ; et puis Resnais avec l'admirable **Chant du Styrène** (ces deux derniers films publiés naguère dans *L'Avant-Scène*), et enfin – surprise – Jean-Luc Godard avec **Opération Béton**...

Bien d'autres aussi, qui se nomment Robert Enrico, Jean Guglielmi, Sidney Jézéquel, Robert Ménégoz, Carlos Vilardebo. C'est à quelques-uns de ceux-là que nous rendons hommage ici. Leurs films sont à peu près invisibles en circuit commercial, et c'est un grand dommage. Les circuits de distribution leur sont (curieusement) fermés. Raison de plus pour les accueillir et, peut-être, conjurer cette malédiction. En espérant que le lecteur nous saura gré d'avoir ainsi entrouvert une fenêtre sur « un autre cinéma ».

Dans lequel l'industrie, à y bien regarder, n'est jamais parvenue à étouffer tout à fait l'art.

L'Avant-Scène

où en est votre abonnement ?

Fig. 11.2. Editorial in *L'avant-scène cinéma* with an image of Charlie Chaplin.

the American magazine *Business Screen* – fittingly described as their "world community of filmmakers."[5] This world community had just assembled at the international festival in West Berlin in October 1962, which had been "no 'kulturfest' of film art lovers, nor a tourist attraction such as those at Cannes, Edinburgh or Venice. The European executives who attended [this festival] were concerned with *industrial* film potentialities."[6] The *Business Screen* journalist recognized that these festivals could play a critical role in more general and pressing efforts to promote industrial film not just as a new direction for film culture but also, and more importantly, as a motor for Europe's post-WWII economic recovery and for global capitalism writ large. In other words, these festivals were not only becoming critical sites – much like the art film showcases in Cannes and elsewhere – for the formation of cosmopolitan film culture, but also, and in ways that extended the business side of Cannes, were fostering a cosmopolitanism that went to the very heart of global capitalism. Much as Cannes helped drive the post-war globalization of capitalist culture by promoting the universalism of entertainment and film language, industrial film festivals helped drive the

5 OHC. "The World Community of Film Makers." *Business Screen*, vol. 23, no. 6, October 31, 1962, p. 45.
6 Ibid.

globalization of capitalism itself by creating an unusual forum for capitalist exchange and by using film's universal language to sell universalizing ideas about technology, industrial development and modernization.[7]

This chapter focuses on the first in a cycle of industrial film festivals that began as a European event held in France in 1958, expanded to include international entries in 1959, and proclaimed itself the world's first international industrial film festival in 1960. Although preceded by similar festivals in locations including Belgium, Britain, Italy and Hungary, this cycle became the world's preeminent industrial film festival thanks to the backing of the Council of European Industrial Federations, and it continued at sites across Europe through the 1960s and into the 1970s.

But why did this French-based festival become the one that mattered for European industry? Was there something about the emergence of French industrial film, about post-WWII French industry or about French cinema that helped ensure that industrial film would be celebrated in this way? What approaches to industrial film did France help institute in its initial conception of the festival and in the norms it established that helped define the festivals that followed? Finally, what value did these festivals have for the executives who attended them and for the filmmakers who formed what that *Business Screen* journalist described as industrial film's "world community of filmmakers?"

French Industrial Film and *Les Trente Glorieuses*

That France would, in 1958, initiate the world's leading series of industrial film festivals had as much to do with developments in French industry as in its famed film industry. The festivals' origins were shaped by developments in the French economy during the first decade of the period from 1946 to 1975 that many historians today still term *Les Trente Glorieuses*. Coined in 1979 by French economist Jean Fourastié, the label "thirty glorious years" encapsulates the considerable changes to French society during and after the nation's post-World War II reconstruction, a period in which France changed more, Fourastié argues, than in the hundred years previous.[8]

7 On Cannes and cosmopolitanism, see Schwartz, Vanessa R. *It's So French! Hollywood, Paris, and the Making of Cosmopolitan Film Culture.* University of Chicago Press, 2007, pp. 57–58.

8 Fourastié, Jean. *Les Trente Glorieuses: ou la révolution invisible de 1946 à 1975*, rev. ed. Hachette, 2000, p. 47.

During these decades, the French population, which had remained relatively static since the 1840s, grew by twenty-five per cent from 40 million to 53 million. The French mechanized agriculture with new industrial fertilizers and tractors, which grew in number by a factor of ten between 1946 and 1958. Food production increased dramatically even as the number of farm workers declined rapidly. The movement of agricultural workers to cities helped drive new patterns of urbanization and a shift in the French economy toward heavy industry, signalled in particular by the production of automobiles, which increased from 750,000 in 1951 to four million in 1958. As a result, the French economy grew at its most dynamic rate ever, averaging five per cent growth each year and reaching ten per cent in the late 1950s and early 1960s. During this period, only Japan's economy could boast higher rates of expansion.[9] As one historian has summarized it: "The economic boom that started in the late 1940s provided a definitive model of the nation's future. Under its impetus [...] France leaped headlong into modernity."[10]

Industrial film production surged with it. Pathé and Gaumont continued producing industrial films, as they had since the 1910s. Active in-house units produced films at the Ministry of Agriculture, the Société national de chemins de fer (SNCF), the automaker Renault and the French postal service.[11] Independent producers, some of which had existed before the war, and others that helped meet market demand after it, served smaller companies and major industries, including coal and petroleum.[12]

These companies took advantage of the same conditions that helped give rise to the Nouvelle Vague and cinéma vérité, especially the portability and

9 Ibid., pp. 35–46.

10 Stovall, Tyler. *France since the Second World War*. Longman, 2002, p. 28.

11 On the Ministry of Agriculture's post-World War II work, see Jacobson, Brian R. "Midcentury Rural Modern: French Agricultural Cinema and the Art of Persuasion." *Screen*, vol. 58, no. 2, 2017; on the SNCF, see Ionascu, Michel. "Le Cinéma des cheminots ou l'introuvable parole ouvrière." *Les cahiers de la cinémathèque*, no. 71, December 2000, pp. 41–46; Farcy, Jean. "Renault et le cinéma d'entreprise." *Renault histoire*, vol. 3, June 1991, pp. 21–68; Michel, Alain P. "Corporate Films of Industrial Work: Renault (1916–1939)." *Films That Work: Industrial Film and the Productivity of Media*, edited by Vinzenz Hediger and Patrick Vonderau, Amsterdam University Press, 2009, pp. 167–86; Hatzfeld, Nicolas, et al. "Filming Work on Behalf of the Automobile Firm: The Case of Renault (1950–2002)." *Films That Work: Industrial Film and the Productivity of Media*, edited by Vinzenz Hediger and Patrick Vonderau. Amsterdam University Press, 2009, pp. 187–209.

12 Independent producers included: Cinétest, Les Films du Compas, Paul de Roubaix's Je Vois Tout, Pierre Long's Son et Lumière, and Jean Mineur, the company still known today for its screen advertising work and the *petit garçon* logo. On the coal industry's use of film, see Caron, Estelle. "Images de la mine et des mineurs." *Les cahiers de la cinémathèque*, no. 71, December 2000, pp. 47–52.

increasing quality of 16 mm cameras and film stock. Beyond the world of cinema proper, they built on the expansion of French industry and increasing interest among French business owners in the value of film as a tool for training workers, educating potential buyers, and creating good feeling among the public at large in an ethos that celebrated modernization and productivity. Public interest in their films was great enough to prompt the companies and government agencies to form their own cinémathèques for distributing industrial films, which were requested directly by schools, churches, ciné-clubs, and other corporations. The films also screened in French theatres as part of regular programming in accordance with rules requiring theatres to show short films before each feature.[13]

In this context, film festivals began to address the importance of short films in general and industrial films in particular. Cannes had featured short films since 1946, and shorts also competed in Edinburgh and at Venice, but no festival focused solely on shorts until 1955. That September, the first such festival began in Tours.[14] There, Alain Resnais' *Toute la mémoire du monde* (1956) – a stunning study of the Bibliothèque nationale in Paris – took home a prize, and the festival quickly became well known. By 1959, Jean-Luc Godard could write in *Cahiers du cinéma* that "we rarely speak of short films in the *Cahiers*, and certain readers have offered their friendly reproaches. The 1958 festival in Tours has encouraged us to fix that gap."[15] Later that year, a new event attempted to cultivate a similar sentiment for industrial film.

The First "European Technical and Industrial Film Days," 1958

Although organized not as a "festival," but rather with the title "Les journées Européennes du film technique et industriel," or the "European Technical and Industrial Film Days," the event came to be known as the nation's first

13 For more about the history of short film exhibition after World War II, see Williams, Alan. *The Republic of Images: A History of French Filmmaking*. Harvard University Press, 1992, p. 362; Mary, Nathalie. "Diffusion." *Une encyclopédie du court métrage français*, edited by Jacky Evrard and Jacques Kermabon. Festival Côté court, 2004, pp. 133–37; and Jeancolas, Jean-Pierre. "Structures du court métrage français, 1945–1958." *Le court métrage français: de l'âge d'or aux contrebandiers*, edited by Dominique Bluher and François Thomas, Presses universitaires de Rennes, 2005, pp. 29–46.

14 On the Festival de Tours, see Mary. "Diffusion," p. 135; Vigo, Luce. "Festivals." *Une encyclopédie du court métrage français*, edited by Jacky Evrard and Jacques Kermabon. Festival Côté court, 2004, p. 179; and Jeancolas. "Structures du court métrage français, 1945–1958," pp. 43–44.

15 Godard, Jean-Luc. "Chacun son tours." *Cahiers du cinema*, vol. 16, no. 92, February 1959, pp. 31–38. Qtd. in Vigo. "Festivals," p. 180.

industrial film festival. Its organizers later embraced that title, but they initially rejected the idea that this was simply a "festival," which suggested a scope that could not encapsulate their broader mission or the activities of the event's five-day programme. To be sure, the event had all the trappings of a film festival: premier screenings, parties, photo ops and fashion. But behind the gloss and glamour, these film days – like many a film festival – ran in the service of business.

The Journées took place in a seemingly unlikely setting: Rouen. Located in north-western France and better known as the Ville Musée, or museum city – or as what Victor Hugo called the "city of a hundred steeples" – Rouen previously had few obvious claims to a cinematic identity of any kind. Even those filmmakers who *had* found their way to north-western France in the 1930s preferred the mysteries and politics found in the port city of Le Havre, the setting, for instance, of Marcel Carné's *Quai des brumes* (1938). Similarly, in Jean Renoir's *La bête humaine* (1938), the doomed train conductor played by Jean Gabin drives the rails from Paris to Le Havre by way of Rouen, but the Ville Musée merits not even a brief stop along the way.

From October 8 to 12, 1958, however, Rouen earned its place on the French film map thanks to its growing importance not for cinema but in the French industrial economy. The Journées were the brainchild of a local Rouen government official named Edmond Loraillère. Under the banner of the city's planning and tourism board, COMET, Loraillère was charged with planning a variety of events, including the city's yearly industrial and agricultural festival.[16] In 1957, he developed the idea for an event featuring industrial films with the intention of bringing greater attention to the films and promoting better distribution.[17] He took inspiration from the short festival in Tours, Britain's first "Festival of Films in the Service of Industry," which took place in Harrogate in October 1957, and similar festivals in Belgium in 1953, Italy in 1956, and another planned for February 1958 in Budapest.[18] In short, Loraillère identified broad European interest in

16 Created in 1951, COMET, the Comité pour l'organisation de manifestations économiques et touristiques, was one means by which Rouen sought to promote its importance in the regional economy.

17 Lorraillère's plan is summarized in an undated document that was presented to Rouen officials and discussed at a COMET meeting on March 25, 1958. See "Présentation des Journées Européennes du film d'entreprise" and "Réunion du C.O.M.E.T." March 25, 1958. Archives départementales de la Seine-Maritime, Rouen (ADSM), Box 137J 87–88.

18 On the 1957 Harrogate festival, see Russell, Patrick, and James Piers Taylor. "Documentary Culture: Groupings, Gatherings and Writings." *Shadows of Progress: Documentary Film in Post-war Britain*, edited by Patrick Russell and James Piers Taylor. Palgrave Macmillan, 2010, pp. 14–19.

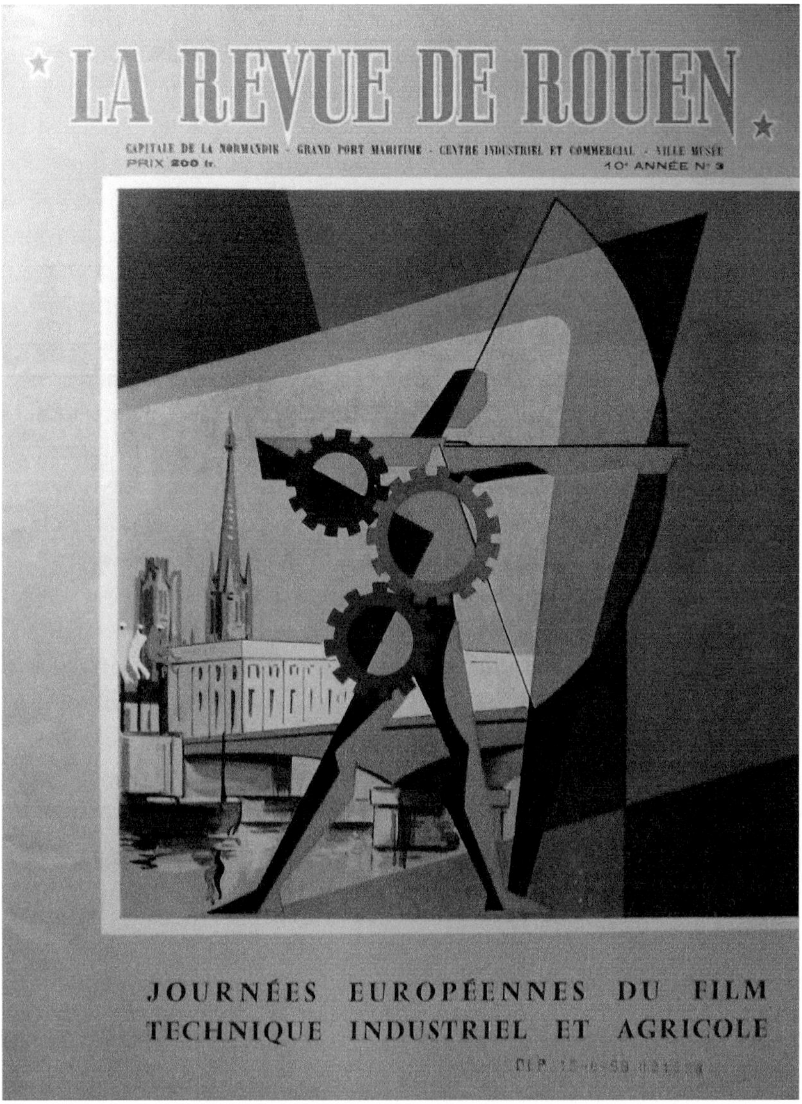

Fig. 11.3. The cover of a special issue of *La revue de Rouen* devoted to the festival (1958).

industrial film and sought to capitalize on it before anyone in France could beat him to it.

Loraillère sold the event to the Rouen municipal council on a calculated mix of economic rewards for the city, the potential to bring glory to France and the artistic value of the films themselves, which he compared to magnificent diamonds. Paris, he reasoned, was already too busy for something like this, so why not Rouen? Such a festival, he argued, would generate local commerce, but even more than just short-term profit, it would also put the city in contact with all the nation's leading corporations, trade federations and a number of government agencies as well as European and international organizations such as UNESCO.[19] In sum, Lorraillère hoped to use industrial film to make his city a key node in a broad network of leading European industrialists.

Loraillère's plans aligned with city leaders' more general aspirations for promoting the local and regional economies. In the context of growth-centred French state planning, Rouen hoped to shed its identity as the "museum city" and redefine itself as a regional hub of industry. A 1955 report commissioned by COMET for the annual city festival had summed up this vision by arguing that the old image of Rouen as a museum city "should not mask this other reality: Rouen is a very important industrial centre."[20] Citing the city's importance as a maritime port that moved the weight of up to four Eiffel Towers each day, its large train station and power plant, and its growing production of petroleum derivatives, chemicals, metals and fabrics, the report claimed for Rouen a place in the top five of great industrial cities in France and no small player in the regional economy.[21]

These efforts to secure an industrial identity reflected broader trends in French state planning, which promoted industrial development beyond the Paris region beginning in the early 1950s.[22] Cities like Rouen could hope to achieve a greater place on the national scene by expanding local and regional economies through industrial development supported by favourable government initiatives. That idea reproduced, on a local scale, what was also happening at the national level, where French planners took advantage of

19 "Présentation des Journées Européennes du film d'entreprise"; and "Réunion du C.O.M.E.T."

20 The report was written by Jacques Leboullenger, secretary of the Centre d'études d'interêt public de la région Rouennaise, in response to COMET's request that the group's contribution to the annual festival catalogue "develop the idea that Rouen is really an industrial centre of very great importance in the national framework." See "Rouen, centre industriel vivant." May 26, 1955; and J. Lagarde to M. Etienne. May 16, 1955. ADSM, Box 137J 87–88.

21 Leboullenger, Jacques. "Rouen, centre industriel vivant."

22 Decentralization of the national economy began in the early 1950s, coordinated by the Ministère de la Reconstruction et de l'Urbanisme and the Commissariat Générale au Plan, and included subsidies for businesses to establish plants in provincial cities. In the mid-1950s, the commissariat pushed forward plans for further regional development. See Wakeman, Rosemary. *The Heroic City: Paris, 1945–1958*. University of Chicago Press, 2009, p. 59; and idem. *Modernizing the Provincial City: Toulouse, 1945–1975*. Harvard University Press, 1997, p. 105; for more about French planning after World War II, see Gildea, Robert. *France Since 1945*, 2nd ed. Oxford University Press, 2002, pp. 96–101; Nord, Philip J. *France's New Deal: From the Thirties to the Postwar Era*. Princeton University Press, 2010, pp. 152–67; and Stovall. *France since World War II*, pp. 29–31.

Marshall Plan funding to redefine France's place in post-WWII Europe by strengthening its industrial economy. In short, Rouen's vision for a new role within France might be understood as a microcosm of France's vision for its new place in the European economy as a whole. In this context, Loraillère's plan seized upon industrial film's potential role in securing these visions. What better symbolic event than a series of industrial film days – a ripe mix of culture and capital – to promote Rouen as an industrial centre?[23] What better way for France to assert its place in European industry than by reasserting its cultural importance and taking a controlling role in future events around industrial film?

The 1958 "Technical and Industrial Film Days" attempted both to promote these ideas about France and its industrial film industry's importance and to cultivate the industrial film industry itself. The event drew an estimated thousand visitors and screened 220 films from nine nations. The festival jury handed out almost fifty awards – the highest of which were termed "Oscars" – in seventeen categories. The prize–winning films' sponsors reached the heights of French industry and government, from Air France, Renault and the SNCF to Saint-Gobain, the Compagnie française des pétroles, the national coal companies and the Ministry of Agriculture. From outside France, prize-winning sponsors included British Petroleum, Shell, Mobil Oil, Edison Volta, Kodak, Nestlé, Olivetti and Fiat.

But film competition was only part of the attraction. For some attendees, the real draw must have been the Film Days' nights, which importantly served not simply to entertain but also to foster the networking that Lorraillère had envisioned. The evening entertainment included premier screenings of three feature films, with stars, writers and directors on hand to give Rouen celebrity sheen, as well as concerts, fashion shows and cocktail after cocktail. On the opening night, Luis Mariano and Richard Pottier attended the world premiere of their *Serenade au Texas* (1958), a musical Western fittingly set in the Texas oil industry. The same evening attendees were treated to a fashion show of the latest Nina Ricci collection. The following night saw the arrival of Marina Vlady, Odile Versois and Robert Hossein for a screening of *Toi, le venin* (*Blonde in a White Car*, 1958), followed by a performance by jazz musician Geo Daly. Friday evening was

23 For all their emphasis on Rouen's industrial qualities, city officials did not forget to highlight its more traditional charms. At the COMET meeting preceding the first Industrial Film Days, for instance, one committee member, Mme. Simon, proposed (successfully) that "the lights on the monuments be extended until 1 am and that a bus be made available to show attendees different views of Rouen illuminated." See "Réunion du C.O.M.E.T." September 25, 1958. ADSM, Box 137J 87–88.

for industry-sponsored parties, first a post-screening cocktail presented by Pathé, followed by a party hosted by Kodak at the Hotel de la Poste.

Saturday evening returned to film with another fitting world premier: Georges Franju's first feature, *La tête contre les murs* (1959). Fitting because Franju had gotten his start making shorts, including numerous notable sponsored films. For at least some of the filmmakers in attendance, his feature was something to root for and emulate as the next step in a career that might take them beyond the red carpets of Rouen, perhaps to those of Cannes. Although Franju did not attend, the author of the 1949 novel that Franju adapted, Hervé Bazin, who had recently become a member of the Académie Goncourt, brought intellectual heft and name recognition.

For the ambitious, or just hardy attendees, the daytime schedule was packed with workshops and a trade fair organized by independent film producers and film technology companies. These components, which were advertised in the film trade press months in advance, were again consistent with Lorraillère's idea of creating a network of business leaders.[24] They made the festival an important site of exchange between film producers and industrialists as well as an educational forum, especially for would-be producers seeking guidance about how film might serve their business interests.

The trade fair took place at the Rouen Chamber of Commerce with the support of the local branch of the Conservatoire national des arts et métiers, the French engineering and continuing education school. It focused primarily on film equipment and demonstrated the range of new devices and services available for the growing industrial film sector. The exhibitors included manufacturers of cameras, lenses, projectors and other film equipment from at least eight nations, including Cinéric, Debrie, and Éclair from France, Cinemeccanica and Ferrania from Italy, English firms including Vinten, Simplex-Ampro, and Rank Precision (the technology branch of the Rank Organisation), the Swiss Omnilux and Perfectone companies, and the Dutch firm Philips. American companies included RCA, the screen maker Radiant, Wisconsin-based Projectograph, Chicago sound recording firm Webcor, and Kodak, while the German contingent included Agfa, Askania, Klangfilm, Perutz, Steenbeck, and Zeiss.

24 The full list of daily conference themes appeared in the May and September issues of *Le technicien du film*. See "Thèmes des Journées Européennes du film technique et industriel de Rouen." *Le technicien du film et le cahier de l'exploitation*, vol. 4, no. 39, May 15, 1958, pp. 12–13; "Les journées Européennes de Rouen." *Le technicien du film et le cahier de l'exploitation*, vol. 4, no. 42, September 15, 1958, p. 11.

Many exhibitors focused on small format equipment, but numerous 35 mm cameras and projectors were also displayed, and one special event included a demonstration of the 70 mm Todd AO. The sheer range of companies highlighted industrial film's potential contribution to the film industry's fiscal health, and the juxtaposition of competing firms underscored the potential national rivalries that also played out in the festival competition itself. Finally, the exhibition included a smaller number of French industrial film producers, including Ana-Ciné, Fred Jeannot, whose exhibition highlighted his firm's 16 mm sound film work, and Tadié-Cinéma, which was selling a new system, "schématographie."[25]

If the exhibitions showed that industrial film had a firm infrastructural base, the workshops explained how art and, more importantly, business would sprout from it. These workshops served one of the event's broader goals of "helping sponsors and filmmakers improve the quality and effectiveness of their films" by giving definition to a young but burgeoning field about which the business owners in attendance were reportedly intrigued but still skeptical.[26] The workshops thus focused on developing the field. They defined industrial film for the uninitiated, instructed owners about how to make them and insisted upon the potential rewards of doing so.

Organizers planned the workshop schedule according to a series of themes that roughly coincided with the daily screenings. The first day focused on personnel issues, where films might help improve internal business functioning, from recruitment and employee orientation to on-the-job training and continuing education, including teaching good sales techniques and training managers, as well as promoting health and workplace safety. Day two continued to focus on internal issues, including how films might contribute to setting up and organizing a business and promote greater efficiency through means such as the introduction of timekeeping devices and better management of supply stocks.

The next two days shifted to a kind of "how-to" guide for those who wanted to make better use of industrial film in their own business. Day three outlined the production process, from how to establish reasonable goals for films to managing the relationship with filmmakers (which was always a point of concern for business owners who were worried about costs

25 See "Après Rouen." *La technique cinématographique*, vol. 27, no. 191, October 1958, pp. 266–67; Gallon, Michel. "Rouen: Utiles journées pour le film technique et industriel." *France Film International*, vol. 11, no. 17, November 1958, p. 6.

26 "Journées Européennes du film technique et industriel." *Le technicien du film et le cahier de l'exploitation*, vol. 4, no. 41, July 15, 1958, p. 12.

and maintaining control of the finished product). Other workshops offered details about distribution through existing national and international networks. The fourth day examined more specific questions about the types of films one might produce, including sales films, prestige films, and different types of propaganda for different audiences, from the mass public to other businesses and professional organizations.

Finally, on the closing day, a symposium brought together speakers from across French industry and the industrial film industry.[27] Producer and industry veteran Etienne Lallier, prefiguring the historical discourse that *L'avant-scène cinéma*'s editors later voiced, situated industrial film in a lineage beginning with Robert J. Flaherty and French director Jean Tedesco. Other directors and producers, including Jean Farcy, Henri Fabiani, Jean Image and André Tadié, showcased industrial film's varying forms and selling points, no doubt with the idea of promoting their own work for the executives in attendance. Other talks, finally, reinforced the overarching theme of the previous days' workshops: why and how to make industrial films and what to do with them.

The goal, put simply, was to make the festival a training ground for culture-minded capitalists. For the industry leaders who attended, the week in Rouen offered the unique opportunity to watch and learn from the best industrial films, to find out how their corporation's or nation's films stacked up against the competition, to see and acquire the newest gadgets that made those films possible, and to learn from film professionals and experienced executives about what it took to plan, create, promote, and distribute such films for one's own business. Here, the culture industry met industrial culture; here, business learned new ways – through an "other cinema," a cinema of its own – to present itself as art.

Conclusion: Art, Industry, Capital

The final day of Rouen 1958 concluded with an awards gala at the local Omnia cinema that highlighted the event's regional importance, its national and international stakes, and industrial film's potential for spectacle, glamour, art, and, above all, business. The gala brought local and regional politicians and representatives of French and European industry together once more with filmmakers and film industry leaders, notably including CNC

27 Texts from the event were reproduced in the next issue of *La revue du Rouen*, vol. 10, no. 1, 1958.

director Jacques Flaud and Fred Orain, president of France's industrial and educational film union. It began with a speech about Normandy and the Seine-Maritime, an emphasis on the region that once more highlighted Lorraillère's intention of promoting Rouen's place on the national scene. Finally, before handing out the prizes, organizers unveiled one final film premier that celebrated French industry and once more underscored the blend of – and tensions between – art and industry that defined the event and its films.

The film, René Zuber's *Le grand oeuvre: panorama de l'industrie française* (p. Roger Leenhardt, 1958), presented just that: a broad, celebratory overview of the many industries that had come to flourish in France in the decade since the early years of post-WWII reconstruction. From steelworks and coal mines, gas wells and oil refineries, cooling towers and electrical works to dams and canals, ports and pipelines, and automobile factories, airports and tanker ships, the film presented France as an industrial wonderland and a thriving example of the heights of modernization. For the film's thirty-six minutes, festival attendees – or the French ones, at least – could revel in the greatest glories of *Les Trente Glorieuses* or, for the non-French, compare it with their own modernizing work. The film offered one final example of industrial film's power to make business and industry spectacular, awe-inspiring, and perhaps even entertaining.

Finally, with an opening that situated French industry in the history of science, technology, and art dating to Louis XIV and the mirrored halls of the Château de Versailles, *Le grand oeuvre* implicitly flattered its viewers – the executives who ran such industries and sponsored such films, and the filmmakers who made them – by reminding them that industry and industrial film were part of a venerable tradition of Enlightenment ideals and the highest of arts. Film, Zuber's "panorama" suggested, had the unique power to make these ideas palpable and to glorify business and industry in ways that could have real consequences, whether for worker training and morale, for promoting sales, for cultivating good feeling among broad publics, or for establishing and maintaining a worldview oriented toward the values of technological progress and industrial modernization.

In the same way, the 1958 Journées Européennes du film technique et industriel in Rouen and the many similar versions that followed helped clarify, consolidate and encourage a process that was already well underway: the process by which the leaders of the new capitalist order appropriated and adapted film art and aesthetics to promote a mid-century modern worldview. These same leaders, after all, were the ones sharing cocktails and watching each other's films in screening rooms; sharing ideas about the aesthetic ideals of capitalist exchange in workshops; and promoting and celebrating a "world community of filmmakers" who they hoped would put the propaganda machine in motion.

In the days following the 1958 event, observers celebrated Rouen as what one journalist later described as "a launching ramp" for industrial film.[28] Perhaps most concretely, the event helped prompt the national French commerce association, the Conseil national du patronat français (CNPF), to create an organization devoted to industrial film distribution with a central cinémathèque in Paris serving regional networks.[29] Its creation represented both success and failure for Lorraillère, who had envisaged just such an institution, but for Rouen, not Paris. Within weeks of the first festival's conclusion, Lorraillère had already overseen a meeting at which local officials outlined plans for a "centre du cinéma," a permanent storage and distribution site for industrial and commercial films in Rouen that would be built at an estimated cost approaching eighteen million francs.[30] Municipal representatives expressed interest in moving forward with definitive plans for this centre so that Rouen would not lose its momentary advantage, but Paris ultimately won out.[31]

Meanwhile, officials in Rouen quickly began planning a follow-up event for 1959.[32] That version followed a similar format, again featuring seventeen categories of 167 competing films. The new Journées expanded in scope from nine to nineteen nations and reportedly doubled to two thousand visitors. The inclusion of films from the United States and Israel made the event international. Meanwhile, the participating nations also included representatives from the Eastern Bloc, a fact that underscored France's

28 Ozanne, Jean-Louis. "Rouen: Rampe de lancement du film industriel." *Présence Normande: Revue de Haute-Normandie*, vol. 13, no. 6, June 1962, pp. 24–28.

29 The organization was named CEFILM (Connaissance de l'économie par le film). For more about its creation, see CEFILM. "Diffusion du film d'information économique et industrielle." Paris. Archives nationales du monde du travail, Roubaix (ANMT), Box 72 AS 816; see also "La diffusion des films d'information économique et industrielle." *Le technicien du film*, vol. 7, no. 74, July 15, 1961, p. 11.

30 "Assemblée générale du C.O.M.E.T." November 12, 1958. ADSM, Box 137J 87–88.

31 Aspects of CEFILM's initial conception suggest that Rouen might have had a leading role. The president of Rouen's Chamber of Commerce and Industry, Georges Lanfry, was appointed as one of two CEFILM vice presidents. Of the seventeen regional distributors for the first CEFILM season (scheduled to run from October 1960 to July 1961), two were based in Rouen: COMET and the Fédération des syndicats patronaux de Rouen. See "Diffusion du film d'information économique et industrielle."

32 At their February 1959 meeting, COMET officials highlighted the need to follow up on the 1958 event's success in 1959 in order to avoid competition from other organizations. "Réunion du C.O.M.E.T." February 24, 1959. ADSM, Box 137J 87–88.

attempts to carve out a Cold War middle ground but which also put future festival organizers on guard against the dangers that communist propaganda posed to the liberal capitalist worldview.

The success of the 1958 and 1959 events helped generate widespread enthusiasm and plans for similar festivals across Europe. In an effort to avoid what its members saw as unnecessary competition, in January 1960 a subcommittee of the Council of European Industrial Federations focused on industrial film festivals finalized plans to consolidate these events into an annual, Council-sponsored festival that would pass between host nations.[33] Despite objections from Belgium, which had already planned its own international festival in Anvers for March 1960, the first officially sanctioned festival was held in October 1960, once again in Rouen, which continued to host the French national festival through most of the 1960s and hosted the international edition again in 1965.

France's central place in these events' origins should be emphasized. Its role in the promotion of industrial film shared much with its role, as home to the Cannes Film Festival, in the contemporaneous promotion of global art film. In her work on Cannes, Vanessa Schwartz argues that the festival helped forge global film culture by providing a collaborative setting for the international exchange of producers, players and filmmakers, and that part of its particular "Frenchness" was the festival's place in French hopes to challenge the domination of American cinema by defying tidy Cold War distinctions between East and West.[34] Something similar could be said of Rouen, where the leaders of French industry sought both to place France at the heart of industrial film and also to give it a guiding role in pan-European and broader global networks of capitalist and industrial film development by making it the site of similar forms of exchange among an international network of industrial filmmakers and their producers. If, as Schwartz argues, the corporations who drove globalization in the years following World War II did so thanks to the creation of a Western society connected by images and to the ideals of universalism and internationalism that were celebrated at Cannes, we should be mindful of the similar work taking place on the less heralded, less glamorous red carpets of Rouen. In other words, this story of capitalist film culture needs to include the culture produced by the non-film corporations that always put the emphasis on capital more than culture. These were the capitalists who gathered at Rouen to learn how to use film

33 "Réunion de la Sous-Commission des Festivals Internationaux du Film Industriel." January 25, 1960. ANMT, Box 72 AS 815.
34 Schwartz. *It's So French!*, p. 203.

culture to promote the industry-centric, growth-focused worldview we're still living with today.

Put another way, the history of industrial film and its festivals is one important aspect of capitalism's appropriation of art's sensory modes and practices. Film allowed corporations to produce and promote their image so successfully – to create the conditions, that is, for what Jacques Rancière terms *aisthesis*, "the sensible fabric of experience within which [works of art] are produced" and the "conditions [that] make it possible for words, shapes, movements and rhythms to be felt and thought as art."[35] If, as the editors of *L'avant-scène cinéma* later insisted, film art had managed to avoid "suffocation," was it not, at least in part, because it had been put in the service of breathing life into business and industry? If so, what would be left for film art in a world in which industry drove so much of its production? What would it mean for our ways of seeing, experiencing, and interpreting art if businesses had so much control over its aims and forms?

These questions go to the heart of today's media-dominated landscape. In a review of *Aisthesis*, Hal Foster poses this challenge to Rancière's optimistic and, in Foster's estimation, naïve idea that art might pose a counterweight to systems of power by creating new sensory experiences to promote political action. For Foster, Rancière misses the point that art simply cannot compete with mass image culture. Rancière's idea of "aesthetic acts," Foster argues, "gives art an agency that it does not possess at present. That it can intervene effectively today in 'the distribution of the sensible' is far from clear; certainly art is no match for the image and information industries that control and concentrate 'the sensible' with such ease and efficiency."[36]

Industrial film festivals offer one critical site for seeing this creation and appropriation in action. They built themselves precisely on the model of film's emerging art worlds – those "culture fests" in Cannes and Venice – and through the work of many of the same would-be film artists who went there. For filmmakers, the festivals were a place to share the art of functionality. For capitalists, they were a place to learn how to sell ideas, to cut deals, and to present business through and even as art – in sum, to create "aesthetic acts" for business action.

In their screening rooms, at their banquets, and in the bars and cafes in between, festivals cultivated the kinds of artistic and capitalistic exchanges

35 Rancière, Jacques. *Aisthesis: Scenes from the Aesthetic Regime of Art*. Translated by Zakir Paul. Verso, 2013, p. x.

36 Foster, Hal. "What's the Problem with Critical Art?" [Review of *Aisthesis* by Rancière]. *London Review of Books*, vol. 35, no. 19, 2013, pp. 14–15.

needed to make the world in the image of post-45 global capital. While this might be described as the industrialization of Cannes – as the transformation of the Cannes model into an industrial film version – it would be better to reverse that formula. The festivals in Rouen and those that followed over the next decade across Europe were, instead, the *Cannes-ification* of industry, the appropriation of film's universal language, its glamour and its appeal, not just for a cinematic view of modern industry, but also, and more importantly, for an industrial capitalist view – projected through film – of the modern world.

Works Cited

"Après Rouen." *La technique cinématographique*, vol. 27, no. 191, October 1958, pp. 266–67.

"Assemblée générale du C.O.M.E.T." November 12, 1958. Archives départementales de la Seine-Maritime, Rouen, Box 137J 87–88.

Caron, Estelle. "Images de la mine et des mineurs." *Les cahiers de la cinémathèque*, no. 71, December 2000, pp. 47–52.

CEFILM. "Diffusion du film d'information économique et industrielle." Paris. Archives nationales du monde du travail, Roubaix, Box 72 AS 816.

Farcy, Jean. "Renault et le cinéma d'entreprise." *Renault histoire*, vol. 3, June 1991, pp. 21–68.

Foster, Hal. "What's the Problem with Critical Art?" [Review of *Aisthesis* by Rancière]. *London Review of Books*, vol. 35, no. 19, 2013, pp. 14–15.

Fourastié, Jean. *Les Trente Glorieuses: ou la révolution invisible de 1946 à 1975*, rev. ed. Hachette, 2000.

Gildea, Robert. *France since 1945*, 2nd ed. Oxford University Press, 2002.

Godard, Jean-Luc. "Chacun son tours." *Cahiers du cinéma*, vol. 16, no. 92, February 1959, pp. 31–38.

Hatzfeld, Nicolas, Gwenaële Rot, and Alain P. Michel. "Filming Work on Behalf of the Automobile Firm: The Case of Renault (1950–2002)." *Films That Work: Industrial Film and the Productivity of Media*, edited by Vinzenz Hediger and Patrick Vonderau. Amsterdam University Press, 2009, pp. 187–209.

Ionascu, Michel. "Le cinéma des cheminots ou l'introuvable parole ouvrière." *Les cahiers de la cinémathèque*, no. 71, December 2000, pp. 41–46.

Jacobson, Brian R. "Midcentury Rural Modern: French Agricultural Cinema and the Art of Persuasion." *Screen*, vol. 58, no. 2, 2017, pp. 141–62.

Jeancolas, Jean-Pierre. "Structures du court métrage Français, 1945–1958." *Le court métrage Français: de l'âge d'or aux contrebandiers*, edited by Dominique Bluher and François Thomas, Presses universitaires de Rennes, 2005, pp. 29–46.

J. Lagarde to M. Etienne. May 16, 1955. Archives départementales de la Seine-Maritime, Rouen, Box 137J 87–88.

"Journées Européennes du film technique et industriel." *Le technicien du film et le cahier de l'exploitation*, vol. 4, no. 41, July 15, 1958, p. 12.

"La diffusion des films d'information économique et industrielle." *Le technicien du film*, vol. 7, no. 74, July 15, 1961, p. 11.

Leboullenger, Jacques. "Rouen, centre industriel vivant." May 26, 1955. Archives départementales de la Seine-Maritime, Rouen, Box 137J 87–88.

"Les Journées Européennes de Rouen." *Le technicien du film et le cahier de l'exploitation*, vol. 4, no. 42, September 15, 1958, p. 11.

Mary, Nathalie. "Diffusion." *Une encyclopédie du court métrage français*, edited by Jacky Evrard and Jacques Kermabon. Festival Côté court, 2004, pp. 133–37.

Michel, Alain P. "Corporate Films of Industrial Work: Renault (1916–1939)." *Films That Work: Industrial Film and the Productivity of Media*, edited by Vinzenz Hediger and Patrick Vonderau. Amsterdam University Press, 2009, pp. 167–86.

Nord, Philip G. *France's New Deal: From the Thirties to the Postwar Era*. Princeton University Press, 2010.

Nowell-Smith, Geoffrey. "General Introduction." *The Oxford History of World Cinema*, edited by Geoffrey Nowell-Smith. Oxford University Press, 1996, pp. xix–xxii.

OHC. "The World Community of Film Makers." *Business Screen*, vol. 23, no. 6, October 1962, p. 45.

Ozanne, Jean-Louis. "Rouen: Rampe de lancement du film industriel." *Présence Normande: Revue de Haute-Normandie*, vol. 13, no. 6, June 1962, pp. 24–28.

"Présentation des Journées Européennes du film d'entreprise." 25 March 1958. Archives départementales de la Seine-Maritime, Rouen, Box 137J 87–88.

Rancière, Jacques. *Aisthesis: Scenes from the Aesthetic Regime of Art*. Translated by Zakir Paul. Verso, 2013.

"Réunion de la Sous-Commission des Festivals Internationaux du Film Industriel." January 25, 1960. Archives nationales du monde du travail, Roubaix, Box 72 AS 815.

"Réunion du C.O.M.E.T." March 25, 1958. Archives départementales de la Seine-Maritime, Rouen, Box 137J 87–88.

"Réunion du C.O.M.E.T." September 25, 1958. Archives départementales de la Seine-Maritime, Rouen, Box 137J 87–88.

"Réunion du C.O.M.E.T." February 24, 1959. Archives départementales de la Seine-Maritime, Rouen, Box 137J 87–88.

Russell, Patrick, and James Piers Taylor. "Documentary Culture: Groupings, Gatherings and Writings." *Shadows of Progress: Documentary Film in Post-war Britain*, edited by Patrick Russell and James Piers Taylor. Palgrave Macmillan, 2010, pp. 11–19.

Schwartz, Vanessa R. *It's So French! Hollywood, Paris, and the Making of Cosmopolitan Film Culture.* University of Chicago Press, 2007.
Stovall, Tyler. *France since the Second World War.* Longman, 2002.
"Thèmes des Journées Européennes du film technique et industriel de Rouen." *Le technicien du film et le cahier de l'exploitation*, vol. 4, no. 39, May 15, 1958, pp. 12–13.
"Un autre cinéma." *L'avant-scène cinéma*, no. 199–200, January 1978, p. 3.
Vigo, Luce. "Festivals." *Une encyclopédie du court métrage français*, edited by Jacky Evrard and Jacques Kermabon. Festival Côté court, 2004, pp. 179–85.
Wakeman, Rosemary. *The Heroic City: Paris, 1945–1958.* University of Chicago Press, 2009.
Wakeman, Rosemary. *Modernizing the Provincial City: Toulouse, 1945–1975.* Harvard University Press, 1997.
Williams, Alan. *The Republic of Images: A History of French Filmmaking.* Harvard University Press, 1992.

About the Author

Brian R. Jacobson is professor of visual culture at the California Institute of Technology. He is the author *Studios Before the System: Architecture, Technology, and the Emergence of Cinematic Space* (Columbia University Press, 2015) and editor of *In the Studio: Visual Creation and Its Material Environments* (University of California Press, 2020), winner of the 2021 SCMS award for Best Edited Collection and the 2021 Limina Prize for Best International Cinema Studies Book.

12 Cinema and Industrial Design

Showmanship, Fairs and the Exhibition Film

Haidee Wasson

Abstract

This chapter documents and examines the use of portable film projectors at the New York World's Fair, 1939–1940, and its surprising range of industrial display practices, including government, public relations and advertising films shown in big and small temporary theatres, and other applications like the use of film projectors to direct pedestrian traffic and to create multimedial and "immersive" experiences and events. The fair is seen as a culmination of decades of industrial film use, multimedial exhibition and display for advertising strategies, while the role of industrial designers and their interest in film is highlighted. The importance of contextualizing "industrial film" within technologies, practices and aesthetics, especially in the fields of industrial design, exhibition and architecture, will be demonstrated.

Keywords: industrial exhibitions; portable film technologies; film projectors; industrial design, Futurama

As we understand more and more about the many kinds of films that fall under the broad umbrella of "industrial film," it is clear that we need more refined categories to help us signpost and differentiate amongst what is clearly a vast body of films. Take but one possible category: the exhibition film. There is great variation and change across the well over hundred-year history of films shown and film technologies used at industrial exhibitions. With only a few projections at fairs at the turn of the century, there were as many as 3,000 films shown by the time that Montreal hosted the World Exposition of 1967, with its gigantic and advanced multi-screen, audiovisual

Hediger, V., F. Hoof, Y. Zimmermann, with S. Anthony (eds), *Films That Work Harder: The Circulation of Industrial Film*. Taylor & Francis Group 2024
DOI: 10.5117/9789462986534_CH12

installations.[1] Earlier work by Ray and Charles Eames, Shirley Clarke, D.A. Pennebaker and Walt Disney at the Brussels World Fair (1958) and then subsequent cinema at the New York World's Fair (1964) mark earlier and well-known examples of immersive and multi-screen environments as they were enacted by artist-industry collaboration.[2] Grand industrial exhibitions present one kind of context for thinking about film, with their event logics, sizable budgets, industrial and state imperatives, technological enthusiasm, rhetorics of innovation and popular (often didactic) address. But in the United States and throughout Europe and elsewhere, there were numerous local and regional fairs that also served as display outposts for ready-made travelling shows constituted by titles held in sizable industrial film libraries. Both types of industrial events, the small and large scale, the local and the international, the fleeting and the epic, showed many of the same kinds of films. Indeed, the voluminous workaday industrial, public relations and advertising films constituted the majority of films made, circulated and seen at most fairs, distinct from the showy premieres and artist-designed spectacles we tend to focus on. Many of the films shown at fairs had small budgets and short life spans; others benefitted from large budgets and had duration built in, circulating for years. Some were architectural, site-specific and unique events, while many more were highly portable, circulating in a can by mail. Most were linked to other industrial communications practices and diversified media use typical not just at fairs but also within organizational operations.[3] While some of these films and installations were

1 For discussion of one of the earliest uses of film at an exposition, see Gunning, Tom. "The World as Object Lesson: Cinema Audiences, Visual Culture and the St. Louis World's Fair, 1904." *Film History*, vol. 6, no. 4, 1994, pp. 422–44; for more on the expanded cinema experiments some 65 years later, see Gagnon, Monika Kin, and Janine Marchessault. *Reimagining Cinema: Film at Expo 67*. McGill-Queen's University Press, 2014.

2 For a study of films at German industrial fairs, see Cowan, Michael. "From the Astonished Spectator to the Spectator in Movement: Exhibition Advertisements in 1920s Germany and Austria." *Canadian Journal of Film Studies*, vol. 23, no. 1, 2014, pp. 2–29; and idem. *Walter Ruttman and the Cinema of Multiplicity: Avant-garde – Advertising – Modernity*. Amsterdam University Press, 2016; for a focused examination of film at the Brussels World's Fair, which merged industry interests with Cold War dynamics, see Nilsen, Sarah. *Projecting America, 1958: Film and Cultural Diplomacy at the Brussels World's Fair*. McFarland, 2011; for a reconsideration of the Charles and Ray Eames film work, see Nieland, Justus. "Midcentury Futurisms: Expanded Cinema, Design, and the Modernist Sensorium." *Affirmations of the Modern*, vol. 2, no. 1, 2014, pp. 46–84; and idem. *Happiness by Design: Modernism, Film, and Media in the Eames Era*. University of Minnesota Press, 2020; for recent considerations of multimedia at fairs, see Turner, Fred. *The Democratic Surround: Multimedia and American Liberalism from World War II to the Psychedelic Sixties*. University of Chicago Press, 2013.

3 Rick Prelinger and others make this point in their examinations of educational and other so-called ephemeral genres. Prelinger, Rick. "Smoothing the Contours of Didacticism." *Learning*

made by noteworthy artists and experimenters, still others were made by nameless technicians and unrecognized labourers, complicating questions of creative agency and intent.[4] These kinds of differentiating factors will be key as we gradually propose new ways of assessing this vast, interconnected area of film, media and exhibitionary activity.[5]

One of the reasons looking at fairs is crucial for thinking about industrial film and media is that they represent one of the most persistent, resource-intensive, spectacular and voluble public interfaces for industrial and sponsored media, supported by what was a highly diversified exhibitionary context, in which – it should be added – film was but one of many presentational techniques. Cinema at fairs became something rather specific, less a stand-alone text and more part of a particular kind of exhibitionary event. Millions of people attended industrial fairs, the largest of which served often as vast monuments to the achievements of state and industry. As such they were a kind of sanctioned leisure, frequently working to assure those without resources about the benevolence and power of those with resources. Film and other media have long served as elements of the many presentational and performance techniques used to promulgate messages celebrating industrial might and technological innovation, and to naturalize the idea of fairs, exhibitions and spectacular technological events in the most general sense. In addition, the largest expositions have throughout the nineteenth and twentieth century served as catalysts for innovation in exhibitionary and display practices beyond fairs proper. As such, they are generators of both technique and content as well as unique venues for display and performance. Let's consider an example.

with the Lights Off, edited by Devin Orgeron et al. Oxford University Press, 2012, p. 351. Prelinger argues that when discussing Jam Handy was a full-service industrial filmmaking company that often made, distributed and exhibited films in the context of coordinated cross-media campaigns, including live shows, press articles, slide films, study guides, and training materials. Ibid., p. 351.

4 Yvonne Zimmerman discusses this and other points related to established methods of analysis in the field of film studies in her essay "'What Hollywood Is to America, the Corporate Film Is to Switzerland': Remarks on Industrial Film as Utility Film." *Films That Work: Industrial Film and the Productivity of Media*, edited by Vinzenz Hediger and Patrick Vonderau. Amsterdam University Press, 2009, pp. 101–18.

5 There is an emerging consensus that heavily contextual analysis will be crucial for developing new vocabularies and categories. For a key early call to think carefully about context, see Elsaesser, Thomas. "Archives and Archaeologies: The Place of Non-fiction Film in Contemporary Media." *Films That Work: Industrial Film and the Productivity of Media*, edited by Vinzenz Hediger and Patrick Vonderau. Amsterdam University Press, 2009, pp. 19–34; context is also fundamental by the concept of "useful cinema," as elaborated in the book of the same name. See Acland, Charles R., and Haidee Wasson, editors. *Useful Cinema*. Duke University Press, 2011.

At the 1939 World's Fair in New York City, the languages and techniques of industrial exhibitions had grown to include a full range of new film technologies. This included 3D, Technicolor and Kodachrome colour, portable projectors with synchronized sound, amateur filmmaking equipment and a range of innovative exhibition techniques that employed films using stroboscopy, microscopy and all manner of animation. Collectively, these languages of cinema illustrated scientific principles, extolled the virtues of corporate engineers and company R & D. They told humanizing tales about imperialist procurement of raw materials from around the world to service American industrial might. And, they offered amusing portraits of anthropomorphic dancing oil drops, self-assembling cars and talking robots. Films were also integrated into live performances to augment and amplify other presentation techniques. Cameras and celluloid extensively documented and made records of these same performances, and also recorded the near-daily speeches by corporate executives, government officials and other dignitaries. Film technologies promulgated an industrial form that was at times entertaining and at others edifying and that consistently invited visitors to marvel at American innovation, achievement and ambition. All of this happened without the direct involvement of Hollywood, which elected to largely bypass the fair, notwithstanding the newsreel coverage generated by the two-year event.[6]

Of the well over five hundred films that were screened at the fair, *To New Horizons* (Jam Handy/General Motors, 1940) stands as a particularly compelling case. It is a twenty-three-minute industrial film best characterized as a work of speculative non-fiction. It borrowed less from previous industrial film conventions and more from those typical of live exhibits, dioramas, travelogues and works of science fiction. It documented, built upon and extended a monumental industrial exhibit called Futurama, the cornerstone of the General Motors (GM) pavilion. The film showed at the fair but it also appeared widely beyond it as part of the growing exhibition circuit constituted occasionally by movie theatres but more so by schools, civic clubs, factories, retail outlets, business associations and itinerant events. *To New Horizons* averred a particular relationship to industrial film and industrial discourses generally. It was about, for, and in a sense constituted by the aims and logics of industrial exhibitions and the modernist design work of Norman Bel Geddes, who is credited as the

6 For the one exception to this statement, see Wagner, Phil. "'A Particularly Effective Argument': *Land of Liberty* (1939) and Hollywood Image (Crisis)." *Film and History*, vol. 41, no. 1, Spring 2011, pp. 7–25.

creative force responsible for the GM Building, and its main attraction, Futurama. With lengthy exterior shots of the GM Building and the waiting fair crowds, and extensive footage of the elaborate diorama and fair ride that operated inside of the building, the film is inextricably linked not just to the visions of a particular industrial designer, but also to a specific architectural and multimedia fair installation, one predicated on a constant, heavy flow of people. Crucially, the same film was also made to function independently of and exist well beyond that fair. Hundreds of 16 mm and 35 mm prints circulated widely, operating as stand-alone, distinct and mobile events – exhibits in a can – which could be recreated wherever the film was projected, addressing a dispersed audience of changing size, shape and constitution for years after the New York attraction closed.

Futurama and *To New Horizons* engaged spectators by asking them to navigate a sizable but highly detailed, miniaturized world using shared structuring devices: a futurist travel narrative, constant movement and authoritative voice-over. Synched, pre-recorded instructions to "look at that," and "marvel at this," directed viewers to focus on particular elements for a specific amount of time as they travelled through the larger whole, helping to break down the vast future into manageable, timed segments. Futurama and *To New Horizons* asked visitor-viewers less to admire a factory or marvel at industrial machinery but more to enter and share a benevolent, streamlined world, one made all the more appealing because of its innovative use of lighting, recorded sound, aerial views and calls to imagine a particular kind of future.[7] In what follows, I will argue that *To New Horizons* serves as a telling index to an emergent phase of industrial film, one in which select films functioned as integral elements of broader, multimedia efforts that used a broadening array of exhibitionary techniques to create complex, innovative, corporate dreamworlds. These films were elements in major industrial design projects which themselves enacted specific exhibitionary logics. In 1939, these dreamworlds had deep links to national and international events, modernist architecture and industrial design practices. They grew across an expanding media infrastructure for disseminating industrial and corporate messages. This infrastructure predates television and grew alongside radio, and made use of movie

7 I am borrowing the phrasing here from Roland Marchand, who notes the importance of Geddes' work at the fair as it marks a shift away from illustrating industrial process toward more generalizable invitations to share a wider social and technological vision, an invitation to "share our world." "The Designers Go to the Fair II: Norman Bel Geddes, the General Motors 'Futurama,' and the Visit to the Factory Transformed." *Design Issues*, vol. 8, no. 2, 1992, p. 25.

theatres as well as an expanding network of portable film projectors, which had by 1939 become familiar elements within industrial media practices generally. This chapter contributes to recent work demonstrating that industrial and sponsored films evinced a complex relationship to some of the received categories that tend to oppose beauty and information, entertainment and didacticism, function with formal play. By examining the cognate disciplines and practices that were shaping film use, namely industrial and exhibition design, as well as public relations, this chapter considers the relationship between film and what was called in the 1930s "industrial showmanship." What follows documents a compelling case study for situating industrial film within a broad aesthetic and technological arena, wherein film must be understood as integral to a highly mediated environment predicated on performance, display and ontological complexity, with links to developments across multiple spheres, from the highly corporate to the avant-garde.

By 1939 when the fair doors opened, the American industrial film was being irretrievably shaped by the ascendant practices of public relations, which were not so much replacing but complementing industrial process films and the folksy parables of modernizing roads and farming techniques that had long been offered by companies such as Ford.[8] The new language of public relations increasingly positioned American corporations at the centre of American identity. Rather than selling excellent products or pioneering efficient factories, public relations work emphasized more general messages about industrial and corporate contributions to civic life, favouring terms like progress, innovation, safety, freedom, and prosperity as prominent elements of company messaging. In addition, public relations practices also adopted new modes and methods of content delivery. There was an increasing use of drama, personal narratives and colloquial language which pervaded outreach efforts found in magazines, radio, travelling exhibits, national and regional fairs, comics and film. Multimediated messages from the edifying to the comedic and cartoonish struck a common chord, expressing the ways in which industry was ostensibly contributing to moral, prosperous and inclusive ways of living.[9]

8 Grieveson, Lee. "The Work of Film in the Age of Fordist Mechanization." *Cinema Journal*, vol. 51, no. 3, 2012, pp. 25–51.

9 Bird, William M., Jr. "Enterprise and Meaning: Sponsored Film, 1939–1941." *History Today*, vol. 39, December 1989, pp. 24–30; see also idem. *"Better Living": Advertising, Media and the New Vocabulary of Business Leadership, 1935–1955*. Northwestern University Press, 1999; and Marchand, Roland. *Creating the Corporate Soul: The Rise of Public Relations and Corporate Imagery in American Big Business*. University of California Press, 1998.

With this shift of address firmly in place, the New York fair presented opportunities to interpret and express these terms by using the tools and techniques of grand industrial expositions. Increasing and unprecedented expenditures on promotional displays and fair exhibits throughout the decade reached a climax in New York City in 1939, facilitating the design and construction of signature buildings and monumental installations.[10] Dubbed "The World of Tomorrow," the optimistic, technocratic and futurist fair stood in stark defiance of the preceding decade of economic strife, and government efforts to intervene. Its general silence on the rising conflict in Europe requires a fulsome and separate treatment. Plainly eager to assert the capacities of unbridled corporate capitalism, the fair became a staging ground to prop up not only abundant products and the processes that produced them but to promulgate very particular and American worldviews. As part of this shift, Roland Marchand has argued that the American fairs of the 1930s, especially New York's, demonstrated a marked move away from industrial didacticism and toward entertainment, what he terms "attractions in tabloid doses."[11] But, it is crucial to clarify that fair planners did not wholly reject inherited ways of talking or exhibiting the fruits of industry. The fair offered plenty of conventional didacticism, with charts, maps, illustrations, working labs and moving models demonstrating industrial innovation and process. Yet, these familiar recreations of industrial reason and might were now also paired with comedies, staged contests, live rocket launches, cartoons and talking robots. Some of these attractions made recognizable histories about industry growth or stories about manufacturing more lively. But others simply replaced such narratives with magical scenarios and dramatic spectacle. At their most advanced, new exhibition techniques invited fair visitors to embrace a future-oriented, technologically utopian worldview, one that manifested in designer-led, multimedial, all encompassing displays that fully left the factory and often the worker behind.[12]

10 Marchand. "The Designers Go to the Fair II," p. 29. Initially budgeted at $2 million, the GM pavillion eventually cost over $7 million, which is roughly equal to $120,540,000 in 2016 (p. 23). This dollar conversion was completed using Bureau of Labor Statistics and government data tracking Consumer Price Index over the last century. "Consumer Price Index." US Bureau of Labor Statistics, www.bls.gov/cpi/home.htm. Accessed February 20, 2018.

11 Marchand. "Corporate Imagery and Popular Education: World's Fairs and Expositions in the United States, 1893–1940." *Consumption and American Culture*, edited by David Nye and Carl Pederson. Amsterdam Vrije University Press, 1991, pp. 18–33.

12 Marchand lays out this case most directly in "The Designers Go to the Fair II." See also Marchand. "The Designers Go to the Fair: Walter Dorwin Teague and the Professionalization of Corporate Industrial Exhibits, 1933–1940." *Design Issues*, vol. 8, no. 1, 1991, pp. 4–17.

With the changing exhibition practices at the fair, the tools and techniques of cinema were playing a part. This fact did not go unnoticed by contemporaneous film advocates. Richard Griffith, for instance, was particularly interested in the potential the fair offered to documentary cinema. Yet he conceded that the "most successful motion pictures at the World's Fair are those which see the Fair itself as an opportunity to use new ways of challenging popular attention."[13] Among these he included discussions of "zoom films," "cinematic murals" and even "poster films," small, short rear-projected looping films. He also noted hybrid, performative uses of film segments, like the short film devoted to the history of transportation that was used to prepare audiences for an actual on-stage rocket launch.[14] Indeed, films and film technologies served many purposes at the fair. There were well over five hundred films shown on the fairgrounds in its first year alone.[15] Some of these films were commissioned for the fair, and many others were part of the growing industrial film libraries. Exhibitors presented these films in small pavilions and large, in dedicated theatres as well as in transient and provisional spaces. In addition to official World Fair and government buildings, many of the largest industrial exhibits included film: Ford, GM, Westinghouse, RCA, Coca-Cola, among others. Some used eye-popping tricks. Chrysler's 3D, stop-motion, colour film featured a self-assembling car, whose parts seemingly flew out from the screen and dangled over the heads of the audience.[16] Indeed, American car companies were highly prominent exhibitors at the fair. Chrysler, Ford and GM each had elaborate, hi-tech pavilions.

GM serves as a telling example of the diverse practices enacted by but one single exhibitor. GM was one of the largest automobile manufacturers in the world in 1939. As such, the company had an elaborate and well-established communications practice, which included a longstanding and active use of film, as well as exalting techno-futurist gadgets and processes emanating from their research labs. Emblematic here was GM's "Caravan of Progress" which toured the United States beginning in 1936, itself an outgrowth of

13 Griffith, Richard. "Films at the Fair." *Films: A Quarterly Discussion and Analysis*, vol. 1, no. 1, November 1939, p. 62.

14 Ibid., pp. 61–72. This report is an interesting one as Griffith is clearly wrestling with his ideals for documentary as a new tool for civic life and the corporate propaganda and sales pitches that predominated.

15 Richard Griffith catalogued and reviewed over five hundred films for the American Film Center in 1939. See ibid.

16 For more on this, see Wasson, Haidee. "Industrial Magic and Light: 3D at the World's Fair." *Public: 3D Cinema and Beyond*, no. 47, 2013, pp. 123–30, https://doi.org/10.1386/public.24.47.123_1.

an earlier fair. The travelling show offered glimpses of GM products and samplings of future products, including early televisions, microwave oven prototypes and stereophonic sound devices. GM even made a film of the caravan, which documented and promoted it.[17] In addition to such caravans, GM also eagerly embraced film throughout the decade. Jam Handy, but one of the production companies with extensive GM contracts, completed as many as forty films for the company in 1937 alone.[18] One estimate suggests that millions of people saw these films, which were plainly part of a more elaborate and multi-modal address orchestrated by GM and addressed to people across the United States and beyond.[19]

At the New York fair, GM's pavilion hosted a theatre where films played regularly, usually along with a live lecture. The benefits of scientific research for industry and employment were a prominent theme."[20] Through a combination of live action, animated images and a range of synch-sound techniques, GM films amplified, narrated and otherwise transformed these themes into short movies. Some were straightforward films about such topics as driver safety. Three of the nine regularly shown films were animated, using surprise and humour woven through light and entertaining narratives that integrated or referred to aspects of GM's products or operations.[21] In other words, by 1939, American industrial films included films for children, with fairy tales being repopulated with GM products (*A Coach for Cinderella*, 1937) or small toys standing in for workers which demonstrated production

17 *Our American Crossroads* (1940). This film documents the diorama that was part of the Caravan of Progress, which itself illustrates the difference between an American town before cars and then afterward. The voice-over for the film is a folksy older man narrative the story of Pleasant Corners. This film can be viewed at http://gmheritagecenter.com and on the Internet Archive. See also "A Film Story of the General Motors Parade of Progress" (Soundmasters), which documents the 1941 travelling version of the caravan (General Motors. "1941: Parade of Progress: GM Takes Its Story to Small-Town America," www.gm.com/mol/3-1941-parade-of-progress.html. Accessed February 20, 2018). Here research, science and industry heroically prepare a prosperous path to the future.

18 Sklar, Paul, and S.H. Walker. *Business Finds Its Voice*. Harper and Brothers, 1938, p. 43.

19 All stats drawn from ibid., ch. 6.

20 Griffith. "Films at the Fair," p. 4.

21 All films shown by GM in the fair's first year were: *A Coach for Cinderella* (1937, 35 mm and 16 mm sound, 10 min, Jam Handy); *Color Harmony* (1938, 35 mm and 16 mm, 10 min, Jam Handy); *Heritage of the Years* (nd. 35 mm and 16 mm, sound, 10 min, Wilding Picture Productions); *Jumping Beans* (1937, 35 mm and 16 mm sound, 10 min, Jam Handy); *King Cotton* (1939, 35 mm and 16 mm, sound, 10 min, Jam Handy); *Modes and Motors* (1938, 35 mm and 16 mm, sound, 10 min, Jam Handy [a film about industrial design]); *Nicky Rides Again* (1937, 35 mm and 16 mm sound, 10 min, Jam Handy); *Quiet, Please!* (nd. 35 mm and 16 mm sound, 10 min); and *We Drivers* (1935, 35 mm and 16 mm sound, 10 min, Jam Handy).

and consumption cycles (*'Round and 'Round*, 1939), a technique used by the Ford Motor Company as well (*Symphony in "F"*, 1940). It is worth noting that GM also made its films available through its Public Relations Department for loan during and after the fair to schools, libraries, YMCAs and other community groups. A good number of these films were plainly addressed to children.

It is in this context that *To New Horizons* takes on special interest as it emerged from within a diversified industrial film programme. In contrast to the children's films described above, *To New Horizons* evinced a notably adult and more conventionally didactic tone. But like the children's films, it elicited a similar sense of wonder through the use of miniatures and models and the creation of magical worlds emptied of conflict and purged of complexity. In some of these films, the factory and worker completely disappeared. New, happy, wondrous things happened in their stead. In addition to these plain connections to other industrial films, the deep formal links between *To New Horizons* and industrial and exhibitionary design are equally important. The film is a spatially dynamic exploration where the marvel of production took backstage to a new world articulated to cars and motorways, celebrating the privileges that individualized, speeding and motorized movement would soon entail. As a fair exhibit, Futurama's goal was to merge a middlebrow didacticism with a wondrous worldview, using architectural models, intricate miniatures and a showman's flair, elements of the exhibit and the film that trace directly back to Bel Geddes and his career.[22] *To New Horizons* sought to capture that, adapting it into a fair film.

Norman Bel Geddes began his career as a stage and theatre designer, crafting sets for several films as well.[23] He was an advocate of new theatrical design, wherein American designers adapted elements of European avant-garde practices, working toward dissolving stark boundaries between audience and stage. Key design features included the simplification of stage

22 For an excellent overview of the multidisciplinary influences on Geddes and the ways in which he then exerted his own influences across the fields of mapping, military strategy, journalism, household appliances, architecture, advertising, interior design, and theatre, see Albrecht, Donald, editor. *Norman Bel Geddes Designs America*. Abrams, 2012.

23 For the most thorough examination of Bel Geddes' relationship to film, including familiarity with the ideas of Sergei Eisenstein see: Dimenberg, Ed. "The Will to Motorization: Cinema, Highways, and Modernity." *October*, vol. 73, Summer 1995, pp. 90–137. Bel Geddes also experimented with microphotography and highspeed filmmaking, as befitting his interest in scale. His films are collected along with his papers in the University of Texas. Designers working with film was not unusual, particularly as filmmaking became less expensive and practical beyond the professional film industry throughout mid-century. The work of Charles and Ray Eames provides a well known case.

scenery and theatre interiors, and the use of inventive lighting and sound to create all-encompassing experiences.[24] Notably, he designed the proscenium out of his theatres, crafting spaces that allowed for more interaction between audience and performers. His avant-gardist contemporaries in Europe were envisioning stages that circumscribed the audience, and that could be moved like modular platforms, higher or lower, or rotated, with ramps and steps that divided the arena and extended into sections of the audience.[25] Some of Bel Geddes' contemporaries also conceptualized theatres that incorporated film, projections and other media into live theatre.[26] Parallel ideas which disrupted traditions of art exhibition by reconceptualizing exhibitions themselves as dynamic and engaged spaces suited to mass education were also hallmarks of Bauhaus experiments throughout the 1920s and early 1930s.[27] There was a rich context of debate to inform Bel Geddes' highly theatrical exhibition design for Futurama, which not only roped the moving audience around the stage, but set them in flight above it. In designing Futurama, Geddes displaced the proscenium stage and the seated audience, replacing both with a cockpit.

In addition to Bel Geddes' work in theatre, he is best known as a leading member of the new profession of American industrial design, which coalesced in the 1930s. Bel Geddes designed industrial objects (stoves, radios) and spaces, which included offices, theatres, gas stations, and shop windows. He also performed public relations and advertising work. Across these areas, Bel Geddes averred an unapologetic embrace of the modern, promulgating a streamlined aesthetic which disavowed ornamentation and foregrounded materials and function. Curved lines, perpetual movement and machine efficiency were key. Bel Geddes declared that the work of the designer should be understood in the same way as the research and development in corporate laboratories, always "peering over the horizon" into the future.[28] He was also known as a showman, with most scholarship that addresses his work highlighting the place of his training in theatre. Donald Albrecht described his building designs as "theatricalized architecture," for his attention to movement around and within spaces, and for the plain will to dynamically

24 For more on Bel Geddes' relationship to theatre design see: McGuire, Laura. "Theaters." *Norman Bel Geddes Designs America*, edited by Donald Albrecht. Abrams, 2012, pp. 173–83.

25 Aronson, Arnold. "Theatres of the Future." *Theatre Journal*, vol. 33, no. 4, 1981, pp. 489–503.

26 For more on thisl, see ibid.

27 For an excellent overview of these activities, see Staniszewski, Mary Anne. *The Power of Display: A History of Exhibition Installations at the Museum of Modern Art*. MIT Press, 1998, esp. ch. 1: "Framing Installation Design: The International Avant-gardes," pp. 1–58.

28 Bel Geddes, Norman. *Horizons*. Little, Brown and Co., 1932, p. 248.

engage people within those spaces.[29] Bel Geddes also evinced a determined fascination with models and miniatures, using them throughout his career not just as part of his working method or as salesmanship but as a unique mode of expression, employing them in his advertising, exhibition design, map designs and his later war work. The latter of these securing recognition from the art world in 1944 when his war and battle miniatures were on display at the Museum of Modern Art in New York.

As an industrial designer, Bel Geddes was not entirely alone in his taste for putting on a show. He, was in fact, joined by several other designers who held prominent fair commissions in New York and indeed throughout the preceding decade.[30] Walter Dorwin Teague is perhaps the most prolific of these with seven major New York fair exhibits to his name, including work for Ford, Du Pont, United States Steel, Kodak, Texaco, and Consolidated Edison. These pavilions integrated films and various moving projections. Teague himself called his exhibits "hit shows," providing a contemporaneous marker of the ways in which architecture, industrial design, exhibition design and their media were increasingly talked about with the same language as popular entertainment.[31] Similarly, Donald Deskey and Raymond Loewy also performed significant work at the fair. Deskey had designed the interiors for Radio City Music Hall several years earlier, and was responsible for the Communications Building at the fair. He shared the idea of performative industrialism that was ascendant during the period. Deskey used the phrase "industrial showmanship." According to him, film technologies played a role in the new language being forged. In 1938 he wrote:

> Every device for the dramatic presentation of products and ideas is being probed. The motion picture is being used in many cases as an important part of the display. However, the use of the sound film alone in a standard theater setting is nothing new to the visitor from the crossroads. But as an instrument for the visualization of ideas, it is being incorporated into

29 For a general overview of Bel Geddes' contributions, see Albrecht, Donald. "Introduction." *Norman Bel Geddes Designs America*, edited by Donald Albrecht. Abrams, 2012, pp. 10–39. The reference to "architecture in motion" is on p. 26 of this essay; for an excellent discussion of Bel Geddes' war work, see Cohen, Jean-Louis. "On the Information Front." *Architecture in Uniform*. Éditions Hazan, 2011, pp. 320–45.

30 For one of the best summaries of the fair as it relates to the field of design, see Meikle, Jeffrey L. *Twentieth Century Limited: Industrial Design in America, 1925–1939*. Temple University Press, 1979, esp. ch. 9.

31 For a closer study of Teague's role at the fair, see Marchand. "The Designers Go to the Fair"; for a focus on Bel Geddes, see idem. "The Designers Go the Fair II," pp. 22–40.

more elaborate mechanical devices; stage presentations for industry with the motion picture as an integral part.[32]

For Deskey, film became part of an experimental ethos in exhibition design and industrial performance, expanding extant industrial techniques as much as expanding previous uses of the medium.

To be sure, Futurama was industrial showmanship writ large. Its scale and scope, detail and drama were unique. It captured the attention of fair visitors and contemporaneous commentators alike who used terms like "ingenuity," "daring" and "genius." Studies at the time deemed it by far the fair's most successful exhibit.[33] It was in essence an automated, theatrical diorama which spectators did not walk through or stand and gaze upon but were carried around, through, and over on moving chairs, strung together like a miniature lounge train. As visitors travelled over and above the planned, efficient, highway-laden future of 1960, a scripted voice-over, choreographed lighting and thousands of moving parts created a living map wherein the whole of the United States was magnificently transformed by efficiency, prosperity and the steady flow of all things and people by the automobile. The scientists and engineers at GM were, in a sense, the invisible protagonists, responsible for this marvellous vision.

GM Chairman Alfred Sloan introduced the pamphlet publicizing the ride, framing Futurama as a dramatic demonstration of a world "hardly yet begun," a world of great opportunities built on the fact that where there are "new and better means of moving people, [...] new communities, new enterprises and new opportunities have everywhere followed."[34] The drama of this message of unbridled progress began with the GM building that housed Futurama – a windowless, curved, structure that featured a winding pedestrian ramp on the outside leading up to an entry way that was partly concealed by the curvature of the building. The ramps facilitated the long lines, but also made waiting a part of the show.[35] In other words, waiting and the display of human anticipation was designed into the architecture. Once inside, visitors entered a large, darkened hall with a giant map (60 x 100 feet) that appeared to be floating. A soft blue light framed the map, which was also illuminated with green and red-light strips that transformed it, one

32 Deskey, Donald. "Industrial Showmanship in '39." *Business Screen*, no. 2, 1938, p. 17.

33 Marchand. "Designers Go to the Fair II," p. 24.

34 Sloan, Alfred P. "A Greeting to Our Guests." General Motors. *Futurama*. 1939, p. 1. Pamphlet. New York Public Library, Manuscripts, World's Fair, Box 2.

35 See, for instance, General Motors. *Futurama*. 1939. Pamphlet. New York Public Library, Manuscripts, World's Fair, Box 2.

moment illustrating current roadways across the United States and another future roadways; at one moment the lighting indicated present traffic flow and the next future traffic flow. The map espoused an up-to-the-minute aesthetic. A pre-recorded voice-over welcomed visitors directing them down the interior spiralling ramp to the fifteen-minute "carry-go-round" ride waiting below, helping to automate and direct human traffic. Once seated, riders transitioned through a dark tunnel – equivalent to dimming the house lights – emerging in the expansive and well-lit, miniaturized diorama.

Bel Geddes' ride used upholstered wing-backed chairs that brought spectators through – and importantly over and above – the working diorama below, itself animated by moving cars, trucks and planes, as well as changing lights.[36] While the chairs were tethered together and operated like a train, the exhibit itself encouraged riders to think of themselves as flying, using words like "gliding," inviting riders to a "magic Aladdin-like flight through time and space" likening the ride to a "transcontinential flight over America."[37] Scripted and pre-recorded sound was a crucial element of the ride, emanating from individual speakers housed in each of the chairs. Commentators at the time noted the soft immediacy and personalization of the pre-recorded audio elements of the tour, itself a recent marvel of innovations in recording and amplification generated by the field of electroacoustics, the same ones that facilitated the concurrent rise of crooning in American cultures of popular music.[38]

Futurama illustrated the highway-saturated future of 1960. Seated spectators travelled for one-third a mile surveying a 35,000 square foot area. Seen largely from above, were "half-million buildings and houses, thousands of multi-lane highways and more than a million trees." Fifty thousand miniscule cars, ten thousand of which zipped around the expansive installation. Official literature boasted of the ride's capacity to carry 28,000 people per day.[39] A key innovation here in terms of exhibition design was the fact that Bel Geddes had seated the spectators on a mechanical chair train that could be moved at a programmed pace, and coordinated with an audio

36 For a technical description of the Futurama, see Van Duyne, Schuyler. "Talking Train Tour, World's Fair Exhibit." *Popular Science*, vol. 135, no. 1, July 1939, pp. 102–05.

37 General Motors. *Futurama*, p. 5.

38 An article in the *New York Telegram* described the reproduced voice as an "a quiet, intimate voice, tensely dramatic, yet direct and almost confidential." Qtd. in Marchand. "Designers Go to the Fair II," p. 38. In contrast to these recollections, the souvenir pamphlet emphasized the technological heft of the sound system, describing it as "twenty-tons-of-voice" (p. 2), suggesting that the elaborate nature of the apparatus was itself presented as a feature of the experience.

39 General Motors. *Futurama*, p. 2.

tour that effectively directed participant attention as they moved through and across and, in a sense, towered over the diorama. The show happened all around them but the narration and audio instructions told them where and how to look.[40] And, while the highways observed from above were designed to support speeding vehicles up to one hundred miles an hour, the slow-moving chairs prioritized "REST and COMFORT"[41] (original emphasis), a small reward after standing so long in line. The climax of the ride actually occurred as visitors disembarked from their chairs and entered a full-sized intersection of the future city, one they could wander around. The suddenly full-size model became an inhabitable rather than merely observable marvel. The buildings that visitors could enter included showrooms for GM cars, as well as an exhibit devoted to the company's research labs, which featured wonders of modern science, including a "talking flashlight" that transmitted speech over a light beam and the "Frig-O-therm," which apparently could cook and freeze food "at the same time," presumably in separate chambers of the same appliance.[42]

Private enterprise, planned highways, cars and consumerism became the foundations of this design-savvy future. Commentators and critics, including Walter Lippmann, noticed the seeming contradiction between unbridled free enterprise and the plain necessity of government policy and spending to build the infrastructure that was elemental to the new vision.[43] Government, public policy, taxes, social safety nets, worker strife were written out of GM's script. Nevertheless, Futurama was a dramatic "total environment" that brokered in a complex play with scale: the human scale of the embodied rider who is first small and part of a waiting mass, then gigantic atop the shrunken future and then right again in the corporate town.[44] It combined dynamic lighting, mechanized spectator movement, directed lines of sight and coordinated pre-recorded sound to help hold it all together. Its extensive plays on scale and use of miniatures pulled forward and in a sense modernized much older panoramic and dioramic

40 Marchand supplies the best and most detailed and elaborate analysis of Futurama. See Marchand. "Designers Go to the Fair II," esp, pp. 31–34.

41 Qtd. in Marchand. "Designers Go to the Fair II," p. 37; Bel Geddes' commitment to comfort is made plain in the book which emerged from Futurama. See Bel Geddes, Norman. *Magic Motorways*. Random House, 1940, esp. pp. 3–10.

42 General Motors. *Futurama*, p. 21.

43 See Lippman, Walter. "A Day at the World's Fair." *New York Herald Tribune*, June 6, 1939, p. 25.

44 Marchand argues persuasively that Bel Geddes' training in theatre and dramatic use of lighting, voice-over, and intense emulation of detailed aerial views created a "total environment" and a distinct kind of corporate experience. See Marchand. "Designers Go to the Fair II."

forms. Also important were new uses of intimate, recorded and amplified sounds, the very same ones transforming and expanding soundscapes and cinema elsewhere at the fair and beyond. Fair exhibition here became a highly choreographed and controlled journey, shaped irretrievably by a highly theatrical, mediated, spatially and temporally complex ride.[45]

Jeffrey Meikle and Roland Marchand have both argued that Futurama successfully inverted industrial exhibition techniques, marking a transition from plainly illustrating industrial assembly lines to enchanting and ultimately emulating them, creating consumers in an exhibitionary assembly line. Following from this, one can easily understand *To New Horizons* as a compelling extension of Futurama's exhibitionary and ultimately productive principles.[46] Like in previous uses of film by museums, film's technical ability to regiment, pace and standardize the relatively free and wandering view of exhibition-goers through vast exhibition space takes on fuller clarity.[47] The film has two distinct parts. The first is a kind of driverless road movie, which proceeds from past to present, observing an identifiable geographic movement beginning with the primordial crashing waves of the Pacific Ocean and travelling eastward across the United States by car, arriving at the advanced engineering marvel of the fair in New York City. This half of the film is also plainly structured as a teleological timeline, charting inevitable and desirable technological progress. The earliest scenes depict sluggish methods of travel such as mule and cart, while later sequences exclusively feature zipping cars and soaring planes. Throughout this segment, log cabins become modern homes and dirt roads become paved motorways. The voice-over declares: "Today engineers are always leading us higher, widening trails, while our men of science are broadening all our mental avenues with new activities, activities based on modern pioneering into new fields of research." A camera is in near constant movement, and often mounted on a speeding automobile, providing a phantom ride across smooth roads that slice through the landscape. The first section of this film ends at the New York fair in front of the seemingly endless lines of fairgoers queued to see what lay within the windowless, curved walls of the GM building. These

45 The fair environment was then further augmented by the multiple discourses about the ride that circulated widely and well beyond the site of the fair. This includes a book entitled *Magic Motorways*, which also provided multiple illustrations of Futurama's models, functioning roadways, detailed drawings, as well as the extensive research that supported Bel Geddes' plans. Bel Geddes. *Magic Motorways*.

46 Meikle. *Twentieth Century Limited*, p. 197.

47 See my "Big, Fast Museums/Small, Slow Movies: Film, Scale and the Art Museum." ***Useful Cinema***, edited by Charles R. Acland and Haidee Wasson. Duke University Press, 2011, pp. 178–204.

shots emulated a key part of the Futurama's iconography as both the building and its structural conceit to ordered crowds became a recurring motif in photographs, postcards, brochures and pamphlets that addressed the exhibit. An on-screen text announces: "We are now approaching Futurama." While the camera lingers on the crowds outside of the building, the black-and-white film transitions into Technicolor, rhetorically inviting viewers into GM's not just colourful but cinematically forward future. Much like *The Wizard of Oz* (1939) released one year before, *To New Horizons* used colour to accentuate the wonder and appeal of the other-worldly.

The second part of the film roughly shadows the first, but continues its journey from past to present, taking it forward from present to future. It too is built with continuous camera motion, edited together as a journey, which similarly culminates at the fair. But, this second phase of the film's trip is conducted through the rabbit hole of Bel Geddes' diorama and is constituted entirely by simulated aerial shots of his miniature models. The voice-over directs the spectator to particular elements of each scene, describing the salient design features of a skyscraper, a greenspace, an airport, an interchange, in each case narrowing the spectator's view by way of coordinating shots which begin wide and then close in on key elements with directional voice-over cues helping to further direct the eye. The tour guide speaks continuously throughout this portion of the film, his script largely the same one as accompanied riders of the Futurama.[48] There is a degree of performative flourish to the voice-over of the guide, who speaks calmly, giving clear instructions and asking big, open-ended questions: Look here. Look there. What will we see? The camera moves continuously if slowly over an expansive landscape entirely remade according to scientific and industrial innovation and insight, or so we are constantly reminded. Viewers are invited to marvel at the harmonious integration of smooth highways with nature's mountainous contours and to admire the ways in which engineering allows us to leap over waterways. In this future world, physics and chemistry have eliminated pests and disease, joining with the farmer in "helpful friendship." Rural areas have been fully electrified; regional roads feed smoothly into busy and multi-laned highways that allow traffic to safely travel at a hundred miles per hour. Cities have also been reimagined, with skyscrapers featuring landing decks for "helicopters and auto-gyros." Residential areas are separated from commercial and industrial spaces with ample parks and green spaces replacing slums and

48 This script also constituted the prose that was detailed in the heavily illustrated souvenir programme handed out at the exhibit to fairgoers.

dated industrial works. As the camera moves across this future, thousands of moving parts, notably all manner of transportation, further animate the highly detailed model. Design magically eliminates structural inequities with a curving overpass here and a smooth roundabout there.

The denouement of the film mirrors precisely that of the ride: a transition to a full-scale actualization of a city intersection in 1960, branded with a large GM logo that sits atop the scene. The film shows people, presumably fairgoers, strolling the elevated sidewalks, while cars whisk safely by one level below. Pedestrians window shop and wander through GM's branded city. In the film, this full-scale environment is surveyed by the camera largely from above. The voice-over reminds us: this is "a world that has hardly begun," resonating with the ride's promotional literature. The camera continues to pan up to the clouds. The film concludes with an abstract animation sequence in colour, with steady organ music giving way to soaring orchestral music, heavy on the strings. The final incantation:

> True each of us may have different ideas as to what that future might be. But every forward outlook reminds us that all the highways of all research and all communications, all the activities of science lead us onward to better methods of doing things, with new opportunity for employment and better ways of living. As we go on, determined to unfold the constantly greater possibilities of the world of tomorrow, as we move more and more rapidly forward, penetrating new horizons, in the spirit of individual enterprise, in the great American way....

The film concludes with the fair's iconic trylon and perisphere, thrusting up into the sky in animated form, returning the film to the fair.

To New Horizons portrays a future in miniature, its cinematic realization in part presented as itself an advancement and sign of what is to come. Camera framing and movement are such that the diorama is used to create a view to a full-sized world rendered photo-realistically; there are no moments where the diorama is revealed in the film as such. The film brokers in scaled illusions, a mainstay of special effects and fair exhibition techniques. Throughout *To New Horizons* the camera is in near-constant motion. The first half features a steadily sometimes quickly moving camera, often mounted on the front of a car, tightly and literally aligning camera and automobile with progress; the phantom frame sometimes shifts rapidly over twisting and winding roads and at others on a smooth and straight path, adding extra-diegetic flourish. In both ride and film, the future is kinetic yet controlled. Editing patterns ensure a smooth and steady view of this new world in motion, with changes of scale and framing managed carefully and gradually.

Roger Luckhurst has suggested that the futurism of the New York Fair can be thought of as a kind of vernacular science fiction, whose discordant play with time and space can be found readily in previous world's fairs, from the Great Exhibition of 1851 (London) onward. Fair exhibitors have long designed and created environments that represented and spectacularized the geopolitics of industrial empires for audiences, representing the world in select and abstract ways, scaling expansionist practices into comprehensible displays, presentations and exhibits. Maps and miniatures, for instance, have long been prominent elements of fair displays, modelling in seemingly benign terms what was an expansive industrial, capitalist and colonial geopolitical system.[49] In this general sense, Futurama was perfectly concordant with the tradition of preceding expositions. Yet it accomplished this with socio-historically specific design sensibilities and media practices that articulated film directly to designed corporate environments that could be recorded and then live again. Its future-oriented utopianism was a design-savvy but politically complex vision that remained silent on the persistent social problems that had characterized the decade, including institutionalized racism, rampant poverty, urban slums, unemployment and labour unrest. And, also Europe.[50] *To New Horizons* continued this highly abstract play of form, translating new techniques of industrial exhibition into reproducible events, making them lasting elements of an enduring exhibitionary ethos not just in the form of films but in amusement parks, subsequent world expositions, densely screened urban sites (such as Times Square), corporate headquarters and shopping malls.

To New Horizons demonstrates that the category of industrial film includes titles that were inextricably linked to educational, entertaining, design-savvy experimental environments undergirded by changing practices of industrial display and corporate expression. The use of cinema at the 1939 New York World's Fair marks an exceptional event that incorporated extant display and exhibition activity and also signalled the way towards future trends. Film shaped these new environments, extending them but also normalizing them as elements of industrial and post-industrial life. In the post-war period, the use of film at exhibitions and expositions continued

49 For an excellent discussion of fairs and their relationship to models, maps, and dioramas, see Luckhurst, Roger. "Laboratories for Global Space-Time: Science-Fictionality and the World's Fairs, 1851–1939." *Science Fiction Studies*, vol. 39, no. 3, pp. 385–400.

50 For such critical assessments of Futurama, see Codgell, Christina. "The Futurama Recontextualised: Norman Bel Geddes's Eugenic 'World of Tomorrow'." *American Quarterly*, vol. 52, no. 2, 2000, pp. 193–245.

to grow and diversify. Indeed, the use of small cinema machines and big ones in 1939 indexes a crucial way in which moving images and the many technologies that facilitated their performance were being harnessed to industrial imperatives. This entailed techniques specific to what I am calling exhibitionary film and film's unique technological and expressive capacities (repeatable spectacle, automated direct address, controlled pacing, visual effects). It also involved articulating cinema to hybrid media spaces and experiments in industrial design. In some instances, this plainly requires disarticulating what we commonly think of as the cinematic apparatus, and rearticulating it to pursue many other display and performance scenarios which plainly thrived far beyond the movie theatre and were shaped by cognate disciplines of visuality, performance and display.

The "World of Tomorrow" was a particularly corporate, governmental and technological utopia wherein alliances across machines came to aid and abet a distinct industrial form. This iteration of industrial media was multidimensional, highly adaptable to a range of spaces and purposes, and announced a common sense for a new kind of cinema in sometimes fantastical but also banal terms. This other kind of cinema provided a new set of materials not just for artists or the entertainment industry but for all industries. These other contexts, and their technological and performative articulations, will help us to identify more fully the specificities of the stylistic and formal properties of these films. Films like *To New Horizons* held a particular relationship to events; they were generated by them but also carried their marks forward in time and well beyond, working to integrate other modes of site-specific display and exhibition (dioramas, live presentations, technological demonstrations) into known and new film types. Overall, the elasticity of this exhibitionary cinema and the films it yielded point to the need for further thinking about the ways in which a new generation of advertising and public relations experts, as well as industrial designers, determined to explore and exploit what they deemed a productive interface between moving images, sounds and spaces that directed millions of people to look, listen to and enjoy corporate dreamworlds.

Works Cited

Acland, Charles R., and Haidee Wasson, editors. *Useful Cinema.* Duke University Press, 2011.

Albrecht, Donald. "Introduction." *Norman Bel Geddes Designs America*, edited by Donald Albrecht. Abrams, 2012, pp. 10–39.

Albrecht, Donald, editor. *Norman Bel Geddes Designs America.* Abrams, 2012.

Aronson, Arnold. "Theatres of the Future." *Theatre Journal,* vol. 33, no. 4, 1981, pp. 489–503.

Bel Geddes, Norman. *Horizons.* Little, Brown and Co., 1932.

Bel Geddes, Norman. *Magic Motorways.* Random House, 1940.

Bird, William M., Jr. *"Better Living": Advertising, Media and the New Vocabulary of Business Leadership, 1935–1955.* Northwestern University Press, 1999.

Bird, William M., Jr. "Enterprise and Meaning: Sponsored Film, 1939–1941." *History Today,* vol. 39, December 1989, pp. 24–30.

Codgell, Christina. "The Futurama Recontextualized: Norman Bel Geddes's Eugenic 'World of Tomorrow'." *American Quarterly,* vol. 52, no. 2, 2000, pp. 193–245.

Cohen, Jean-Louis. "On the Information Front." *Architecture in Uniform.* Éditions Hazan, 2011, pp. 320–45.

"Consumer Price Index." *US Bureau of Labor Statistics,* www.bls.gov/cpi/home.htm. Accessed February 20, 2018.

Cowan, Michael. "From the Astonished Spectator to the Spectator in Movement: Exhibition Advertisements in 1920s Germany and Austria." *Canadian Journal of Film Studies,* vol. 23, no. 1, 2014, pp. 2–29.

Cowan, Michael. *Walter Ruttmann and the Cinema of Multiplicity: Avant-garde – Advertising – Modernity.* Amsterdam University Press, 2016.

Deskey, Donald. "Industrial Showmanship in '39." *Business Screen,* vol. 1, no. 2, 1938, p. 17.

Dimenberg, Ed. "The Will to Motorization: Cinema, Highways, and Modernity." *October,* vol. 73, Summer 1995, pp. 90–137.

Elsaesser, Thomas. "Archives and Archaeologies: The Place of Non-fiction Film in Contemporary Media." *Films That Work: Industrial Film and the Productivity of Media,* edited by Vinzenz Hediger and Patrick Vonderau. Amsterdam University Press, 2009, pp. 19–34.

Gagnon, Monika Kin, and Janine Marchessault. *Reimagining Cinema: Film at Expo 67.* McGill-Queen's University Press, 2014.

General Motors. *Futurama.* 1939. Pamphlet. New York Public Library, Manuscripts, World's Fair, Box 2.

General Motors. "1941: Parade of Progress: GM Takes Its Story to Small-Town America," www.gm.com/mol/3-1941-parade-of-progress.html. Accessed February 20, 2018.

Grieveson, Lee. "The Work of Film in the Age of Fordist Mechanization." *Cinema Journal,* vol. 51, no. 3, 2012, pp. 25–51.

Griffith, Richard. "Films at the Fair." *Films: A Quarterly Discussion and Analysis,* vol. 1, no. 1, November 1939, pp. 61–75.

Gunning, Tom. "The World as Object Lesson: Cinema Audiences, Visual Culture and the St. Louis World's Fair, 1904." *Film History,* vol. 6, no. 4, 1994, pp. 422–44.

Lippmann, Walter. "A Day at the World's Fair." *New York Herald Tribune*, June 6, 1939, p. 25.

Luckhurst, Roger. "Laboratories for Global Space-Time: Science-Fictionality and the World's Fairs, 1851–1939." *Science Fiction Studies*, vol. 39, no. 3, pp. 385–400.

Marchand, Roland. "Corporate Imagery and Popular Education: World's Fairs and Expositions in the United States, 1893–1940." *Consumption and American Culture*, edited by David Nye and Carl Pederson, Amsterdam Vrije University Press, 1991, pp. 18–33.

Marchand, Roland. *Creating the Corporate Soul: The Rise of Public Relations and Corporate Imagery in American Big Business*. University of California Press, 1998.

Marchand, Roland. "The Designers Go to the Fair: Walter Dorwin Teague and the Professionalization of Corporate Industrial Exhibits, 1933–1940." *Design Issues*, vol. 8, no. 1, 1991, pp. 4–17.

Marchand, Roland. "The Designers Go to the Fair II: Norman Bel Geddes, the General Motors 'Futurama,' and the Visit to the Factory Transformed." *Design Issues*, vol. 8, no. 2, 1992, pp. 22–40.

McGuire, Laura. "Theatres." *Norman Bel Geddes Designs America*, edited by Donald Albrecht. Abrams, 2012, pp. 173–88.

Meikle, Jeffrey L. *Twentieth Century Limited: Industrial Design in America, 1925–1939*. Temple University Press, 1979.

Nieland, Justus. *Happiness by Design: Modernism, Film, and Media in the Eames Era*. University of Minnesota Press. 2020.

Nieland, Justus. "Midcentury Futurisms: Expanded Cinema, Design, and the Modernist Sensorium." *Affirmations of the Modern*, vol. 2, no. 1, 2014, pp. 46–84.

Nilsen, Sarah. *Projecting America, 1958: Film and Cultural Diplomacy at the Brussels World's Fair*. McFarland, 2011.

Prelinger, Rick. "Smoothing the Contours of Didacticism: Jam Handy and His Organization." *Learning with the Lights Off: Educational Film in the United States*, edited by Devin Orgeron, Marsha Orgeron, and Dan Streible, Oxford University Press, 2012, pp. 338–55.

Sklar, Paul, and S.H. Walker. *Business Finds Its Voice*. Harper and Brothers, 1938.

Sloan, Alfred P. "A Greeting to Our Guests." General Motors. *Futurama*. 1939, p. 1. Pamphlet. New York Public Library, Manuscripts, World's Fair, Box 2.

Staniszewski, Mary Anne. *The Power of Display: A History of Exhibition Installations at the Museum of Modern Art*. MIT Press, 1998.

Turner, Fred. *The Democratic Surround: Multimedia and American Liberalism from World War II to the Psychedelic Sixties*. University of Chicago Press, 2013.

Van Duyne, Schulyer. "Talking Train Tour, World's Fair Exhibit." *Popular Science*, vol. 135, no. 1, July 1939, pp. 102–105.

Wagner, Phil. "'A Particularly Effective Argument': *Land of Liberty* (1939) and Hollywood Image (Crisis)." *Film and History*, vol. 41, no. 1, Spring 2011, pp. 7–25.
Wasson, Haidee. "Big, Fast Museums/Small, Slow Movies: Film, Scale and the Art Museum." *Useful Cinema*, edited by Charles R. Acland and Haidee Wasson. Duke University Press, 2011, pp. 178–204.
Wasson, Haidee. "Industrial Magic and Light: 3D at the World's Fair." *Public: 3D Cinema and Beyond*, no. 47, 2013, pp. 123–30, https://doi.org/10.1386/public.24.47.123_1.
Zimmermann, Yvonne. "'What Hollywood Is to America, the Corporate Film Is to Switzerland': Remarks on Industrial Film as Utility Film." *Films That Work: Industrial Film and the Productivity of Media*, edited by Vinzenz Hediger and Patrick Vonderau. Amsterdam University Press, 2009, pp. 101–17.

About the Author

Haidee Wasson is Distinguished University Research Professor of film and media at Concordia University, Montreal. Her publications include *Everyday Movies: Portability and the Transformation of American Culture* (UC Press, 2021), *Museum Movies: The Museum of Modern Art and the Birth of Art Cinema* (UC Press, 2005), as well as numerous edited collections, including *Useful Cinema* (with Charles R. Acland) and *Cinema's Military Industrial Complex* (with Lee Grieveson).

Section 4

Teaching Oneself and Others

13 Putting Films to Work

System, the Magazine for Business

Gregory A. Waller

Abstract

Focusing on the 1910s, this chapter examines the role of motion pictures for American retailers and manufacturers, as seen from the perspective of the period's most prominent business publication, *System, the Magazine for Business*. Published in Chicago by the A.B. Shaw Company, *System* paid attention to the managerial practices of the commercial film industry but more often surveyed and promoted the manifold utility of motion pictures for the purposes of sales, public relations, and advertising, as well as for training and overseeing employees. According to *System*, key to the successful use of film by business was both the availability of portable 35 mm projectors and new strategies for creating and distributing sponsored business films for theatrical exhibition.

Keywords: periodical literature; sales; portable projector; distribution; exhibition; industrial film

> *The "movie" has long ceased to be merely the plaything of a public seeking amusement. It has become a force in the world of business, serving a useful end and filling a real need.*
>
> – David Lay, "Drawing the Crowds to Your Films," *System, the Magazine for Business* (September 1915)

In this chapter I examine the history of films that work during the 1910s not in terms of early corporate adopters or entrepreneurial schemes but rather from the contemporaneous perspective of the most profitable and prominent American business periodical of its day, *System, the Magazine for Business*.

Hediger, V., F. Hoof, Y. Zimmermann, with S. Anthony (eds), *Films That Work Harder: The Circulation of Industrial Film*. Taylor & Francis Group 2024
DOI: 10.5117/9789462986534_CH13

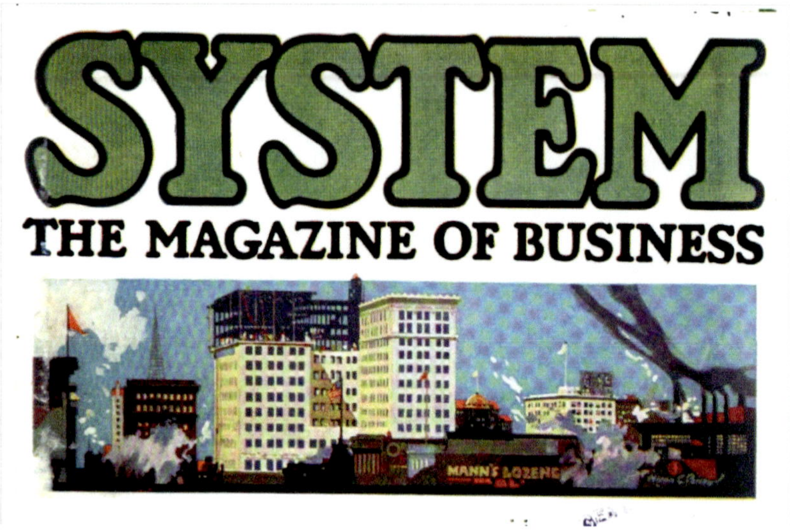

Fig. 13.1. System colour logo, July 1915.

In a number of feature articles and briefer items, *System* described how both manufacturers and retailers could utilize moving pictures to boost sales and increase efficiency and managerial oversight in the workplace. In so doing, *System* – much more fully than the motion picture trade press of the period – at this formative historical moment articulated the parameters and the possibilities of film as an emerging and still largely untapped "force in the world of business," well beyond the limited opportunities for theatrically screening industrial and advertising films in the early 1910s.

In December 1910, *System* published its first substantial article on how motion pictures have been "taken out of the amusement category and set to work."[1] By this date, as the nickelodeon boom in North America showed few signs of abating, exhibitors and moviegoers alike were becoming fully dependent on the regular delivery of one-thousand-foot reels (which might be comprised of two different films). The increasingly systematized and standardized commercial production and distribution of motion pictures in the United States left little room for non-fiction titles, particularly for films intended to serve more than immediate box-office ends. December of that year, for instance, saw 203 films released in the United States, classified in the trade press (and often in advertisements as well) by production company, length (measured in feet), and genre. According to the trade magazine, *The*

1 Mitchell, Henry W. "The Camera as a Salesman." *System*, vol. 18, no. 6, December 1910, pp. 580–87.

Nickelodeon, sixteen of the new films that month were non-fiction (that is, not identified as Comedy or Drama), including four "Industrials": *Cocoanut Plantation* (Pathé), *The Life of a Salmon* (Edison), *An Old Silver Mine in Peru* (Edison), and *The Police Force of New York City* (Edison).[2]

At the level of the local exhibitor this prioritizing of fiction over non-fiction was equally evident. A venue like the Auditorium Theater in San Bernardino, California, which changed its programme every three days and typically offered five or six films (along with two live acts), booked only one self-styled industrial – *Life of a Salmon* – among the forty-seven different films it screened during December 1910.[3] The situation was much the same at The Victor in Allentown, Pennsylvania, which screened two industrials out of the fifty-one films it presented that month: *The Police Force of New York* (said to show "all the branches of the force, including practice work of stopping runaways and the police dogs") and *Life of a Salmon* (billed as a "scenic picture of great photographic beauty").[4]

The situation in December 1910 was typical of industry practice. The previous five months had seen a total of eighteen industrials reach the American market, according to the listings of current releases in *The Nickelodeon*. Even this number might have been overly generous – or at least open to debate. *The Film Index* and *Moving Picture World* only identified six industrials released between June and December 1910, suggesting that the category itself was hardly stable or uniformly understood.[5] In this period for the US market, an industrial was most commonly taken to be a film that followed the process of making or manufacturing a product. *Variety*, for example, praised *Marble Quarrying in Tennessee* (Lubin, 1910) as "an excellent 'industrial.' It shows in detail the process of taking marble out

2 *The Nickelodeon*. Vol. 4, no. 11, December 1, 1910, pp. 326–27.

3 This information comes from the daily advertisements for this theatre in the *San Bernardino [CA] County Sun*, December 1–31, 1910.

4 Advertisement for *The Police Force of New York*. *Allentown [PA] Democrat*, December 21, 1910, p. 8; Advertisement for *Life of a Salmon*. *Allentown [PA] Democrat*, December 30, 1910, p. 8. This sort of tabulation and listing of titles does not do justice to the historical and interpretive questions raised by the fact that these industrials were included with fiction films in programmes that also featured live performances. See, in this regard, the still illuminating discussion of programming at the 1994 Amsterdam Film Workshop as transcribed in Togs, Daan Her, and Nico de Klerk, editors. *Nonfiction from the Teens*. Stichting Nederlands Filmmuseum, 1997, pp. 49–57.

5 In fact, probably none of the "industrials" released in December 1910 are clear examples of the "industrial moving pictures" that Martin Loiperdinger covers in his extremely helpful survey, "Early Industrial Moving Pictures in Germany." *Films That Work: Industrial Film and the Productivity of Media*, edited by Vinzenz Hediger and Patrick Vonderau. Amsterdam University Press, 2009, pp. 66–73.

of the earth" and then the "manufacturing process" that turns marble into a "finished product."[6] Yet Edison's promotional material did not describe *The Police Force of New York* as the recounting of a process (or the telling of a continuous narrative) but rather as an eye-opening survey of several different police "operations" shot in various locations around New York City.[7] Was this film, then, an industrial or an "educational" or even a "scenic," as the *Moving Picture World* indicated? This shifting terminology is especially noteworthy not only in relation to the argument by Frank Kessler and Eef Masson concerning "generic overlap in early non-fiction films," but also in the more immediate context of an editorial in *The Nickelodeon* in December 1910 that declared: "[T]he subject of industrial motion pictures and their use for advertising and general publicity purposes is becoming one of great importance."[8] What then to make of the paucity of films labelled as "industrials" in American moving picture shows? Perhaps the industrials exhibited theatrically did not exhaust or even best fulfil the promise *The Nickelodeon* saw in "industrial motion pictures" designed for "advertising and general publicity purposes."

Given the undeniable draw of moving pictures for a business intent on improving public relations, for an enterprising moving picture entrepreneur, or for an advertiser on the lookout for a potentially viable new medium, the configuration of the commercial film industry in the early 1910s could

6 Review of *Marble Quarrying in Tennessee*. *Variety*, vol. 17, no. 9, February 1910, p. 16; Yvonne Zimmermann claims that the "manufacturing or process film (*Fabrikationsfilm*), which shows the step-by-step manufacturing of goods from raw material to consumable product [...] is the earliest type of industrial film to represent a separate category" ("'What Hollywood Is to America, the Corporate Film Is to Switzerland': Remarks on Industrial Film as Utility Film." *Films That Work: Industrial Film and the Productivity of Media*, edited by Vinzenz Hediger and Patrick Vonderau. Amsterdam University Press, 2009, p. 105.) In an article for *System*, H.B. Vanderblue explained the success of industrials as follows: "[T]he average man and woman likes to see wheels go around and witness the intricate working of ingenious machinery. The pictures, therefore, met with immediate popular favor. Having a high educational value the pictures make palatable publicity. The detailed processes of making goods when placed before the ultimate consumer implies that the maker has nothing to hide. The effect is that of a demonstration by the manufacturer. The buyer no longer takes the word of another regarding values. He knows" ("Bringing the Factory to the Clerk: Motion Pictures of Making Processes Visualize Sales Arguments for Counter Use." *System*, vol. 22, no. 6, December 1912, p. 578).

7 Edison promotional notice for the Police Force of New York City. *The Film Index*, vol. 6, no. 24, December 24, 1910, p. 14.

8 Kessler, Frank, and Eef Masson. "Layers of Cheese: Generic Overlap in Early Non-fiction Films on Production Processes." *Films That Work: Industrial Film and the Productivity of Media*, edited by Vinzenz Hediger and Patrick Vonderau. Amsterdam University Press, 2009, pp. 75–84; *The Nickelodeon*. Vol. 4, no. 12, December 15, 1910, p. 329.

have seemed enticing if difficult to access. On the one hand, there was the relatively still novel power of projected moving images as a form of communication; the overwhelming fact of audiences filling thousands of theatres, day in and day out; and the need by exhibitors for a constant supply of new product. On the other hand, there was a production and distribution system firmly in place that seemed to have little use or need for non-fiction film. What would it take to make film cost-effectively useful for American business? How and where might moving pictures be set to work?

As I have indicated elsewhere, the International Harvester Corporation during the early 1910s had an answer to these questions or at least made a concerted effort to deploy motion pictures, in part by sponsoring *Back to the Old Farm*, a dramatic one-reeler designed for theatrical distribution.[9] A different and more complete answer was offered on the pages of *System*, which contributed significantly to the discursive construction of American business in this period.[10] Introduced in December 1900 by the Shaw-Walker Company of Muskegon, Michigan, *System* was taken over fully by publisher-editor A.B. Shaw and moved to Chicago in 1903, where it had by 1910 become the cornerstone of a burgeoning publishing empire. The System Company premiered its companion monthly, *Factory, the Magazine of Management*, in 1907 and published a host of mass-marketed pamphlets and books, including the *Business Man's Library* (1907). By 1917, when Shaw introduced another monthly, *System on the Farm*, he could claim in an advertisement that the System Company with its periodicals and more than sixty books on business topics "was by far the largest publisher of business literature in the world."[11] *System* established the priorities and set the tone for all of Shaw's publications. While his flagship monthly rarely took notice of what historian Michael McGerr calls the antitrust and regulatory measures that attempted to "take hold of big business in the progressive era," *System* was ever on the lookout for anything that

9 Waller, Gregory A. "International Harvester, *Business Screen* and the History of Advertising Film." *Films That Sell: Moving Pictures and Advertising*, edited by Bo Florin et al. Palgrave, 2016, pp. 40–53; idem. "Advertising with Moving Pictures: International Harvester's *the Romance of the Reaper* (1910–1913)." *The Image in Early Cinema: Form and Material*, edited by Scott Curtis. Indiana University Press, 2018.

10 In paying close attention to the actual material that appeared in a popular magazine like *System*, my approach is quite distinct from a study such as Galambos' *The Public Image of Big Business in America, 1880–1940*, which relies on content analysis of select periodicals to measure the "social perception" of large corporations over decades. Galambos, Louis. *The Public Image of Big Business in America, 1880–1940: A Quantitative Study in Social Change.* Johns Hopkins University Press, 1975

11 Advertisement for *System on the Farm. Chicago Daily Tribune*, September 10, 1917, p. 7.

might increase efficiency, effectiveness and earnings of American "offices, factories, and stores," large and small.[12]

Systematizing the Film Business

Given *System*'s inclusive and optimistic vision of American business, from the captain of industry to the small-town shopkeeper, it is somewhat surprising that the magazine offered only scattered coverage of the commercial motion picture business, which became an increasingly visible part of the American cultural and economic landscape during the first two decades of the century. While the contracts of Mary Pickford and Charlie Chaplin or the box-office returns of *The Birth of a Nation* might not have captured this magazine's attention, accounting methods that had made the manufacture of motion pictures more profitable warranted an article.[13] So did certain particularly effective movie industry executives. For example, Carl Laemmle, the head of Universal, merited a brief profile in 1912 because he was a millionaire whose profits were supposed to have increased substantially after he instituted a meticulous system to keep track of raw materials (film stock and chemicals) and the employees who handled them. Two years later, Harry E. Aitken was featured since his effective branding of the Mutual Film Corporation proved to be not only lucrative but also evidence that the film industry had increasingly emerged from "chaos [...] into a stature of definite and organized business."[14] This maturation was driven, *System* would propose in 1918, by executives like Famous Players-Paramount chief Adolph Zukor, who exemplified how far a "hard-headed business sense" and a hands-on managerial style can take a businessman – even in a field of endeavour as risky and volatile as the "movie game."[15] An article attributed to Zukor elaborates in some detail on the significant innovations he claimed to have brought to the

12 McGerr, Michael. *A Fierce Discontent: The Rise and Fall of the Progressive Movement in America, 1870–1920*. Free Press, 2003, p. 155; early covers of the magazine announce its concern with "Offices, Factories, and Stores." See, for instance, the cover of *System*, vol. 4, no. 4, August 1903.

13 Pavant, James A. "Why These Costs Agree with Estimates." *System*, vol. 31, no. 1, January 1917, pp. 90–94.

14 Corey, Herbert. "Carl Laemmle." *System*, vol. 22, no. 6, December 1912, pp. 609–10; Thomas, Hugh. "Harry E. Aitken." *System*, vol. 26, no. 1, July 1914, pp. 75–76.

15 Crowther, Samuel. "How They Did It: Adolph Zukor." *System*, vol. 33, no. 3, March 1918, pp. 375–76; Sayford, Irving S. "How They Did It: John R. Freuler." *System*, vol. 33, no. 2, February 1918, pp. 202–03.

film industry, including collecting information directly from consumers, monitoring and improving the practices of exhibitors, and putting into place a multi-tiered distribution policy. In Zukor, *System* clearly saw a forward-thinking, aggressive modern business executive, who earned his success by developing new strategies that enabled theatre owners to become more profitable retailers, adjusting manufacturing to suit the demands of the market, introducing a new data-based approach to the study of his clientele, and bringing about fundamental changes in the operation of own corporation and the film industry at large.

Readers of *System* were encouraged to take inspiration – and even borrow directly – from Zukor's example. But could moving pictures as a medium become a tool in the service of achieving the aims this magazine associated with Zukor (and a host of other successful executives): increased efficiency, effective managerial oversight, systematization of production and distribution and innovative strategies for increasing retail sales? Simply put, beyond Hollywood studios and theatre owners, how could a business profit from using moving pictures? This question informed much of *System*'s coverage of film.

The Camera as a Tool of Management

In contrast to magazines like *Scientific American* or *Popular Mechanics*, *System* was not interested in motion picture technology for its own sake or in research and experiments that made use of film, but rather in practical – and, therefore, profitable – deployment of the medium by businesses of all sorts. Given *System*'s enthusiastic endorsement of what it called "the national movements for greater efficiency in manufacturing and distribution,"[16] it is not surprising that a 1915 article by H.F. Porter identified the moving picture camera as one of several types of "equipment that mechanically multiples the efficiency of the executive" by "extending" his "eyes and presence" in the factory. Porter reasoned that while telephone and pneumatic tube systems (and even whistles and gongs) facilitate the transmission of information in the workplace, the moving picture camera (like the time-clock) can "automatically register or record the activity of men and equipment" – most obviously, "the motions of workmen." Thus Porter claimed that time-motion studies based on moving pictures could be used

16 "What's Coming in Business?" *System*, vol. 27, no. 3, March 1915, p. 226.

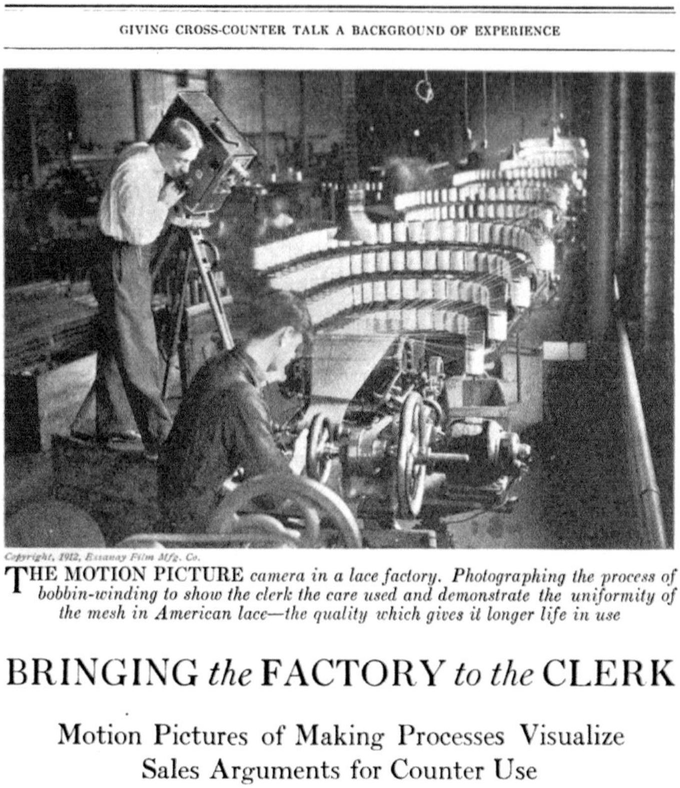

GIVING CROSS-COUNTER TALK A BACKGROUND OF EXPERIENCE

Copyright, 1912, Essanay Film Mfg. Co.

THE MOTION PICTURE *camera in a lace factory. Photographing the process of bobbin-winding to show the clerk the care used and demonstrate the uniformity of the mesh in American lace—the quality which gives it longer life in use*

BRINGING *the* FACTORY *to the* CLERK

Motion Pictures of Making Processes Visualize Sales Arguments for Counter Use

Fig. 13.2. H.B. Vanderblue, "Bringing the Factory to the Clerk: Motion Pictures of Making Processes Visualize Sales Arguments for Counter Use," *System* (December 1912).

by a manager to "raise the skill of workmen."[17] This was hardly a new idea. A 1911 article illustrated with images of strips of film in *Factory* – *System*'s companion publication – broached the possibility that moving pictures might be of great value when it came to "scientific management" since the film of a specific stage of a work process could be useful in showing workers "how to do better work more carefully by graphic teaching of detailed operations."[18]

Two years later W.S. Ball described at some length for *Factory*'s readers the process of "taking a motion apart" by filming it and then examining the

17 Porter, H.E. "The Long Arm of Management." *System*, vol. 27, no. 1, January 1915, pp. 50–56; see Hoof, Florian. *Engel der Effizienz: Eine Mediengeschichte der Unternehmensberatung*. Konstanz University Press, 2015, pp. 120–28.

18 "Explaining How Work is Done with Films of Moving Pictures." *Factory*, vol. 7, no. 6, December 1911, pp. 400–01.

footage with a magnifying glass – in effect rendering the "moving picture" as a "photographic record" comprised of a series of still images. Following this investigation, the examiner – and the manufacturer who hired him – would be able to distinguish "useful" from "useless motions in a piece of work," which, in turn, would dictate how a task might be carried out more efficiently. Ball goes so far as to claim – without any hint of irony or misgivings – that by utilizing the moving picture machine in this manner, "every process can be studied and improved, and the best way of doing it discovered and standardized."[19]

For Porter, the "automatic" recording of workmen on the job held a somewhat different promise than Ball's dream of ubiquitous standardization: the opportunity to provide (and empower) the manager with a broad "perspective of the activity of his factory," an otherwise unattainable point of view. Porter even imagines the rich possibilities of an "apparatus coupling the principle of the moving-picture machine with that of telephotography, whereby a panoramic action-picture of any part of the shop can be produced in the office at the manager's will."[20] This vision of a surveillance apparatus using moving pictures nicely foreshadows *Metropolis* and reflects the "technocratic utopianism" that historian Elspeth H. Brown sees as informing corporate uses of photography during the Progressive Era.[21]

Like *Factory, System* also identified photography as a valuable tool for business, publishing an article in 1908 that highlighted the efforts in this field by Frank B. Gilbreth, soon to become famous as what Scott Curtis calls "one of the most prolific popularizers of scientific management."[22] In this instance, however, it was Gilbreth's use of the photograph as "record – automatic, accurate, incontrovertible" – that garnered *System*'s praise. Photography was lauded as a medium for creating advertising material and, more important, for producing easily filed and readily accessible visual records of construction

19 Ball, W.S. "Taking a Motion Apart." *Factory*, vol. 10, no. 2, February 1913, pp. 131–32, 159–60, 162.

20 Porter. "Long Arm of Management," pp. 52–53.

21 Brown, Elspeth H. *The Corporate Eye: Photography and the Rationalization of American Commercial Culture, 1884–1929*. Johns Hopkins University Press, 2005, p. 23.

22 Hanson, E.S. "The Camera on the Job." *System*, vol. 13, no. 5, May 1908, pp. 479–87; Curtis. "Images of Efficiency: The Films of Frank B. Gilbreth." *Films That Work: Industrial Film and the Productivity of Media*, edited by Vinzenz Hediger and Patrick Vonderau. Amsterdam University Press, 2009, p. 85; see also Brown's chapter on Gilbreth, "The Corporate Eye," pp. 65–118; and Hoof. *Angels of Efficiency: A Media History of Consulting*. Oxford University Press. 2020; for broader background on the efficiency movement and the cultural rhetoric concerning what he calls "liberation and limitation," see Lears, Jackson. *Rebirth of a Nation: The Making of Modern America, 1877–1920*. Harper, 2009, pp. 222–61.

work, equipment installation and factory accident sites.[23] Here the camera takes its place as one more useful tool among the array of business machines that were prominently advertised in each issue of *System* – accurate, time-saving devices that, for example, could take dictation, automatically write checks, perform calculations or make duplicate copies.

Factory reiterated these points in a two-part article in 1917 that enumerated more than thirty ways the still camera could be "applied" by businesses. In addition to preserving all manner of data in a convenient, durable format, photography figured for *Factory* as an extremely useful and flexible medium for creating visual aids for workers, documenting work done well or poorly and producing specialized micro-photographic, stereoscopic and X-ray images.[24] In these articles, moving pictures are only mentioned in passing, not as a medium for what Vinzenz Hediger and Patrick Vonderau call preserving "institutional memory" or "optimizing process," but as a way of providing "valuable safety instruction" and "lessons in citizenship and Americanization."[25] In effect, moving pictures are here simply one iteration of photography as a multipurpose "tool of management."[26]

So, too, is the lantern slide produced from a photographic negative and then incorporated into an illustrated lecture as a means of instructing employees on work practices and safety concerns. Given the widespread use of lantern slides, it is not surprising to find references in *System* to this practice deployed by concerns like New York Edison and US Steel.[27] Most

23 Porter. "The Long Arm of Management," pp. 50–56; similar claims for the value of photography appear in Rice, Martin P. "Where the Picture Tells the Story." *System*, vol. 16, no. 2, August 1909, pp. 166–71; and "Unique Uses of the Camera in Business." *System*, vol. 31, no. 4, April 1917, pp. 441–42; an early statement is idem. "The Camera and Cutting Costs," *System*, vol. 14, no. 4, October 1908, pp. 407–08, on the use of photographs to evaluate railroad operation. Rarely, if ever, would a source like *System* take into account issues involved in storing and accessing a significant number of moving pictures, as opposed to photographic negatives and prints.

24 Gooding, Frank E. "Using the Camera as a Tool of Management." *Factory*, vol. 18, no. 6, June 1917, pp. 836–40; idem. "Using the Camera as a Tool of Management II: Thirty-seven Other Ways the Camera Can Help You." *Factory*, vol. 19, no. 1, July 1917, pp. 29–34.

25 Gooding. "Using the Camera as a Tool of Management II," p. 33; Hediger, Vinzenz, and Patrick Vonderau. "Record, Rhetoric, Rationalization: Industrial Organization and Film." *Films That Work: Industrial Film and the Productivity of Media*, edited by Vinzenz Hediger and Patrick Vonderau. Amsterdam University Press, 2009, p. 11.

26 Conversely, an arrangement of sixteen captioned photographs in sequence to show how waste could be put "to practical use – at a profit to the middle man" was called a "motion-picture sequence" in Banning, Kendall. "Where the 'Waste' Goes." *System*, vol. 18, no. 1, July 1910, pp. 35–40.

27 Banning. "Getting Employees in Line." *System*, vol. 15, no. 4, April 1909, pp. 379–82; see also Sumner, Lynn. "Putting Details in the Spotlight." *System*, vol. 17, no. 2, February 1910, pp. 115–24; idem. "Teaching the Employee." *System*, vol. 20, no. 2, August 1911, p. 221.

often, however, *System* mentioned instructional sessions involving both lantern slides and moving pictures, as in "Training Workers to Be Careful," a 1913 article by then associate editor, poet Carl Sandburg.[28] The combination of projected slides, screened moving pictures, and (most likely) spoken commentary created a multiple-media presentation, typically offered to a captive audience inside the workplace. Such presentations could also be mounted in the service of broader public relations ends by trade organizations like the Portland Cement Association, which, according to *System*, was by 1916 funding a full-service Extension Division that "prepares and delivers lectures, technical and popular; prepares lecture outlines; trains speakers, arranges data; cooperates with independently paid lecturers; and prepares the motion picture films, lantern slides, charts and models which are needed to show the use of concrete."[29]

The Camera as a Salesman

Even given the dividends that illustrated lectures, time-motion studies and factory-floor cameras might pay as instructional tools, productivity boosters or surveillance devices, *System* was most interested in moving pictures as a way to market goods and services and establish brand identity – in other words, as a medium for delivering advertising via moving pictures that included but was not limited to the sort of industrials then in theatrical release. As early as 1909, *System* noted the promise of a "novel method" of advertising products by theatrical screenings of a specially scripted film (with accompanying "talking-machine record") that drew on "all the dramatic force – humor, pathos and action – that has made the motion picture show so popular."[30] *System* likewise reported favourably in 1910 on a bank's successful efforts to attract new first-generation immigrant depositors by renting out a local picture show on Monday evenings to present a free, ticketed "entertaining program" that included slides and motion pictures detailing bank operations, with explanatory talks given in a different foreign language each week.[31] Such articles were hardly anomalous. Over the first decade of the century, *System* became increasingly concerned

28 Sandburg, Carl. "Training Workers to Be Careful." *System*, vol. 24, no. 2, August 1913, pp. 124–32.

29 Husband, Joseph. "Putting a Big Crimp in 'Overhead'." *System*, vol. 30, no. 6, December 1916, p. 665.

30 "Pictures That Pay Dividends." *System*, vol. 15, no. 5, May 1909, pp. 517–18.

31 Loomis, William W. "How Banks Reach Out for Business." *System*, vol. 18, no. 2, August 1910, pp. 185–87.

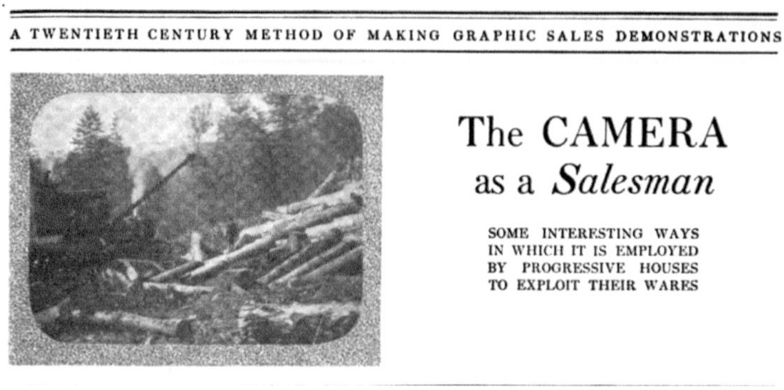

A TWENTIETH CENTURY METHOD OF MAKING GRAPHIC SALES DEMONSTRATIONS

The CAMERA as a *Salesman*

SOME INTERESTING WAYS IN WHICH IT IS EMPLOYED BY PROGRESSIVE HOUSES TO EXPLOIT THEIR WARES

Fig. 13.3. Henry W. Mitchell, "The Camera as a Salesman," *System* (December 1910).

with the importance of advertising for all modern business enterprises, leading to a six-part series of articles in 1908 to 1909 (called "Advertising in Operation") based on the premise that "in America we believe that the people should select. And advertising is the direct primary of the people. Through advertising the people need not take the selections others make for them; they can take into their own hands the choice and determination of our commodities."[32] The role that moving pictures might play in this all-important democratic marketplace as a "distinctly modern American method" of "enforcing selling arguments" was the subject of *System*'s first feature article directly concerned with film, Henry W. Mitchell's "The Camera as a Salesman" (December 1910).[33]

The attention *System* paid to "the possibilities of the camera" and to what a 1911 article would call "new applications of the [moving] picture idea to business" reflects this magazine's own commitment to providing useful information to its readers, as evidenced by the material included in the 110 pages of the December 1910 issue that included Mitchell's article.[34] Except for one editorial-styled piece considering whether individuals should be held accountable for actions taken in the name of a corporation, this issue was almost entirely devoted to three types of material: (1) anecdotal "real stories" about how relatively small-scale and short-term problems were solved or novel strategies successfully implemented, such as how a

32 Balmer, Edwin. "Advertising in Operation – Its Results." *System*, vol. 15, no. 2, February 1909, p. 153.

33 Mitchell. "The Camera as a Salesman," pp. 580–87.

34 The phrase comes from "A 'Magic Carpet' for Salesmen." *System*, vol. 19, no. 5, May 1911, p. 557.

"credit man" devised ingenious schemes to collect overdue payments or how a pen manufacturer quickly adapted to meet demand with supply; (2) practical advice (often complete with forms and diagrams) about how to run a more efficient and profitable business by, for instance, making smart use of floor space or developing a new system for tracking customers; and (3) descriptions of innovative equipment, material and technology, like motor trucks in contrast to horse-drawn wagons and concrete in contrast to wood and stone. In effect, the version of the progressive business world (or perhaps simply America) offered in *System* is filled with inevitable yet solvable complications. This is a world responsive to all manner of individual initiative, ready to adopt new technologies if they prove useful, and always capable of being more fully systematized and efficiently managed. Here was a domain primed to make use of or at least test out the practical possibilities of moving pictures.

While Mitchell notes the continuing relevance of "stereopticon lectures" (sometimes employing opaque projectors) as a "method of instructing employees," "The Camera as a Salesman" highlights several successful examples of what he calls "graphic sales demonstrations." And the article itself makes extensive use of visual evidence, illustrating its text with photographic images of lantern slides, strips of film, and exhibition sites. Virtually all the ingenious "sales talks" that Mitchell describes combine slides and moving pictures accompanied by a company representative, with the slides allowing for "magnified" details and the moving pictures providing a dramatic view, typically of machines in motion. Such usage thus stands as an early and small-scale example of what Yvonne Zimmermann among others has called the "orchestrated media mix (or *Medienverbund*)," which would subsequently become the hallmark of "corporate communication."[35]

According to Mitchell, the sites for these multiple-media presentations and the advertising strategies that deployed moving pictures varied considerably, including:

- Branch offices of a manufacturing concern equipped with a permanently installed, "light-proof" screening room, seating only a few potential customers
- A large tent show that travelled the circuit of major agricultural fairs, capable of entertaining "thousands of daylight visitors" with a "fluent lecturer and a succession of colored slides and motion pictures"

35 Zimmermann. "'What Hollywood Is to America, the Corporate Film Is to Switzerland'," p. 102.

- A touring "advertising show" using motion pictures and slides, presented for free in rented moving picture theatres as an added attraction to the regular programme
- Specially produced films about the manufacturing of a product designed to be shown in moving picture theatres as well as non-theatrical sites like "commercial shows, clubs, and at exhibits at fairs and expositions."

Most obviously missing from this list is a non-theatrical practice that would figure prominently until at least mid-century in the American discourse concerning the use of motion pictures by and for business: the sales representative armed with a portable projector and a reel of film ready to pitch a product virtually anywhere at any time.[36] This option took centre stage as early as August 1912 in a *System* article written by erstwhile humorist Homer Croy. Croy lauded the capacity of what he called "industrial pictures" to offer striking "pictorial evidence" of complex production processes and huge machines in action. While Croy noted that such films have been successfully circulated for "general exhibition" in theatres and other public venues by manufacturers of borax, automobiles and agricultural equipment, he claimed that an equally promising opportunity for businesses to take advantage of this "new phase of salesmanship made possible by modern invention" hinged on the availability of a more mobile, private and flexible version of exhibition. "For the purposes of displaying these [industrial] pictures," Croy explained,

> a firm has just put out a small projecting machine for the special use of salesmen. The whole machine may be carried in a compact little case. When the salesman brings it into a prospect's office, he merely draws the shades on the windows, attaches a socket to the electric light, and begins to turn a crank – while the machine he is selling leaps out onto a wall and goes through its workings as noiselessly as a shadow. The prospect does

36 Almost two decades later, Frank Presbrey argues basically the same thing: that the "main place" of motion pictures "in advertising is probably as a visualizer of industrial equipment in action. Here it has become an important aid in the selling of machinery and other articles of which samples cannot be carried. The product can with the motion picture be demonstrated in a vivid and highly interesting way in the prospect's office, perhaps even better than if the machine itself were at hand. If necessary the projector can be slowed down so that the eye can follow the action of an intricate and fast machine." Presbrey, Frank. *The History and Development of Advertising*. Doubleday, 1929, p. 579. I discuss this image in relation to the history of 16 mm in "Projecting the Power of 16 mm, 1935–1945." *Useful Cinema*, edited by Charles R. Acland and Haidee Wasson. Duke University Press, 2011, pp. 125–48.

> not have to study a catalogue. He merely leans back in his swivel chair and watches the evolutions of the apparatus while the salesman explains each detail as it comes or stops or pauses at his command.[37]

Unlike a theatrically screened industrial, the advertising film for Croy resembles a catalogue brought to life, fully dependent on a portable projector that has the capacity to stop or freeze the film image – a break in screened motion that would never be anything but a distracting mistake under theatrical conditions of exhibition. Furthermore, this particular use of moving pictures requires the presence of a salesman-projectionist to manage the private screening and provide necessary explanation. This company representative delivers his sales pitch not to some targeted demographic at a county fair or to an even more random collection of spectators at the picture show, but to one attentive and "busy "executive, for "the motion picture brings the factory, the mine, or the big ditch into his private office."[38]

Portable projectors suitable for the business purposes that Croy imagines began to be advertised in *System* as early as 1915, if not sooner. For example, the Standard Film Service claimed in its ad that "our portable projecting machine enables your salesmen to bring your plant – your product – before the *very eyes* of customers. Carried like a suitcase – may be used in the prospect's office, in the hotel – wherever there is electric current."[39] An ad in the June 1916 issue for the DeVry Portable Projector – "simple, efficient, durable" and weighing only 19½ pounds – actually shows a man carrying this suitcase unit and promises "increased selling ability for every salesman."[40] And a notice in 1918 announcing the American Projectoscope, a "portable commercial type of motion picture machine [...] intended particularly for the use of salesmen," emphasized that this projector "can be run both forward and backward and the picture held

37 Croy, Homer. "Sales Demonstrations by Moving Pictures." *System*, vol. 22, no. 2, August 1912, p. 136.

38 Ibid., pp. 129–36. Much of this article was reprinted under the title "Films a Success as Salesmen" in *Motography*, vol. 7, no. 10, November 9, 1912, pp. 376–77. Croy would soon begin a long, if intermittent, connection with the commercial film industry when he was hired by Universal to travel through the "Orient" collecting "humorous happenings and curious customs" for a series of travelogues. "Croy Off on Long Trip." *Motography*, vol. 11, no. 7, April 4, 1914, p. 246.

39 Advertisement for Standard Film Service. *System*, vol. 28, no. 6, December 1915, p. 756, advertising section. One problem with ascertaining the earliest appearance of such advertisements is that libraries typically bound issues of periodicals like *System* without including the separate advertising section found in each issue, which could cover as many pages as the non-advertising matter. These bound copies often become the source for digitization.

40 Advertisement for DeVry Projector. *System*, vol. 29, no. 6, June 1916, p. 783.

stationary on the screen when desired" – features that would have been of no use in theatrical situations.[41] It is impossible to tell the extent to which these and other portable projectors were actually purchased and used by businesses, though the advertisements themselves functioned discursively to help imagine how and where businesses might effectively deploy moving pictures.

"Movies" That Find Customers

As these portable projectors entered the market, *System* in the mid-1910s continued to keep its readers aware of a range of tasks that motion pictures could fulfil, including training retail clerks and travelling salesmen by providing "a new and live point of view on the merchandise" they sold.[42] Advertising goods and services, however, remained for *System* the primary benefit of motion pictures, as became quite evident in four articles by regular contributor, David Lay, which draw from material previously published in *System* and constitute the magazine's fullest statement on the topic.

Lay's series, appearing in July, August, and September 1915 and January 1916, begins with a description of a "present-day salesman" opening his grip and setting up a projector for a private screening, this text accompanied by two soon-to-be iconic illustrations: a well-dressed man holding a suitcase projector and a film being screened to a small group in a non-theatrical space. It is as if the potential opportunity for "sales demonstrations by moving pictures" glimpsed by Croy three years earlier had by 1915 become an accomplished fact. In his suitcase Lay's salesman "may carry anything, from railroad trains and printing presses in motion, to dynamite explosions." Frequently framing the written words in Lay's articles are images of strips of celluloid, drawn from widely screened films by well-established firms to market a range of products, from dynamite (DuPont Powder), paint (Sherwin-Williams), and firearms (Winchester) to automobiles (Studebaker) and made-to-order suits (Royal Tailors). The obvious implication is that "the motion-picture film as a salesman is no longer a novelty."[43]

41 Advertisement for American Projectoscope. *System*, vol. 33, no. 5, May 1918, p. 771.

42 Vanderblue. "Bringing the Factory to the Clerk," pp. 573–78; see "Increasing Sales by Motion Pictures." *System*, vol. 19, no. 3, March 1911, p. 321; and "A 'Magic Carpet' for Salesmen," p. 557.

43 Lay, David. "Putting 'Movies' on the Sales Force." *System*, vol. 28, no. 1, July 1915, pp. 73–77.

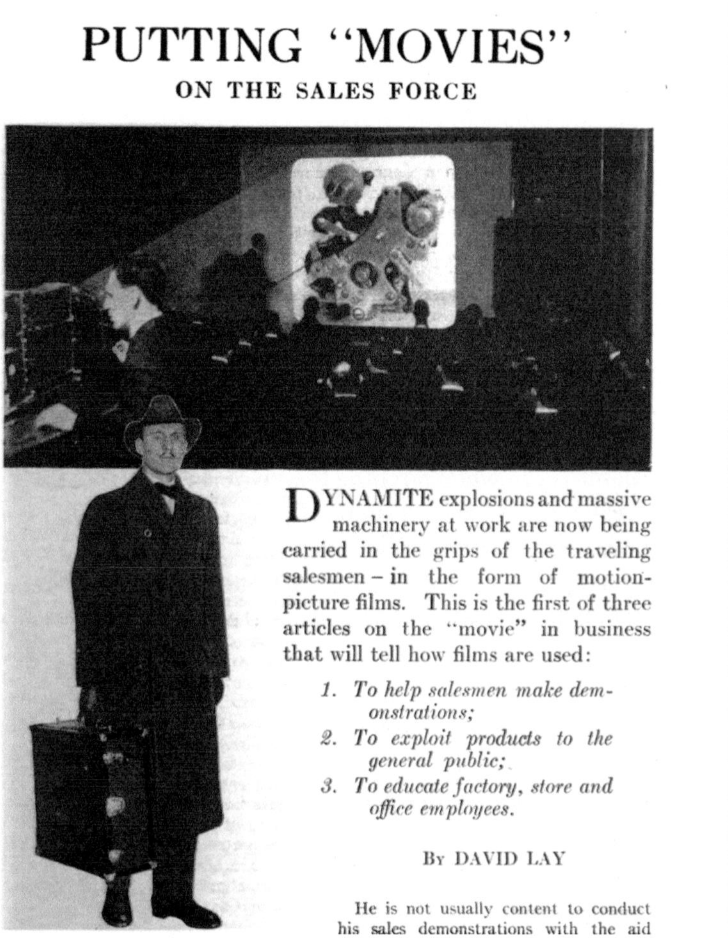

PUTTING "MOVIES"
ON THE SALES FORCE

DYNAMITE explosions and massive machinery at work are now being carried in the grips of the traveling salesmen – in the form of motion-picture films. This is the first of three articles on the "movie" in business that will tell how films are used:

1. *To help salesmen make demonstrations;*
2. *To exploit products to the general public;*
3. *To educate factory, store and office employees.*

By DAVID LAY

He is not usually content to conduct his sales demonstrations with the aid

Fig. 13.4. David Lay, "Putting 'Movies' on the Sales Force," *System* (July 1915).

Yet unlike virtually all of the previous discussions of moving pictures in *System*, Lay focuses much more directly on successful strategies for "making interesting 'movie' advertisements, and getting them on the programs of 'movie' theaters." The goal was, in his words, "to exploit products to the general public," by tapping the "vast purchasing power" represented by the "millions of men, women, and children who attend movie theaters." This enticing prospect requires particular strategies for producing, distributing and exhibiting motion picture advertising: it calls for thinking of these films, in effect, as "'movies,'" a term that for *System* in 1915 still required scare quotes, presumably because it was slang or perhaps because it seemed audacious to claim that the "movies" (for Lay, a thrilling, dramatic entertainment with a devoted audience of ten million daily) could have anything to do with "motion picture campaigns" carefully orchestrated by modern businesses.[44]

Keeping the model of the movies in mind, Lay argues, means that the producer of films for business must revivify the "industrial film" by "inject[ing] 'human interest' – that intangible but essential element into a brief series of pictures intended to attain some practical end that eventually touches a man's pocketbook."[45] While trick shots, recreated historical scenes, images of a factory in operation, or simply any sort of "action" could enliven sales films, according to Lay the prime strategy for injecting human interest involved what we might call narrativizing the sales film. Thus Lay highlights pictures that, for instance, effectively weave a "little comedy-drama" into an advertisement for ready-made suits or introduce a "love theme" even when pitching the virtues of cement as a building material.[46] Readers of *System* would have been quite familiar with this sort of narrativization, since the magazine itself depended heavily on first-person testimonials from successful businessmen and had since 1907 frequently published fictional "tales of business strategy." By the 1910s a regular feature of *System* was Daniel Louis Hanson's series of Moses Irons stories, self-styled "business fiction" that put the acumen and organizational skills of a wise modern capitalist to the test in dramatically rendered situations.[47]

In addition to this emphasis on narrativization, what's particularly revealing about Lay's articles as a survey of the state of moving pictures in the service of business is that he devotes much attention not only to how sales films should be produced (including costs and the process of developing a scenario) but also to successful and economically viable

44 Idem. "'Movies' That Find Customers." *System*, vol. 28, no. 2, August 1915, pp. 190–94.

45 Idem. "Putting Your Sales Arguments into 'Movies'." *System*, vol. 29, no. 2, February 1916, p. 169; see Clark, Neil M. "Adding 'Human Interest' to the Goods You Sell." *System*, vol. 19, no. 3, March 1916, pp. 328–32, which includes discussion of efforts to train salespeople by using motion pictures that include "picturesque or striking facts" about the products they sell. Lay, of course, had not invented this strategy. The Essanay Company ran ads in *System* in 1912 offering to "dramatize your selling points with moving pictures," thereby tapping the "novel and human interest appeal in motion-picture advertising." Advertisement for Essanay. *System*, vol. 22, no. 6, December 1912, advertising section.

46 Lay. "'Movies' That Find Customers," pp. 192–93.

47 See, for instance, Hanson, Daniel Louis. "Moses Irons and Company Collects a Mexican Account." *System*, vol. 29, no. 2, February 1916, pp. 139–45; "tales of business strategy" comes from one of the earliest of these stories: Cartwright, Henry. "Chance and Circumstance – F.O.B." *System*, vol. 11, no. 2, February 1907, p. 139.

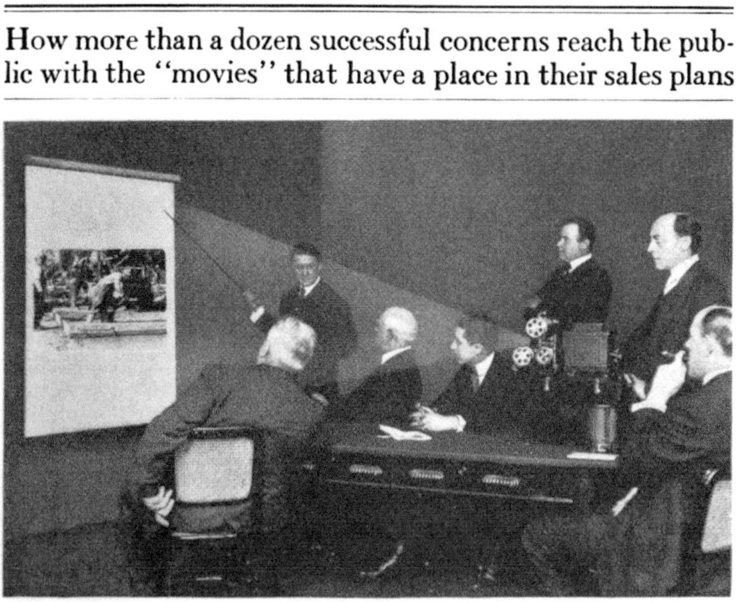

Fig. 13.5. David Lay, "Drawing the Crowds to Your Films," *System* (September 1915).

models of distribution and exhibition. "Drawing the Crowds to Your Films," Lay's contribution to the September 1915 issue of *System,* considers several methods of reaching a broader public, with the Studebaker automobile company serving as his prime example of "securing a wide distribution at a very small expense." He quotes at length from Studebaker's instructions to its agents concerning how to get a one-reel film showing the production and road-testing of a car screened in a movie theatre. The local exhibitor ought to be pleased to have the opportunity to screen the film, "which cannot but prove of absorbing interest to every person in the audience," declared the company, but if the exhibitor is recalcitrant, the agent should first offer to pay for newspaper advertising for the film and the rest of the program; then, upping the ante, agree to purchase a 100 or 150 tickets to the screening and distribute them to prospective customers; or, in the worst case scenario, simply pay the exhibitor to screen the film a certain number of times. Studebaker further advised local dealers to make sure to purchase lantern slides from the company to be projected immediately after the conclusion of the film. This multi-tiered plan underscores Studebaker's commitment to getting its sales film on the programme at movie theatres, with no concern for whatever other films might share the bill. The plan also assumes that while exhibitors might have resistance to screening advertising films, they also have some leeway in arranging their programming and, with the right incentives, can make room for a sponsored one-reeler.[48]

Conclusion

In 1917, *System* marvelled at the Ford Motor Company's modern, highly efficient factory that contained a 10,000 square foot "complete up-to-date motion picture" production facility, and a year later the magazine noted the full commitment to slides and motion pictures by another very prominent early corporate adopter, the National Cash Register Company.[49] But by and large, *System* after 1916 devoted considerably less copy to how businesses might make use of moving pictures. Late in the decade, as I noted earlier, it was executives in the theatrical film industry whom *System* deemed worthy of attention, like Adolph Zukor, as well as Hiram Abrams, the president of United Artists.[50] And when the publisher of *System* released *Making More Out of Advertising* in 1919, this volume in The Shaw Retailing Series, which largely dealt with newspapers, considered the value of direct mailings, street car signs, and billboards, but made no mention of advertising using moving pictures (or lantern slides).[51]

Yet earlier in the decade and culminating in David Lay's four articles, *System* fully bought into the notion that a range of American businesses could profitably utilize moving pictures for different purposes at different sites on different sorts of occasions. In the material I have considered in this chapter, six options drew the most attention in *System*:

- Time-motion studies to improve worker efficiency, not designed to be screened
- Instructional films intended for employees, typically screened to captive audiences at the workplace

48 Lay, David. "Drawing the Crowds to Your Films." *System*, vol. 28, no. 3, September 1915, pp. 327–32.

49 Porter, Harry Franklin. "Sales Methods That Net $1,000,000 a Week." *System*, vol. 31, no. 5, May 1917, p. 510; Patterson, John H. "How I Get My Ideas Across." *System*, vol. 33, no. 6, June 1918, pp. 875–79.

50 Abrams, Hiram. "How I Keep 12,000,000 Customers Sold." *System*, vol. 36, no. 2, August 1919, pp. 199–204; see also Schaefer, Fred. "Builders of Business: Albert E. Smith." *System*, vol. 36, no. 1, July 1919, pp. 59–60.

51 *Making More Out of Advertising*. Edited by Wheeler Sammons, chairman of the Shaw Publications editorial board. A.B. Shaw Company, 1919. Part of the Shaw Retailing Series of books.

- Films of products screened by salesmen in private demonstrations for prospective customers:
- Programmes combining moving pictures, slides, and a lecture presented outside of theatres to the general public
- Programmes combining moving pictures, slides, and a lecture presented in rented theatres to the general public
- One-reel "commercial films" with sufficient "human interest" screened as part of regular programming at movie theatres to audiences of moviegoers[52]

This list, I propose, is a fair approximation of how "the 'movie' in business" was deployed and constructed discursively in the United States during the early and mid-1910s. Of course, the relative importance of these options would come to vary considerably over time, depending on exhibition and distribution possibilities, projection technology, corporate priorities, state regulation, and a host of other factors – not least of all the vast discourse comprising articles, news items, promotional notices, editorials, and advertisements published in a wide range of specialized trade and general interest periodicals that reported on, gave practical advice about, predicted the future of, and championed the use of moving pictures by business.

The menu of options derived from *System*'s various articles hinged on a set of sometimes overlapping – and, again, quite historically specific – distinctions related to the sites, functions, presentation strategies and priorities of useful cinema:

52 *System* offered very little even in the way of passing references to the advertising possibilities of product placement in movies, which Lay treats as a very preliminary stage in what would become full-fledged "motion-picture campaigns [...] conducted by many enterprising business concerns" ("'Movies' That Find Customers," pp. 190–91). Even by early in 1910, a *Printers' Ink* article (subsequently reprinted in *Moving Picture World*) could identify three distinct possibilities: "pictures with semi-advertising themes" produced for screening in movie theatres; films specifically shot for general "sales work," circulated via "picture machines and lecturers [...] sent out over the country, exhibiting in halls hired for the purpose"; and films designed to be shown by salesmen in private screenings to "one or two prospective purchasers" (Collins, James H. "Advertising via the Moving Picture." *Printers' Ink*, vol. 70, no. 8, February 23, 1910, pp. 24–26, 28. Reprinted with the same title in *Moving Picture World*, vol. 6, no. 11, March 19, 1910, pp. 42–43). Three years later, *Advertising & Selling* found basically the same options in play, citing examples of each strategy for utilizing what it called "commercial 'movies,'" including a handful of titles that had made it into wide theatrical release, like the Waterman Company's *The Story of the Fountain Pen* (King, A. Rowden. "The Commercial 'Movies': How the Manufacturer Is Using Moving Pictures to Help Business." *Advertising & Selling*, vol. 22, no. 8, January 1913, pp. 50–54).

- Restricted as distinct from unrestricted exhibition
- Narrowly targeted viewers as distinct from general audiences
- Non-theatrical sites as distinct from movie theatres
- The movie theatre as rented space as distinct from the movie theatre in normal operation
- Moving pictures as distinct from multiple-media presentations
- Films infused with "human interest" as distinct from films without such "human interest"
- Instructional as distinct from advertising aims
- Movie-like business films as distinct from the movies

A few of these distinctions – like the relation between multiple-media performances and self-contained films – are relatively clear-cut and easily charted. Others are better understood as paired variables, very much open to debate and to recalibration – like the relation between instruction and advertising, between the movies and movie-like sponsored films, and even between theatrical and non-theatrical exhibition. Taken together, this set of distinctions and the options listed above provide, I would propose, a framework for examining specific films, the efforts of individual companies that produced or employed moving pictures, and the broader prospects for and the implementation of not only moving pictures as what *System* calls a "practical tool of modern economic life,"[53] but also multi-sited and multipurpose cinema more generally during the 1910s and beyond.

Works Cited

Abrams, Hiram. "How I Keep 12,000,000 Customers Sold." *System*, vol. 36, no. 2, August 1919, pp. 199–204.

Advertisement for American Projectoscope. *System*, vol. 33, no. 5, May 1918, p. 771.

Advertisement for DeVry Projector. *System*, vol. 29, no. 6, June 1916, p. 783.

Advertisement for Essanay. *System*, vol. 22, no. 6, 1912, advertising section.

Advertisement for *Life of a Salmon*. *Allentown [PA] Democrat*, December 30, 1910, p. 8.

Advertisement for *The Police Force of New York*. *Allentown [PA] Democrat*, December 21, 1910, p. 8.

Advertisement for Standard Film Service. *System*, vol. 28, no. 6, December 1915, p. 756, advertising section.

Advertisement for *System on the Farm*. *Chicago Daily Tribune*, September 10, 1917, p. 7.

53 Lay. "Putting Your Sales Arguments into 'Movies'," p. 170.

Advertisements for Auditorium Theater. *San Bernardino [CA] County Sun*, December 1–31, 1910.
Ball, W.S. "Taking a Motion Apart." *Factory*, vol. 10, no. 2, 1913, pp. 131–32, 159–60, 162.
Balmer, Edwin. "Advertising in Operation – Its Results." *System*, vol. 15, no. 2, February 1909, pp. 151–53.
Banning, Kendall. "Getting Employees in Line." *System*, vol. 15, no. 4, 1909, pp. 379–82.
Banning, Kendall. "Where the 'Waste' Goes." *System*, vol. 18, no. 1, 1910, pp. 35–40.
Brown, Elspeth H. *The Corporate Eye: Photography and the Rationalization of American Commercial Culture, 1884–1929*. Johns Hopkins University Press, 2005.
"The Camera and Cutting Costs." *System*, vol. 14, no. 4, 1908, pp. 427–28.
Cartwright, Henry. "Chance and Circumstance – F.O.B." *System*, vol. 11, no. 2, 1907, pp. 133–39.
Collins, James H. "Advertising via the Moving Picture." *Printers' Ink*, February 23, 1910, pp. 24–26, 28. Reprinted with the same title in *Moving Picture World*, vol. 6, no. 11, March 19, 1910, pp. 42–43.
Corey, Herbert. "Carl Laemmle." *System*, vol. 22, no. 6, December 1912, pp. 609–10.
Crowther, Samuel. "How They Did It: Adolph Zukor." *System*, vol. 33, no. 3, March 1918, pp. 375–76.
Croy, Homer. "Films a Success as Salesmen." *Motography*, vol. 7, no. 10, 1912, pp. 376–77.
Croy, Homer. "Sales Demonstrations by Moving Pictures." *System*, vol. 22, no. 2, August 1912, pp. 129–36.
"Croy Off on Long Trip." *Motography*, vol. 11, no. 7, 1914, p. 246.
Curtis, Scott. "Images of Efficiency: The Films of Frank B. Gilbreth." *Films That Work: Industrial Film and the Productivity of Media*, edited by Vinzenz Hediger and Patrick Vonderau. Amsterdam University Press, 2009, pp. 85–99.
"Explaining How Work Is Done with Films of Moving Pictures." *Factory*, vol. 7, no. 6, 1911, pp. 400–01.
Galambos, Louis. *The Public Image of Big Business in America, 1880–1940: A Quantitative Study in Social Change*. Johns Hopkins University Press, 1975.
Gooding, Frank E. "Using the Camera as a Tool of Management." *Factory*, vol. 18, no. 6, June 1917, pp. 836–40.
Gooding, Frank E. "Using the Camera as a Tool of Management II: Thirty-seven Other Ways the Camera Can Help You." *Factory*, vol. 19, no. 1, 1917, pp. 29–34.
Hanson, David Louis. "Moses Irons and Company Collects a Mexican Account." *System*, vol. 29, no. 2, February 1916, pp. 139–45.
Hanson, E.S. "The Camera on the Job." *System*, vol. 13, no. 5, May 1908, pp. 479–87.
Hediger, Vinzenz, and Patrick Vonderau. "Record, Rhetoric, Rationalization: Industrial Organization and Film." *Films That Work: Industrial Film and the*

Productivity of Media, edited by Vinzenz Hediger and Patrick Vonderau. Amsterdam University Press, 2009, pp. 35–49.

Hediger, Vinzenz, and Patrick Vonderau, editors. *Films That Work: Industrial Film and the Productivity of Media*. Amsterdam University Press, 2009.

Hoof, Florian. *Angels of Efficiency: A Media History of Consulting*. Oxford University Press. 2020.

Hoof, Florian. *Engel der Effizienz: Eine Mediengeschichte der Unternehmensberatung*. Konstanz University Press, 2015.

Husband, Joseph. "Putting a Big Crimp in 'Overhead'." *System*, vol. 30, no. 6, December 1916, pp. 660–65.

"Increasing Sales by Motion Pictures." *System*, vol. 19, no. 3, March 1911, p. 321.

Kessler, Frank, and Eef Masson. "Layers of Cheese: Generic Overlap in Early Nonfiction Films on Production Processes." *Films That Work: Industrial Film and the Productivity of Media*, edited by Vinzenz Hediger and Patrick Vonderau. Amsterdam University Press, 2009, pp. 75–84.

King, Rowden. "The Commercial 'Movies': How the Manufacturer Is Using Moving Pictures to Help Business." *Advertising & Selling*, January 1913, pp. 50–54.

Lay, David. "Drawing the Crowds to Your Films." *System*, vol. 28, no. 3, 1915, pp. 327–32.

Lay, David. "'Movies' That Find Customers." *System*, vol. 28, no. 2, September 1915, pp. 190–94.

Lay, David. "Putting 'Movies' on the Sales Force." *System*, vol. 28, no. 1, July 1915, pp. 73–77.

Lay, David. "Putting Your Sales Arguments into 'Movies'." *System*, vol. 29, no. 2, February 1916, pp. 168–72.

Lears, Jackson. *Rebirth of a Nation: The Making of Modern America, 1877–1920*. Harper, 2009.

Loiperdinger, Martin. "Early Industrial Moving Pictures in Germany." *Films That Work: Industrial Film and the Productivity of Media*, edited by Vinzenz Hediger and Patrick Vonderau. Amsterdam University Press, 2009, pp. 66–73.

Loomis, William W. "How Banks Reach Out for Business." *System*, vol. 18, no. 2, August 1910, pp. 185–87.

"A 'Magic Carpet' for Salesmen." *System*, vol. 19, no. 5, May 1911, p. 557.

McGerr, Michael. *A Fierce Discontent: The Rise and Fall of the Progressive Movement in America, 1870–1920*. Free Press, 2003.

Mitchell, Henry W. "The Camera as a Salesman." *System*, vol. 18, no. 6, December 1910, pp. 580–87.

The Nickelodeon. Vol. 4, no. 11, December 1, 1910, pp. 326–27.

The Nickelodeon. Vol. 4, no. 12, December 15, 1910, p. 329.

Patterson, John H. "How I Get My Ideas Across." *System*, vol. 33, no. 6, June 1918, pp. 875–79.
Payant, James A. "Why These Costs Agree with Estimates." *System*, vol. 31, no. 1, January 1917, pp. 90–94.
"Pictures That Pay Dividends." *System*, vol. 15, no. 5, May 1909, pp. 517–18.
Porter, Harry Franklin. "Sales Methods That Net $1,000,000 a Week." *System*, vol. 31, no. 5 May 1917, pp. 509–17.
Porter, H. E. "The Long Arm of Management." *System*, vol. 27, no. 1, January 1915, pp. 50–56.
Presbrey, Frank. *The History and Development of Advertising*. Doubleday, 1929.
Promotional notice for *The Police Force of New York City*. *The Film Index*, December 24, 1910, p. 14.
"Record of Current Films." *The Nickelodeon*, vol. 4, no. 11, 1910, pp. 326–27.
Review of *Marble Quarrying in Tennessee*. *Variety*, vol. 17, no. 9, February 1910, p. 16.
Rice, Martin P. "Where the Picture Tells the Story." *System*, vol. 16, no. 2, August 1909, pp. 166–71.
Sammons, Wheeler, editor. *Making More Out of Advertising*. A.B. Shaw, 1919.
Sandburg, Carl. "Training Workers to Be Careful." *System*, vol. 24, no. 2, August 1913, pp. 124–32.
Sayford, Irving S. "How They Did It: John R. Freuler." *System*, vol. 33, no. 2, February 1918, pp. 202–03.
Schaefer, Fred. "Builders of Business: Albert E. Smith." *System*, vol. 36, no. 1, July 1919, pp. 59–60.
Sumner, Lynn. "Putting Details in the Spotlight." *System*, vol. 17, no. 2, February 1910, pp. 115–24.
"Teaching the Employee." *System*, vol. 20, no. 2, August 1911, p. 221.
Thomas, Hugh. "Harry E. Aitken." *System*, vol. 26, no. 1, July 1914, pp. 75–76.
Togs, Daan Her, and Nico de Klerk, editors. *Nonfiction from the Teens*. Stichting Nederlands Filmmuseum, 1997.
"Unique Uses of the Camera in Business." *System*, vol. 31, no. 4, April 1917, pp. 441–43.
Vanderblue, H.B. "Bringing the Factory to the Clerk: Motion Pictures of Making Processes Visualize Sales Arguments for Counter Use." *System*, vol. 22, no. 6, December 1912, pp. 573–78.
Waller, Gregory A. "Advertising with Moving Pictures: International Harvester's *The Romance of the Reaper* (1910–1913)." *The Image in Early Cinema: Form and Material*, edited by Scott Curtis. Indiana University Press. 2018.
Waller, Gregory A. "International Harvester, *Business Screen* and the History of Advertising Film." *Films That Sell: Moving Pictures and Advertising*, edited by Bo Florin, Nico de Klerk, and Patrick Vonderau. Palgrave, 2016, pp. 40–53.

Waller, Gregory A. "Projecting the Power of 16 mm, 1935–1945." *Useful Cinema*, edited by Charles R. Acland and Haidee Wasson. Duke University Press, 2011, pp. 125–48.

"What's Coming in Business?" *System*, vol. 27, no. 3, March 1915, p. 226.

Zimmermann, Yvonne. "'What Hollywood Is to America, the Corporate Film Is to Switzerland': Remarks on Industrial Film as Utility Film." *Films That Work: Industrial Film and the Productivity of Media*, edited by Vinzenz Hediger and Patrick Vonderau. Amsterdam University Press, 2009, pp. 101–17.

About the Author

Gregory A. Waller is Provost Professor of Cinema and Media Studies in the Media School at Indiana University. Since 2013 he has been editor of *Film History: An International Journal.* Most recently, his research and publications have focused on non-fiction film, the discourse concerning motion pictures in American periodicals, and the production and exhibition of non-theatrical cinema in the United States during the first decades of the twentieth century.

14 New Media for the Schools of Tomorrow

The AV Instructional Films of Robert W. Wagner

Charles R. Acland

Abstract
During the mid-twentieth century, the Ohio State University's Bureau of Educational Research was an influential venue for the development of ideas about new media in classrooms, advancing forms of audiovisual pedagogy and influencing today's teaching methods. One of the ways its views and findings circulated was through the production of instructional films about audiovisual pedagogy, largely directed toward the training of teachers. Through an examination of the production and distribution circumstances of such Robert W. Wagner-produced films, this chapter reveals the dominant ideological frames for the advancement of new technological conditions. These frames show the way audiovisual advocates and experts sought to establish "schools for tomorrow" as a flexible media environment that featured media making. Doing so, they fortified conditions and discourses about the new technological society and the technocratic citizenry appropriate to those conditions.

Keywords: audiovisual instruction, multimedia systems, National Defense Education Act of 1958, modularity, technological society

Distinctive in American culture of the post-WWII period was a sense of the immersive and pervasive presence of media. In every realm of life, it seemed that a new environment of images and sounds was upon us. For all of Marshall McLuhan's analytical shortcomings, he was absolutely in tune with the times when he oriented critical and popular attention to a global media environment and to the sensory realm media produced.

Hediger, V., F. Hoof, Y. Zimmermann, with S. Anthony (eds), *Films That Work Harder: The Circulation of Industrial Film*. Taylor & Francis Group 2024
DOI: 10.5117/9789462986534_CH14

Hitting the peak of his influence in the mid-1960s, McLuhan's success is seen in the lasting language he provided for how we talk about new media and technological change, and this language zeroed in on the newly developing atmosphere of multimedia experience. But the language and the media experience he sought to explain had its origins in many different orbits. Education was one especially important, and underappreciated, discursive sphere that made foundational contributions to how we have come to understand media society. This chapter documents this influence through the contributions of educational scholar and practitioner, Robert W. Wagner, who produced films about teaching and other subjects at Ohio State University (OSU). I place his activities to promote multimedia education in relation to the currents in audiovisual pedagogy in the 1950s and 1960s but focus on Wagner's series of films about new educational media, what I'll call the Communication Series.

There is a growing store of research assessing the Cold War media ecology: Lynn Spigel's work on taste and media spaces, Anna McCarthy's analysis of American television's mode of governance, Vanessa Schwartz's examination of a new phase of cosmopolitanism and cultural hybridity, and Christina Klein's study of the reinvigoration of colonial power, *Cold War Orientalism*.[1] Seeking to contribute to these and other studies of post-WWII American media history, this chapter addresses the then novel idea that the electronic media environment was one in which people experienced different media as an ensemble. Media may have operated synchronously or asynchronously, coordinated or chaotic, but overall Cold War media featured agendas to establish the category of *multimedia*. Interest in this topic I share with Fred Turner, whose important book *The Democratic Surround* reads these conditions as an expansion of a liberal experiential realm.[2] Significantly, in period, multimedia became more than an environment of diverse media forms; it was characterized by built, systemic, and purposeful relationships among media.

Historical accounts of multimedia tend to focus on innovations introduced at mid-century expositions and by artists. But one principle

1 Spigel, Lynn. *Welcome to the Dreamhouse: Popular Media and Postwar Suburbs*. Duke University Press, 2001; McCarthy, Anna. *The Citizen Machine: Governing by Television in 1950s America*. Duke University Press, 2010; Schwartz, Vanessa R. *It's So French! Hollywood, Paris and the Making of Cosmopolitan Film Culture*. University of Chicago Press, 2007; and Klein, Christina. *Cold War Orientalism: Asia in the Middlebrow Imagination, 1945–1961*. University of California Press, 2003.

2 Turner, Fred. *The Democratic Surround: Multimedia and American Liberalism from World War II to the Psychedelic Sixties*. University of Chicago Press, 2013.

institutional site for multimedia experiments, and a major force responsible for integrating multimedia systems and skills into contemporary quotidian life, were schools and training facilities. The rise of American audiovisual education is a continuous trajectory from the Payne Fund Studies on motion pictures and children of the 1930s to "media literacy" decades later. In this sweep, a major reorientation of pedagogy transpired, one that refigured the sites, goals, and procedures of US education. As early as the 1940s a generation of educational technologists thought of schools as a testing ground for a new vision of multimedia, and during the years that followed, they transformed classrooms into multichannel environments. Edgar Dale, one of the most prominent educational technologists, introduced the new term "audiovisual" to teachers via his influential book *Audio-visual Methods in Teaching* (1946).[3] This teaching manual argued that the main orientation of audiovisual instruction was to "make experiences," coordinated across devices. Though each device made its own contributions, the category for pedagogical decisions was the *media system*. The multimedia aspect is even captured in the multi-sensory appeal implied by the term "audiovisual," which replaced the older term "visual education" in the mid-1940s. There were economic and demographic forces encouraging the turn to mediatized classrooms, and among some educational technologists there was a sincere sense that the benefits to teachers and students of multimedia teaching would be substantial. Yet, without question, the technological refiguration of education was to produce the appropriate citizen for the coming media and information economy, and to prepare a generation for a new environment of images, sounds, and sensations. At root, the post-war educational technologists understood that a new media epoch was arriving and that it must be met with enhanced technological materials and skills. Consequently, their pedagogical agendas hastened the normalization of our age as inescapably one of technology, information, and ongoing disruptive change.[4]

Paul Edwards reflects on the "closed-world discourse" of Cold War information technology. Describing the emergence of the "electronic battlefield," Edwards found that "it is a vision of a closed world, a chaotic and dangerous space rendered orderly and controllable by the powers of rationality and

3 Dale, Edgar. *Audio-visual Methods in Teaching*. Dryden Press, 1946.

4 C.f. my "Curtains, Carts and the Mobile Screen." *Screen*, vol. 50, no. 1, 2009, pp. 148–66; *Swift Viewing: The Popular Life of Subliminal Influence*. Duke University Press, 2012; and "American AV: Edgar Dale and the Information Age Classroom." *Technology & Culture*, vol. 58, no. 2, 2017, pp. 392–421. For an excellent history of the expansion of technology in America schools, see Cain, Victoria. *Schools and Screens: A Watchful History*. MIT Press, 2021.

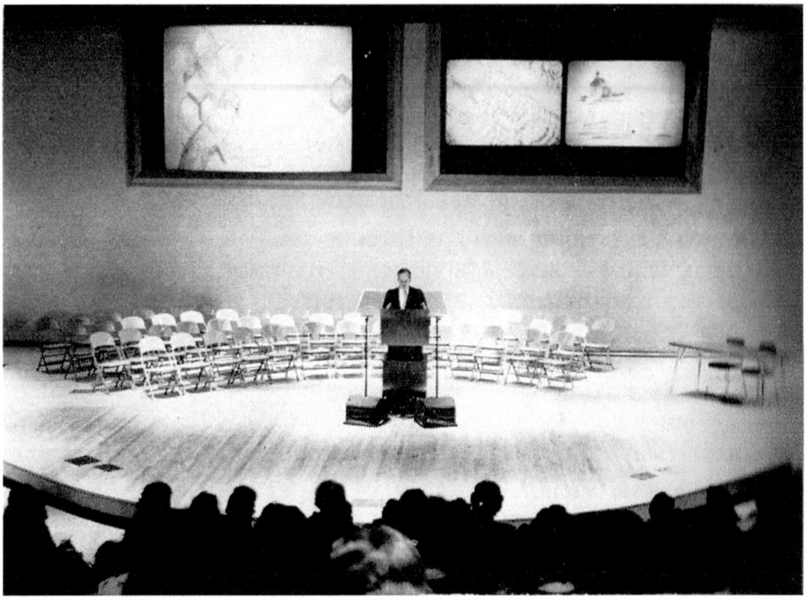

Fig. 14.1. Chicago Teachers College, experimental multimedia teaching environment, in *Business Screen* (1962).

technology."[5] Edwards covers human-machine integration of post-WWII military research, but it easily parallels the closed-world discourse of the "electronic classroom," where similarly a media system would "render orderly and controllable" the chaos of modern education. Why was modern education understood as chaotic? Because, as educational technologists reasoned, it had to contend with a fast-approaching future of information, media, and message abundance; this would create pressures to know more and to learn faster; and a swelling population pushed the infrastructural limits of existing facilities. Audiovisual education, then, prepared students for a media age that was imagined before it existed.

Elaborate efforts transpired to install multimedia as part of "the school of tomorrow," including intellectual labour, budget allocations, government lobbying, industrial and market formation, policy actions, and performance measures. Few institutions contributed to this as extensively in the US, and over several decades, as OSU, with notable figures Edgar Dale, head of OSU's Bureau of Educational Research (BER), Samuel Renshaw and his famed tachistoscopic training of WWII naval personnel, Keith Tyler and

5 Edwards, Paul N. *The Closed World: Computers and the Politics of Discourse in Cold War America*. MIT Press, p. 72.

his educational radio research, and Sidney Pressey, who sent shockwaves with the original teaching machine in the 1920s.

Another massively influential individual at OSU was Robert W. Wagner (1918–2011). He was first and foremost a filmmaker – a documentarian, screenwriter, producer and director.[6] He spent roughly five decades at OSU, with only brief time away as the head of the Cinema School at the University of Southern California in 1958, retiring in 1991. Wagner had made films at the Office of War Information (OWI) (1942–1943) and the Office of Coordinator of Inter-American Affairs (1943–1944). As chief of information for the Division of Mental Hygiene of the Ohio Department of Public Welfare, Wagner produced mental hygiene films, including *City of the Sick* (c. 1946) and *Problem Children* (c. 1946). He also helped launch the International Congress of Schools in Cinema and Television (CILECT) at the 1955 Cannes Film Festival, and reportedly assisted in bringing Czechoslovakia and the USSR into the organization, a significant achievement during the Cold War. Wagner served as the vice president of CILECT for eight years (1974–1982).[7] He held other key positions, most notably his long-time service on the executive of the University Film Producers Association (UFPA), a premier professional venue for university production units in the US established in 1947, and he was the editor of their publication, the *Journal of the University Film Producers Association*, for nearly two decades (1956–1975), which exists today as the *Journal of Film and Video*.[8]

His service to CILECT and UFPA mirrored his life-long production of films for university and government purposes, sometimes crossing paths with notable creative talent. In Turkey and Iran in 1951 for the US State Department, he wrote *Saydie, Village Midwife* (1951) that was eventually directed by Irvin Kershner, later of *The Empire Strikes Back* fame. In the 1960s, Wagner produced films about, and with the participation of, Hollywood luminaries: *George Stevens, Filmmaker* and *Frank Capra, and Why We Fight*, footage of which appears in the Netflix documentary *Five Came Back*.[9]

6 Wagner began at OSU as a research assistant in 1941–1942, became an assistant professor in 1944 at the BER, and was the director of the motion picture division of the Department of Photography and Cinema (1956–1966) and then chair (1966–1974).

7 Wagner, Robert W. "Oral History." Interviewed by Virginia Tiefel. OSU Archives, February 1, 2001; "Robert W. Wagner: Obituary." *Columbus Dispatch*, February 9, 2011; and Rose, Ernie. "UFVA Mourns the Loss of Robert W. Wagner." University Film & Video Association, March 9, 2011.

8 The publication becomes *Journal of the University Film Association* (1968–1981), *Journal of the University Film and Video Association* (1982–1983), then adopted its current title.

9 All broadcast locally, the first is not to be confused with *George Stevens: A Filmmaker's Journey* (1984).

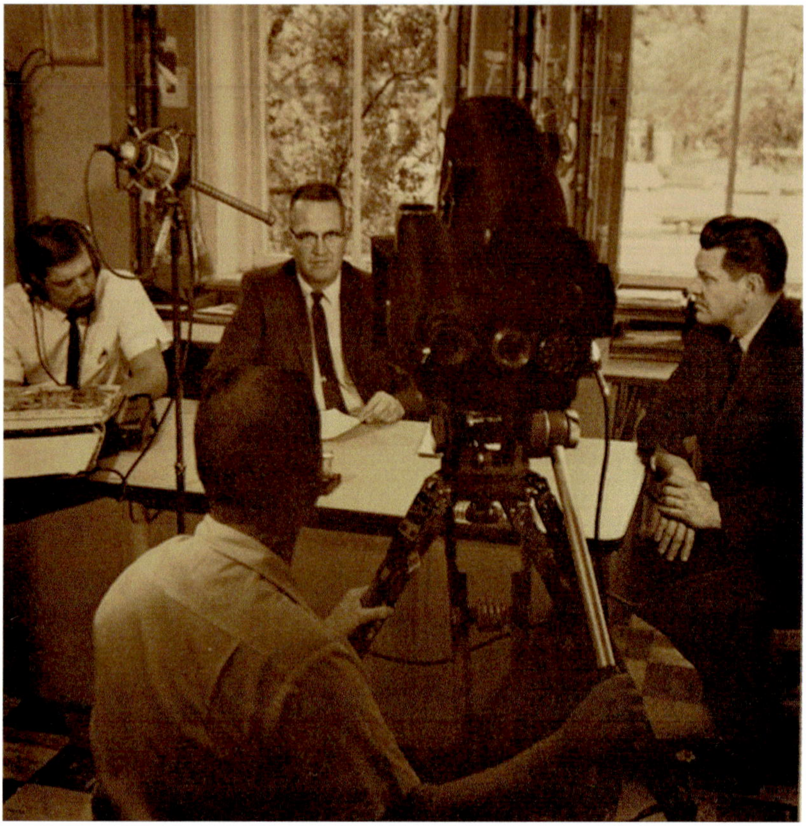

Fig. 14.2. Production Still, "Communication Series," interim report, January 1965. Robert Wagner is on the right.

And actor James Cromwell narrated one of his more successful television documentary productions, *The View from Malabar* (2000), on urban development, co-produced with WOSU-TV (PBS). But most of Wagner's films were made as part of course work and student training, with some commissioned by outside bodies. With Wagner, students and faculty shot sporting events and made films for organizations like the Red Cross, Seeing Eye Dogs, the American Automobile Association, and the US Department of Health, Education, and Welfare.[10] Wagner's films are not systematically archived, and the production roles are not always clearly credited, so no comprehensive filmography is possible. Take note that in this respect Wagner's career is a portrait of the foundational importance of universities to the history of filmmaking and media making, where we find a high level of fluidity

10 Wagner. "Oral History." pp. 4–8.

between making, critiquing, and experiencing. Moreover, this fluidity was often presented as an ideal form of engagement for AV pedagogy.

We don't often take account of how schools, especially universities and colleges, were production entities with an enormous impact upon the genres of instructional and sponsored media. Our emerging histories of "useful cinema" tend to focus on schools as consumers of films. But school-based film, television and radio units were also sites of media practice.[11] Importantly, once established, such units enjoyed relative stability, and they were outlasted only by the most successful of private production entities. University media making involved teams of staff and faculty and incorporated revolving pools of students. Some works were for university-specific promotional, research or historical/documentary purposes. Some were used in a single university setting; others circulated widely. Others still fulfilled course requirements, destined to be seen once in class, if that, and then be consigned to basements and storage lockers. In other words, university media productions were training exercises and institutional service works. Looking at this material now, it is not always apparent if a work was process-oriented, with training through media making being the objective, or a singular production for a university client. Consequently, as we develop our investigations of "useful" media making, we need to attend to overlapping pedagogical, research and institutional agendas.

AV Pedagogy and Airborne Television

Wagner wrote many essays that sought to advance the place of media in teaching and the importance of production. He contributed to the newly founded *Audio Visual Communication Review*, which became the leading American scholarly journal on education technology. His essays included coverage of the UFPA's eighth annual conference held at OSU in 1954 as well as reviews of films.[12] His advocacy included consideration of contemporary media systems. While in the 1950s he primarily focused on film, in 1961, he

11 Mike Zryd similarly points out the interrelationship between experimental filmmakers and the university classroom in "Experimental Film as Useless Cinema." *Useful Cinema*, edited by Charles R. Acland and Haidee Wasson. Duke University Press, 2011, pp. 315–36.

12 Wagner, Robert W. "The University Film Producers Association: A Report on the Organization and Its Recent Convention." *Audio Visual Communication Review* (*AVCR*), vol. 2, no. 4, Fall 1954, pp. 299–303; idem. "Films without Endings." *AVCR*, vol. 3, no. 2, Fall 1955, pp. 138–43; idem. "Thirty New University-Produced Films." *AVCR*, vol. 4, no. 4, Fall 1956, pp. 334–39; and idem. "Three Films Worth Seeing." *AVCR*, vol. 7, no. 2, Spring 1959, pp. 138–41.

became the *Audio Visual Communication Review*'s editor of a new section, "Media." Even earlier in 1952, Wagner argued that the boom in television would simultaneously be a boom in film, and that university training should respond accordingly. Television meant that university producers of instructional and educational films should now understand their audience as much broader, beyond the classroom and into the living rooms of viewers.[13] He discussed materials that presented media possibilities and modes of informational exchange, as with his essay "Films about Communication," and introduced readers to fresh innovations when television and film share personnel and skill sets, as with "New Dimension in Audio-visual Design."[14] He was a voice for creative approaches to the "idea" film, as seen in "The Formula Film," in which he decried that non-theatrical film fell just as easily into imitative and conventionalized treatments as Hollywood.[15] For the entirety of this career, Wagner studied and disseminated programs of pedagogical advantage made possible by new audiovisual materials, even writing some historical overviews.[16] The intellectual labour of setting out the terms of discernment, and helping scholars and teachers navigate a new sea of media texts and media making, was a significant part of Wagner's long-term project. His turn to producing a series of films about audiovisual teaching experiments was a logical extension of his demonstrated research priorities.

An elaborate example of educational experimentation, one that melded with a defence agenda, was the Midwest Council on Airborne Television Instruction Project, based at Purdue University. Called Stratovision, videotaped courses were to be broadcast via DC-7 airplanes to 13,000 schools, though in fact this ambition was never fully realized and actual participation was about 1,800 schools. With Ford Foundation funding, the project continued until 1968, and was later adopted for US military propaganda overseas.[17] Wagner made a film to document and promote the experiment. *Airborne*

13 Idem. "TV and University Motion Picture Production." *Journal of the University Film Producers Association*, vol. 4, no. 3, 1952, pp. 7–9.

14 Idem. "Films about Communication." *AVCR*, vol. 3, no. 3, 1955, pp. 213–19; and idem. "New Dimensions in Audio-visual Design." *AVCR*, vol. 4, no. 2, 1956, pp. 142–47.

15 Idem. "The Formula Film." *AVCR*, vol. 3, no. 1, Winter 1955, pp. 53–57.

16 Idem. "The Educational Film in Transition." *Audiovisual Instruction*, vol. 9, 1964, p. 172.

17 A brief discussion of Stratovision appears in James Schwoch, *Global TV: New Media and the Cold War, 1946–69*. University of Illinois Press, 2009, pp. 84–85. See also Perlman, Alison. "Television Up in the Air: The Midwest Program on Airborne Television Instruction, 1959-1971." *Critical Studies in Media Communication*, vol. 27, no. 5, 2010, pp. 477-497; Tracey, Monica W., and Jill E. Stefaniak. "MPATI: The Midwest Program on Airborne Television Instruction (1959-1971)." *International Journal of Designs for Learning*, vol. 5, no. 2, 2014, pp.29-33.

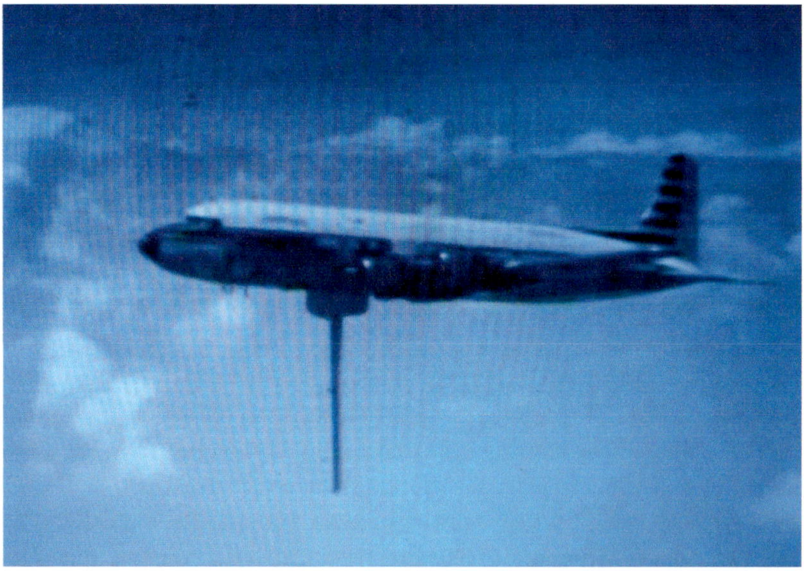

Fig. 14.3. Still from *Airborne Television: Profile of a School* (1962).

Television: Profile of a School (1962, 17 min) was notable for the way that it set a template for Wagner's films to follow, and at times Wagner would remember it as part of his Communication Series.

Airborne Television presents an elementary school in Dublin, Ohio, where this specialized broadcast system enhances conventional lessons and helps teachers confront an increasing number of students and subjects to cover. Teachers testify to the ease of integration, administrators confirm efficiencies and heightened student enthusiasm is repeated throughout. In a Spanish-language class, as a female instructor leads the class, the film cuts to an airplane broadcasting a pre-recorded lecture by a male teacher, which students see on a classroom television set and to which they respond by repeating phrases. In advanced math, a female teacher guides the class following the direction of a televised male authority; exploring science, a broadcasted male expert supplements a female teacher's lesson. *Airborne Television* depicts several examples of male experts broadcast to classrooms run by female teachers. In this way, the film presents functional integration of the new educational media, and covers pedagogical materials like film, field trips and in-class demonstrations in addition to Stratovision broadcasts. Further, it illustrates that standardization of lessons across schools reasserted a gendered hierarchy of authority. The implication was that the complexity and rapid evolution of subject knowledge require more centralized operations, ones that marginalize and devalue the workforce of mostly female teachers. This model of centralization of pedagogical control would later be critiqued and supplemented by an emphasis on local adaptability and media making, as is described below.

Airborne Television and the Communication Series to come are records of a developing closed-world discourse of the "electronic classroom." They show teachers and students engaging in and benefiting from a new constellation of media. Other films of Wagner provocatively present recent technological change as making rapid educational restructuring necessary. This is best illustrated by *Crisis and the University* (1964, 28 min), which earned a Local Emmy nomination for Documentary Film. His most direct statements about the future of educational technology, however, are found in the Communication Series. The films highlight the core features of individual and group instruction, and the coming information revolution, but they are more pointedly arguments to persuade adoption of new media instructional ideas. They circulated as evidence of the new media system's efficient delivery to large numbers of students, of its expertly constructed content, and of its fit with the technological environment, assuring that the media classroom matched future work and societal needs.

Funding the School of Tomorrow

The Communication Series followed soon after *Airborne Television*, taking about three years to make.[18] Funds came from the US Office of Education via Title VII of the National Defence Education Act of 1958 (NDEA), which was President Eisenhower's response to the Sputnik crisis of 1957. Just as there was a space race and an arms race, there was a Cold War education race, and the act was an American effort to best its competitor nations. It resulted in fundamentally changing the federal government's relationship to education. The act established student loan programs; increased investments in science, math, languages, and area studies; and built new procedures for counselling and testing students. The NDEA also did considerable work to remake the modern American classroom as a multimedia audiovisual space. While visual education predates 1958 by decades, classroom utilization soared after this turning point settlement in 1958. The act provided substantial funding

18 Wagner, Robert W. "A Galaxy of Motion Picture Documents on Communication Theory and the New Educational Media." Department of Health, Education, and Welfare, US Office of Education, Washington, DC, 1966, p. iii; and idem. "A Series of Motion Picture Documents on Communication Theory and the New Educational Media. Final Scripts." Ohio State University, 1966.

for experiments in new educational media, reports on new AV methods, catalogues and aids to utilization, and, most uniquely, Title III provided state grants with matching federal funds "to help equip and remodel laboratories and classrooms," funding conversion as well as storage and maintenance.[19] By 1967, federal funds given to the states for AV equipment, media content and materials, and classroom remodelling, totalled $514,982,462. With the state's own contributions, this investment reached $930,911,000.[20] Oriented toward knowledge production and dissemination, Title VII had funded over six hundred projects by 1968 for the more modest sum of $40.3 million.[21]

In 1960, the Appropriations Committee of the US Congress heard the following rationales for Title VII and its focus on "new educational media":

> The exploration of development of all types of communication media as aids to teaching are moving ahead under Title VII. The shortage of teachers, the vast amounts of knowledge instructors must impart to students, and the excellence of scholarship our modern world demands of graduates, place great strain on conventional teaching resources.[22]

Accompanying claims about resource shortages were statements about the sheer volume of material that needed to be taught. And the expectation that media would solve these problems was taken for granted. Title VII was "to lay a foundation for wide use of newer media."[23] Such interpretations of a new learning environment, characterized by information abundance, infrastructural limits and rapidly changing technological means infused the operations of the NDEA.

19 "NDEA Title III Guidelines." US Department of Health, Education, and Welfare, Office of Education, Washington, DC, January 1965 (revised April 1967), p. v. Materials eligible under Title III included darkening shades, blinds, or curtains; filmstrip or slide previewers; flannel and magnetic boards; microfilm reader printers; reading pacers; tachistoscopes; portable stands with sound amplification for large-group teaching; projection screens, stands, and tables; projectors: 16 mm and 8 mm, sound or silent, filmstrip, microfilm, opaque, overhead, slide; record or transcription players; stereoscopes; tape recorders and playbacks; teaching machines; television receivers (with amplifiers, antennae, coaxial cable, converters); storage equipment (cabinets, shelves, carts, racks); media content (books, charts, dioramas, discs, films, filmstrips, microfilms, photographs, programmed learning materials); and devices for media preparation (cameras, tape splicers). Ibid., pp. 56–68.

20 "NDEA Title III: Fiscal Years 1959–67: A Management View." US Department of Health, Education, and Welfare, Office of Education, Washington, DC, May 1969, p. 109, p. 111.

21 C.f. Barksdale Clowse, Barbara. *Brainpower for the Cold War: The Sputnik Crisis and National Defense Education Act of 1958*. Greenwood Press, 1981.

22 "Review of Activities under National Defense Education Act." Government Printing Office, 1960, p. 6. Hearings before the Subcommittee of the Committee on Appropriations House of Representatives. 86th Congress, 2nd Session.

23 Ibid.

The five-year official assessment of the NDEA reported, "No real instructional technology could be said to exist in the United States in 1958" though "[m]ore than 40 years of applied instructional research had suggested that the multisensory experiences provided by motion pictures, radio, television, and other media of communication could be used effectively in the schools." Despite that certainty "relatively little use was being made of such resources in the schools." Teachers did not have adequate training with AV, they used them as substitutes for teaching, and school administrators treated AV materials as extras or as a way to increase class size.[24] The report summarized that with the NDEA "the newer media have become more widely and more effectively used."[25]

Title VII consisted of two parts. Part A was for research grants, focusing on new studies of instructional media, teacher training in "newer media" – this being the term used prior to "new media" becoming the most common term – and academic uses of "new educational media." Part B made funds available through the Office of Education to disseminate information on new educational media. Part B was the source of the funds for Wagner's Communication Series, officially titled "A Series of Motion Picture Documents on Communication Theory and the New Educational Media."[26] Wagner's grant was for $6,000 (December 31, 1961–August 31, 1962) then $159,461 (May 28, 1963–March 31, 1965) (or, taking them together as 1963 dollars, worth about $1.65 million in 2023). Of the first four hundred plus Title VII project contracts signed between 1958 and 1963, only fifteen had higher amounts awarded.[27] The goal of promoting educational media included steps to assure that attitudes toward new media teaching improved following screenings. At the preparation stage, a pool of 719 students saw four test versions of each film in 1964. A questionnaire called the "New Educational Media Attitude Scale," which had been used on *Airborne Television*, was administered. It showed a significantly favourable impression of new educational media following a screening of the test films, and favourable impression was the objective.[28]

To provide context for Wagner's contract, consider another Title VII B contract (#SAE-8758). This went to the Society of Motion Picture and

24 "Report on the National Defense Education Act, Fiscal Year 1963." US Department of Health, Education, and Welfare, Government Printing Office, 1964, p. 55.

25 Ibid., p. 59.

26 The grant for this project was B-131 and B-131-A (contract OE-3-16-020).

27 "Projects Initiated under Title VII, Part B, National Defense Education Act, September 2, 1958–June 30, 1963." US Department of Health, Education, and Welfare, US Office of Education, Washington, DC, 1963.

28 Wagner. "A Galaxy of Motion Picture Documents," pp. 78f.

Television Engineers (SMPTE) to coordinate technical standards for AV educational media. It was to do for education what their report "Recommended Procedure and Equipment Specifications for Educational 16-mm Projection" (May 1941) did for the coordination of efforts between US military usage and film equipment manufacturers.[29] A conference was held in Princeton, New Jersey, June 4–7, 1961. The backdrop for their discussions was that "American education is in violent ferment," the result of "four explosive influences:" population growth, knowledge growth, psychology-of-learning advances and instructional-technology developments.[30] The participants took note of the underutilization of AV materials and of teacher resistance to new media instruction.[31] They called for "a professional breed of highly competent, teaching-learning systems experts," and clearly advocated for the "systems approach" and the production of integrated packages of materials.[32] The Midwest Program on Airborne Television Instruction "was singled out for adverse criticism, largely because local control over the curriculum was being removed. In addition, they felt that large numbers of children, with varying degrees of ability, were being forced, despite these differences, to proceed at an arbitrarily uniform class rate."[33] In many respects, Stratovision represented a pedagogical idea about fixed packages arriving from a central production and distribution agent. In response, there was a growing agreement that multimedia systems were advanced through flexible and programmable materials.[34] Though Wagner did not attend this conference, the ideas presented about AV education, the claims about the nature of the educational crisis, and the understanding about the future direction of educational media echo with great fidelity those represented in Wagner's Communication Series.

Wagner's Communication Series

The four central films in the Communication Series were *The Information Explosion* (1966, 34 min), *Perception and Communication* (1966, 32 min), *The*

29 Allen, William H., and John Flory. *Designing New Apparatus for Learning: A Survey of Needs for, and Developments in, Audiovisual Devices Suitable for Educational Uses.* Society of Motion Picture and Television Engineers, 1963, p. v.

30 Ibid., p. 3.

31 Ibid., pp. 4–5.

32 Ibid., p. 14.

33 Ibid., p. 86.

34 Ibid., p. 136.

Teacher and Technology (1966, 49 min) and *The Process of Communication* (1966, 46 min). This last title begins with a colour animated sequence about communication that shows how "we move in a world of symbols, statements and events." The soundtrack consists of electronic sounds and the editing is fast-paced. Information theory is the ostensible topic, but the film focuses on AV instruction. One sequence shows new ways to create simulated situations for flight attendant training at United Airlines (UA). At Fort Leavenworth, Kansas, we see military decision-making being studied. The stable information transmitted in the UA example is something more complex in the military setting. A dramatic gender division is apparent with the exclusively female UA example contrasting the male military one. In all illustrations, AV materials include a range of demonstrations, TV, film, overhead projectors, radio, daylight projections, as well as more traditional blackboards and books. AV-facilitated simulations appear several times, and in one sequence, teachers in training interact with film projections that change depending upon the response they give. A filmed role-playing exercise halts as a student turns to the camera to ask the teacher in training a question about dictatorship. The female teacher's inadequate response generates an unsatisfying shrug from the filmed students. After consultation about her performance with her male supervisor, she tries a different tact and elicits a more productive response from the filmed students. This plays out similarly in other films in the series, in which we see teacher training through mediated classroom situations with specifically political and gendered content. Moreover, orientation is toward technology, and students and teachers alike speak directly to screens to simulate dialogue. With the multimedia materials presented, there is nearly always classroom discussion before and after usage.

The most advanced AV examples in *The Process of Communication* come at the end, with computer-based teaching machine SOCRATES at the University of Illinois. An individual student encounters a lesson at a terminal, where the computer advances frames, makes assignments, and places answers on a memory disk. The system highlights individual accommodation. As this sequence explains, SOCRATES can identify patterns and respond uniquely to dozens of students at once. Another University of Illinois computer-based teaching machine, PLATO, contains lessons on slides, scanned and seen on TV monitors attached to keyboards. A student watches a film and then navigates responses, corrected by the computer. The automation of lessons is challenged by one PLATO researcher, who comments, these computer systems are "not talking about removing the teacher from the teaching situation," but about taking care of mundane tasks. Communication scholar

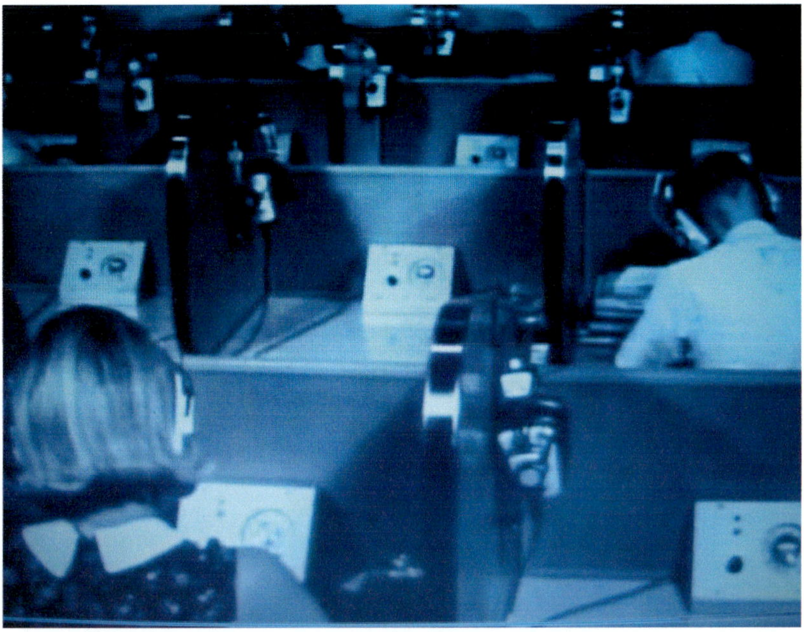

Fig. 14.4. Dial-up language instruction, in *The Teacher and Technology* (1966).

George Gerbner underlines this point in a closing statement, emphasizing the "human end of a technological revolution in communication and in teaching." The certainty and inevitability of that revolution, though, is taken for granted by all the film's participants.

The Teacher and Technology covers similar ground and again presents PLATO computer learning. Commentary comes from educational technology specialist James Finn (USC). Segments present various forms of simulated environments to train teachers, flight attendants and military personnel. Among the experiments covered is a dial-up language-learning and information system at OSU where "accessibility is the key value of the system" and the University of Miami's octagonal instructional resource centre at Coral Gables, which includes three-hundred-seat classrooms, rear projection, a central projection deck and feedback mechanisms for students. Architectural designs of new media classrooms create the "school of tomorrow" as "an experience centre" and "multimedia environment." The film demonstrates that schools don't only consume pre-packaged media plans but also produce their own materials. This is best illustrated in a sequence on the US Air Force Academy at Colorado Springs, included in *The Teacher and Technology* but taken from another film, *Teacher-Centered Television* (c. 1964). It shows the Director of AV Services saying they see the "teacher as manager of instructional

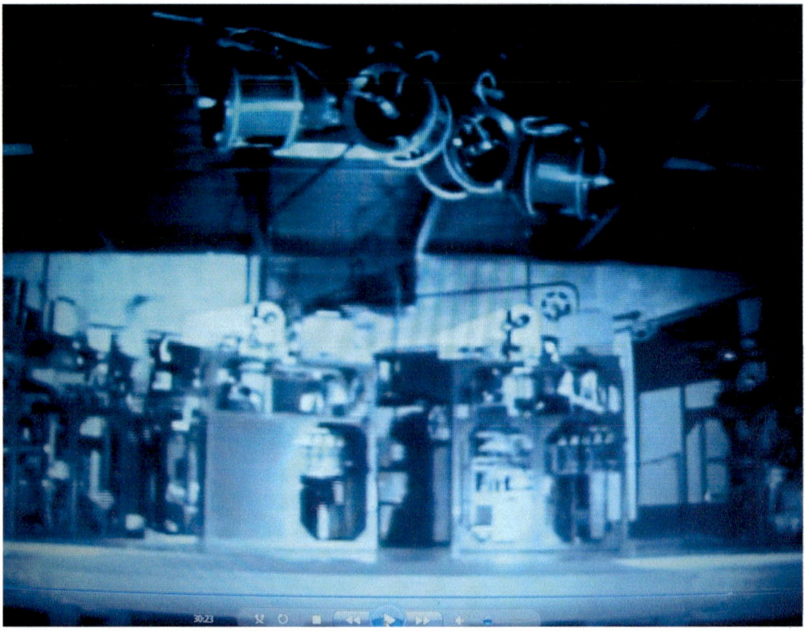

Fig. 14.5. Central control room, U of Miami's octagonal AV instructional facility, in *The Teacher and Technology* (1966).

resources or as a manager of the student's environment." This sequence features the "instructor-centred" emphasis, showing how media making is part of training for the complex decision-making of military operations. The full development of the budding officer includes exposure to poetry, and a teacher leads students through a discussion of Beat poet Lawrence Ferlinghetti's "Dog" (1958), partly presented on television. A sequence at Stephen's College documents one response to "today's world of exploding knowledge," as the narrator puts it. Access to the world of literature is opened up with live radio broadcasts of an author's talk, a telephone hook-up for student questions and a projected image of the figure, thus connecting several classrooms across the state to Vance Bourjaily, John dos Passos and Ralph Ellison. *The Teacher and Technology* concludes with an image from another film by Wagner, *Music Research* (1964), in which a girl with headphones guides her own lesson, with the narration declaring "This is a teacher of tomorrow. Born in a time of communication revolution, the child of today is already at home with instructional technology."

The Information Explosion begins with a colour montage of advertisements and Times Square, followed by Edgar Dale asking how the information revolution might be controlled. The film switches to black and white and has communication studies luminary Wilbur Schramm introduce a sequence

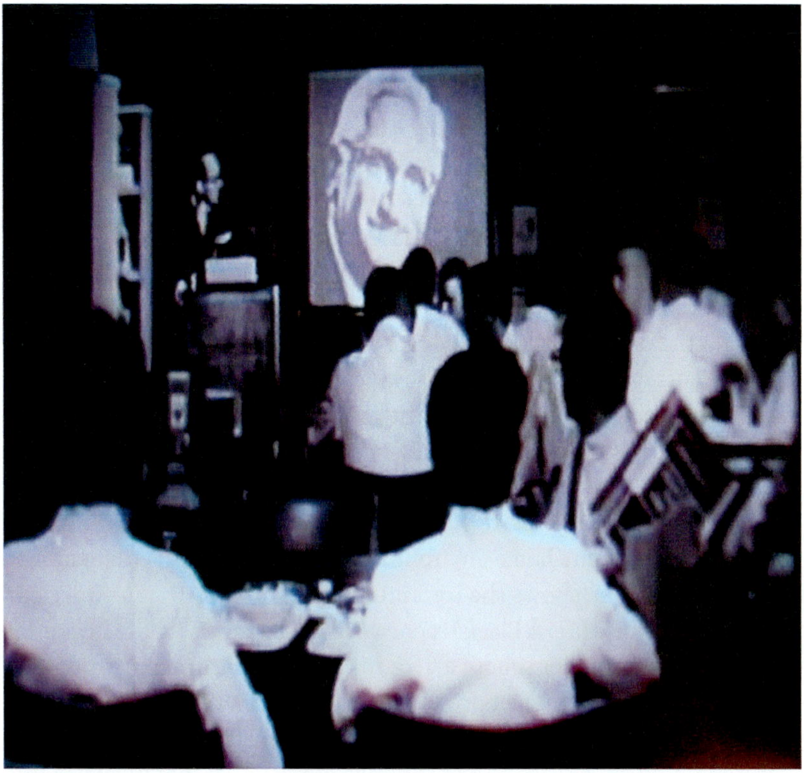

Fig. 14.6. Mulitmedia classroom usage, in *The Information Explosion* (1966).

about evolutionary changes in communication from cave drawings to satellites, accompanied by eerie electronic music. Returning to Dale in a bookstore, who emphasizes the need for the free flow of information, the sequence ends with a televised news conference by President Kennedy. The film presents several examples of coping with information abundance. There is hi-tech hourly decision-making at UA using film, lantern slides, and transparencies, with links to statistical analysis offices in different cities. Another sequence shows a micro-broadcast to Grant Hospital in Columbus, Ohio, of new medical advances. The film includes portraits of the media consumption of a newspaper-reading white boy and a Black teenage girl who reads classic novels and listens to classical, jazz and rock 'n' roll music. Dale wonders, with all this informal learning, whether they will "learn to cope independently and critically with the world's assault on their senses?" A final sequence, from Wagner's *Communication Revolution* (1960), opens with Dale saying, "We need teachers who are sensitively aware of the nature and impact of the communications revolution."

Perception and Communication is an aesthetic departure from the other films. The opening is more observational than the other didactically scripted films, with soft-focus colour cinematography, overlapping audio, and zooms. Light music accompanies images of kittens, turtles and water sprinklers, all part of a summer camp for visually impaired children. This first sequence looks like a film from the 1970s, where the other films seem from the 1950s. The didacticism returns as the film shifts to black and white, and as male experts begin to narrate. Still, *Perception and Communication* retains its forward-looking tone, and it treats civil rights era classrooms and the training and expertise of women more elaborately than the other three films. It depicts group discussions and role play and advances the importance of international experience. International university students appear, and one segment follows an anthropology student in Borneo and in the Peace Corps. A section about an integrated urban class of five-year-old children, in a predominantly poor Black neighbourhood, includes white teacher Shirley Duncan emphasizing the need to know the children's home environment. A segment, shot with handheld cameras on a farm, shows the unfamiliarity of urban African-American children with rural life. A Black teacher, Brenda R. Taylor, tells of becoming aware that many in her school arrive in the morning hungry, prompting teachers to start a breakfast programme. *Perception and Communication* aims to address the diversity of experience that constitutes a contemporary classroom and encourage respectful incorporation of that diversity into lesson plans. A curious illustration of self-perception includes an exercise in which children document themselves with photographs and movies, telling of a white kid, among his predominantly black classmates, not realizing he was white until seeing these photos. Kenneth Norberg (Sacramento State College) summarizes the film's position saying, "This means that the teacher must not only be a subject matter specialist, but also a transactional agent, a professional designer and arranger of learning experiences."

Wagner's films document new educational media and promote a systemic approach to instructional media, showing the integration of material purchase, rented, and produced by schools. While some sequences include interaction among students and teachers, most are orientated toward a unified system of media technology. This might be to produce simulated work situations or to master basic knowledge through self-guided instruction. Teachers become managers of information delivery via the new multimedia systems, rather than experts themselves. The films reassuringly highlight the learned advocates for this approach, presenting several of the most impactful communication and education scholars of the time: Wilbur Schramm, George Gerbner, Hadley Cantril, Edgar Dale, James Finn, Kenneth

Norberg, Keith Tyler, Marshall McLuhan and Gilbert Seldes. A dominant current underlying the films is that the overwhelming world of media and information abundance requires an embrace of educational media. This embrace involves the rapid imparting of information, but also the making of experiences, which include various forms of media practice. And while the films reassert male authority in narration, on-camera experts and occupational hierarchies, there are some hints of the advancements generated by civil rights struggles.

A Galaxy of Modularity and Media Making

Responding to concerns about inflexible AV content, Wagner inventively built media making into the Communication Series, inviting users to see the films as works to be adapted, cut up and reassembled. Wagner's final report presented an argument for the modular design of the films in this series. It included information on how to cut a new film using the series as source footage. The guide offered suggestions for thematic reassembly. Emblazoned with "A Galaxy of Motion Picture Documents on Communication Theory and the New Educational Media" – a McLuhanesque flourish substituting "galaxy" for "series," which appeared in the original project title – Wagner's eighty-two-page report included a user's manual for the films, a usage study, and an essay on communication theory by Harbans Singh Bhola, who had received his doctorate at OSU in 1965 under Edgar Dale's supervision.

Extending the astronomical analogy, the report called the four main films "planetary films." Each could be broken apart into smaller "asteroid films," which were self-contained sequences, numbering thirty-one in total. Supplementing these works were five "satellite films": *The Communication Revolution* (1960, 22 min), *Communication Conference* (1965, 31 min), *Teaching Machines and Sidney Pressey* (1964, 17 min), *Music Research* (1964, 17 min), and *Teacher-Centered Television* (c. 1964, 23 min). These were all made by Wagner outside the parameters of his Office of Education contract and were repackaged as part of the Communication Series.[35]

Teaching Machines and Sidney Pressey includes footage of Pressey's device from the 1920s, demonstrations of the machine's reward system

35 *The Communication Revolution*, released in 1960, is often misidentified as 1966 or later. For example, *Educational Films 1973* lists all the planetary and satellite films as 1967. *Educational Films 1973*. University of Michigan, Media Resources Center, Audio-visual Education Center, Ann Arbor, 1973.

and automatic grading, and a sequence with B.F. Skinner. Where the teaching machine was "ignored by mainstream education," it was now adopted due to the increased amount of material people must learn today, articulating the "information explosion" idea that rested behind AV instructional agendas at the time. Skinner appears again in *Music Research*, which introduces some of the technological advances in music education, especially self-instruction. In fact, all the technology featured in this film is for individual lessons, including touch screens. As the narrator states, "[t]hrough the use of the media [educators have] more control of the learning process."

Teacher-Centered Television was originally a videotape document of multimedia instruction at the US Air Force Academy in Colorado Springs. *The Communication Revolution* and *Communication Conference* are both literally discussion films. They are depictions of panel discussions among experts. The first film has Edgar Dale, Keith Tyler, Marshall McLuhan, and Gilbert Seldes as participants. Filmed in 1960, this film captures McLuhan prior to the peak of his fame a few years later, but shows that many of his ideas about media were in place. We see the other panellists clearly taken by his imaginative suggestions and his charismatic presence – for instance, his distinction between hot and cool media – and for the most part their positions are proximate and in accord. *Communication Conference* includes Dale joined by educational technology specialists Charles Hoban, Franklin Knower, Kenneth Norbert, James Finn, and George Gerbner. Where *The Communication Revolution* is a more free-wheeling discussion about the technological society, *Communication Conference* more pointedly addresses the role of teachers and new educational media.

Wagner explained that the "galaxy of film" concept responded to a widespread sense that pre-packaged teaching materials were too standardized for general classroom use. Without flexibility and programmability, works could not be adapted to changing pedagogical situations. With Wagner's innovation, the film topics would educate teachers about new developments in media and communication theory, and their "galaxy" design would promote application of those concepts in their step toward media making.[36]

Asteroid films – that is, the sequences that made up the four main films in the series – tend to feature single institutionally specific examples, often with a particular technological application depicted. Never do any of the efforts and experiments depicted fail or eliminate other advantages. Every single one is a contribution to some facet of learning, whether by individuals

36 Wagner. "A Galaxy of Motion Picture Documents," pp. 6–7.

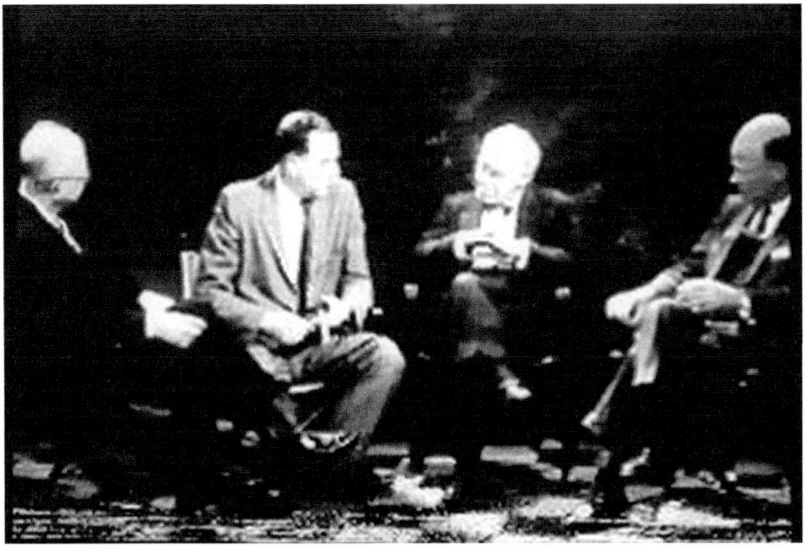

Fig. 14.7. Panel discussion, *The Communication Revolution* (1960), with, from the right, Keith Tyler, Marshall McLuhan, Gilbert Seldes, and Edgar Dale.

or in groups, and each provides a way to navigate what they accept as modern technological upheaval. Oddly, these educational media and materials are not described as part of the troubling condition of upheaval. Bhola expressed it this way in the "Galaxy" user's manual: "[T]he teacher must be interested in [information technology] because we *are* living in a world of information industries and universal information consumption. We are in the midst of the communication revolution, and the nation's schools must respond to the social, economic and cultural realities of life."[37] In a neat bit of self-serving calculation, designed to ease teacher concerns that their influence on teaching decisions will be lost to new educational media but no doubt producing more anxiousness, Bhola proposed that "Teaching = (Teacher x Information Technology)." Therefore, if IT is zero, the formula tells us that teaching too would be zero.[38]

Wagner later remembered that the Communication Series included eight films.[39] There were, in actuality, nine different films (five satellites and four planetary films). The bibliographic information for the series characterizes the project as forty films (four planetary, thirty-one asteroid, and five satellite). Wagner made this claim, too, double counting asteroid

37 Ibid., p. 71

38 Ibid., p. 73.

39 Wagner. "Oral History," p. 13.

A. Media in Instruction (Continued)

DESCRIPTION	PLANETARY	ASTEROID	SATELLITE
Overhead Projector	INFO EXPL PROC COM TEACH TECH	4 4 3,5,6	
Printed Materials, Libraries	INFO EXPL TEACH TECH	1,2,6 3,4,6,7,10	S-2
Programmed Instruction	PERCEPTION TEACH TECH PROCES COM	3 8,9 7	S-4; S-5
Radio	INFO EXPL	5,6	S-2
Simulation	PROCES COM	3,4,6	
Slides	PERCEPTION TEACH TECH	4 4,9	
Teaching Machines	PROCES COM	6,7	S-4
Television	INFO EXPL PROCES COM TEACH TECH	3,6 7 3,7,9,10,11	S-2; S-3

Fig. 14.8. Index for modular possibilities, "A Galaxy of Motion Picture Documents on Communication Theory and the New Educational Media" (1966).

segments and planetary films. He then expanded that all the combinations represented 1,600 films (an erroneous calculation, simply forty times forty, which is not even close to the mathematically accurate figure for all possible permutations).[40]

In theory, the modular design facilitated uses in different lessons. Wagner thought that it would allow for segments to be removed if they didn't work with an audience or if the information became dated. The self-contained segments varied in length, making them adaptable for different formats, including broadcast television and cartridges for 8 mm projectors.[41] Short instructional films had been developed prior to this date to facilitate adaptability to different teaching contexts. Wagner, himself, had only recently produced *Single-Concepts Physics Clips* (1963), which consisted of nineteen self-contained clips of physics processes, such as inertial forces, temperature waves and radioactive decay. Each clip was between three and four minutes long. The idea of the "single concept" film circulated elsewhere, and Wagner was a featured speaker at the National Conference on Cartridged Films, at the University of Michigan, February 22–24, 1967, that was part of the "Single Concept Film Clip Project." At this event, Wagner explained in detail

40 Wagner. "A Galaxy of Motion Picture Documents." p. 40.

41 Ibid., p. 35.

the modular design of his Communication Series.[42] Wagner wrote of the planetary films:

> The structure of each [...] is more than a sequence of single-concept films or conventional filmic episodes. Each is not only related to others in the film by a single, unifying theme, but also repeats, extends, and sophisticates the ideas contained in the previous segment. Each is, at the same time, designed to be used as a self-contained filmic message with a simple technique for detaching and using it separately, as well as for replacing it in the original sequence.[43]

To this end, the four main films in the Communication Series came with their title and segment printed along the sprocket perforations to make identification easier for those looking to splice and extract a particular topic. The user's manual indicated possible uses of these components, compiling an index to the films of institutions, topics and individuals presented. A film unit about audiovisual specialists might consist of *The Process of Communication*, *The Teacher and Technology*, or asteroid 4 of the first film, or asteroids 3, 6, and 7 of the latter. A unit about teaching with television might include *The Process of Communication* (or just asteroid 7), *The Teacher and Technology* (or just asteroids 3, 7, 9, 10, and 11), *The Information Explosion* (or asteroids 3 and 6), and satellite films *The Communication Revolution* or *Teacher-Centered Television*.[44]

The modular idea directs film usage toward random access, where the galaxy is a database navigated through tags that link components and themes. Overall, Wagner's project presented an early enactment of what today we call "database film," a form of non-linear film consisting of a set of segments that can be assembled and navigated variously by different users. Database cinema, today, is often erroneously understood as particular to digital media. Wagner's series is evidence against this ahistorical assumption. Wagner claimed a follow-up study showed that many teachers used this modularity to cut and re-edit films.[45] Three usage surveys recorded that approximately one-third had recut these films at some point. They also noted

42 Wagner, Robert W. "Modular Design for a Series of Films on Communication Theory and the New Educational Media for Use in Teacher Education." *Single Concept Film Clip Project – Final Report*, edited by Elwood E. Miller. Office of Education, Bureau of Research, Washington, DC, 1967.

43 Wagner. "A Galaxy of Motion Picture Documents." pp. 7, 9.

44 Ibid., p. 44, p. 49.

45 Wagner. "Oral History," p. 19.

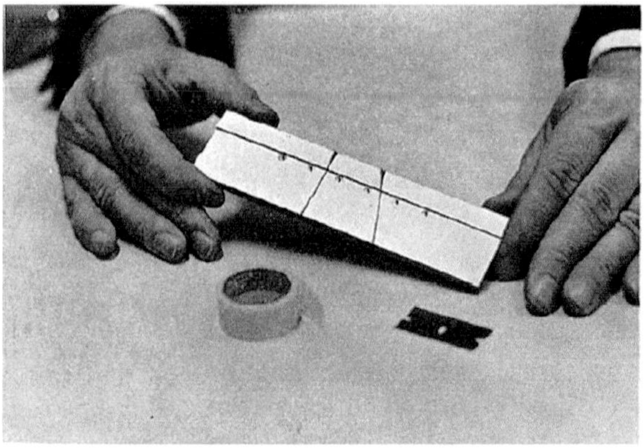

Above: The equipment for tape-splicing segments splicing block; perforated 16 mm tape; and a razor blade. Below: The film is cut either at right angles or on the diagonal. The tape is slipped over pins and adheres to film. Film may be spliced without loss of frame.

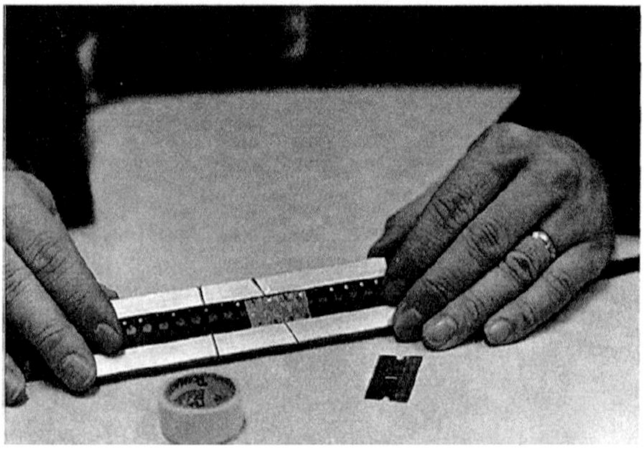

Fig. 14.9. Splicing illustrated, "A Galaxy of Motion Picture Documents on Communication Theory and the New Educational Media" (1966).

the extreme discomfort most had with doing so. *The Teacher and Technology* received the most favourable responses in early surveys but was surpassed in 1981 by *Perception and Communication*.[46] Reviews of the films thought they were informational, suited for high schools, but technically only "fair," "pedestrian," and "designed to assuage the fears of teachers in a technological age facing the awesome challenge of mechanized instruction."[47]

46 Wagner. "The Teacher and Technology: The Bottom Line." *Techtrends*, January–February 1998, pp. 37–42.

47 Ragsdale, Donald J. "Review: *Perception and Communication*." *Speech Teacher*, vol. 18, no. 4, November 1969, p. 329; Keyes, Elizabeth A. "Review: *Communication Conference* and *The Process of Communication*." *Speech Teacher*, vol. 18, no. 4, November 1969, p. 328; and Erlich, Michael G.

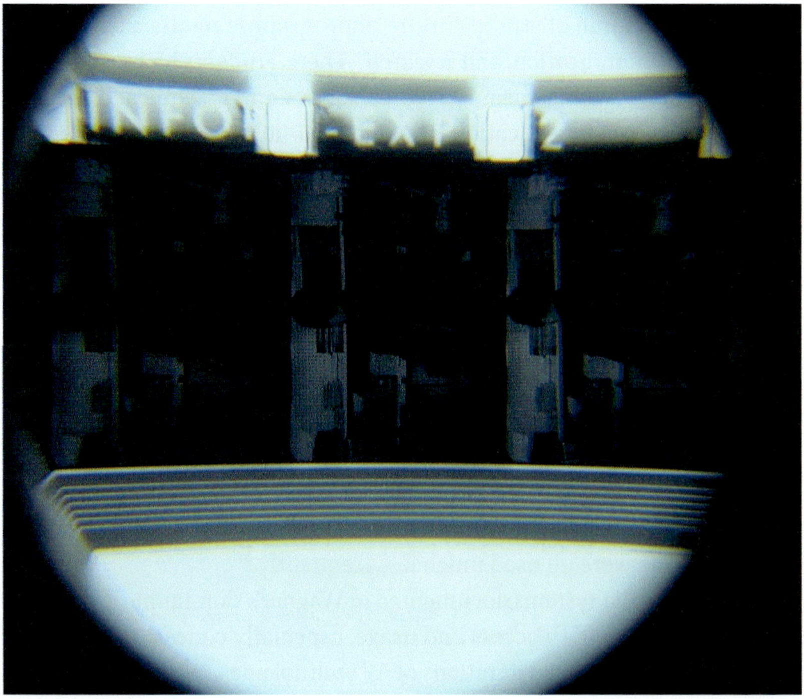

Fig. 14.10. Identification of asteroid 2 on print of *The Information Explosion* (author photo, 2014).

Multimedia and the Common Sense of Technological Society

The *Journal of the University Film Producers Association* devoted an issue to film and photography at OSU in 1967, the cover of which featured two stills from the production of the Communication Series. Wagner wrote in the issue, "In a time of information explosion, the photographic image not only serves as an organizing agent, but also as a storage and retrieval system in situations where visual evidence is vital to the interpretation of events."[48] Images were seen as solutions, as mechanisms to interrupt the flow of information excess and to provide tools for judgement. Just as the Communication Series did not present its depiction of multimedia classrooms as contributing to the chaos of modern information, neither was image abundance part of the problem. "The university," Wagner continued,

"Review: *The Communication Revolution* and *The Information Explosion*." *Speech Teacher*, vol. 18, no. 4, November 1969, p. 329.

48 Wagner, Robert W. "A University Department of Photography and Cinema: An Environment for Seeing." *Journal of the University Film Producers Association*, vol. 19, no. 1, 1967, p. 4.

"is the logical home of studies in the photographic media as art forms, education, communication, and research." There, students had access to "an environment for seeing," meaning critical analysis and technical training.[49]

The enthusiasm for educational technology – and that "environment for seeing" – swelled through the 1960s. In 1971, AV purchases in the US hit $1.9 billion, an unprecedented sum, with substantial sales increases in filmstrips, non-theatrical film, tape cassettes, multimedia kits, AV programme control hardware, film production equipment, rear projection screens, and 8 mm and 16 mm motion picture projectors.[50] The National Education Association surveyed teachers in 1967 and showed that more than 80% had access to a range of AV materials, though only about 27% had access to 8 mm film projectors, 11% to closed-circuit television, and only 3% to computers. Actual use reported was about 75% for 16mm film projectors and more than that for maps, records, and silent filmstrips. Overhead projectors, sound filmstrips, programmed instruction materials, 8 mm film projectors, closed-circuit tv, and computers were all used much less frequently.[51]

The multimedia systems documented in Wagner's Communication Series sought to advance both access and usage, especially concerning the most technologically advanced iterations of AV technologies, teaching machines, closed-circuit television and computers. Educational technologists were not the only voices calling for the purchase and deployment of more AV materials. The corporations that developed, built, and sold these devices contributed immeasurably and banked on a significant expansion of the educational market. RCA, to take one example, constructed the "classroom of tomorrow" at the HemisFair '68 International Exposition, in San Antonio, Texas. Build around its Spectra 70 computer, the pavilion displayed the coordination and monitoring of computer learning, with individual student performance assessed. Not only were lessons delivered to consoles, and data gathered for analysis, but administrative tasks, like scheduling and payroll, could be handled. While RCA hoped technological wonders would sell its product, such pavilions equally promoted the dominant ideas about educational technology that were advanced following the NDEA of 1958. These ideas included corporate participation in schools, education via technological interface, central control of large student numbers, adaptability to

49 Ibid., p. 5.

50 "AV Market Reaches $1.9 Billion." *Journal of the University Film Association*, vol. 24, no. 4, 1972, p. 91.

51 "Instructional Resources in the Classroom." *NEA Research Bulletin*, October 1967. University of Maryland, College Park, National Public Broadcasting Archives, Series 5, Box 13, File 7, AECT Papers. Hereafter AECT Papers.

individual learning needs and pedagogical contexts, and, most importantly, inevitability of the information society.[52]

The merits of technologically enhanced teaching were, and continue to be, debated. Certainly, many wrong turns litter the field of educational technology. Reviewing the previous twenty-five years of educational experiments, Fred M. Hechinger in 1975 did not see significant change, but saw "entrenched power of the status quo" in which "Teaching machines are dead," open classrooms and voucher systems did not deliver their promises, and performance contracting drew for-profit content and hardware delivery schemes into classrooms. He commented, "[c]losed-circuit television failed to make a lasting impact" and cited the Airborne Television Project as little more than "a giant intellectual crop duster." He quoted an assessment from the Ford Foundation pamphlet *A Foundation Goes to School*: "[G]reat numbers of teachers were exposed to, and encouraged to use, overhead projectors, tape recorders, filmstrips etc., and in many cases these practices continue. In far too many instances, however, equipment of all kinds is gathering dust."[53]

Cold War technologically driven teaching, as organized by such advocates as Wagner, involved programmable multimedia systems, experience as a teaching objective, ideas that technological change was rapid and inevitable, and media as something students and teachers interacted with and made rather than passively received. Many of these decades-old ideas continue to be celebrated today as new, whether as concepts of interactivity, the naturalization of change, the inherent value of media making, or interoperability of media forms. In many ways, what was the "closed-world discourse" of the "electronic classroom" has since travelled to all corners of our contemporary context to become an ideologically laden common sense. When we study these discourses about educational media we see the intellectual and institutional labour that produced and solidified our hegemony of technological inevitability. Other forces contributed to what Toby Miller calls our "cybertarianism," especially economic ones, but those forces would never have achieved hegemonic status without the work of those advocating and circulating arguments that made this closed world imaginable and viable.[54] Indisputably, discursive

52 "RCA to Demonstrate Computer-Based System for 'Classroom of Tomorrow' at HemisFair '68." *RCA News*, January 23, 1968. Press release. University of Maryland, College Park, National Public Broadcasting Archives, series 5, box 13, file 7, 1–4, AECT Papers.

53 Hechinger, Fred M. "Where Have All the Innovations Gone?" *New York Times*, November 16, 1975, Section 11, p. 30.

54 Miller, Toby. "Cybertarian Flexibility – When Prosumers Join the Cognitariat, All That Is Scholarship Melts into Air." *Precarious Creativity: Global Media, Local Labor*, edited by Michael Curtin and Kevin Sanson. University of California Press, 2016, pp. 19–32.

and institution-building work of the post-WWII education technologists tilled the ground for the multimedia classrooms of today. Our educational era entwines technological imperatives, teaching and student performance measures, individual and self-guided lessons, mass course delivery, curricula driven by industrial sector labour demands, and for-profit corporate service of educational hardware and software. Many teachers and administrators find their work days oriented toward the management of technological systems, adapting to technological demands rather than having technological systems adapt to their work needs and capabilities. And the current stable place of these technologized educational practices is a direct product of the Cold War and its figuring of the school of tomorrow.

Wagner's truly extraordinary career deserves more consideration than has been afforded here. He drew attention to film and media preservation, hosting a turning point conference on the topic in 1968 and writing essays to alert us to the issue, well before it was the pressing issue it became.[55] His work on the education and media included a filmography of 170 films about film.[56] A career milestone, Wagner received the Eastman Kodak Gold Medal from SMPTE in 1981. This prestigious honour explicitly recognized his contributions to AV education, preservation, and the innovative modular design of the Communication Series.[57]

Wagner's series is but one outcome of the NDEA's Title VII, in the service of making the classroom media system, and is but one illustration of the dominant ideologies that circulated about the "electronic classroom." It shows us something very important: in our current research, we need to consider that *it isn't just films that work but multimedia systems*. Moveover, the integrated design of the teaching technologies we navigate today still privileges media making, media experience, and *both* individual and mass group learning. In Wagner's endeavours, we witness an expanding common sense about media and education, one whose priorities and biases ring through our digital world. For the generations that followed the work of the post-WWII educational technologists, the classroom media system and the acceptance of perpetual technological upheaval have only become more solidly hegemonic.

55 Wagner. "Oral History," p. 14; idem. "Preservation and Restoration of Photographic Materials through Nuclear and Ultra-Sonic Methods." *Illinois Public Libraries*, vol. 53, no. 1, January 1971; idem. "Motion Picture Preservation." *The American Archivist*, vol. 32, no. 2, Apr. 1969; idem. "Restoration of the Newsreels of President Harding." *Ohio History*, vol. 78, no. 2, Spring 1969.
56 Wagner, Robert W., and David L. Parker. *A Filmography of Films about Movies and Moviemaking*. Eastman Kodak, 1968.
57 Wagner. "Oral History," p. 19.

Works Cited

"AV Market Reaches $1.9 Billion." *Journal of the University Film Association*, vol. 24, no. 4, 1972, p. 91.

Acland, Charles R. "American AV: Edgar Dale and the Information Age Classroom." *Technology & Culture*, vol. 58, no. 2, Apr. 2017, pp. 392–421.

Acland, Charles R. "Curtains, Carts and the Mobile Screen." *Screen*, vol. 50, no. 1, 2009, pp. 148–66.

Acland, Charles R. *Swift Viewing: The Popular Life of Subliminal Influence*. Duke University Press, 2012.

Allen, William H., and John Flory. *Designing New Apparatus for Learning: A Survey of Needs for, and Developments in, Audiovisual Devices Suitable for Educational Uses*. Society of Motion Picture and Television Engineers, 1963.

Barksdale Clowse, Barbara. *Brainpower for the Cold War: The Sputnik Crisis and National Defense Education Act of 1958*. Greenwood Press, 1981.

Cain, Victoria. *Schools and Screens: A Watchful History*. MIT Press, 2021.

Dale, Edgar. *Audio-visual Methods in Teaching*. Dryden Press, 1946.

Eucational Films. University of Michigan, Media Resources Center, Audio-visual Education Center, Ann Arbor, 1973.

Edwards, Paul N. *The Closed World: Computers and the Politics of Discourse in Cold War America*. MIT Press, 1996.

Erlich, Michael G. "Review: *The Communication Revolution* and *The Information Explosion*." *Speech Teacher*, vol. 18, no. 4, 1969, p. 329.

Hechinger, Fred M. "Where Have All the Innovations Gone?" *New York Times*, November 16, 1975, Section 11, p. 30.

"Instructional Resources in the Classroom." *NEA Research Bulletin*, October 1967. University of Maryland, College Park, National Public Broadcasting Archives, series 5, box 13, file 7, AECT Papers.

Keyes, Elizabeth A. "Review: *Communication Conference* and *The Process of Communication*." *Speech Teacher*, vol. 18, no. 4, 1969, p. 328.

Klein, Christina. *Cold War Orientalism: Asia in the Middlebrow Imagination, 1945–1961*. University of California Press, 2003.

McCarthy, Anna. *The Citizen Machine: Governing by Television in 1950s America*. Duke University Press, 2010.

Miller, Toby. "Cybertarian Flexibility – When Prosumers Join the Cognitariat, All That Is Scholarship Melts into Air." *Precarious Creativity: Global Media, Local Labor*, edited by Michael Curtin and Kevin Sanson. University of California Press, 2016, pp. 19–32.

"NDEA Title III: Fiscal Years 1959–67: A Management View." US Department of Health, Education, and Welfare, Office of Education, Washington, DC, May 1969, p. 109–11.

"NDEA Title III Guidelines." US Department of Health, Education, and Welfare, Office of Education, Washington, DC, January 1965 (rev. April 1967), p. v.

"Projects Initiated under Title VII, Part B, National Defense Education Act, September 2, 1958–June 30, 1963." US Department of Health, Education, and Welfare. US Office of Education, 1963.

Ragsdale, Donald J. "Review: *Perception and Communication*." *Speech Teacher*, vol. 18, no. 4, 1969, p. 329.

"RCA to Demonstrate Computer-Based System for 'Classroom of Tomorrow' at HemisFair '68." *RCA News*, January 23, 1968. Press release. University of Maryland, College Park, National Public Broadcasting Archives, series 5, box 13, file 7, 1–4, AECT Papers.

"Report on the National Defense Education Act, Fiscal Year 1963." US Department of Health, Education, and Welfare. Government Printing Office, 1964, p. 55.

"Review of Activities under National Defense Education Act." Government Printing Office, 1960, p. 6.

"Robert W. Wagner: Obituary." *Columbus Dispatch*, February 9, 2011.

Rose, Ernie. "UFVA Mourns the Loss of Robert W. Wagner." University Film & Video Association, March 9, 2011.

Schwartz, Vanessa R. *It's So French! Hollywood, Paris, and the Making of Cosmopolitan Film Culture*. University of Chicago Press, 2007.

Schwoch, James. *Global TV: New Media and the Cold War, 1946–69*. University of Illinois Press, 2009.

Spigel, Lynn. *Welcome to the Dreamhouse: Popular Media and Postwar Suburbs*. Duke University Press, 2001.

Tracey, Monica W., and Jill E. Stefaniak. "MPATI: The Midwest Profram on Airborne Television Instruction (1959-1971)." *International Journal of Designs for Learning*, vol. 5, no. 2, 2014, pp.29-33.

Turner, Fred. *The Democratic Surround: Multimedia and American Liberalism from World War II to the Psychedelic Sixties*. University of Chicago Press, 2013.

Wagner, Robert W. "The Educational Film in Transition." *Audiovisual Instruction*, vol. 9, 1964, p. 172.

Wagner, Robert W. "Films about Communication." *Audio Visual Communication Review*, vol. 3, no. 3, 1955, pp. 213–19.

Wagner, Robert W. "Films without Endings." *Audio Visual Communication Review*, vol. 3, no. 2, Fall 1955, pp. 138–43.

Wagner, Robert W. "The Formula Film." *Audio Visual Communication Review*, vol. 3, no. 1, Winter 1955, pp. 53–57.

Wagner, Robert W. "A Galaxy of Motion Picture Documents on Communication Theory and the New Educational Media." Department of Health, Education, and Welfare, US Office of Education, Washington, DC. 1966.

Wagner, Robert W. "Modular Design for a Series of Films on Communication Theory and the New Educational Media for Use in Teacher Education." *Single Concept*

Film Clip Project – Final Report, edited by Elwood E. Miller. Office of Education, Bureau of Research, Washington, DC, 1967.

Wagner, Robert W. "Motion Picture Preservation." *American Archivist*, vol. 32, no. 2, 1969, pp. 125–32.

Wagner, Robert W. "New Dimensions in Audio-visual Design." *Audio Visual Communication Review*, vol. 4, no. 2, 1956, pp. 142–47.

Wagner, Robert W. "Oral History." Interviewed by Virginia Tiefel. Ohio State University Archives, February 1, 2001.

Wagner, Robert W. "Preservation and Restoration of Photographic Materials through Nuclear and Ultra-Sonic Methods." *Illinois Public Libraries*, vol. 53, no. 1, 1971.

Wagner, Robert W. "Restoration of the Newsreels of President Harding." *Ohio History*, vol. 78, no. 2, 1969.

Wagner, Robert W. "A Series of Motion Picture Documents on Communication Theory and the New Educational Media. Final Scripts." Ohio State University, 1966.

Wagner, Robert W. "The Teacher and Technology: The Bottom Line." *Techtrends*, January–February 1998, pp. 37–42.

Wagner, Robert W. "Thirty New University-Produced Films." *Audio Visual Communication Review*, vol. 4, no. 4, 1956, pp. 334–39.

Wagner, Robert W. "Three Films Worth Seeing." *Audio Visual Communication Review*, vol. 7, no. 2, 1959, pp. 138–41.

Wagner, Robert W. "TV and University Motion Picture Production." *Journal of the University Film Producers Association*, vol. 4, no. 3, 1952, pp. 7–9.

Wagner, Robert W. "A University Department of Photography and Cinema: An Environment for Seeing." *Journal of the University Film Producers Association*, vol. 19, no. 1, 1967, p. 4.

Wagner, Robert W. "The University Film Producers Association: A Report on the Organization and Its Recent Convention." *Audio Visual Communication Review*, vol. 2, no. 4, 1954, pp. 299–303.

Wagner, Robert W., and David L. Parker. *A Filmography of Films about Movies and Moviemaking*. Eastman Kodak, 1968.

Zryd, Mike. "Experimental Film as Useless Cinema." *Useful Cinema*, edited by Charles R. Acland and Haidee Wasson. Duke University Press, 2011, pp. 315–36.

About the Author

Charles R. Acland is Distinguished University Research Professor in the Department of Communication Studies, Concordia University, Montreal. His most recent book is *American Blockbuster: Movies, Technology, and Wonder* (Duke University Press 2020).

15 We Must Know More Than We Can See

Images for Vocational Training and the Emergence of Cognitive Ergonomics

Guilherme Machado

Abstract
When labour moves out of the visible realm, new visual techniques for transmitting labour-related knowledge appear to keep production systems running. This chapter examines some corporate practices of knowledge transmission in the era of post-industrial labour in France. The emergence of cognitive ergonomics in the last decades of the twentieth century led to a reconfiguration of visual techniques for vocational education, which have gradually replaced systems of education based on visual prescriptions. Henceforth, learners no longer stand before the images of labour to endure the constraint of their truth. Educational and organisational images of labour now frequently have the function of *situating* workers in learning environments with carefully planned aesthetic affordances, so that the design of efficiency becomes the workers' own daily concern. The chapter presents a few examples of such dispositives for transmitting labour-related knowledge since the 1990s, particularly in the French energy production company Électricité de France.

Keywords: knowledge; transmission; film; vocational education; post-industrial discipline

The Inscription of Labour

The emblematic project of the European Enlightenment, Denis Diderot's *Encyclopédie*, includes an early visual and literary collection of artisanal knowledge. The books lay out hundreds of texts and engravings depicting the procedures, tools and gestures of various trades. Along with the Royal

Hediger, V., F. Hoof, Y. Zimmermann, with S. Anthony (eds), *Films That Work Harder: The Circulation of Industrial Film*. Taylor & Francis Group 2024
DOI: 10.5117/9789462986534_CH15

Academy of Sciences' *Description des arts et métiers*, a similar project initiated at the end of the seventeenth century and published nearly at the same time as the *Encyclopédie* in the 1760s,[1] Diderot's enterprise was among the first significant scientific programmes against the secrets of artisanal expertise: a successful attempt to collect and store labour-related knowledge on easily transportable and technically reproducible media. Through the methodical analysis and transcription of aspects of labour that were worthy of interest, the *Encyclopédie* aimed to provide an educated audience with a didactic means of understanding labour.

Diderot's endeavour to record the knowledge involved in the procedures, machines and performance of artisanal trades stands at the threshold of the industrial age. It was not exempt of political intent. The forces that led to the completion of the *Dictionary of Arts and Crafts* were not unlike those behind Adam Smith's plea for the release of artisanal knowledge from the custody of guilds. Smith extolled the importance of knowledge circulation beyond the associations of artisans. Such corporate hogging of knowledge, he contended, restricted access to the labour market, promoted idleness and slowed down innovation within the trades.[2] A great admirer of the *Encyclopédie*, Smith shared the conviction that the secrets of labour should be revealed in lightweight materials which could make the rounds of the reading rooms of enlightened nations.

In fact, the methods of technical instruction that sprang up from the early days of industrialisation are seen by historians of education as the institutional heirs to the first encyclopaedic projects. The Conservatoire National des Arts et Métiers, for instance, a French establishment for the promotion of education and research into industrial technical innovation, was inaugurated in 1794 by Abbé Grégoire as a bank of tool and machinery models, with a collection of 3,335 drawings and replicas related to a profusion of trades.[3] As in other engineering schools emerging in the period, instruction in industrial activities was carried out with collections of reliable depictions

1 On the controversial relationship between the plates of the *Encyclopédie* and those of the Academy of Sciences, see Huard, Georges. "Les planches de l'*Encyclopédie* et celles de la *Description des Arts et Métiers* de l'Académie des Sciences." *Revue d'histoire des sciences*, vol. 4, no. 3–4, 1951, pp. 238–49. See also Proust, Jacques. *Diderot et l'Encyclopédie*. Armand Colin, 1967, especially the second chapter.

2 See Smith, Adam. *An Inquiry into the Nature and Causes of the Wealth of Nations* [1776]. The Modern Library, 1937, especially pp. 118–23.

3 Picon, Antoine. "Connaissances techniques sous les Lumières." *Le conservatoire national des arts et métiers au cœur de Paris: 1794–1994*, edited by Claudine Fontanon et al. CNAM, 1994, pp. 69–73.

of good labour practices and machines, the *dictionnaires raisonnés* that made labour explicit through textual and visual inscriptions.[4]

In the twentieth century, with the generalisation of vocational education, such stable depictions of labour became crucial in the institution of industrial discipline. The work efficiency engineer Frederick W. Taylor became known for paving the way for teams of non-skilled outsiders to acquire theoretical control over the practical knowledge needed to keep industrial production running. His management system still relied on paper inscriptions for assembling, storing and communicating labour-related knowledge: "The managers assume [...] the burden of gathering together all of the traditional knowledge which in the past has been possessed by the workmen and then of classifying, tabulating, and reducing this knowledge to rules, laws, and formulae which are immensely helpful to the workmen in doing their daily work."[5] For Taylor, the production of educational material was essential to palliate the scarcity (and the cost) of expertise, as it provided a pedagogical control over the workforce renewal process. Current lean strategies of industrial organisation continue to use visual "labour standards" in vocational training, i.e. collections of photographic and textual depictions of tasks, including gestures, tool positioning and machine handling. In order to make production levels programmable, such detailed depictions of the *modi operandi* of industrial work have proved to be effective and sustainable tools of management.[6]

Throughout most of the last century, textual and visual inscriptions remained sites where good labour practice was made explicit for apprentices in industrial work. They required from trainees mimetic adaptation to repeat the displayed truth of safe and efficient performance. Indeed, the evolution of visual technologies since the eighteenth century has enabled an ever-more comprehensive *visual display* of labour. When, in 1750, Diderot justified the methods of his titanic enterprise, saying, "one would never finish

4 The term "explicit" must be understood here not only as "explained," but also in a phenomenological sense, as the enlightened appearance of labour. For a recent definition of this term ("*Explizit*") with a critical focus on phenomenology, see Sloterdijk, Peter. *Sphères III: Écumes*. Maren Sell, 2005, esp. pp. 57–78, 184–202, with a focus on Bruno Latour's concept of *articulation*. Bruno Latour took up Sloterdijk's concept of *Explikation* (hereafter translated as *explication*) in his essay "A Plea for Earthly Sciences." *New Social Connections: Sociology's Subjects and Objects*, edited by Judith Burnett, Syd Jeffers, and Graham Thomas. Palgrave Macmillan, 2010, pp. 72–84.

5 Taylor, Frederick W. *The Principles of Scientific Management*. Harper & Brothers, 1911, p. 36.

6 On the application of the lean manufacturing system at the Peugeot-Citroën group and the place of the "pictorial ranges" ("*gammes imagées*") in this system, see Morais, Alexandre, and Raynald Aubineau. "Articulation entre l'ergonomie et le lean manufacturing chez PSA." *Activités*, vol. 9, no. 2, 2012, pp. 179–97.

if one proposed to render in figures all the states through which a piece of iron passes before being transformed into a needle,"[7] he was certainly not trying to convince an audience acquainted with image technologies operating at a rate of twenty-four "figures" per second. The evolution of labour inscription technologies would culminate with the appearance of film media, a set of advanced techniques which greatly facilitated the recording of and access to significant amounts of data at a time when even the subtlest details of labour – understood as bodily performance in tandem with machines – had to be made visually explicit in some way, if they were to be effectively communicated as collective references of performance.

In the following pages, the eventual erosion of the encyclopaedic system to enhance the transmission of labour-related knowledge by using collections of enlightening inscriptions will be tackled in the context of the emergence of cognitive ergonomics and its impact on vocational education. The recent obsolescence of the body and its muscles as means of production due to the generalisation of cybernetic systems in industry has rendered the encyclopaedic didactics ineffective for transmitting knowledge that cannot be recorded through descriptive geometry or photosensitive apparatuses. As industrial productivity began to demand cognitive rather than physical performance, knowledge transfer techniques were adapted to ensure new flows of skills and innovation. The circulation of labour-related knowledge in the cybernetic age is beginning to rely on collections of images that are quite different from those of the vocational schools of previous centuries. These images are part of a diverse, otherwise effective strategy for making labour explicit, another educational solution for administering the skills that keep industrial production running.

Thermodynamic Labour vs Activity in Situation

The encyclopaedic method of making labour explicit is characterised by the extent of its presence in the visible field. For the encyclopaedist, explaining labour means opening it up to sight. There is no explication that is not at the same time an assent that truth can fit on paper (later on celluloid or electronic screens) to be exposed to the intelligent eyes of readers. For Roland Barthes, the picture in the *Encyclopédie* is "a sort of rational synopsis: it not only illustrates the object or its path [in the production process], but also

7 Qtd. in Pannabecker, John R. "Representing Mechanical Arts in Diderot's 'Encyclopédie'." *Technology and Culture*, vol. 39, no. 1, 1998, pp. 33–73.

the very mind that conceives it."[8] In the spirit of the encyclopaedia, the ideal reader is the one whose mind espouses visual information. Similarly, in the twentieth-century workshop, apprentices are successful when they reproduce predetermined labour standards with minimal deviation. In other words, workshop apprentices, like the readers of the *Encyclopédie*, are engaged in the practice of acquiring knowledge that is complete in the inscription. Readers and apprentices are expected to see real practice as a field of faithful reproduction of the intelligible systems they have previously observed on paper or other stable, authentic inscription.[9]

Encyclopaedic didactics aligned well with the thermodynamic conception of labour as the transformation of energy into visible performance. This has been a predominant conception since the generalisation of the combustion engine in industrial processes in the nineteenth century. As Anson Rabinbach has shown, the thermodynamic conceptualisation of energy as an abstract universal force, which explained the functioning of the motor as well as that of the planets and the human body, led the sciences and politics of the nineteenth and twentieth centuries to a "transcendental materialism,"[10] according to which labour denoted any material manifestation of the universal substance that is energy. Labour is based on the fundamental laws of energy conservation and dissipation; it can therefore be calculated and managed independently of contextual idiosyncrasies, social situations or political issues, insofar as it is considered a physical, "natural" event.

Since the 1950s, however, the effects of automation on industrial production have considerably extended the category of labour beyond the visible spectrum. The generalisation of robots and computers in industrial processes resulted in the replacement of physical performance in production by abstract systems. Rabinbach concludes his essay by recalling the various accounts on the disappearance of human labour as conceived by thermodynamics:

> Organized around the substantiality of the muscles and nerves, and around the material objects that the body in tandem with machines

8 Barthes, Roland. "Image, raison, déraison." *L'univers de l'Encyclopédie*, edited by idem et al. Les libraires associés, 1964, pp. 10–16. My trans.

9 In Foucault's words, in the *épistémè* of the classical age, "l'être du sens est tout entier du côté du signe." Foucault, Michel. *Les mots et les choses. Une archéologie des sciences humaines*. Gallimard, 1966, p. 80.

10 Rabinbach, Anson. *The Human Motor: Energy, Fatigue, and the Origins of Modernity*. University of California Press, 1990, see especially the first two chapters.

> transformed into products, nineteenth-century labour was being rapidly superseded by the work of technologies operating on and proliferating through abstract systems of knowledge. Technology was making corporal work obsolete.[11]

In his analysis of two corporate films where the visual representation of labour as data-processing seemed to have been a real problem, Vinzenz Hediger showed in the first volume of *Films That Work* how the persistence of traditional instruments of corporate communication in the cybernetic age might be condemned to the reproduction of "thermodynamic kitsch". Since the computer introduced a new paradigm of productivity, according to which labour and its outputs can no longer be seized in the visible realm, the typical thermodynamic rhetoric based on light, motion, volume and other photographic attractions lost its relevance. As Hediger says, "[t]he emergence of information technology provokes a shift in what you might call the visual rhetoric of industrial production."[12]

That shift affected vocational training practices and industrial disciplinary systems. When labour is essentially the transformation of energy into visible performance according to laws that not everybody can know, one can still count on visual standards of efficiency and the endurance of workers to coordinate labour. But what happens when labour leaves the province of the observable? How to prescribe and monitor operations that lurk in the inner reaches of the brain and the hardware? What is the use of training techniques such as photography and film when they cannot capture and open up to the sight of novices the performances that ensure productivity?

One of the alternatives that emerged at the turn of the century in an attempt to avoid the thermodynamic kitsch in vocational education is the reconfiguration of audiovisual-based training methods, and the consequent shift of sites where labour is made explicit in the knowledge transmission process. For the sake of instructional efficiency, visual inscriptions have been denied the onto-administrative role of revealing to workers what good labour is. Instead, they have been assigned the role of stimulants in

11 Ibid., p. 297.

12 Hediger, Vinzenz. "Thermodynamic Kitsch: Computing in German Industrial Films, 1928/1963." *Films That Work: Industrial Film and the Productivity of Media*, edited by Vinzenz Hediger and Patrick Vonderau. Amsterdam University Press, 2009, p. 128. For another account on the obsolescence of the body as the visible vehicle of labour, see Vinzenz Hediger. "Body Rebuilding. Körper und Arbeit an der Schwelle zum kybernetischen Zeitalter." *Prometheische Kultur. Wo kommen unsere Energien her?*, edited by Claus Leggewie et al. Wilhelm Fink Verlag, 2013, pp. 195–222.

endless processes in which the workers themselves are committed to design effective cognitive performance.

The attention paid to the invisible realm of mind in contemporary practices of vocational training in France stems from recent labour sciences, especially ergonomics and work psychology. To understand the role of visual inscriptions in vocational education today, it is therefore necessary to take into consideration the advance of the concept of *activity* over that of *behaviour* in labour analysis. Since the 1970s, Western psychologists have lifted the embargo on Soviet theories of activity. Authors such as Lev Vygotsky and Alexei Leontiev became common references among work psychologists in Europe and America. Their theories offered relevant analytical frameworks for the problems posed by the invasion of information technology in all industrial sectors. Regarding the problem under consideration here, suffice it to say that activity, according to these theories, is not a phenomenon devoid of situational idiosyncrasies. The activity of a supervisor in a nuclear power plant monitoring screens and light indicators is hardly comparable to that of an accountant updating her files in a private company. Instead of manifestations of the same universal laws, activities are always the result of subjective interpretations of singular situations. In the terms of a French work psychologist, it is not enough to think of activity as observable behaviour, for it is also "a field of possibilities and impossibilities, where what one cannot do, what one would have liked to do, what one wishes to do, counts as much as what one actually does."[13] Ergonomists like to define activity as what is *subjectively put into play* in order to perform tasks. The analysis of activity must then be carried out on the "actual work situations," where the analyst is able to pinpoint the specific interactions that operators have with technical objects and colleagues – so as to identify through the visible aspects of activity "the unobservable mechanisms"[14] of behavioural regulation.

Analyses of activity do not, in any way, lead to behavioural prescriptions. Where the latter exist, they are part of the object of analysis, given that activities are nexuses of objects, prescribed procedures and goals, assigned values, motives, inhibitions, worries and patterns of attention in the arena

13 Clot, Yves. "Qu'est-ce que l'activité dans l'analyse du travail?" *Performances humaines et techniques* [special issue], 1995, pp. 2–6. My trans.; see also idem. *La fonction psychologique du travail*. Presses Universitaires de France, 1999, pp. 119–29.

14 Leplat, Jacques. "A propos des compétences incorporées." *Education permanente*, no. 123, 1995, pp. 101–14; on the ergonomic approach to activity, see also idem, and Jean-Michel Hoc. "Tâche et activité dans l'analyse psychologique des situations." *Cahiers de psychologie cognitive*, vol. 3, no. 1, 1983, pp. 49–63.

of work in which actors are physically as well as psychologically situated. The analyst of activity takes any workplace as a setting of precepts, assumptions and sensations. This setting includes most of the conditions for the accomplishment of productive labour.[15] Accordingly, activity theory has led mainstream labour analysis to consider that labour-related knowledge is distributed in the specific play of interactions one maintains in a given milieu, which is in constant variation, but which also has its generic features for each specific occupation. For ergonomists and psychologists, the study of the observable part of these interactions is central to the diagnosis of work problems and deficiencies, as it provides access to the knowledge that is "encapsulated in action,"[16] i.e. tacit and context-related knowledge incorporated by experience. Whether one calls it skill, expertise, know-how or aptitude-to-do-the-job is at this level just a matter of terminological taste. The fact that the psychological analysis of activity has defined *situations* as means of access to the control rooms of cognitive performance should be hailed as an important inaugural act of post-industrial labour governance. Instead of making labour instruction and control viable through the production of behavioural standards, these theories opened the way to the production of *situational standards* for transmitting cognitive performance as individual adaptations to well-designed labour environments.

From the point of view of vocational training, this means that efficient knowledge transfer techniques can be conceived as systems of *mise en situation*. Occupations have been envisaged by occupational didacticians in France as classes of generic situations, while competent professionals are those capable of calculating relevant, safe and efficient actions autonomously for as many situational variants as possible within their trades.[17] The more situations shop sellers experience – in terms of sales targets, clients' moods and demands, environmental sets, stress conditions etc. – the more resourceful they are in their activity. Expertise grows as one develops cognitive and even emotional means of managing increasingly rich and complex labour situations.

In this context, one important reason why film techniques have been assigned to capture and communicate situational standards of labour is

15 Activity's situation as a "setting" is conceptualised by Jean Lave in *Cognition and Practice: Mind, Mathematics and Culture in Everyday Life*. Cambridge University Press, 1988, pp. 149–52. Lave's concept has been used in work psychology; see, for example, Clot. *La fonction psychologique du travail*.

16 Leplat. "A propos des compétences incorporées."

17 See Pastré, Pierre, et al. "La didactique professionnelle." ***Revue française de pédagogie***, no. 154, 2006, pp. 145–98.

the fact that they provide access to *actual* labour situations. While for work psychologists film techniques are used to analyse "real activity,"[18] for trainers, they help to reduce the mismatch between actual occupations and training. They satisfy corporate requirements for the design of vocational training in authentic work environments. It would be wrong, however, to assert that the current effectiveness of audiovisual techniques in transmitting labour-related knowledge relies essentially on the situations they depict. While collectively displaying common situations of production, they demonstrate their virtuosity rather by regulating dynamics of verbal enunciation.

Labour, from the Horse's Mouth

One of the major sponsors of the audiovisual avant-garde of vocational education in France is the country's largest electricity supplier. Since the 1990s, when the mass retirement of the baby-boomer generation posed the problem of the transfer of highly qualified workers' knowledge, Électricité de France (EDF) commissioned research into techniques of analysis of its activities.[19] The challenge of preserving and communicating experts' "intangible cultural heritage,"[20] which started within the company, has since been stimulating a group of international researchers.[21] Outlining a successful method of labour analysis developed in this context should help give a clear idea of the role played by audiovisual techniques in some contemporary training practices.

EDF's investment in new instruments and protocols for the capture and transmission of labour-related knowledge can be understood as an example of the above-mentioned shift in the visual rhetoric of industrial

18 One of the principles of cognitive ergonomics and post-behaviourist work psychology is the attitude of starting the analysis from "real activities," instead of observing the effects of *stimuli* introduced by the analysts in the work process. Video cameras are then often used as tools for capturing "real activity," as it occurs before any intervention by the analyst. See, for instance, Clot et al. "Entretiens en autoconfrontation croisée: une méthode en clinique de l'activité." *Education permanente*, no. 146, 2001, pp. 17–25.

19 See Lahlou, Saadi. "Observing Cognitive Work in Offices." *Cooperative Buildings: Integrating Information, Organizations and Architecture*, edited by Norbert A. Streitz et al. Springer, 1999, pp. 150–63.

20 Le Bellu, Sophie, et al. "Studying Activity in Manual Work: A Framework for Analysis and Training." *Le travail humain*, vol. 79, no. 1, 2016, pp. 7–30.

21 The research on dispositives developed in EDF was later carried out by a group of researchers at the London School of Economics under the title Subjective Evidence Based Ethnography (SEBE).

production. As one of the researchers engaged in the project to collect, study and transfer expert know-how asserted: "The classic observation of behaviours is not enough anymore. It becomes necessary to come up with new techniques to analyse the flow of activities."[22] In the research project, the goal was to succeed in "externaliz[ing] both tacit and explicit know-how" while giving a "suitable educational format"[23] to the results of the research. The analysis of activity runs through the entire educational dispositif set up by the researchers. It even enters the training rooms, where activities find their final analysts in those looking to make their way into the labour arena, i.e. the trainees themselves.

The chief instrument of activity analysis developed in EDF is the subcam, a miniature subjective-camera with a wide-angle lens and a stereo microphone. It is designed to capture the perceptual situation of expert operators, who do not shy away from recording their own activities when it comes to aiding in the education of young apprentices. The researchers and designers of the educational dispositif assert that contemplating an operational milieu from an expert's point of view is an experience with a high referential identification value: "empathy occurs based on *sharing similar phenomenological experience*."[24] To ensure accuracy, the camera is usually placed next to the expert's eyes (normally attached to a pair of spectacles), so that the resulting images provide an optical and auditory point of view as close as possible to that of the expert operator.[25]

Subcam images provide a sensorial immersion in actual labour situations. In contrast to the classic prescriptive display of tasks to be performed, they reveal in detail the operators' spatial and temporal organisation, their solutions to problems that are not covered by formal instructions, and especially their attentional dynamics at work.[26] When expert operators follow a procedure, subcam images display not only the buttons, levers

22 Le Bellu, Sophie. *Capitalisation des savoir-faire et des gestes professionnels dans le milieu industriel. Mise en place d'une aide numérique au compagnonnage métier dans le secteur de l'énergie*. PhD thesis, Université Bordeaux 2, 2011, p. 104. My trans.

23 Le Bellu et al. "Studying Activity in Manual Work," p. 8.

24 Lahlou, Saadi. "Digitization and Transmission of Human Experience." *Social Science Information*, vol. 49 [special issue: "Digitize and Transfer"], 2010, pp. 291–327. Author's emphasis.

25 Since the 1990s, there have been several variants of this type of apparatus, as well as a number of uses in industrial, sporting and military contexts. See, for instance, the studies in decision-making in medical environments: Omodei, Mary M., and Jim McLennan. "Studying Complex Decision Making in Natural Settings: Using a Head-Mounted Video Camera to Study Competitive Orienteering." *Perceptual and Motor Skills*, vol. 79, no. 3, 1994, pp. 1411–25.

26 Sophie Le Bellu says that for research on manual work, subcam images provide sufficient information on the user's focus of attention. However, when it comes to record interactions with

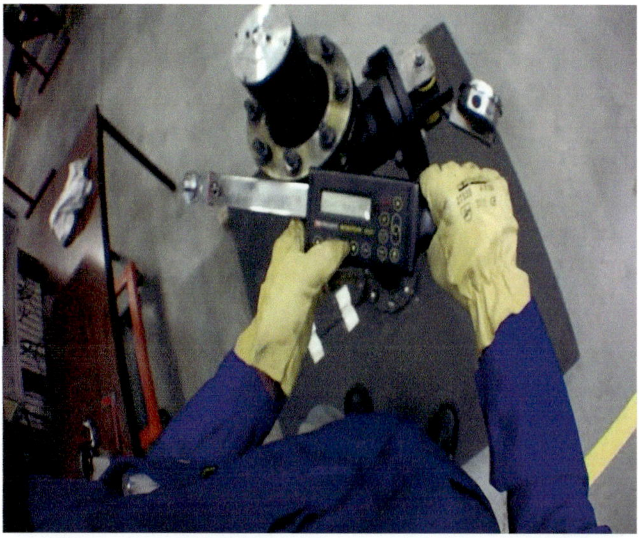

Fig. 15.1. Example of a subcam user's point of view. Image provided by Saadi Lahlou.

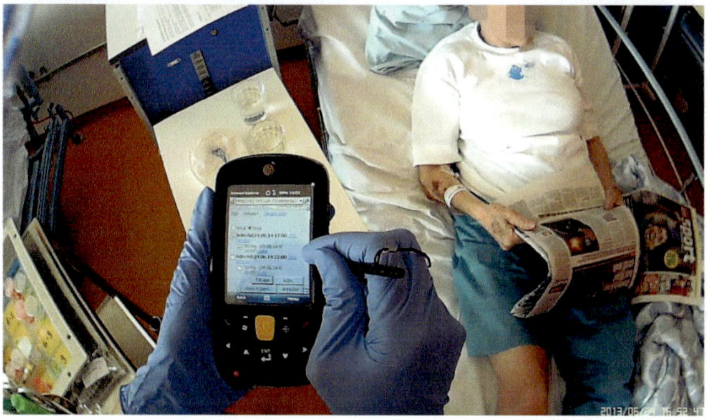

Fig. 15.2. Example of a subcam user's point of view. Image provided by Saadi Lahlou.

and valves that have to be pushed, pulled and turned, but also secondary elements of the operational milieux, which help operators to find their gestural bearings. While the prescribed procedure is limited to stating which files must be opened and which modifications made, the subcam captures in real-time the difficulties posed by inopportune phone calls, the accumulation of files, and the unexpected behaviour of software. Crucially, subcam images reveal the experts' tricks to get around the problems, their inventiveness in coping with the unexpected, and their time management strategies to accomplish their tasks. In a word, subcam images provide access to what work psychologists call "real activity," which differs from

computer screens, the eye-tracker may be necessary. See Le Bellu. *Capitalisation des savoir-faire et des gestes professionnels dans le milieu industriel*, p. 123.

procedural formulae in that it refers to a series of contingent, conscious and unconscious phenomena in the work environment that contribute to the subjective organisation of activity.

But the virtue of this labour archiving tool is not only that it gives the unique opportunity to experience workplaces through the eyes of experts. What makes it so valuable to researchers seeking to solve the problem of post-industrial knowledge transmission appears in a second stage of the analysis, after the first data collection. The "replay interview" is a phase in which a first cut of the subcam's footage – retaining only "the best moments" of one or several days of recorded activities for future instruction – is screened to the expert whose activities were recorded. Here, subcam images bring operators face to face with their own operations in progress, and are deemed to manifest a "remarkable capacity [to remind the experts of their] actual mental states during activity."[27]

Audiovisual self-confrontation is a technique for providing awareness of the streams of consciousness during activity, too subtle for the operators to consciously be aware of when immersed in their labour routines. Generally, it is claimed, workers are "unaware of the amazing psycho-motor expertise they display."[28] In the course of activity, they deploy complex networks of goals and cognitive processes to deal with sequences of tasks conducted often simultaneously. These networks of goals and cognitive processes result from the memory of past experiences and the ability to predict situations; they do not always have a reference in the visible field, although they play a major role in supporting action: "The aircraft pilot is aware of the position of the undercarriage; the train driver is aware of his passengers – even though he does not see them; and in fact, these are important elements in their decisions."[29]

The replay interview is therefore a moment of detachment from the pressing concerns of being-at-work. It offers an occasion for self-examination. It is designed to induce verbal enunciation of "the various layers of goals and motives that lie behind the observed action."[30] The key role of the interviewer in this phase is to shift the focus of the expert operators' attention and prompt them to use their words to exhaust the meanings and affects behind their actions on screen. The interview is recorded by a fixed camera focusing

27 Lahlou. "Digitization and Transmission of Human Experience," p. 308.

28 Idem. "How Can We Capture the Subject's Perspective? An Evidence-Based Approach for the Social Scientist." *Social Science Information*, no. 50, 2011, p. 633.

29 Idem. "Digitization and Transmission of Human Experience," p. 304.

30 Idem. "How Can We Capture the Subject's Perspective?" p. 621.

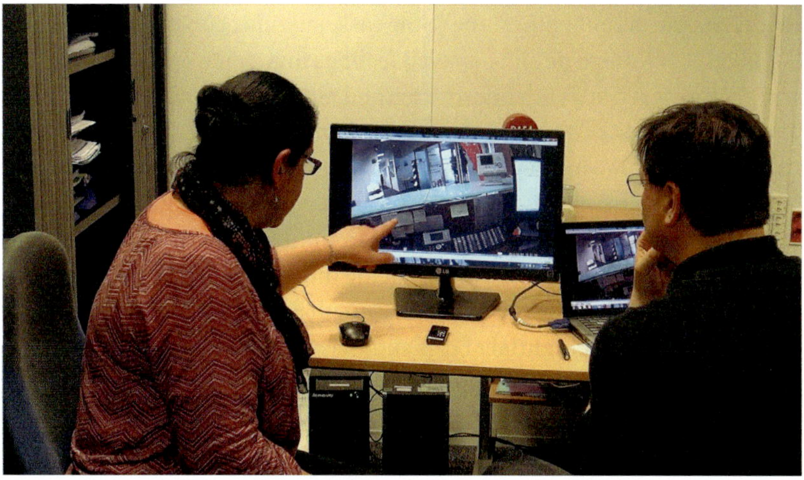

Fig. 15.3. The replay interview. Image provided by Saadi Lahlou.

the expert operator and the screen as the first describe their reasoning, or by a subcam carried by the interviewed expert. This record is intended to help the researchers retrospectively identify the visible actions referred to by the operators while they were reporting their streams of consciousness. New editing of the video may follow, with the operators' voice-over taken from the interviews as additional sound layers – adding to the explication of their labour.

Replay interviews, as subcam footage, can be assembled and disassembled, gathered in a single multi-layered audiovisual document showing the entire deployment of a complex procedure, or isolated in short scenes to form different stages of instruction. Images coming from the record of different expert operators' activities can be compared or combined, so one scene can complete another. The Multimedia Platform for Apprenticeship is a digital platform developed for an organised storage of data from this image production system. It aims to facilitate access by apprentices to large amounts of pedagogical data in nuclear power plants, where the platform was originally launched. A digital interface allows the users to navigate the platform. Ordinary operations in the plants are broken down into goals and subgoals, as they were made explicit by experts, with detailed comments "from the horse's mouth"[31] on how different sorts of events were interpreted and the emotions they provoked. Users interested in specific operations can compare different qualified operating modes, time-saving and stress management strategies and other expert tips stored in the platform. The abundance of operating

31 Idem. "Digitization and Transmission of Human Experience," p. 309.

modes suggests that behind the same task, singular subjective activities, distinct sets of subgoals and personal feelings may come up that reflect the way each expert deals with information and manages to accomplish the task. The accumulation of data allows a thorough investigation of work situations, with the aim of raising awareness about effective ways to organise action: "We extracted and analysed systematically 101 interruptions during office work by a visitor or by phone, retrieved from fifty hours of subfilms. This allowed us to better understand how being interrupted was handled by the subjects and how it affected their subsequent activity."[32]

The educational potential of this audiovisual pedagogy is praised by its designers, who state their ambition to reform traditional training methods: "The aim of [the Multimedia Platform for Apprenticeship] is not to offer a single, correct way of performing an action, but to provide a new way of representing the professional experience by enabling a novice to share the real-time experience of an expert through the first-person perspective [...] and accompanying explanations."[33] In this way, the project brings the good news that workers have moved from their unpleasant status of pure research objects to that of participants in labour analysis and design: "[Participants] intensely contribute to the interpretation of collected data"; as "research collaborators," they are "given the choice of staying anonymous or being credited" in the scientific publications.[34] This cooperative knowledge-building principle, closely akin to contemporary management practices, bears witness to a conception of labour analysis that no longer leads to fixed behavioural inscriptions, hierarchical structures of power, and traditional media of knowledge communication: "Indeed the construction, maintenance and use process of the new information tools seem to involve inputs from all the stake-holders [sic] in continuous work in progress, in sharp contrast to the former 'book-publication' mode."[35]

Moreover, in such educational systems of image production and visualisation, not only do expert workers state what good labour is in the process of knowledge transmission, but apprentices are also actively involved in the labour design process via the accumulation and comparison of audiovisual content. The exploration of image databases for training, whether in video, game, or other digital formats, is usually accompanied by debriefing

32 Idem. "How Can We Capture the Subject's Perspective?" p. 631.

33 Le Bellu et al. "Studying Activity in Manual Work," p. 25.

34 Quotes from Lahlou, Le Bellu, and Sabine Boesen-Mariani. "Subjective Evidence Based Ethnography: Method and Applications." *Integrative Psychological and Behavioural Science*, vol. 49, no. 2, 2015, p. 218.

35 Lahlou. "Digitization and Transmission of Human Experience," p. 298.

interviews designed to encourage apprentices to formulate their own alternatives for dealing with common labour situations. Indeed, watching images of labour appears to be an opportunity for personal expression, for making judgments in the light of extraneous experiences, and for proposing potentially better alternatives for action. Occupational trainers are familiar today with methods of "conceptualisation of action"[36] drawn from clinical psychology, which advocate the verbalisation of lived situations – whether real or simulated – in order to facilitate collective understanding, assessment, insight, and commitment to the development of original forms of action and decision-making. Since situations involving "subjective mechanisms" and the processing of information are never exactly the same, it becomes convenient that labour as cognitive performance evolves according to continuous analyses to make it explicit, i.e. intelligible, negotiable, and amenable to change. Regular updates of situational-images databases for frequent analyses of activities and collective accounts of good labour have thus proved to be effective systems of performance improvement and self-regulation.

Images Make People Talk

The fact that contemporary systems for transmitting labour-related knowledge have assumed the form of chat rooms facilitated by audiovisual input is not trivial. For a long time, the secrets of intimacy, the tricky machinery of mind, the emotions and other hidden "subjective mechanisms" have found in speech a trusted means of revelation. In the history of the psychological analysis of activity, visual inscriptions have sometimes played the role of auxiliaries to men and women who were called on to make their thoughts explicit by speaking. Yves Clot, an influential labour psychologist in France, designed his so-called clinic of activity based on "self-confrontational interviews". His use of video sequences to confront workers with their own activities aims to modify labour situations thanks to the dynamics of verbal interactions that the videos provoke between workers in the context of collective interviews. For Clot, when interviews are facilitated by the screening of "real activities[37]" carried out by the interviewees, cognitive

36 Pastré, Pierre. "L'analyse du travail en didactique professionnelle." *Revue française de pédagogie*, no. 138, 2002, pp. 9–17.

37 Clot asserts that the analysis of real activity explains the intrinsic relation that cognitive mechanisms maintain with emotions: "By posing the problem from the point of view of real

and emotional processes tend to surface by way of speech. The activity then becomes transformative, insofar as its actors get verbally involved in its analysis: "The intersubjective exchange sets in motion the existing activity."[38]

It is thus not surprising that the moment when the sciences of subjectivity began to devise techniques to stimulate workers to verbalise their mental states and to help them articulate their implicit affects and know-how coincided with the moment when managers began to claim the status of "engineers of discussion".[39] Managers in the post-industrial era have realised that the best way to increase efficiency at work is to let it be formulated by those who do the work. Therefore, any process aimed at making good labour explicit is likely to take the form of a personal conclusion based on the analysis of reliable data. The "spaces for discussion" that have proliferated in contemporary organisations are spaces where groups of workers engage in analyses of their own situations in order to come up with credible solutions for their transformation. These spaces require "management tools of a new kind, less geared towards conforming behaviour than towards supporting the discursive exploration of novelties."[40] Managerial intervention must focus, according to the expert manager Mathieu Detchessahar, on the configuration of frameworks of exchange through formal, infrastructural investments: "These formal investments must allow, mainly, to inform the opinion of the operators (distribution and geography of information) and to support discursive mechanisms." Feeding spaces for discussion with constantly new problematic situations and soliciting workers to interpret and classify data in order to achieve together regulatory solutions for their own activities have thus become common practices for creating contexts of exchange in which, as Detchessahar says, "'saying' is the essential framework for 'doing.'"[41]

This may partly explain why, in today's organisations, educational images are not confined to training rooms. More than just a means of controlling

activity [...] we can avoid the ruinous dualism of the cognitive and the emotional." Clot. *Travail et pouvoir d'agir*. Presses Universitaires de France, 2008, p. 4. My trans.

38 Idem. "La formation par l'analyse du travail: pour une troisième voie." *Manières de penser, manières d'agir en éducation et en formation*, edited by Bruno Maggi. Presses Universitaires de France, 2000, pp. 133–56. My trans.

39 See Detchessahar, Mathieu. "Santé au travail. Quand le management n'est pas le problème, mais la solution...." *Revue française de gestion*, vol. 5, no. 214, 2011, pp. 89–105; see also idem. "Quand discuter, c'est produire. Pour une théorie de l'espace de discussion en situation de gestion." *Revue française de gestion*, no. 132, 2001, pp. 32–43.

40 Idem. "Santé au travail." My translation.

41 The last quotes of this paragraph are from idem. "Quand discuter, c'est produire." My translation.

the renewal of the workforce, the production of educational content became a key means of constantly regulating activities and ensuring innovation. Weekly meetings in the screening room – while expressly motivated by the fact that they place the workers' subjectivities at the centre of organisational design[42] – could be defined as dispositifs for the permanent solicitation of verbal formulations of attitudes that commit their authors. As discussion engineering products, these meeting rooms equipped with inquiring moderators and screens showing new labour situations every week are powerful dispositifs that prompt workers' to voice their views on core organisational issues, which usually revolve around safety, productivity and efficiency. The images that animate these meetings have the function of steering the debates towards context-related themes and problems of daily work, often too circumstantial and conjunctural to constitute a theme for conversation in the absence of a visual reference, or an audiovisual support bringing specific, unresolved or perfectible situations back into focus.

Audiovisual techniques have been increasingly mobilised to fight against "organizational silence."[43] In the same French company where, as noted above, labour has been made explicit through the discourse of workers seeking to report more than what images can show, management methods have been developed which use audiovisual inscriptions to support exchanges between workers with the aim of having them regulate their own activities. In EDF's distribution sector, a group of ergonomists has implemented a knowledge-sharing system in the early 2010s to create an efficient "safety culture". Field operators are asked to record, photographically or on video, abnormal situations in their daily activities that force them to deviate from established safety conventions. The images are collected by line managers who select the most relevant ones for weekly discussions with the workers. Procedures are then developed collectively to deal safely with situations such as those depicted in the images.[44] Here too, images are thus used to pass the tiny world of everyday life through the endless mill of verbal expression, engaging workers in a revision of their own limits, prodding them to reframe their action within new norms of performance, new conditions and criteria for assessing risk, identifying hazards, and making decisions.

42 "Unlike the shaping of the Taylorist organisation, which aimed to exclude the subjectivity of the operators from the production system, the shaping of the discussion space aims to place it at the heart of the system." Ibid., p. 37. My trans.

43 Rocha, Raoni. *Du silence organisationnel au développement du débat structuré sur le travail: les effets sur la sécurité et sur l'organisation*. PhD thesis, Université de Bordeaux, 2014.

44 On the implementation of this procedure, see Rocha et al. "Work Debate Spaces: A Tool for Developing a Participatory Safety Management." *Applied Ergonomics*, no. 46, 2015, pp. 107–14.

The use of images as agents in a dispositif of explication of labour by the workers themselves, which relies on the ability of images to situate observers within a given framework of enunciation endowed with controllable aesthetic affordances, was also extensively practiced in France, even before EDF's research experiments, by the French *ovnimaker* René Baratta, a labour documentarist specialised in ergonomic sciences.[45] What Baratta calls his OVNI (*objets vidéo non-identifiés*)[46] are moving pictures which lack the status of "films" because they are devoid of "personal narrative construction." He claims that his video records of labour situations, which he uses to confront teams of workers with their own activities in order "to create the conditions for the emergence of a collective speech," are not the "expression of [his] personal point of view," but impersonal testimonies of real activities conveying the details of their specific milieux and with a remarkable power to interpellate workers on the forms of their activities.[47] Indeed, pictures of real work situations present the advantage of not carrying any scent of purpose, authorial creation, or hierarchical instruction. In contrast to prescriptive labour standards, audiovisual testimonies of everyday activity demonstrate significant capacity to stimulate those who find in the pictures motives for self-justification, transient events that cause difficulties, daily challenges of unstable interactions with the social and material environment, and opportunities to promote personal experiences as solutions for collective problems. These "unidentified" images (i.e. images that do not "express" any point of view or intentional conception) do not reveal the truth of labour on their readable, illuminating surface. Rather, they evidence problematic situations capable of triggering discursive acts of enlightenment.

The Adventure of Making Labour Explicit

Labour entered a modern era when the activities that belonged to this category were put into circulation in the form of inscriptions to disseminate, beyond the associations of artisans and other restricted clusters, the knowledge involved in the production of wealth. One can speak of labour

45 See *Aucun risque! Parole de compagnons* (René Baratta, 1992), winner of the European Club of Health award in 1993.

46 The term "OVNI" (*objet volant non-identifié*), in the usual sense of the term, is French for UFO.

47 Quotes from Baratta, René. "Du film à l'Ovni en passant par le travail." *Communications*, no. 80, 2006, pp. 235–54. My trans.

modernity by referring to the end of a period – which lasted roughly until the eighteenth century – in which the body was the exclusive medium for the transmission of labour-related knowledge. Projects such as the *Encyclopédie*, which have sought to make this knowledge accessible in the form of iconographic and scriptural collections, attempted to place it in the category of *explicit knowledge*. Their aim was to replace the contingent, obstinate bodily movements of master craftsmen with stable and reproducible signs, thanks to scientific methods for analysing the fleeting, thoughtless, and spontaneous nature of the productive exercise. The new material conditions of labour-related knowledge greatly contributed to the institutional organisation of apprenticeship and the consequent new rhythm of circulation and innovation of production techniques.[48] Explicit knowledge must here be understood as opposed to implicit knowledge, which prevailed in craft apprenticeship until the late eighteenth century (and beyond) in Western Europe, when it was not uncommon for training practices to last – as Adam Smith decried – up to ten years, during which time apprentices were supposed to assimilate through observation and practice the intuitive knowledge of master craftsmen.[49]

Modern instruction, by contrast, refuted the tacit dimension of labour. Institutional techniques of knowledge communication in the industrial age have always been linked to procedures to display healthy and productive labour practice such as truths of which a significant characteristic is the lack of discretion. The modern pedagogy of work consists of adapting elusive and erratic aspects of labour practice – previously subject to the contingency of the body at work – into short episodes of explicitness, which can be brought to the attention of apprentices in orderly fashion as information that makes a real difference. This is what makes modern workplaces to be animated by *movements of explication*, i.e. programmes of controlled manifestation, by waves of attractional evidence, of knowledge previously slumbered in the tacit domain and held by only a select few.

Offices and control rooms in the cybernetic age do not escape this modern dynamics of labour explication. In the present-day organisations, however, cognitive performance and procedural knowledge became key factors of productivity. Retirement and the increasing mobility of experts might be

48 See Simondon's remarks on this matter. Simondon, Gilbert. *On the Mode of Existence of Technical Objects*. Univocal Publishing, 2017 [1958], pp. 103-121.

49 See Smith. *An Inquiry*; on apprenticeship in craft guilds, see also Epstein, S.R. "Craft Guilds, Apprenticeship, and Technological Change in Preindustrial Europe." *Journal of Economic History*, vol. 58, no. 3, 1998, pp. 684–713.

economically threatening for the organisations devoid of tools to externalise experts' "intangible cultural heritage," as it was put by EDF's ergonomic researchers. The risk of a loss of knowledge and innovation capacity and the urgency of organisations' independence from knowledge holders maintain the relevance of knowledge transmission techniques.[50] In this light, the emergence of cognitive ergonomics can be seen as the condition for an update of the modern dispositives of labour explication, so that they are adapted to the formalisation and communication of performance currently required by post-industrial activities.

The real novelty in the post-industrial systems of knowledge transmission seems to be the successful involvement of workers in the frenzy of labour explication. Their innovation consists in making appear as spontaneous and expedient the obligation for workers to establish a permanent attitude of self-exposure; an attitude of unveiling by way of speech their cognitive performance and the intimate conditions of their daily exercise. This happens in part when images have ceased to be vehicles of explicit labour; when they have been released from the sporadic role of coercive announcers of good labour in the educational process in order to be employed systematically as informers of decision-makers; when they have lost their utility as capturers of productive performance and have been reconfigured into techniques of *mise en situation*; when they have stopped composing a machine of control functioning on the idea of prescription and obedience to be part of a more fashionable engine operating "spaces of discussion," where each worker commits to respecting the discipline they are stating for themselves.

All this points to a somewhat new mode of labour governance through labour explication, which continually dictates to workers – but without really saying it, so as not to disrupt the magic of its operation – that *they must know more than they can see*. This formula reflects a specific reading of Michael Polanyi's famous statement: "we can know more than we can tell."[51] Polanyi summarised in this sentence the result of his humanistic study of a realm of tacit knowledge underlying human experience and communication practice. He thus stated the existence of a pervasive implicit, prelinguistic, and discrete effort indispensable to all human action and understanding.

50 See for instance the plea by organisational theorists Ikujiro Nonaka and Hirotaka Takeuchi for the externalisation of knowledge in the "knowledge-creating company": "externalization […] holds the key to knowledge creation. It is in this mode that tacit knowledge, which is personal, context-specific, and therefore hard to formalize and communicate to others, is converted into knowledge that is transmittable and articulable," Nonaka, Ikujiro, and Hirotaka Takeuchi. *The Knowledge-Creating Company*. Oxford University Press, 1995, pp. 237–38.

51 Polanyi, Michael. *The Tacit Dimension* [1966]. Peter Smith, 1983, p. 4.

Contemporary apprentices of labour, however, are virtually compelled to read Polanyi's statement as an epistemological challenge involving the struggling verbalisation of personal knowledge that is *circumstantially not yet* formulated; as if the fate of being able to know more than one is able to speak could be redeemed by an additional effort of spectatorial speech. This is the additional effort required from workers in their contemporary environments densely populated with images. An effort that has been widely celebrated along with their recent opportunity to participate in the modern adventure of making labour explicit.

Works Cited

Baratta, René. "Du film à l'Ovni en passant par le travail." *Communications*, no. 80, 2006, pp. 235–54.

Barthes, Roland. "Image, raison, déraison." *L'univers de l'Encyclopédie*, edited by Roland Barthes, Robert Mauzi, and Jean-Pierre Seguin. Les libraires associés, 1964, pp. 10–16.

Clot, Yves. *La fonction psychologique du travail*. Presses Universitaires de France, 1999.

Clot, Yves. "La formation par l'analyse du travail: pour une troisième voie." *Manières de penser, manières d'agir en éducation et en formation*, edited by Bruno Maggi. Presses Universitaires de France, 2000, pp. 133–56.

Clot, Yves. "Qu'est-ce que l'activité dans l'analyse du travail?" *Performances humaines et techniques* [special issue], 1995, pp. 2–6.

Clot, Yves. *Travail et pouvoir d'agir*. Presses Universitaires de France, 2008.

Clot, Yves, Daneil Faïta, Gabriel Fernandez, and Livia Scheller. "Entretiens en autoconfrontation croisée: une méthode en clinique de l'activité." *Education permanente*, no. 146, 2001, pp. 17–25.

Detchessahar, Mathieu. "Quand discuter, c'est produire. Pour une théorie de l'espace de discussion en situation de gestion." *Revue française de gestion*, no. 132, 2001, pp. 32–43.

Detchessahar, Mathieu. "Santé au travail. Quand le management n'est pas le problème, mais la solution…." *Revue française de gestion*, vol. 5, no. 214, 2011, pp. 89–105.

Epstein, S.R. "Craft Guilds, Apprenticeship, and Technological Change in Preindustrial Europe." *Journal of Economic History*, vol. 58, no. 3, 1998, pp. 684–713.

Foucault, Michel. *Du gouvernement des vivants. Cours au Collège de France. 1979–1980*. Seuil/Gallimard, 2012.

Foucault, Michel. *Les mots et les choses. Une archéologie des sciences humaines*. Gallimard, 1966.

Hediger, Vinzenz. "Body Rebuilding: Körper und Arbeit an der Schwelle zum kybernetischen Zeitalter." *Prometheische Kultur. Wo kommen unsere Energien her?*, edited by Claus Leggewie, Ursula Renner, and Peter Risthaus. Wilhelm Fink Verlag, 2013, pp. 195–222.

Hediger, Vinzenz. "Thermodynamic Kitsch: Computing in German Industrial Films, 1928/1963." *Films That Work: Industrial Film and the Productivity of Media*, edited by Vinzenz Hediger and Patrick Vonderau. Amsterdam University Press, 2009, pp. 127–49.

Huard, Georges. "Les planches de l'*Encyclopédie* et celles de la *Description des Arts et Métiers* de l'Académie des Sciences." *Revue d'histoire des sciences*, vol. 4, no. 3–4, 1951, pp. 238–49.

Lahlou, Saadi. "Digitization and Transmission of Human Experience." *Social Science Information*, vol. 49 [special issue: "Digitize and Transfer"], 2010, pp. 291–327.

Lahlou, Saadi. "How Can We Capture the Subject's Perspective? An Evidence-Based Approach for the Social Scientist." *Social Science Information*, no. 50, 2011, pp. 607–55.

Lahlou, Saadi. "Observing Cognitive Work in Offices." *Cooperative Buildings: Integrating Information, Organizations and Architecture*, edited by Norbert A. Streitz, Jane Siegel, Volker Hartkopf and Shin'ichi Konomi. Springer, 1999, pp. 150–63.

Lahlou, Saadi, Sophie Le Bellu, and Sabine Boesen-Mariani. "Subjective Evidence Based Ethnography: Method and Applications." *Integrative Psychological and Behavioral Science*, vol. 49, no. 2, 2015, pp. 216–38.

Latour, Bruno. "A Plea for Earthly Sciences." *New Social Connections: Sociology's Subjects and Objects*, edited by Judith Burnett, Syd Jeffers, and Graham Thomas. Palgrave Macmillan, 2010, pp. 72–84.

Lave, Jean. *Cognition and Practice: Mind, Mathematics and Culture in Everyday Life*. Cambridge University Press, 1988.

Le Bellu, Sophie. *Capitalisation des savoir-faire et des gestes professionnels dans le milieu industriel. Mise en place d'une aide numérique au compagnonnage métier dans le secteur de l'énergie*. Université Bordeaux 2, 2011. PhD thesis.

Le Bellu, Sophie, Saadi Lahlou, Valery Nosulenko, and Elena Samoylenko. "Studying Activity in Manual Work: A Framework for Analysis and Training." *Le travail humain*, vol. 79, no. 1, 2016, pp. 7–30.

Leplat, Jacques. "A propos des compétences incorporées." *Education permanente*, no. 123, 1995, pp. 101–14.

Leplat, Jacques, and Jean-Michel Hoc. "Tâche et activité dans l'analyse psychologique des situations." *Cahiers de psychologie cognitive*, vol. 3, no. 1, 1983, pp. 49–63.

Morais, Alexandre, and Raynald Aubineau. "Articulation entre l'ergonomie et le lean manufacturing chez PSA." *Activités*, vol. 9, no. 2, 2012, pp. 179–97.

Omodei, Mary M., and Jim McLennan. "Studying Complex Decision Making in Natural Settings: Using a Head-Mounted Video Camera to Study Competitive Orienteering." *Perceptual and Motor Skills*, vol. 79, no. 3, 1994, pp. 1411–25.

Nonaka, Ikujiro, and Hirotaka Takeuchi. *The Knowledge-Creating Company*. Oxford University Press, 1995.

Pannabecker, John R. "Representing Mechanical Arts in Diderot's 'Encyclopédie'." *Technology and Culture*, vol. 39, no. 1, 1998, pp. 33–73.

Pastré, Pierre. "L'analyse du travail en didactique professionnelle." *Revue française de pédagogie*, no. 138, 2002, pp. 9–17.

Pastré, Pierre, Patrick Mayen, and Gérard Vergnaud. "La didactique professionnelle." *Revue française de pédagogie*, no. 154, 2006, pp. 145–98.

Picon, Antoine. "Connaissances techniques sous les Lumières." *Le conservatoire national des arts et métiers au cœur de Paris: 1794–1994*, edited by Claudine Fontanon, Michel Le Moël, Raymond Saint-Paul, and Béatrice de Andia. CNAM, 1994, pp. 69–73.

Polanyi, Michael. *The Tacit Dimension* [1966]. Peter Smith, 1983.

Rabinbach, Anson. *The Human Motor: Energy, Fatigue, and the Origins of Modernity*. University of California Press, 1990.

Rocha, Raoni. *Du silence organisationnel au développement du débat structuré sur le travail: les effets sur la sécurité et sur l'organisation*. PhD thesis, Université de Bordeaux, 2014.

Rocha, Raoni, Vanina Mollo, and François Daniellou. "Work Debate Spaces: A Tool for Developing a Participatory Safety Management." *Applied Ergonomics*, no. 46, 2015, pp. 107–14.

Simondon, Gilbert. *On the Mode of Existence of Technical Objects*. Univocal Publishing, 2017 [1958].

Sloterdijk, Peter. *Sphères III: Écumes*. Maren Sell, 2005.

Smith, Adam. *An Inquiry into the Nature and Causes of the Wealth of Nations* [1776]. The Modern Library, 1937.

Taylor, Frederick W. *The Principles of Scientific Management*. Harper & Brothers, 1911.

About the Author

Guilherme Machado holds a doctorate in Cinema Studies. He is a former member of the Graduate Research Training Program "Configurations of Film" at Goethe University Frankfurt (2019–21). His research focuses on the visual culture of labour, the media theory and anthropology of labour environments. He recently co-edited *Tacit Cinematic Knowledge* (Meson Press, forthcoming).

16 Free Enterprise Film

Aims of Industry, Economic Propaganda and the Development of a Neoliberal Cinema

Scott Anthony

Abstract

In the aftermath of World War II, Aims of Industry emerged as a significant propaganda organization. Through films, cartoons and public discussions it combined the interests of corporations, local businessmen and consumer groups in opposition to austerity. While sharing many rhetorical and visual idioms of the films produced by the British state, Aims of Industry's propaganda favoured the international over the domestic, exchange over production and individualism against restraint. Beside its role in the longer-term resurgence of economic liberalism, Aims' story also illustrates how the pedagogic norms of the Documentary Film Movement were elbowed aside by hidden persuading, how privacy and consumption began to alter the UK's media culture, and how de-industrialization eroded the modernist foundations of British documentary.

Keywords: economics; propaganda; neoliberal; documentary; Britain; empire

In *Films That Work*, Vinzenz Hediger and Patrick Vonderau argued that the study of industrial film offered film scholars a way to move beyond the representational power of images to address the mediatization of social phenomena that has traditionally been the preserve of social scientists.[1]

Research for this chapter was made possible by Start-up Grant no. M4081571 from Nanyang Technological University, Singapore, and the "Economics in the Public Sphere" (ECONPUBLIC) project funded by the European Research Council FP7 (2007–2013)/ERC grant agreement 283754. Many thanks to Aled Davies, Florian Hoof, Kit Kowol, Peter Mandler, Patrick Russell, and Martin Stollery for feedback on an earlier draft.

1 Hediger, Vinzenz, and Patrick Vonderau. "Introduction." *Films That Work: Industrial Film and the Productivity of Media*, edited by Vinzenz Hediger and Patrick Vonderau. Amsterdam University Press, 2009, pp. 9–18.

Hediger, V., F. Hoof, Y. Zimmermann, with S. Anthony (eds), *Films That Work Harder: The Circulation of Industrial Film*. Taylor & Francis Group 2024
DOI: 10.5117/9789462986534_CH16

They posit – entirely convincingly – that industrial films have provided the sine qua non for new social practices such as large-scale industrial production and globalized financial markets.

In a British context, the Second World War provides the textbook study of the ways visuality, power and economic organization are tied together by industrial and commissioned film. In *Britain's War Machine*, David Edgerton showed how the conflict marked the zenith of British coordination of transglobal technology, production and finance.[2] Not uncoincidentally, the war also marked the cultural apogee of what is often called the "Griersonian tradition" of government-sponsored industrial film.[3] Wartime saw the creation of an enormous information infrastructure gathering together and evangelizing technological practices, political discourses and forms of social discipline.[4] By the close of the war, film could be identified as "the main highway of future social progress."[5]

However, the end of the war saw British coordination of this transglobal system of production begin to wane and with it the conviction that industrial film was an engine of social progress. The election of democratic socialists coupled with war debts, patriotic pride and the spectre of decolonization nudged the nation's political economy in the direction of economic nationalism. Although the new Labour government attempted to re-mobilize the nation's industries towards production and export, freed from the totalizing experience of war the national informational infrastructure sagged, shrivelled and then shrunk. One consequence of this is that after the Conservative Party's victory in 1951, the public sector "Griersonian tradition" increasingly sought to attach itself to the project of European modernization, of colonial development and the building of post-war international organizations.[6] Meanwhile, in the first ten years or so after the war, privately funded film libraries (such as Aims of Industry's) would increasingly move onto Griersonian territory at home. Aims of Industry (Aims) was part of an emerging

2 Edgerton, David. *Britain's War Machine: Weapons, Resources, and Experts in the Second World War.* Oxford University Press, 2011.

3 See Sussex, Elizabeth. *The Rise and Fall of British Documentary.* University of California Press, 1975; and Swann, Paul. *The British Documentary Film Movement, 1926–1946.* Cambridge University Press, 1989; see, for instance, Stollery, Martin, and John Corner. "The Postwar British Documentary Movement – Decline or Transition?" *Journal of British Cinema and Television*, vol. 10, no. 3, 2013, pp. 387–94.

4 See McLaine, Ian. *Ministry of Morale: Home Front Morale and the Ministry of Information in World War II.* Allen & Unwin, 1979.

5 Editorial. *Manchester Guardian*, March 31, 1944.

6 See, for example, Druick, Zoë, and Deane Williams, editors. *The Grierson Effect: Tracing Documentary's International Movement.* Palgrave BFI, 2014.

movement clearing the ground for the emergence of a Western imaginary of consumption that depended on both the might and the managerial independence of transnational corporations.[7]

This is the context in which Aims, a "non-political" educational organization, emerged as both an influential and prescient political and cultural force in Britain. Although it produced propaganda in a range of cultural forms, this chapter will mostly consider Aims' work in the immediate post-war years when its interest in film production and investment in its film library were at their peak. Aims' library not only circulated a range of industrial and sponsored film but it curated them – self-consciously privileging films that focused on private consumption, personal growth and international trade.

Industrial film in the Second World War had often been addressed to specific problems, working as a mechanism for the cultural redistribution of esteem as well as the sharing of information. Although utilizing the motifs, formats and social practices of the wartime industrial film, Aims' use of film was a small part of a wider effort to scramble, subvert or steal the meanings of those idioms.[8] Rather than operating in the tight industrial and organizational feedback loops described by Hediger and Vonderau, the production and circulation of British industrial film in the post-war world would become a site for the blurring of this knowledge.[9] While wartime film privileged civic understanding of interconnected specialisms, industrial film as utilized by Aims dealt with foundational if not universal issues. It preferred to deal in sweeping ideological vistas although it operated under the tight control of a small, socially connected, political subculture.

In that the advent of Aims coincided with the rise of McCarthyism and the formation of Gladio networks, telling the story of Aims is telling the story of the immediate impact of the Cold War on British industrial film.[10] Aims might even be thought of as a kind of cultural Gladio network. Considered over a longer period, Aims' activities illustrate how industrial film in Britain began its fractious and non-linear journey into the post-industrial economy.

7 Here it's significant that the progressive elements of the "Griersonian tradition" that found their way into the films of large corporations – such as Shell, ICI, and Unilever – tended to be strongest during the process of decolonization and the attendant need to manage "international issues" such as environmental degradation. See Russell, Patrick, and James Piers Taylor, editors. *Shadows of Progress: Documentary Film in Post-war Britain*. Palgrave Macmillan, 2010.

8 See Kisch, Richard. *The Private life of Public Relations*. MacGibbon & Kee, 1964.

9 Hediger, and Vonderau. "Introduction."

10 See, for example, Ganser, Daniele. *NATO's Secret Armies: Operation Gladio and Terrorism in Western Europe*. Frank Cass, 2005; and Roulin, Stéphanie, et al., editors. *Transnational Anti-Communism and the Cold War: Agents, Activities, and Networks*. Palgrave Macmillan, 2014.

This chapter argues that Aims' film programmes serve as a significant and politically complex staging post on the road to a new culture of de-industrial film. In the consumerist, privatizing and post-industrial mode of cinema patronized by Aims some of the cultural origins of neoliberalism can be discerned.

Introducing Aims

Aims was a campaigning organization created in 1942 to promote the cause of free enterprise.[11] During the war the British state had become central to the management of the economy; Aims' task was to ensure that after the conflict had ended control of industry was handed back to the private sector. Founded by H.G. Starley of Champion Sparking Plugs, it was ostensibly a "non-party" and "non-political" educational organization; it defended the larger principle of private enterprise and promised to protect small and medium-sized businesses from government interference.[12] Aims began to invest in industrial film in an effort to shunt post-war political debate away from state intervention.

The cultural activism that bought Aims to public prominence was its anti-nationalization campaigns. Labour had proposed nationalizing the cement industry, the water supply and sugar refining as part of its commitment to break up private monopolies and grow the size of the public sector (despite the rapid growth of the state during the war, eighty per cent of the UK's economy remained in private hands). Labour's intention to nationalize the sugar industry led Lord Lyle of Westbourne, head of Tate & Lyle, to call an extraordinary general meeting at which he successfully sought permission "to fight nationalisation with every weapon at our command."[13] Aims became the vehicle through which Tate & Lyle – which had built a near monopoly of the British sugar market in the 1930s – opposed the Labour government. Under the rubric of promoting free enterprise Aims built a

11 See Cockett, Richard. *Thinking the Unthinkable: Think-tanks and the Economic Counter-revolution, 1931–1983*. HarperCollins, 1994.

12 Starley was chairman of Society of Motor Manufacturers and Traders. Other American companies associated with the motor industry, including Ford and Standard Oil, would also be long-term sponsors of Aims of Industry. Contemporaries picked up on this very quickly. Letter from Honey to Andrew. May 1, 1946. University of Warwick, Modern Records Centre (MRC), British Employers Confederation Papers, MSS 200/B/3/2/C822.

13 Wilson, H.H. "Techniques of Pressure – Anti-nationalization Propaganda in Britain." *Public Opinion Quarterly*, vol. 15, no. 2, Summer 1951, p. 227.

broad coalition with the Federation of British Industries (FBI) by stressing their collective commitment to private ownership.[14] By working through Aims, Tate & Lyle was able to turn a sectarian battle into something more akin to a universal principle. Over the next fifty years Aims' campaigns against state run businesses would occupy a distinctive niche in the British political landscape.[15]

It would be easy to draw a line from the norms established by the hucksters of the Edwardian factory gate film to Aims' initial approach to cinema.[16] But while Aims worked through existing traditions that tied together cinematic showmanship, local industries and their workers, it also shifted its attention from civic spaces towards the emerging infrastructure of organized post-war consumerism. For instance, Aims held its events in Butlin's holiday camps and at consumer goods fairs.[17] It staged competitions – such as the chance to win a television – alongside discussions about the virtues of private enterprise. Aims made its presentations in entertainment venues; leisure settings could serve as lecterns to encourage dissatisfaction with post-war austerity and rationing. The campaign against the nationalization of Tate & Lyle was to be inserted into a campaign against austerity. Beyond the immediate context of the anti-sugar nationalization campaign, Aims' focus on consumers was its default strategy during its first decades. Touring consumer shows were used as a hook to draw people into debates, exhibitions and film screenings. By 1955, Aims claimed to be placing four hundred stories a month in the British press, each of its mobile film seasons had reached more than 180,000 people while the films in the Aims library had been watched by more than 500,000 people.[18]

Alongside Aims' debt to the practices of Edwardian industrial film was a reliance on the collective habits encouraged by the Ministry of Information during the Second World War and propagated by the Central Office of Information (COI) in the years immediately after 1945. During its six years in office the post-war Labour administration funded more than 100,000 lectures, five hundred films, 170 exhibitions and thirty national advertising

14 Letter from Sherwood to Sinclair. December 22, 1950. University of Warwick, MRC, *Papers* of the *Federation of British Industries* (*PFBI*), MSS 200/F/3/P1/1/12.

15 It was eventually wound up in the 1990s after the Labour Party abandoned its commitment to "Clause IV" and the collapse of the Soviet Union.

16 See Russell, Patrick. "From Acorn to Oak: Industrial and Corporate Films in Britain." *Business Archives*, no. 107, 2011, pp. 53–76.

17 "Winter Programme 1954/55." *PFBI*, MSS 200/F/3/04/13/4.

18 "Progress Report – January, 1949"; "Progress Report – February, 1955"; "Progress Report – April, 1955"; "Progress Report – August, 1955." PFBI, MSS 200/F/3/04/13/4.

campaigns alongside the publication of innumerable pamphlets and books.[19] Aims' campaigns were designed to operate through the informational networks sponsored by the state.

Ironically, although it had been founded to fight against the political consensus for state intervention that had gathered momentum with the wartime publication of the Beveridge report, the success of Aims' campaigns depended upon the liberal administrative machinery the war had done so much to create. The publication of Beveridge's report in 1942 provided the blueprint for the introduction of a wide-ranging social insurance programme that established national minimums of welfare. However, while it provided the rationale for state expansion it mostly eschewed coercion and stressed the need for cooperation, voluntarism and individual initiative.[20] By offering its services to schools, churches, working men's clubs, and the study groups of the armed forces, Aims was arguably attempting to appropriate the idiom of "consensus" as expressed by the Beveridge report as well as its infrastructure.[21] The industrial films it screened were less ideological arguments against nationalization and more celebratory expositions of existing industrial concessions. Tate & Lyle hired BBC broadcaster Richard Dimbleby to record a Q&A outlining how state control would damage labour relations, pensions and marriage payments in the sugar industry. Aims published leaflets by Labour movement figures such as J.R. Clynes and Alfred Edwards, as well as by trade unionists such as Florence Hancock, arguing against further nationalization.[22] Fundamental if not irreconcilable differences of political economy were being contested in constrained social, linguistic and aesthetic spaces. This was intense ideological hostility to the social democratic project that was nevertheless able to assume its garb and parrot its language.

The Free Enterprise Film Library

Although Tate & Lyle underwrote the financial costs of Aims' entry into film, its film library was not biased towards the sugar industry in an immediately

19 See Moore, Martin. *The Origins of Modern Spin: Democratic Government and the Media in Britain, 1945–51*. Palgrave Macmillan, 2006.

20 See Harris, José. *William Beveridge: A Biography*. Clarendon Press, 1977.

21 For more on Aims' energetic efforts to work through a variety of clubs and organizations, see Letters from Boyfield to Jago, April 13, 1949; Baker to Tewson, May 18, 1949; Adamson to Tewson, June 17, 1949; Pople to Fletcher, December 22, 1949. University of Warwick, MRC, Trade Unions Congress Papers (TUCP), MSS 292/500/6.

22 For instance, Clynes, J.R. "The Danger of the Unofficial Strike." TUCP, MSS 292/500/6.

obvious way. Instead Aims' agenda became apparent by the films it would not lend. Nationalization had turned public utilities, transport and the Coal Board into major sponsors of industrial film in the post-war era, but Aims circulated no films produced by any of the major publicly owned industries despite their apparent popularity and high aesthetic qualities.[23] Instead, films produced by private industry were supplemented by work from Gaumont British, the British Council, the US Information Service and very occasionally the COI.

Alongside this bias of omission Aims was cautious not to overpopulate its programmes with films produced by private industry. In a typical screening films celebrating the contributions of British industry were interspersed with programmes about the production of sherry, the architectural history of provincial cities, cookery shorts, new developments in medicine and anti-vivisection films. In addition to economics, Aims' catalogues invited film groups to programme titles across categories such as careers, "men of enterprise," sport, holidays and travel.[24]

Nevertheless, by examining the industrial categories of the Aims film library we can discern an agenda. Special attention is paid to chemical industries, mining industries and various agricultural concerns. These industrial categories map onto the major sponsors of private sector industrial films: films commissioned by companies such as Dunlop, ICI and Shell sat alongside those made for the tea, tobacco and sugar industries. Aside from the line drawn between public and private industries, a further dividing line in the Aims catalogue is between those industries primarily concerned with domestic production and those dependent on overseas trade. The finer distinction is between films made that emphasize primary or extractive industries and those that deal with secondary or manufacturing industries. This distinction marked a significant juncture in the history of the British industrial film form.

From its beginning, the idea of the industrial film in Britain was intertwined with the process film format. Over time the evolution of films such as Cricks and Sharp's *A Visit to Peek Frean and Co.'s Biscuit Works* (1906), Charles Urban's *A Day in the Life of a Coal Miner* (1910) and Lever Brothers' *Port Sunlight* (1919) established the expectation that industrial

23 See Russell and Taylor. *Shadows of Progress*.

24 "Subject Headings: Aims of Industry Speaker's Bureau." Pamphlet. 1956. University of Warwick, Modern Records Centre, British Employers Confederation Papers, MSS 200/B/3/2/C82; "Aims of Industry: Public Relations Division." Pamphlet. Undated. University of Warwick, Modern Records Centre, British Employers Confederation Papers, MSS 200/B/3/2/C82.

films illustrate how raw materials are transformed into packaged products, distributed, delivered and consumed.[25] During the 1930s this pattern had been reformulated by Stephen Tallents and John Grierson's Documentary Film Movement, which bought together film criticism, artistic endeavour and the idea of social and political service in an effort to imbue the "process film" with a symbolic purposefulness.[26] Canonic films such as *Drifters*, *Night Mail* and *Spare Time* used the mechanics of the "process film" to hang a political vision of an interconnected society.[27] The Documentary Film Movement's political achievement can be understood as the imaginative embedding of the idea that the UK's extractive, manufacturing and consumer industries should be configured to the benefit of all. From the UK's departure from the gold standard in 1931 until the end of Bretton Woods, the impact of the Documentary Film Movement was to rebalance the national socio-economic imaginary away from consumption towards production.[28] Ultimately, it was the collapse of the nationalized industries – and with it their film units – that ended this totemic period of British industrial film, but it was an imaginative shibboleth of British political life that had been under attack since the Conservatives' abolition of the government's film unit in 1952 and the post-war rise of organizations like Aims.

Arguably the signature art, film and information series produced by the post-war Labour government was the *Britain Can Make It!* series, which explained why production for export was needed to pay off war debts, re-equip the military and build a welfare state. By contrast Tate & Lyle films such as *From Cane to Cube* (1950) and *A Family Affair* (1959) focused on the benefits of production to the development of the local economies in the West Indies as well as to the consumer in Britain. One immediate way of understanding the Labour government's conflict with Tate & Lyle is as a battle on behalf of low-paid producers of domestic sugar beet against the profitable importers of sugar cane. By contrast, the industrial films of private companies whose success rested on international exchange used

25 Sargeant, Amy. "GPO Films: American and European Models of Advertising in the Projection of a Nation." *Twentieth Century British History*, vol. 23, no. 1, 2012, pp. 38–56.

26 See Anthony, Scott. *Public Relations and the Making of Modern Britain*. Manchester University Press, 2012.

27 Idem. "The GPO Film Unit and *'Britishness'* in the 1930s." *The Projection of Britain: A History of the GPO Film Unit*, edited by Scott Anthony and James G. Mansell. Palgrave Macmillan, 2011, pp. 10–17.

28 For some post-GFC reflections on the trajectory of the British national economic imaginary, see Bowman, Andrew, et al. *The End of the Experiment? From Competition to the Foundational Economy*. Manchester University Press, 2014.

the Aims film library to challenge government deference towards national production and the nationalized industries.[29]

The broader nature of Aims' critique of the nationally directed mixed economy can be ascertained through the film titles listed in its "economics" section. The following descriptions are typical of the economic entries in the Aims catalogue:

> **Balance 1950** is a coloured cartoon made to explain ICI's balance sheet to the Company's employees. The film tells its story by means of animated diagrams and a rhyming commentary with sung choruses.
>
> **Get off the Hose** – another ICI film – shows how international trade has been hampered since the war by documentation and controls. Gilbert Harding takes the lead.
>
> **Can We Be Rich?** – sponsored by the British Electrical Development Association – has Mr Geoffrey Crowther, Editor of *The Economist*, discussing with an unseen interviewer the question contained in the title. The conditions governing the nation's standard of living are defined, and it is made clear by ingenious analogy that we at present consume more than we produce.[30]

These were all films underlining the precariousness of the mixed economy; films that did not directly contradict the broad thrust of the government's economic thinking but instead emphasized the myriad of risks of intervention.[31] Three themes jump out of the film descriptions in the Aims catalogue: demonstrations of the power of private enterprise to foster wealth and prosperity, an entreaty to identify with the personalities, terminology and tools that private industry used to describe the world and warnings that further state interference could limit if not undermine future prosperity.

The film library of Aims had been purposely configured to stress the importance of the international over the domestic, the importance of exchange over production and of individual agency against state restraint.

29 Ussishkin, Daniel. "Morale and the Postwar Politics of Consensus." *Journal of British Studies*, vol. 52, no. 3, 2013, pp. 722–43.

30 "Industrial Films Distributed by Aims of Industry." Pamphlet. University of Warwick, MRC, Aims of Industry: Ephemera, MSS.21/2116.

31 For a definitive statement of industrial resistance to Labour's productivity drives, see Tomlinson, Jim. *Democratic Socialism and Economic Policy: The Attlee Years, 1945–1951*. Cambridge University Press, 1997.

Aims and the Party Politics of Industrial Film

While the informational apparatus of wartime was gradually wound down, no previous peacetime government had allocated so great a proportion of its resources to the tasks of informing and cajoling its citizens than the post-war Labour administration.[32] One reason for this was the Labour Party's determination to strike its own ideological path and resist American efforts to shape the global economic order in its own image.[33] British industrial films stressed the importance of British workers and British production because only by rebuilding exports could Britain free itself from dependence on American finance and construct a welfare state.[34] One impulse behind the post-war production of British industrial films was the effort to stave off Americanization and with it the prospect of being "Europeanized."[35]

Additionally, the Labour government was fighting a long-running battle against the private commercial news media that had sought to undermine the party during the interwar period.[36] However, the Labour government's decision to launch a royal commission investigating the private ownership (and bias of) the press occurred at the same time as independent journalism in Eastern Europe was being suppressed.[37] Aims' campaigns skilfully tapped into this spirit of embattled defiance among journalists.[38] In one month of the 1951 general election campaign Aims claimed to have generated 11,269 column inches across 362 newspapers and magazines. The topics addressed included water, sugar, coal, meat, bulk purchasing, profits, bureaucracy, steel, cement, housing and "the achievements of free enterprise."[39] This media and cultural activism apparently made an impact in knitting together a range

32 Crofts, William. *Coercion or Persuasion? Propaganda in Britain after 1945*. Routledge, 1989, p. 250.

33 See Mercer, Helen, et al., editors. *Labour Governments and Private Industry: The Experience of 1945–51*. Edinburgh University Press, 1992.

34 Rogow, Arnold A. "The Public Relations Program of the Labour Government and British Industry." *Public Opinion Quarterly*, vol. 16, no. 2, Summer 1952, pp. 201–24.

35 Later Aims of Industry were apparently involved in efforts to get the UK to join the EEC. Lashmar, Paul, and James Oliver. "How MI6 Pushed Britain to Join Europe." *Sunday Telegraph*, April 27, 1997, p. 10.

36 For more on this, see Beers, Laura. *Your Britain: Media and the Making of the Labour Party*. Harvard University Press, 2010.

37 See *Royal Commission on the Press, 1947–49: Report*. His Majesty's Stationery Office (HMSO), Cmd. 7700, 1949.

38 An alternative explanation is that press magnate Lord Beaverbook may have been involved in the creation of Aims. "Aiming Lower." *Daily Telegraph*, February 3, 1996, p. 15.

39 Wilson. "Techniques of Pressure," p. 236.

of constituencies either ideologically opposed to, commercially frustrated by, or simply bored with, continuing austerity.

Whatever their misdemeanours during the interwar years in the immediate post-war period, privately owned newspapers were constrained by paper rationing. Geopolitical Cold War anxieties were pervasive, everyday thrift inescapable. These ideologically and materially fraught circumstances encouraged brisk, pithy writing styles that in turn lent themselves to impatient and pointedly defiant editorials that Aims' campaigns were well placed to exploit.[40] Commercially stretched newspapers were targeted with cartoons and pre-written features and these newsprint campaigns were followed up with events, exhibitions and leaflet drops as well as films. For example, the script for the Aims-commissioned film *Bringing Home the Bacon* (1950) ran:

> George sits with a good free enterprise cup of good free enterprise tea while he sits back in his free enterprise chair at his free enterprise table, covered with a free enterprise cloth and free enterprise crockery while he listens to his free enterprise radio and rests his feet in free enterprise slippers on a free enterprise carpet and at the same time puffs contentedly at his free enterprise cigarette. It sounds as if there is a moral to all this somewhere.[41]

The release of this SADFAS-produced film was to be accompanied by the mass production of leaflets that were sized to be slipped into ration books by grocers, butchers and local shop keepers. One leaflet read: "Don't shoot your Butcher – HE's not to blame for – tough meat, poor quality, tiny rations. THE cause is – state bulk purchase of meat, state control of distribution."[42] Whereas the films of governmental agencies conscientiously attempted to develop accurate, convincing and dramatic representational models able to illustrate the importance of economic and social cooperation, Aims faced no such restraints. At the centre of Aims' publicity was "Mr Cube," who became a fixture of both Tate & Lyle's prestige publicity and a mouthpiece for Aims during it "Tate not State" anti-nationalization campaign.[43] As Aims

40 See Curran, James, and Jean Seaton. *Power without Responsibility*. Routledge, 2010.

41 It is unclear whether this film was produced. "Bringing Home the Bacon." July 13, 1950. Script. *PFBI*, MSS 200/F/3/04/13/2.

42 Wilson. "Techniques of Pressure," p. 237.

43 It's likely that Mr Cube's existence was prompted by the popularity (and success) of the Gas Light and Coke Company's Mr Therm in the interwar period. In opposition to Mr Cube was the COI's "Charley" (and to a lesser extent the films of actor Richard Massingham). Clampin, David, and Ron Noon. "The Maverick Mr Cube: The Resurgence of Commercial Marketing in Postwar Britain." *Journal of Macromarketing*, vol. 31, no. 1, 2011, pp. 174–75.

admitted, Mr Cube could say things that might have appeared exaggerated or overly aggressive in the mouth of one of its spokesmen.

Perhaps most effectively, Tate & Lyle printed anti-nationalization messages on the ten million bags of sugar it sold in the UK each week. Pointed messages on domestic products usually delivered by local hands had the further advantages of both knitting together a politically and socially diverse constituency of discontents while also giving a highly sophisticated media campaign the appearance of a spontaneous rebellion. Rationing and sugar had proved perfect vehicles for Aims to politically canalize popular resentment against government control.[44] The democratic multitude were protesting against the governmental monolith.

By employing these methods Aims' ideological concerns found purchase in a world of shortages and government control of supply and prices. Its campaigns targeted families, women and children in an attempt to turn dissatisfaction with material hardship to scepticism about government intervention in industry. Large industrialists and small businesses, conservatives, liberals and the anti-Marxist labour movement as well as women and former fascists were drawn together in significant numbers by Aims to help defeat the Labour government in 1951 and by doing so stave off nationalization of the sugar industry. The end of rationing in 1954 saw a flurry in consumption of fats and sugars: in the aftermath British sugar consumption – which had always been exceptionally high – rose to an average of 500g per person per week.[45]

From a contemporary perspective Aims' activities – especially the production, distribution and exhibition of industrial films – can appear rather sinister. Although often focusing on local enterprises like the butcher, the baker and the grocer, Aims' campaign for free enterprise swiftly became a vehicle for big business. Companies such as Ford, Imperial Chemical Industries (ICI) and Rolls-Royce sat alongside Tate & Lyle as major funders, while Aims also operated as a conduit for a variety of Cold War propaganda produced by the American government.[46]

However, Aims' success in achieving particular sectarian goals was also a barrier to its long-term aims of promoting (in longer historical terms you

44 See Zweiniger-Bargielowska, Ina. *Austerity in Britain: Rationing, Controls, and Consumption, 1939–1955*. Oxford University Press, 2000.

45 Collingham, Lizzie. *The Taste of War: World War Two and the Battle for Food*. Allen Lane, 2011, p. 495.

46 For more on the US propaganda at this time, see Cull, Nicholas. *The Decline and Fall of the United States Information Agency: American Public Diplomacy, 1989–2001*. Palgrave Macmillan, 2012, pp. 81–133.

might argue resurrecting) a popular culture of liberal free enterprise.[47] Fear of further nationalizations dissipated with the defeat of the Labour government and with it the broader coalition of private industry that Aims had been able to draw on. The FBI had become involved with Aims because it believed in protecting the reputation of private industry, but it was alienated by Aims' further attempts to propagate a particular ideological agenda and feared that pursuing strident campaigns would bring it into political disrepute.[48] The FBI disassociated itself from Aims and as a result the organization fell more tightly into the financial grip of its big business funders, eventually becoming something more akin to a lobbying organization operating at the fringes of the Conservative Party.[49] No longer animated by shared experiences (such as, for example, the flavourless, stodgy wartime food and the "National Loaf"), Aims' arguments lost their immediate explanative appeal. It is also not a coincidence that Aims' mid-century successes occurred at the popular zenith of the British industrial film in the aftermath of the war, before the form fragmented into the more specialized fields of the art, advertising, education and training film and before the post-war demographic shift that would fracture the cultural, political and sociological bedrock that the golden age of the British industrial film had been built upon.[50]

Before the 1951 election James Chuter Ede, then home secretary, had complained that Aims' Luton office was actually the side entrance of the local Conservative Party headquarters and that they shared the same telephone number.[51] Furthermore, Aims' initiatives, such as the anti-nationalization newspaper *Workers Forum*, would eventually erode its "educational" authority and enable Labour to dismiss its activities. Film clubs were encouraged to borrow films from the Co-Operative Wholesale Society Film Unit, the Scientific Films Association or the COI instead.[52] Trade union monitoring of Aims would eventually reveal the organization's links with strike breaking,

47 See Trentmann, Frank. *Free Trade Nation: Commerce, Consumption, and Civil Society in Modern Britain*. Oxford University Press, 2008.

48 For one of many examples of the FBI's displeasure, see Letter from Edwards to Sewell. February 23, 1954. *PFBI*, MSS 200/F/3/04/13/4.

49 Although acolytes of Aims apparently also remained on the fringes of the FBI. Rollings, Neil. "Cracks in the Post-war Keynesian Settlement? The Role of Organised Business in Britain in the Rise of Neoliberalism before Margaret Thatcher." *Twentieth Century British History*, vol. 24, no. 4, 2013, pp. 637–59.

50 See Anthony, Scott. "Governing for Happiness: Mark Abrams, Subjective Social Indicators and the Post-war Explosion of 'Middle-Opinion'." *Scientific Governance in Britain, 1914–79*, edited by Don Leggett and Charlotte Sleigh. Manchester University Press, 2016, pp. 274–94.

51 Wilson. "Techniques of Pressure," p. 232.

52 Letter from Tewson to Millar. November 28, 1952. TUCP, MSS 292/500/6.

intimidation and the Economic League. "Aims [was] responsible for the 'Mr Cube' campaign on behalf of Messrs Tate & Lyle," ran one confidential pitch to potential subscribers. "[R]ecently we were mainly responsible for Breaking the Bristol Dock Strike and the Briggs Ford Strikes. [Our] services are increasingly being used for this type of job."[53] By the 1970s, Aims focused its propaganda efforts on white collar workers and management schools; its efforts to build a mass popular culture for free enterprise were abandoned.

Aims and the Post-war Order

In the mid-1920s the British state began to experiment with film in an attempt to stimulate imperial trade, encourage pan-imperial networks of intellectual exchange and build "Empire-mindedness."[54] The production, distribution and theorizing of documentary cinema developed by John Grierson, whose work initially depended on the patronage of the Empire Marketing Board (EMB), is a key example of this.[55] During the interwar years, however, the vision of global development propounded by the EMB existed at the margins of domestic politics. Speculative imperial investment was impossible to sell domestically in the context of mass unemployment, while not all Colonial administrators believed that British-style capitalist development could (or should) be rolled out around the globe. Resistance to new technologies could be justified by imperialist paternalism and respect for native cultures in addition to the enormous expense.[56] The poorly funded EMB was disbanded after just seven years in existence; its cinematic wings shunted to the General Post Office, where it was put to work propagating the need for a new domestic communications infrastructure appropriate to an era of mediatized mass democracy.[57]

53 "Scottish Council." August 6, 1952. University of Warwick, MRC, TUCP, MSS 292/500/6.

54 See Anthony. *Public Relations and the Making of Modern Britain*; or for a less neutral interpretation, see Grieveson, Lee. "The Cinema and the (Common) Wealth of Nations." *Film and the End of Empire*, edited by idem and Colin MacCabe. Palgrave Macmillan, pp. 73–113.

55 See Anthony, Scott. "John Grierson: From the Poster Movement to the Informational State." *The Major Realist Film Theorists*, edited by Ian Aitken. Edinburgh University Press, 2016, pp. 203–18.

56 Half of the respondents to a 1948 poll could not name a single colony, a fact which goes some way to demonstrating both the ineffectiveness and complete irrelevance of imperial propaganda in Britain during the interwar period. See Porter, Bernard. *The Lion's Share: A History of British Imperialism, 1850 to the Present*. Pearson, 2012, p. 254.

57 See Constantine, Stephen. "'Bringing the Empire Alive': The Empire Marketing Board and Imperial Propaganda, 1926–1933." *Imperialism and Popular Culture*, edited by John M. Mackenzie. Manchester University Press, 1986, pp. 191–231.

Ironically, the Second World War – more usually understood in British historiography as "the people's war" – saw aspects of the EMB's development-led vision resurrected, as colonial development became an existential priority. Defeating Nazi Germany depended on the coordination of global production: interventionist investment in, and direction of, the resources of the British Empire was essential to military success. By the end of the war an influential strand of technocratic opinion (as much internationalist as imperialist) argued for further reconfiguration of the world's political economy towards human need: international coordination was understood as the answer to the nationalist economic policies of the interwar period that had led to war. For example, the nutritional scientist John Boyd Orr, whose research had been sponsored by the EMB and whose work was the focus of Paul Rotha's *World of Plenty*, became the head of the United Nations Relief and Rehabilitation Administration (UNRRA).[58] Orr advocated for a World Food Board and the expert-led management of worldwide agricultural production.[59] John Grierson became director of mass communications and public information of UNESCO where he wrote *The Voice of the State*, a visionary history-cum-manifesto of the role of communications technology in ordering human civilization. The idealization of world government helps explain why the Labour government passed the Colonial Development and Welfare Act in 1945 – this was investment in imperial infrastructure, healthcare and education that could be self-consciously presented as considerate rather than exploitative and necessitated the continuing existence of the British Empire.[60]

This was the global context that Aims operated in and it helps explain why the Aims film library would come to prize propaganda produced by the

58 See MacDonald, Richard. "Evasive Enlightenment: World without End and the Internationalism of Postwar Documentary." *Journal of British Cinema and Television*, vol. 10, no. 3, 2013, pp. 452–74; and Druick, Zoë. "Visualising the World: The British Documentary at UNESCO." *The Projection of Britain: A History of the GPO Film Unit*, edited by Scott Anthony and James G. Mansell. Palgrave Macmillan, 2011, pp. 272–80.

59 See Collingham. *The Taste of War*, pp. 476–501; and Vernon, James. *Hunger: A Modern History*. Harvard University Press, 2007.

60 One reason for the Labour government's enthusiasm for the empire was the ideological influence of the Fabian socialists. Fabians had long argued that the empire could provide the organizational basis for a world welfare state, and the Colonial Bureau they set up in 1940 operated as a forum for bringing together socialist imperialists from the Labour Party with enthusiasts for development-orientated imperialism from the political right. Porter. *The Lion's Share*, pp. 254–65; see also Constantine. *The Making of British Colonial Development Policy, 1914–1940*. Cass, 1984.

Marshall Plan.[61] Aims was a vehicle for private business to strike out against technocratic consensus: its media campaigns enlisted anti-imperialists, entrepreneurs and weary consumers to its cause. While industrial films sponsored by the British state focused on production, effort and imperial cooperation, Marshall Plan propaganda posited citizens as consumers rather than workers and producers. American know-how, competition and mass consumerism were offered as a blueprint for both prosperity and political stability. While the British government fell into disrepute for mismanaging colonial resources, most famously a disastrous attempt to cultivate peanuts in Tanganyika, the Marshall Plan was illustrating the ability of American finance to power European reconstruction. Aims both offered trade union members scholarships to study in America and produced pamphlets warning that ideas like nationalization threatened the long-term future of US investment.[62]

At a more granular level it's noticeable how abruptly Aims' campaigns deal with production processes. While nationalized industries produced films that stressed multiple threads of connection, Aims' campaigns were more likely to focus on either the end of the chain or its very beginning. They zoom in on domestic consumers or how the benefits of that consumption empower poorer overseas producers. The fact that the buyer and seller are unaware that they are joined together in this propaganda mystifies rather than clarifies the historical, commercial and social forces that have connected them. In this way Aims' work promoted an understanding of the world that placed concepts such as agency, choice and individual identity at its heart. Associational national groupings would be superseded by the decisions of individual citizens.

As outlined previously Aims can be understood as using film to propagate a media culture that favoured the international over the domestic, exchange over production and individualism against restraint. However, it can also be understood as an attempt to create a cinema that worked against the idea of industrial and social intervention. In Aims' campaigns unexceptional private individuals are shown as being brought together by their own volition in stark contrast to government films where cooperation is directed by the state.

61 Financing provided by the Marshall Plan was crucial to the agricultural modernization that enabled European farming to recover and thrive over the short to medium term. Alongside the economic aid, however, was a political and ideological project that required European nations to set aside a proportion of the Marshall Plan loans to produce film propaganda that demonstrated the benefits of the American way of life.

62 Letter from Storrs to Sewill. August 9, 1955. TUCP, MSS 292/500/6.

Towards the De-industrial Film

Film developed in the world's wealthy urban centres, its development condensed, restructured and bought together aspects of economic, social and intellectual life as well as art and communication. The so-called Second Industrial Revolution of the early twentieth century – a series of revolutions in the applied sciences, transport and telecommunications – further implanted film's role as a technology for understanding both the mechanical and psychological configuration of the modern world.[63] By the 1930s an influential cross-section of activists, educationalists and government officials had become interested in using industrial film to help shape, negotiate and facilitate the demands of a modern mass democracy in Britain. By the Second World War the British industrial film had become a national mode of production, representation and circulation that supported a political economy gradually feeling its way from liberalism towards something more statist.

Aims can be understood as a story about the origins of the liberal resurgence, a story whose origins are still more usually sited in 1970s America. Conversely, it is a story about the ways in which even when social democracy was at the peak of its electoral powers in the UK it remained a surprisingly beleaguered and embattled force. In addition to this, Aims' campaigns can also be understood as exemplifying three concurrent changes that were reshaping the field of industrial film during the early post-war period.

Firstly, by the 1930s industrial film had evolved into a self-consciously pedagogical form. The applied art – including industrial film – produced by the Empire Marketing Board, for example, worked in an idiom that was forever announcing and explaining what it was trying to do. In EMB propaganda elements of emotion, humour and aesthetic beauty are invariably tugged back into expositional modes. During the interwar years the term "propaganda" had become politically loaded, while commercial media interests complained about state encroachment into their businesses. The result had been a "Griersonian" mode of industrial film that wore its social and political purposes extremely self-consciously and to the point were they could become objects of parody. By contrast, Aims' appeals are part of a wave of industrial filmmaking that begins to prune back this pedagogical apparatus. Aims' direction of creative travel was towards the "hidden

63 Boon, Timothy. "Old Industry, New Science? The GPO Film Unit between Palaeotechnology and Neotechnology." *The Projection of Britain: A History of the GPO Film Unit*, edited by Scott Anthony and James G. Mansell. Palgrave Macmillan, 2011, pp. 28–37.

persuading" of post-war advertising. Comics for children, records and a dice game in which players won by throwing "E N T E R P R I S E" and lost by throwing "S T A T E" were among Aims' early innovations.[64]

Secondly, British industrial films had not only evolved into a form that made a virtue out of their overtly pedagogical format; they also operated on the principle that they had to operate in socially useful contexts. Films were to be screened to activist community associations, trade unions and study groups. Cinema was conceptualized as a way of strengthening the operation of pre-existing social movements. By contrast, Aims' campaigns were increasingly designed to work in a domestic or leisure context. Whereas sponsors such as Stephen Tallents understood modern industrial film as an addition to the existing public sphere, Aims campaigns increasingly pushed the discussion into private spheres.[65] The difference was between those that addressed new media technology to revitalizing civic activism and those that saw the existing public sphere as increasingly contained by (and accessed through) new media technology.

Lastly, the shift away from production-focused industrial film was underpinned by shifting patterns of employment. Peak employment in industry in the UK occurred in 1955 as compared to West Germany in 1970, Italy in 1971 and France in 1973.[66] This helps explain why the cultural reputation of the British industrial film would fall so rapidly and why there was no cinematic analogue to the industrial films of *Les Trente Glorieuses*.[67] The point where standard historiographical arguments mark the decline of the British industrial film correlate exactly with the decline of employment in industrial production. The early onset of de-industrialization illustrates why the imaginative resonance of the British industrial film faded so quickly and by the 1980s was chiefly associated with the corporate training video. New opportunities in the post-war media were being made in advertising and television. Aims favourites such as *Balance 1950* anticipate this shift: whereas the industrial films of the early post-war period celebrate combined effort, ICI's short represents the world through a corporate balance sheet. The production-centred "process film" had been replaced by a disaggregated world that is measured in cost centres. The industrial film had moved from

64 Wilson. "Techniques of Pressure," p. 237.

65 See Anthony. *Public Relations and the Making of Modern Britain*; and Tallents, Stephen. *The Projection of England*. Faber, 1932.

66 Tomlinson, Jim. "De-industrialization Not Decline: A New Meta-narrative for Post-war British History." *Twentieth Century British History*, vol. 27, no. 1, 2016, p. 78.

67 See Jacobson, Brian. "On the Red Carpet in Rouen: Industrial Film Festivals and a World Community of Filmmakers." This volume.

a world where things were modernistically aggregated (it was claimed) for the benefit of everyone to a post-modern world where (it was claimed) the world was shaped by the autonomous decisions of individuals.

Works Cited

"Aiming Lower." *Daily Telegraph*, February 3, 1996, p. 15.

"Aims of Industry: Public Relations Division." Pamphlet. Undated. University of Warwick, Modern Records Centre, British Employers Confederation Papers, MSS 200/B/3/2/C82.

Aitken, Ian. *Film and Reform: John Grierson and the Documentary Film Movement*. Routledge, 1990.

Anthony, Scott. "Governing for Happiness: Mark Abrams, Subjective Social Indicators and the Post-war Explosion of 'Middle-Opinion'." *Scientific Governance in Britain, 1914–79*, edited by Don Leggett and Charlotte Sleigh. Manchester University Press, 2016, pp. 274–94.

Anthony, Scott. "The GPO Film Unit and '*Britishness*' in the 1930s." *The Projection of Britain: A History of the GPO Film Unit*, edited by Scott Anthony and James G. Mansell. Palgrave Macmillan, 2011, pp. 10–17.

Anthony, Scott. "John Grierson: From the Poster Movement to the Informational State." *The Major Realist Film Theorists*, edited by Ian Aitken. Edinburgh University Press, 2016, pp. 203–18.

Anthony, Scott. *Public Relations and the Making of Modern Britain*. Manchester University Press, 2012.

Anthony, Scott, and James G. Mansell, editors. *The Projection of Britain: A History of the GPO Film Unit*. Palgrave Macmillan, 2011.

Beers, Laura. *Your Britain: Media and the Making of the Labour Party*. Harvard University Press, 2010.

Boon, Timothy. "Old Industry, New Science? The GPO Film Unit between Palaeotechnology and Neotechnology." *The Projection of Britain: A History of the GPO Film Unit*, edited by Scott Anthony and James G. Mansell. Palgrave Macmillan, 2011, pp. 28–37.

Bowman, Andrew, et al. *The End of the Experiment? From Competition to the Foundational Economy*. Manchester University Press, 2014.

"Bringing Home the Bacon." July 13, 1950. Script. University of Warwick, Modern Records Centre, Papers of the Federation of British Industries, MSS 200/F/3/04/13/2.

Clampin, David, and Ron Noon. "The Maverick Mr Cube: The Resurgence of Commercial Marketing in Postwar Britain." *Journal of Macromarketing*, vol. 31, no. 1, 2011, pp. 174–75.

Clynes, J.R. "The Danger of the Unofficial Strike." University of Warwick, Modern Records Centre, Trade Unions Congress Papers, MSS 292/500/6.

Cockett, Richard. *Thinking the Unthinkable: Think-tanks and the Economic Counter-revolution, 1931–1983*. HarperCollins, 1994.

Collingham, Lizzie. *The Taste of War: World War Two and the Battle for Food.* Allen Lane, 2011.

Constantine, Stephen. "'Bringing the Empire Alive': The Empire Marketing Board and Imperial Propaganda, 1926–1933." *Imperialism and Popular Culture*, edited by John M. Mackenzie. Manchester University Press, 1986.

Constantine, Stephen. The Making of British Colonial Development Policy, 1914–1940. Cass, 1984.

Crofts, William. *Coercion or Persuasion? Propaganda in Britain after 1945*. Routledge, 1989.

Cull, Nicholas. *The Decline and Fall of the United States Information Agency: American Public Diplomacy, 1989–2001*. Palgrave Macmillan, 2012.

Curran, James, and Jean Seaton. *Power without Responsibility*. Routledge, 2010.

Druick, Zoë. "Visualising the World: The British Documentary at UNESCO." *The Projection of Britain: A History of the GPO Film Unit*, edited by Scott Anthony and James G. Mansell. Palgrave Macmillan, 2011, pp. 272–80.

Druick, Zoë, and Deane Williams, editors. *The Grierson Effect: Tracing Documentary's International Movement*. Palgrave BFI, 2014.

Edgerton, David. *Britain's War Machine: Weapons, Resources, and Experts in the Second World War*. Oxford University Press, 2011.

Editorial. *Manchester Guardian*, March 31, 1944.

Ganser, Daniele. *NATO's Secret Armies: Operation* Gladio *and Terrorism in Western Europe*. Frank Cass, 2005.

Grieveson, Lee. "The Cinema and the (Common) Wealth of Nations." *Film and the End of Empire*, edited by Grieveson and Colon MacCabe. Palgrave Macmillan, 2011, pp. 73–113.

Hagener, Malte. *Moving Forward, Looking Back: The European Avant-garde and the Invention of Film Culture 1919–1939*. Amsterdam University Press, 2007.

Harris, José. *William Beveridge: A Biography*. Clarendon Press, 1977.

Hediger, Vinzenz, and Patrick Vonderau. "Introduction." *Films That Work: Industrial Film and the Productivity of Media*, edited by Vinzenz Hediger and Patrick Vonderau. Amsterdam University Press, 2009, pp. 9–18.

"Industrial Films Distributed by Aims of Industry." Pamphlet. University of Warwick, Modern Records Centre, Aims of Industry: Ephemera, MSS.21/2116.

Jacobson, Brian. "On the Red Carpet in Rouen: Industrial Film Festivals and a World Community of Filmmakers." This volume.

Kisch, Richard. *The Private life of Public Relations*. MacGibbon & Kee, 1964.

Lebas, Elizabeth. *Forgotten Futures: British Municipal Cinema 1920–1980*. Black Dog, 2011.

Lashmar, Paul, and James Oliver. "How MI6 Pushed Britain to Join Europe." *Sunday Telegraph*, April 27, 1997, p. 10.

Letter from Adamson to Tewson. June 17, 1949. University of Warwick, Modern Records Centre, Trade Unions Congress Papers, MSS 292/500/6.

Letter from Baker to Tewson. May 18, 1949. University of Warwick, Modern Records Centre, Trade Unions Congress Papers, MSS 292/500/6.

Letter from Boyfield to Jago. April 13, 1949. University of Warwick, Modern Records Centre, Trade Unions Congress Papers, MSS 292/500/6.

Letter from Edwards to Sewell. February 23, 1954. University of Warwick, Modern Records Centre, Papers of the Federation of British Industries, MSS 200/F/3/04/13/4.

Letter from Honey to Andrew. May 1, 1946. University of Warwick, Modern Records Centre, British Employers Confederation Papers, MSS 200/B/3/2/C822.

Letter from Pople to Fletcher. December 22, 1949. University of Warwick, Modern Records Centre, Trade Unions Congress Papers, MSS 292/500/6.

Letter from Sherwood to Sinclair. December 22, 1950. University of Warwick, Modern Records Centre, Papers of the Federation of British Industries, MSS 200/F/3/P1/1/12.

Letter from Storrs to Sewill. August 9, 1955. University of Warwick, Modern Records Centre, Trade Unions Congress Papers, MSS 292/500/6.

Letter from Tewson to Millar. November 28, 1952. University of Warwick, Modern Records Centre, Trade Unions Congress Papers, MSS 292/500/6.

MacDonald, Richard. "Evasive Enlightenment: World without End and the Internationalism of Postwar Documentary." *Journal of British Cinema and Television*, vol. 10, no. 3, 2013, pp. 452–74.

Marwick, Arthur. "Middle Opinion in the Thirties: Planning, Progress and Political 'Agreement'." *English Historical Review*, vol. 79, no. 311, 1964, pp. 285–98.

McLaine, Ian. *Ministry of Morale: Home Front Morale and the Ministry of Information in World War II*. Allen & Unwin, 1979.

Mercer, Helen, Neil Rollings, and Jim Tomlinson, editors. *Labour Governments and Private Industry: The Experience of 1945–51*. Edinburgh University Press, 1992.

Moore, Martin. *The Origins of Modern Spin: Democratic Government and the Media in Britain, 1945–51*. Palgrave Macmillan, 2006.

Porter, Bernard. *The Lion's Share: A History of British Imperialism, 1850 to the Present*. Pearson, 2012.

Purdon, James. *Modernist Informatics: British Literature and the Government of Information, 1900–1950*. Oxford University Press, 2015.

"Progress Report – April, 1955." University of Warwick, Modern Records Centre, Papers of the Federation of British Industries, MSS 200/F/3/04/13/4.

"Progress Report – August, 1955." University of Warwick, Modern Records Centre, Papers of the Federation of British Industries, MSS 200/F/3/04/13/4.

"Progress Report – February, 1955." University of Warwick, Modern Records Centre, Papers of the Federation of British Industries, MSS 200/F/3/04/13/4.

"Progress Report – January, 1949." University of Warwick, Modern Records Centre, Papers of the Federation of British Industries, MSS 200/F/3/04/13/4.

Rogow, Arnold A. "The Public Relations Program of the Labour Government and British Industry." *Public Opinion Quarterly*, vol. 16, no. 2, Summer 1952, pp. 201–24.

Rollings, Neil. "Cracks in the Post-war Keynesian Settlement? The Role of Organised Business in Britain in the Rise of Neoliberalism before Margaret Thatcher." *Twentieth Century British History*, vol. 24, no. 4, 2013, pp. 637–59.

Roulin, Stéphanie, Giles Scott-Smith, and Luc van Dongen, editors. *Transnational Anti-communism and the Cold War: Agents, Activities, and Networks*. Palgrave Macmillan, 2014.

Royal Commission on the Press, 1947–49: Report. His Majesty's Stationery Office, Cmd. 7700, 1949.

Russell, Patrick. "From Acorn to Oak: Industrial and Corporate Films in Britain." *Business Archives*, no. 107, 2011, pp. 53–76.

Russell, Patrick, and James Piers Taylor, editors. *Shadows of Progress: Documentary Film in Post-war Britain*. Palgrave Macmillan, 2010.

Sargeant, Amy. "GPO Films: American and European Models of Advertising in the Projection of a Nation." *Twentieth Century British History*, vol. 23, no. 1, 2012, pp. 38–56.

"Scottish Council." August 6, 1952. University of Warwick, Modern Records Centre, Trade Unions Congress Papers, MSS 292/500/6.

Sissons, Michael, and Philip French, editors. *The Age of Austerity 1945–1951*. Penguin, 1964.

Stollery, Martin. "Grierson's 'First Principles' as Beginning and Origin: The Documentary Tradition in the Field of Non-fiction Film." *Screen*, vol. 58, no. 3, September 1, 2017, pp. 309–31.

Stollery, Martin, and John Corner. "The Postwar British Documentary Movement – Decline or Transition?" *Journal of British Cinema and Television*, vol. 10, no. 3, 2013, pp. 387–94.

"Subject Headings: Aims of Industry Speaker's Bureau." Pamphlet. 1956. University of Warwick, Modern Records Centre, British Employers Confederation Papers, MSS 200/B/3/2/C82.

Sussex, Elizabeth. *The Rise and Fall of British Documentary*. University of California Press, 1975.

Swann, Paul. *The British Documentary Film Movement, 1926–1946*. Cambridge University Press, 1989.

Tomlinson, Jim. "De-industrialization Not Decline: A New Meta-narrative for Post-war British History." *Twentieth Century British History*, vol. 27, no. 1, 2016, pp. 76–99.

Tomlinson, Jim. *Democratic Socialism and Economic Policy: The Attlee Years, 1945–1951*. Cambridge University Press, 1997.

Trentmann, Frank. *Free Trade Nation: Commerce, Consumption, and Civil Society in Modern Britain*. Oxford University Press, 2008.

Ussishkin, Daniel. "Morale and the Postwar Politics of Consensus." *Journal of British Studies*, vol. 52, no. 3, 2013, pp. 722–43.

Vernon, James. *Hunger: A Modern History*. Harvard University Press, 2007.

Wilson, H.H. "Techniques of Pressure – Anti-nationalization Propaganda in Britain." *Public Opinion Quarterly*, vol. 15, no. 2, Summer 1951, p. 227.

Winston, Brian. *Claiming the Real: The Documentary Film Revisited*. British Film Institute, 1995.

"Winter Programme 1954/55." University of Warwick, Modern Records Centre, Papers of the Federation of British Industries, MSS 200/F/3/04/13/4.

Zweiniger-Bargielowska, Ina. *Austerity in Britain: Rationing, Controls, and Consumption, 1939–1955*. Oxford University Press, 2000.

About the Author

Scott Anthony is Deputy Head of Research and Public History at the Science Museum Group in the UK. His books include *Shell: Art and Advertising* (2021), *Public Relations and the Making of Modern Britain* (2012), *The Projection of Britain: A History of the GPO Film Unit* (2011, co-edited with James G. Mansell) and a forthcoming history of film propaganda for BFI/Bloomsbury. He is currently PI of the 'Museums and Industry: Long Histories of Collaboration' (MaILHoC) project, funded by the Joint Programming Initiative on Cultural Heritage, which examines the ethics of industrial patronage in historical perspective.

Section 5

Post/Colonial Industries and Third Industrial Cinemas

17 Framing Local and International Sentiments and Sounds

Unilever and Royal Dutch Shell in a Changing Nigeria

Rudmer Canjels

Abstract
The period after World War II was one of flux and change, during which many countries became independent of colonial rule. This chapter examines the use of locally made films by Unilever and Shell in relation to one of these new independent countries – Nigeria. It will highlight how these international companies began using film in Nigeria as a strategic medium, trying to create a bond between the companies and the films' viewers – especially through sound – and showing that British and Dutch companies were actively involved in the decolonization process. This chapter also underlines the significance of the international processes and dynamics of industrial film, which not only serve as advertisement for products, processes or company images, but are also shot through with, and transmit, important technical, political, cultural and social discourses.

Keywords: documentary; crude oil; palm oil; post-colonial film; Nigeria; industrial film

While advertising is designed to sell a product or service through media exposure, public relations is more concerned with the creation and maintenance of a company's positive image and reputation within a society. According to public relations director D.H. Buckle from Unilever's subsidiary the United Africa Company (UAC), public relations is "a continuous effort […] to promote the best possible understanding of its policy, activities and

Hediger, V., F. Hoof, Y. Zimmermann, with S. Anthony (eds), *Films That Work Harder: The Circulation of Industrial Film*. Taylor & Francis Group 2024
DOI: 10.5117/9789462986534_CH17

social attitudes among those outside who influence its development."[1] Rather than requiring immediate results with advertising, long-term means and strategies were important for public relations. This kind of promotion could, according to Buckle, be done by producing press releases, booklets, exhibitions, company magazines, speeches and films.

As indigenous nationalist movements grew in the 1950s, expatriate companies sought ways to represent their interests to local Nigerian audiences as well as the global market. The measured use of public relations was very important to direct the public carefully to better understand the company and its relation to the changing socio-political context, seeing the company as a good citizen playing an important role in economic and social development.[2] In this chapter I will show that during the difficult transitional period around the time of the independence of Nigeria (October 1960), Unilever/UAC and Royal Dutch Shell began using film in Nigeria as a strategic medium to try to create a bond with the viewer, actively addressing the viewer and making links to the company, especially through sound.[3] The films produced in Nigeria by these two companies were meant not only for a local Nigerian audience, but were distributed internationally as well, promoting both the local and global corporate image. This research thus opens up perspectives on the complex international, national and local construction of a corporate image, the dynamics of differentiated marketing communications, the cultural, social, economic and political dimensions of media forms, especially in the transition to a post-colonial setting.

Unilever and the United Africa Company

In 1930 the British soap-making company Lever Brothers merged with the Dutch Margarine Union to form Unilever, which remains until today one of the world's largest multinational consumer goods companies. At the time of its formation, Unilever's most important subsidiary was UAC, which

1 Buckle, D.H. *Newsletter*, no. 4, October 1967, pp. 1–2. Unilever Archives & Records Management (UARM), United Africa Company collection (UAC), 1/11/1/7/2.

2 Idem. *Public Relations and Advertising Services*. October 1967. UARM, UAC/1/11/1/7/2.

3 Both firms mostly gloss over the diverse ethnic groups and linguistic diversity that are present in Nigeria (such as belonging to Igbo, Hausa, Fulani, and Yoruba). This research is part of a larger study on Unilever and Shell films in Nigeria during Independence; see Canjels, Rudmer. *The Dynamics of Celluloid on the Road to Independence: Unilever and Shell in Nigeria*. Netherlands Institute for Sound and Vision, 2017. Sponsored by the (HERA) EU-funded project *Technology, Exchange & Flow*.

concentrated on West African trade. Palm oil was a major raw material for both margarine and soap, and so supplies from British West Africa (where Lever Brothers was already trading in the 1910s) were an important export product for Unilever. Unilever sold its merchandise – from food products to clothing, from bicycles to building materials – in West Africa through UAC, and the subsidiary ensured that Unilever (as well as other traders) would get supplies of raw materials such as palm oil, cocoa or groundnuts. Modern-day Nigeria originated in 1914 as the British formally merged the British Southern and Northern Nigerian protectorates. It was the most populous country in Africa and formed a very important import and export market for UAC.[4]

Between the 1900s and the 1920s Unilever had occasionally commissioned a number of factory, informational and promotional films (including some filmed in Nigeria).[5] In order to professionalize and coordinate the company's production of these films, an administrative department called the Film Section was set up in London's Watergate House in 1949 with a small staff.[6] Rather than producing its own films, the Film Section commissioned films from independent British production companies. After two films on the history of the company, other subjects were also undertaken, such as Eskimo life and their teeth, the harvesting of peas, and the importance of the forklift truck. The 16 mm or 35 mm films were distributed free of charge for educational, industrial or household use. From the late 1950s production picked up and the films' subjects became more scientific. In 1954, the Unilever film catalogue contained only twelve films; ten years later forty-two films could be rented. Some of these educational films never mention the Unilever brand name or show Unilever products. From the early 1960s onwards, a greater variety of multiple language versions were made. Unfortunately, not much is known about actual distribution and audience numbers. According to Unilever itself, in the early 1960s seventy per cent of the UK bookings from the Unilever film library were for primary and secondary schools. At that time the UK library had around 28,000 rentals a year with an estimated audience of 2.8 million. International rentals with copies on permanent loan to other organizations accounted for an additional two million viewers.[7]

4 Fieldhouse, David Kenneth. *Merchant Capital and Economic Decolonization: The United Africa Company, 1929–1987*. Oxford University Press, 1994, pp. 3–9.

5 More information on this period, see Canjels. *The Dynamics of Celluloid*.

6 "Unilever Presents...." *Unilever House Magazine*, vol. 8, no. 5, October–November 1955, p. 3.

7 "When You Borrow a Unilever Film." *Film User*, February 1963, p. 77.

Filming UAC

It was in the period of initial growth of the film catalogue that Unilever made UAC's first film as well as Unilever's first colour film, *The Oil Rivers* (1956). *The Oil Rivers* deals with the harvesting of palm fruit and the production of palm oil in Nigeria. The film was quickly followed by two other films concerning Nigerian trade: *The Twilight Forest* on timber and *Traders in Leather* on hides and skins.[8] All three films were made by the British film company Editorial Film Productions, which had worked for Unilever as well as for the Colonial Office. All technical personal came from the UK, with the same director, Sydney Latter, and the same editor and producer. Each film cost between £18,000 and £23,000, a large sum for non-theatrical subjects.[9]

The three films showing three different UAC products in Nigeria deal with the positioning of old versus new, of old habits and crafts versus newly introduced technology and knowledge. As was common in colonial films, the UAC films underline the idea that through the help of the West (in the guise of an international company), an internal, progressive change was made possible, whereby the colony and its independent successor state could develop economically and modernize quickly along Western lines. As was also noted in the British press, the films hardly explained any scientific ideas, but functioned rather as "interest film" – an informational documentary giving an overview of a trade.[10]

There are no local Nigerian-language versions known for the UAC films, so translation by a commentator would sometimes have been necessary, creating the risk of transforming a film.[11] The films make use of a typical British-sounding male voice-over with clean and crisp diction, sustaining a decidedly Western point of view. In *Traders in Leather*, for instance, the film voice-over introduces the Fulani herdsmen as "a strange nomadic people," who live their "simple pastoral life," while nobody knows much about their origins or customs which "are buried in the ages before history." As the Fulani village is introduced, Western string music is played, which turns into to the

8 Buckle, D.H. "UAC Public Relations Director." *Newsletter*, no. 3, September 1967, p. 1. UARM, UAC/1/11/1/7/2. In the 1950s in cooperation with UAC, *Design for Africa* within *Unilever Magazine* 2 (1956) and *The Surf Boats of Accra* (1958) were also made. As these films do not focus on Nigeria specifically, these films will not be dealt with.

9 An equivalent of around £330,000 and £400,000 in today's money.

10 "The Seeds of Change." *Monthly Film Bulletin*, no. 276-277, 1957, p. 36.

11 Rice, Tom. "Working through the End of Empire." This volume.

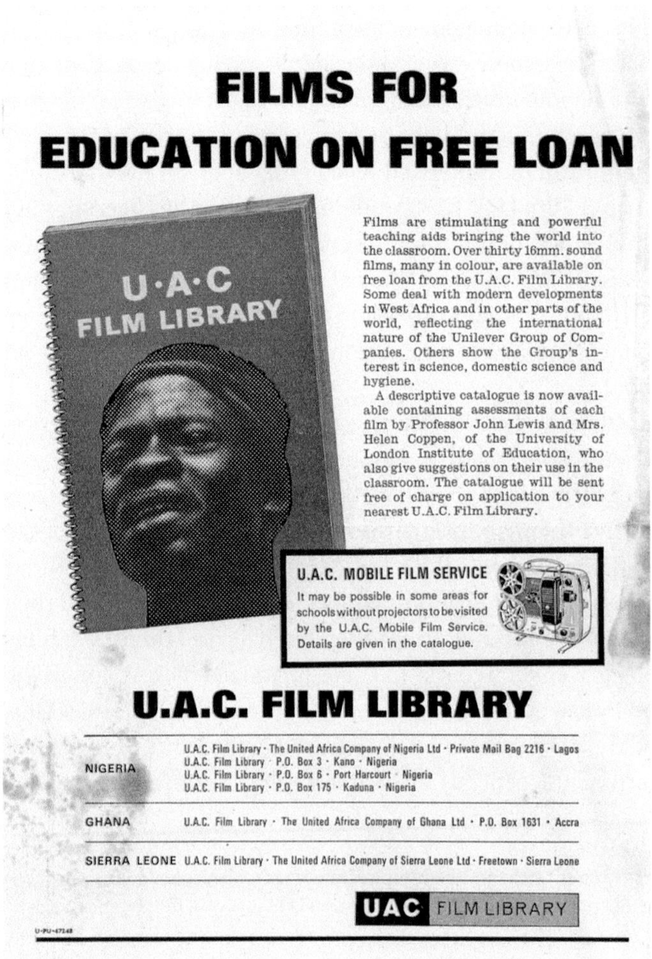

Fig. 17.1. Over thirty Unilever films were available in the Nigerian UAC film library in the early 1960s, suggestions on their classroom use were supplied. Reproduced with kind permission of Unilever from an original in Unilever Archives. Advertisement file, UARM UAC/1/11/21/8.

sound of joyous plucking violins as a woman carries a bowl on her head.[12] It places the woman in a humorous aural surrounding which, together with the condescending voice-over, seems to be done for the Western viewer. One wonders if and how the film would be used within Nigeria itself as it uses mostly stereotypical images. UAC/Unilever did not make use of mobile cinema vans touring the country yet, which were a common and popular method of showing films in Nigeria, also used for government-sponsored

12 There were two versions available for UK schools, one "virtually free from musical background"; it is unclear if the violins were still present in this version. ("Traders in Leather." *Visual Education*, August 1958, p. 10). The music was composed by Elizabeth Lutyens.

colonial films.[13] A UAC mobile cinema system would only be established in 1960; until that time expenses were considered too high to send cinema vans to visit more remote villages and schools.[14]

However, UAC probably did benefit from the familiarity of the non-theatrical presentation method used by colonial sponsored film units, such as the Colonial Film Unit (CFU) and its successor, the Nigerian Film Unit (NFU). UAC public relation managers were equipped with 16 mm projectors, enabling them to show films at staff clubs and travel to their audience.[15] Each public relations office and trading branch, of which there were several around Nigeria, had a collection of films which could be loaned to Nigerian schools.

Politicians, traders, and teachers were seen as a key section of the community to win over, especially to make them realize "that our business is an integral part of the life of their constituency and that our Managers are available to give them an insight into the functioning of much of the economic life of their country."[16] Both *The Oil Rivers* and *The Twilight Forest* were shown to Nigerian state commissioners in December 1956 and then went into general release in January 1957.[17] In March 1958 all three UAC films were shown during a cocktail party in Lagos organized by UAC for all the members of the House of Representatives.[18] Commercial companies thus clearly followed the path created by the NFU, whereby films were used to connect to governmental officials.[19] These screenings for officials would be

13 Brian Larkin writes more on the fascinating use of media technologies and the use of mobile cinema in colonial and post-colonial Nigeria. Commercial cinema was, for instance, seen by mainstream Hausa society as a lower-class and un-Islamic activity; mobile cinema was accepted, however. Larkin, Brian. *Signal and Noise: Media, Infrastructure, and Urban Culture in Nigeria*. Duke University Press, 2008, pp. 123–24.

14 *Summary of Public Relations: Proposals Previously Considered and Discarded*. C. 1959. UARM, UAC/1/11/4/1/2; "Film Rover for Nigeria." *UAC News*, vol. 10, no. 1, January 1961, p. 2. *UAC News* is located at UARM, UAC/2/19/3/6/1/1.

15 "Film Shows at Kano." *UAC News*, vol. 5, no. 6, June 1956, p. 9.

16 *Summary of Public Relations: Summary of Company's Public Relations Activities in Commonwealth West Africa*. C. 1959, pp. 6–7. UARM, UAC/1/11/4/1/2.

17 Technical and Distribution Records of *The Oil Rivers* and *The Twilight Forest*. UARM.

18 "Film Show for Federal Legislators." *West African Pilot*, March 12, 1958, p. 3.

19 The films were also used by government, as the Eastern Nigeria Ministry of Information ordered the Unilever films to be screened by the governmental mobile units in 1961. Robson. *Report on Public Relations 1961/62*. 1961, p. 7. UARM, UAC/1/2/3/6/1; for more on the NFU, see Rice, Tom. "From the Inside: The Colonial Film Unit and the Beginning of the End." *Film and the End of Empire*, edited by Lee Grieveson and Colin MacCabe. Palgrave Macmillan, 2011, pp. 135–53; and Smyth, Rosaleen. "Images of Empires on Shifting Sands: The Colonial Film Unit in West Africa in the Post-war Period." *Film and the End of Empire*, pp. 155–75.

progressively used by UAC and other companies in the years after Nigeria's independence.[20] Compared to Unilever's public relations in the UK, UAC in Nigeria thus used film on a more political level from the onset. To contact these groups, press releases, goodwill advertisements, staff magazines and economic reviews also were used.[21]

Trading for Nostalgia

While the opening credits claim that the films were originally made for Nigeria, they had an extended international distribution life. However, when the films were advertised for non-theatrical release in the UK, they were framed mostly as a farewell, a salute to the country's "primitive" way of life, a colonial image of a nostalgic time.

An advertisement for the three Nigerian films together illustrates the complicated issues around the reuse of Nigerian films in Britain. In October 1958 an advertisement entitled "The New Face of Africa" showed a drawn image of one of the most famous images in African art, the sixteenth-century bronze head of Queen Mother Idia, a powerful monarch at the Benin court at that time.[22]

The Benin Empire was located in what is now southern Nigeria and was one of the oldest and most developed regions in West Africa, dating back to the twelfth century. The head is slightly turned, showing off the naturalistic details of the face against a clean white background. The adjacent text tells of the "recently discovered Benin bronzes" showing a "highly developed culture" with "superb technical and artistic qualities." Functioning as a link to the films that also showcase these qualities, the text explains how "today, West Africa is emerging from a long period of obscurity, with British ideas and techniques contributing to a new social and cultural order."

The advertising image of the head is in fact a reversed drawn copy of an often-used photograph (with some added lines across the face). The anonymous photograph was first used in the British Museum's *Handbook to the Ethnographical Collections* (1910) and would be reprinted in several books during the 1910s and 1920s, such as Carl Einstein's famous

20 "Premiere of 'Enterprise in Nigeria'." *UAC News*, vol. 11, no. 1, January 1962, p. 3; "'African Awakening' Launched in Nigeria." *UAC News*, vol. 11, no. 8, September 1962, p. 3.

21 Stanton, J.L. *Group Public Relations in Commonwealth West Africa*. December 1, 1960, p. 4. UARM, UAC/1/11/6/1.

22 Mack, John. *Africa, Arts and Cultures*. British Museum Press, 2000, p. 14.

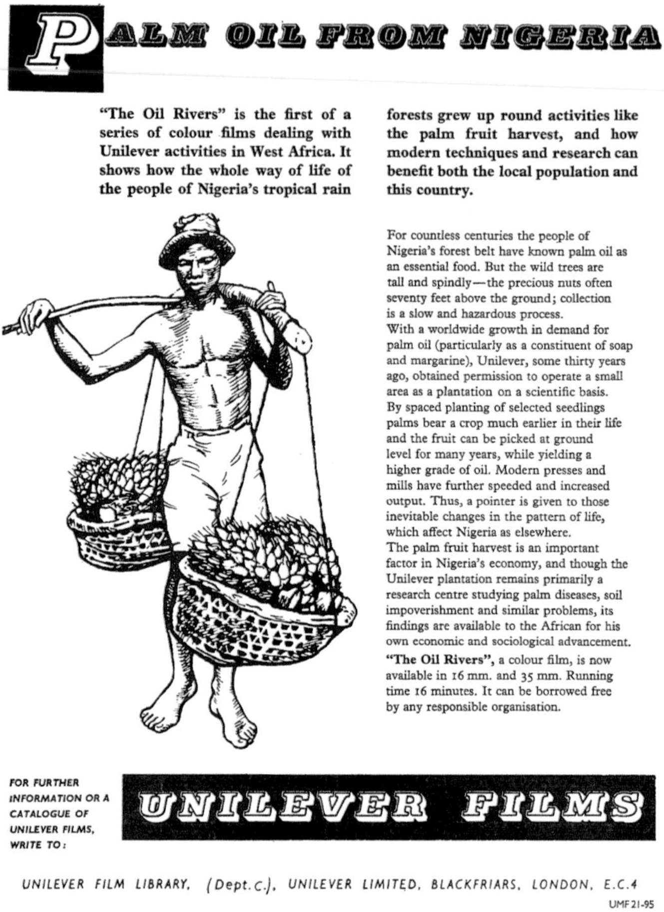

Fig. 17.2. The UK advertisements for the first three UAC films conjure up a pre-industrial world, in line with traditional colonial images. Advertisement, *Film User*, March 1957, p. 93.

Negerplastik (1915). This picture would, together with other images of African statues, help to promote new ways of seeing non-Western images in a modern and avant-garde light, and influence cubism and German expressionism, in the years after its publication.[23] One has to realize, however, that within the photographed and drawn image is embedded an important colonial history, one that also relates directly to the trading of palm oil. The bronze head was given to the British Museum by Sir William Ingram in 1897 after a "punitive expedition" was formed to capture the king and destroy the city of Benin in the Benin Empire, which had until

23 Grossman, Wendy A. "From Ethnographic Object to Modernist Icon: Photographs of African and Oceanic Sculpture and the Rhetoric of the Image." *Visual Resources*, vol. 23, no. 4, 2007, pp. 293–94.

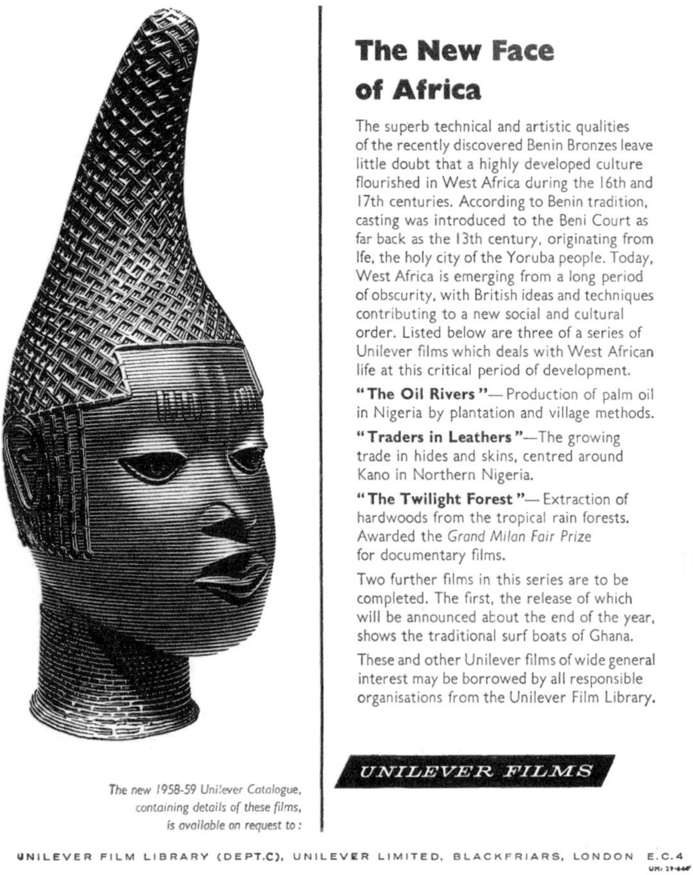

Fig. 17.3 The new face of Africa is represented by the sixteenth-century bronze head of Queen Mother Idia, emerging from a long period of obscurity. Advertisement, *Film User*, October 1958, n.p.

then managed to retain its independence. The expedition was in response to the defeat of a previous British invasion in 1896, brought about by a Benin trade embargo of palm oil against the British.[24] As a result, the city and the palace were looted and much of it burned down. Some two thousand artefacts were sent to Europe, where they were auctioned to the most prominent world museums, galleries and private collections in order to finance the expedition.[25] The image of the bronze head is embedded, therefore, in a very troubled and violent colonial past. Both the photograph and the drawing represent not just the bronze head, but refer to and reiterate the manner in which African artefacts were reused

24 Home, Robert. *City of Blood Revisited: A New Look at the Benin Expedition of 1897*. Rex Collings, 1982, p. 27.

25 Ibid., pp. 88–89 and 100–01.

and repurposed within a modern Western society, away from its original place and use.

Not long after the release of the three films in Nigeria and the UK, the UAC started to withdraw from trading its central commodities, cocoa, groundnuts and palm products, and would quit these and other produce businesses by 1961.[26] However, as late as 1966 the films were still used in schools to teach geography and current affairs.[27] According to the 1967 Unilever catalogue, *Traders in Leather* was an "example of the way in which British knowledge and resources have helped a young and recently developed territory towards a greater prosperity and self-reliance."[28] The films continued to be shown in the West, solidifying the assertion that Unilever had provided knowledge and opportunities to Nigeria, even though the country's descent into civil war in 1967 quickly rendered this representation obsolete.

The Shell Film Unit in Nigeria

Since 1903, several international oil companies had tried to locate and exploit mineral oil in on the territory of what was later to become Nigeria, largely to no avail. Only in 1956 were commercially viable quantities of oil found by a joint venture of Royal Dutch Shell and British Petroleum, at that time the sole remaining oil company in Nigeria. The Nigerian Shell-BP films produced in relation to these activities were produced by the Shell film unit rather than BP's.

Shell management recognized film as an ideal medium for reaching out and building public support for its activities and started the Shell Film Unit (SFU) in London in 1933 after consulting John Grierson. For several decades, Shell would produce hundreds of documentaries, often dealing with scientific and technological subjects, distributing them mainly non-theatrically in cooperation with educational and cultural organizations.[29] Rather than through outright propaganda, Shell films sought to increase the company's prestige and standing through association. The company name appeared only in the credit titles. As the company's global reach expanded, additional film units were set up around the world. According to Shell's own

26 Fieldhouse. *Merchant Capital and Economic Decolonization*, p. 391.

27 "Three Million at School." *Film User*, May 1966, p. 215.

28 *Unilever Films, Catalogue 1967–1968*. Unilever Ltd., 1967, p. 8.

29 For a historical overview of the SFU, see Canjels, Rudmer. "From Oil to Celluloid: A History of Shell Films." *A History of Royal Dutch Shell, Vol. 4*, edited by Jan Luiten van Zanden. Oxford University Press, 2007, pp. 6–33.

public relations research at the time, in 1960 the international audience for the company's films had grown to 45 million, with films showing on a regular basis in around thirty countries.[30]

The Nigerian Shell film unit was set up in 1959, initially for two years, but it would last for four, centred in Lagos. Douglas Gordon had joined the SFU as a trainee in 1954 and had been assistant director to Bert Haanstra's *The Rival World* before directing several Shell films himself, such as *The Ruthless One* (1956) and *Oil in Asia* (1958). Gordon was sent with a small crew to Nigeria to train Nigerian technicians, mostly students, and act as manager and producer.[31] The films would be made by a self-contained unit in Nigeria. Only the final stages of the film post-production and sound editing would, together with adding the commentary and music recording, be done in London.[32] Though more localized compared to the set-up by Unilever, where the whole crew was from the UK, the "local" Nigerian films still had to travel to London before they could be screened in Nigeria (similar to the governmental CFU/NFU productions). The unit would consist of an expatriate producer as well as a director, cameraman and editor, each having a Nigerian assistant. The expatriate staff would train the assistants in directing, camera work and editing; perhaps inspired by the CFU training schools.[33] Originally it was thought that Nigerian assistants might eventually take over their jobs, but it seems to have never gotten that far.[34] Other staff members were a projectionist/handyman, a clerk and a driver, all from Africa. The cost of the Nigerian film unit was estimated at around £50,000 a year (around £750,000 today).[35]

The unit was set up to secure and propel Shell-BP's image as a benefactor that brought vital economic assets to the country and had a wealth of experience which, according to the company, were both needed by Nigeria. As Douglas Gordon wrote: "The successful 'projection' of this image may well prove an essential factor in the Company's unhampered prosecution

30 Drummond, John. *Shell Film Operations*. December 16, 1960, p. 2. Shell Film & Video Unit (SFVU) archive, PAC 21 B SFU 14.

31 His initial crew was editor Roy Ayton, director Frank Nesbitt, and cameraman Maurice Picot. Director Richard Taylor and Philip Owtram came later. Among the Nigerian assistants was the later accomplished film producer and documentary maker Levi Ezeasor.

32 Gordon, Douglas. *Films in Shell-BP: Interim Report*. June 2, 1959, p. 5. EYE Filmmuseum Amsterdam, Bert Haanstra Archive (Haanstra Archive), 136/A 6-4.

33 Rice, Tom. "Working through the End of Empire." This volume.

34 Gordon. *Films in Shell-BP: Interim Report*, p. 6.

35 "Minutes of 18th meeting, Group Film Committee." The Hague, July 2, 1959, p. 6. Haanstra Archive.

of its business."[36] The fact that Nigerians were working on these films was seen as beneficial to Shell-BP's public image as well. Film was, according to Gordon, particularly suited for this job. The first films would be made "to win general approval rather than [provide] detailed understanding. [...] The emphasis should be on movement, drama, human interest, and technical intricacies [should be] studiously avoided." The films had to be made in English, but spoken by a Nigerian, avoiding "detailed technical exposition and concentrate rather on dramatic action and human interest, but otherwise make no special allowances which might give the appearance of condescension."[37] This personal touch and dramatic focus was rather different from the traditional Shell film made by the unit in London, which were indeed more technical overall. In its four years of existence, the Nigerian unit would make around ten films. The first two, made by the same team of a British director, cameraman, and editor, each having a Nigerian assistant, were *The Search for Oil in Nigeria* and *Oilman's Move*, both released in 1961. The two films were first shown, according to Gordon, at several Independence Festival celebrations in Nigeria.[38]

Local Voices and Global Views

Compared to films made by the SFU in London, the two films indeed focus a lot more on the human interaction, instead of various technical and scientific methods of discovering oil. The use of technique and equipment is in *The Search for Oil in Nigeria* and *Oil Man's Move* mostly linked with human endurance, of crossing swamps, drilling, setting off explosives, driving through muddy roads, constructing and tearing down oil rigs, and of battling nature with its wet and humid climate. Like all Shell films from that period, there is usually no synchronized speech within the frame, and commentators are only heard, not seen. This might have been a consequence of the many international language versions that were needed, making it much easier to change the language of the unseen voice-overs. Most Shell commentaries explained scientific and technological advances to the audience by using a voice-over in "received pronunciation," the standard English accent spoken in the south of England. *The Search for Oil in Nigeria* is not narrated by a native British speaker but by the Nigerian writer and political activist Wole

36 Gordon. *Films in Shell-BP: Interim Report*, p. 1.

37 Ibid., pp. 2–3.

38 Gordon. Transcript interview recorded May 18, 1983, p. 8. SFVU archive, PAC/21 B SFU 11.

Soyinka (who later received the Nobel Prize in Literature).[39] Using local voice talent for internationally released films was a very recent development. From the early 1960s, however, it became common practice within Shell to use local voice-over talent for the English version, communicating for viewers that they were seeing an authentic portrayal of Nigerians working in the oil industry, lessening the colonial feel.[40]

Like the voice-over, the music used in these Shell films was related to Nigeria. The music for *Oilman's Move* uses a varied musical palette, performed by the highlife bands of Ambrose Campbell and Fitzroy Coleman, who had both worked in London and greatly influenced London's highlife and jazz community in the 1950s. The term "highlife" came into use when prestige entertainment bands added versions of local dance tunes to their repertoire of European music, blending traditional West African rhythms and Caribbean calypso melodies with European musical elements, and thus transcended class and cultural markers. In Nigeria the very popular highlife music symbolized progress, modernity, internationalism and the independence of the country, thereby making it an interesting, but fitting, choice for Shell to use in its promotional campaign.[41] During the film the up-tempo music with its rhythmic jazz tunes returns every now and then, mostly when showing the slow-moving transport of trucks, making the work and move of the oil rig seem more fluent. In the middle of the film a calypso sung by Nigerian-born musician Samuel Akpabot, explaining the troubles the trucks are having travelling on the watery roads, even getting stuck. A calypso is a type of music that originated in the West Indies, and is characterized by improvised lyrics on topical, political or broadly humorous subjects. Like Campbell and Coleman, Akpabot had studied and played in London.[42]

The music in *Oilman's Move* (as well as *The Search for Oil in Nigeria*) thus weaves the sounds and music of immigrants together with those influenced by British and Western culture, tapping into a rich cultural mixture of high

39 *African Awakening* (1962) from UAC would also use Soyinka's voice and show scenes from one of his plays performed in Ibadan. Soyinka lived and worked in London during 1954–1959.

40 One exception is the films from Shell-Burmah. In the 1950s this local unit used Indian voice talent, such as in *The Weavers of Maindargi* (1956) and *Maharajah Meets a Challenge* (1959). These films were not released internationally, however.

41 Collins, John, and Paul Richards. "Popular Music in West Africa." *World Music, Politics and Social Change*, edited by Simon Frith. Manchester University Press, 1991, pp. 12–46.

42 Akpabot would continue to make calypsos for industry. In UAC's *The New Traders* (1962) Akpabot's prominently placed calypso comments (with comical elements) on the new opportunities of trading in West Africa.

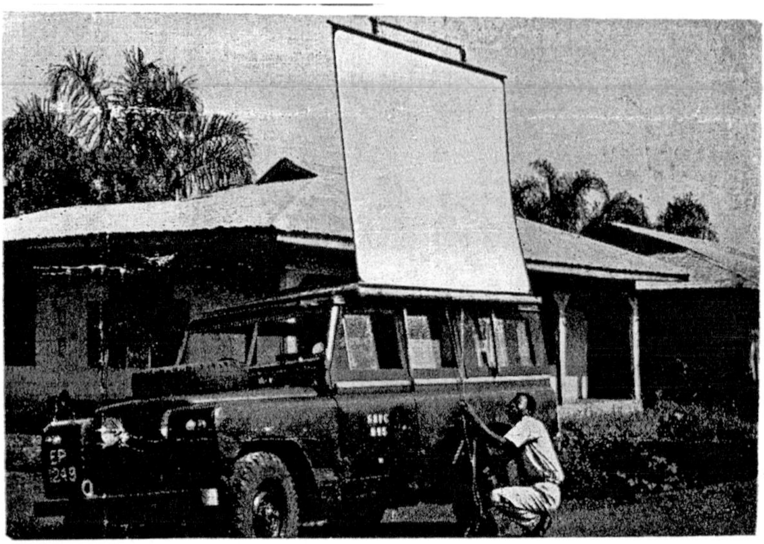

Fig. 17.4. One of Shell-BP's "cinerovers" that would tour various districts and regions of Nigeria with a 16 mm projector, a generator, speakers and a collapsible screen on the roof. "More Requests for Films," *Shell-BP Bulletin*, no. 1, May 1963, p. 3.

and low, the old and new, urban progress and the rural, the local and the foreign. The use of non-received pronunciation, using dialects, even the heavy accents in the calypso, must also have related to the idea of intimacy and authenticity, qualities Grierson advocated as the ideal aim of a good commentary.[43] Thus, just by listening to the commentary track and music, Nigerian Shell films signalled they were not something from the typical Shell mould.

There exist two versions for *The Search for Oil in Nigeria*, a Shell-BP version as well as an adjusted Shell version for international release, trimmed by eight minutes. The international version does not refer to BP at all, and to Shell only in the credits (as was common for most international Shell films at the time). The original thirty-minute Shell-BP film is much clearer about its sponsorship, mentioning Shell-BP several times in the voice-over as well in the calypso song. The original version also references local points of interest and specific towns, showing and introducing more Nigerian workers and villagers. So, there was a clear distinction between the use of and the rhetoric attached to the films for national and international distribution. Both *The Search for Oil in Nigeria* and *Oilman's Move* had a

43 Stollery, Martin. "Voiceover/Commentary." *The Projection of Britain: A History of the GPO Film Unit*, edited by Scott Anthony and James G. Mansell. Palgrave Macmillan, 2011, p. 172.

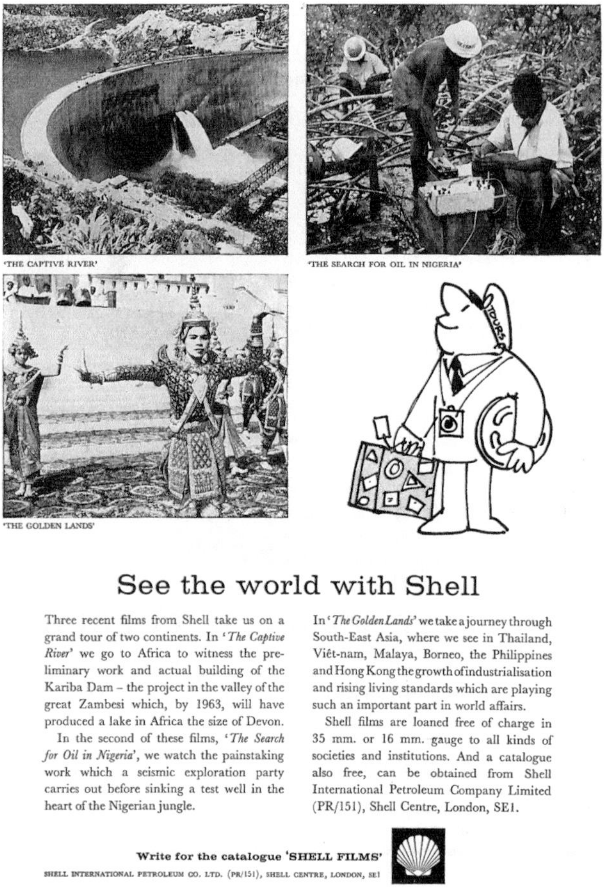

Fig. 17.5. Shell's tour operator invites everybody to take a journey to far away exotic lands. Advertisement, *Film User*, November 1962, p. 567.

long international distribution history; in 1971 they were still listed in Shell's international film catalogue.

Unfortunately, little is known about the initial release of the first Shell-BP films in Nigeria, but it is safe to assume that they travelled on mobile cinema vans and were offered for distribution.

Neither the *Shell-BP Bulletin* nor regional Nigerian newspapers contain advertisements for these films, and they are barely mentioned. In the UK, on the other hand, *The Search for Oil in Nigeria* was advertised in specialized film magazines, with ads showing a drawing of a Shell tour operator inviting everybody to see the world with Shell, highlighting three stills from exotic looking films made by various local Shell film units.[44]

44 Advertisement 1962. Advertisement file, UARM, UAC/1/11/21/8.

The advertisement thus positions the locally made films within an exotic framework, suggesting that they allow the viewer to travel in the cinema to experience unknown cultures and different people. The critical reaction to the films in the UK was limited to a few general comments, rather than extended reviews.[45]

Framing Nation and Enterprise

At the end of 1961 UAC released the first of three new films on West Africa. The three films had a different focus than those of 1957. *Enterprise in Nigeria* was the first to be released and was the only one of these three that dealt exclusively with Nigeria.[46] The film was in colour, twenty-one minutes long, and was available in 16 mm and 35 mm. It was made by a different team and production company than the first three UAC films, though Sydney Latter returned as director. *Enterprise in Nigeria* was previewed in November 1961 at the Unilever House in London. The high commissioner for the federation of Nigeria in the UK was present, as were representatives from Northern and Western Nigeria, and Sierra Leone.[47] A month later the film was shown several times in Nigeria at private and public showings for state rulers, ministers, business delegates and industry leaders.[48]

Part of the change that can be seen in the three new UAC films must be found with a different focus that the public relations department was taking up. According to internal communications in 1960, Nigerian independence necessitated a re-examination of the public relations programme, retuning its activities to present events and the future of an independent Africa. By focusing more on the present MPs, ministers, schoolteachers, but also those who would make the leaders of tomorrow (students, business men, members of trade unions), the department thought the importance of the operations of the company would become more known and prevent the company from being characterized as "an unhealthy anachronism in an independent African Country."[49] Thus, much more than in 1957, film was

45 "The Search for Oil in Nigeria" [review]. *Film User*, September 1961, p. 487; "The Search for Oil in Nigeria" [review]. *Industrial Screen*, January–February 1962, p. 282.

46 Because of this, the other two, *The New Traders* and *African Awakening*, will not be dealt with in this research.

47 "London Preview of Enterprise in Nigeria." *UAC News*, vol. 11, no. 1, January 1962, p. 2.

48 "Premiere of 'Enterprise in Nigeria'," p. 3; "Enterprise in Nigeria Launched in Ibadan." *UAC News*, vol. 11, no. 1, January 1962, p. 10; "Enterprise in Nigeria." *UAC News*, vol. 11, no. 2, February 1962, p. 10.

49 Stanton. *Group Public Relations in Commonwealth West Africa*, pp. 1–4.

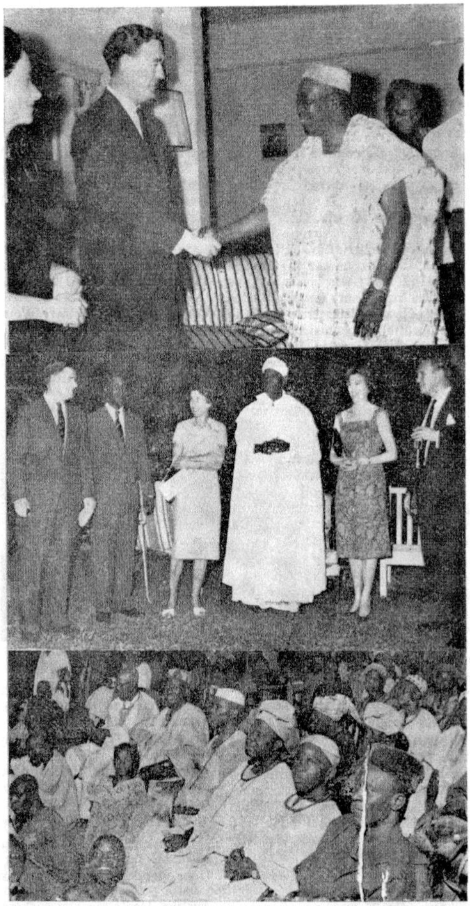

"ENTERPRISE IN NIGERIA"

The Company's latest film "Enterprise in Nigeria" has been shown in the three districts of The United Africa Company of Nigeria Limited in Western Nigeria — Ibadan, Ilesha and Benin/Warri. In Ibadan the film was shown to His Excellency the Governor, the Premier and other distinguished personalities. Above are some of the pictures taken during the shows: Top: the Regional Director — Mr. W. Hardy being received by His Excellency the Governor Sir Adesoji Aderemi at a private showing at Government House. Middle: The Oba of Benin, Akenzua II, was one of the guests at the party which preceded the showing of the film at Benin. Bottom: The Owa of Ilesha, Oba Biladu III, and President of the Western House of Chiefs, watching the film at a special showing in Ilesha with other distinguished personalities in the town.

Fig. 17.6. After the premiere, *Enterprise in Nigeria* toured several districts in western Nigeria and was shown to state rulers, governmental officials and other distinguished personalities. "Enterprise in Nigeria." *UAC News*, vol. 11, no. 2, February 1962, p. 10. Reproduced with kind permission of Unilever from an original in Unilever Archives, UARM UAC/2/19/3/6/1/1.

treated actively within public relations, with premieres for relatively small groups, while the mobile cinema and the film library were used to reach larger numbers.

Compared to the earlier UAC films (as well as the Shell films), *Enterprise in Nigeria* clearly takes place in modern Nigeria and does not take place in a small town or in the tropical forest. Taking centre stage now is the city of

Lagos, which had become an attractive symbol of the new, modern Nigeria, a place far away from traditional rural life. Signifying progress through speed and modern architecture, *Enterprise in Nigeria* starts with a moving shot of a Nigerian on a motorcycle driving through the city. The shot lasts one minute during which the credits are shown, and relaxed jazzy music is heard, written by the London-based, Jamaican-born jazz pianist Yorke de Souza. The man on the motorcycle stops and enters the Niger house, headquarters of the UAC, and the voice-over by Eldred Fiberesima begins.[50] While images of new buildings such as a school, a university and a hospital are shown, the voice-over explains that the predecessors of UAC had been in Nigeria from the seventeenth century (trading with "their friends") and have helped Nigeria to prosper. Then a long section, which takes up most of the film, introduces the viewer to various factories.

One specific mode of address in the film becomes strikingly clear when one compares *Enterprise in Nigeria* to the Dutch version, *Nigeria bouwt aan zijn toekomst* (Nigeria Is Building Its Future). The English version seems especially keen on speaking to the viewer on an intimate level, connecting the government, citizens and workers to the company; this is much less so in the Dutch version.[51] The voice-over is by a white Dutch person, speaking standard Dutch, and instead of "we" and "you" he uses "them" and "they." While the voice-over in the English version tells about the history of Nigeria and the country's relation with UAC, of growing up together, of learning the value to youth of being young in a young land, in the Dutch version the accompanying images of new buildings are shown without any commentary. In the English version, images of plastic water pipes being laid by Nigerians, are accompanied by the following commentary: "Made by Nigerians, laid by Nigerians, used by Nigerians. Look as you ride by, the new age might be under your feet." The Dutch voice-over condenses this into the observation that the pipes are "made in the country itself." Where the English version claims that "every tree that falls brings prosperity for everyone: government, company and worker," in the Dutch version the tree just brings "economic prosperity." And, finally, when the English version claims that the skills learnt will make the trainees "an asset to the nation," no such mention occurs in the Dutch version. Much more than the earlier UAC films or the Dutch version, the English version of *Enterprise in Nigeria* strives to forge a bond with the African viewer, actively addressing her and making connections to the company, creating positive feelings for a unified

50 Fiberesima worked at Radio Lagos and would later produce several plays.

51 Apart from a Dutch voice-over and translated credits, no other alterations were made.

national identity where post-colonial companies were a welcome part of society. The film was released in the UK in January 1962, but did not receive much advertisement or notice.

Conclusion

This chapter has looked into the use of industrial films by two major multinational corporations, UAC/Unilever and Shell, and how they responded to the uncertainty surrounding Nigeria's transition to independence. Though nowhere near present-day advertising budgets, both companies spent several millions of pounds in today's currency making films that were produced in Nigeria and intended primarily for local use. Though it remains unclear how much these films contributed to achieving UAC/Unilever and Shell's goal of seamless continued operations in Nigeria, my research has shown how invested these companies were in creating and adapting their corporate images to the shifting political and cultural landscape. In Nigeria, UAC/Unilever and Shell films could be rented for free by interested parties. They were actively used by the companies' public relation managers and shown with their own mobile film units, but they were also used by the government's mobile film units. The films reached different audiences, from schoolchildren, students and (future) employers, to highly sought after traders, teachers and high-level government officials. Some films were also screened abroad by the companies themselves, as international company film libraries offered them for free for educational, industrial or household use. They existed in various languages and sometimes in re-edited versions.

From my research it becomes clear that British and Dutch companies were much more than passive observers during the decolonization process. Their films were planned and executed with explicit public relations objectives in mind and reveal a strong desire to forge a close bond between multinational corporations and the newly independent nation. After independence, the use of sound becomes a key element in this strategy. Local voice talent and (highlife) music with a connection to both Africa and London were intended to authenticate the corporate image and support the claims of cooperation and joint growth made in the films. Rather than focus on the technical exposition that had been perfected by the Shell Film Unit in London, the film unit in Nigeria also more explicitly showcased human interest stories and created a dramatic focus to underline the connection of the company with Nigerians. The rhetoric framing and distribution strategies vary accordingly for national and international distribution. Some films were only meant for distribution in Nigeria, while others were produced in several different versions and launched with a different advertising campaign for international distribution. My research thus underlines the significance of the processes and dynamics of industrial film. While serving as an advertisement or promotional spot for a product, process or company image, industrial films are shot through with, and transmit, important technical, political, cultural and social discourses.

Works Cited

Advertisement. 1962. Advertisement file, Unilever Archives & Records Management United, Africa Company Collection/1/11/21/8.

"'African Awakening' Launched in Nigeria." *UAC News*, vol. 11, no. 8, September 1962, p. 3.

Buckle, D.H. *Newsletter*, no. 4, October 1967. Unilever Archives & Records Management, United Africa Company collection/1/11/1/7/2.

Buckle, D.H. *Public Relations and Advertising Services*. October 1967. Unilever Archives & Records Management, United Africa Company collection/1/11/1/7/2.

Buckle, D.H. "UAC Public Relations Director." *Newsletter*, no. 3, September 1967, p. 1. Unilever Archives & Records Management, United Africa Company collection/1/11/1/7/2.

Canjels, Rudmer. *The Dynamics of Celluloid on the Road to Independence: Unilever and Shell in Nigeria*. Netherlands Institute for Sound and Vision, 2017.

Canjels, Rudmer. "From Oil to Celluloid: A History of Shell Films." *A History of Royal Dutch Shell, Vol. 4*, edited by Jan Luiten van Zanden. Oxford University Press, 2007, pp. 6–33.

Collins, John, and Paul Richards. "Popular Music in West Africa." *World Music, Politics and Social Change*, edited by Simon Frith. Manchester University Press, 1991, pp. 12–46.

Drummond, John. *Shell Film Operations*. December 16, 1960. Shell Film & Video Unit Archive, PAC 21 B SFU 14.

"Enterprise in Nigeria." *UAC News*, vol. 11, no. 2, February 1962, p. 10.

"Enterprise in Nigeria Launched in Ibadan." *UAC News*, vol. 11, no. 1, January 1962, p. 10.

Fieldhouse, David Kenneth. *Merchant Capital and Economic Decolonization: The United Africa Company, 1929–1987*. Oxford University Press, 1994.

"Film Rover for Nigeria." *UAC News*, vol. 10, no. 1, January 1961, p. 2.

"Film Show for Federal Legislators." *West African Pilot*, March 12, 1958, p. 3.

"Film Shows at Kano." *UAC News*, vol. 5, no. 6, June 1956, p. 9.

Gordon, Douglas. *Films in Shell-BP: Interim Report.* June 2, 1959. EYE Filmmuseum Amsterdam, Bert Haanstra Archive, 136/A 6-4.

Gordon, Douglas. Transcript interview recorded May 18, 1983. Shell Film & Video Unit Archive, PAC/21 B SFU 11.

Grossman, Wendy A. "From Ethnographic Object to Modernist Icon: Photographs of African and Oceanic Sculpture and the Rhetoric of the Image." *Visual Resources*, vol. 23, no. 4, 2007, pp. 293–94.

Home, Robert. *City of Blood Revisited: A New Look at the Benin Expedition of 1897.* Rex Collings, 1982.

Larkin, Brian. *Signal and Noise: Media, Infrastructure, and Urban Culture in Nigeria.* Duke University Press, 2008.

"London Preview of Enterprise in Nigeria." *UAC News*, vol. 11, no. 1, January 1962, p. 2.

Mack, John. *Africa, Arts and Cultures.* British Museum Press, 2000.

"Minutes of 18th meeting, Group Film Committee." The Hague, July 2, 1959. EYE Filmmuseum Amsterdam, Bert Haanstra Archive.

"Premiere of 'Enterprise in Nigeria'." *UAC News*, vol. 11, no. 1, January 1962, p. 3.

Rice, Tom. "From the Inside: The Colonial Film Unit and the Beginning of the End." *Film and the End of Empire*, edited by Lee Grieveson and Colin MacCabe. Palgrave Macmillan, 2011, pp. 135–53.

Robson. *Report on Public Relations 1961/62.* 1961. Unilever Archives & Records Management, United Africa Company collection/1/2/3/6/1.

"The Search for Oil in Nigeria" [review]. *Film User*, September 1961, p. 487.

"The Search for Oil in Nigeria" [review]. *Industrial Screen*, January–February 1962, p. 282.

"The Seeds of Change." *Monthly Film Bulletin*, no. 276–277, 1957, p. 36.

Smyth, Rosaleen. "Images of Empires on Shifting Sands: The Colonial Film Unit in West Africa in the Post-war Period." *Film and the End of Empire*, edited by Lee Grieveson and Colon MacCabe. Palgrave Macmillan, 2011, pp. 155–75.

Stanton, J.L. "Group Public Relations in Commonwealth West Africa." December 1, 1960. Unilever Archives & Records Management, United Africa Company collection/1/11/6/1.

Stollery, Martin. "Voiceover/Commentary." *The Projection of Britain: A History of the GPO Film Unit*, edited by Scott Anthony and James G. Mansell. Palgrave Macmillan, 2011, pp. 168–78.

Summary of Public Relations: Proposals Previously Considered and Discarded. C. 1959. Unilever Archives & Records Management, United Africa Company collection/1/11/4/1/2.

Summary of Public Relations: Summary of Company's Public Relations Activities in Commonwealth West Africa. C. 1959. Unilever Archives & Records Management, United Africa Company collection/1/11/4/1/2.

"Three Million at School." *Film User*, May 1966, pp. 214–15.
"Traders in Leather." *Visual Education*, August 1958, p. 10.
UAC News. Unilever Archives & Records Management, United Africa Company collection 2/19/3/6/1/1.
Unilever Films, Catalogue 1967–1968. Unilever Ltd., 1967.
"Unilever Presents…." *Unilever House Magazine*, vol. 8, no. 5, October–November 1955, p. 3.
"When You Borrow a Unilever Film." *Film User*, February 1963, pp. 76–77.

About the Author

Rudmer Canjels is an independent media researcher fascinated with seriality, ephemeral media, fandom, and industrial film. He is the author of *Distributing Silent Film Serials* (Routledge, 2011), a study on the international distribution and cultural transformation of silent film serials. He has collaborated on the production of several documentaries for *A History of Royal Dutch Shell* (Oxford University Press, 2007) as well as researched Shell's own cinematic history in *Films That Work* (Amsterdam University Press, 2009). His monograph *The Dynamics of Celluloid* (beeldengeluid.nl, 2017) deals with the industrial films by Unilever and Shell made in Nigeria while it became an independent country.

18 Working through the End of Empire

Tom Rice

Abstract

This chapter examines how industrial film was representing, negotiating and managing the loss of the British Empire to colonial audiences. It highlights the centrality of industry and argues that the colonial industrial film was defined, and enacting change, by a specific set of aesthetic values. Therefore, it foregrounds the work of government officials, and subject experts, within industrial film histories. Through the example of different government film units, the chapter foregrounds the performance of work and industry, both on, and off, screen, in the nation-building process. In the immediate aftermath of war film both represents and embodies a new model of industry and economic partnership for colonial audiences, revealing the informal economies of cinema that would often operate beyond independence.

Keywords: Colonial Film Unit (CFU); industrial film; government film-making; training schools; Africa; Jamaica

This chapter examines the *work* of the Colonial Film Unit, a government organization established at the outbreak of war in 1939. Over its sixteen year career, the CFU produced, distributed and, through a fleet of mobile cinemas, exhibited films for local audiences across the British Empire. Film became an increasingly important part of government work, used to define and shape productive citizens and to formalize economic ties between colonizer and colonized. Throughout this tumultuous period, film both represents and embodies a new model of industry and economic partnership for colonial audiences, revealing the informal economies of cinema that would often operate beyond independence.

The chapter asks a series of questions to understand the notion of "work" within this colonial cinema. How did the CFU represent work and industry, whether validating manual labour or foregrounding "hard work" as part of

Hediger, V., F. Hoof, Y. Zimmermann, with S. Anthony (eds), *Films That Work Harder: The Circulation of Industrial Film.* Taylor & Francis Group 2024
DOI: 10.5117/9789462986534_CH18

the nation-building process? How did film itself work within the colonies, developing a language for "illiterate" audiences that perpetuated division and also shaped the ways in which industry was represented? How does this, in turn, redefine the industrial film for colonial audiences? Finally, how did film function as work and develop as an industry, producing, training and developing local units and filmmakers? In this way the CFU tentatively visualized a post-imperial identity not only on screen but also through its operating practices, establishing training schools and local units which, at least ostensibly, reflect and shape these broader moves towards self-government.

Within the context of this collection, it is important to note that this cinema for colonial audiences is, to an extent, a cinema of expertise. The founder and producer of the CFU, William Sellers, was formerly a sanitary inspector and civil servant, using film as part of the government health unit in Nigeria. One of his initial innovations with the CFU, the Raw Stock Scheme – which provided film stock for the colonies – was specifically intended to allow "experts," whether on hygiene or industry, to make films that "adhere to the instructions given from time to time in the CFU's quarterly magazine, *Colonial Cinema*."[1] In this way, Sellers was encouraging colonial officials to make films as part of government administration, inviting us to consider not simply what film brought to these areas of government work, but also what these government workers brought to film and its use within the colonies. While the CFU did employ highly accomplished filmmakers, one of the arguments when piloting CFU training schools later in the 1940s was that it was easier to train someone knowledgeable in the culture to make a film than it was to teach a professional filmmaker about a local culture.[2] This cinema of expertise, in this case functioning alongside other media – including filmstrips, posters, talks, broadcasts and plays – remains somewhat marginalized within film history as value judgements based primarily on aesthetics have privileged documentarians that seek social

I would like to thank James Kearney at the AP Archives in London and Karl Magee at the Grierson Archive, University of Stirling. A number of the films mentioned within this article can be viewed online at www.colonialfilm.org.uk.

1 "Filming in Africa." *Colonial Cinema*, vol. 1, no. 5, July 1943, pp. 1–2.

2 Sellers wrote in 1958 that "local people can be trained as film technicians more quickly, and more effectively, than expatriate technicians are able to acquire anything except a superficial knowledge of the anthropological and sociological complexities of African habit and custom and traditional way of life." Sellers, William. "The Production and Use of Films for Public Informational and Educational Purposes in British African Territories." Colloquy of the Official Information Services in Black Africa, Brussels, 1958.

change through film at the expense of department officials and subject experts, who apply their expertise to film.

Films of Work

When the CFU produced its first film in 1940, *Mr English at Home*, it depicted a lower-middle-class family living in Surbiton with the husband, an artisan, shown at work on a building site. The example was useful for the CFU, particularly during war, as it depicted hard-working British citizens relatable to the African audience. Audience reports commented on the work of Mrs English – "Do many English women have to work as hard as that?" "Do they not have servants?" – while a colonial films officer further remarked that it came as a revelation to many Africans that not all English households had a "complement of servants."[3] The CFU sought to challenge perceptions in subsequent films both by foregrounding the value of manual labour and by showing British men and women undertaking it. This served, as a report from Sierra Leone noted, to "disabuse their minds of the idea that the British are a race of rulers and overlords who, lily-like, neither toil nor spin."[4]

The earliest experiments in filmmaking in the colonies had recognized and propagated this notion of "hard work." When Julian Huxley took three films out to East Africa in 1929 to test audience responses to film, he included *Black Cotton*, a film of the cotton industry in Nigeria that had first played at the Empire Exhibition of 1924 to 1925, a massive celebration and promotion of imperial trade.[5] Huxley noted how audiences grew "wildly excited" when they saw "natives" working on the screen. "They applauded, stomped with their feet, laughed, shouted, explained to each other," he continued, "the noise was particularly deafening when anyone was seen on the film doing a hard job of work."[6] In promoting hard work and industrial labour, these films (and the discussions around them) were defining and showcasing film's

3 "Kumasi Kumbunga and John English." *Colonial Cinema*, vol. 3, no. 2, June 1945, p. 47; Smyth, Rosaleen. "The British Colonial Film Unit and Sub-Saharan Africa, 1939–1945." *Historical Journal of Film, Radio and Television*, vol. 8, no. 3, 1988, pp. 285–98.

4 "Sierra Leone." *Colonial Cinema*, vol. 4, no. 1, March 1946, p. 13.

5 See Rice, Tom. "Exhibiting Africa: British Instructional Films and the Empire Series (1925–8)." *Empire and Film*, edited by Lee Grieveson and Colin MacCabe. Palgrave Macmillan, 2011, pp. 115–33.

6 Huxley, Julian S. *African View*. Chatto and Windus, 1931, p. 292; idem. "Report on the Use of Films for Educational Purposes in East Africa." 1930. The National Archives (TNA), London, Colonial Office (CO) 323/1252/15.

function in the creation of *productive* colonial citizens. Indeed films often spoke directly to workers in the colonies, playing, for example, on mining circuits. A wartime lecturer with Jamaica Welfare Limited's travelling cinema recalled the ways in which its shows sought to shape the colonial worker by privileging "hard work" as part of the nation-building process. The shows highlighted "that hard work is enabling and not a curse," he wrote, "that efficiency will receive reward even if long delayed at times, that building a nation means each individual, being and doing his best at all times."[7]

These earliest films for colonial audiences sought to further the economic productivity of the colonies by modernizing industries and their workforce. Huxley defined the success of one of the earliest experiments in health education in the 1920s, conducted by Dr. Paterson, a medical officer in Kenya, by stating that "the white settlers report an increase in the efficiency of their labourers."[8] Many of these early experiments – most notably the early work of William Sellers, then a sanitary inspector in Nigeria – were ostensibly preoccupied with improving health, but the undercurrent of productivity was never far away. In 1934 Paterson suggested a series of films for East African natives, which were designed to "make the native peoples happier, healthier and more useful." He placed most emphasis on "useful." In trying to create productive citizens – as defined by their relationship to Europeans – Paterson explained that an African might never be healthy or happy unless he could provide the European, American and Eastern manufacturers "with those products of heavy industry and with such hardware and textiles as he cannot produce for himself."[9]

In the post-war period, against a backdrop of civil unrest, Cold War politics and mass decolonization, films of trade and industry help to imagine a new model of economic partnership for colonial audiences. The CFU, now funded through the Welfare and Development Act, sent units to Africa to produce films that would, as its own magazine reported, develop self-reliance and ensure that "the seeds of progress in health, industry and agriculture could be planted." Films like *Weaving in Togoland* (1946), *Better Homes* (1948), *Why Not You?* (1950) and *Good Business* (1947), sought to instruct local African audiences in modern agricultural and building methods.[10]

7 Rennalls, Martin A. "Development of the Documentary Film in Jamaica." MS thesis, Boston University, 1967, p. 35.

8 Huxley. *African View*, p. 161.

9 Reynolds, Glenn. *Colonial Cinema in Africa: Origins, Images, Audiences*. McFarland, 2015, p. 168.

10 Sellers, William. "Address to the British Kinematograph Society." *Colonial Cinema*, vol. 6, no. 1, March 1948, p. 9.

The example of the 1947 film *Good Business* reveals the ways in which the CFU films sought not to show, but to do, to create these productive citizens through film. *Good Business* was made with the help of a local farming cooperative to "encourage Yoruba cocoa farmers in the development of co-operative marketing societies." William Sellers decided to adapt the film for British and international audiences, under the title *Nigerian Cocoa Farmer*, to show "not the dark continent of the picture-postcard travelogue, but [...] a little-known aspect of colonial development in which African initiative, self-reliance and self-government play a significant part."[11]

There are two functions at play here. In *Good Business* the emphasis on cooperative societies is part of a membership drive for African viewers, while in *Nigerian Cocoa Farmer* it is intended to generate domestic and international support for continued expenditure on the colonies. Similarly the CFU's home unit depicted events, such as Colonial Month in 1949, which were intended to showcase to British audiences the *value* of the colonies. The CFU films of these events then sought to connect the African viewer to Britain, to show them this British support and, in turn, to encourage Africans to modernize and develop their industrial methods.

This process is evident in the 1948 film *An African Conference in London*, which represents African workers welcomed to Britain, working with, and learning from, their British counterparts. The economic ties between the areas are shown at the Bourneville factory, as the African delegates watch the export of "good African cocoa," seeing the "process through from beginning to end." The film articulates a gradual shift in this colonial relationship. It includes sequences in London at official events (meeting the king and visiting London landmarks) which foreground a traditional imperial relationship and highlight difference (for example, through costume). It then shows the African delegates learning first-hand from a demonstration of British farming in Hertfordshire, and visiting a car factory in Coventry, now moving away from the centre to celebrate industrial Britain ("it's the modern way to progress") and promote social partnership. However, this movement of personnel is very carefully articulated. *An African Conference* emphasizes that the Africans depicted here are "visitors." They are shown waving goodbye from the car factory, while the commentator notes that the

11 Morton-Williams, Peter. *Cinema in Rural Nigeria: A Field Study of the Impact of Fundamental-Education Films on Rural Audiences in Nigeria*. Federal Information Services, Lagos, 1952, p. 20; Sellers. "Address to the British Kinematograph Society," p. 10.

skills they learn will be taken back "to their own country."[12] This notion of taking modern British-taught skills back to the colonies would now extend to the practice of filmmaking.

How Film Works

On May 24, 1950, a series of Colonial Film Unit productions played "to a large and appreciative audience" at a British Film Institute event intended to highlight "film production in the colonies" at the Institut Français Theatre in London.[13] The programme notes included two essays. The first, written by Harry Franklin, the director of information in Northern Rhodesia, discussed the effect of cinema on African audiences. In a now-familiar rhetoric, Franklin outlined the particular requirements for producing films "which are likely to be understood by the very backward peasant audiences." Franklin noted that cartoons, maps, diagrams or "any type of trick filming" are not understood and extolled a specialized film technique that had been propagated by the Colonial Film Unit over the previous decade. "The ideal film for the villager," he determined, "should be of slow tempo, on a subject with which he is familiar, with a soundtrack for music and effects only, and a commentary in the local vernacular given through the microphone."[14]

Franklin's writing sought to perpetuate a division between colonizer and colonized, applying an ideology of film form, popularized by William Sellers, that served to define and differentiate the colonial worker. This is evident in the films shown at the Institut Français in 1950, such as *Pamba*, a CFU production made by Norman Spurr in Uganda. Spurr ostensibly followed Sellers' theories on film form, arguing that the experience of film for an "illiterate peasant" in Africa was "nearer 1910 than 1950." He believed that "shots needed to be left on the screen for an appreciably longer time than with films for European audiences," and that "there must be a simple aim."[15]

12 Rice, Tom. "From the Inside: The Colonial Film Unit and the Beginning of the End." *Film and the End of Empire*, edited by Lee Grieveson and Colin MacCabe. Palgrave Macmillan, 2011, pp. 135–53.

13 "The British Film Institute, Commonwealth Film Production, Summer 1950." University of Stirling Library Archives and Special Collections, John Grierson Archive, G5: 20:3; "Editorial." *Colonial Cinema*, vol. 8, no. 3, September 1950, p. 51.

14 "The British Film Institute, Commonwealth Film Production, Summer 1950"; Franklin, Harry. "The Central African Screen." *Colonial Cinema*, vol. 8, no. 4, December 1950, pp. 85–88.

15 Spurr, Norman F. "Films for Africans – 1910 or 1950?" *British Kinematography*, vol. 16, no. 6, June 1950, pp. 185–88.

Pamba is a "narrative-teaching" film, which follows, as Spurr himself noted, the "hoary old formula of the good versus the bad," by showing the correct (translated as modern, British) way and the bad methods. The film was imagined and functioned as part of government administration. Spurr worked very closely throughout with the local agricultural officer who, he noted, recognized that visual aids "were not to replace him, but to be his servants," and incorporated a comedy character, Kapere ("a victim of his own stubbornness, conceit or disobedience"), previously established and popularized by the Public Relations and Welfare Department through a comic strip and in the plays of its touring demonstration teams.[16] Indeed Kapere features prominently in Spurr's *A Challenge to Ignorance* (1950), a record of the demonstration teams' work in Uganda. The film shows a wide range of performances – music, a Mr Wise and Mr Foolish play showing the "proper" way to plant cotton, demonstrations on soil erosion, banana growing and bicycle safety – and highlights how the government units use different media and organize the local space "to bring all the people together." The film shows almost exclusively Africans on screen – performing and watching – except for the British welfare officer (always alone in the frame) who watches and advises on particular problems. This model of colonial supervision – a mediated move towards self-governance – is also evident in the film itself, which while depicting African government work is directed by Spurr and with a British voice of God narration.

Spurr recognized the potential value of film here. "The problem in our colonies," he explained, "is the problem of a multitude who have much to learn as a changing world attacks the very roots of living, and yet there are too few teachers." "Film, properly used," he suggested, "may well become an important answer to this problem of how to teach quickly yet surely."[17] Spurr acknowledged that film does only part of the work here, working alongside other government materials and, most crucially, local personnel at the site of exhibition. "The expert should always be in attendance to answer questions, stimulate discussion, encourage endeavour," Spurr argued, "then the film becomes an instrument capable of injecting new ideas into the very blood stream of the people."[18]

We can glimpse this distinctive mode of address in the commentary of *Good Business*. In *Nigerian Cocoa Farmer*, the commentator identifies himself with British and American interests, noting that the Americans, "like

16 Spurr, Norman F. "Pamba." *Empire Cotton Growing Review*, vol. 27, June 1950, pp. 172–76.
17 Ibid.
18 Ibid.

ourselves, must import cocoa." In *Good Business* the commentator identifies with the African protagonist and now directly involves the audience, by asking questions like: "Where is Lawani going in the truck?"

This shift in address acknowledges the ways in which these CFU films were intended for exhibition, as part of a larger political event, alongside talks, music and demonstrations, that invariably culminating with the British national anthem. Indeed Spurr argued that for illiterate audiences film cannot be a "teacher on its own" and presented the local commentator as "the most vital link between the film and the audience."[19] This figure might offer call and responses, ask questions of the audience, provide additional talks and direct where the audience looks on screen. He would organize the film space and explain and rework these films at the site of exhibition.

Audiences often then understood and experienced a film in ways that were unimagined when the film was processed in London. Spurr noted the parochial responses that ensured that even a film on "How to Wash" was severely criticized in Uganda because the Nigerian on screen washed his head last, while the local Baganda would wash their heads first. This was not, however, simply a question of whether a film was understood or even "useful" to an audience, but also in some cases whether it could be directly damaging to the colonial authority.[20]

At the height of the Emergency in Malaya (Malaysia) in the 1950s, the government cancelled screenings of a propaganda film made by the Malayan Film Unit after reports that cinemagoers had cheered the on-screen appearance of communist leader Chin Peng.[21] In Malta, the governor suggested it would be "little short of a disaster" to show *A Queen Is Crowned*, the biggest box-office hit of 1953 in the UK, given that it depicts an Anglican ceremony, and all the more so "at a time when there are signs that certain elements wish to attack the British connection, even by going to the length of disparaging the Monarchy."[22] What we see here are examples of *films that don't work*.

Looking back on the CFU's work in 1953, Sellers surprisingly suggested that the majority of its films might fall into this category. Although he argued that "the technical and pictorial quality" of CFU films was "of a high standard," few could be classed as "successful and right for purpose." Sellers suggested

19 "Commentary and Commentators." *Colonial Cinema*, vol. 6, no. 1, March 1948, pp. 13–14.

20 Spurr. "Films for Africans – 1910 or 1950?" pp. 186–87; Rice, Tom. "'Are You Proud to Be British?': Mobile Film Shows, Local Voices and the Demise of the British Empire in Africa." *Historical Journal of Film, Radio and Television*, vol. 36, no. 3, 2016, pp. 331–51.

21 "Chin Peng Cheered as KL Audiences See Peace Talks Newsreel." *The Straits Times*, January 6, 1956, p. 4.

22 "A Queen Is Crowned in Malta." TNA, FCO 141/10508.

that the cultural and language barriers were the problem here – foreign technicians had a tendency to relate a subject to their own culture, while language barriers placed a huge onus on local commentators. He saw the solution in local units, arguing that "if successful films were to be made *for* the people and *with* the people, they must be made *by* the people."[23]

Film as Work

While the CFU had looked to film to instruct productive colonial workers, after the war it sought to create filmmakers to continue, and embody, this work. "But the production of films is only one side of the Unit's activities," the Institut Français programme outlined in its second essay on the Colonial Film Unit, "and is perhaps ultimately less important than the task of training Africans and West Indians to make films for themselves." Film was presented here as an industry, creating not only a product (film) but also the workers. The programme described a "great experiment: an experiment in film production, an experiment in education, and an experiment in the training of colonial people as filmmakers who will understand the problems and needs of their own countries, and who, with the help of the Unit, will be able to interpret these problems to their own people."[24] Whether idealism or a practical response to the shifting political climate, this was extending a long-held view that government ideology was better received when coming from local voices – an earlier CFU report had claimed that audiences "believe much more readily what is told them by other Africans," and that "their jokes went down better than ours."[25] The CFU was now looking to shape the producers as well as consumers of film, training values, ideals and a dominant colonial ideology as much as technical skills.

The first CFU training school took place in Accra in 1948 to 1949 and was the subject of the first film shown at the Institut Français event. *A Film School in West Africa* is, in some respects, an industrial process film, following a familiar CFU trajectory in showing the creation, completion (and often export) of a product. The film opens at Accra airport with the arrival of the Nigerian trainees and then proceeds to show the different

23 Sellers, William. "Film Use and Production in British Colonial Territories." *Report on the Seminar on Visual Aids in Fundamental Education*. UNESCO, 1954.

24 "The British Film Institute, Commonwealth Film Production, Summer 1950."

25 Film Centre. *The Use of Mobile Cinema and Radio Vans in Fundamental Education*. UNESCO, 1949, p. 60.

stages of the training school (technical training, still camera work, loading film, script writing, filming, editing and finally, "viewing results"). The product at the end here is film and, in particular, filmmakers, who would over the next decade be tasked with developing local film productions in their own countries. This is a fantasy of production, of citizens as much as celluloid, with the structure positioning film alongside the other established industries that it would record.

The film is, however, also imagining a new model of empire here. It shows the moderated handing over of film apparatus. The instructor is very rarely out of shot, often framed over the shoulder of the six trainees, while the commentator reminds us that all film travels to London for final editing and processing. In so doing, it proposes an idealised partnership between colonizer and colonized. When first arriving in the classroom, the commentator states that "The students are welcomed cheerfully by the instructor and they, in turn, already feel that he is their friend." The film's message of cooperation is inescapable. They are, we are repeatedly told, a "team," filmmaking is "above all a test of teamwork" and their success depends on "true cooperation with him [the British instructor] and with each other." Later the commentator reaffirms that the school is training character and values, as much as technique: "The development of the spirit of co-operation was one of the main purposes of the training for picture making is no one man job, it is teamwork throughout." This message was foregrounded in reports of subsequent training schools. "It cannot be too strongly stressed," the CFU wrote of the West Indies film school, "that this spirit of friendliness and co-operation went a very long way to make the school initially a great success."[26]

As with all CFU productions, the film was intended for colonial (and primarily African) audiences. This may be evident in the film form, which largely follows the simplified, specialized technique espoused by Franklin and, in particular by Sellers. The film also talks directly to its African audience – the students are "writing what is called the film script" – and in this way seeks to train and create film literate viewers, to make them aware of the processes and techniques of film as well as the British efforts to mobilize African workers. Indeed the training and development of filmmakers was

26 "Colonial Film Unit Training School in the West Indies." *Colonial Cinema*, vol. 9, no. 2, June 1951, p. 43. As so often with these government films, there are repressed histories within these shots. The scriptwriting takes place away from Accra on the battlements of the old castle at Anomabu. While presented as a picturesque coastal setting, a relaxed environment for informal discussion around a table, the fort represented the centre of British slave trading into the nineteenth century.

closely aligned to the "development" of film audiences. "Trainees are turning out a type of straight-forward film," wrote the Colonial Office's C.Y. Carstairs in 1952, "eschewing thrills, but strong in content and local touch, which very closely fits the stage of film education which their audience have reached."[27]

Yet, as the screening at the Institut Français attests, *A Film School in West Africa* also served an ulterior motive, to promote and redefine the work of the CFU to British authorities at a moment when its funding was being cut. The CFU was seeking to retain a foothold in the colonial territories as moves towards independence gathered pace, offering local access to the camera while still moderating and mediating its use.

As well as playing *A Film School in West Africa*, the screening showed extracts from two films produced by the trainees at the Accra film school. *The Good Samaritan* from March 1949 is introduced as "[a] film story made by the Nigerian students of the Accra Film Training School at the end of their six month course – entirely without guidance." The central characters are Kafi, "an unfortunate beggar" who cannot find work, and Osei, the titular good Samaritan, who offers him food, clothing and, most crucially, work with a builder. The film shows Kofi at work, highlighting how the building is done, but focuses on the results of this labour. He is shown being paid and, as "Kafi is wise," uses the money to buy a blanket. The moral of the film is explicitly revealed in the titles. "Osei tells Kafi that he must work hard and help others as he has been helped," finally concluding that "Happiness comes from giving – not taking." The film defines worth through hard work, but is not instructing in specific building methods (as in many CFU instructional films), but instead in values, in the importance of working, taking instruction and helping others. The goals of the film school are embedded in the story, offering a message of ongoing colonial partnership, but one in which the colonial figure, who has been given a helping hand, is now expected to take greater responsibility ("giving not taking").

The screening also showed an extract from *Basket Making*, a film made by the Gold Coast students, which shows the process from cutting branches to selling at market. Ostensibly this is a typical CFU industrial process film, but significantly the process is conducted on screen by students from a village school. This is, once more, not simply an instructional film for a particular industry, but also a film about education and training, showing the development and construction of "useful" workers. Titles discuss both the specifics of construction – "Peeling the Adobe Branches for Interlacing" – and also the wider skills developed

27 Carstairs, C.Y. "Edinburgh Film Festival." *Colonial Cinema*, vol. 10, no. 4, December 1952, p. 78.

through work ("work demands concentration and skills"). It is evident from the films produced during the courses that particular emphasis was placed on industry – another example from the Gold Coast students shows the process of extracting oil from Copra – and these films followed a well-established model, which presents the product from the land to market or export.

While these films showcase, and in their subject matter promote, the training and development of African filmmakers, the discussions surrounding these training programmes often reveal broader tensions. H.M.K. Howson, who worked at the training school and wrote the script for *A Film School in West Africa*, corresponded with Sellers about employing the trainees in Nigeria as regional films officers. Howson had spoken with H. Cooper, a member of the Nigerian Public Relations office, who argued that the trainees were not of "high enough educational standard" and expressed concerns about their "limited ability." He further argued that they would need "one year of trial at least" to acquire "sufficient experience."[28]

Attitudes were worse still in East Africa. The CFU abandoned plans to host the next training school at Makerere College in Uganda in 1949 after East African officials argued that it would be difficult to find suitable students as the "proposed training course may be overloaded beyond the capacity of African trainees."[29] Instead the next course took place in Jamaica at the recently established University of the West Indies, but the proposals for this were also adapted. The convener now suggested that the approach adopted at Accra – described as "extreme simplicity" to cater for the "ignorance of cinema convention" – would be unsuitable for cosmopolitan West Indies, which was culturally exposed to "sophisticated" American and British influences.[30]

This slightly different approach is evident in some of the work produced by the nine trainees at the West Indian school, although once again it is notable that films of industry feature prominently here. The first film made by graduates in British Guiana was on the rice industry, celebrating new methods and machinery introduced through a cooperative scheme, while in Trinidad it was on cocoa farming. *Cocoa Rehabilitation* was directed by trainee Wilfred Lee and made in collaboration with the Department of Agriculture. The film follows the Mr Wise and Mr Foolish format, with John adhering to government directives and gaining a bumper crop and Tom,

28 Howson, H.M.K., to Sellers. "Future Film Section Organisation in Nigeria in Relation to the Accra Training School." TNA, INF 12/285.

29 Smyth, Rosaleen. "The Post-war Career of the Colonial Film Unit in Africa: 1946–1955." *Historical Journal of Film, Radio and Television*, vol. 12, no. 2, 1992, p. 170.

30 Evans, Gareth. "The Colonial Film Unit's West Indian Training Course in Jamaica." *Visual Aids in Fundamental Education: Some Personal Experiences*. UNESCO, 1952, pp. 130–39.

whose "scepticism is mainly responsible for his lowly position," ignoring instruction and ultimately realizing the error of his ways. The film is notably more ambitious in form – there are dissolves, close-ups, panning, the use of a chart, shorter shot lengths and even a match cut – but retains a familiar rhetoric around work. On receiving his pay, the commentator notes, "What a tidy sum it is indeed, high dividends for hard and honest labour." The film emphasizes the economic importance of industry to the individual ("Every bag produced means more dollars") but also to the nation, as the film concludes with shots of export (the well-worn image of the boat being loaded). "To win this battle and restore our export trade," the commentator explains, "every farmer must help."

The first production from the Jamaican trainees, *Farmer Brown Learns Good Dairying* was presented in the local press as a film "made in Jamaica by Jamaicans" but inevitably bears the marks of the CFU instruction. The film appeared alongside a brochure and a filmstrip – a form used particularly extensively in the West Indies – as part of the Livestock Division's advisory work.[31] It promotes the need for continued British assistance as the British instructor – who "knew very well [...] they have never followed his advice" – demonstrates modern farming methods to Farmer Brown. The film establishes a clear contrast between old and new, between the modern urban environment in which a group of young Europeans enjoy a drink in their car, and the old country district where locals transport goods on their heads and use horse and cart. While the film depicts the Jamaican workers preparing and delivering the milk, the Europeans are seen in two roles – as government officials modernizing the industry and as consumers.

This disparity in roles is often more explicitly outlined in sponsored documentaries for British audiences, such as *From Cane to Cube*, a film from 1950 produced for the British sugar-refining company, Tate and Lyle, which shows the movement of sugar from the fields of Jamaica to the factories of London. There is a clear division between the first half set in Jamaica and the second half set in London. In this example, the conventions of the industrial film help to privilege the British role within this imperial relationship.

From Cane to Cube initially follows a West Indian protagonist, who also provides the commentary for the first half of the film. The film projects familiar British interpretations of West Indian culture ("You know, we cane cutters are happy people when we are cutting"). The second half of the film in London is introduced by a smart British voice, claiming to work in

31 "Home-made Documentary for the Screen." *Daily Gleaner* [Jamaica], February 15, 1952, p. 5; Government of Jamaica. *Annual Report of the Department of Agriculture*. 1952.

the Tate and Lyle Thames refinery. The British commentator explains the industrial processes using microscopic close-ups, scientific drawings and shots of a British schoolboy in a laboratory. The British worker states that "centrifugal force throws the mixture against the rotating basket" in order to remove the molasses. The Jamaican commentator? "No I never was one for knowing just what that machine does with that cane," merely noting that that the cane is squeezed until "it just doesn't look like cane no more."

While this is a sponsored documentary and not a CFU film, this exaggerated shift is helpful in understanding the CFU's representation of industry to colonial audiences. Sellers' ideology of film form, while modified and adapted in the 1950s, shaped a colonial cinema that offered lengthy sequences without scientific explanation, with limited camera movement and minimal editing. The films officer in Nigeria even suggested that the laboured message on cooperative schemes seen within *Good Business* was not picked up by Nigeria peasants as "it may be that the whole construction and tempo of this film is rather in advance of the film education of these audiences."[32] In this way, the film form was used to reinforce dominant colonial ideologies not only by creating a division between the European and colonial spectator, but also in representing a largely static, traditional model of local industry. This is a cinema defined, and enacting change by, a different set of aesthetic values and this evidently shapes the ways in which industry is presented to colonial audiences.

The Work Continues

The CFU's training programme continued into the 1950s. A film school at Cyprus in 1951 involved nine students from Cyprus, Mauritius, Hong Kong and the Sudan, while by 1953, Sellers reported that seventy students "from various countries" had been trained at the CFU's London headquarters.[33] There were trainees with scholarships from Iraq and Haiti in London in 1954 and, significantly, such training did not end when the Colonial Film Unit closed its doors in 1955. In 1958 when George Pearson wrote about six young Nigerians who could potentially "devote their lives" to film and "grow old in its service," he asked: "Where could they obtain the needed training?" The answer was the Overseas Film and Television Centre (OFTVC), which had been set up by the CFU's chief editor Vic Gover and included prominent CFU alumni like Dennis Bowden. The centre operated as a commercial enterprise

32 "Showing Films in the Villages." *Colonial Cinema*. September 1948, p. 63.

33 Sellers. "Film Use and Production in British Colonial Territories."

to continue the work of the CFU and to connect these "comparatively 'small voices'" to a "technical agent working from a main centre." As it expanded, serving as the London agents for more territories, OFTVC provided not only post-production services but also training work, initially offering the six Nigerians – "well chosen, eager and anxious to serve" – a six-month course run by George Pearson in 1958, which was housed at the OFTVC studio.[34]

The CFU officially disbanded in 1955 on the cusp of widespread independence, by which time Sellers estimated that films were being made in thirty-two of the British territories.[35] In some cases, these units sought to shake off the traces of the CFU – pointedly rejecting Sellers' philosophies – but we can also often see continuity in exhibition practices, equipment, technologies and personnel. In reviewing the work of the West Indian school, *Colonial Cinema* had stated that "the pump has been primed" for future work and that the CFU was "most anxious" to retain close contact with the students.[36] So it proved as the head of the post-independence Jamaica Film Unit, the Trinidad Film Unit and the managing director of the Ghana Film Industry Corporation were all veterans of the CFU's first two training schools in the late 1940s. The director of *A Film School in West Africa*, Lionel Snazelle, would head the Nigerian Film Unit until independence, training numerous students in ways articulated at the original Accra school, while other figures return beyond independence. Barely a year after independence, Norman Spurr moved out to Ghana, working with the audiovisual unit of the Department of Social Welfare and Community Development. On his return from Ghana in 1962, Spurr became principal of the OFTV Training School in London. For film, like so many industries across the empire, work would often continue even after the British flag was lowered.

Works Cited

"The British Film Institute, Commonwealth Film Production, Summer 1950." University of Stirling Library Archives and Special Collections, John Grierson Archive, G5: 20:3.

34 Pearson, George. "Overseas Film and Television Centre, Film Training School." Colloquy of the Official Information Services in Black Africa, Brussels, 1958; Gover, Vic. "A Brief Review of the Services Provided by the Overseas Film and Television Centre for Film Units Working in Africa." Colloquy of the Official Information Services in Black Africa/Universal and International Exhibition, Brussels, 1958.

35 Sellers. "The Production and Use of Films."

36 "Colonial Film Unit Training School in the West Indies."

Carstairs, C.Y. "Edinburgh Film Festival." *Colonial Cinema*, vol. 10, no. 4, December 1952, p. 78.
"Chin Peng Cheered as KL Audiences see Peace Talks Newsreel." *The Straits Times*, Singapore, January 6, 1956, p. 4.
"Colonial Film Unit Training School in the West Indies." *Colonial Cinema*, vol. 9, no. 2, June 1951, pp. 40–44.
"Commentary and Commentators." *Colonial Cinema*, vol. 6, no. 1, March 1948, pp. 13–14.
"Editorial." *Colonial Cinema*, vol. 8, no. 3, September 1950, pp. 50-51.
Evans, Gareth. "The Colonial Film Unit's West Indian Training Course in Jamaica." *Visual Aids in Fundamental Education: Some Personal Experiences*. UNESCO, 1952, pp. 130–39.
Film Centre. *The Use of Mobile Cinema and Radio Vans in Fundamental Education*. UNESCO, 1949.
"Filming in Africa." *Colonial Cinema*, vol. 1, no. 5, July 1943, pp. 1–2.
Franklin, Harry. "The Central African Screen." *Colonial Cinema*, December 1950, pp. 85–88.
Gover, Vic. "A Brief Review of the Services Provided by the Overseas Film and Television Centre for Film Units Working in Africa." Colloquy of the Official Information Services in Black Africa/Universal and International Exhibition, Brussels, 1958.
Government of Jamaica. *Annual Report of the Department of Agriculture*. 1952.
"Home-made Documentary for the Screen." *Daily Gleaner* [Jamaica], February 15, 1952, p. 5.
Howson, H.M.K., to Sellers. "Future Film Section Organisation in Nigeria in Relation to the Accra Training School." The National Archives (TNA), London, INF 12/285.
Huxley, Julian S. *African View*. Chatto and Windus, 1931.
Huxley, Julian S. "Report on the Use of Films for Educational Purposes in East Africa." 1930. The National Archives (TNA), London, Colonial Office 323/1252/15.
"Kumasi Kumbunga and John English." *Colonial Cinema*, vol. 3, no. 2, June 1945, pp. 46-48.
Morton-Williams, Peter. *Cinema in Rural Nigeria: A Field Study of the Impact of Fundamental-Education Films on Rural Audiences in Nigeria*. Federal Information Services, Lagos, 1952.
Pearson, George. "Overseas Film and Television Centre, Film Training School." Colloquy of the Official Information Services in Black Africa, Brussels, 1958.
"A Queen Is Crowned in Malta." The National Archives (TNA), London, FCO 141/10508.
Rennalls, Martin A. "Development of the Documentary Film in Jamaica." MS thesis, Boston University, 1967.

Reynolds, Glenn. *Colonial Cinema in Africa: Origins, Images, Audiences*. McFarland, 2015.
Rice, Tom. "'Are You Proud to Be British?': Mobile Film Shows, Local Voices and the Demise of the British Empire in Africa." *Historical Journal of Film, Radio and Television*, vol. 36, no. 3, 2016, pp. 331–51.
Rice, Tom. "Exhibiting Africa: British Instructional Films and the Empire Series (1925–8)." *Empire and Film*, edited by Lee Grieveson and Colin MacCabe. Palgrave Macmillan, 2011, pp. 115–33.
Rice, Tom. "From the Inside: The Colonial Film Unit and the Beginning of the End." *Film and the End of Empire*, edited by Lee Grieveson and Colin MacCabe. Palgrave Macmillan, 2011, pp. 135–53.
Sellers, William. "Address to the British Kinematograph Society." *Colonial Cinema*, vol. 6, no. 1, March 1948, pp. 9-12.
Sellers, William. "Film Use and Production in British Colonial Territories." *Report on the Seminar on Visual Aids in Fundamental Education*. UNESCO, 1954, pp. 42–46.
Sellers, William. "The Production and Use of Films for Public Informational and Educational Purposes in British African Territories." Colloquy of the Official Information Services in Black Africa, Brussels, 1958.
"Showing Films in the Villages." *Colonial Cinema*. September 1948, pp. 62-64.
"Sierra Leone." *Colonial Cinema*, vol. 4, no. 1, March 1946, p. 13.
Smyth, Rosaleen. "The British Colonial Film Unit and Sub-Saharan Africa, 1939–1945." *Historical Journal of Film, Radio and Television*, vol. 8, no. 3, 1988, pp. 285–98.
Smyth, Rosaleen. "The Post-war Career of the Colonial Film Unit in Africa: 1946–1955." *Historical Journal of Film, Radio and Television*, vol. 12, no. 2, 1992, 163–77.
Spurr, Norman F. "Films for Africans – 1910 or 1950?" *British Kinematography*, vol. 16, no. 6, June 1950, pp. 185–88.
Spurr, Norman F. "Pamba." *Empire Cotton Growing Review*, vol. 27, June 1950, pp. 172–76.

About the Author

Tom Rice is a professor in film studies at the University of St. Andrews. He is the author of *White Robes, Silver Screens: Movies and the Making of the Ku Klux Klan* (Indiana University Press, 2015) and *Films for the Colonies: Cinema and the Preservation of the British Empire* (University of California Press, 2019). He previously worked as the senior postdoctoral researcher on the project "Colonial Film: Moving Images of the British Empire" (www.colonialfilm.org.uk).

19 Cinema-going on the Railway Tracks

Transportation, Circulation and Exhibition of Information Film in Colonial India

Ravi Vasudevan

Abstract

This chapter seeks to explore the impact of so-called useful cinema as it emerges through processes of transportation, circulation and exhibition, mainly in relation to the railways. This instructional cinema related to agricultural improvement, animal husbandry and public health, and it targeted rural and small-town populations in India from the 1920s. The braiding of media forms and transportation technologies was key to how audiences were invited to position themselves in order to access a wider universe of images, spaces, temporalities, technologies, production practices and commodity life. This film culture was driven by governmental and business enterprise, but also by more dispersed commercial networks in which an avowedly useful object and communication medium segued into a complex film, media and commodity ecology.

Keywords: instructional film; demonstration; film programme; railways; motor vans; bazaar; travel

It is a remarkable and little-known fact that the Indian railway was a key vehicle for the production, circulation and exhibition of films. These were largely for audiences in small towns and the rural world which did not have ready access to built cinema theatres or even travelling cinema facilities. The films made and shown were mostly of the information/instruction genre – nowadays called "useful cinema" in film studies research[1] – though travel

1 Acland, Charles R., and Haidee Wasson, editors. *Useful Cinema*. Duke University Press, 2011

Hediger, V., F. Hoof, Y. Zimmermann, with S. Anthony (eds), *Films That Work Harder: The Circulation of Industrial Film*. Taylor & Francis Group 2024
DOI: 10.5117/9789462986534_CH19

films also constituted a sizeable segment of the repertoire. The railways were the most spectacular transport infrastructure and symbolic edifice to be involved in film circulation and exhibition; other, more modest agencies included motor vans and bullock carts. This chapter seeks to explore the logic of so-called useful cinema as it emerges through processes of transportation, circulation and exhibition. Specifically, we are considering here "films that work," industrial films produced to impart instruction, to record manufacturing processes, to generate a rhetoric of governance and address different publics, and a means of reflecting on and rationalizing production.[2] This chapter argues that to track this agenda we need to research the mode of cinematic delivery and presentation, the actual occasion of the exhibition of useful cinema, in order to capture a sense of the impact and meaning of such film enterprises.[3] I argue that this braiding of technologies and circuits was key to how audiences were being invited to position themselves in order to access a wider universe of images, spaces, temporalities, technologies, work practices and commodity life. Crucial coordinates of this film culture were governmental and business enterprises, but significant too was a more dispersed, less instrumental way in which an avowedly useful object and communication medium segued into a complex film, media and commodity ecology.

Railways and Cinema

In exploring the relationship between film exhibition, transportation and overlapping circuits of goods and commodities, I focus on the railways, only briefly referring to the wider transportation ecology within which it functioned. If we trawl through the international history of cinema, we will observe that it is not unusual for railways to be used to circulate and exhibit cinema. The best-known instance of this was in the agit-trains of the Soviet Union which went out to propagandize for the revolution amongst peasant populations. The Soviet case did not use

2 Hediger, Vinzenz, and Patrick Vonderau. "Record, Rhetoric, Rationalization: Industrial Organisation and Film." *Films That Work: Industrial Film and the Productivity of Media*, edited by Vinzenz Hediger and Patrick Vonderau. Amsterdam University Press, 2009, pp. 35–50.

3 This follows Elsaesser's argument that the analysis of industrial film requires inquiry into the occasion, the purpose and the addressee of the film. Elsaesser, Thomas. "Archive and Archaeologies: The Place of Industrial Film in Contemporary Media." *Films That Work: Industrial Film and the Productivity of Media*, edited by Vinzenz Hediger and Patrick Vonderau. Amsterdam University Press, 2009, pp. 19–34.

railways to exhibit instructional films, it appears, though film was used to capture distressed conditions to inform and raise resources from the wider population. Filmmakers such as Dziga Vertov, who would later be a key figure of the avant-garde, were involved.[4] Exhibition would take place through stationary cinema-wagons. Later, the film train of the 1930s would also house a film studio for making, editing and processing films, in a bid to create a live link and feedback loop between agitational film crews and peasant audiences, and featured important filmmakers such as Alexander Medvedkin.[5] In colonial India railways were used to exhibit films in rural situations to impart information and instruction about improved techniques and technologies. Film use was similar to that in the wider colonial film setting, in which mobile cinemas using vans exhibited instructional films. The best known of these instances was the Bantu Educational Kinema Experiment in East Africa, and there has also been substantial research on South Africa by James Burns, by Tom Rice on the Colonial Film Unit, and by Brian Larkin on colonial Nigeria; there is also good scholarship on European travelling cinemas.[6] This broad "developmental" swathe lies in counterpoint to another way of relating railways and cinema. In the British case a railway cinema car exhibited films to passengers on board the train, to relay information in newsreels, as part of the experience of combining modern technologies of travel and representation to embody and imagine space-time transformation, and as an experience of leisure.[7] In the cases I refer to, the transport vehicle is not

4 Tode, Thomas. "Agit-trains, Agit-steamers, Cinema Trucks: Dziga Vertov and Travelling Cinema in the early 1920s in the Soviet Union." *Travelling Cinema in Europe*, edited by Martin Loiperdinger, KINtop Schriften 10, 2008, pp. 143–56; Kinez, Peter. *The Birth of the Propaganda State: Soviet Methods of Mass Mobilisation 1917–1929*. Cambridge University Press, 1985; Loiperdinger, *Travelling Cinema*, for more articles on European uses of travelling cinema.

5 Widdis, Emma. *Alexander Medvedkin*. I.B. Tauris, 2005; see also Heftberger, Adelheid. "Dziga Vertov's and Alexander Medvedkin's Film Trains and Agit-Steamers of the 1920s and 1930s." *Apparatus: Film, Media and Digital Cultures in Central and Eastern Europe*, no. 1, 2015, www.apparatusjournal.net/index.php/apparatus/article/view/2/75.

6 Windel, Aaron. "The Bantu Educational Kinema Experiment and the Political Economy of Community Development." *Film and the End of Empire*, edited by Colin MacCabe and Lee Grieveson. Palgrave Macmillan, 2011, pp. 207–26; Sanogo, Aboubakar. "Colonialism, Visuality and the Cinema: Revisiting the Bantu Educational Kinema Experiment." *Film and the End of Empire*, pp. 227–46; Burns, James. *Flickering Shadows: Cinema and Identity in Colonial Zimbabwe*. Ohio University Press, 2002; Larkin, Brian. *Signal and Noise: Media, Infrastructure, and Urban Culture in Nigeria*. Duke University Press, 2008; Rice, Tom. "Working through the End of Empire." This volume.

7 Harrison, Rebecca. "Inside the Cinema Train: Britain Empire and Modernity in the 20th Century." *Film History*, vol. 26, no. 4, 2014, pp. 32–57.

itself the exhibition space, but is a facilitator, bringing the material into the place; importantly for my argument, its purpose does not end there, for it also provides a background to the exhibition, and is thus crucial to the mise en scène of how the film was seen, how it was positioned in a configuration of objects and people in space, and in relationship to possible and imaginary destinations.

Why did railways assume a filmmaking and exhibiting function in India? The answer is not readily clear in any formal policy statement, but may lie in the fact that as infrastructure it had a critical and penetrative outreach function that spread geographically and in the depth of social formations; that is, it extended in its network from urban forms through to peasant cultures. Ian Kerr and other historians of Indian railways have outlined the colonial operations fulfilled by the railways, including the mobilization of military force, the transportation of labour and of commodities, and the integration and reshaping of so-called primary production to address the needs of industrial manufacture in the imperial metropolis.[8] If the railways was integral to these functions of colonial power and economic restructuring, to this was now added a media/communication function to spread information about the government's development agenda. We will see that, to make such films useful, the railways publicity saw two genres, agricultural improvement and travel films, working in tandem to motivate peasant audiences to try new methods, enhance productivity and earning capacity, and encourage travel, thereby augmenting railway revenues. To travel not only fitted the railway's immediate economic objectives, but would induct the peasant population into new circuits of consumption that the railways would in turn target by recourse to advertising and retailing commodities in a budding consumer economy.

Government-sponsored film and commercial entertainment film supply came together in the programme offered to targeted audiences, the mix ensuring that audiences would not find instruction boring. It also brought two different film distribution cultures together, with commercial travelling companies also taking government contracts to show public health and agricultural improvement films while continuing to be involved in the lower end of the entertainment film market. The possibility of the government film arriving in front of target audiences and pandering to them then depended on a complex film ecology.

8 Kerr, Ian. "Introduction." *Railways in Modern India*, edited by idem. Oxford University Press, 2001.

Railway Film Publicity: Agrarian Improvement, Travel Films and the Mobilization of Peasant Populations

A substantial component of railway film production were instructional films, all of which related to agricultural matters, including livestock and cattle rearing. There were also films catering to tourist consumption, such as shorts on Indian tourist attractions, including sacred towns, rural scenes, fairs and festivals, indigenous sports, and actualities (such as films relating to the Great Indian Peninsula Railways' own regiment and its infrastructural work). A film in the 1927 programme on the subject of Wembley was likely to refer to the Imperial Exhibition of 1924 to 1925, which, apart from exhibiting the empire was also probably used to encourage the tourist trade in various colonies of the empire. Alongside agricultural improvement and travel films, the railways also made and circulated so-called topical films, and also films of a technical sort which were for the railway staff.

The central publicity officer of the GIP Railways, A.E. Tylden-Pattenson, outlined the types of audiences the films aimed at:

> [A]mongst the non-railway audience the range is intended to appeal to the cultivators of cotton, wheat and sugar and the agriculturists generally, or to the pilgrim and potential traveller. The films are also intended to broaden the outlook of the railway staff and to stimulate their interest in the primary products and industries, and in general travel with which they are intimately involved.[9]

The picture we have here is of a filmmaking and exhibition practice which addressed railway employees as well as quite specific publics who were to be drawn into a world of improvement and travel. Programming rationale made films of travel speak to films of agricultural improvement. Tylden-Pattenson presented this particular economic argument linking travel and improvement, in which film instruction would lead to enhanced productivity, purchasing power and the circulation of commodities and passengers. Thus, referring to films on cattle breeding, wheat and cotton the publicity officer noted that they showed

> the necessity of farming and fertilising the ground in order to get a better crop and again – in the cycle of improvement – the re-action in the

9 "Memorandum: Film Production and Display on the Indian State Railways." September 1927, p. 3. National Archives of India (NAI), GOI Railway Department/Traffic 1750T/38B. Hereafter, "Memorandum."

> superior fodder becoming available for breeding of better cattle. The finer the harvest and the greater the output, the earlier the general uplift and the wider the circulation of money through the movement of people with a higher purchasing power. From the Railway point of view this means greatly enhanced traffic, again benefiting the producer in the shape of lower rates.[10]

To focus on genres of travel and improvement would be to misrepresent the combination of genres that composed the film screening programme, which we should take as a critical vector of audience address. As Tylden-Pattenson noted, an entertainment component was also crucial, otherwise audiences, whether of railway employees or a more general public, would find unrelenting instruction boring. When asked what kind of film was most appreciated in railway film programmes, Tylden-Pattenson noted, "The Serio-comic 'The Thief of Baghdad' has been immensely popular. We have had it on the lines three times and it will still draw enormous audiences. Then there are 'The Gold Rush,' 'The Lost World,' and 'Safety Last' (Harold Lloyd). They are very popular indeed." When asked about the response of "the purely agricultural classes," he declared them "just as amused. Tremendous appreciation is exhibited."[11] This reply was corroborated by other railway publicity officers. The East Bengal Railway publicity officer mentioned similar titles to those shown regularly by the cinema car on Great Indian Peninsular service and described a typical film programme (table 19.1).[12]

Tylden-Pattenson indicated that he had been in negotiation with leading film distributors, Madan and Pathé, in order to secure good-quality films for railway cinema exhibition, and a major hit such as *The Thief of Baghdad* had been acquired through Madans. There was a plan to develop an agreement to sustain such quality content in the film programme.[13] However, he also noted that films were secured from film circulating libraries, and it is likely these were supplied by the distributors of cheaper films.[14] Here we are entering a different network, that of film as commodity rather than as useful resource, and such a terrain was quite differentiated in terms of the

10 "Memorandum," p. 2.

11 Tylden-Pattenson, A.E., Chief Publicity Officer, Railway Board. Oral evidence, *Indian Cinematograph Committee, Evidences* (hereafter ICCE), Government of India, Central Publications Branch, 1928, vol. 1, pp. 70–71.

12 Written Statement by Publicity Officer, East Bengal Railway. March 19, 1928. *ICCE*, vol. 4, p. 192.

13 "Memorandum," p. 5.

14 Ibid., p. 3.

Table 19.1. Typical Film Programme on the Great Indian Peninsular Service

Film Titles	Number of Reels
Agra and Fatehpur Sikri (Indian State Railways scenic film)	1
Cattle Improvement in the Central Provinces (Indian State Railways film)	3
Cry of the Children (Maternity and child welfare film with vernacular titles produced by the Public Health Department)	3
Big Game Shooting in Central India (Indian State Railways film)	1
A comic film	2
Total	10

supply chain, as I have noted. It could be argued that such a commodity function was neutralized by the fact that government shows tended to be free in order to attract audiences; however, the hiring of commercial entertainments still drew government into the commodity network. The work of instruction was never severed from that of commodity circulation and the entertainment business.

Railways, Commodity Networks and Imaginaries

Advertising and canvassing to encourage railway travel amongst new travellers, came together with the cultivation of a new commodity imagination, as railways sought to draw businessmen and an expanding constituency of consumers into a new sense of trade possibilities and markets. For example, the Bengal Nagpur Railway published a directory of commodities that went into a second edition in 1931. It shows the commodities in alphabetical order, the names and addresses of individual merchants dealing in them, the names of the nearest stations and the headquarters of the commercial districts in which the stations concerned are situated. The directory was compiled with a view to bringing producers and buyers in close touch with one another in the hope that internal trade would thereby be stimulated and railway traffic increased.[15]

Key to this building of a commodity network through the railways was the phenomenon of bazaar trains, which incorporated a cinema facility, and now started featuring regularly on many lines, including the Assam

15 India Railway Board. *Report by the Railway Board on Indian Railways*. 1928–1929. British Library, India Office Records, 1928/29–1931/32 IOR/V/24/3550, p. 90.

Bengal, Eastern Bengal, Great Indian Peninsula and Burma Railways from 1927. “These specials are an innovation arranged for the first time by the Eastern Bengal Rly in September 1927, when a train of seven third class carriages specially adapted for use as a travelling shop, a catering car, an electric charging van and a cinema car was run.”[16] For the GIP railways in 1929, it was noted that, after

> much preliminary advertising 10 firms were induced to take part in the [...] Diwali special train. [...] The train was on tour for 4 weeks and halted at 25 important stations for approximately 12 hours each. The object in view was to attract as many people as possible to the halting stations to see the train. This was carried out by means of suitably coloured posters and handbills and also by beat of drums. As a result of this, approximately 4.5 lakhs of people visited the train and exhibitors made extensive cash sales in addition to booking substantial orders.[17]

From available information, the bazaar train featured infrastructural products, such as cement, fertilizer and farm machinery, along with everyday consumer products, soap, tea, clothes, shoes, matches, watches. International firms such as Lever Brothers were involved, and a significant managing agency, Killick Nixon, which apart from managing cement firms also happened to manage several railway lines as well.[18]

A Wider Information and Commodity Imaginary

We need to situate the circulation, advertising and publicizing of commodities through railways not only in terms of its own history, but also in relation to other circuits through which commodities were displayed and retailed, and in terms of a variety of media technologies, of which cinema was only one. To take commodity circulation within the agrarian sphere as the general frame to situate media use, we can see particular emphasis on key goods such as new seeds, fertilizer and technologies ranging from rice mills to tractors. To promote these improving commodities there developed a particular

16 Ibid.

17 Ibid. 1929–1930, p. 92.

18 Narrative Report by Agents of Class I Railways for the Year 1929–1930. National Archives of India, Defence Department, Public Relations Section, file no. 59, Railway Department, 1253-statistical/166-200, B (K).

collaborative logic, linking government and business corporation. Indeed, in the cases we refer to, the public utility of a new technology emerged at the junction of government and corporate collaboration, something which was not highlighted in the proliferating discourses and policy statements that developed at this time about improved agriculture and health. Ian Petrie's excellent work on science and technology in the Bengali countryside shows how international business set up local circuits to promote their merchandise, with companies taking recourse to travelling salesmen, and using exhibitions to inculcate knowledge and interest in new technologies.[19] In Bengal there was a bid to introduce new varieties of rice and jute, the former less successful, the second much more effective. Fertilizer was also a major object of promotion, with the Chilean Nitrate Company promoting natural nitrogen fertilizer in Egypt and India as major markets with relatively nascent chemical industries; subsequently, other companies arrived, and Imperial Chemical Industry's sales of sulphate of ammonia, a soil fertilizer, in Dhaka and Mymensingh districts proved successful. The incidence of cultivators actually taking to industrially produced fertilizer was relatively limited.[20] Their preference was to use the less expensive option of its domestic production by women. But the fertilizer expert C.M. Hutchinson's testimony to the Indian Cinematograph Committee indicates that there was considerable initiative in the promotion of chemical fertilizer, as he moved from the state-run Imperial Agricultural Institute at Pusa to the ICI-supported Fertiliser Propaganda of India Company. Hutchinson's testimony captures the relationship between government and business enterprise to produce and publicize commodity as public utility very well:

> My actual employers are the Imperial Chemical Industries in England. That is a very large combine with about 60 millions as capital. Their idea is to promote the use of fertilisers in Indian agriculture. There is no doubt at all that it would be a very good thing for Indian agriculture. We want to do that and the greatest difficulty in doing that is to get the conservative *ryot* [cultivator] to use something which has been proved to be profitable. As a first step towards that, we are running this fertiliser propaganda company which is a non-profit earning company and which is directed solely for the purpose of introducing the idea of the use of

19 Petrie, Ian. *Village Visions: Science and Technology in the Bengal Countryside, c. 1860–1947*. PhD diss., University of Pennsylvania, 2004. Petrie gives the example of German salesmen in the early twentieth century, and also the influx of Japanese goods into Bengal. Ibid., pp. 102–03.
20 Ibid., pp. 174–78. For fertiliser strategy.

> fertilisers in India. We are working in collaboration with the Agricultural Departments all over India. And one of the ways we want to advance is by the use of the cinema.[21]

Political Imperatives

In all this, the central government's use of agencies and technologies to promote rural reconstruction suggest a distinctive political imperative. Media presentations took place within a political and administrative framework, a framework involving health, agricultural and publicity officers and police, and often the local landlord as well. This was a political and administrative apparatus which not only tried to ensure that messages were communicated by specialists who could use and interpret the media, but was also a local power which could ensure audience presence.[22] After the First World War, rural publicity for government-development initiatives were supported by landlord classes anxious to undermine the growing influence of anti-colonial agitators amongst the peasantry. The Rural Reconstruction report of 1935 to 1939 devotes considerable space to the reproduction of a lecture on "The Government and Rural Development: United Provinces Experiments" (file pages 269–277) by Sir Edward Blunt, a civil servant. Blunt's opening remarks highlight the political context and motivation for rural reconstruction. He noted that the peasant was confronted with a psychological problem:

> [T]he old conservatism shifted with war, and then even more strongly with depression.
> [He] had never encountered this kind of trouble, and displayed, along with old reserves of patience, a new resourcefulness; it puzzled him, it filled him with apprehension, and made him ready to listen to anybody who could suggest a way out. At first, he was inclined to take short cuts proposed by Congress demagogues, such as refusing to pay his rent. But when, largely through the activities of the Publicity Department, he

21 Hutchinson, C.M., Chief Scientific Adviser, Fertiliser Propaganda of India. Oral evidence. December 16, 1927. *ICCE*, vol. 2, p. 735. Without any irony intended, Hutchinson said he told his employers, "I think the cinema is the most fertile method of changing the idea of the people." Ibid., p. 735.

22 The incidence of official mediation of instructional film was common across colonies. Some of the reportage on film shows, lectures, and audience reaction was quite elaborate, as James Burns (*Flickering Shadows*) shows for colonial Zimbabwe. See also the work on the Bantu Educational Kinematograph Experiment cited earlier; Larkin. *Signal and Noise*.

> realised what Government and his own landlords were actually doing for him, he soon decided that their performances were preferable to Congress promises. He now knows exactly what he wants, and under the new constitution will see that he gets it.

This was a fairly representative statement of the rationale and techniques of rural reconstruction, and Blunt's account about the importance of publicity went into details about the mobile vans and media apparatuses which were used to engage the peasant audience in the villages.[23]

By the 1930s, India was going through an important transition. Blunt's lecture refers to the new constitution, enacted in the Government of India Act (1935), which enabled some ten per cent of the population to vote in provincial elections, thus mobilizing the upper ranks of peasant households into the electoral process. The elections were to take place in 1937, and it is advisable to read successive central reports on rural reconstruction, running from 1935 through 1939, and detailing media strategies to engage audiences, as an index of a specifically colonial political understanding that aimed to neutralize agrarian mobilization and the influence of anti-colonial nationalism. The colonial central government here exceeded the writ of nationalist governments elected in several key provinces, including the United Provinces, Bombay and Madras Presidency, Bihar, Central Provinces and Assam. In these reports, while motor vans were the key vehicles of information for the government, the railways were also part of the overall communication design.[24]

Commercial Film Networks

I have tried to suggest that we need to situate the government's film initiatives in a series of circuits. These relate to an overall economic logic which involved key businesses, and shaped the circulation of information about new materials, techniques and technologies. Such initiatives circulated through the agency of salesmen, agricultural exhibitions, and health weeks, and involved media use in which film had a significant, if not dominant presence.

23 Blunt, Edward, Sir, KCIE, OBE. "Government and Rural Development: United Provinces Experiments." Paper to be read before the East India Association at the Caxton Hall, Westminster, SW1, on May 13, 1936, with Sir Atul Chatterjee, GCIE, KCSI, in the Chair. IOL L/I/1/596 Rural Development in India, p. 2.

24 Government of India Grant for Rural Development in 1935–36. Ibid., p. 7.

Most reports on rural communication indicate that slides, magic lanterns, gramophone, radio transmitters and cinema composed the spectrum of media used. The circuits of commodity information were augmented by film and media circuits involving government sponsorship, and sometimes actual filmmaking by government through railways, and by a more dispersed field of professional and amateur filmmakers.

Crucially, such film production and exhibition also drew upon the travelling cinema circuit. Companies such as Aurora in Bengal provide a significant example. This travelling cinema company took contracts to make public health films, and also showed films provided by government in that area. Testimony by company representatives to the Indian Cinematograph Committee in 1928 indicated that the company screened films from the second-hand market in foreign imports to audiences for their living, and also produced and exhibited government information films. The government-sponsored film programme included entertainment sourced from major film distributors, but also probably drew upon commercial travelling film companies. It was thus knowingly implicated in the flow of entertainment cinema as various levels of its existence, from relatively high-end films to what is often referred to as "junk" cinema. For the travelling cinema company, while the returns on entertainment films were better, they took time to be realized, and the government commission provided a staple source of earning.[25]

On the Horizon: Layering Transport and Media Imaginaries

In this concluding section I want to return to the key thematic of transportation, and the varied ways in which the cinema, and the film audience, was implicated within a media experience which was inextricably tied to the backdrop of road, rail, mechanized and non-mechanized forms of transportation, against which screenings took place. The travelling cinemas perhaps exemplify this motif best. Entertainment films were sent by railway parcel, and the company would receive it on the day of the performance and send it back the next day.[26] However, the government-

25 Bose, M.M., Professor, and Others of the Aurora Cinema Company. Oral evidence. December 15, 1927. *ICCE*, vol. 2, pp. 665–77.

26 Bose, A.N., Aurora Cinema, Das, H.K., Wellington Bioscope and Variety Co., Ghose, J.N., Monarch Bioscope Co., Banerjea, B.B., Temple Cinema, Dey, H.K., Wellington Cinema. Oral evidence. December 23, 1927. *ICCE*, vol. 2, p. 1089.

sponsored film would be retained for a longer period, as indicated by the information around the exhibition of the Aurora film on maternity. The Aurora testimony described how the company had to carry the entire apparatus of cinema with them. They had to carry equipment and tents and also the seating for those who paid more than two annas. Generators also had to be carried for film screening in the villages. All of this meant employing a big staff.[27] The mode of local transportation from the railhead was by bullock cart.

This process, of the cinema reaching into the countryside, was part of a wider bid to make the rural accessible through transport infrastructure. A key business here was Burmah Shell, which, while primarily invested in the sale of petroleum-related products, such as kerosene, petrol fuel, lubrication oils and insecticide, was also involved in marketing road surfacing material and developing a market for tractor retail and rental.[28] The company also worked with the Bengal government to sponsor experiments at government facilities in irrigation trials.[29] The second key business player in terms of road infrastructure were the tyre manufacturers Goodyear and Dunlop. Here, the material changes in the bullock cart, a key delivery device not only of goods but of publicity, were the main object of transformation. As Ian Petrie and Stefan Tetzlaff have shown, company research was devoted to the improvement of carts, with a bid by Goodyear to encourage the introduction of pneumatic tyres.[30]

The challenge of thinking through transportation, circulation and exhibition is highlighted when the technology of transportation becomes not only the crucial vehicle for the delivery of media technology, but also the background, the horizon as it were, against which the technology not only operated but to which it was imaginatively and materially aligned. At one level, the technology itself has a representational function, articulating, in its own artefactual existence a distinctive form that generates a spectacle of transformation. As Ian Kerr points out, the railway not only undertook the work of speedy transportation, it conjured up a formidable image of steam, sound and movement, and of how its bridges, tracks, embankments

27 Ibid., p. 1095.

28 Vasudevan, Ravi. "From Advertising to Public Education: Notes on Burmah Shell in India." *Marg*, vol. 68, no. 3 [special issue: "The Story of Early Indian Advertising"], March–June 2017, pp. 114–17.

29 Petrie. *Village Visions*, p. 188.

30 Ibid., pp. 166–67; Tetzlaff, Stefan. *The Motorisation of the "Mofussil": Automobile Traffic and Social Change in Rural and Small-Town India c. 1915–1940*. DPhil thesis, University of Göttingen, 2015.

and tunnels reorder nature.[31] As I have suggested, motorized transport not only relays new ideas of speed, but it also implies the new metalled and cemented road networks in counterpoint to the bullock cart, even if that too was being remodelled to prevent damage to the new roadways. These material infrastructures of transportation therefore are part of the space-time horizon within which the vehicular form arrives, that annihilation of space by time which Wolfgang Schivelbusch and others have written of.[32] These vehicles and their infrastructure are both material and semiotic, embodying the network in their very presence, and thereby signifying the possibilities of travel and movement.

Transportation can mean more than moving from one place to another. As the Merriam-Webster dictionary informs us, transport can mean carriage of something or someone from one place to another; to cause (someone) to imagine that he or she is in a different place or time; to cause (someone) to feel very happy, interested, or excited. It is in these additional imaginative and affective terms that we may consider not only the semiotic work of the artefacts of transportation, but also of the media technologies which were brought into place by these travel networks. Here, apart from the attractions of the medium itself, the prominent film genres, the agricultural improvement films and the travel and pilgrimage films generated interest in other spaces, the latter for tourism or religious experience, or a mixture of both. Arguably, a standard and repeated concern expressed by officials about how films were received by peasant audiences, the alienness of peoples, customs and places which the imported instructional film brought along may also have elicited curiosity rather than estrangement. While officials believed that peasant viewers did not like seeing things they could not readily recognize, perhaps their anxiety stemmed from the possibility that such films would distract the peasant and undermine the instructional imperative. We will recall that it was exactly a transportation of peasant viewers that the railways wanted to bring into being. This was a figure who would imbibe new techniques, improve production, earn more, undertake new acts of consumption and travel, and engage in new commodity networks through bazaar trains. In the process there would emerge, in the railway publicity officer's design, a

31 Kerr, Ian J. "Representation and Representations of the Railways of Colonial and Post-colonial South Asia." *Modern Asian Studies*, vol. 37, no. 2, 2003, pp. 287–326; see also Prasad, Ritika. *Tracks of Change: Railways and Everyday Life in Colonial India*. Cambridge University Press, 2015, ch. 3, for the physical impact of embankments, and ch. 5, for the growing anxiety about railways' association with the spread of contagion.

32 Schivelbusch, Wolfgang. *The Railway Journey: The Industrialization of Time and Space in the 19th Century*. University of California Press, 1986.

mutually beneficial, looping effect that brought improvement and enhanced railway traffic together. As Petrie warns us in the case of rural Bengal, neither was the peasant as sedentary as this model suggests – short-term and seasonal migration was quite common – nor was he so receptive to the seeds and fertilizer and machinery retailed through this circuit. Nevertheless, people did buy from bazaar trains, and further, the event of viewing emerges as an absorbing material and imaginative engagement for the way it was profoundly framed by the wider horizon of transportation, a horizon which the viewer was invited to move into. Key here was the transportation artefact as an iconic register of the traversal of vast space.

I draw attention to the information available on the material forms of the actual exhibition context to suggest how this horizon was put together. There was a mise en scène here involving separate sidings away from the station, creating a distinct space for the exhibition train, the cinema car, and an arena for demonstrations of different sorts. These come from a slightly later date, the early 1940s and the era of World War II. Nevertheless, they suggest the types of viewing arrangements that were put in place.

> Exhibition Train
>
> The Train will spend one day at each station. It will arrive at approximately 6 a.m. each day and leave at 11 p.m. The exhibition will open at 12 noon.
>
> [...] The exhibition is divided into four parts:
>
> (a) Six coaches containing exhibits, photographs, etc.
>
> (b) Two trucks on the train on which demonstrations will be given, i.e. gun drill and explanation of an aeroplane fuselage.
>
> (c) A demonstration in an arena near the Train.
>
> (d) A cinema shows after dark.
>
> [...] The train will be "dazzle" painted and will be well lit at night.
>
> [...] It will be necessary to make arrangements at each station prior to the arrival of the train. The [...] arrangements required:
>
> (a) The length of train is 1,167 feet excluding the engine.
>
> (b) A siding is required for the train at each station. This must not be a line in normal use by trains. It should be as far away as possible from the working lines, provided that it is easily accessible to the public. The train should not be berthed at any platform. The crowd will be kept one side of the train only, and there should therefore be plenty of space on the side to be opened.
>
> (c) The arena will be one hundred yards long by eighty yards wide. The site must not be crossed by railway lines, telegraph wires or cables.

> It should be not nearer than fifty yards from the train and not more than eight hundred yards away. It may be possible to find a suitable area amongst sidings or shunting yards at the bigger stations, or outside the railway precincts altogether at country stations.
>
> (d) It does not matter if the train and arena are some way from the station provided there is easy access for the public without interfering with the working of the railway.[33]

We may observe here how the railway exhibition train is placed on a siding to constitute an arena for demonstration away from the main station. It has an aspect of spectacle in the dazzle painting which would make it visible at night, and it creates a horizon line, uninterrupted by wires or cables, for the crowd assembled on one side of it who would be witness to the spectacle. Another note recommends that the "ideal site for the Arena is a clear piece of ground with the Train forming a background." After visits by the public to the exhibition coaches during the day, one of the preparatory notes schedules that a "Pipe Band will play 'Retreat' until the Cinema Show opens at approx. 1815 hrs. Cinema continues until 2045 hrs. [...] News Bulletins will be broadcast in the vernacular at 1815 & 2110 hrs. [...] Music (Indian) will be amplified when no demonstrations are taking place in the Arena."[34]

The notes also work out the location of a *shamiana* (tented enclosure) for the elites, and the number and volume of police required to regulate inflow to the exhibition coaches and to take care of the Arena and later the "Cinema Show."[35] We thus have in these notes an outline of the multiple sensory dimensions of the exhibition, involving band, recorded music, radio news bulletins, and the cinema; the document also indicates the layered audience space that surrounds the train, and the policing arrangements made to keep order. Other notes on a Defence Services Exhibition train will provide much more focused exhibition, including physical demonstrations and film exhibitions of armed warfare as well as exhibits of the artefacts of war.

All of this contrasts with the arrangement of film spectatorship that places us to view the film in darkened auditorium with the apparatus of projection concealed; the question of the mode of delivery is not an issue in this scenario of classical, absorptive spectatorship (even if that construct

33 "Exhibition Train" Note, including Appendix. Mysore Residency, Bangalore, 1941. NAI, Defence Department, Public Relations Section, File no. 59.

34 "Defence Services Exhibition Train" arrangements to be made in advance of the train's arrival at the station. Ibid.

35 Ibid.

too has had to be historically qualified). In the viewing situation we are looking at now, there are several levels at which the infrastructure that enables media experience is arranged. The first is the horizon line of the train and, in a more modest register, the motor van. The second is the multiple media that define the experience, rather than a singular one, so that radio, gramophone, magic lantern, film and live performance were part of the attractions, and the loudspeaker appears as a key technology as well. Such media presentations were never purely instructional, and, apart from involving travel films, also included entertainment cinema culled from various levels of commercial film distribution. Thirdly, we have the phenomenon of railway and motor van-organized exhibition showcasing not only state approved techniques and technologies, but exhibitions and retail undertaken by business corporations. The latter, scholarship on rural colonial India has shown, was also involved in developing the material infrastructure of roadways, remodelled bullock carts and tractors, that were changing the networks of transportation which were also those of media publicity. Whether messages were actually communicated, whether media actually "worked" in terms of eliciting the desired cognitive engagement is, of course, another matter. And I have suggested that we need to place such a bid to create useful film and media experience as part of overlapping circuits, from what Ian Petrie refers to as the agrarian exhibition complex involving district administrations and itinerant salesmen to the work of travelling cinemas who would show government films and low-cost cinematic amusements.

Brian Larkin, exploring similar media use in colonial Nigeria, has argued that films highlighting health and agricultural improvement addressing peasant and tribal populations constituted a species of political cinema, separated out from the commodity circuits of film as entertainment. The category seems ironic in its intent, in the face of how the political in cinema has mainly been associated with left-wing agitprop, avant-garde, and, later, what has come to be known as post-colonial cinema. He uses the term to identify a cinema which relays state to its subjects as a form of tutelage, without the mediation of the market. He also takes this address as a vector of impending citizenship discourse, where the citizen is the beneficiary of welfare, the swathe of techniques and technologies which would help develop the economy, health, and thereby productivity and taxability of subjects.[36] While his analysis may reflect a particular way in which the circuit of the state-sponsored public information film operated separately from other circuits, my research suggests that film and media use, and the audiences

36 Larkin. *Signal and Noise*, ch. 3.

solicited by their circulation, inhabited much more complicated terrain. This was open to complex weaves of exposure to the world of commodities, the pedagogy of instruction, the affective and visceral attractions of entertainment, and the imaginative realm of travel.

Works Cited

Acland, Charles R., and Haidee Wasson, editors. *Useful Cinema*. Duke University Press, 2011.

Blunt, Edward, Sir, KCIE, OBE. "Government and Rural Development: United Provinces Experiments." Paper to be read before the East India Association at the Caxton Hall, Westminster, SW1, on 13 May 1936, with Sir Atul Chatterjee, GCIE, KCSI, in the Chair. British Library, India Office Library, L/I/1/596 Rural Development in India, p. 2.

Bose, A.N., Aurora Cinema, Das, H.K., Wellington Bioscope and Variety Co., Ghose, J.N., Monarch Bioscope Co., Banerjea, B.B., Temple Cinema, Dey, H.K., Wellington Cinema. Oral evidence. December 23, 1927. *Indian Cinematograph Committee*, vol. 2, p. 1089.

Bose, M.M., Professor, and Others of the Aurora Cinema Company. Oral evidence. December 15, 1927. *Indian Cinematograph Committee, Evidences*, vol. 2, pp. 665–77.

Burns, James. *Flickering Shadows: Cinema and Identity in Colonial Zimbabwe*. Ohio University Press, 2002.

Elsaesser, Thomas. "Archive and Archaeologies: The Place of Industrial Film in Contemporary Media." *Films That Work: Industrial Film and the Productivity of Media*, edited by Vinzenz Hediger and Patrick Vonderau. Amsterdam University Press, 2009, pp. 19–34.

Harrison, Rebecca. "Inside the Cinema Train: Britain Empire and Modernity in the 20th Century." *Film History*, vol. 26, no. 4, 2014, pp. 32–57.

Hediger, Vinzenz, and Patrick Vonderau. "Record, Rhetoric, Rationalization: Industrial Organisation and Film." *Films That Work: Industrial Film and the Productivity of Media*, edited by Vinzenz Hediger and Patrick Vonderau. Amsterdam University Press, 2009, pp. 35–50.

Heftberger, Adelheid. "Dziga Vertov's and Alexander Medvedkin's Film Trains and Agit-Steamers of the 1920s and 1930s." *Apparatus: Film, Media and Digital Cultures in Central and Eastern Europe*, no. 1, 2015, www.apparatusjournal.net/index.php/apparatwindelus/article/view/2/75.

Hutchinson, C.M., Chief Scientific Adviser, Fertiliser Propaganda of India. Oral evidence. December 16, 1927. *Indian Cinematograph Committee, Evidences*, vol. 2, p. 735.

Indian Cinematograph Committee, Evidences. Government of India, Central Publications Branch, 1928.

India Railway Board. *Report by the Railway Board on Indian Railways.* 1928–29. British Library, India Office Records, 1928/29-1931/32 IOR/V/24/3550.

Kerr, Ian. "Introduction." *Railways in Modern India,* edited by Ian Kerr. Oxford University Press, 2001.

Kerr, Ian J. "Representation and Representations of the Railways of Colonial and Post-colonial South Asia." *Modern Asian Studies,* vol. 37, no. 2, 2003, pp. 287–326.

Kinez, Peter. *The Birth of the Propaganda State: Soviet Methods of Mass Mobilisation 1917–1929.* Cambridge University Press, 1985.

Larkin, Brian. *Signal and Noise: Media, Infrastructure, and Urban Culture in Nigeria.* Duke University Press, 2008.

"Memorandum: Film Production and Display on the Indian State Railways." September 1927, pp. 2–5. National Archives of India, Defence Department, Public Relations Section, file no. 59, GOI Railway Department/Traffic 1750T/38B.

Narrative Report by Agents of Class I Railways for the Year 1929–1930. National Archives of India, Defence Department, Public Relations Section, file no. 59, Railway Department, 1253-statistical/166-200, B (K).

Petrie, Ian. *Village Visions: Science and Technology in the Bengal Countryside, c. 1860–1947.* PhD diss., University of Pennsylvania, 2004.

Prasad, Ritika. *Tracks of Change: Railways and Everyday Life in Colonial India.* Cambridge University Press, 2015.

Rice, Tom. "Working through the End of Empire." This volume.

Sanogo, Aboubakar. "Colonialism, Visuality and the Cinema: Revisiting the Bantu Educational Kinema Experiment." *Film and the End of Empire,* edited by Colin MacCabe and Lee Grieveson. Palgrave Macmillan, 2011, pp. 227–46.

Schivelbusch, Wolfgang. *The Railway Journey: The Industrialization of Time and Space in the 19th Century.* University of California Press, 1986.

Tetzlaff, Stefan. *The Motorisation of the "Mofussil": Automobile Traffic and Social Change in Rural and Small-Town India c. 1915–1940.* DPhil thesis, University of Göttingen, 2015.

Tode, Thomas. "Agit-trains, Agit-steamers, Cinema Trucks: Dziga Vertov and Travelling Cinema in the early 1920s in the Soviet Union." *Travelling Cinema in Europe,* edited by Martin Loiperdinger, KINtop Schriften 10, 2008, pp. 143–56.

Tylden-Pattenson, A.E., Chief Publicity Officer, Railway Board. Oral evidence. *Indian Cinematograph Committee, Evidences,* vol. 1, pp. 70–71.

Vasudevan, Ravi. "From Advertising to Public Education: Notes on Burmah Shell in India." *Marg,* vol. 68, no. 3 [special issue: "The Story of Early Indian Advertising"], March–June 2017, pp. 114–17.

Widdis, Emma. *Alexander Medvedkin.* I.B. Tauris, 2005.

Windel, Aaron. "The Bantu Educational Kinema Experiment and the Political Economy of Community Development." *Film and the End of Empire*, edited by Colin MacCabe and Lee Grieveson. Palgrave Macmillan, 2011, pp. 207–26.

Written Statement by Publicity Officer, East Bengal Railway. March 19, 1928. *Indian Cinematograph Committee, Evidences*, vol. 4, p. 192.

About the Author

Ravi Vasudevan is Honorary Fellow, Centre for the Study of Developing Societies (CSDS), Delhi (www.csds.in). With Ravi Sundaram, he directs Sarai, the CSDS media research programme (www.sarai.net). Vasudevan coordinates the media module for the International Centre for Advanced Studies: Metamorphoses of the Political (https://micasmp.hypotheses.org/), a German-Indian collaboration. He has edited *Making Meaning in Indian Cinema* (2000), co-founded the journal *Bioscope: South Asian Screen Studies* (2010–), and is author of *The Melodramatic Public: Film Form and Spectatorship in Indian Cinema* (2010, 2015).

20 The Latin American Process Film

Salomé Aguilera Skvirsky

Abstract

This chapter proposes an alternative reading of the early films of the New Latin American Cinema (NLAC). Rather than focusing on the representation of underdevelopment in these films, the chapter reframes this production through the prism of the process genre. The chapter asks, "Why do process films play such a key role in the history of political filmmaking in the region?" The chapter goes on to address this question largely by exploring the Chilean case, and particularly the work of Sergio Bravo.[1]

Keywords: Sergio Bravo; *Mimbre*; process genre; New Latin American Cinema; Fernando Balmaceda; *Tejidos Chilenos*

The historiography of the New Latin American Cinema (NLAC), the continental film movement responsible for one of the most sophisticated bodies of left political filmmaking, usually begins in the late 1950s with mostly short documentaries about the strenuous life ways of mixed-race peasants living in rural areas; the movement achieves its height in the late 1960s with such masterworks as *Black God, White Devil* (Glauber Rocha, Brazil, 1964), *The Hour of the Furnaces* (Fernando Solanas and Octavio Getino, Argentina, 1968), *Memories of Underdevelopment* (Tomás Gutiérrez Alea, Cuba, 1968), *The Jackal of Nahueltoro* (Miguel Littín, Chile, 1969) and *The Battle of Chile: Part I* (Patricio Guzmán, Chile, 1975); and begins to fracture and dissolve in the late 1970s as the revolutionary fervour of the late 1960s is met with violent reaction in Argentina, Chile and Brazil or as revolution gives way to state capitalism in the case of Cuba.

1 A version of this chapter was published as chapter 4, "Nation Building," in *The Process Genre: Cinema and the Aesthetic of Labor* (Duke University Press, 2020).

Hediger, V., F. Hoof, Y. Zimmermann, with S. Anthony (eds), *Films That Work Harder: The Circulation of Industrial Film*. Taylor & Francis Group 2024
DOI: 10.5117/9789462986534_CH20

The standard scholarly account of this first inaugural moment of the NLAC has emphasized three aspects of this production: the representation of "underdevelopment"; the influence of Italian Neorealism and Griersonian documentary; and the extent to which these early works were a developmental "stage" along the path toward a militant, autochthonous (rather than derivative) production that would be emblematized by the masterworks of the late 1960s.[2] It is certainly true that several of these films fix – for the first time in this region – a new visual lexicon to accompany the emergent discourse of underdevelopment. It is also undeniable that Neorealism and Griersonian documentary were important referents for the cosmopolitan filmmakers involved in the NLAC, especially the Argentine Fernando Birri and the Cubans Tomás Gutiérrez Alea and Julio García Espinosa, who all studied filmmaking in Italy in the early 1950s at the Centro Sperimentale di Cinematografía in Rome. But, the scholarly focus on the depiction of poverty and "underdevelopment," and on the works' European antecedents has obscured what stands out today as the most striking commonality of several of these most emblematic and canonical early works of the NLAC and their difference from later production: they are mostly process films.[3] That is, they are films organized by the sequential, and usually chronological, representation of a process of production that systematically follows the trajectory of raw material from unformed matter to manufactured commodity. The manufacturing process is often artisanal rather than industrial and the final commodity is intended for local consumption rather than international export.

The acme of Latin American political new wave cinemas – the NLAC – was inaugurated by the process film. The classic process films of the NLAC include *Arraial do Cabo* (Mário Carneiro and Paulo Cezar Saraceni, 1959) about traditional fishing and fish salting in a small municipality on the coast of the state of Rio de Janeiro; *Aruanda* (Linduarte Noronha, 1960) about the fabrication of ceramic housewares in a remote part of Paraíba, Brazil; *El mégano* (Tomás Gutiérrez Alea and Julio García Espinosa, 1954)

2 For an elaboration of this argument, see Aguilera Skvirsky, Salomé. "Realism, Documentary, and the Process Genre in Early New Latin American Cinema." *The Routledge Companion to Latin American Cinema*, edited by Marvin D'Lugo et al. Routledge, 2018.

3 This term was coined by Tom Gunning to refer to a genre of early films made between 1906 and 1917. See *The Process Genre: Cinema and the Aesthetic of Labor* (Skvirsky, Duke University Press, 2020), which extends the category, proposing a conceptualization of the "process genre" across media and across time. See also Gunning, Tom. "Before Documentary: Early Nonfiction Films and the 'View' Aesthetic." *Uncharted Territory: Essays on Early Nonfiction Film*, edited by Daan Hertogs and Nico de Klerk. Stichting Nederlands Filmmuseum, 1997, pp. 9–24.

about charcoal burning in the Zapata swamps of Cuba; *Araya* (Margot Benacerraf, 1958) about the process of salt mining on the Arayan peninsula in Venezuela; *Mimbre* (Sergio Bravo, 1957), *Trilla* (Sergio Bravo, 1959) and *Día de organillos* (Sergio Bravo, 1959) are, respectively, about craft in wicker, traditional wheat threshing, and organ making in Chile; *Faena* (Humberto Ríos, 1960) and *Ceramiqueros de Traslasierra* (Raymundo Gleyzer, 1965) are, respectively, about industrial slaughter and the production of ceramic figurines in Argentina; *Chircales* (Marta Rodríguez and Jorge Silva, 1967) is about brick making in Colombia. All of these films to varying degrees make use of processual representation – that is, the sequential how-to syntax that is the defining feature of what I call "the process genre." The processual character of these films has never been remarked in the literature despite the fact that, I would argue, it opens up a new horizon for the analysis of their aesthetic and political power. In this chapter, I aim to address this aporia. Why do process films play such a key role in the history of political filmmaking in the region? Why does processual syntax seem aesthetically and politically appropriate for the early NLAC cineastes? How do they transform its more usual manifestations to fit their ends? While the Latin American process film was especially significant in three national contexts – Brazil, Argentina and Chile – that are well-represented among the radical process films of the New Latin American Cinema, I will focus here on the Chilean case, in which the early work of Sergio Bravo emerged as an exceptionally compelling example of processual representation.

Processual representation is a familiar and well-used structuring principle in films from around the world and from different time periods. Some examples are *A Visit to Peek Frean and Co.'s Biscuit Works* (Cricks & Martin, UK, 1906), *Nanook of the North* (Robert J. Flaherty, USA, 1922), *Master Hands* (Jam Handy, USA, 1936), *A Man Escaped* (Robert Bresson, France, 1956), *Jeanne Dielman, 23 Commerce Quay, 1080 Brussels* (Chantal Akerman, Belgium, 1975), *The Ice Miners of Chimborazo* (Igor and Gustavo Guayasamín, Ecuador, 1980), *Red Persimmons* (Shinsuke Ogawa and Peng Xiaolian, Japan, 2001), *In Comparison* (Harun Farocki, Austria, 2009), *Leviathan* (Lucien Castaing-Taylor and Verena Paravel, USA, 2012) and *Raw Herring* (Hetty Naaijkens-Retel Helmrich and Leonard Retel Helmrich, Netherlands, 2013). Moreover, across the history of world cinema, process films are prominently represented in a range of (generic) categories of filmmaking from industrial film to educational film to ethnographic film to art cinema.

The process films of the New Latin American Cinema had antecedents in the local non-fiction, useful filmmaking scene. They emerged in a context in which non-fiction filmmaking was dominated by state and corporate institutions and initiatives and in which processual representation was a standard organizing convention. Of course, each national context had its own particular characteristics. In Chile and Brazil, the non-fiction filmmaking scene through the 1950s was dominated by sponsored films (educational as well as industrial films); in Argentina, there was a broader spectrum of production circumstances and an independent sector. It is in the context of the ubiquitous deployment of processual representation that several NLAC filmmakers adopted processual syntax in their early films. But they transformed it. In their experimentations, processual representation is not an incidental feature of the films' style; the processual syntax of the films is central to their power and aims. These experimentations ultimately reveal something about the process film as a genre.

Chile

In Chile, it was the prominence of industrial filmmaking that provides the relevant backdrop for the emergence of the NLAC process films. Chile had a tradition of mostly industrial – corporate and institutional – filmmaking going back to *Cemento Melón* (Carlos Eckardt Zúñiga, 1917) about the newly constructed cement factory in Valparaíso; and the film that scholars consider to be the first Chilean documentary, *El mineral "El Teniente"* (1919), by the Italian-Argentinian Salvador Giambastiani, about copper mining activities at the U.S.-owned Braden Copper company. While *El mineral "El Teniente"*, the only one of the two that survives in some form, is better read as an "album film"[4] featuring a collection of discrete views of the geography of the mining town and some disparate mining activities than as a process film, –by the early 1940s process films dominated non-fiction filmmaking in Chile. Of the approximately fifty-seven documentaries produced between 1941 and 1957 (the year Sergio Bravo made *Mimbre*), at least twenty-four could be considered process films; many others depicted processes within an album film format.[5] The remainder are mostly travelogues. The process films had titles like *La pesca en alta mar* (*Fishing on the High Seas*, Armando Rojas Castro, 1941), *Central de leche* (*Milk Station*, Guillermo Yánquez, 1944), *Carbón Chileno* (*Chilean Coal*, Pablo Petrowitsch, 1944), *Salitre* (*Salt Mine*,

4 "Album film" comes from Peterson, Jennifer Lynn. *Education in the School of Dreams: Travelogues and Early Nonfiction Film.* Duke University Press, 2013.

5 This estimation is mine based on data compiled in Vega, Alicia. *Itinerario del cine documental chileno, 1900–1990.* Universidad Alberto Hurtado, Estudios y Artes de la Comunicación, 2006.

Pablo Petrowitsch, 1946), *Cobre* (*Copper*, Pablo Petrowitsch, 1950), *Acero* (*Iron*, Patricio Kaulen, 1950), *Chile y su explotación ovina* (*Chile and Its Sheep Industry*, Vinicio Valdivia, 1951), *Petróleo chileno* (*Chilean Oil*, Armando Parot and Fernando Balmaceda, 1955) and *Pãnos de Tomé* (*Cloth of Tomé*, Boris Hardy, Alejandro Wehner and Jorge Di Lauro, 1955). Most of these films depicted industrial processes and factory production, and celebrated Chilean modernization. A few of the album documentaries, such as *Tarapacá, un pasado y un porvenir* (Egidio Heiss, 1943), made for the local provincial government, depicted non-industrial production. Almost all of these films – the process films and the album films – were funded by private companies like Carbonífera Lota-Schwager, Braden Copper and Nestlé, or by public entities like the Instituto de Cinematografía Educativa de la Universidad de Chile and the Ministry of Agriculture. Many were produced by private publicity companies like Cinep, founded by Fernando Balmaceda and Armando Parot, and the Argentinian company Emelco. The films were produced by the same cast of filmmakers, including the cinematographer Egidio Heiss and the directors Guillermo Yánquez, Armando Rojas Castro, Pablo Petrowitsch, Naúm Kramarenco, Patricio Kaulen and Fernando Balmaceda.[6]

Against this backdrop, Sergio Bravo's process films – which were made between 1957 and 1962 and are among the inaugural films of the NLAC – stand out *not* because they are the first process films. As we have seen, processual representation was a standard film rhetoric of the period in Chile. Bravo used the same genre conventions, but transformed them. He was the founder, with Pedro Chaskel, of the Centro de Cine Experimental in 1957. The Centro de Cine Experimental – which would, in 1961, be incorporated into the Audiovisual Department of the University of Chile – and the Instituto Fílmico de la Universidad Católica would revolutionize Chilean national cinema, constituting a break with previous production both in fiction and non-fiction.[7] In the first phase of its existence, from 1957 to 1963, the Centro produced ten short films, the most outstanding of which are five process documentaries directed by Bravo – *Mimbre* (*Wicker*, 1957), *Día de organillos* (*Barrel Organ Day*, 1959), *Trilla* (*Wheat*, 1959), *Láminas de Almahue* (*Plates from Almahue*, 1961), *Parkinsonismo y cirugía* (*Parkinsons and Surgery*, 1962) – and the city symphony film, *A Valparaíso* (1963), attributed to Joris Ivens, who spent several months at the University of Chile between 1962 and 1963. For his part, Bravo (at the time) understood his work to be opposed

6 I discuss some of this material in Aguilera Skvirsky. "Realism, Documentary."

7 See Salinas Muñoz, Claudio, et al. *Historia del cine experimental en la Universidad de Chile, 1957–1973*. Uqbar Editores, 2008.

to that of his contemporaries – filmmakers like Fernando Balmaceda, for whom he worked as a cameraman in 1957 and 1958; Pablo Petrowitsch; and Jorge Di Lauro.[8] These were filmmakers making publicity process films for industry. While their films were almost universally nationalist celebrations of machine-driven extractive industries like copper, coal and limestone, Bravo's films were celebrations of artisanal, manual labour: the craft in wicker in *Mimbre*; wheat threshing by hand in *Trilla*; barrel organ fabrication in *Día de organillos*; traditional wheelwright craftsmanship in *Láminas de Almahue*; the surgeon's craft in *Parkinsonismo y cirugía*.

Bravo's difference from his contemporaries should not be sought only in the fact that these contemporaries were producing films for paying patrons – whether corporate or governmental – with directly useful objectives. After all, several of these sponsored films circulated nationally and internationally as works of cinematic art, some winning awards like *Salitre*, which, in 1947, won a prize from the Chilean film and theatre magazine *Ecran* for best short film of the year. Fernando Balmaceda, perhaps the most productive process filmmaker of the late 1950s and 1960s and the recipient of several awards and honours, reports with surprise the positive reception of his films "if one considers that all of them were commissioned films that were either public service announcements or publicity films."[9] Balmaceda's documentaries were exhibited in theatres and written up in the local newspapers. He even won an annual prize from the Critics of the Art. His work circulated in international film festivals and received honourable mentions in Montevideo and Karlovy-Vary; a gold medal in Moscow; a Golden Dove in Leipzig's documentary festival; the medal for best short national film at the fourth Festival Internacional de Cine de Viña del Mar for the sponsored film *Carbón* (1965). In addition, television stations in Poland, the Soviet Union, Switzerland and France acquired the rights to exhibition of some of them, such as *Tejidos Chilenos* (1965).[10]

8 Ibid., p. 43. There is an interesting story to tell about Sergio Bravo's educational background. He had often said that he was more influenced by his training in architecture than by his participation in the cine-club that he started at the University of Chile. Bravo studied in the School of Architecture at the University of Chile in the late 1940s, after the university reform of 1946. Bravo studied under the direction of the Hungarian Bauhaus architect, Tibor Weiner, who had who arrived in Chile in 1939 and became a major force in the curricular revamping of the School of Architecture between 1946 and 1947 along functionalist lines. See Talesnik Y., Daniel. "Tibor Weiner y su role en la reforma: una re-introducción." *Revista de arquitectura*, vol. 12, no. 14, 2006, pp. 64–70.

9 Balmaceda del Río, Fernando. *De zorros, amores y palomas: memorias*. El Mercurio/Aguilar, 2002, p. 343.

10 See ibid., p. 343; Pino, Diego. "Fernando Balmaceda del Río." *Revista séptimo arte*, www.r7a.cl/article/fernando-balmaceda-del-rio/. Accessed August 31, 2015.

Nor are the differences between these industrials and Bravo's films to be sought in their nationalist orientation: they are all – at least, on the face of it – paeans to the intelligence of the Chilean worker. For example, in *La metalurgia del cobre* (Patricio Kaulen, 1960), sponsored by the Braden Copper company, the voice-of-god narrator ends his step-by-step account of the process of copper refining with a crystallization of the explicit message peddled by the film:

> The cycle is complete. The blind furies of nature were dominated by man. His intelligence made possible the exploitation of her riches, and in a process that appears unreal, fantastic, hallucinatory. Here what began as a simple rock became a refined product of an indefatigable labour and an irrefutable illustration of what man is capable of making with what nature proffers.[11]

The narration's celebration of the sophistication of the design of the process and of the machinery is displaced from the American engineers of an American-owned copper mine to the Chilean workers employed by the company in a naked bid to attract more workers.

Nor are the differences between industrial process films of this period and Bravo's process films to be found in the former's neglect of artisanal production. Some of Fernando Balmaceda's best-received films, *Manos creadoras* (1961) and *Tejidos Chilenos* (1965), centrally involve craft production. *Manos creadoras* is a short film depicting crafts (including work in wood, textiles, ceramics, engraving, tapestry) practiced at the School of Applied Arts, University of Chile. *Tejidos Chilenos* – commissioned by one of the major textile companies in Latin America, Bellavista-Tomé Cloth Factory, to celebrate its one hundred year anniversary, and produced by Balmaceda's own publicity company, Cinep – pridefully traces the history of Chilean textiles from the pre-Columbian practices of the Mapuche people that survive into the present to the practices of the Chilean cowboy, "*el huaso*," to the machine production of the Bellavista-Tomé factory. It is a particularly instructive example.

The final voice-over narration of *Tejidos Chilenos* declares:

> The Chilean textile tradition saw itself incorporated into the movement of progress and its slow and patient flow became an agile and powerful industrial impulse. Chilean textiles since then have achieved a new

11 Translation mine.

> dimension that may be traced back to their traditional beauty and quality. The art of textiles – whether it is the kind that sprouts spontaneously near a dry stone wall in Atacama or in the smoky, warm interior of a hut in Temuco or beneath a tile roof on some patio in Doñihue – is an art with one single origin and one single destiny, to which industrialized art is not foreign. The same love of material binds them since they have in common the spinning and the dyeing and the warping and the knitting. The one – the craft – sprouts pleasantly and lovingly like the quiet flowering of nature. The industrial art is, on the other hand, manmade machine and power, ingenious and rhythmic, unlimited in his reach, as if he returns to himself for the delight and satisfaction of men. This is what Bellavista-Tomé has accomplished in its hundred years and what it will continue doing: evolving in a textile tradition that is also its own tradition.[12]

The argument of *Tejidos Chilenos*' narration is unmistakable: the industrial production of cloth borrows everything that is good and natural from the artisanal production of the autochthonous craftsman and develops and improves it. While the world "evolves new techniques and designs, here [in the countryside among artisans], [...] these silent centres of creative activity flower in their own eternity." The dichotomy between city and country, factory and artisanal workshop, indigenous craftsperson and creole machine operator mobilizes familiar evolutionary scripts from anthropological discourse: the city, the factory, the industrial worker are dynamic, always evolving and innovating; the country, the small-scale workshop, "the Indian" are eternal, traditional, natural.[13] Yet, in Balmaceda's narration, both the

12 Translation mine.

13 Johannes Fabian has argued that this allochronism or denial of co-evalness – that is, "a persistent and systematic tendency to place the referent(s) of anthropology in a Time other than the present of the producer of anthropological discourse" (p. 31) – was a constitutive feature of the emergent discipline of anthropology. See Fabian, Johannes, and Matti Bunzl. *Time and the Other: How Anthropology Makes Its Object*. Columbia University Press, 2014. Undergirding the denial of co-evalness is a broadly shared embrace of cultural evolutionism that may be traced back through the work of Immanuel Kant, Georg Hegel, Herbert Spencer, Edward B. Tylor, Lewis Morgan, but also Friedrich Engels and Karl Marx. Ethnographic process films – even contemporary ones – *often* exhibit some variation of cultural evolutionism. *In Comparison* (Harun Farocki, Austria, 2009), for example, is a film that plays with this close relation between ethnographic film, the process film, and cultural evolutionism. For an application of Fabian's theory to film, see Fatimah Tobing Rony, *The Third Eye: Race, Cinema, and Ethnographic Spectacle*. Duke University Press, 1996; for more on cultural evolutionism, see Robert L. Carneiro, *Evolutionism in Cultural Anthropology: A Critical History*. Westview Press, 2003; Harris, Marvin. *The Rise of Anthropological Theory: A History of Theories of Culture*. Crowell, 1968.

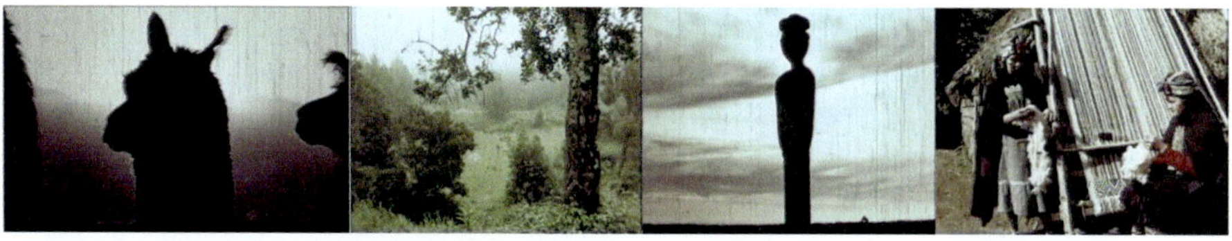

Fig. 20.1. From *Tejidos Chilenos* (Fernando Balmaceda, 1965). From the opening.

dynamic factory and the enduring autochthonous craftsman have their place in the story of Chilean textiles: the latter gives rise to the former before branching off into a dead end.

This approach to Bellavista-Tomé's centennial, which devotes half its screen time to a positive assessment of artisanal craft production, was initially of concern to the company's executives – until they saw the final film.[14] Despite the glowing review of traditional, autochthonous Chilean textile practices in which, according to the narration, these practitioners "appear limitless in their creative imagination, in the good taste of their designs, in their weaving practice which is pure mastery" – despite this explicit praise, the film's visual treatment of machines and industrial production has quite a different character from its treatment of artisanal practices. The portion of the film devoted to artisanal production features several postcard-perfect landscape shots, what one might expect from a tourist brochure or a travelogue: of alpacas at sunrise; of misty virgin forests; of manicured meadows; of pre-Columbian totems at dusk (fig. 20.1). When the camera does turn to Mapuche and "*huaso*" weaving processes, the images are almost continuously accompanied by a voice-over narration except for a one-minute interlude in which a Mapuche woman works on a handmade wood loom. Several of the shots that make up this section are artfully framed and photographed; still, the particulars of her labour and the sequence of her work is difficult to make out.

By contrast, in the portion of the film devoted to industrial production, Balmaceda constructs a particularly striking 00:02:20 narrationless sequence in which machines spin and ply and then pick and gill and dye and dry and card and comb and warp and weave (fig. 20.2). One contemporary critic for the Chilean magazine *Ercilla* praised the sequence for illustrating the industrial process "in a *precise way*, without the habitual excesses of narration. The cinematography ably takes advantage of the colour and texture of the different elements, while the music of Sergio Ortega lends an

14 Vega. *Itinerario del cine documental chileno, 1900–1990*.

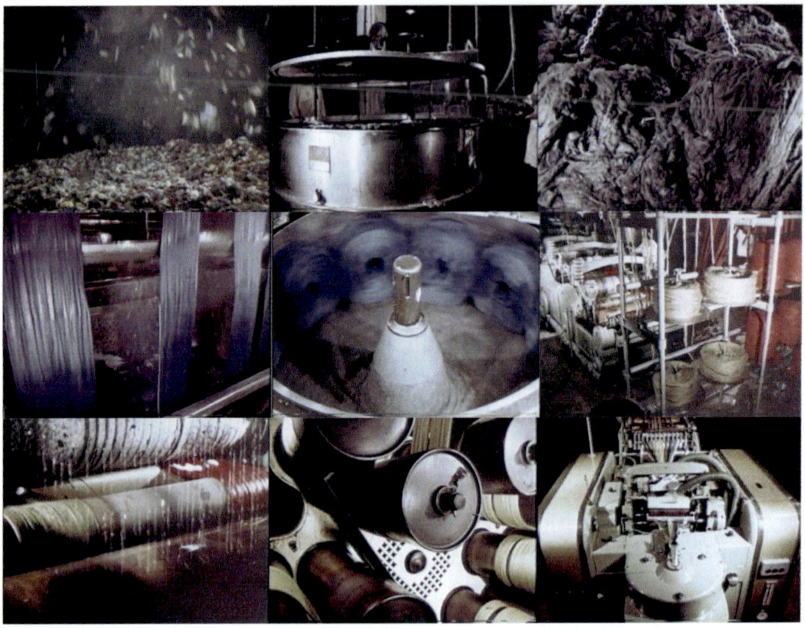

Fig. 20.2. From *Tejidos Chilenos* (Fernando Balmaceda, 1965). Part of the 2:20 narrationless sequence, which is not in sequential order. There are shots missing from this figure, but not steps (as there are multiple views of the same step).

important auditory complement to the visual rhythm of this sequence."[15] The stars of this lyrical sequence are unquestionably the dazzling automated machines – filmed in dramatic close-up – that spin and roll and twist and transform shorn wool into colourful, patterned fabric. If the narration gives both industrial and artisanal production a favourable treatment, the image track and the score tell another story about the unlikely technological modernity of the Chilean textile industry.

This divergent approach in the representation of indigenous craft production and industrial mass production is not atypical in colonial or we might say, internal colonial, contexts. As Rianne Siebenga has shown in the case of the representation of Indian crafts in early British films, processual representation was reserved for the depiction of industrial production while the album film format was applied to the representation of traditional Indian crafts. Siebenga concludes that in the Indian case an album approach to craft followed colonial discourse in underscoring a lack of modernity. The point could be elaborated: processual representation as a structuring principle suggests modernity, and in a context like the Latin American context in which

15 Qtd. in ibid., p. 230. Translation mine. Emphasis mine.

it is ideologically paramount that the indigenous element's "backwardness" be folded into and overcome in a dialectical synthesis emblematized by the mestizo mixed-race national subject – in such a context any suggestion of the modernity and dynamism of the indigenous national subject poses a problem for any state's project of acculturation (and whitening).[16]

The ideology of *mestizaje* in which the indigenous national element is not expelled from the nation (as in U.S. ideology) but incorporated – swallowed and absorbed – maps well onto the ideology of reactionary modernism in which the nation's non-industrial, rural, folk past leads inevitably to a modern, industrial, urban present.[17] In an adaptation of reactionary modernism to the Latin American context, *Tejidos Chilenos*' solution to the rurality, indigeneity, "archaicism" and "underdevelopment" of the Chilean countryside is to propose a triumphalist cultural evolutionism that passes unilinearly from tradition to modernity, from country to city, from artisanal labour to machine automation, from "Indian peasant" to mestizo worker all in an inexorable march toward progress.

What will set Bravo's films apart from Balmaceda's best exemplars are not that the former employs poetic photographic and editing strategies. After all, *Tejidos Chilenos* uses those very resources in its treatment of machines – their rollers, their axles, their gears. The major difference is in their respective treatments of artisanal craft and the artisan. If *Tejidos Chilenos* reserves processual syntax for industrial production, Bravo's films apply it to artisanal production performed by peasant-artisans.

Mimbre (Sergio Bravo, 1957)

Mimbre was Bravo's first process film. He must have been working on it at the same time as he was working as a cameraperson for Balmaceda

16 For more on the discourse of mestizaje in Latin American, see Vasconcelos, José. *The Cosmic Race/La raza cosmica*. Translated by Didier T. Jaén, 2nd ed. Johns Hopkins University Press, 1997; Freyre, Gilberto. *The Masters and the Slaves: A Study in the Development of Brazilian Civilization*. Translated by Samuel Putnam, 2nd rev. ed. Alfred A. Knopf, 1956; Miller, Marilyn Grace. *Rise and Fall of the Cosmic Race: The Cult of Mestizaje in Latin America*. University of Texas Press, 2004; Martínez-Echazábal, Lourdes. "Mestizaje and the Discourse of National/Cultural Identity in Latin America, 1845–1959." *Latin American Perspectives*, vol. 25, no. 3, 1998, pp. 21–42.

17 For the classic text on "reactionary modernism," see Jeffrey Herf, *Reactionary Modernism: Technology, Culture, and Politics in Weimar and the Third Reich*. Cambridge University Press, 1986; Michael Cowan has written about reactionary modernism in relation to industrial films in the German context. See Cowan, Michael. *Walter Ruttmann and the Cinema of Multiplicity: Avant-garde – Advertising – Modernity*. Amsterdam University Press, 2016.

on publicity projects for Cinep. *Mimbre* is a ten-minute film that depicts the work of Manzanito, a master wicker craftsman, living in the Quinta Normal commune of the metropolitan city of Santiago. The film opens on a medium long shot of Manzanito seated outside, concentrated on the task of splitting and preparing wicker fibres for weaving. The figure absorbed in work, facing the camera, shares the frame with a leaning tree trunk in a composition that undeniably borrows the conventions of the picturesque. Manzanito labours to a guitar score composed especially for the film by the Chilean musician and folklorist Violeta Parra. The film will depict – in a loose process structure – the steps in Manzanito's production of a wicker sea animal beginning with the preparation of the wicker fibres and moving to the weaving of fibres around vibrating upright stakes then to the bending of fibres then to the creation of the animal's body cavity then to the construction of the fish's head. The sequences of Manzanito labouring are interrupted twice, each time is marked by a distinct musical thread that introduces two visual interludes in which the meandering camera surveys, from near and far, the marvels of Manzanito's completed wicker animals: the simulation of the undulating wing of a dove in flight; the approximation of bird feathers; the open mouth of a fish as if gasping for water in a fisherman's net; the contrast of the fish's tail fin with the scales-effect created by the wicker weave; the plasticity of a horse's wicker mane; cow eyes and pointy, curving wicker horns; arched chicken beaks and wicker wattles and combs etc. (fig. 20.3). The interludes highlight, at once, the versatility of the wicker material that can just as well approximate fish scales as horse manes and chicken wattles and the ingenuity and flourish of the craftsman who has vivified his birds by his crafting of the feather wing in flight and enlivened his fish, paradoxically, by freezing it in its last open-mouthed gasp for water. The displayed wicker animals are further animated by the breeze that stirs the wicker dove hanging from a string or by the children in the film that commandeer the life-sized wicker animals, putting them to use as masks and play costumes.

The animation of inanimate matter extends to raw materials as the camera films vibrating wicker stakes. In one shot, the upright vibrating stakes occupy the foreground, Manzanito's face obscured behind them and his working hands – the cause of their vibration – out of the frame entirely. In another shot, the fibre stakes extend horizontally out of the leftmost edge of the frame, the cause of their movement wholly hidden from view (fig. 20.4). Matter – wicker animals and wicker fibres – seemingly moving itself is not so different from Balmaceda's automated machines that miraculously card and comb and weave seemingly without the aid

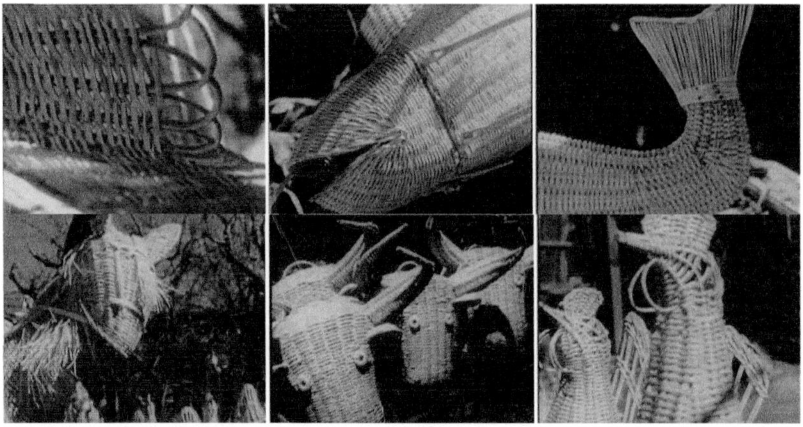

Fig. 20.3. From *Mimbre* (Sergio Bravo, 1957). Finished creations.

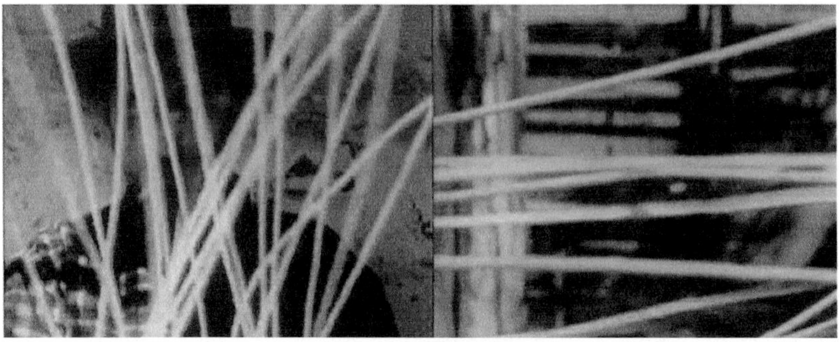

Fig. 20.4. From *Mimbre* (Sergio Bravo, 1957). Vibrating fibers.

of a human agent. But whereas Balmaceda reserved this dynamism for industrial textile production and relegated "*huaso*" and Mapuche artisanal textile production to the realm of an inert, unchanging, eternal (though aesthetically pleasing) tradition – Bravo's craftsman, his workshop, his materials, and his artefacts teem with life, dynamism, movement, and an evolving creative ferment. Each interlude presents a mystery about how Manzanito was able to manipulate wicker in such a way as to approximate a flowing mane or a tapered horn or a fish's tail fin. And because the interludes punctuate the processual representation, each time we are thrown back to observing Manzanito at work, manipulating his materials, our apprehension of the craftsman's dexterity at the level of execution and his ingenuity at the level of conception (after all, how did he think to solve the problem of the dove's feathers with wicker loops rather than something else?) is enhanced. It is both a sense of dexterity and ingenuity that is missing from Balmaceda's Mapuche and "*huaso*" weavers whose labours look peripatetic

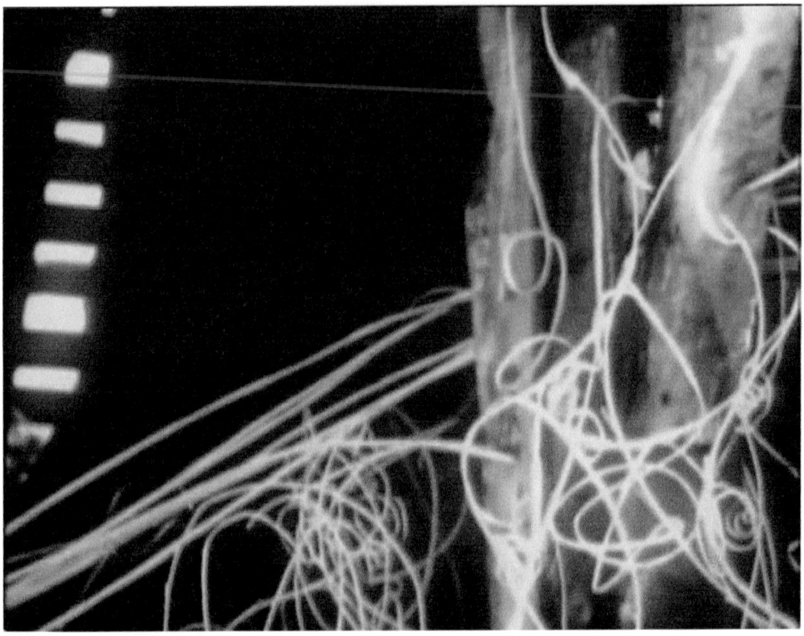

Fig. 20.5. From *Mimbre* (Sergio Bravo, 1957). Film strip—two ways.

and unsystematic. But if a sense of another's dexterity can be achieved by close, patient and sustained observation, the representation of ingenuity (which is an element of technological development) is more elusive.

Mimbre achieves an acute sense of the craftsman as a technologist. Defamiliarization is the governing aesthetic conceit of the film. In *Mimbre*, the defamiliarization that is evidenced in the enlivening of wicker animals and wicker fibres also manifests in the proliferation of visual and aural analogies. The self-moving, dry fibres catch the light in such a way that their reality is uncertain – are they tangible matter, fibres, or the illusion of matter created by light rays? Is it a light show that we are seeing? Or, are the fibres the liberated strings of Violeta Parra's guitar that plays over the images? The discarded fibre curlicues produced by Manzanito's wicker fibre-shredding process look like discarded celluloid film reels (fig. 20.5).

The body cavity of Manzanito's wicker whale is filmed from an aerial perspective, then visually analogized, through associative montage, to a daisy, a field of daisies, and to a spider spinning its web. In this segment of the film, we are presented with four shots: a field of daisies, differentially illuminated by natural light; the wicker animal's body cavity, filmed from above; a close-up shot of a spider, washed out, spinning a web; and a close-up image of Manzanito working on a dramatic wicker hourglass shape. The first two shot transitions are form dissolves: the high angle shot of the bunch of daisies leaches into the shot of the hollow centre of the whale's body, the round darkness of the daisy field in the left third of the frame becomes the round darkness of the hollow centre of the animal. For ten seconds the camera records the changing play of light inside the animal's body cavity as the dark round centre shrinks and throbs with the sun's inconstant illumination. The shot suggests two other visual analogies – to a single daisy with its dark disk the hollow centre and with its light-coloured petals the wicker weave, and to the eye of a living creature whose pupils dance with the reflection of light. In the next transition, a quick form dissolve superimposes a round, washed-out, white spider on the dark centre of the whale cavity – the juxtaposition both similar (in shape) and distinct (in colour); the spider's patterned silk threads are like the wicker fibres, which, once woven, are solid enough to hold together but permeable enough to allow air and light to pass. The spider shot cuts to a close up of Manzanito at work, the round white spider in the top left of the frame is displaced by a round, dark hollow cavity and Manzanito's hand at work weaving through splayed wicker stakes – the spider's labour, thus, explicitly likened to the craftsman's labour (fig. 20.6).

Bravo's films produce an acute material consciousness as one is confronted by a comparison of forms: the wicker dying fish is like and unlike the real dying fish; the vibrating wicker fibres are like and unlike the strings of the guitar, light rays, reels of discarded celluloid; the wicker weave is like and unlike the cobweb; the spider is like and unlike the artisan; the wicker cavity is like and unlike a daisy or an eye, etc. The sequence of a poetic cinematography that defamiliarizes and abstracts the work of the machines in *Tejidos Chilenos* functions similarly, but the formal comparisons are to non-representational, abstract, modernist shapes and patterns. Still, in both cases, these analogies of forms seem to deinstrumentalize the steps in an instrumental process with a determinate, useful end.

Surely in the case of *Mimbre* one might think that Bravo's defamiliarization of the materials of wicker craftsmanship deinstrumentalizes them, as it emphasizes their physical properties – treating those properties as ends in themselves – rather than emphasizing their function in the creation of use objects (e.g. wicker animals, for example). Such a reading would suggest that *Mimbre* just inverts the industrial process films' instrumentalism: means displace ends as the isolable beauty of each step in the process of wicker animal production risks rendering the final object superfluous. On this view, *Mimbre* just extends what in *Tejidos Chilenos* is an interregnum, a flourish, to the film as a whole.

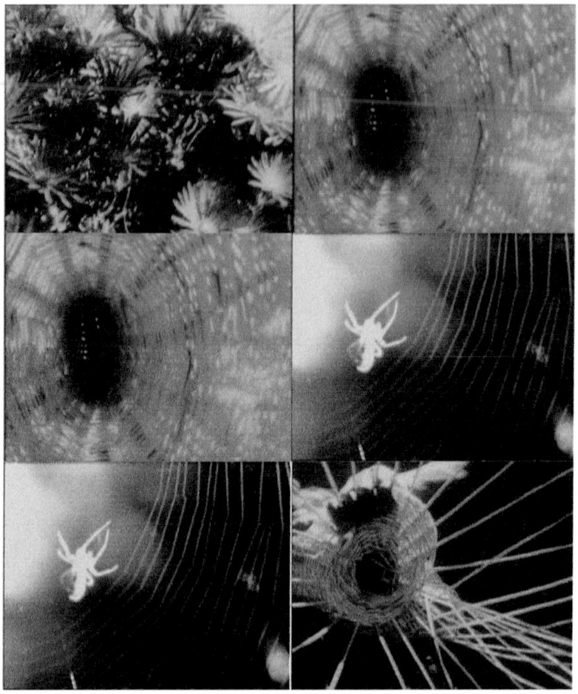

Fig. 20.6. From *Mimbre* (Sergio Bravo, 1957). Four consecutive shots. The first two transitions are dissolves; the third is a straight cut. All the transitions are establish graphic matches.

But, consider the *usefulness* of defamiliarization to technological innovation. Does not innovation often sprout from a fresh perception of the possibilities of matter that then leads to repurposing? Where does wicker come from? How did wicker first become the basis for basket making? One source of wicker is tree branches. When the branches are attached to trees, growing leaves, shading the ground, it is difficult to imagine that those very same branches could be "repurposed," used for basket-making or – as in the case of Manzanito's work – for life-sized wicker animals. In order to acquire such an insight, one would have to know something about those branches, the character of their fibres – when wet, when dry, when living, when dead etc. The possibility of using those branches for making wicker depends on defamiliarizing their usual use, which requires acquaintance with their properties. Making strange is a phase in much technological innovation; deliberate defamiliarization is, thus, a strategy.

In the case of *Mimbre*, it is striking that the film is bookended by a tree. While the first shot juxtaposes the tree's trunk with long wicker fibres, culled and piled on the ground awaiting Manzanito's treatment, the film ends with a more explicit survey of the tree. An aerial medium shot looks down on a dark bucket of water that Manzanito uses to make the wicker

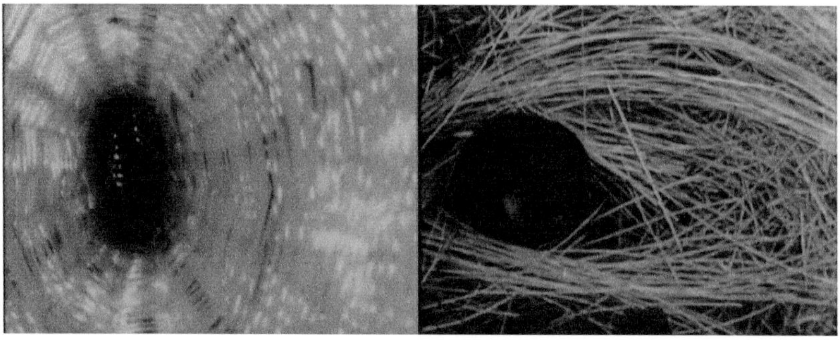

Fig. 20.7. From *Mimbre* (Sergio Bravo, 1957). Compare the visual eye metaphor of the earlier moment (left) to the eye of the penultimate shot of the film (right), an aerial view of the outdoor work space. In both, the metaphoric pupil uses light (from the bottom of the cavity in the first case and from the reflection of the sky in the second) to suggest life.

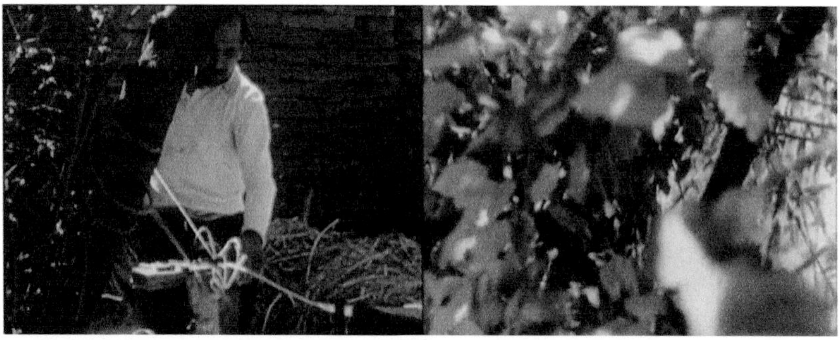

Fig. 20.8. From *Mimbre* (Sergio Bravo, 1957). The opening and close shots of the film—of a tree.

fibres more pliable; the water in the bucket catches the light presumably from the moving clouds overhead in such a way that for a moment it seems that an animal – a rabbit, perhaps, like in a magician's trick – is stirring in the bucket, about to emerge. The bucket is surrounded by long strands of wicker fibre hugging its rounded contours; the image reprises the visual metaphor of the daisy with its round, dark centre and of the spider in the eye of its web and of the eye with its reflective pupil (fig. 20.7). The image cuts to a slightly longer shot of the same scene, but now the camera tilts and tracks upwards to reveal the tree of the opening shot with all its diverging branches and its fluttering leaves. This final moment of the film as it lingers over a view of leaves quivering in a gentle zephyr recalls the Lumière brothers' early film *Baby's Meal* (1895), in which the wind in the trees came to betoken, for contemporary audiences, the miracle of the cinema (fig. 20.8). From the tree Manzanito has derived wicker fibres and from wicker fibres wicker animal figures. Manzanito's craft labour has repurposed the stuff of nature, and the film suggests the miracle was achieved not despite an investigation of form(s), but because of it. Moreover, the kinesis that permeates every shot and all matter in the film underscores the dynamism of life where nothing stands still, where everything is subject to constant change.

But there is more: Bravo's material consciousness includes the careful observation of how matter, instruments and labour work. His is a dialectic of anti-instrumentalism and instrumentalism. It is not enough to bring out the dryness of wicker as a material (i.e. a sense brought out by analogizing the wicker to beams of light); to understand wicker and its potential one needs to understand how it bends, how it threads, how it knots, how it cuts, how it ravels and unravels, how to manipulate it, how to shape it, etc. In other words, one needs to see it in action; one needs to see what it causes and what it effects. And so, at the centre of this picture is Manzanito, the craftsman, making wicker imitations of life out of trees – making something from "nothing."

The most fascinating, compelling examples of processual representation deliver this kind of knowledge, and often depend on defamiliarizing cinematographic techniques as one strategy. Sergio Bravo's films are captivating process films; and they provide this knowledge. This is how we can make sense of critics' treatments of his films. In a short description of *Trilla*, Alicia Vega writes that the camera follows the wheat-threshing activity of the campesino "meticulously," "operating at times from the perspective of the axis of a [cart's] wheel."[18] Of *Mimbre* and *Día de organillos*, Corro et al. explain that these films prioritize the close up, "thereby conferring dramatic and plastic centrality to the artisan's hands that thread and intertwine wicker [and] to the adjustments that the barrel organ fabricator makes to his instrument in his miserable ranch."[19] Moreover, the attention in *Mimbre*, for example, is focused on the "material *manipulations* that precede and constitute the object [...] and not so much on the person of the artisan."[20] Unlike Balmaceda's *Tejidos Chilenos*, Bravo's artisans are not analogized to an inert nature, resigned to an unchanging, eternal biological life. They are paragons of dynamism, human intelligence. They are cast as men of action, makers of worlds.

The human capacity for technological development (for that is what we are talking about) has a political significance in this context. For, just as the human being can repurpose nature and make some useful object from mere

18 Vega. *Itinerario del cine documental chileno, 1900–1990*, p. 165. Translations mine.

19 Corro Pemjean, Pablo. *Teorías del cine documental chileno, 1957–1973*. Pontificia Universidad Católica de Chile, Facultad de Filosofía, Instituto de Estética, 2007, p. 96. My translation.

20 Ibid., p. 101. Translation mine. Emphasis mine.

matter, surely she is also capable of transforming the world, of organizing it according to a different set of principles. In the case of Bravo's process films, the agent of world-making and remaking is neither the industrialist of Balmaceda's *Tejidos Chilenos* nor "the Indian," but rather a hybrid national subject – the Chilean *volk* – constituted by the realities and ideology of racial and cultural mixing. The processes that Bravo poeticizes – from the wicker work in *Mimbre* to wheelwrighting in *Láminas del Almahue* to barrel organ construction in *Día de organillos* to threshing in *Trilla* – are not autochthonous in the sense that they are unique to Chile, invented there; they are "autochthonous" only in the sense that they were practiced there at the time Bravo filmed them. The important point is not that Chile has a unique, singular culture; it is, rather, that its symbolic national subject has skill, capacity, intelligence and agency. Bravo's achievement was to marshal processual representation to the ends of perceiving – viscerally, bodily, not merely intellectually – the significance of the practical intelligence of the folk subject.

The process films of the first phase of the NLAC vary. Some, like *Mimbre*, place more emphasis on the ingenuity and technicity of the artisan's conception. Others, like *Aruanda*, on her dexterity. And still others on the complexity or efficiency of the steps of the process. All of them – as a direct consequence of the adoption of processual representation – transmit a sensuous experience of the labour of their national subjects as skilled, intelligent, and systematic.

Conclusion

In one of the landmark films of the NLAC, Glauber Rocha's *Barravento* (*The Turning Wind*, 1962), the impetus to revolutionary struggle is allegorized by an act that interrupts a process of production.[21] Firmino returns to the fishing village of his birth after an unsuccessful stint in the city. There, he finds his community stuck in a repeating cycle of bare subsistence: they survive by selling the fish they have fished collaboratively with a rented net to the distributer who has rented them the net; they pray to the saints of the syncretic religion of Candomblé to keep them safe and alive. Firmino, finding his village just as he left it, cuts the net on which the community's survival depends; he does this in order to provoke resistance, to invite the "turning

21 A strikingly similar moment occurs in *Redes* (Emilio Gómez Muriel and Fred Zinnemann, Mexico, 1936).

wind," he says. Without a net, the community faces either starvation or it must change its mode of producing and reproducing life – its way of life.

The process films of the first phase of the NLAC are organized around processes of production that have a cyclical character: production proceeds in a series of steps – with a beginning, middle, and end – passing from unformed matter to saleable commodity. But the processes presented are not singular, unrepeatable events; at the same time that a process of production is being presented, a process of reproduction is being implied. In every case, the process of (re)production is cast as the basis of a way of life. The films, thus, are largely allegories of a parallel national community with a distinct mode of production. As such, it is little wonder that many of the titles – *Aruanda, Arraial do Cabo, Araya, Ceramiqueros de Traslasierra* – refer to geographic places.

It is common to think that in these process films a rustic "primitive," "underdeveloped" way of life is the object of criticism and the occasion for intervention. On this view, it is the untenable cycle of "underdevelopment" from which the national peasant subject needs liberation. This is what is, in effect, proposed in *Barravento*. Of course, such a reading makes the choice of a processual syntax (missing from *Barravento*) in the films I have been discussing here odd. This is especially so given that processual syntax had been a staple in non-fiction industrial and educational films, films that were strenuously making the case for Chilean, Brazilian and Argentinian *technological modernity*. Thus, it would be surprising if the processual representation of craft production was functioning as a vehicle for making a case for the *technological underdevelopment* of these communities. Such a reading of these films (i.e. as documents of "underdevelopment") chafes against the fascination with skilled craft labour evinced by many of them – enabled precisely by their adoption of a processual syntax.

The play of instrumentalism and anti-instrumentalism in these films suggests something else. On the one hand, processual syntax is the most instrumentalist of organizing structures; after all, the process is a process in virtue of the fact that it produces a result, a use value. On the other hand, when the materials involved in that process are defamiliarized and metaphorized (whether as abstract art or as some other analogous phenomenon in the world like a spider web) – or when the craftsman is presented as absorbed by his labours – an anti-instrumentalism is at play; means (as opposed to ends) take on a new significance and centrality. The manifold pleasures and the potential for innovation evinced in artisanal production are thus placed at the centre of an alternative, parallel, utopian nation. In effect, these early process films were offering up craft labour as a component of a flourishing and meaningful human life, and proposing the craftsman as a true modernizer. And processual syntax was crucial to apprehending the labour of the artisan as consummately skilled, intelligent – and technological.

Works Cited

Aguilera Skvirsky, Salomé. "Realism, Documentary, and the Process Genre in Early New Latin American Cinema." *The Routledge Companion to Latin American Cinema*, edited by Marvin D'Lugo, Ana M. López, and Laura Podalsky. Routledge, 2018, pp. 119–32.

Balmaceda del Río, Fernando. *De zorros, amores y palomas: memorias*. El Mercurio/Aguilar, 2002.

Carneiro, Robert L. *Evolutionism in Cultural Anthropology: A Critical History*. Westview Press, 2003.

Corro Pemjean, Pablo. *Teorías del cine documental chileno, 1957–1973*. Pontificia Universidad Católica de Chile, Facultad de Filosofía, Instituto de Estética, 2007.

Cowan, Michael. *Walter Ruttmann and the Cinema of Multiplicity: Avant-garde – Advertising – Modernity*. Amsterdam University Press, 2016.

Fabian, Johannes, and Matti Bunzl. *Time and the Other: How Anthropology Makes Its Object*. Columbia University Press, 2014.

Freyre, Gilberto. *The Masters and the Slaves: A Study in the Development of Brazilian Civilization*. Translated by Samuel Putnam, 2nd rev. ed. Alfred A. Knopf, 1956.

Gunning, Tom. "Before Documentary: Early Nonfiction Films and the 'View' Aesthetic." *Uncharted Territory: Essays on Early Nonfiction Film*, edited by Daan Hertogs and Nico de Klerk. Stichting Nederlands Filmmuseum, 1997, pp. 9–24.

Harris, Marvin. *The Rise of Anthropological Theory: A History of Theories of Culture*. Crowell, 1968.

Herf, Jeffrey. *Reactionary Modernism: Technology, Culture, and Politics in Weimar and the Third Reich*. Cambridge University Press, 1986.

Martínez-Echazábal, Lourdes. "*Mestizaje* and the Discourse of National/Cultural Identity in Latin America, 1845–1959." *Latin American Perspectives*, vol. 25, no. 3, 1998, pp. 21–42.

Miller, Marilyn Grace. *Rise and Fall of the Cosmic Race: The Cult of Mestizaje in Latin America*. University of Texas Press, 2004.

Peterson, Jennifer Lynn. *Education in the School of Dreams: Travelogues and Early Nonfiction Film*. Duke University Press, 2013.

Pino, Diego. "Fernando Balmaceda del Río." *Revista séptimo arte*, June 7, 2021, https://www.r7a.cl/2021/06/fernando-balmaceda-del-rio/.

Rony, Fatimah Tobing. *The Third Eye: Race, Cinema, and Ethnographic Spectacle.* Duke University Press, 1996.
Salinas Muñoz, Claudio, Hans Stange Marcus, and Sergio Salinas Roco. *Historia del cine experimental en la Universidad de Chile, 1957–1973.* Uqbar Editores, 2008.
Talesnik Y., Daniel. "Tibor Weiner y su role en la reforma: una re-introducción." *Revista de arquitectura*, vol. 12, no. 14, 2006, pp. 64–70.
Vasconcelos, José. *The Cosmic Race/La raza cosmica.* Translated by Didier T. Jaén, 2nd ed. Johns Hopkins University Press, 1997.
Vega, Alicia. *Itinerario del cine documental chileno, 1900–1990.* Universidad Alberto Hurtado, Estudios y Artes de la Comunicación, 2006.

About the Author

Salomé Aguilera Skvirsky is associate professor in the Department of Cinema and Media Studies at the University of Chicago. She works on transnational political cinema, with a special focus on Latin America. Her first book, *The Process Genre: Cinema and the Aesthetic of Labor* (Duke University Press, 2020), won the 2021 Best First Book Award from the Society for Cinema and Media. Her essays have appeared in *Cinema Journal*, *Journal of Latin American Cultural Studies*, *Social Identities*, *Hispanófila*, and in *The Routledge Companion to Latin American Cinema*.

Section 6

Production Cultures and/of the Industrial Film: Amateurs and Professionals

21 Soviet Industrial Film across Categories

Negotiating between Utility, Art and Science

Maria Vinogradova

Abstract

Due to the centrality of the industrial working class to the Marxist project and its Leninist interpretations, industry as a theme had a significant presence in Soviet cultural production. "Films on an industrial topic," the term favoured within the Soviet film industry, presented a thematic category rather than a type. The topic's treatment featured a strong aesthetic component that manifested itself on par with the more utilitarian functions, such as disseminating technical literacy. This chapter maps the industrial film across the whole spectrum of the Soviet non-fiction film production, including the *kulturfilms* of the 1920s, technical films of the 1930s, as well as the popular science, educational and amateur films of the post-World War II period.

Keywords: Soviet industrial film; *kulturfilm*; educational film; popular science film; amateur film; mobile projection

Soviet typologies of non-fiction film rarely mentioned industrial films as a category. The term *proizvodstvennyi fil'm* (from *proizvodstvo* [industrial production]) was more frequently associated with industrial dramas, a genre that developed in the 1930s. Such films featured formulaic plots and usually involved a conflict between a young, enlightened and enthusiastic worker willing to promote industrial development at his factory, and the old "reactionary forces" that opposed innovation or actively disrupted socialist construction. During the later period, in the 1960s to 1980s, these dramas evolved towards a greater psychological and moral complexity. Such *proizvodstvennye fil'my* were aimed for theatrical distribution alongside other fiction feature films and formed a genre (or sub-genre) rather than

Hediger, V., F. Hoof, Y. Zimmermann, with S. Anthony (eds), *Films That Work Harder: The Circulation of Industrial Film*. Taylor & Francis Group 2024
DOI: 10.5117/9789462986534_CH21

type, although one directly responding to the general line of Soviet state propaganda. A Soviet equivalent of the term "industrial film" as accepted in Western academic scholarship would be "films on an industrial topic" (*"fil'my na industrial'nuiu temu"*), and such films were created across the whole spectrum of the Soviet non-fiction film production. Mapping this spectrum is the central goal of this chapter.

Western academic studies of industrial films that have developed in the last two decades mostly agree that an industrial film is usually a sponsored film meant for non-theatrical, non-commercial dissemination.[1] The main argument in favour of their inclusion as an object of media studies is the necessity to shift attention from studying film as a work of art towards a broader set of categories. As Thomas Elsaesser has suggested, "there are many histories of the moving image, only some of which belong to the movies."[2] In industrial films communication is more important than aesthetics, and it is this communicative aspect, rather than industrial history, that interests media scholars in the first place. At the same time, in capitalist contexts industrial corporations were some of the biggest commissioners of utility films, and it is largely for this reason that industrial films appear as a category of their own.

In the Soviet socialist context both the structuring of the industrial sector and the organization of film production, as well as the vision of their public purpose, account for the differences in the production, dissemination and typology of industrial films. First of all, the industry was nationalized, centrally governed and viewed as a public resource. Moreover, it was consistently propagated as the central pillar of the Soviet modernity. Therefore, all communication on topics related to the industry was infused with the vision of its significance for the nation and every individual within it. Secondly, film production was also nationalized and structured to accommodate the broadest possible range of audiovisual products beyond theatrical films. Utility films, as elsewhere, were the most numerous and, produced within

I express my gratitude to American Council of Learned Societies for its generous support through the Mellon/ACLS Dissertation Completion Fellowship (2015–2016) that enabled me to work on this chapter.

1 See, for instance, Zimmermann, Yvonne. "'What Hollywood Is to America, the Corporate Film Is to Switzerland': Remarks on Industrial Film as Utility Film." *Films That Work: Industrial Film and the Productivity of Media*, edited by Vinzenz Hediger and Patrick Vonderau. Amsterdam University Press, 2009, p. 109.

2 Elsaesser, Thomas. "The Place of Non-fiction Film in Contemporary Media." *Films That Work: Industrial Film and the Productivity of Media*, edited by Vinzenz Hediger and Patrick Vonderau. Amsterdam University Press, 2009, p. 19.

the nationalized context, had to negotiate the interests of commissioners with broader visions of the public good. The latter was largely, but not exclusively, defined by state propaganda.

Throughout the history of Soviet utility films we can see a search for an ideal film that would combine the communicative function with an aesthetic entrenched in socialist ideals. Quite representatively, a 1959 essay by Aleksandr Zguridi, a major director and theorist of popular science films, discusses the meaning of scientific and artistic vision of the world that such films would ideally have to profess. Zguridi argues that "[s]cience reflects the world in the form of scientific notions. Art – in the form of artistic images. This way, the way of artistic interpretation [*poznanie*] of the world has its specific features, but, in principle, does not differ from the scientific interpretation."[3] Scientific interpretation of the world presupposed its study in the broadest possible range of aspects, and the Soviet concept of the popular science film was based on such a broad interpretation. Echoing this sentiment, writer Mikhail Arlazorov criticized subject matter-based fragmentation of genres, such as "'the genre of biological film,' 'the genre of medical film,' 'the genre of technical film,' 'art historical' etc. [...] It is doubtful that expressive means used by a director explaining issues in biology are different from those used by a film popularizer [*kinopopuliarizator*] of technology."[4]

These examples help to explain why the term "films on an industrial topic" was preferable to "industrial films." Films on an industrial topic were produced by studios of different specializations, including popular science, documentary, educational and, during the earlier periods, by those specializing on *kulturfilms* and technical films. This chapter aims to trace the development of such films in the Soviet Union across these contexts. I will refer to them as "industrial films," mirroring the already well-established category in international media scholarship. Despite the differences in terminology, as well as those that stem from the conditions of state socialism, these films share many important features with their international counterparts. While Soviet theorists aspired to the status of art for them, in practice their bulk consisted of utility films, and this utilitarian function, in the vast majority of cases, defined them in more

3 Zguridi, Aleksandr. "O nekotorykh voprosakh teorii i praktiki nauchno-populiarnoi kinematografii." *Nauchno-populiarnyi fil'm*, vol. 1, edited by Boris Al'tshuler et al. 1959, p. 29. My translation.

4 Arlazorov, Mikhail. "Fil'my o tekhnike – eto ochen' interesno." *Nauchno-populiarnyi fil'm*, vol. 1, edited by Boris Al'tshuler, Igor' Vasil'kov, Aleksandr Zguridi, Vladimir Razumnyi, Vladimir Nikolai, M.V. Tikhonov, Vladimir Shneiderov, and David Iashin. Iskusstvo, 1959, pp. 84–85.

significant ways than aesthetic. Just as elsewhere, industrial films in the Soviet Union were some of the most ubiquitous types of film production, but this ubiquity has always gone side by side with their marginality. This quality of marginality has been remarked by Michel de Certeau, who stated, in the early 1980s, that "marginality is today no longer limited to minority groups, but is rather massive and pervasive."[5]

Bringing the Soviet industrial film into the spotlight of attention is a step to uncover the history of a significant but largely unknown phenomenon. Including this body of films not only challenges but also complements the auteur-centred film historiography, since their utilitarian purpose was never completely separated from the same aesthetic motivations and sensibilities.

Industrial Topic in the Soviet Discourse

Since the earliest days of the Soviet state, industrialization was seen as a central vehicle to deliver the promises of the revolution, and it was broadly understood that coverage of the topic, while serving the interests of the industry, simultaneously operated within the field of public interest on many levels. In the 1920s, bringing industrial topics to the forefront of the public attention was more than a matter of spreading awareness of the issue, training the workforce, or disseminating the desired attitudes. Discourse around its cultural significance culminated in productivist theories of the 1920s, in which the idea of industrial production as a guiding principle of organization of social relations was a crucial topic.

Productivism was a branch of constructivism that developed one of the most radical theories of proletarian culture by the mid-1920s, according to which art had to be merged with industrial production.[6] In the most significant way, these ideas were articulated by the art historian and critic Boris Arvatov. In his theory of proletarian culture, Arvatov identified and elaborated four central aspects of artistic practice that had to be transformed on the way from bourgeois to proletarian art: "1) artistic technique, 2) cooperation in art, 3) artists' ideology, 4) art and the everyday [*byt*]."[7] Throughout his discussion, Arvatov stressed the importance of the social dimension of

5 De Certeau, Michel. *The Practice of Everyday Life*. Translated by Steven Rendall. University of California Press, 2011, p. xvii.

6 On the productivist ideas in the 1920s avant-garde, see Zalambani, Maria. *Iskusstvo v proizvodstve: Avangard i revoliutsiia v sovetskoi Rossii 20-kh godov*. Translated by Natalia Kardanova. IMLI RAN/Nasledie, 2003.

7 Arvatov, Boris. *Iskusstvo i proizvodstvo*. Proletkul't, 1926, pp. 94–120.

art, in which collective production had to replace the individualism of a lone artist. He proclaimed the artisanal mode of production (*remeslennichestvo*) obsolete, arguing that industrial approach had to be extended to artistic practice, eliminating the separation between labour and creativity.

Echoing the points of Marxist theory, theorists of the Soviet left stressed that in the bourgeois practice, individualism of the sole creator led to fetishization, social irrelevance and alienation from his work process, products of his creativity, and their consumption by the public. Arvatov attacked the traditional separation between the high and low, or high and applied art – in essence, "art" and "non-art," arguing that working class could and had to "put artistic labour into any deed." The separation between art and non-art had to be abolished in favour of a classification with regards to specificity of each medium, viewing the material that it organized as a basis for classification. Thus, poetry and journalism would both be arts of speech, theatre play and street celebration would be arts of performance, or "stage forms of organization of human action," and the "art of paint as a specific branch of painting industry" would make no difference between easel painting and house painting. Abolishing the distinction between the aesthetic and purely utilitarian would overthrow the hierarchy of modes of production, individual perspectives of creators rooted in social origins, and eliminate the fetishism of materials and tools. If the use of bronze and marble in traditional sculpture had to accentuate its status as a luxury object, a proletarian artist would find inspiration in aluminium and steel, create music with noises and poetry with street slang.[8] The same would apply to tools, or means of production. A bourgeois artist rejects industrial tools as an antithesis of creative freedom, however, "the question of tool is a social one: it is only in an individualistic society that (paint) brush, violin etc. are monopolized and fetishized tools of creativity. For the proletariat, a class of conscious-collective producers, this limitation ceases."[9] In Arvatov's theory, art organizes the raw materials of daily life, which is its main difference from decoration that is dispensable and redundant.[10]

8 Ibid., pp. 96–98.

9 Ibid., p. 103.

10 Ibid., p. 112. To be sure, such functionalist approach was shared not only by Soviet theorists of the 1920s but also was one of the preoccupations of the international avant-garde at the time. Le Corbusier's architectural theory presents one of the best examples: having famously defined the house as "a machine for living in," he defined architecture as a medium of mass, surface, and plan, whereas "the 'styles' are a lie." Le Corbusier. *Towards a New Architecture.* Translated by Frederick Etchells. Dover Publications, 1986, p. 3. [Originally published in 1923 as *Vers une architecture.*]

Industrial work, as an aspect of daily life, would thus be a valid and desired subject to be reflected in art, including cinema, while production of this art would have to be closely connected with industrial production.[11] Many of the synthetic new art forms that emerged in the 1920s were preoccupied with engaging the material of daily life. This is most evident in Proletkult theatre, in particular, the genre of the "living newspaper," in which participants acted out sketches related to recent news.[12] At the same time, news media proper developed networks of worker-correspondents, or *rabkory* (from *rabochii korrespondent*), who would report on events of their daily life, focusing, in particular, on the life of their factory. Most newspapers had an editor or section that managed work with such worker-correspondents who contributed not only texts but also photographs, and were called *fotokory* and *khudozhkory*, respectively.

Dziga Vertov's Kino-Eye movement that developed by the mid-1920s from the small group working on the *Kino-Pravda* newsreels (1922 to 1925),[13] largely shared the principles of the worker-correspondent movement, where *kinoks* were frequently called *kinokorrespondenty*, or film-correspondents. While Vertov's idea to create a "factory of facts"[14] never materialized, a later organization, Society of Friends of Soviet Cinema (Obshchestvo druzei sovetskogo kino, ODSK), took active steps towards developing a network of nonprofessional film-correspondents nationwide, peaking in 1926 to 1930.

Soviet *Kulturfilm* in the 1920s

While understanding productivist theories is important for understanding of industrial topics, films that dealt with it were, for the most part, more practically oriented. Educating the masses about the importance of technological

11 Within the Constructivist circles, organization of industrial work was at times approached as a work of art in and of itself. Alexei Gastev (1882–1939), a poet, writer and revolutionary who eventually dedicated himself to the work as a director of Central Institute of Labour, expressed these ideas in his writings, some of which were compiled in the volume *Poeziia rabochego udara*. The book underwent six editions during his lifetime between 1918 and 1926, the final one reprinted in 1971 (Khudozhestvennaia literatura).

12 On the form of "living newspaper," see Mally, Lynn. "The Americanization of the Soviet Living Newspaper." Carl Beck Papers in Russian & East European Studies, no. 1903. University of Pittsburgh, Center for Russian and East European Studies, February 2008.

13 The group included Vertov himself, cinematographer Mikhail Kaufman, his brother, and film editor Elizaveta Svilova, Vertov's wife.

14 Vertov, Dziga. "The Factory of Facts" [1926]. *Kino-Eye: The Writings of Dziga Vertov*, edited by Annette Michelson, translated by Kevin O'Brien. University of California Press, 1984, pp. 58–60.

and industrial development was one of the cornerstones of the early Soviet cultural policy, and film, unsurprisingly, was assigned a central role in this process. In 1922 Goskino, the state body that regulated film production, signed a few contracts with the biggest industrial trusts in Moscow that bound these trusts to fund short films about specific factories.[15] For film industry that was in a deep crisis such commissions were a significant source of income, so that various film companies, such as Goskino, Sevzapkino, Kino-Moskva, Proletkino, and other, began to get actively involved in what filmmaker and the chief historiographer of Soviet *kulturfilm* Nikolai Lebedev called "a lucrative and irresponsible affair":

> Irresponsible because the films could be made as badly as possible. Managers of the trusts that funded their production, who understood nothing about cinematography, accepted these films without any evaluation. In any case, these pictures were never subjected to the judgement of the public opinion, they were never shown at theatre screens: printed in one or two copies, they were given to the patron and disappeared without a trace inside fire-resistant safes.[16]

As a result of this "film production fever," as Lebedev called it, it became common for industrial trusts, big or small, in the centre or at the periphery, to have at least one film made about each of its factories.

Lenin's course towards the New Economic Policy (1921 to 1928) eased the restrictions of nationalization and collectivization, allowing for creation of private enterprises. In filmmaking this resulted in creation of a large number of small film companies. As noted by film historian Peter Bagrov, the second half of the 1920s saw a proliferation of "peripheral" film collectives staffed by hastily educated graduates of filmmaking courses at vocational schools, and many of these collectives dispersed after producing only one film. Analysing the situation in Leningrad, Bagrov argues that one of the reasons that such small companies flourished there was that the local film training college produced dozens of certified filmmakers, frequently at a semi-amateur level of preparation, who were not able to find employment at Leningrad state film factory.[17] According to Lebedev, by the second half

15 Lebedev, Nikolai. "Kul'turfil'ma na Zapade i u nas." *Kinovedcheskie zapiski*, vol. 58, 2002. Excerpts from unpublished book manuscript, www.kinozapiski.ru/ru/article/sendvalues/315. My translation.
16 Ibid.
17 Bagrov, Peter. "On proletel po zhizni … Alfred Dobbel't: kinematografist iz Riazani." *Riazanskaia starina*, vol. 2–3, edited by Aleksandr Nikitin and Pavel Tribunskii. Krai, 2006, pp. 218–19.

of the 1920s, at least twenty-four factories in Leningrad stored industrial films.[18]

The "film production fever" began to wane around 1926 when funds for production of commissioned films were reduced, while film factories began to turn towards producing more feature films. Production of non-fiction and commissioned films was relegated to special departments within these factories, such as the scientific section created at Mezhrabpom-Rus'.[19] In the catalogue of the section's *kulturfilms,* published in 1928, at least twenty appear to be industrial films, with straightforward titles, such as *From Cotton to Fabric, How to Build a Locomotive, Production of Axes, Production of Nails, Production of Shoes* or *How a Newspaper Is Made.*[20] Lebedev viewed this development as evolution of *kulturfilm* from its earlier, "primitive" stage, but still regarded Soviet domestic production of *kulturfilms* as a missed opportunity, plagued by poor management, shortages of positive film stock to print circulation copies and, most importantly, the inability to overcome the broad perception of *kulturfilm* as a third-rate type of film production.[21]

Another prominent advocate of *kulturfilm* was Vladimir Erofeev, a documentary filmmaker who also spent a few years in Germany working for the film branch of the Soviet chamber of commerce. Erofeev argued that *kulturfilm* had a strong potential to be a spectacle for a mass audience, and pointed to the box-office success of *kulturfilm* in Germany.[22] Among these films, Erofeev particularly stressed the achievements of German industrial, science and expedition films.[23] Erofeev's dreams of a theatrical *kulturfilm* culture were paralleled by discussions in the press and targeted efforts to open commercial theatres specializing in exhibition of *kulturfilms.*[24] Lebedev himself was among the cofounders of one such theatre, *Artes,* that opened in 1927 in Moscow. However, *Artes* continued to screen popular feature films three days a week, which Lebedev criticized, calling it a "half-measure."[25]

18 Lebedev. "Kul'turfil'ma na Zapade i u nas."

19 Ibid.

20 Mezhrabpom-Fil'm. *Kul'turfil'my proizvodstva nauchnogo otdela aktsionernogo obshchestva "Mezhrabpom-fil'm".* Leninskaia tipografiia Upromtorga, 1928.

21 Lebedev. "Kul'turfil'ma na Zapade i u nas."

22 See Deriabin, Aleksandr. "'Nasha psikhologiia i ikh psikhologiaa – sovershenno raznye veshchi'. 'Afganistan' Vladimira Erofeeva i sovetskii kul'turfil'm dvadtsatykh godov." *Kinovedcheskie zapiski,* vol. 54, 2001, www.kinozapiski.ru/ru/article/sendvalues/617; and Schlegel, Hans-Joachim. "Nemetskie impul'sy dlia sovetskikh kul'turfil'mov 20-kh godov." *Kinovedcheskie zapiski,* vol. 58, 2002, www.kinozapiski.ru/ru/print/sendvalues/313.

23 Schlegel. "Nemetskie impul'sy dlia sovetskikh kul'turfil'mov 20-kh godov."

24 See, for instance, Boltianskii, Grigorii. "Teatr kul'turfil'my." *Sovetskoe kino,* no. 1, 1927, pp. 2–3.

25 Lebedev. "Zadachi ekranov kul'turfil'my." *Kino-Front,* vol. 4, 1927, pp. 6–9.

The majority of screenings of non-fiction films were organized at schools, worker clubs, and delivered through mobile film screening units (*kinoperedvizhka*) in rural areas. By the late 1920s, there was a wealth of publications in film pedagogy reviewing educational, technical and sanitary films, with suggestions on their best uses.[26] A number of available films were imported from abroad, especially sanitary films. Many of these films were subjected to re-editing, a common practice in the Soviet film industry of the 1920s, when both documentary and fiction films were changed not only to make them ideologically suitable but also to make them more appealing to the Soviet audience, the way the editor imagined it. At the end of the 1920s, however, non-fiction films were rarely discussed as an attraction per se, while a growing body of publications offered methodological recommendations and practical advice on educational uses of non-fiction film.[27]

Technical Film in the 1930s

From around 1929, appeals to produce better educational and technical films are seen in many publications. One of the most frequent points of criticism for the existing films was that they did not make clear for whom they were made and what exactly they aimed to teach. Thus, Nikolai Kudriavtsev cites audience feedback for a few films, such as *At the Steering Wheel* (*Za rulem*) and *Diesel Construction* (*Dizelestroienie*), where the audience members complained that the main message of these films was inarticulate.[28] Kudriavtsev also criticizes the overall design of many films that give too many or too few details, or do not sufficiently organize and systematize the material.[29] He especially attacks what he regards as excessive and pointless aesthetization of images in many industrial films:

26 In addition to the aforementioned Mezhrabpomfilm catalogue, other notable titles include Nesturkh, Mikhail. *Kino i obshchedostupnye lektsii*. Novaia Moskva, 1925, and Sukharebskii, Lazar'. *Obzor sanprosvetitel'nykh kinofil'm za 10 let proletarskoi revoliutsii (1917–1927)*. Izdatel'stvo Moszdravotdela, 1928.

27 For more on the rise and fall of Soviet *kulturfilm* in the 1920s, see Sarkisova, Oksana. "The Adventures of the *Kulturfilm* in Soviet Russia." *A Companion to Russian Cinema*, edited by Birgit Beumers. John Wiley & Sons, 2016, pp. 92–115.

28 Kudriavtsev, Nikolai. "Bor'ba za kachestvo uchebnykh fil'mov." *Kino v pomoshch osvoeniiu tekhniki*, edited by Alexander Anoshchenko. Sektor vedomstvennoi i zaochnoi literatury, 1935, p. 10.

29 Ibid., pp. 14–15.

> In some films the viewer's attention is led away from the most important. In this respect, the film about milled peat (made in 1933) is especially representative, where the machines are shown against a background of such big and beautiful clouds that in comparison to them the machines appear as shapeless dark spots. This is the result of chasing after pretty clouds! Had the author accentuated the machines rather than clouds, then the lighting and clouds would have helped to show more impressively those very machines that the author rightfully attempts to admire.[30]

The example with the clouds is so ubiquitous in the literature of the time that it seems that they became a frequent joke in film pedagogical circles. The criticism of infatuation with "photogenie" of the machines and industrial processes[31] is consistent with the persecution of the "formalist trends" in art in the 1930s, but it also reflects the decade's course towards intensive industrial development and efficiency. Assessing Soviet educational and *kulturfilms* produced in the 1920s, filmmaker and theorist Iurii Zheliabuzhskii observed that there were almost no proper technical films: the existing ones were mainly commissioned advertising films. Thus, the film *Severoles,* commissioned by the forestry trust of the same name, was made with the goal to demonstrate its products to the potential consumer of timber abroad, and persuade to buy timber from this specific manufacturer, states Zheliabuzhskii. Many such films, he observes, "usually included demonstration of broadly smiling or thoughtfully frowning directors and managers of a given enterprise."[32]

A large number of books on non-fiction film published in the 1930s contain practical recommendations on how to make better technical films, along with extensive advice on their pedagogical uses. One of the most common modes of working with educational films of various kinds was the so-called "film-lecture" that included an introduction, commentary and discussion led by a trained educational worker. This mode of work had already been widespread in the 1920s, which was largely necessitated by the absence of sound in the films. Even when sound films became available in the 1930s, an

30 Ibid., p. 16. Here and below, unless otherwise noted, translation from Russian is mine.

31 In the same anthology, Iurii Zheliabuzhskii, criticizing Soviet 1920s films, argued that "[t]he proverbial 'photogenie' of machines, at times, carried the authors away so much that behind it, they completely failed to see the essence of processes being filmed, and apparently, in order to make this less evident, they ultimately confused the spectator, spicing up the picture with a vast abundance of 'beautiful' shots – running clouds, running water or oil etc." Zheliabuzhskii, Iurii. "Sovetskii tekhnicheskii fil'm." *Kino v pomoshch osvoeniiu tekhniki*, p. 25.

32 Ibid.

active intervention and interactive presentation by a live commentator was still regarded as necessary, in part, "to correct possible errors in the film," which could be an easier alternative to sending the film back to the factory for re-editing.[33] The commentator, or lecturer, was also expected to read all the captions aloud, assuming that a part of the audience may be illiterate.[34] A film-lecture was an event in several parts. First of all, a thematic exhibition had to be organized in the foyer to prepare the viewers for the material in the film. The screening, together with the introductions and comments by the lecturer, concluded with a discussion and audience feedback.

A number of annotated catalogues of industrial films were published in the 1930s. One such book from 1931, entitled *Industrial Films*,[35] is arranged into multiple sections according to the focus of films, such as "Rationalization, Standardization, Mechanization," "Vaporescence, Production of Steam Engines, Work of Steam Machines," "Radio and Radio Industry," "Construction of Factories and Plants," or "On Worker Innovations." Each section contains several film titles with descriptions, and is concluded by suggestions on how to lead the conversation, along with a one- or two-page bibliography. A few of the films in the book are fiction films on industrial topics, such as *Dva druga, model' i podruga* (*Two Friends, a Model and a Girlfriend*, 1928). The model here has nothing to do with fashion – it is a prototype of a machine for packing soap bars made by two workers of a provincial soap factory. The film is an adventure comedy, in which the friends have to overcome a lot of obstacles on their way to implementing their invention. The antagonists that they fight against are the so-called *nepmen*, or private entrepreneurs who were benefitting from the New Economic Policy (NEP) that allowed private enterprise – and allegedly were the reactionary proto-capitalist elements in the Soviet society.

There were broader treatises on the methodology of film education that provided detailed guidance for lecturers on selection of films, organizational questions, and advertising of screenings and film-lectures. The latter was understood as a part of the projectionist's work, and a projectionist's kit often included supplies and tools for advertising. "Film advertising should be economical, eye-catching, and artistic," states one such book from 1934, recommending that "a film projectionist should have two posters, one for

33 Sukharebskii, Lazar'. "Metodika pokaza nauchno-prosvetitel'noi fil'my." *Kulturfil'ma*, edited by Kirill Shutko. Tea-Kino-Pechat', 1929, pp. 178–79.

34 Ibid., p. 187.

35 Shagin, Vasilii. *Proizvodstvennye fil'my*. Gosudarstvennoe izdatel'stvo khudozhestvennoi literatury, 1931.

the current screening and another for the following one." In rural areas it was common for projectionists to create handmade posters, and the book recommends that "a village film projectionist should always have two brushes (a thick and a thin one) for writing, a few bottles of paint, and ink." However, the most efficient form of in-advance advertising is "on-film advertising," states the author – essentially a film trailer – that would be projected in a foyer or, if there was none, in the screening room before the main film.[36]

Apart from the "film-lecture," other suggested types of working with film were the "kino-dispute," "kino-court" and "kino-meeting."[37] The audience was shown a film (frequently a fiction film featuring two types of characters, a modern progressive worker willing to embrace the new social and industrial relations, and an old-style "reactionary element"), and asked to make their judgement of the characters. The recommendations on how to moderate this discussion contained clear suggestions on how to lead the audience to making a "correct" judgement without stating it openly.

Later, during the 1930s and early 1940s, we see more targeted recommendations on how to work with film in specific industries, such as the railroad transport, for instance.[38] After a break during the war, publication of annotated filmographies of industrial, technical, agricultural and sanitary films resumed during the immediate post-war years. Besides the existing films, these publications also included plans for production of films for the following year, arranged by theme, and stating the number of films in the plan for each theme. Such books were meant as a resource for educators and published by a variety of departments that oversaw film distribution, arts and education.

Developing the Art of Popular Science Film

The next period in question begins around the mid-1950s. Khrushchev's course taken towards de-Stalinization began a new wave of modernization. In contrast to breakneck speed industrialization pressed by Stalin in the 1930s, Khrushchev's new vision of modernity emphasized the development

36 Vronskii, Boris, and Robert Vartazarian. *Kino i tekhpropaganda*. Izdatel'stvo upravleniia kinofikatsii pri SNK RSFSR, 1934, pp. 29–31.

37 See Sukharebskii. *Kino-disput*. Roskino, 1932; and idem. *Kino-sud*. Roskino, 1933; Zel'manov, Mark. *Kino-miting*. Roskino, 1932.

38 *Opyt ispol'zovaniia tekhnicheskikh kinofil'mov na zheleznodorozhnom transporte*. Upravleniie uchebnymi zavedeniiami NKPS, 1941.

of advanced technologies within the framework of science-technological revolution (STR), which included a general improvement of standards of living. This period was marked by liberalization of all spheres of life, and in the cultural sphere these changes resonated by making authenticity into one of its central preoccupations. Authenticity meant paying greater attention to the individual, a romantic fascination with lived experiences, and a yearning for proven facts as opposed to ideological dogmas. By and large, it meant a return to an earlier, purer vision of socialism untainted by the cult of personality – a socialism based on a scientific understanding of the world that would harmonize the relationships between the individual, society and the natural environment.

In the world of cinema, profound social changes at the time were reflected in a true "explosion of documentarism." In addition to the increased interest in documentary cinema as such, a significant part of debates within Soviet film circles since the second half of the 1950s was concerned with developing a greater scope of non-theatrical filmmaking. It is evident that this interest, in part, reflected an increased familiarity with the cinematographic process abroad: Soviet delegations were frequently invited and travelled to international festivals of science, industrial, educational, ethnographic, tourist and other non-theatrical films, and sought membership in international film associations of various kinds, such as the International Scientific Film Association (ISFA), or AICS in the French version.[39] Soviet filmmakers and theorists at the time were particularly interested in developing the art of popular science film, distinct from purely educational films.

In the Soviet context, the definition of popular science film was very broad and included not only films on the topics of science, technology and industry, but also films on broader topics in social sciences and humanities. Unclear boundaries of this category caused a lot of debates on what exactly popular science film was. Alexander Zguridi, a major science filmmaker

39 The first ISFA Congress was held in Paris in 1947, and after 1992, the association was succeeded by the International Association for Media in Science (IAMS). In his speech at the opening of the 20th IAMS Congress in 2012, Jan T. Goldschmeding, its first president, called IAMS "a descendant of a famous family of scientific film associations," where the first one was the Association pour la Documentation Photographique et Cinématographique dans les Sciences (ADPCS, 1933–1938) founded by Charles Claoué and Jean Painlevé. Goldschmeding, Jan T. "IAMS, Past, Present and Future." International Association for Media and Science, www.media-in-science.org/wp-content/uploads/2014/10/IAMS-past-present-future-2-0.pdf. Accessed September 20, 2017. Soviet participation in the international science film community is reflected in the archival collection of the Union of Filmmakers at Russian State Archive of Literature and Art (Rossiiskii gosudarstvennyi arkhiv literatury iskusstva – RGALI, Fonds 2936, Union of Filmmakers/Soiuz kinematografistov SSSR, 1957–1973).

and subsequently teacher and theorist, defended this broad definition in his 1958 article written together with another veteran filmmaker, Boris Al'tshuler. The authors stated that popular science as a category was suitable for films on topics from all fields of knowledge aimed at a broad audience. In contrast, films categorized as educational had to be tied to school curriculum at secondary or vocational schools, or at universities, and thus target an audience of a specific age and level of education.[40]

Such definitions formed the basis of organization of Soviet films production, which comprised several types of studios according to the type of target audience and channels of distribution. This structuring was hierarchical, and the top of this hierarchy belonged to theatrical feature films. Studios in major cities across the country, such as Mosfilm in Moscow or Lenfilm in Leningrad, that produced theatrical features, enjoyed the most lavish funding. They were also highly selective in their staffing, and for key creative positions, such as directors and cinematographers, a degree from the All-Union State Institute of Cinematography (VGIK), the national film school founded in 1919 in Moscow, was a prerequisite. Documentary, television and children's film studios stood somewhat lower in the hierarchy. Popular science film studios, such as Tsentrnauchfilm in Moscow, Lennauchfilm in Leningrad, or Kievnauchfilm in Kiev, stood another step down, followed by educational film studios, such as Shkolfilm (School Film) and Soiuzvuzfilm (Union Higher Education Film), that were closer to the bottom of the hierarchy. For aspiring filmmakers the choice to specialize in popular science and educational film was frequently motivated by their greater accessibility: admissions at film school were less competitive, while work within these contexts allowed a greater degree of experimentation.[41]

This structuring could be quite fluid in reality. Thus, the Gorky Film Studio in Moscow, which specialized in theatrical films for children and youth, frequently made films for a general adult audience. Kievnauchfilm, in addition to its main specialization in popular science, developed a strong section of animation, including animation films for children. Outside of

40 Al'tshuler, Boris, and Aleksandr Zguridi. "O klassifikatsii nauchnykh fil'mov." *Iskusstvo kino*, vol. 8, August 1958, pp. 142–48.

41 Such was the case of Vladimir Kobrin (1942–1999), a well-known Russian experimental filmmaker whose career began at Tsentrnauchfilm in the late 1970s. In Kobrin's own words, "I deliberately decided to choose a [...] niche for myself in cinema, the exact one that had the lowest prestige at the time, and for that reason [...] [was] the most free from the state's paternalizing." Hanegen, Sabine, and Vladimir Kobrin. "Voprosy Vladimiru Kobrinu, zadannye Sabinoi Hanegen." *Kobrin: Svetiashschiisia sled*, edited by Andrei Gerasimov et al., Nizhnii Novgorod, ROF "Fond Iuriia Norshteina," 2005, p. 84. My translation.

major centres, such as Moscow, Leningrad and Kiev, local film studios frequently created films of all types, as it was the case with Sverdlovsk Film Studio in Sverdlovsk or Uzbekfilm in Tashkent.

Industrial films, or films on industrial topics, were not assigned a separate category, but could be produced by studios of all types, depending on their target audience and channel of distribution. Studios' work was subject to thematic planning, but for those that specialized in non-fiction film, commissions from factories, industrial departments and other organizations made a significant part. It can be implied that their standing in the hierarchy of film production was proportionate to the share of commissioned work in their output. Commissioned films presuppose a more specific target audience, and the narrower the audience, the lower was the studios' place in the hierarchy. It can be argued that educational films made for schools stood at the bottom for this reason: created for a clearly defined purpose, they were defined by utility that narrowed down the space for ambiguity and abstraction. Popular science film studios presented a borderline category. Much of the theoretical debates in the 1960s centred around developing popular science as a form of art, stressing broader enlightenment, rather than specialized knowledge, as their main objective. In a 1964 essay, writer and theorist of popular science Igor' Vasil'kov criticized the increasing number of commissioned films in the output of popular science film studios, lamenting that their "endless stream drowns the clear understanding of characteristic features of a popular science film."[42] Concurrently, an increasing number of large organizations created their own media departments. Such was, for instance, the Dnepropetrovsk Experimental Laboratory for Production of Study Aids of the Ministry of Iron Industry (Dnepropetrovskaia eksperimental'naia laboratoriia po proizvodstvu sredstv obucheniia Minchermeta SSSR). A number of manuals published by educators provided clear guidelines for creation of in-house film units at organizations of various sizes, arguing for broader application and greater accessibility of audiovisual means in instruction and technical documentation.[43]

These patterns parallel those observed by European and American media scholars in relation to industrial films produced, disseminated and retrospectively evaluated in capitalist contexts, specifically, their placement

42 Vasil'kov, Igor'. "Sovremennye problemy kinopopuliarizatsii." *Nauchno-populiarnyi fil'm*, vol. 2, edited by Boris Al'tshuler et al. Iskusstvo, 1964, p. 10.

43 Such as Miasnikov, Vitalii. *Opyt sozdaniia tekhnicheskikh kinofil'mov na predpriiatiiakh i organizatsiiakh*. Leningradskii dom nauchno-tekhnicheskoi propagandy, 1962; and Brysin, Pavel. *Sozdanie ekrannykh nagliadnykh uchebnukh posobii v proftekhuchilishche*. Vysshaia shkola, 1982.

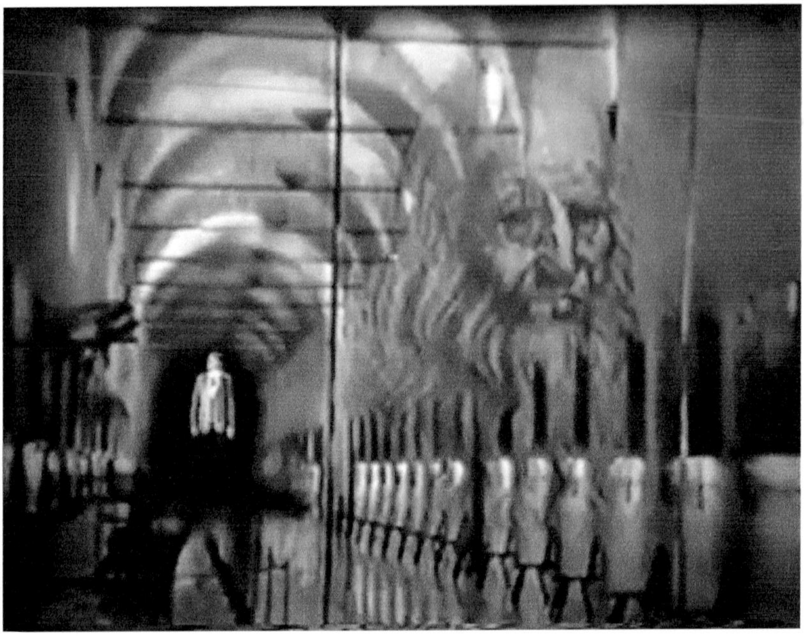

Fig. 21.1. Soviet delegate walking through the Leonardo Da Vinci Gallery at the Museum of Science and Technology in Milan. Leonardo was one of the early researchers of the process of formation of cracks in metals.

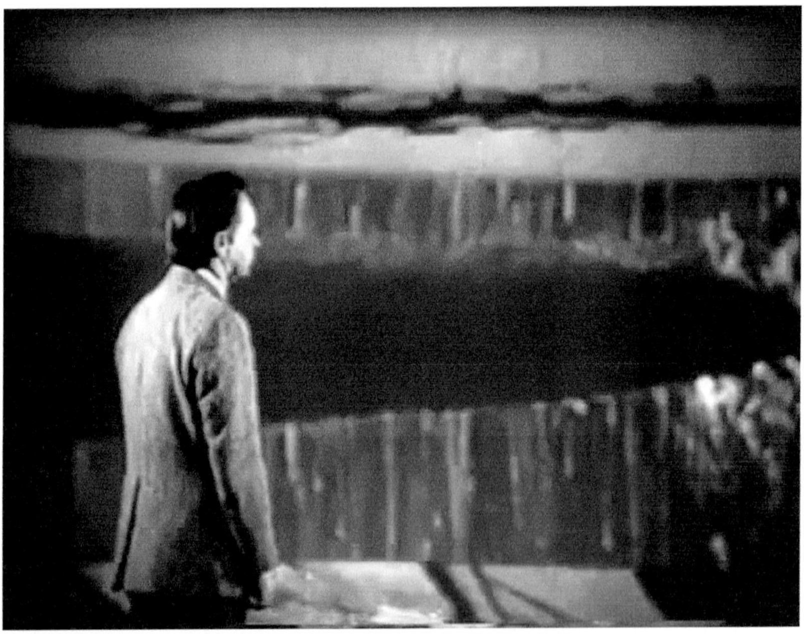

Fig. 21.2. University professor using film to explain the process of formation of cracks in metals.

within the hierarchy of culturally valorized objects. As Yvonne Zimmermann has noted in her discussion of industrial films in Switzerland, "sponsored and auteur films seemed to be two incompatible concepts."[44] As an attempt to bridge this divide, Soviet visions of the popular science film after the late 1950s were guided by an avant-garde impulse to infuse function with an aesthetic sense. Industrial films did not fit easily into this project, although there are a number of examples of popular science films on industrial topics that neatly blend enlightenment and art. *Treshchiny v metalle* (*Cracks in Metals*, 1984) directed by Iurii Ivanov and produced by Kievnauchfilm is one such example. The film explains the physical foundations of the process of formation of cracks in metals, putting these processes in perspective by outlining the history of human understanding of the process of decay. The vast majority of Soviet industrial films, however, stayed clear of philosophical abstractions, and were made with more pragmatic goals in mind.

Industrial Films by Amateur Film Studios

While creation of in-house film studios staffed with professional filmmakers was a growing trend at Soviet organizations after the 1960s, this was not the only type of in-house film production units at the time. Simultaneously, amateur film collectives, or studios, were becoming more common at industrial plants and other organizations, and in some cases studios of both types could coexist in the same organization. Driven by enthusiasm of their participants, amateur film studios relied on material support from the state, trade unions and their host organizations.

Organized amateur filmmaking was not a new phenomenon in the 1960s, but a series of initiatives undertaken with advocacy of film professionals helped to turn it into a truly widespread phenomenon at the time. These initiatives targeted the main impediment, the lack of available small-gauge film equipment: domestic production of compact and small-gauge cameras had not begun in the Soviet Union until 1956. In 1957, upon the founding of the Union of Filmmakers, a section for work with film amateurs was established within it. The chief advocate of amateur filmmaking was Grigorii Roshal', a professional film director who had begun his career as a Proletkult theatre director in the 1920s. Other enthusiasts within the professional community were the likes of Iakov Tolchan, a prominent documentary

44 Zimmermann. "'What Hollywood Is to America, the Corporate Film Is to Switzerland'," p. 101.

cinematographer whose career had started in 1924 in Vertov's *Kino-Eye*. In order to establish domestic production of small-gauge cameras, the section regularly organized conferences that gathered both professional and amateur filmmakers, as well as representatives of various industrial factories, such as Moscow Gramophone Factory, for instance, as well as the Ministry of Trade and the military. The latter was the most important carrier of technological innovation, and its presence was therefore crucial.

Amateur film section at the Union of Filmmakers took a complex of measures to stimulate the development of amateur filmmaking, such as organizing national and local film festivals and exhibitions of handmade filmmaking devices, regularly covering amateur cinema in its monthly journal *Iskusstvo Kino* (Film art), representing Soviet amateurs at international amateur film festivals, and initiating cultural exchange with amateur film organizations in various countries on both sides of the iron curtain. Transcripts of debates within the Union of Filmmakers[45] indicate that there was a great deal of enthusiasm within film professionals about the most diverse uses of amateur film, such as recording community and family life and generally developing a more personal style of filmmaking; experimenting with film form, and, of course, making practical films. This enthusiasm reflects the increased interest in non-theatrical and non-fiction film that characterized the period.

Many factories and other large and medium-size organizations, such as universities, vocational schools and houses of culture (Soviet community arts and leisure organizations) housed and partially sponsored amateur film collectives that received additional support from trade unions and other bodies. The bigger studios also had paid staff units. Such arrangements puzzled observers from capitalist countries, such as Geoffrey Levy, a member of London's prominent amateur film group Grasshopper, who visited Moscow in 1963. Levy subsequently reported on the trip in an article for the British magazine *Amateur Cine World,* and the article concluded with a description of his conversation with Grigorii Roshal': "Finally, Mr. Rochal had one more question to ask. 'Tell me,' he said, 'What salaries are paid to Secretaries, and other official clubs?' I realised then that we still had quite a lot to learn about one another."[46] While this situation seemed luxurious to a foreign observer, Soviet amateurs still had to work on a shoestring budget and frequently rely on handmade equipment. The same, however,

45 RGALI, F. 2936, Union of Filmmakers.

46 Levy, Geoffrey. "An Amateur in Moscow." *Amateur Cine World,* January 30, 1964, p. 141.

was often true about professionals working at popular science, educational and documentary studios.

In most cases such amateur film collectives were self-started, and there were infinite varieties of the ways in which they interacted with their host organizations and state bodies that coordinated amateur filmmaking. It was usually assumed that an amateur film collective would regularly make films for its host organization, and on some occasions factory directors, in pursuit of prestige associated with festival prizes, interfered into the filmmaking process and attempted to dictate how to make films and on what subjects. However, this practice was frequently criticized in the filmmaking community and occasionally in the press. Some studio directors were quite adamant about separating those commissioned films from the ones that they made "for creativity."[47] For such commissions, many directors kept a few skilled amateurs that made a core of the collective, and they were granted a leave off their main work to make films.

Interviews with former participants of Soviet amateur film collectives at factories indicate that many of them balanced the utilitarian function in their filmmaking with building community around a recreational activity. An undated film, most likely from the early 1980s, by People's Film Studio Izhor-Film at Izhora Factories in the town of Kolpino outside of Leningrad, presents a vivid example of this balance. The piece focuses on the studio itself and includes fragments from its other films, out-takes and working moments. The latter are mostly performed for the camera rather than caught spontaneously. As the filmmakers told me in an interview, they "simply put bits and pieces together." The film was not aimed to be screened at film festivals, does not have a soundtrack or narrative, and is not perceived as significant by its creators, as it became clear from my interview with them.[48] Located in the public sphere of a workplace environment, this fragment is motivated by private intentions, celebrating friendship between co-workers, while at the same time performing and stressing the casual attitude of creativity within a formal structure, a palace of culture supervised by the factory and state organizations.

Work-related films created by such studios often reflect a greater flexibility in the choice of subject matter compared to that allowed by thematic planning that guided professional film production. Thus, Vitalii Pavlov,

47 Interview with Oleg Bazilevich, director of the former People's Film Studio LOMO in Leningrad. St. Petersburg, September 28, 2012.

48 Interview with Maxim Deordiiashchenko and Sergei Kuz'muk, participants of the former People's Film Studio Izhor-Film in Koplino, Leningrad oblast'. Kolpino, December 12, 2012.

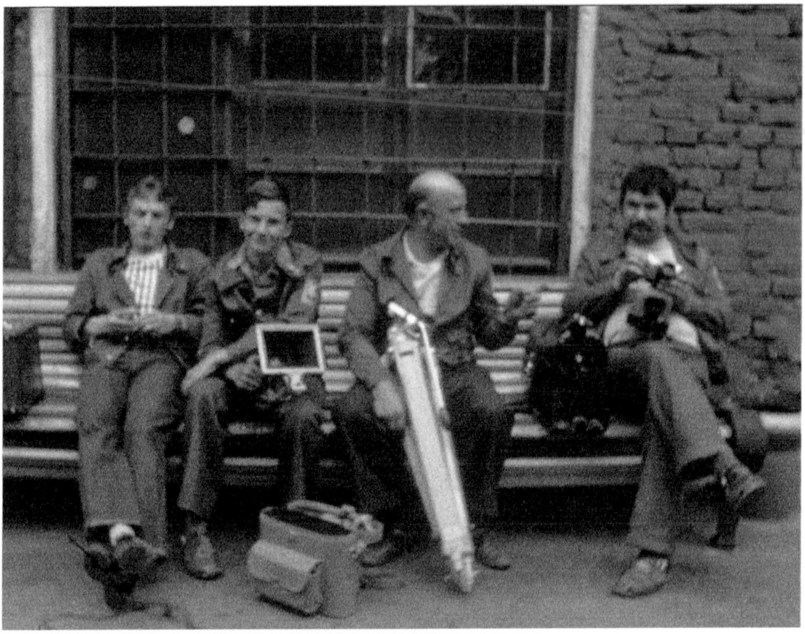

Fig. 21.3. Izhor-Film amateur filmmakers preparing for a shoot.

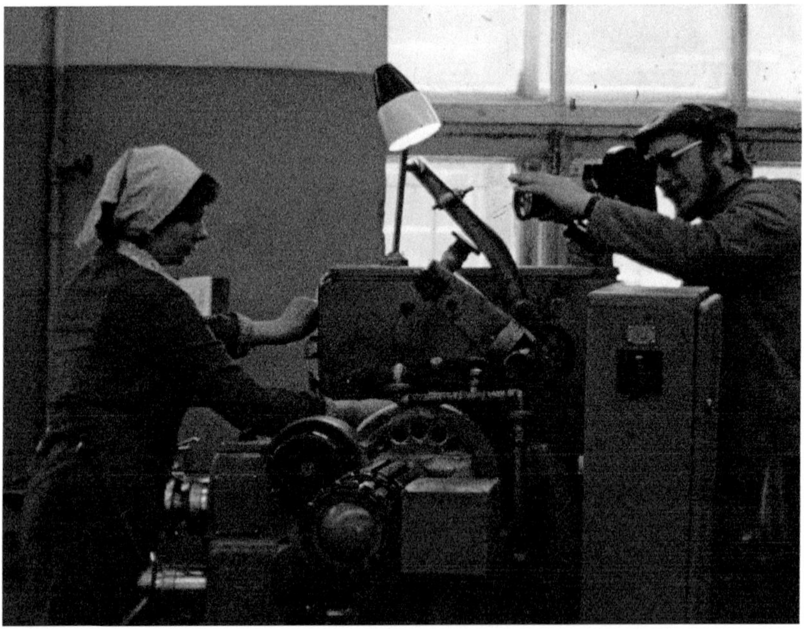

Fig. 21.4. Filming coworkers on the factory floor.

the director of the former People's Film Studio Start-Film collective at Leningrad's Ravenstvo factory that specialized in advanced military and civil technologies, remembers that as more co-workers became aware of the studio, the filmmakers were approached with increasing frequency by employees from a variety of departments asking to document "whenever they created something special, some kind of rationalization or innovation." Some of the inquiries were even more informal, as colleagues from one of the labs called Pavlov and requested, "Come over with your camera, our cactus is blooming!"[49]

***

This brief survey of the history of Soviet production of non-fiction films on industrial topics shows that, while "industrial film" was formally not a category in film production, such films occupied a broad territory across a variety of other categories. What becomes evident from this history is a recurrent aspiration to create a culturally prestigious industrial film, especially pronounced during the periods marked by a high degree of social change and aesthetic experimentation, such as the 1920s and the 1960s. During these decades, Soviet artists and theorists embarked on a project of expanding the territory of art to areas that traditionally did not belong to it. In the 1920s these debates revolved around the binary of art and production, replaced in the 1960s by another binary, that of art and science. This shift evidently reflects the evolution of the vision of modernity that dominated the Soviet discourse, from one driven by industrialization towards a new one defined by advanced technologies within the framework of the science-technological revolution (STR). During other periods, such as the 1930s, utility was prioritized before aesthetics, and, parallel to the experimental wave of the 1960s, broader availability of filmmaking technologies lead to increasing separation of commissioned films from popular science, although this separation was always partial. Despite the differences in the economic and conceptual structuring of film production, industrial films produced in the Soviet socialist context share certain affinities with their counterparts from Western capitalist countries, in particular, the low cultural status of commissioned films, both at the time of their creation and in retrospective evaluation. The difference, however, is that commissioned films made only a part of the corpus of Soviet industrial films, although the overall share of the former was growing over the decades.

49 Interview with Vitalii Pavlov, director of the former People's Film Studio Start-Film in Leningrad. St. Petersburg, December 17, 2012.

Works Cited

Al'tshuler, Boris, and Aleksandr Zguridi. "O klassifikatsii nauchnykh fil'mov." *Iskusstvo kino*, vol. 8, August 1958, pp. 142–48.

Arlazorov, Mikhail. "Fil'my o tekhnike – eto ochen' interesno." *Nauchno-populiarnyi fil'm*, vol. 1, edited by Boris Al'tshuler, Igor' Vasil'kov, Aleksandr Zguridi, Vladimir Razumnyi, Vladimir Nikolai, M.V. Tikhonov, Vladimir Shneiderov, and David Iashin. Iskusstvo, 1959, pp. 81–94.

Arvatov, Boris. *Iskusstvo i proizvodstvo*. Proletkul't, 1926.

Bagrov, Peter. "On proletel po zhizni ... Alfred Dobbel't: kinematografist iz Riazani." *Riazanskaia starina*, vol. 2–3, edited by Aleksandr Nikitin and Pavel Tribunskii. Riazan', Krai, 2006, pp. 216–34.

Boltianskii, Grigorii. "Teatr kul'turfil'my." *Sovetskoe kino*, no. 1, 1927, pp. 2–3.

Brysin, Pavel. *Sozdanie ekrannykh nagliadnykh uchebnukh posobii v proftekhuchilishche*. Vysshaia shkola, 1982.

De Certeau, Michel. *The Practice of Everyday Life*. Translated by Steven Rendall. University of California Press, 2011.

Deriabin, Aleksandr. "'Nasha psikhologiia i ikh psikhologiaa – sovershenno raznye veshchi'. 'Afganistan' Vladimira Erofeeva i sovetskii kul'turfil'm dvadtsatykh godov." *Kinovedcheskie zapiski*, vol. 54, 2001, www.kinozapiski.ru/ru/article/sendvalues/617.

Elsaesser, Thomas. "The Place of Non-fiction Film in Contemporary Media." *Films That Work: Industrial Film and the Productivity of Media*, edited by Vinzenz Hediger and Patrick Vonderau. Amsterdam University Press, 2009, pp. 19–34.

Gastev, Alexei. *Poeziia rabochego udara* [1926], 6th ed. Khudozhestvennaia literatura, 1971.

Goldschmeding, Jan T. "IAMS, Past, Present and Future." International Association for Media and Science, www.media-in-science.org/wp-content/uploads/2014/10/IAMS-past-present-future-2-0.pdf. Accessed 20 September 2017.

Hanegen, Sabine, and Vladimir Kobrin. "Voprosy Vladimiru Kobrinu, zadannye Sabinoi Hanegen." *Kobrin: Svetiashschiisia sled*, edited by Andrei Gerasimov, Mikhail Kamionskii, and Alla Romanenko, Nizhnii Novgorod, ROF "Fond Iuriia Norshteina," 2005, pp. 83–87.

Kudriavtsev, Nikolai. "Bor'ba za kachestvo uchebnykh fil'mov." *Kino v pomoshch osvoeniiu tekhniki*, edited by Aleksandr Anoshenko. Sektor vedomstvennoi i zaochnoi literatury, 1935, pp. 9–23.

Lebedev, Nikolai. "Kul'turfil'ma na Zapade i u nas." *Kinovedcheskie zapiski*, vol. 58, 2002. Excerpts from unpublished book manuscript, www.kinozapiski.ru/ru/article/sendvalues/315.

Lebedev, Nikolai. "Zadachi ekranov kul'turfil'my." *Kino-Front*, vol. 4, 1927, pp. 6–9.

Le Corbusier. *Towards a New Architecture*. Translated by Frederick Etchells. Dover Publications, 1986. [Originally published in 1923 as *Vers une architecture*.]

Levy, Geoffrey. "An Amateur in Moscow." *Amateur Cine World*, January 30, 1964, p. 141.

Mally, Lynn. "The Americanization of the Soviet Living Newspaper." Carl Beck Papers in Russian & East European Studies, no. 1903. University of Pittsburgh, Center for Russian and East European Studies, February 2008.

Mezhrabpom-Fil'm. *Kul'turfil'my proizvodstva nauchnogo otdela aktsionernogo obshchestva "Mezhrabpom-fil'm"*. Leninskaia tipografiia Upromtorga, 1928.

Miasnikov, Vitalii. *Opyt sozdaniia tekhnicheskikh kinofil'mov na predpriiatiiakh i v organizatsiiakh*. Leningradskii dom nauchno-tekhnicheskoi propagandy, 1962.

Nesturkh, Mikhail. *Kino i obshchedostupnye lektsii*. Novaia Moskva, 1925.

Opyt ispol'zovaniia tekhnicheskikh kinofil'mov na zheleznodorozhnom transporte. Upravleniie uchebnymi zavedeniiami NKPS, 1941.

Sarkisova, Oksana. "The Adventures of the *Kulturfilm* in Soviet Russia." *A Companion to Russian Cinema*, edited by Birgit Beumers. John Wiley & Sons, 2016, pp. 92–115.

Schlegel, Hans-Joachim. "Nemetskie impul'sy dlia sovetskikh kul'turfil'mov 20-kh godov." *Kinovedcheskie zapiski*, vol. 58, 2002, www.kinozapiski.ru/ru/print/sendvalues/313.

Shagin, Vasilii. *Proizvodstvennye fil'my*. Gosudarstvennoe izdatel'stvo khudozhestvennoi literatury, 1931.

Sukharebskii, Lazar'. *Kino-disput*. Roskino, 1932.

Sukharebskii, Lazar'. *Kino-sud*. Roskino, 1933.

Sukharebskii, Lazar'. "Metodika pokaza nauchno-prosvetitel'noi fil'my." *Kulturfil'ma*, edited by Kirill Shutko. Tea-Kino-Pechat', 1929, pp. 172–93.

Sukharebskii, Lazar'. *Obzor sanprosvetitel'nykh kinofil'm za 10 let proletarskoi revoliutsii (1917–1927)*. Izdatel'stvo Moszdravotdela, 1928.

Vasil'kov, Igor'. "Sovremennye problemy kinopopuliarizatsii." *Nauchno-populiarnyi fil'm*, vol. 2, edited by Boris Al'tshuler, Leonid Belokurov, Igor' Vasil'kov, Aleksandr Zguridi, Vladimir Nikolai, M.D. Nechaeva, and Vladimir Razumnyi. Iskusstvo, 1964, pp. 5–46.

Vertov, Dziga. "The Factory of Facts" [1926]. *Kino-Eye: The Writings of Dziga Vertov*, edited by Annette Michelson, translated by Kevin O'Brien. University of California Press, 1984, pp. 58–60.

Vronskii, Boris, and Robert Vartazarian. *Kino i tekhpropaganda*. Izdatel'stvo upravleniia kinofikatsii pri SNK RSFSR, 1934.

Zalambani, Maria. *Iskusstvo v proizvodstve: Avangard i revoliutsiia v sovetskoi Rossii 20-kh godov*. Translated by Natalia Kardanova. IMLI RAN/Nasledie, 2003.

Zel'manov, Mark. *Kino-miting*. Roskino, 1932.

Zguridi, Aleksandr. "O nekotorykh voprosakh teorii i praktiki nauchno-populiarnoi kinematografii." *Nauchno-populiarnyi fil'm*, vol. 1, edited by Boris Al'tshuler,

Igor' Vasil'kov, Aleksandr Zguridi, Vladimir Razumnyi, Vladimir Nikolai, M.V. Tikhonov, Vladimir Shneiderov, and David Iashin. Iskusstvo, 1959, pp. 18–33.
Zheliabuzhskii, Iurii. "Sovetskii tekhnicheskii fil'm." *Kino v pomoshch osvoeniiu tekhniki*, edited by Aleksandr Anoshenko. Sektor vedomstvennoi i zaochnoi literatury, 1935, pp. 24–32.
Zimmermann, Yvonne. "'What Hollywood Is to America, the Corporate Film Is to Switzerland': Remarks on Industrial Film as Utility Film." *Films That Work: Industrial Film and the Productivity of Media*, edited by Vinzenz Hediger and Patrick Vonderau. Amsterdam University Press, 2009, pp. 101–17.

About the Author

Maria Vinogradova is a film and media historian specializing in the study of Soviet film culture. She is currently working on her book manuscript "On the Public Rails: A History of Soviet Amateur Filmmaking (1957–1991)." Her research has been supported by fellowships from American Council of Learned Societies (ACLS) and National Endowment for the Humanities (NEH).

22 "There Is No Life More Reckless and Adventuresome Than That of the Oil Prospector"[1]

ENI's Geologist-Filmmakers in Iran[2]

Luca Peretti

Abstract

This chapter analyses the amateur films shot by the employers of the Italian national oil company ENI in Iran between the end of the 1950s and the beginning of the 1960s. At the centre of geopolitical and economic interests, Iran was one of the sites where the Italian company looked for oil, signing agreements with the local government. Several teams of geologists and geophysics worked there at the time, and together with maps and different tools they also brought amateur cameras that they used to shoot at least eighty short films, now preserved in the company's archive. The chapter discusses these films and how they came to be, at the intersection of sponsored, amateur, ethnographic, and travel cinema.

Keywords: amateur cinema; Iran; exploration; ENI; Italian cinema

Italian screen studies have only been marginally touched by an important, recent trend in film and media studies: the study of the macro-field of

1 "Non esiste vita più spericolata e romanzesca di quella del cercatore di petrolio". Spadini, Raffaele. "Visita ai nostri cantieri in Iran, Con i geologi della Mineraria sulle vette e fra le gole dell'Iran (monti Zagros)." *Il gatto selvatico*, no. 4, 1964, p. 8. All the translations are mine, unless otherwise noted.

2 Acknowledgements: I wish to thank Anna Landolfi, Lucia Nardi, Daniela Scamuzzi, Mattia Voltaggio, and all the archivists and employees of the Archivio Storico ENI in Rome and Pomezia; the editors of this book and, in particular, Vinzenz Hediger, who invited me to write this chapter; and Francesco Casetti and Millicent Marcus for their comments and support. This essay was completed in 2018.

Hediger, V., F. Hoof, Y. Zimmermann, with S. Anthony (eds), *Films That Work Harder: The Circulation of Industrial Film*. Taylor & Francis Group 2024
DOI: 10.5117/9789462986534_CH22

non-fiction filmmaking that has to do with ephemeral, non-theatrical, educational, and industrial films.[3] However, there is no shortage of primary material, as a number of films of this kind have been produced in Italy and a critical and theoretical discussion around these films has taken place, especially in the 1960s. While this legacy is largely forgotten in the studies on Italian cinema (at least those in English language), some companies and different institutions are cataloguing, archiving, and in some cases even digitalizing their visual patrimony.[4] The case of ENI (Ente Nazionale Idrocarburi, National Hydrocarbons Authority) is one of the most interesting examples of the companies that made films at the time, one that is worthwhile studying both for the films themselves[5] and for the role that the company had in Italian and Mediterranean history. ENI is a formerly state-owned company (now privatized) founded in 1958, whose core business was and continues to be the extraction, selling, and distributing of gas and oil, but it has operated in a large number of fields, including nuclear power, energy, mining, chemicals and plastics, refining/extraction, and distribution machinery. It had/has a peculiar role in the state – it became a sort of unofficial Ministry of Foreign Affairs for Italy, often cooperating and working with the actual minister or local ambassadors. The company claimed to be not just a commercial firm but a public agency invested in the well-being of Italian citizens and those of the countries it was dealing with. This was the case especially under the leadership of its founder and first president, Enrico Mattei (1906–1962), a

3 See Heymer, Anna, and Patrick Vonderau. "Industrial Films: An Analytical Bibliography." *Films That Work: Industrial Film and the Productivity of Media*, edited by Vinzenz Hediger and Patrick Vonderau. Amsterdam University Press, 2009; Acland, Charles R., and Haidee Wasson, editors. *Useful Cinema*. Duke University Press, 2011; Orgeron, Devin, et al., editors. *Learning with the Lights Off: Educational Film in the United States*. Oxford University Press, 2012; for the Italian context, see Bonifazio, Paola. *Schooling in Modernity: The Politics of Sponsored Films in Postwar Italy*. University of Toronto Press, 2014.

4 Particularly remarking is the work done at the Archivio del Cinema Industriale e della Comunicazione d'Impresa (Industrial Film and Business Communication Archive) in Castellanza, Varese, and the Archivio Nazionale Cinema d'Impresa (National Film Archive for Industrial Film), in Ivrea, Turin, as a branch department of Fondazione Centro Sperimentale di Cinematografia (Italian National Film Archive) in Rome, as well as by private companies like ENI, Olivetti, and the AEM Foundation.

5 Frescani, Elio. *Il cane a sei zampe sullo schermo. La produzione cinematografica dell'Eni di Enrico Mattei*. Liguori Editore, 2014; Latini, Giulio. *L'energia e lo sguardo. Il cinema dell'Eni e i documentari di Gilbert Bovay*. Donzelli, 2011; Bonifazio, Paola. "United We Drill: ENI, Films, and the Culture of Work." *Annali d'Italianistica: From Otium and Occupatio to Work and Labor in Italian Culture*, vol. 32, 2014, pp. 329–50. ENI's vast filmography includes films by Paolo and Vittorio Taviani, Gilbert Bovay, Joris Ivens, Gillo Pontecorvo, Bernardo Bertolucci among others.

former partisan and one of the most influential figures in the history of contemporary Italy.[6]

Between the end of the 1950s and the beginning of the 1960s several teams of geologists and geophysics were sent to Iran: these workers returned with maps, analysis of the grounds, information, scientific evaluations, but also over eighty short films, shot in 8 mm and 16 mm. These films are short (most of them are between two and three minutes long), largely unedited, made by non-professional filmmakers and "raw," not reworked after the filming.[7] Several different conventional types of filmmaking are being mobilized: home movies, expedition and explorational films, travel films. Furthermore, they are conceived within the framework of utility cinema, but their function is unstable and unclear: they clearly have a scientific and utility goal (preparing the ground for future explorations; collecting visual aid for drilling sites), but they also document the daily lives of the ENI experts. Possibly beyond their original intent, they also depict local populations and their lifestyles. According to Charles Tepperman, "in contrast to both popular cinema, which addressed spectators as part of a mass audience, and documentary film, which quickly came to designate a specific kind of government and institutional propaganda, amateur films generally reflected a personal and independent attitude toward practical filmmaking."[8] As these films oscillate between being useful, in one sense, for the information they collect and as travel memos, what happens to the independent attitude Tepperman describes when films are also useful, in another sense, when they are utility films? Finally, these films are ephemeral for three reasons: (1) for scholarly and classificatory reasons, that is, they are part of the growing sub-field concerned with industrial and sponsored films; (2) as artefacts that were never completed, edited or polished, and which live as works of complex and partially unidentifiable

6 Pozzi, Daniele. *Dai gatti selvaggi al cane a sei zampe. Tecnologia, conoscenza e organizzazione nell'Agip e nell'Eni di Enrico Mattei*. Marsilio, 2009; Bini, Elisabetta. *La potente benzina italiana. Guerra fredda e consumi di massa tra Italia, Stati Uniti e Terzo Mondo (1945–1973)*. Carocci Editore, 2013.

7 According to Roger Odin, "the home movie is not or, more exactly, should not be considered a 'film' – that is at least if we call a film a structured production with specific communicative goals, and with a beginning and an end." In this chapter I refer to them as "films" for practical reasons and employing a broad definition of the term that precisely include also films without a beginning and an end. Odin, Roger. "The Home Movie and Space of Communication." *Amateur Filmmaking: The Home Movie, the Archive, the Web*, edited by Laura Rascaroli, Gwenda Young with Barry Monahan. Bloomsbury Academic, 2014, p. 17.

8 Tepperman, Charles. "Mechanical Craftsmanship: Amateurs Making Practical Films." *Useful Cinema*, edited by Charles R. Acland and Haidee Wasson. Duke University Press, 2011, p. 290.

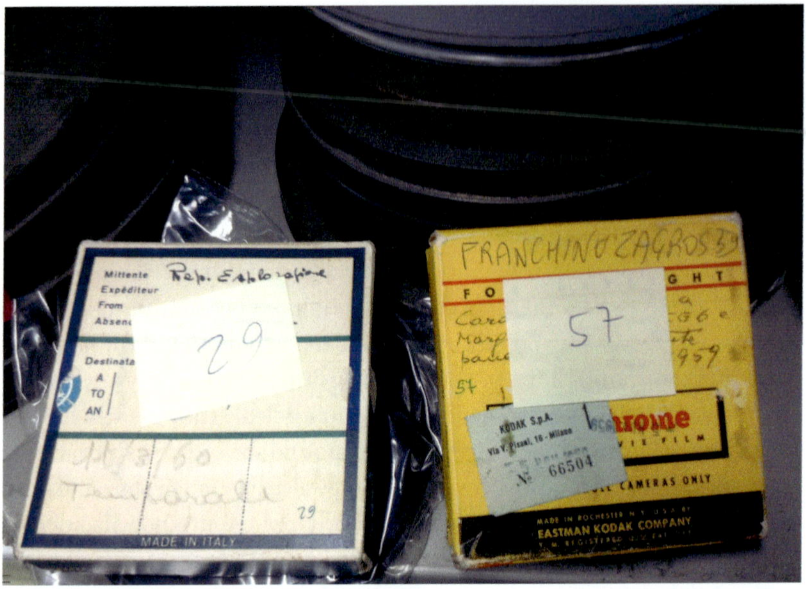

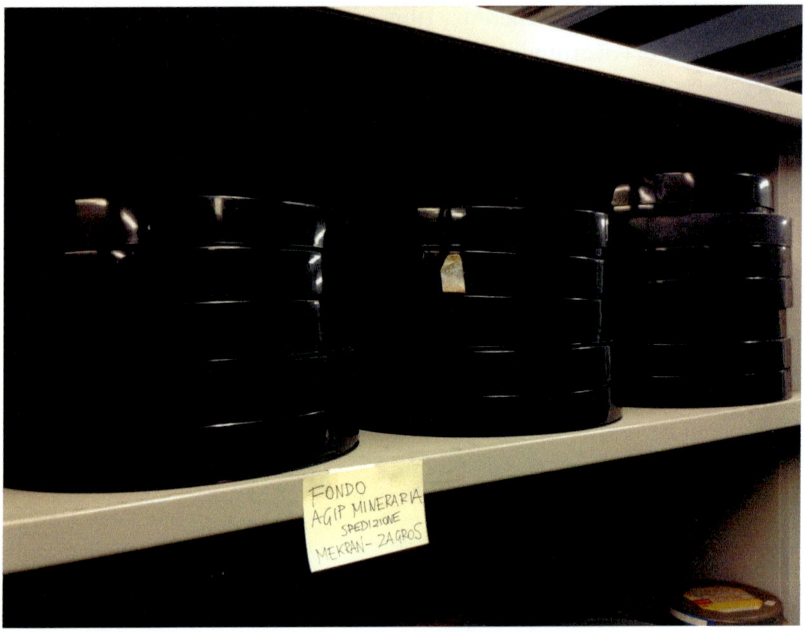

Figs. 22.1 and 22.2: The films in the Archivio Storico ENI. Photographs by Luca Peretti.

status (between amateur and industrial cinema) and, moreover, which never found an audience, diverging from one of the key characteristics of industrial films ("an industrial film can fulfil its function only if it reaches its target audience," as Yvonne Zimmermann wrote[9]); and, finally, (3) the main reason and context for which they came to exist – the search for oil – was also eventually ephemeral. The expeditions were fully successful only offshore.[10] The drilling on Mount Zagros was postponed and never fully took place.[11]

Before starting the analysis of the films and examining their context and the methodological tools needed to study them, a few words are necessary on where they are preserved and who organized them. The ENI archive owns eighty-eight short films made in Iran between 1958 and 1961 (figs. 22.1 and 22.2), which have been digitized, and catalogued by archivist Letizia Cortini.[12] We know that no professional filmmakers were sent with the workers thanks to the documentation on the explorations.[13] The films are all silent, some in black and white and some in colour, a choice that seems to be random more than aesthetic or formal. An array of secondary sources, from interviews with the workers to fiscal and bureaucratic documents regarding the expeditions, can help us understanding the context of these films. As is the case for many industrial films, the filmmakers are nameless,[14] but

9 Zimmermann, Yvonne. "'What Hollywood Is to America, the Corporate Film Is to Switzerland': Remarks on Industrial Film as Utility Film." *Films That Work: Industrial Film and the Productivity of Media*, edited by Vinzenz Hediger and Patrick Vonderau. Amsterdam University Press, 2009, p. 108.

10 See Varvelli, Riccardo. "Il campo di Bahrgan Sar. Le prime attività Agip in Iran (2nda parte)." *NIA*, 1999, p. 29.

11 Eventually ENI, like other foreign companies, was expelled in 1979.

12 In this chapter I follow the way the films were numbered by the archivists at ENI, in progressive number from 1 to 88. These numbers do not follow a specific criteria, chronological or of other kinds, and were randomly assigned when the films were archived. The titles are also attributed, I cite them using the number and the title in Italian followed by the English translation, as in *1. Scenette locali villaggi zona Goxar* (*Local Vignettes from Towns in the Goxar Area*). On Letizia Cortini, see her website: "Visioni dalla Storia. Le fonti audiovisive nella didattica – Audiovisual Sources for Teaching," visionandonellastoria.net/about/. Accessed April 13, 2018. Aristide Franchino, one of the ENI workers of the time, saw these films and helped Cortini to recognize names and places.

13 Such as "Proposal for a Work Plan in Iran for 1957" and "IRAN: Programmi di lavoro e relativa organizzazione" ("Iran: Work plan and organization") for the year 1958 where the list of workers does not include any photographers or cameramen. Archivio Storico ENI, Rome/Pomezia (henceforth ASE), Estero, box (b.) 21, folder (f.) FF5 and FF6. There is however in the list of equipment a very generic "*attrezzatura fotografica*" (photographic equipment), besides two specifically geological photographic equipment. Another similar plan is in ASE, Estero, b. 21, f. FF3.

14 As Heymer and Vonderau wrote, "Industrial films […] are often nameless, have no place and no life other than the task at hand. […] They seem, like memos, company brochures, and other forms of 'grey literature', ephemeral to media studies." "Industrial Films," p. 407.

differently from many utility films, the scope, goals, intended audience, in other words, the reasons for their existence as films, are unclear: no single document owed by the company, in fact, speaks to this.

Explorational Home Movies of Unclear Utility

Roger Odin and Patricia R. Zimmermann distinguish between amateur cinema and home movies: for Odin, the home movie is a sub-category of amateur cinema (in itself a macro-category opposed to professional cinema), and Zimmermann uses *amateur film* as a "covering term for the complex power relations defining amateur filmmaking" and *home movies* "as a descriptive term for actual films produced by families."[15] The films I discuss in this chapter could be classified as home movies, while they also strongly recall the characteristics of amateur travel films and even ethnographic ones. Amateur films "rarely display an autonomous visual esthetic," as Alexandra Schneider noted in a paper on an "amateur film [that] intersects with the travel film."[16] These films are no exception. The filmmakers seem to have some understanding of film language as they try to imitate professional filmmaking, such as in the case of a long pan at the beginning of *4. No title*, or the beginning of *26. Riprese aeree di un vasto territorio dell'Iran, in particolare dell'area del Mekran* (*Aerial View of a Vast Territory in Iran, the Area of Makran, in Particular*).[17] However, the images are usually poorly taken,[18] the quality is low, and, in some cases, because they are taken from a means of transportation, the images are also shaken and blurred: for example, in *28. Difficoltà di avanzamento della spedizione in Mekran e allestimento del campo lungo la costa* (*Travelling Difficulties of the Expedition in Makran and Dismantling the Camp along the Coast*), the difficulties of filming and the harshness of the condition where they work become one and the same. According to Odin, they work well precisely for

15 Odin, Roger. "Rhétorique du film de famille." *Rhétoriques, sémiotiques. Revue d'Esthétique*. Union Générale d'Editions, 1979, pp. 340–73; Zimmermann, Patricia R. *Reel Families: A Social History of Amateur Film*. Indiana University Press, 1995, p. x.

16 Schneider, Alexandra. "*Autosonntag* (Switzerland, 1930) – A Film Safari in Swiss Alps." *Visual Anthropology*, vol. 15, 2002, pp. 115–16.

17 From the memories we also know that they were avid filmgoers as soon as they could reach the cities.

18 As Cortini notes, "Immagini solo in alcuni casi nitide, per lo più sfocate o sovraesposte" ("Images usually not clear, in most of the cases blurred or overexposed"), "Inventario foto-cinematografico di viaggi in Iran per l'Archivio Storico Eni," ASE, Roma 2011, no archival collocation, no page.

Fig. 22.3. Filming. From *1. Scenette locali villaggi zona Goxar* (*Local Vignettes from Towns in the Goxar Area*).

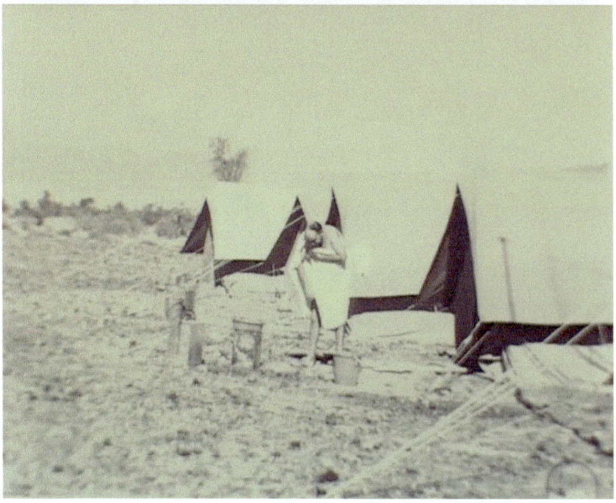

Fig. 22.4. Bath. From *27. No title*.

this reason, as "to work well, the home movie should be made like a random succession of scenes only offering snippets of family life from which each family member [in this case, each worker] might be able to reconstruct the family history in his or her own way."[19] As it happens in many home movies, people look at the camera and in a flash of the home the filmmaker sometimes becomes clearly visible, such as in *23. Momenti della spedizione dell'Agip Mineraria negli Zagros* (*Moments in the Agip Mineraria Expedition in the Zagros Mountains*), where we see an entire figure filming with his small camera (fig. 22.3). Among the motives of home and family movies, we see intimate moments, such as a man having a bath (in *27. No title*) (fig. 22.4)

19 Odin. "The Home Movie," p. 17.

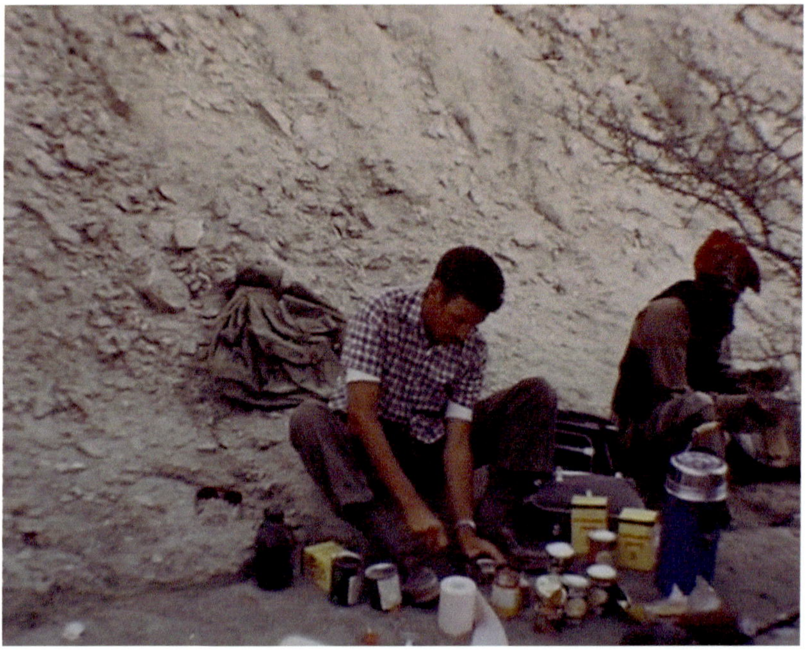

Fig. 22.5. Eating. From *36. Campo attrezzato con tende lungo la costa (Camp with Tents along the Coast).*

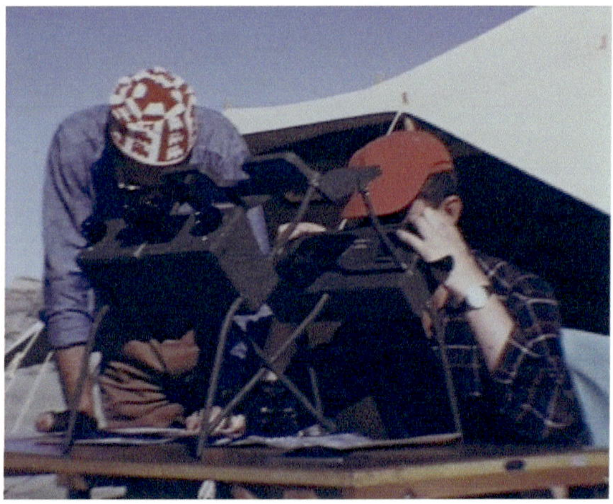

Fig. 22.6. Working. *38. Mezzo di trasporto in movimento, locali al lavoro (Means of Transport in Movement, Locals at Work).*

or daily activities like eating (fig. 22.5) (for example, a lunch break in *19. Sosta della spedizione nei pressi delle rovine di antichi edifici medievali* (*The Expedition Stops in the Vicinity of Ancient Medieval Buildings*). One of the most important of these activities is the setting up and dismantling of the several camps the expeditions built along the trip, which are depicted in several films (for example, *53. Riprese dell'allestimento del campo Husdan a*

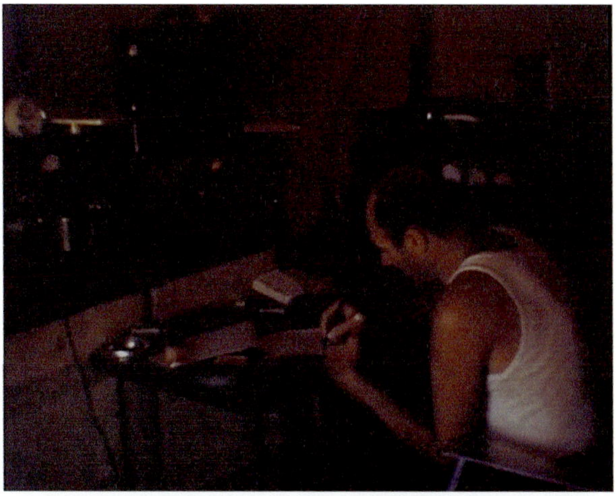

Fig. 22.7. Working. From *36. Campo attrezzato con tende lungo la costa* (*Camp with Tents along the Coast*).

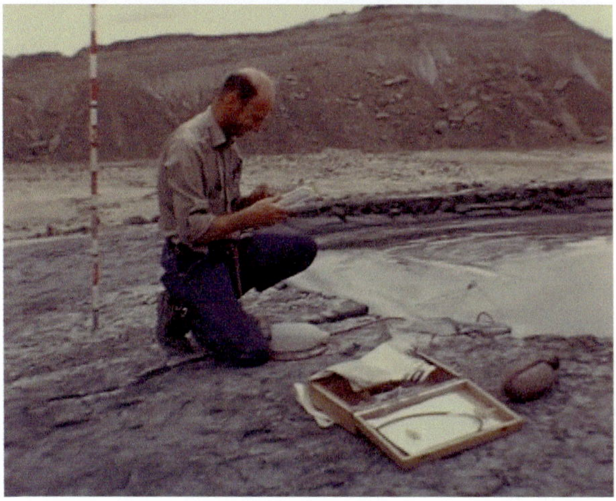

Fig. 22.8. Working. From 24. *Riprese del territorio del Mekran e di alcune attività di rilevamento dei geologi* (*Filming of the Makran Territory and of Some Survey Activities of the Geologists*).

Jask (*Filming the Setting up of the Husdan Camp in Jask*), and, of course, we see men working, busy with a variety of different tasks (figs. 22.6, 22.7 and 22.8). The kind of family and community represented in these films, then, is an unusual one, a comradeship of the type found in military expeditions, of an all-male world of men struggling for survival. The amateur films made in Iran are part of a number of different media that strengthen this sense of community, one that, in this case, is built on group pride, the sense of belonging to a company and engaging in a pioneer experience. In a world with no women, these films also show men doing typically feminine work – at least according to the gender normative conventions common in mid-century Italy – like cleaning dishes or washing clothes. The only women who appear in the film are the exotic subjects to look at, the local women.

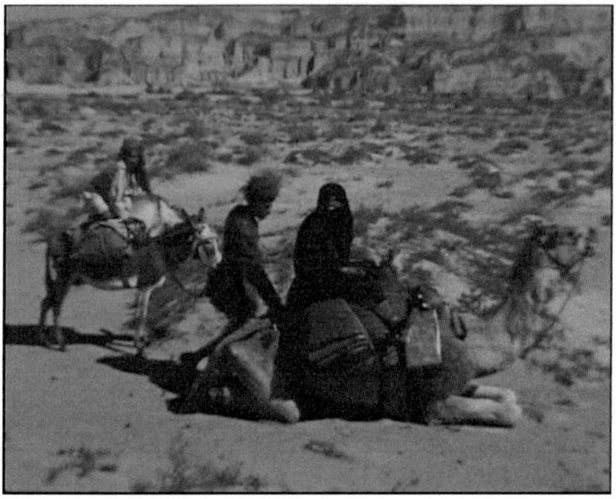

Fig. 22.9. Local people, including a veiled woman. From *38. Mezzo di trasporto in movimento, locali al lavoro* (*Means of Transport in Movement, Locals at Work*).

However, we learn from the memories of the pioneers, they were hard to film: as Franchino states, "Le donne in genere, per ovvii motivi religiosi, non si lasciano fotografare" ("Usually women, for obvious religious reasons, do not let us take pictures of them"[20]) (fig. 22.9).

These films also share characteristics typical of industrial films – those that Friedrich Kittler[21] and several other scholars, including Yvonne Zimmermann, have described as a media mix (or *Medienverbund*):

> As is the case with corporate photography, an industrial film rarely comes alone; it is usually part of a series of films. Furthermore, industrial films are always integrated in an orchestrated media mix (or *Medienverbund*) for corporate communication, and thus not used in isolation, but in correlation with other media.[22]

As already pointed out, reports and memories by the pioneers were written on the same subjects of the films, and a number of articles (in house organ

20 Franchino, Aristide. "Ricordi di un geologo. MEKRAN (Iran sud-orientale) 1958 – Viaggio e rilevamento geologico." Associazione Pionieri e Veterani Eni, February 2010, available on the APVE website. Accessed September 18, 2016.

21 Kittler, Friedrich. *Discourse Networks 1800/1900*. Stanford University Press, 1990; and idem. *Gramophone, Film, Typewriter.* Stanford University Press, 1999.

22 Zimmermann. "'What Hollywood Is to America, the Corporate Film Is to Switzerland'," p. 104. In the same book, see other uses and discussions of the idea of Medienverbund in Elsaesser, Thomas. "Archives and Archaeologies: The Place of Non-fiction Film in Contemporary Media," pp. 19–34; Hediger, and Vonderau. "Record, Rhetoric, Rationalization: Industrial Organization and Film," pp. 35–50; and Szczepanik, Petr. "Modernism, Industry, Film: A Network of Media in the Baťa Corporation and the Town of Zlín," pp. 349–76.

magazines like *Il gatto selvaggio* and *NIA* or in newspapers like *Il tempo*) can be studied together with these films. More importantly, a number of photographic albums accompany these films. There are official ones, like one created on the occasion of an anniversary celebration of the Iranian oil industry in March 1958,[23] composed of a series of photographs of Iranian and Italian personalities, including Enrico Mattei, visiting oil drilling sites, participating to conferences and commemorations events, including traditional games. Others albums (as happens with the films) consist of photographs taken to serve geological or geophysical purposes: pictures of vast lands, of geological surveys, of analysis made on the ground, and similar subjects. This is the case, for example, with a large album completed in February 1961 by L. Anelli and G. Dattilo[24] called "Note sul rilevamento da Chahbahar a Zahedan" ("Notes on the Survey from Chabahar to Zahedan").[25] As in the films, we also have albums that mix working preoccupations, tourist interests and portraits of daily life. This is the case with an album made by three employees (named Mengoli, Pesce and Messori), and also with another large album containing photos shot by several named geologists.[26]

In all these materials, paradoxically, we have names of the photographers but not of the filmmakers. We can suppose that in some cases they are the same people, especially since the points of view, types of images and subjects are extremely similar to those we find in the films. Much like Bernardo Bertolucci, who in 1965 was invited by the company to Iran to film *La via del petrolio* (*The Oil Route*, 1967), an auteur was also invited by ENI to take photographs during the expeditions: Federico Patellani, a world-renowned photographer at the time.[27] His photographs taken in Iran at the beginning of the 1960s are surely more refined, official and canonically artistic than those taken by geologists, even though the subjects are nonetheless similar. Part of this media mix (or *Medienverbund*) is also the presence of ENI at an important exposition that took place in Teheran in 1958 called "Italia Produce" ("Italy Produces").[28] In 1961, ENI also sponsored a film entirely

23 National Iranian Oil Company. "Fiftieth Anniversary Celebrations of the Iranian Oil Industry, March 1958." Archivio Storico ENI, Rome/Pomezia, no archival collocation. The album seems to be a gift of the NIOC to ENI.

24 ASE, no archival collocation.

25 ASE, Album 8, no archival collocation.

26 ASE, Album 265, no archival collocation.

27 On Patellani, see Bolognesi, Kitti, and Giovanna Calvenzi, editors. *Federico Patellani: professione fotoreporter*. Silvana editoriale/Museo fotografia contemporanea, 2015.

28 On this exposition, see Ciucci, Carlo. "'*Affermazione dell'Eni alla mostra di Teheran*' (padiglione dell'Eni alla mostra di ottobre), '*Italia produce*' a Teheran, in Iran." *Il gatto selvatico*, no. 10, 1958, pp. 16–17.

dedicated to its foreign activities, *Panorama attività estero 1961*, which was composed of images shot in different areas of the world where the company was operating. Iran is, of course, included, but the images used to talk about it are that of a newsreel shot on the occasion of the visit of Enrico Mattei and President of the Italian Republic Giovanni Gronchi after the signature of the agreement and not those shot by the geologists. The unclear utility of these films is because we can only guess what purpose they had, while we have more information about the employment of other media – radio, for example, which served to enable communication between the different contingents in Iran,[29] or photographic equipment, which had geological purposes.[30]

In what follows, I will briefly discuss the importance of the deal that ENI struck with Iran, then turn to the discussion of the main themes that emerge in these films, before concluding with a discussion on the nature of these films and their role in the strategy of the company and of Italy at the time.

A Different Deal: ENI in Iran

In August 1953 a US and British intelligence operation overthrew the democratically elected government of Mohammad Mossadeq. The prime minister had nationalized the Anglo-Iranian Oil Company, which, according to journalist Stephen Kinzer, was "the most profitable British business in the world."[31] The UK had historically controlled the region, and Winston Churchill had no interest in giving it up, while the newly elected US government lead by Republican Dwight Eisenhower framed the intervention in anti-communist terms: not because Mossadeq was communist, but because Iran was geopolitically important. The new US-backed government would eventually sign agreements with the Seven Sisters. In January 1954, Mattei wrote a note to Vittorio Zoppi, a high-ranking diplomat within the Italian Ministry of Foreign Affairs, saying, "Finora in questo giuoco noi siamo fuori" ("So far, we are excluded from this game").[32] Merely nine years after the end of

29 "Schema dei collegamenti Radio." ASE, Superfondo Eni, Fondo Estero Serie Assistente del Presidente per i rapporti con l'Estero (1953–1963), b. 21, f. FF3.

30 "Particolare importanza [...] ha rivestito l'uso dell'aereofotografia e 'l'aereofotointerpretazione'" ("Particularly important [...] was aerial photography and 'aerialphotointerpretation'"). Cortini. "Inventario foto-cinematografico."

31 Kinzer, Stephen. *All the Shah's Men: An American Coup and the Roots of Middle East Terror.* John Wiley & Sons, 2003, p. 2.

32 Tremolada, Ilaria. *La via Italiana al petrolio.* Ornitorinco Edizioni, 2011, p. 134.

the Second World War, Italy was slowly recovering socially and economically, and in parallel it was also trying to assert its role in the Mediterranean and in the Middle East. Of the few Italian colonies only one (Libya) had been in the area, leaving Italy excluded of the "game" of the geopolitical strategies in the Middle East and North Africa region; on the other hand, however, Italy could present itself as an ally and not as a former colonial power, a strategy that Mattei pursued vigorously. The concern was not about access to natural gas, which was the first main hydrocarbon extracted by ENI. Italy in fact lacked petroleum, and therefore the company had to look abroad.[33] Iran was the second country with which ENI signed a deal after Egypt, upsetting other oil companies and in general US, French and British interests in the area. Mattei and ENI at the time even sided explicitly with anti-colonial movements.[34]

Historian Ilaria Tremolada notes how Mattei in the above-mentioned letter to Zoppi used the adverb *finora* (so far), as the situation changed when, after some initial contacts between Italian independent oil companies and the Iranian government, in August 1957 ENI signed an agreement with the NIOC (National Iranian Oil Company). The agreement included the exploration in three distinct areas: the Makran (Mekran, in Italian) costal region, Mount Zagros, and Bahrgan Sar, an offshore area in the Persian Gulf.[35] The type of agreement they signed with the Iranian government would have a huge impact on the way Western oil companies dealt with "developing" and decolonizing countries. As Paul Frankel, an important oil consultant, wrote at the time, "for the first time in the Middle East, a State becomes a partner with a foreign company in a concessionary company."[36] This partnership was based on a very simple assumption, well expressed in a note by the General Direction of Economical Affairs of the Minister of Foreign Affairs: "La Persia ha il petrolio che noi non abbiamo, mentre noi abbiamo i cantieri che ad essi mancano" ("Persia has oil, which we do not have, whereas we have the ability to build construction yards that

33 Pozzi. *Dai gatti selvaggi*, pp. 406–12. Mattei himself discussed this topic in several speeches, for example, "L'ENI e il problema italiano dell'energia," Piacenza, September 12, 1957. Mattei, Enrico. *Scritti e Discorsi, 1945–1962*. Rizzoli, 2012, pp. 490–97.

34 ENI and colonialism is discussed in Peretti, Luca. "Before The Battle of Algiers: Sartre, Colonialism, Industrial Cinema, and Unmade Film." *Senses of Cinema*, no. 84, September 2017, http://sensesofcinema.com/2017/sartre-at-the-movies/sartre-colonialism-industrial-cinema-and-an-unmade-film/.

35 Tremolada. *La via Italiana*, p. 189.

36 Frankel, Paul H. Note written to Mattei in February 1958. ASE, Presidenza/Estera, b. 50, f. 1743. As a scholar, Frankel is also author of a book on Mattei and other works on oil. On this agreements, see Pozzi. *Dai gatti selvaggi*, pp. 414–15; and Tremolada. *La via Italiana*, pp. 189–225.

they do not have").[37] The exploration that followed was the first one of this kind that the company undertook abroad. Parallel to the growth of the company domestically and internationally, ENI had started producing films to document its main projects – such as *Il gigante di Ravenna* (*The Giant of Ravenna*, Fernando Cerchio, 1959) – or newsreel-type films that illustrated the work of the company more broadly, such as *Panorama delle attività del gruppo Eni 1959* (*Overview of ENI's Activities in 1959*). The great importance of the deal that the company struck may explain the presence of 8 mm and 16 mm camera during expeditions, while only later did the company sponsor an official documentary in the area (Bertolucci's film). But these ENI workers may have had cameras for another, simpler reason: enthusiasm. Amateur cinema was on the rise in Italy at the time and several manuals had already been published by the beginning of the 1960s to help *cinematori* (amateur directors) film and develop their films.[38]

"We Really Worked Like Pioneers"[39]: Working, Meeting and Mapping in Faraway Lands

In the short films made in Iran two main themes emerge: the recounting of the epic of the pioneers and the scientific explorations and analysis of the ground. The former is articulated in two different aspects: the daily lives of the ENI workers and their encounters with populations that, even discarding the Orientalist pathos of the "untouched savages" (which was otherwise present in some of the contemporary narrations of the expedition[40]), had indeed

37 Qtd. in ibid., p. 66.

38 Among others, Costa, Enrico. *Filmare in 8mm*. Hoepli, 1963; and Staab, Joachim Gerhard. *Coraggio, filmate! La regia nel film amatoriale e professionale*. Translated by L. Moggio. Edizioni del progresso fotografico, no date but probably 1963, original edition *Filmregie für Leinwand und Bildschirm*. Heering-Verlag, 1962.

39 "Lavoravamo in maniera veramente da pionieri." "Intervista a Benito Gasparotto." Summer 2016. Archivio Storico ENI, Rome/Pomezia, no archival collocation. The concept and word *pionieri* (pioneers) is very dear to the company and the people who worked at ENI: the association of former workers is exactly called Associazione Pionieri e Veterani Eni (Veterans and Pioneers Association, APVE). We can find many requests for memories of the pioneers in specific publications edited by the company, like Muzzin, Adriano. "Prefazione. Per salvare la memoria comune." *La memoria raccontata. Storie e ricordi di uomini e donne di eni, Documenti dall'Archivio storico*, vol. 3. ENI, 2013, pp. 7–8; or Guidi, Francesco. "In Iran il Gatto Selvatico arriva nel 1962." *NIA*, no. 107, April 1995.

40 One of the journalists that visited them wrote: "L'autocolonna dei nostri tecnici si spinse in zone mai visitate da europei, in villaggi che mai avevano visto la faccia di un uomo bianco" ("The convoy of our technicians drive there where no Europeans has ever been, in villages that have not seen the face of the white men before"). Salvatori, Sandro. "Faticosa e affascinante avventura dei petrolieri italiani in Persia." *Il Tempo*, October 1958.

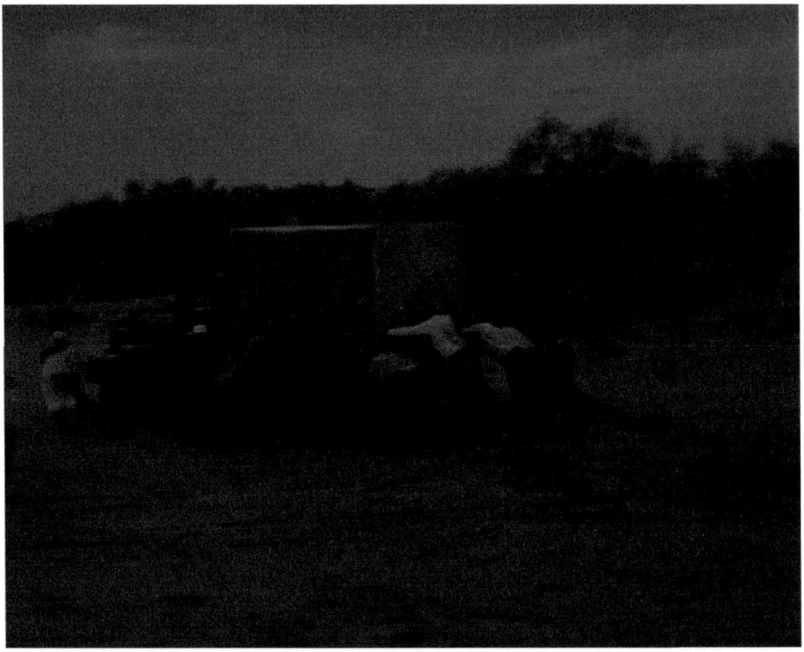

Fig. 22.10: Harsh conditions. From *31. Momenti di difficoltà della spedizione dei tecnici verso la Provincia di Fars* (*Difficult Moments in the Expedition of the Technicians toward the Fars Province*).

had possibly little if any contact with Western people. These encounters are of three different kinds: a mere detached observation of people living in villages; more articulate interactions, often for what concerns food or exchange of different kind of goods; and, finally, the employment of these people as workers for the company. Often these themes are combined in the same films.

In these films we see how the ENI pioneers had to get through the desert, encountering strong winds and other forms of inclement weather (fig. 22.10), and long pans of completely uninhabited lands that give a sense of where the workers were operating. Reports and articles of the time are filled with language such as "Nelle difficili terre dell'Iran i nostri lavorano con entusiasmo" ("In the difficult lands of Iran our men work with enthusiasm"),[41] and similarly in the official documents of the company: "Il rilievo geofisico nel Mekran risulta assai pesante e difficile: soprattutto per le gravi difficoltà di spostamento dei mezzi su estese sabbie mobili o regioni paludose, e difficoltà di rifornimento" ("Geophysical surveying in the Makran is very burdensome and difficult. Above all, it is particularly awkward to travel across quicksand

41 Carey, Jane, and Andrew Carey. "Nelle difficili terre dell'Iran i nostri lavorano con entusiasmo." *Il gatto selvatico*, no. 9, 1959, pp. 11–14. The authors are two American political scientists.

or swampy regions, and refuelling is problematic").[42] Sometimes, this had nationalistic overtone, as in "le circospezioni geologiche non sono state facili. Anzi possiamo dire che soltanto il coraggio e la tenacia dei tecnici italiani hanno potuto avere ragione delle avversità e delle asprezze naturali" ("Geological explorations have not been easy. On the contrary, we could say that only the courage and the skill showed by Italian experts could overcome the adversities and the natural harshness").[43] Most of these pioneers were young (under thirty), had to sustain extremely long stays away from home[44] and often were hired specifically for this job, as the experienced workers from the Po Valley (where ENI work in Italy began) were not able for those conditions.[45] We see them, however, being interviewed by Bertolucci five years later, when the drilling sites have been built and extraction has begun: this shows, on a more general note, how industrial films can help us to understand the evolution and the organization of the work within the company.

The pioneer epic is framed in military terms: Milziade Torelli, in the house organ *Il gatto selvatico*, talks about "Le pattuglie avanzate dell'E.N.I" ("ENI's advanced squads") or "gli uomini impegnati nella dura e appassionante battaglia" ("men engage in a harsh and fascinating battle").[46] Military personnel are also an occasional presence in these films, for example, in *Mezzo di trasporto in movimento, locali al lavoro* (*Means of Transport in Movement, Locals at Work*): their role, we learn thanks to diaries and memorials of the time, was to protect the workers in dangerous areas.[47] But from the films, we also get the sense of how the lure of the camera must have conquered

42 Agip Mineraria. "Attività in Iran. Programma dei lavori e & Preventivo di spese [Programme of works and estimate of expenditure], Anno 1959, Tehran, 27 Novembre 1958." Archivio Storico ENI, Rome/Pomezia, Presidenza/Estera, b. 21, f. FF6.

43 Salvatori. "Faticosa e affascinante." Emphasis added.

44 Franchino talked about twenty-two months, and another pioneer, Giovanni Bozzolato, about two years, without ever going back home. See Franchino, Aristide. "Ricordi di un 'geologo' in Iran 1958–1959: Adventure in the Zagros"; and Bozzolato, Giovanni. "Ricordi iraniani con l'Agip, anni '60." Both available on the Associazione Pionieri e Veterani Eni (APVE) website: www.pionierieni.it/wp/, accessed September 18, 2016.

45 Pozzi. *Dai gatti selvaggi*, p. 422.

46 Torelli, Milziade. "27.000 chilometri sulla via italiana del petrolio." *Il gatto selvatico*, no. 12, 1959, pp. 4–5; Bertoli, Ubaldo. "Seguendo una squadra dell'Agip Mineraria (sulla squadra sismica, il topografo e il delmag)." *Il gatto selvatico*, no. 6, June 1957, pp. 7–9. An article completely articulated through military metaphors.

47 Franchino recounts how the gendarmes helped them in different situations. Franchino, Aristide. "L'esplorazione in Iran, monti Zagros." Associazione Pionieri e Veterani Eni, December 15, 2009, www.pionierieni.it/wp/?p=929. Accessed September 18, 2016. This is an expanded version of the memories published in *Industria Mineraria*.

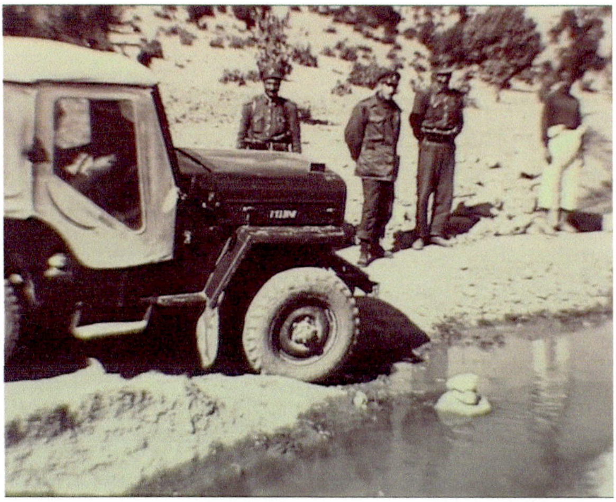

Fig. 22.11. Iranian police. From *23. Momenti della spedizione dell'Agip Mineraria negli Zagros (Moments in the Agip Mineraria Expedition in the Zagros Mountains)*.

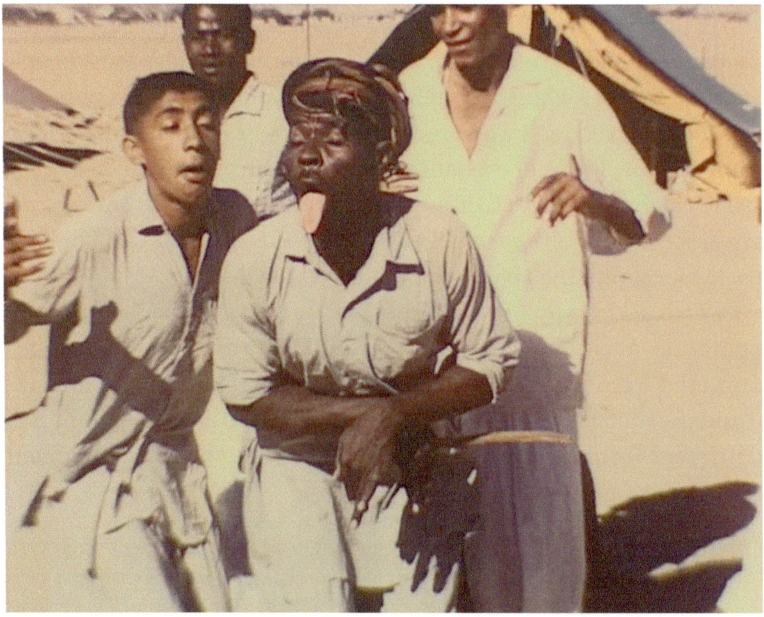

Fig. 22.12. Dance. From *20. Danze e svaghi di baktiari in un campo. Aerei dell'Agip Mineraria sulla pista nei pressi di Jask (Dance and Recreation in a Bakhtiari Camp. Agip Mineraria's Airplanes on a Landing Field Close to Jask)*.

officials as well (fig. 22.11). Dance is also a recurrent topic, such as in 2. *Festeggiamenti per le nozze di una coppia di beluchi. Arrivo dell'ambasciatore italiano in Iran (Wedding Party for a Beluchi Couple. Arrival of the Italian Ambassador in Iran)*. In some cases it has the form of a very ritualistic and complex tone, with imitations of animals and sexualized movements, as in *20. Danze e svaghi di baktiari in un campo. Aerei dell'Agip Mineraria sulla pista nei pressi di Jask* (*Dance and Recreation in a Bakhtiari Camp. Agip Mineraria's Airplanes on a Landing Field Close to Jask*) (fig. 22.12).[48] In these kinds of scenes and many of the scenes that involve local people, there is a constant dynamic of very long shots and close-ups, including people directly looking into the camera in a way typical of home movies and often linked to notions of "authenticity."[49] Often, it is unclear whether or not if the people who are filmed have granted their consent. In *21. Avanzamento della spedizione dalla valle Milak Haimini alla catena del Kuh i Jar, tra difficoltà e attraversamento di villaggi locali* (*The Expedition Passes through the Milak Haimini Valley to the Kuh i Jar Range, Facing Difficulties and Going through Local Villages*) we see children looking at the camera with curiosity and then escaping, while women who are presumably carrying water are not even aware of the camera. Some of these close-ups are staged, a default feature in home movies.[50] Particular physical traits are emphasized, like a big moustache, and the encounter of these filmmakers with local people is never just one of looking at: often, the ENI employers are the observed besides the observers. ENI workers even mimic people dancing, or perform more predictable forms of exchange, such as how to deal with local food, or how to catch or preserve fish, as in *10. Scene di pesca e di lavorazione del pesce a Jask* (*Scene of Fishing and Processing Fish at Jask*). In Homi Bhabha's terms, the locals are the object of both "desire and derision."[51]

It is worth asking what ENI workers knew about the people they were encountering. According to pioneer Franchino, the different populations encountered were "Baktiari, Boyrakma, Teiebi, Bahamei, Dosmanziari, Zilui and the nomads Qashaqai." He also notes how the local population was different from that of the other areas of Iran.[52] A long document

48 While this does not look like a ritual, it is worth pointing to the tradition of dancing and rituals in ethnographic films, recounted among others by Catherine Russell, where she analyses, among others, films by Jean Rouch and Maya Deren. Russell, Catherine. *Experimental Ethnography: The Work of Film in the Age of Video*. Duke University Press, 1999, pp. 193–237.

49 "The direct look into the camera of mothers, children and fathers, their naivety and spontaneity in front of the cameras instill a regime of belief in contemporary viewers." Simoni, Paolo. "Eyewitnesses of History: Italian Amateur Cinema as Cultural Heritage and Source for Audiovisual and Media Production." *VIEW Journal of European Television History and Culture*, vol. 4, no. 8, 2015, p. 2. doi:10.18146/2213-0969.2015.jethc092.

50 Schneider. "*Autosonntag* (Switzerland, 1930)."

51 Bhabha, Homi. *The Location of Culture*. Routledge, 1994, p. 60.

52 Franchino, Aristide. "I ricordi di un geologo in Iran." *Industria Mineraria*, no. 5–6, 1998, p. 24, note p. 34 (I left the names as he wrote them); and idem. "Ricordi di un geologo. MEKRAN (Iran sud-orientale) 1958 – Viaggio e rilevamento geologico."

written by a doctor (not an ENI employer) who worked in Iran for a long time, Vincenzo Bianchini,[53] contains instructions on how to dress, what to eat, what to touch in the deserts. He recommends people who just got there to "Avere un inizio all'orientale" ("To start off as an Oriental type"), that is, spending a lot of time sitting down, drinking tea, smoking as little as possible and avoid drinking alcohol. In this way, it is possible to understand the surrounding environment "non solo per valutarlo dal punto di vista geologico ma cercando di sgorgere i suoi valori estetico musicali" ("not only to evaluate it from a geological point of view but also to catch its aesthetic, musical values").[54] Bianchini also recommends treating local people

> *umananmente ma nessuna confidenza.* [...] *Il concetto deve essere: Lavorate ubbidite, sarete fedeli rispettosi, un salario generoso ma basta, il di più sarà ritenuto debolezza e l'orientale è abilissimo nel profittare e porre infinite e sottili difficoltà.*

> humanly but without getting too intimate. [...] The concept should be: work, obey, be faithful, [and pay] a generous salary, but that's it. Something more will be considered a form of weakness and the Oriental people are very crafty at exploiting this and make everything more difficult.

The relationship with the Iranians is definitely a very complex one. One of the larger goals of ENI was precisely to cooperate equally with the people of the countries where they operated, in order to win their trust and be able to establish long-term relationships – and the official statements of the company, old and recent, films included, emphasized this aspect.[55] At

53 The document is preserved in the ENI archive in one of the albums donated by the pioneers, and it is not an official document of the company or presents as an official position of the company concerning the encounters with local populations. It can be taken as an example of the discourse that such encounters produced in Italy at the time.

54 The report is in one of the albums (number 8) that the ENI pioneers donated to ASE; see a folder with correspondence between Vincenzo Bianchini and Enrico Mattei regarding a possible collaboration in Iran. ASE, Presidenza/Estero, b. 50, f. 1743. Bianchini's name is not in the database of people employed by ENI.

55 See Nardi, Lucia. "La 'Diplomazia Parallela' dell'Eni e il ruolo degli 'uomini di Mattei' nei paesi dell'Africa del Nord." *Enrico Mattei e l'Algeria durante la Guerra di Liberazione Nazionale*. ENI, 2010, pp. 40–46. Among other films, see *Oro nero sul Mar Rosso* (*Black Gold on the Red Sea*, Vittorio Gallo, 1962), filmed in Egypt, and *Questioni d'oggi: il Marocco* (*Today's Questions: Morocco*, Franco Dodi, 1960).

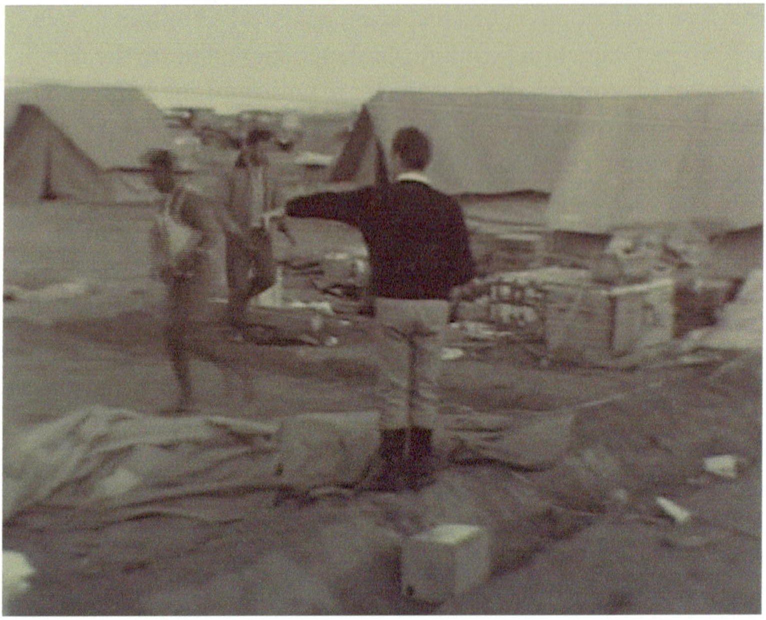

Fig. 22.13. Orders. From *6. Riprese di danze di beluchi, adulti e bambini e delle attività di un campo in disallestimento* (*Dance of the Baloch People, Adults and Children, and Dismantling a Camp*).

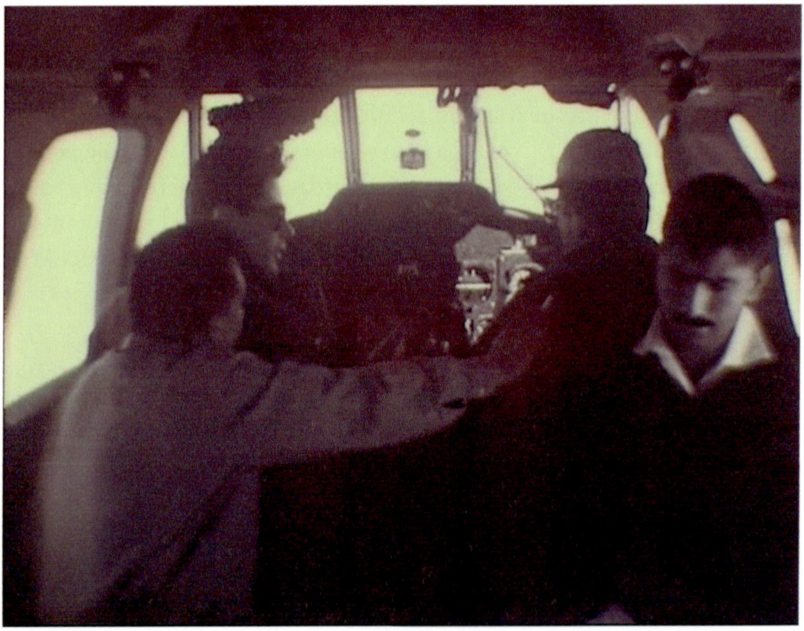

Fig. 22.14. Inside a helicopter. From *26. Riprese aeree di un vasto territorio dell'Iran, in particolare dell'area del Mekran* (*Aerial View of a Vast Territory in Iran, the Area of Makran, in Particular*).

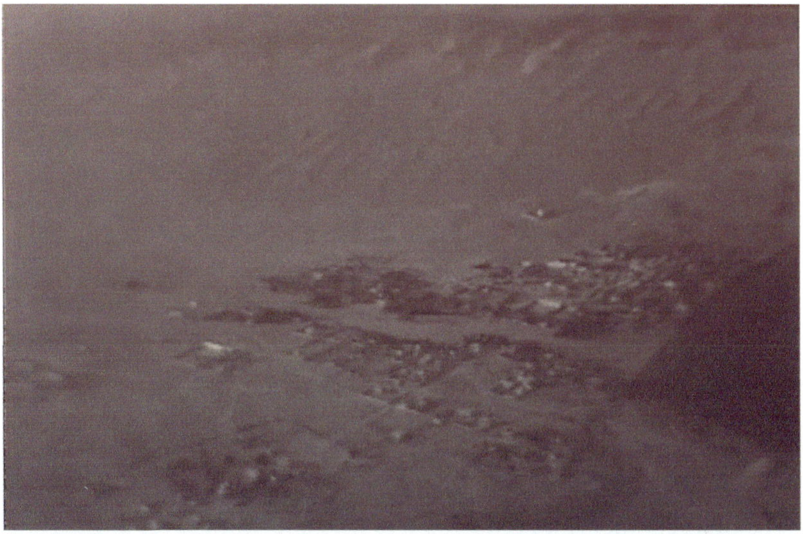

Fig. 22.15. View from above. From *26. Riprese aeree di un vasto territorio dell'Iran, in particolare dell'area del Mekran* (*Aerial View of a Vast Territory in Iran, the Area of Makran, in Particular*).

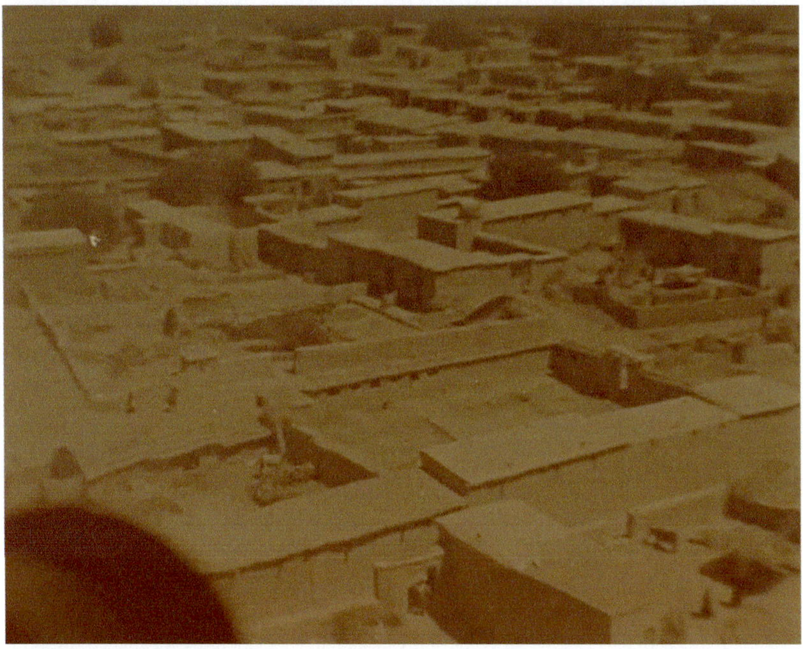

Fig. 22.16. A city from above. From *17. Documentazione filmica che mostra riprese aeree del territorio degli Zagros, dei villaggi e il campo base della spedizione* (*Film Documentation Showing Aerial Shots of the Zagros Territory, the Villages and the Base Camp of the Expedition*).

the same time, there are clearly hierarchical relationships between ENI experts and their local helpers that emerge strongly in these films. For example, in *6. Riprese di danze di beluchi, adulti e bambini e delle attività di un campo in disallestimento* (*Dance of the Baloch People, Adults and Children, and Dismantling a Camp*), where an (Italian) man is showed vigorously giving orders to another (Iranian) man (fig. 22.13), and in *9. Lo sbarco di materiali, automezzi e rifornimenti al campo di Jask* (*Unloading Materials, Vehicles and Supplies at the Jask Camp*) we see a number of Baloch men disembarking from a large barge carrying big wooden boxes while Italians supervise them and give orders. The presence of these moments has to do with the nature of these films as well: in an edited, carefully controlled film such an explicit example of ordering more than of cooperation would probably not have been shown. However, we also see Iranians employed in specialized jobs, such as pilot in *20. Danze e svaghi*, or as cartographers and topographers who appear in several films and were named and remembered by Franchino and other ENI workers. As in many holiday films, in these films as well, the experience of mobility is a recurrent feature, rendered through airplanes, trucks of different kinds, boats, a large barge, horses and donkeys, helicopters.[56] We see them and we see *from* them: in particular, views from above are recurrent then in these films (figs. 22.14, 22.15 and 22.16). As Siegfried Kracauer says, the aerial views typical of mid-century modernity "bring patterns and configurations into focus."[57] The helicopter becomes then the symbol of what is new and modern. In *17. Documentazione filmica che mostra riprese aeree del territorio degli Zagros, dei villaggi e il campo base della spedizione* (*Film Documentation Showing Aerial Shots of the Zagros Territory, the Villages and the Base Camp of the Expedition*) we see how the arrival of the helicopter caused a big runaway of animals, probably scared by the noise – which we cannot hear, given that the film is silent.

To conclude, I want to return to the ambiguous status of these films. If their role in the business strategy of the company remains unclear, it

56 "Holiday film is not only a record of what is seen, but also an impulse to capture the experience of mobility." Neumann, Mark. "Amateur Film, Automobility and the Cinematic Aesthetics of Leisure." *Amateur Filmmaking: The Home Movie, the Archive, the Web*, edited by Laura Rascaroli, Gwenda Young with Barry Monahan. Bloomsbury Academic, 2014, pp. 51–64. Means of transportations that are listed in details in several official documents, such as "Proposta" in ASE, Estero, b.21, FF5.

57 Aerial photographs, he wrote, "are bound to bring normally unseen patterns and configurations into focus. This explains their potential revealing power." Kracauer, Siegfried. *History: The Last Things before the Last*. Markus Wiener Publishers, 1995, p. 42.

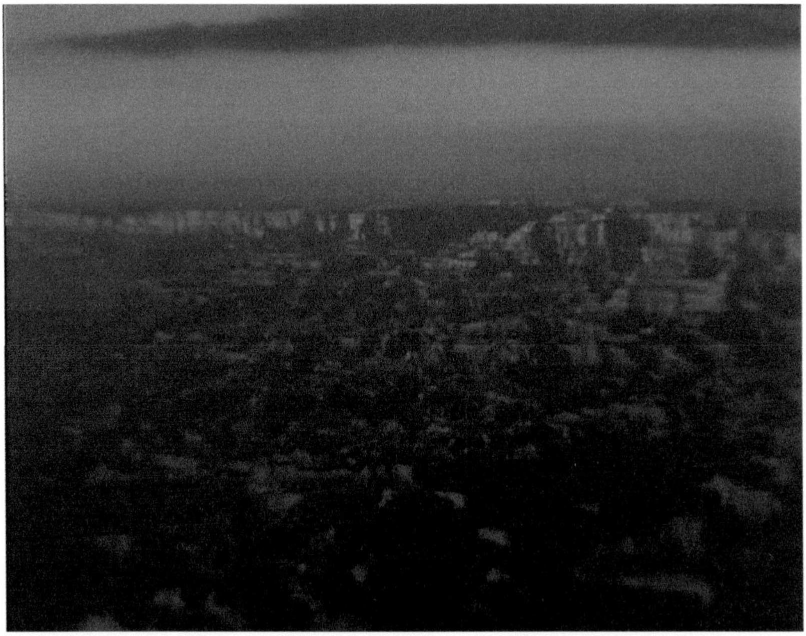

Fig. 22.17. Land. From *26. Riprese aeree di un vasto territorio dell'Iran, in particolare dell'area del Mekran* (*Aerial View of a Vast Territory in Iran, the Area of Makran, in Particular*).

is safe to assure that for the workers of that company they function like tourist photography, that in Susan Sontag's terms "help people to take possession of space in which they are insecure."[58] In this they work together with one of the main working goals of the expeditions, that is, mapping lands previously unmapped – at least, not by Westerners. Filming them is one way to gain a sense of control; mapping is another one,[59] and in the films we see a number of cartographers working on maps, while in the literature of the time this is widely emphasized.[60] And it is precisely in the depiction of the landscapes we see the most clear mixing of naturalist, ethnographic and travel interests with working goals: *24. Riprese del territorio del Mekran e di alcune attività di rilevamento dei geologi* (*Filming of the Makran Territory and of Some Survey Activities of the Geologists*), *11. Immagini di manifestazioni gassose in località Bahikistan* (*Images of*

58 Sontag, Susan. *On Photography.* Farrar, Straus & Giroux, 1977, p. 9.

59 Spadini. "Visita ai nostri cantieri." Nonetheless, it has been impossible to identify quite a few locations, for example, lake "Dilibou" of the river "Rud i Kirsan." On the problems of identifying places, see Cortini, Letizia. "Inventario foto-cinematografico."

60 Spadini. "Visita ai nostri cantieri in Iran," p. 7.

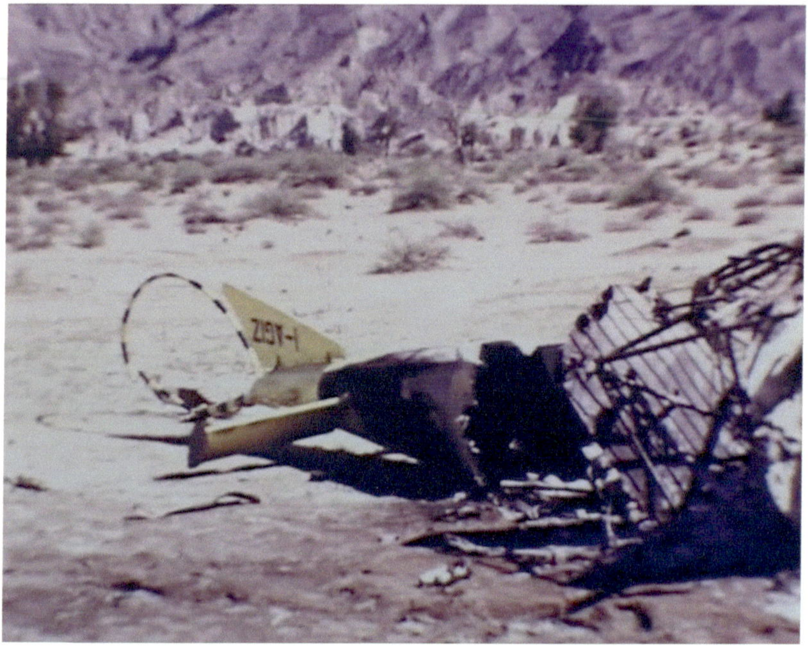

Fig. 22.18. A broken helicopter. From *38. Mezzo di trasporto in movimento, locali al lavoro* (*Means of Transport in Movement, Locals at Work*).

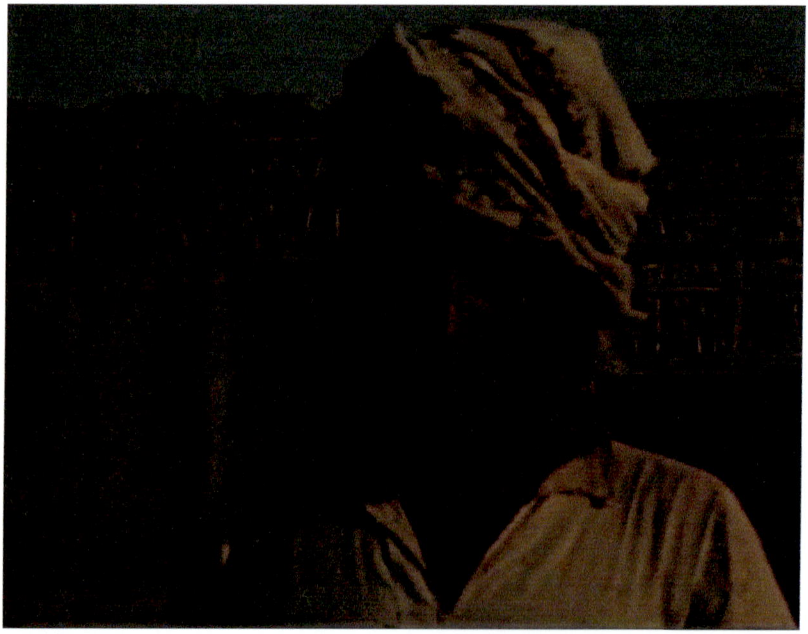

Fig. 22.19. Close up. From *1. Scenette locali villaggi zona Goxar* (*Local Vignettes from Towns in the Goxar Area*).

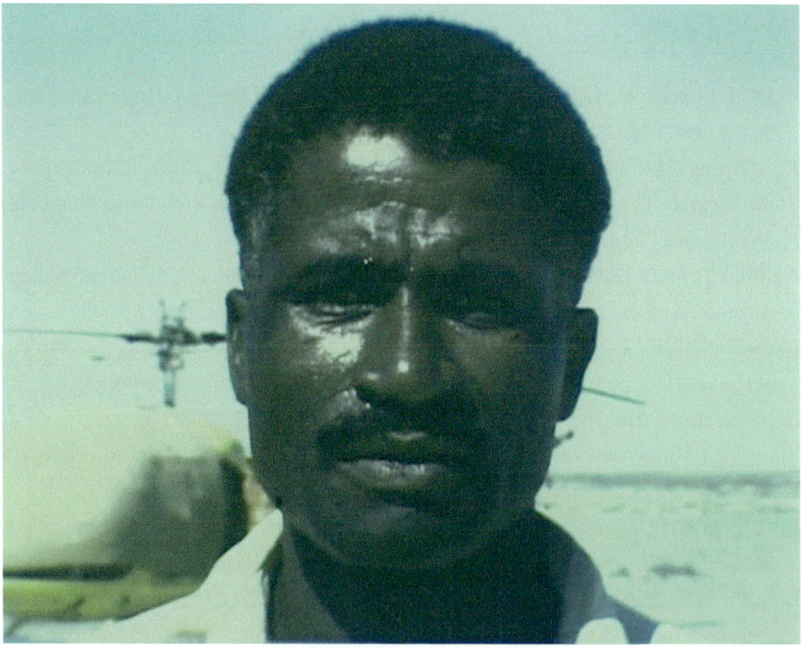

Fig. 22.20. Close up. From *37. Immagini varie da una barca in navigazione, scarico pacchi in spiaggia e atterraggio dell'elicottero al campo base lungo la costa del Mekran* (*Various Images from a Sailing Boat, Unloading Boxes on the Beach and a Helicopter Landing at the Base Camp along the Makran Coast*).

Gaseous Manifestations in Bahikistan), and 12. *Riprese del paesaggio roccioso e desertico da un veivolo che sorvola una vasta area del Mekran* (*Shots of the Rocky Desert Landscape from a Vehicle Flying over a Large Area of the Makran*) (fig. 22.17). One of the main feature of industrial films, their role as means for governing and controlling, is here almost completely missing.[61] These films are literally out of control – they have not been controlled by cinematic means (editing) or company ones. At the same time, filming these unknown landscapes surely helped the workers to gain a sense of control, while their status of home movies is also ambivalent, their filmmakers were amateurs as filmmakers but they were working in a business and not a leisure setting.

61 "As a corporate communication tool the industrial film was particularly efficient at media events that could be governed and controlled." Zimmermann. "'What Hollywood Is to America, the Corporate Film Is to Switzerland'," p. 112.

Works Cited

Acland, Charles R., and Haidee Wasson, editors. *Useful Cinema*. Duke University Press, 2011.

Agip Mineraria. "Attività in Iran. Programma dei lavori e & Preventivo di spese, Anno 1959, Tehran, 27 Novembre 1958." Archivio Storico ENI, Rome/Pomezia, Presidenza/Estera, b. 21, f. FF6.

Anelli, L., and G. Dattilo. "Note sul rilevamento da Chahbahar a Zahedan." February 1961. Archivio Storico ENI, Rome/Pomezia, Album 8, no archival collocation.

Bertoli, Ubaldo. "Seguendo una squadra dell'Agip Mineraria (sulla squadra sismica, il topografo e il delmag)." *Il gatto selvatico*, no. 6, June 1957, pp. 7–9.

Bhabha, Homi. *The Location of Culture*. Routledge, 1994.

Bianchini, Vincenzo. Report. Archivio Storico ENI, Rome/Pomezia, Album 8, no archival collocation.

Bini, Elisabetta. *La potente benzina italiana. Guerra fredda e consumi di massa tra Italia, Stati Uniti e Terzo Mondo (1945–1973)*. Carocci Editore, 2013.

Bolognesi, Kitti, and Giovanna Calvenzi, editors. *Federico Patellani: professione fotoreporter*. Silvana editoriale/Museo fotografia contemporanea, 2015.

Bonifazio, Paola. *Schooling in Modernity: The Politics of Sponsored Films in Postwar Italy*. University of Toronto Press, 2014.

Bonifazio, Paola. "United We Drill: ENI, Films, and the Culture of Work." *Annali d'Italianistica: From Otium and Occupatio to Work and Labor in Italian Culture*, vol. 32, 2014, pp. 329–50.

Bozzolato, Giovanni. "Ricordi iraniani con l'Agip, anni '60." Associazione Pionieri e Veterani Eni, www.pionierieni.it/wp/.

Carey, Jane, and Andrew Carey. "Nelle difficili terre dell'Iran i nostri lavorano con entusiasmo." *Il gatto selvatico*, no. 9, 1959, pp. 11–14.

Ciucci, Carlo. "'*Affermazione dell'Eni alla mostra di Teheran*' (padiglione dell'Eni alla mostra di ottobre), '*Italia produce*' a Teheran, in Iran." *Il gatto selvatico*, no. 10, 1958, pp. 16–17.

Correspondence between Vincenzo Bianchini and Enrico Mattei. Archivio Storico ENI, Rome/Pomezia, Presidenza/Estero, b. 50, f. 1743.

Cortini, Letizia. "Inventario foto-cinematografico di viaggi in Iran per l'Archivio Storico Eni." Archivio Storico ENI, Rome/Pomezia, Roma 2011, no archival collocation.

Cortini, Letizia. "Visioni dalla Storia. Le fonti audiovisive nella didattica – Audiovisual Sources for Teaching." *visioni dalla storia*, https://visionandonellastoria.net/about/.

Costa, Enrico. *Filmare in 8mm*. Hoepli, 1963.

Elsaesser, Thomas. "Archives and Archaeologies: The Place of Non-fiction Film in Contemporary Media." *Films That Work: Industrial Film and the Productivity of*

Media, edited by Vinzenz Hediger and Patrick Vonderau. Amsterdam University Press, 2009, pp. 19–34.

Franchino, Aristide. "I ricordi di un geologo in Iran." *Industria Mineraria*, no. 5–6, 1998.

Franchino, Aristide. "L'esplorazione in Iran, monti Zagros." Associazione Pionieri e Veterani Eni, December 15, 2009, www.pionierieni.it/wp/?p=929. Accessed September 18, 2016.

Franchino, Aristide. "Ricordi di un 'geologo' in Iran 1958–1959: Adventure in the Zagros." Associazione Pionieri e Veterani Eni, www.pionierieni.it/wp/. Accessed 18 September 2016.

Franchino, Aristide. "Ricordi di un geologo. MEKRAN (Iran sud-orientale) 1958 – Viaggio e rilevamento geologico." Associazione Pionieri e Veterani Eni, February 2010, www.pionierieni.it/wp/. Accessed September 18, 2016.

Frankel, Paul H. Note written to Mattei in February 1958. Archivio Storico ENI, Rome/Pomezia, Presidenza/Estera, b. 50, f. 1743.

Frescani, Elio. *Il cane a sei zampe sullo schermo. La produzione cinematografica dell'Eni di Enrico Mattei*. Liguori Editore, 2014.

Guidi, Francesco. "In Iran il Gatto Selvatico arriva nel 1962." *NIA*, no. 107, April 1995.

Hediger, Vinzenz, and Patrick Vonderau. "Record, Rhetoric, Rationalization: Industrial Organization and Film." *Films That Work: Industrial Film and the Productivity of Media*, edited by Vinzenz Hediger and Patrick Vonderau. Amsterdam University Press, 2009, pp. 35–50.

Hediger, Vinzenz, and Patrick Vonderau, editors. *Films That Work: Industrial Film and the Productivity of Media*. Amsterdam University Press, 2009.

Heymer, Anna, and Patrick Vonderau. "Industrial Films: An Analytical Bibliography." *Films That Work: Industrial Film and the Productivity of Media*, edited by Vinzenz Hediger and Patrick Vonderau. Amsterdam University Press, 2009, pp. 405–61.

"Intervista a Benito Gasparotto." Summer 2016. Archivio Storico ENI, Rome/Pomezia, no archival collocation.

IRAN: Programmi di lavoro e relativa organizzazione. Archivio Storico ENI, Rome/Pomezia, Estero, box (b.) 21, folder (f.) FF5 and FF6.

Kinzer, Stephen. *All the Shah's Men: An American Coup and the Roots of Middle East Terror*. John Wiley & Sons, 2003.

Kittler, Friedrich. *Discourse Networks 1800/1900*. Stanford University Press, 1990.

Kittler, Friedrich. *Gramophone, Film, Typewriter*. Stanford University Press, 1999.

Kracauer, Siegfried. *History: The Last Things before the Last*. 1969. Markus Wiener Publishers, 1995.

Latini, Giulio. *L'energia e lo sguardo. Il cinema dell'Eni e i documentari di Gilbert Bovay*. Donzelli, 2011.

Mattei, Enrico. *Scritti e Discorsi, 1945–1962*. Rizzoli, 2012.

Mixed photographs. Archivio Storico ENI, Rome/Pomezia, Album 265, no archival collocation.

Muzzin, Adriano. "Prefazione. Per salvare la memoria comune." *La memoria raccontata. Storie e ricordi di uomini e donne di eni, Documenti dall'Archivio Storico*, vol. 3. ENI, 2013, pp. 7–8.

Nardi, Lucia. "La 'Diplomazia Parallela' dell'Eni e il ruolo degli 'uomini di Mattei' nei paesi dell'Africa del Nord." *Enrico Mattei e l'Algeria durante la Guerra di Liberazione Nazionale*. ENI, 2010, pp. 40–46.

National Iranian Oil Company. "Fiftieth Anniversary Celebrations of the Iranian Oil Industry, March 1958." Archivio Storico ENI, Rome/Pomezia, no archival collocation.

Neumann, Mark. "Amateur Film, Automobility and the Cinematic Aesthetics of Leisure." *Amateur Filmmaking: The Home Movie, the Archive, the Web*, edited by Laura Rascaroli, Gwenda Young, with Barry Monahan. Bloomsbury Academic, 2014, pp. 51–64.

Odin, Roger. "The Home Movie and Space of Communication." *Amateur Filmmaking: The Home Movie, the Archive, the Web*, edited by Laura Rascaroli, Gwenda Young, with Barry Monahan. Bloomsbury Academic, 2014, pp. 15–26.

Odin, Roger. "Rhétorique du film de famille." *Rhétoriques, sémiotiques. Revue d'Esthétique*. Union Générale d'Editions, 1979, pp. 340–73.

Orgeron, Devin, Orgeron, Marsha, and Dan Streible, editors. *Learning with the Lights Off: Educational Film in the United States*. Oxford University Press, 2012.

Peretti, Luca. "Before *The Battle of Algiers*: Sartre, Colonialism, Industrial Cinema, and Unmade Film." *Senses of Cinema*, no. 84, September 2017, http://sensesofcinema.com/2017/sartre-at-the-movies/sartre-colonialism-industrial-cinema-and-an-unmade-film/.

Pozzi, Daniele. *Dai gatti selvaggi al cane a sei zampe. Tecnologia, conoscenza e organizzazione nell'Agip e nell'Eni di Enrico Mattei*. Marsilio, 2009.

"Proposal for a Work Plan in Iran for 1957." Archivio Storico ENI, Rome/Pomezia, Estero, box (b.) 21, folder (f.) FF5 and FF6.

"Proposta." Archivio Storico ENI, Rome/Pomezia, Estero, b.21, FF5.

Russell, Catherine. *Experimental Ethnography: The Work of Film in the Age of Video*. Duke University Press, 1999.

Salvatori, Sandro. "Faticosa e affascinante avventura dei petrolieri italiani in Persia." *Il Tempo*, October 1958.

"Schema dei collegamenti Radio." Archivio Storico ENI, Rome/Pomezia, Superfondo Eni, Fondo Estero Serie Assistente del Presidente per i rapporti con l'Estero (1953–1963), b. 21, f. FF3.

Schneider, Alexandra. "*Autosonntag* (Switzerland, 1930) – A Film Safari in Swiss Alps." *Visual Anthropology*, vol. 15, 2002, pp. 115–28.

Simoni, Paolo. "Eyewitnesses of History: Italian Amateur Cinema as Cultural Heritage and Source for Audiovisual and Media Production." *VIEW Journal of*

European Television History and Culture, vol. 4, no. 8, 2015. doi:10.18146/2213-0969.2015.jethc092.

Sontag, Susan. *On Photography*. Farrar, Straus & Giroux, 1977.

Spadini, Raffaele. "Visita ai nostri cantieri in Iran, Con i geologi della Mineraria sulle vette e fra le gole dell'Iran (monti Zagros)." *Il gatto selvatico*, no. 4, 1964, pp. 7–9.

Staab, Joachim Gerhard. *Coraggio, filmate! La regia nel film amatoriale e professionale*. Translated by L. Moggio. Edizioni del progresso fotografico, [1963?]. Originally published as *Filmregie für Leinwand und Bildschirm*. Heering-Verlag, 1962.

Szczepanik, Petr. "Modernism, Industry, Film: A Network of Media in the Bat'a Corporation and the Town of Zlín." *Films That Work: Industrial Film and the Productivity of Media*, edited by Vinzenz Hediger and Patrick Vonderau. Amsterdam University Press, 2009, pp. 349–76

Tepperman, Charles. "Mechanical Craftsmanship. Amateurs Making Practical Films." *Useful Cinema*, edited by Charles R. Acland and Haidee Wasson. Duke University Press, 2011, pp. 289–314.

Torelli, Milziade. "27.000 chilometri sulla via italiana del petrolio." *Il Gatto Selvatico*, no. 12, 1959, pp. 4–7.

Tremolada, Ilaria. *La via Italiana al petrolio*. Ornitorinco Edizioni, 2011.

Varvelli, Riccardo. "Il campo di Bahrgan Sar. Le prime attività Agip in Iran (2nda parte)." *NIA*, 1999.

Work Plan. Archivio Storico ENI, Rome/Pomezia, Estero, b. 21, f. FF3.

Zimmermann, Patricia R. *Reel Families: A Social History of Amateur Film*. Indiana University Press, 1995.

Zimmermann, Yvonne. "'What Hollywood Is to America, the Corporate Film Is to Switzerland': Remarks on Industrial Film as Utility Film." *Films That Work: Industrial Film and the Productivity of Media*, edited by Vinzenz Hediger and Patrick Vonderau. Amsterdam University Press, 2009, pp. 101–17.

About the Author

Luca Peretti is a British Academy Postdoctoral at the University of Warwick. He co-edited volumes on terrorism and cinema (Postmedia books), Pier Pasolini Pasolini (Bloomsbury Academics), and on Italian cinema and Algeria (AAMOD, 2022). His work has appeared in, among others, *Senses of Cinema, The Italianist: Film Issue, Journal of Italian Cinema and Media Studies, Historical Materialism, Comunicazioni Sociali, Quest. Issues in Contemporary Jewish History*. He is on the editorial board of *Zapruder World, Cinema e Storia,* and *L'Avventura*. He wrote and co-produced the film *Mister Wonderland* (dir. Valerio Ciriaci, 2019). He collaborates with newspapers and magazines.

23 Industrial Film from the Home Studio

Amateur Cinema and Low-Budget Corporate Moving Image Culture in West Germany (1950 to 1977)

Alexander Stark

Abstract
This chapter examines the case of the female West German filmmaker Elisabeth Wilms (1905–1981), who started out as a film amateur and became a successful producer of sponsored films. By exploring her methods in industrial film production, the chapter shifts the focus away from high-value industrial films and their well-known production companies to the lower financial end of film production. The author aims to bring to the fore the historic intersections and interconnections between amateur and industrial filmmaking which have long been overlooked by film studies and exemplifies how filmmaker and client were often amateurs in the film business – a fact that has gained little attention in previous research, but often had a huge impact on the film production process.

Keywords: amateur film; industrial film; low-budget film; useful cinema; female filmmaker; Germany

For god's sake, please don't make a sensation in the press! I regard my work, and the blessings it lies under, as a gift from God, for which I am really thankful. By the way, I am first and foremost the wife of a master baker, and the time that I put into my films is just the amount that I can spare from my duties in the household and shop.
– Elisabeth Wilms[1]

1 My translation. The original quote reads as follows: "Machen Sie doch um Gotteswillen in der Presse kein Aufsehen! Ich sehe in meiner Arbeit und auch in dem Segen, der über dieser Arbeit liegt, ein Geschenk unseres Herrgotts, für das ich ihm zutiefst dankbar bin. Im übrigen bin ich in erster Linie die Frau eines Bäckermeisters, und nur soviel, wie ich an Zeit neben meinen Pflichten im Haushalt und im Ladengeschäft erübrigen kann, gehört meiner Filmarbeit." See

Hediger, V., F. Hoof, Y. Zimmermann, with S. Anthony (eds), *Films That Work Harder: The Circulation of Industrial Film*. Taylor & Francis Group 2024
DOI: 10.5117/9789462986534_CH23

This is how, in 1959, the magazine *Die Frau in der Bäckerei* (The woman in the bakery) quoted a woman named Elisabeth Wilms (1905–1981) from the city of Dortmund in West Germany's Ruhr area. For this niche publication, which focused on a readership consisting of female bakery personnel, Wilms was of interest because she was not only a baker's wife herself and ran a bakery with a grocery store together with her husband full time. On top of that, she worked as a successful producer of sponsored films. Her body of work not only included clerical films and image productions for Dortmund's city administration, but also industrial films. And while Wilms presented this filmic work as a form of divinely blessed leisure activity in the magazine, the vast number of documents she left behind tells another story: as her tax records show, for example, she made an annual profit of DM 13,599.18 (roughly €31,000 today) with this alleged leisure activity in the year the article was published. In other years, her profits were more than twice this amount, and the filmmaker worked on five to ten commissioned films. Her oeuvre grew to around one hundred films by the time of her death in 1981, and includes a wide range of films for a great variety of purposes and audiences. But on closer inspection it was the production of industrial films for various customers that formed the backbone of her film business for many years.

Wilms remained a freelancer throughout her whole life. In the space of twenty-seven years, she produced twenty-eight industrial films for a total of fifteen different clients. The list includes big German industrial corporations like the Rheinstahl AG, regional actors like the electricity supplier Vereinigte Elektrizitätswerke Westfalen AG, and small, local customers like Dortmund's municipal utility company Dortmunder Stadtwerke AG. The fact that these clients, regardless of their scope of action, were mostly based in Dortmund and its surrounds means that the range of subjects of Wilms' industrial films was largely oriented towards many of the industrial activities of the Ruhr area at the time: steel production and processing, mechanical engineering, electricity and water supply. It seems that the filmmaker worked with all types of industrial film in the course of her commercial film career. Her estate includes process films, image films, sales representatives' films, trade fair and exhibition films, educational and current affairs films.[2]

By exploring Elisabeth Wilms' work methods in industrial film production and examining the particularities of her case and her films, this chapter

"Frau Elisabeth Wilms, Dortmunds filmende Bäckersfrau." *Die Frau in der Bäckerei*, vol. 7, no. 3, 1959, p. 5.

2 Typology taken from Zimmermann, Yvonne. "Industriefilme." *Schaufenster Schweiz: Dokumentarische Gebrauchsfilme 1896–1964*, edited by idem. Limmat, 2011, pp. 246–47.

shifts the focus away from high-value industrial films and their well-known production companies to the lower financial end of film production. In doing so, it aims to bring to the fore the interconnections between amateur and industrial filmmaking, to which attention has recently been drawn by scholars such as Charles Tepperman, Brian R. Jacobson and Yvonne Zimmermann.[3] Their studies have shown that historically there have been various overlaps and intersections between the two fields and in doing so, they have underlined that amateur filmmaking was not limited to a mere leisure activity. In studying Wilms' case, it is possible to further pursue this research direction. Corporate and other archives are filled with low-budget industrial films that often have a rather amateurish appearance. But in the most cases they lack further contextual documentation due to their minor importance or local scale within the corporation. Wilms' case makes it possible to shed some light on this gap. Moreover, it exemplifies how both filmmaker and client were often amateurs in the film business – a fact that has gained little attention in previous research, but often had a huge impact on the film production process.

Picking up on Charles Tepperman, Elisabeth Wilms can be regarded as a "serious" or "advanced amateur."[4] Indeed, the documents show that the filmmaker proactively used the amateur label. I will argue that in her efforts both to work as a one-woman company and to assert herself as an inexpensive amateur, she was able to cover a low-budget niche in industrial filmmaking that was not profitable for larger production companies. As a woman, Elisabeth Wilms became engaged with her activity first in a clearly male hobby and later in a professional field that was a male domain as well. However, it is difficult to trace the gender aspect, which must have played a role in this daily work, because it doesn't come up explicitly in the written documents handed down. The filmmaker herself also remained relatively vague in this respect. In a television interview in 1980 she said:

> It was kind of unusual that I was a filmmaker as a woman. Most of the time it was men who made films. And they also wanted to shape me and

3 See Tepperman, Charles. *Amateur Cinema: The Rise of North American Moviemaking, 1923–1960*. University of California Press, 2015, pp. 222–27; and Jacobson, Brian R. "The Boss's Film: Expert Amateurs and Industrial Culture." *Amateur Movie Making: Aesthetics of the Everyday in New England Film, 1915–1960*, edited by Marta J. McNamara and Karan Sheldon. Indiana University Press, 2017, pp. 198–218; as well as Zimmermann, Yvonne. "Amateur Film in the Factory: Forms and Functions of Amateur Cinema in Corporate Media Culture." *Global Perspectives on Amateur Film Histories and Cultures*, edited by Masha Salazkina and Enrique Fibla. Indiana University Press, 2021.

4 See Tepperman. *Amateur Cinema*, p. 6.

> force and encourage me to write scripts and think about what I filmed. But I never did that. I filmed intuitively and the films were quite good, they were appreciated. However, I noticed again and again that as a woman one didn't have the validity you had if you were a man. You felt it again and again.[5]

Unlike other industrial films, Wilms' works left almost no traces in the corporate archives of her clients, which is either because they ceased to exist due to mergers or liquidations or due to the fact that the institutions have discarded the relevant holdings at some point. This makes it difficult to retrace what Thomas Elsaesser called the *Medienverbund* that the films were originally part of.[6] Hence, this chapter will focus primarily on the perspective of the filmmaker, by mainly using documents from the extensive written and filmic estate of the "filming baker's wife," which is kept in the Stadtarchiv Dortmund and the LWL Medienarchiv in Münster.[7]

"The Filming Baker's Wife": The Amateur as a Business Label

Elisabeth Wilms never visited a film school and didn't get any professional film training. In fact, she started making films as an amateur and hobbyist in 1941. But due to some coincidences, her hobby had developed into a business on its own and made her a person of public interest to some degree within a few years. The first of these coincidences was that at the beginning of her filmic activity she had decided on the 16 mm film format (and remained true to it throughout her life). By her own admission, it wasn't possible to get any

5 Own translation of a transcript. The original quote reads as follows: "Es war irgendwie ungewöhnlich, dass ich als Frau Filmerin war. Meistens waren es die Männer, die Filme drehten. Und man wollte mich auch formen und mich dazu zwingen und anzuregen, Drehbücher zu schreiben, und zu überlegen, was ich filmte. Das hab ich aber nie gemacht. Ich habe intuitiv gefilmt und die Filme wurden ganz gut, wurden auch anerkannt. Allerdings, ich hab das immer wieder gemerkt, dass man doch als Frau nicht so die Geltung hatte, als wenn man ein Mann gewesen wäre. Man hat es immer wieder empfunden." See Klauß, Jürgen, and Michael Lentz. *Brot und Filme. Das große Hobby der Elisabeth Wilms*, Westdeutscher Rundfunk, December 5, 1981, b/w and colour, 60 min.

6 See Elsaesser, Thomas. "Archives and Archaeologies: The Place of Non-fiction Film in Contemporary Media." *Films That Work: Industrial Film and the Productivity of Media*, edited by Vinzenz Hediger and Patrick Vonderau. Amsterdam University Press, 2009, p. 22.

7 This chapter is based on findings which were part of my doctoral project and are laid out in more detail in my book *Die „filmende Bäckersfrau" Elisabeth Wilms – Amateurfilmpraktiken und Gebrauchsfilmkultur.* Schüren, 2023, pp. 191–254.

of the cheaper 8 mm equipment she initially wanted, due to the wartime circumstances. By that time, 16 mm was *the* standard for professional film work outside of commercial cinema and had been longer established and more widely distributed than 8 mm. Since copying to 35 mm without great loss of quality was possible since the late 1920s, it met the demands of both amateur and professional filmmakers, making it the format of choice for documentary, industrial and educational film production.[8] Nevertheless, these opportunities did not play a role during the first six years of Wilms' filmmaking, when she concentrated on creating documentary films about everyday life in her surroundings, which she screened in private or at the local chapter of the BDFA (Association of German Film Amateurs).

At the end of the Second World War Wilms hid her camera, films and raw material and subsequently kept on filming scenes of daily life – this time in the heavily bombed and destroyed city centre of Dortmund. In another coincidence, this came to the attention of the clerical German charity organization Hilfswerk der Evangelischen Kirchen in Deutschland (Relief Organization of the Protestant Churches in Germany) in 1947. At this time, the organization needed a fundraising film to support their work both nationally and internationally. Since the filmmaker was assessed as having "the right faith" and as being an amateur without too many political preconceptions, and possessing a good sense of style to boot, they asked her to produce first one and then a second documentary film about the – still quite difficult – living conditions in Wilms' hometown. An observant Christian, the baker's wife gladly agreed.[9]

The films were unable to fulfil their original purpose due to various delays, mainly during the duplication process. When prints were finally ready, the client felt that they presented an outdated picture of Germany and declined to distribute and circulate them.[10] But this fact was not subsequently made public by Wilms, who instead presented them as a great success in the following years, when first the local press, then larger

8 See Eisert, Gelia, and Gerhard Kemner. *Lebende Bilder: Eine Technikgeschichte des Films*. Nicolai, 2000, pp. 149–52; as well as Schneider, Alexandra. *"Die Stars sind wir." Heimkino als filmische Praxis*. Schüren, 2004, pp. 57–58; and Roepke, Martina. *Privat-Vorstellung. Heimkino in Deutschland vor 1945*. Olms, 2006, pp. 59–60.

9 See unpublished document by Pawlowski, Karl. "Vorschläge für die Werbung." November 4, 1945. Landeskirchliches Archiv der Evangelischen Kirche von Westfalen, Bielefeld, Stock 13.110, no. 95/2; as well as Schmidt, Heinrich. Letters to Hauptbüro des Evangelischen Hilfswerkes Westfalen. June 4, 1947, and June 18, 1947. Landeskirchliches Archiv der Evangelischen Kirche von Westfalen, Bielefeld, Stock 13.110, no. 95/2.

10 See Landeskirchliches Archiv der Evangelischen Kirche von Westfalen, Bielefeld, Stock 13.110, no. 95/2.

newspapers published reports about the two fundraising films made by "the filming baker's wife," as they nicknamed her early on. Indeed, it seems that the extensive press attention given to the filmmaker, particularly during the early 1950s, played an important role in her enduring role as a producer of sponsored films. Soon after the earliest reports, the first film orders were placed. In 1950, she began work on a documentary film for Dortmund's city administration about the reconstruction of the city centre, as well as working on her first industrial film, commissioned by the steel producer Westfalenhütte Dortmund AG, which featured the production of wire rod. Thereafter, more engagements with clients based in and around Dortmund followed. In the written estate of the filmmaker there are as many as seventy-one newspaper and magazine articles dating from the 1950s, all of which can be characterized as benevolent and essentially functioning as free advertisements. They praise her, for example, as "Germany's most sought-after small gauge filmmaker," her "one-woman company" as "the smallest and most useful film production in the world" and her films as "small genuine works of art."[11] The filmmaker herself was far from passive towards the media's interest in her person and work – indeed, she seems to have quickly discovered its commercial potential. She therefore soon began to play an active role in shaping the coverage on her by giving interviews, lending the journalists photos and older newspaper articles to write their stories, and promoting the alleged international success of her post-war fundraising films. Especially at the beginning of her career, Wilms could build on the promotional nimbus she had created.

The combination of this with the facts that she not only was a female filmmaker, but also came from the artisanal milieu made her a person of interest for the newspapers of the booming Federal Republic of Germany of the 1950s. The story of her self-made success "out of nowhere" fits well in the upbeat mood of the West German "economic miracle." In this context, it is also important to note that Wilms was a successful woman, but in the media presented herself as someone who only broke with the socially predominant classic gender role to a limited extent and therefore still operated within social conventions. Thus, while many articles emphasized how self-determined and independent she was in her "hobby," they also point out that despite her success, she still worked every day in the family business and did not "neglect her domestic duties."[12] In the course of this

11 See Damrow, Hildegard. "Produktion im Herrenzimmer." *Hamburger Abendblatt*, December 10, 1955.

12 "Binsenweisheit ... und Bäckerkünste." *Heute*, vol. 5, no. 114, 1950.

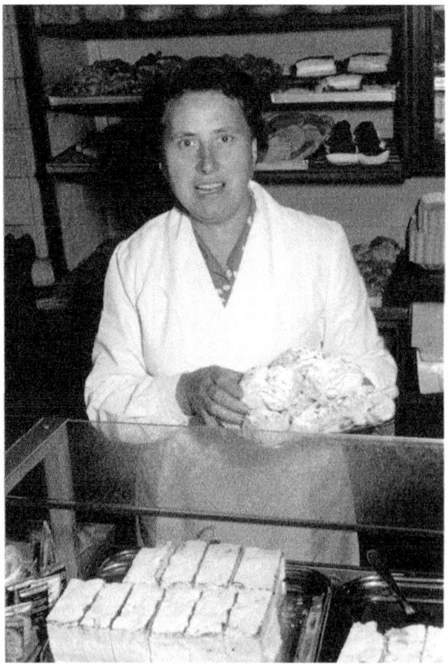

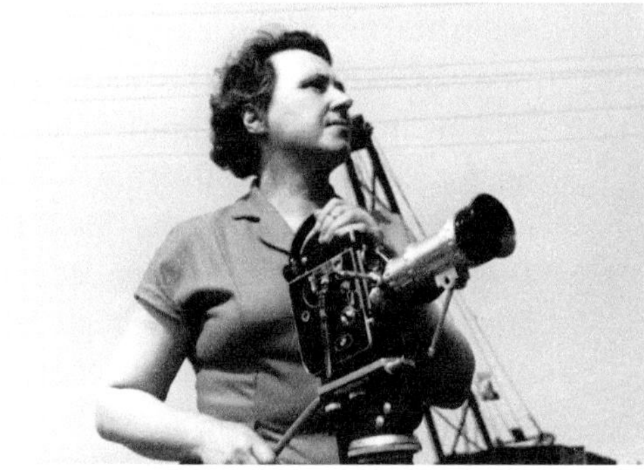

Figs. 23.1 and 23.2: Elisabeth Wilms behind the bakery counter and behind the camera. The photos, often used by the press, point to the two roles that Wilms regularly played: the baker's wife, for whom the family and the family business come first, and the filmmaker who produces several sponsored films per year on the other. Image source: Stadtarchiv Dortmund.

(self-)portrayal, it was repeatedly suggested that she did not pursue her film projects out of commercial interest, but out of devotion for filmmaking. In fact, the documents show that Elisabeth Wilms made some projects as a labour of love, but these were charitable or clerical films and not industrial films. Her pricing may have been cheap (as I will show), but her tax records testify that she made a significant profit on her commercial projects in most years of her career. In turn, this profit was partly used for new equipment for her business or for buying expensive colour film material, which she used for her private projects.

However, although press coverage and public film screenings were the only channels of acquisition actively used by Elisabeth Wilms, it would be insufficient to think that her success was based solely on her fame. In fact, we must also take into account what Manfred Rasch stated in 2003, namely that the major West German film production companies were not particularly interested in the production of industrial films. According to Rasch, instead regional networks developed and a large number of small production companies emerged to fill this gap and to try to become the sole film producer for one of the large industrial corporations.[13] In this context, it should be noted that Wilms lived in the middle of West Germany's

13 See Rasch, Manfred. "'Film ab!' Zum Industriefilm des Ruhrgebiets zwischen 1948 und 1959." *Industriefilm 1948–1959. Filme aus Wirtschaftsarchiven im Ruhrgebiet*, edited by Manfred Rasch et al. Klartext, 2003, p. 28.

industrial heartland, and that, due to the local concentration of the filmmaker's activities on Dortmund and its surrounds, there were personnel and institutional proximities and sometimes overlaps among her clients. So, for example, the Dortmunder Stadtwerke AG was a company on its own, but was owned by the City of Dortmund, one of the first clients of the filmmaker ever. Close connections also existed between the Dortmunder Stadtwerke AG and the Vereinigte Elektrizitätswerke Westfalen AG, another of the filmmaker's clients. In connection with this, it is important to point out that some of her customers became regular clients. Although there were often several years in between the individual commissions, the number of regular customers kept the business of Elisabeth Wilms running. It is therefore also not surprising that she worked on several films on the same or very similar subjects over the years, such as four films on Dortmund's water supply for the Dortmunder Stadtwerke AG, four films about the construction or expansion of regional power plants, three of them made for the Vereinigte Elektrizitätswerke Westfalen AG, or two films for the Rheinstahl AG on convertible steel highways. Last but not least it is also important to mention that the filmmaker repeatedly presented her works to the public in a wide variety of contexts in and around Dortmund, which may have contributed to her local fame and garnered her new commissions. I will return to this later.

Low-Budget Industrial Filmmaking

Another decisive factor in Wilms' success as a producer of sponsored films was that she not only presented herself as a potentially inexpensive amateur filmmaker, but also actually produced relatively low-priced films. It is difficult to find precise statements on the production costs of industrial films. Contemporary guidebooks often remain vague in terms of total costs and refer to the amount of money which the client is willing to pay per viewer as being the relevant factor from which the total costs are to be calculated. So, as an example of this calculation and without making statements about film gauge, material and length, Friedrich Mörtzsch mentions the sums of DM 100,000 and DM 150,000 for industrial films in 1959 in his well-known guidebook *Die Industrie auf Zelluloid* (Industry on celluloid).[14] In the 1978 Kodak guide *Industrie-, Informations-, Wirtschafts-Film im Auftrag* (Industrial, information, and business film on commission),

14 See Mörtzsch, Friedrich. *Die Industrie auf Zelluloid. Filme für die Wirtschaft.* Econ, 1959, p. 113.

the question of costs is raised in two chapters: Film producer Hans J. Motzkus claims that the practical values for the costs of PR films would range "from DM 200,000 to amounts in the millions."[15] Volker Bergmann, also a film producer, states in the same publication: "There are utility films, product information for example, with high efficiency (running time 8 to 12 minutes, 16 mm format, colour), whose production costs are DM 30,000. Requirements are few locations and short production time. On the other hand, there are elaborate PR and image films that cost several hundred thousand."[16] Scholarly literature in turn mostly focuses on industrial films that were flagship productions of big corporations and therefore justified high costs. Historian and archivist Manfred Rasch mentions two of these rather prominent examples in his texts. One of them is the 1955 film *Von der Bramme zum Breitband* (sound, b/w), of which a twenty-eight-minute long and a twelve-minute short version exist. According to Rasch, it had been produced "very inexpensively" for a total of DM 64,209.[17] In turn, the film *Werkstatt für Europa – Feuer an der Ruhr* (1956–1957, 71 min, sound, colour), which was created as a joint project of several major clients with the aim of producing a full-length colour film about the Ruhr area, was budgeted at DM 360,000.[18] However, it must pointed out that both films were shot in 35 mm, which meant that the costs for the film material alone were significantly greater than for a comparable 16 mm production. Hans Schaller noted in 1997 in his publication *Der Industriefilm schrieb Geschichte* (The industrial film wrote history), which unfortunately lacks references in many places, that the question of production costs can only be answered on a case-by-case basis. Without going into further details, he states that the price scale could range from DM 40,000 up to half a million DM.[19] What

15 Motzkus, Hans J. "Kosten/Nutzen-Relation des Industriefilms." *Industrie-, Informations-, Wirtschafts-Film im Auftrag. Ein Wegweiser für alle, die mit einem der wirksamsten Kommunikationsmedien arbeiten wollen*, edited by Wilfried von Tresckow. Kodak, 1978, p. 143.

16 My own translation. The original quote reads as follows: "Es werden Gebrauchsfilme hergestellt, Produktinformationen zum Beispiel, mit hoher Effizienz (Laufzeit 8 bis 12 Minuten, Format 16mm Farbe), deren Herstellungskosten bei 30.000,- liegen. Voraussetzung sind wenig Drehorte und kurze Herstellungszeit. Andererseits gibt es aufwendige PR- und Imagefilme, die einige Hunderttausend kosteten." See Bergmann, Volker. "Kalkulation." *Industrie-, Informations-, Wirtschafts-Film im Auftrag. Ein Wegweiser für alle, die mit einem der wirksamsten Kommunikationsmedien arbeiten wollen*, edited by Wilfried von Tresckow. Kodak, 1978, pp. 124–25.

17 See Rasch, Manfred. "Von der Kinoempfehlung zum firmeneigenen, prämiierten Dokumentarfilm. Über die Anfänge des Industriefilms auf der August Thyssen-Hütte AG." *Industriefilm – Medium und Quelle. Beispiele aus der Eisen- und Stahlindustrie*, edited by Rasch et al. Klartext, 1997, p. 88.

18 See idem. "'Film ab!'," pp. 37–38.

19 Schaller, Hans. *Der Industriefilm schrieb Geschichte. 1895–1995, 100 Jahre Industrie- und Wirtschaftsfilm*. PR Trend Agentur, 1997, p. 109.

Table 23.1. Elisabeth Wilms' Pricing for Industrial Films

Production	Film Title	Features	Length	End Price
1951	*Aufgleisgeräte für Schienenfahrzeuge System Maschinenfabrik Deutschland* (*Rerailing Equipment for the Maschinenfabrik Deutschland Rolling Stock System*)	silent, b/w	≈25 min	2,000 DM
1952	*Wasserversorgung der Industriestadt Dortmund* (*The Water Supply for the Industrial City of Dortmund*)	sound, b/w	37 min	3,900 DM
1952–1955	*Dokumentarfilm von der Erweiterung des Kraftwerks Gersteinwerk 1952/55* (*Documentary about the Extension of the Gersteinwerk Power Plant, 1952–1955*)	sound, b/w	39 min	25,000 DM
1957–1959	*GWH Gemeinschaftswerk Hattingen Erweiterung 1957/59* (*The Extension of Gemeinschaftswerk Hattingen (GWH), 1957–1959*)	silent, colour	35 min	30,000 DM
1959–1963	*Schiffshebewerk Henrichenburg in Waltrop* (*The Henrichenburg Shiplift in Waltrop*)	sound, colour	≈36 min	40,650 DM
1960–1963	*Bau "Kraftwerk Westfalen" 1960–1963* (*Construction of the Westfalen Power Plant, 1960–1963*)	sound, colour	38 min	40,000 DM
1961	*P&H-Raupenbagger in interessanten und schwierigen Einsätzen* (*P&H Crawler Excavators in Interesting and Difficult Operations*)	silent, colour	≈35 min	29,318 DM
1962–1963	*Kühlwasserkanal in Vakuumbeton für das Kraftwerk "Westfalen"* (*The Cooling Water Channel in Vacuum Concrete for the Westfalen Power Plant*)	silent, colour	≈12 min	12,960 DM
1966	*Umsetzbare Stahlhochstraße Schnellbauweise System Rheinstahl* (*The Rheinstahl Movable Steel Rapid Highway Construction System*)	sound, colour	≈25 min	13,072 DM
1970–1972	*Stahlbau zum Küstenschütz* (*Steel Construction for Coastal Protection*)	sound, colour	17 min	22,000 DM
1976–1977	*Thyssen-Klönne Stahlbrückenbau* (*The Construction* of the *Thyssen-Klönne Steel Bridge*)	sound, colour	≈21 min	48,840 DM

all these examples have in common is that they illustrate the wide price range in industrial film production, which was unquestionably dependent on factors such as film length, format, material and prestige of the project.

In addition, they testify that the prices of Elisabeth Wilms' industrial films are situated within the range of potential prices (see table 23.1).

These low prices were possible due to the fact that, unlike most of her competitors, Wilms was not financially dependent on her labour as a filmmaker, but still had the family-run bakery with grocery store in the background. To add to that, she was also the only employee of her film business, and she was willing to work on a project over long periods of time, in some cases a few years, and realize follow-up projects as cost-effective "updates" of existing films, i.e. adding only new material to the existing film and doing minor revisions. In combination with her prices, this made Wilms particularly interesting for customers with a low advertising budget, such as municipal companies like the Dortmunder Stadtwerke AG. One could even go further and say that Wilms' low pricing gave such institutions, which otherwise would not have had the financial means to do so, the opportunity to have their own films made in the first place.

Amateur Sponsors

In many cases, Wilms' clientele themselves had no experience at all in the film business, which often had a direct effect on the contracting and further process of filmmaking. Although it is an example from early on in her career, the case of the film *Dortmund's neue Westfalenhalle – Der Gigant unter den Sportpalästen* (1950–1952) illustrates very well how inexperienced both sides could be: both the filmmaker and her client, the communal Westfalenhalle AG, failed to have their agreement on the scope of the film project fixed in a written contract. This, as well as Wilms' enthusiasm for this nearly two-year long project (which was evident in the fact that she visited the building site almost daily), and her belief in being financially remunerated for also documenting the manufacturing processes of individual construction parts, were the ingredients for a prolonged conflict. Due to the filmmaker's way of working, production costs exploded, and her client refused to cover them, referring to a verbal agreement on a total of DM 2,500 (approximately €6,000 today) for the finished film. In addition, the companies involved in the construction did not want to participate financially in Elisabeth Wilms' expenses, as she had imagined. For their part, they had not commissioned a film and were not interested in the filmmaker incorporating the company-specific recordings made during production into the film against payment. *Dortmund's neue Westfalenhalle* finally became a financial disaster for the filmmaker, but it did not stop her from accepting orders from local municipal companies regularly later on, which involved repeated conflicts, since lessons were clearly not learned from the mistakes of previous film projects. In turn, there are indications that the smaller film sponsors tended to prefer working regularly with film amateurs. As such, some letters in Wilms' estate show that her initial contact with the Dortmunder Stadtwerke AG apparently came about after another amateur filmmaker had not completed a commissioned film for the company, who thus had to look for a replacement.[20]

As mentioned earlier, Elisabeth Wilms herself stated that she had never written scripts for her film projects, but had instead "filmed intuitively." The archival documentation supports this statement: according to it, she mainly worked with rudimentary, sometimes only handwritten scene lists that often describe the content of a scene in just one word. This does not exclusively apply to her industrial film projects. In post-production, she also drew up editing lists in which she noted the length of shots or entire scenes and, if necessary, the off-commentary text spoken over them. In particular, her smaller and less experienced clients often gave her a lot of leeway to work in such a way by not producing scripts or treatments for the film projects themselves, largely because they had no experience in filmmaking. While the municipally owned company Dortmunder Stadtwerke AG, for example, delegated a great deal of decision-making power to her in the first place, the obviously more experienced Rheinstahl AG provided detailed and well planned scripts and treatments for the two films on convertible steel highways (1966–1967 and 1972) which Wilms shot for them.

The procedure of screening work samples, which was also common elsewhere in sponsored filmmaking, was also practised by Wilms. But in her case, these screenings happened at home in her living room, and later in her home studio. More importantly, this replaced the development of scripts in most cases. On these occasions, the filmmaker tried to understand the wishes of her clients. However, a retrospective examination shows that it was precisely these vaguely outlined projects (for which in some cases even no written contracts existed) that often caused disagreements on both sides as the work progressed. This could, for example, lead to Wilms producing a film that deviated content-wise from what the customer had originally intended and only vaguely expressed. This usually only became apparent during the process of approval.

20 See Wilms, Elisabeth. Letter to Dortmunder Stadtwerke AG. January 23, 1953. Stadtarchiv Dortmund, Stock 624, no. 13.

On the one hand, Wilms' enthusiasm about a project, which obviously increased the fewer specifications were made by the client, is often also partly responsible for these conflicts. This enthusiasm sometimes resulted in films that were twice as long in the first cut as originally agreed. On the other hand, there were also disagreements if a client, in the absence of a script accepted by both sides in advance, wanted to expand the length or content of the film more or less spontaneously during the production or post-production process (e.g. by recording new equipment at work). In the case of various industrial film productions, this led to a massive delay in their completion, as innovations often required to be incorporated again and again, which in turn brought up discussion about the price agreed upon in advance. In the case of the Dortmunder Stadtwerke AG, in particular, one can also speak of a general lack of interest in the actual production process, which manifested itself in the absence of a contact person or in not answering or answering very late Wilms' inquiries. These cases often were accompanied by even more serious differences of opinion during the approval of the film. Another extreme, which occurred at least once, is that the completion of a film could also be brought to a standstill for an indefinite period by the merger of the client with another company.[21]

Professional Production Equipment

In 1941, Elisabeth Wilms had started making films with a used Zeiss Ikon Movikon camera and a used film projector. But according to her inventory lists she started buying new equipment soon after the war and further expanded these investments, especially in the course of the 1950s. In doing so, she already had built up an extensive stock by 1953, which included a Bolex H16 camera besides the Movikon, four projectors and three screens, a group of speakers and a mixer, an AEG tape recorder and a device to create rolling titles. Within a few years, two more Bolex H16 cameras and two zoom lenses followed among other things. The acquisition and parallel use of three to four cameras is unusual in the context of amateur filmmaking and must be considered in the context both of her efforts at professionalization and the financial security of the Wilms family, which was based on the operation of the bakery and grocery store. While for the vast majority of amateur filmmakers there was simply no practical need to own that many

21 See Wilms. Letter to Thyssen Industrie AG Klönne. September 9, 1977. Stadtarchiv Dortmund, Stock 624, no. 42.

cameras, this was not the case with Wilms. During the 1950s she produced several commissioned films per year, which means that she used her cameras more than was usual for amateur filmmakers, and that the equipment wore out faster. Numerous repair documents in the filmmaker's estate testify to this. On the other hand, she could hardly afford longer breakdowns of her equipment during a shoot, as she herself repeatedly emphasized in her letters to the repair departments of the camera manufacturers. This was due to the fact that she often had to capture work processes that could be filmed only at certain points in time or only once (e.g. industrial production processes or the construction phases of buildings). All this made the existence of replacement equipment necessary.

In fact, Wilms herself confirmed the parallel use of three cameras for her productions in a 1957 letter. In this context, it should not be overlooked that the filming capacity of each reel was limited to only a few minutes. Similarly, in certain situations it was logical to keep further cameras ready for immediate use, as a reel with exposed film had to be changed laboriously against an unexposed one under shielding from daylight. It is possible, but not verifiable, that Wilms used the different cameras, which could have been equipped with different lenses she owned, for specific environments and shooting situations as well as for film material with different light intensities. Among the things she bought from the profit of her film projects, however, was not only film equipment, but also other auxiliary equipment. In 1955–1956 she invested in a car and a caravan with which she could reach locations outside Dortmund and even stay overnight, but also go on vacations. But going on vacation did not exclusively mean leisure time. Rather, the filmmaker began at some point to shoot her own stock footage on her travels, which was used, for example, in her films about Dortmund's water supply to illustrate the water cycle.[22] Aside from that these particular films also illustrate very well that she recycled older footage in new films, such as older steel mill shots, which were then used to illustrate water consumption.

A large part of the post-production process of Wilms' films took place in her home in Dortmund. In her early creative years, she did the film editing at the living room table with the help of a glue press and a clothesline. The latter was stretched through the room and hung with the material of individual shots and scenes in its later order. After just a few years (1953 at the latest), the filmmaker purchased a viewing and cutting table for these works, which shows once again how early she was already pursuing professionalization. But those were not the only efforts in this direction: in the course of a

22 See Klauß, and Lentz. *Brot und Filme.*

few years several rooms of the Wilms' house were completely or partially repurposed for film production. In 1957 for example, the filmmaker invested DM 3,500 in the electrical installations and the design of these rooms, before she eventually had an architect design and build a fifty-six square metre screening room with separate projectionist cabin in an already existing annex to the family's house in 1961–1962.

Film Form as a Matter of Financial Means

In the course of her almost forty years of work, Wilms produced both black-and-white and colour films, as well as both silent and sound films (the latter being films which were accompanied with music and a voice-over commentary). Concerning the use of colour film material in her industrial films, there is a clear moment of change in her career. While the filmmaker used to shoot her first industrial films only in black and white, she switched to colour in 1956, which means that eighteen of her twenty-eight industrial films were made with colour film. It is not possible to say for certain why the filmmaker changed this particular year, but it can be said that she began working with colour film material much earlier in her private life and preferred it. Maybe it was because she had established a business relationship with the London-based company Colour Film Services Ltd. in 1955, where she mainly had her colour films processed and copied until 1961. So far, the role of film laboratories has received little attention in research. However, the available literature on the subject suggests that it would have been at least extremely difficult for Wilms to find a company in West Germany in 1956 that could not only develop 16 mm colour films, but also copy them. If the films were set with sound, this added to the difficulty.[23] What is remarkable besides that about the business relationship with Colour Film Services Ltd. is that in urgent cases the filmmaker herself brought her films to London and back again. In other cases, friends were apparently used as couriers to speed up the delivery of the material.

By contrast, there can be no question of a "moment of change" when it comes to silent or sound film in the sense that Wilms exclusively produced

23 See Read, Paul. "A Short History of Cinema Film Post-Production." *Zur Geschichte des Filmkopierwerks*, edited by Joachim Polzer. Polzer Media Group, 2006, p. 87, as well as the unfortunately incomplete listing of film laboratories and their technical capabilities in Bell, Frank. "Kopier- und Entwicklungsanstalten in der BRD seit 1945." *Zur Geschichte des Filmkopierwerks*, pp. 231–44.

the latter from a certain point in time on. In the case of her commissioned productions, it was rather the combination of the purpose of the film, the financial capabilities of the client as well as the client's infrastructure for the projection of the finished film, which were fundamental to the question of whether a film should be produced as silent or sound film. Of all her twenty-eight industrial films, half were obviously produced as silent films while the other half featured music and a voice-over commentary.

In any case, the development of a text explaining the film images was part of the production process. How the text development took place in the making of her industrial films, however, cannot be reconstructed today. But there are indications that the filmmaker herself often played a relatively small part in this process. Several draft texts and final versions that can be clearly assigned to clients testify to this. It seems logical and not at all limited to amateurish production processes that the responsibility for the text lay predominantly on the side of the client, as they probably knew best how to explain the content of the pictures.

In the case of a silent film, the text could either be a visual part of the film itself in the form of short intertitles (as used, for example, in *Aufgleisgeräte für Schienenfahrzeuge System Maschinenfabrik Deutschland*), or it could be a more extensive text which would have been read out during the screenings. It seems that the latter model was used more often in Wilms' case. This practice also involved an economic aspect, since it left the possibility for the customer to write, translate and modify the commentary text even after the completion of the film, in order to save money during the production process and to be able to utilize it in more diverse contexts with a smaller number of copies.

In contrast, the development of a spoken text for sound films and its recording in post-production was a far more complex process. In addition, Wilms herself apparently had no great pleasure in the work steps connected to the soundtrack. As her preserved inventory lists show, she was technically able to do this early on, yet she regularly outsourced these works. In some cases, scoring was done by local sound studios in Dortmund. Another possibility, apparently used on a regular basis, was to outsource the work to another member of the Dortmund Amateur Film Club, named Werner Smacka, who apparently specialized in scoring and is mentioned in one of Wilms' letters as being the owner of an "amateur studio."[24] In other respects as well, the soundtrack was kept at a low amateur level. In some

24 See Wilms. Letter to Aberbach GmbH. August 1, 1973. Stadtarchiv Dortmund, Stock 624, no. 10.

cases, especially when the client was short of funds, local theatre actors or people with a work-related connection to the theme of the film were hired (such as a master baker who spoke the off-text for a film about a bakery exhibition). It seems that also in this aspect the (often limited) financial possibilities of the customer determined the form of the films. Aside from the soundtrack, Smacka also took over the public screening of Wilms' city image films several times on behalf of the Dortmund city administration.[25]

The Circulation of Niche Films

While aspects such as editing and sound took place mainly locally in Dortmund, the opposite applies to film development and copying. As was already mentioned regarding Colour Film Services Ltd. In the course of her career, the filmmaker was very keen to experiment this context and called on a total of fourteen service providers from West Germany, England and France. However, she worked particularly frequently with the company KINAX from Dillenburg, Germany, with whose founder she was connected by the fact that he also came from the bakery trade and had then started his own business in the film industry.[26]

As Friedrich Mörtzsch pointed out in 1959, the producer of an industrial film usually did not transfer the copyright to the client, but retained the right of duplication and distribution and the possibility to make further copies of the negative and sell them to the client.[27] It is mainly due to the fact that Elisabeth Wilms also oriented herself to these professional procedures, which helps to understand the modalities of copying and circulation at least for some of her industrial films. Furthermore, this fact also points to one of the success factors for Wilms' film production: By keeping the right to duplicate and thus sell additional film copies to the customer, a production company was able to create extra profits in addition to the income from the actual film production. Now, taking a look at Wilms' industrial films it becomes evident that these works were not only to be found at the lower end of the price scale (as mentioned before), but also circulated only in a

25 See, for example, Stadt Dortmund. Letter to Werner Smacka. August 20, 1969. Stadtarchiv Dortmund, Stock 113, no. 124. For further remarks on the usage of sound and music in industrial films, see Donhauser, Peter. "Musik, Ton und Sprache im Industriefilm." *Ferrum. Nachrichten aus der Eisenbibliothek*, vol. 76, 2004, pp. 14–23.

26 See Bell, Frank. "Unsere Welt ist der Film: Wilhelm Ax und KINAX in Dillenburg." *Zur Geschichte des Filmkopierwerks*, edited by Joachim Polzer. Polzer Media Group, 2006, pp. 205–16.

27 See Mörtzsch. *Die Industrie auf Zelluloid*, p. 123.

Table 23.2. Film Copies, Including First Contract Copy

Film Title	Production	No. of Positive Film Copies
Aufgleisgeräte für Schienenfahrzeuge System Maschinenfabrik Deutschland (*Rerailing Equipment for the Maschinenfabrik Deutschland Rolling Stock System*)	1951	**13** (2 German, 3 Spanish, 8 English/French)
Harnischfeger Auto- u. Raupenkrane aus Dortmunder Fertigung (*Harnischfeger Truck Cranes and Track Cranes Manufactured in Dortmund*)	1959	**4** (2 German, 1 French, 1 Italian)
P&H-Raupenbagger in interessanten und schwierigen Einsätzen (*P&H Crawler Excavators in Interesting and Difficult Operations*)	1961	**1**
Wasserversorgung der Industriegroßstadt Dortmund (*The Water Supply for the Major Industrial City of Dortmund*)	1962	**2**
Bau "Kraftwerk Westfalen" 1960–1963 (*Construction of the Westfalen Power Plant, 1960–1963*)	1961–1963	**4**
Kraftwerk im grünen Land (*Power Plant in the Countryside*)	1958–1966	**1**
Umsetzbare Stahlhochstraße Schnellbauweise System Rheinstahl (*The Rheinstahl Movable Steel Rapid Highway Construction System*)	1966	**2**
Wasser für Dortmund (*Water for Dortmund*)	1971–1972	**7**
Wasser – Mehr als H_2O (*Water – More Than H_2O*)	1972	**4**
Umsetzbare Stahlhochstraße Schnellbauweise System Rheinstahl (*The Rheinstahl Movable Steel Rapid Highway Construction System*)	1972	**8** (4 German, 4 English)

very small number of copies in almost all cases (see table 23.2). Of course, this number always has to be seen in the context of the intended purpose of the individual film: It is therefore understandable that more copies were needed of a film for an internationally operating client, which was intended to explain the working method and the advantages of a product to potential customers, than of a film which was only shown during company tours or at in-house events.

Some clients not only had their films made by Wilms, but also had them screened by her on various occasions. Best documented are the screenings of the films she produced for the Vereinigte Elektrizitätswerke Westfalen AG (VEW) about the firm's power plants. On behalf of this client, Wilms screened one or more of the three "power plant films" twenty-one times between 1960 and 1969, always in North Rhine-Westphalia. A number of these screenings can be reconstructed: many of them took place in the vicinity of the power plants and obviously served the purpose of bringing the residents closer to their new "neighbours." However, there is also evidence of screenings within the framework of naval club evenings, advisory board meetings and anniversary celebrations. Besides that the "power plant films" were also projected directly on-site by VEW employees for visitors to the plants. Here again Wilms was asked to instruct the employees in the use of the projection equipment. But not only her regional clients like the VEW, who presumably had little or no projection infrastructure and personnel of their own, made use of this, but also the Rheinstahl AG. On behalf of this client, the filmmaker, for example, screened her 1967 film on movable steel highways in Munich, Brussels and Antwerp, probably to potential customers.

However, besides this, some of Wilms' industrial films were also shown outside their original usage context. Also this time the filmmaker herself played an important role, because it was she who lent copies to interested institutions or who presented the films herself, sometimes in combination with other of her sponsored films. In this way, her films made it to schools, technical colleges, engineering associations or even conferences, promoting her person and work as well as the initial client or his product. Finally, it can be said that Elisabeth Wilms' initiative in screening her industrial films was probably not due to direct financial interests, which is suggested by her invoices. They usually display amounts between DM 50 and DM 80 per screening plus travel costs. On top of that, the filmmaker had to bring and use her own projection equipment. It rather seems that the "filming baker's wife" was keen to show her films to the public and to receive recognition for her cinematic talent, which, of course, does not exclude the possibility that such screening events might result in further orders.

The Films

As already mentioned, Elisabeth Wilms' complete filmic oeuvre as such, but also the corpus of her industrial films is characterized by a great diversity in content, which is the result of the various clients, different purposes and occasions the films were made for, as well as the fact that the filmmaker was active for several decades. Nevertheless, some generalizing statements about the films and their form can be made: No matter whether they were shot in black and white or colour, left silent or set with sound, all of them can be described as conventional and conservative in their appearance. At the same time, they are, apart from Wilms' very first industrial film, lacking classic attributions for amateur films, such as unsteady camera work. In fact, the films show stylistically how the filmmaker professionalized over the years and how she adopted professional standards in filmmaking, which is why her works can be regarded as conventional industrial films.

Elisabeth Wilms was, as her other films also show, only open to filmic experiments to a small extent, her joy in experimenting apparently being limited to finding new perspectives. Therefore, things like a diagonal low-angle view, single panoramic shots from high points or from an airplane, an overhead shot from a moving hall crane or a shot filmed through a power plant turbine are the visual highlights of her industrial films. Expensive animation is the exception. Aside from this, the films are first and foremost committed to the same cinematic conventions as most representatives of this film category, with pans and long shots being some of their standard elements. This in turn shows that although Wilms came from the amateur film context, she quickly adopted the conventions of professional industrial filmmaking.

As far as the stylistic sound design of her scored industrial films is concerned, it can be said that, like her visual work, it is consistently very conventional. The sound level essentially contains a maximum of two elements: Classical or contemporary instrumental music (in the latter case most likely often arranged by Werner Smacka or coming from a music publisher) and the voice-over commentary, usually read out by a male voice. The narration of Wilms' industrial films is, just as most of the visual work, always kept simple. With topics like industrial building projects and Dortmund's water supply being recurring subjects of the films, they all feature straightforward step-by-step storytelling. However, the visual duration of many films can probably be explained by the fact that the live commentary had to be given space, at least in part of the cases.

Conclusion

Elisabeth Wilms' "amateur" industrial films were never the flagship films of big corporations. Likewise, they did not circulate in the "*kulturfilm*" context at all and were the opposite of stylistically innovative industrial films made by well-known filmmakers.[28] Nor were they "amateur" films in the conventional sense. Rather, the films were either niche products on niche subjects, or low-budgeted image films, befitting the purpose of their

28 Like, for example, *Nur der Nebel ist grau* by Robert Menegoz, produced in 1965 for Thyssen. See Hediger, Vinzenz, and Patrick Vonderau. "Einleitung." *Filmische Mittel, industrielle Zwecke. Das Werk des Industriefilms*, edited by idem. Vorwerk 8, 2007, pp. 9–10.

commission and the financial capabilities of the customer. It seems indeed that the financial aspect must be regarded the most important factor in Wilms' case. The "filming baker's wife" was able to cover a low-budget niche in industrial filmmaking that wasn't profitable for larger production companies while making use of the amateur label to promote herself. Her work methods show a mixture of amateur and professional practices, as she repeatedly worked without a written contract, and often without a script, but, on the other hand, produced works that complied with the professional conventions of industrial films and (at least in some cases) circulated internationally, produced her own stock footage, recycled old material and applied to common copyright standards. Above all, however, what I have sought to emphasize in this chapter is the fact that some of Wilms' customers can also be described as amateurs when it came to filmmaking, which had direct effects on the production process. They possessed meagre financial resources, seldom had employees with experience in film, and hardly any of their own infrastructure for projection. It was for this reason that they apparently regularly made use of local film amateurs to produce and screen their films.

Works Cited

Bell, Frank. "Kopier- und Entwicklungsanstalten in der BRD seit 1945." *Zur Geschichte des Filmkopierwerks*, edited by Joachim Polzer. Polzer Media Group, 2006, pp. 231–44.

Bell, Frank. "Unsere Welt ist der Film: Wilhelm Ax und KINAX in Dillenburg." *Zur Geschichte des Filmkopierwerks*, edited by Joachim Polzer. Polzer Media Group, 2006, pp. 205–16.

Bergmann, Volker. "Kalkulation." *Industrie-, Informations-, Wirtschafts-Film im Auftrag. Ein Wegweiser für alle, die mit einem der wirksamsten Kommunikationsmedien arbeiten wollen*, edited by Wilfried von Tresckow. Kodak, 1978, pp. 122–25.

"Binsenweisheit ... und Bäckerkünste." *Heute*, vol. 5, no. 114, 1950.

Damrow, Hildegard. "Produktion im Herrenzimmer." *Hamburger Abendblatt*, December 10, 1955.

Donhauser, Peter. "Musik, Ton und Sprache im Industriefilm." *Ferrum. Nachrichten aus der Eisenbibliothek*, vol. 76, 2004, pp. 14–23.

Eisert, Gelia, and Gerhard Kemner. *Lebende Bilder: Eine Technikgeschichte des Films*. Nicolai, 2000.

Elsaesser, Thomas. "Archives and Archaeologies: The Place of Non-fiction Film in Contemporary Media." *Films That Work: Industrial Film and the Productivity of*

Media, edited by Vinzenz Hediger and Patrick Vonderau. Amsterdam University Press, 2009, pp. 19–34.

"Frau Elisabeth Wilms, Dortmunds filmende Bäckersfrau." *Die Frau in der Bäckerei*, vol. 7, no. 3, 1959, pp. 5–7.

Hediger, Vinzenz, and Patrick Vonderau. "Einleitung." *Filmische Mittel, industrielle Zwecke. Das Werk des Industriefilms*, edited by Vinzenz Hediger and Patrick Vonderau. Vorwerk 8, 2007, pp. 8–15.

Jacobson, Brian R. "The Boss's Film: Expert Amateurs and Industrial Culture." *Amateur Movie Making: Aesthetics of the Everyday in New England Film, 1915–1960*, edited by Marta J. McNamara and Karan Sheldon. Indiana University Press, 2017, pp. 198–218.

Landeskirchliches Archiv der Evangelischen Kirche von Westfalen, Bielefeld, Stock 13.110, no. 95/2.

Mörtzsch, Friedrich. *Die Industrie auf Zelluloid. Filme für die Wirtschaft*. Econ, 1959.

Motzkus, Hans J. "Kosten/Nutzen-Relation des Industriefilms." *Industrie-, Informations-, Wirtschafts-Film im Auftrag. Ein Wegweiser für alle, die mit einem der wirksamsten Kommunikationsmedien arbeiten wollen*, edited by Wilfried von Tresckow. Kodak, 1978, pp. 142–45.

Pawlowski, Karl. "Vorschläge für die Werbung." November 4, 1945. Landeskirchliches Archiv der Evangelischen Kirche von Westfalen, Bielefeld, Stock 13.110, no. 95/2.

Polzer, Joachim, editor. *Zur Geschichte des Filmkopierwerks*. Polzer Media Group, 2006.

Rasch, Manfred. "'Film ab!' Zum Industriefilm des Ruhrgebiets zwischen 1948 und 1959." *Industriefilm 1948–1959. Filme aus Wirtschaftsarchiven im Ruhrgebiet*, edited by Manfred Rasch, Hans Ulrich Berendes, Peter Döring, Karl-Peter Ellerbrock, Michael Farrenkopf, Renate Köhne-Lindenlaub, Birgit Siekmann, Hans-Georg Thomas, Manfred Toncourt, Horst A. Wessel, and Burkhard Zeppenfeld. Klartext, 2003, pp. 11–50.

Rasch, Manfred. "Von der Kinoempfehlung zum firmeneigenen, prämiierten Dokumentarfilm. Über die Anfänge des Industriefilms auf der August Thyssen-Hütte AG." *Industriefilm – Medium und Quelle. Beispiele aus der Eisen- und Stahlindustrie*, edited by Manfred Rasch, Karl-Peter Ellerbrock, Renate Köhne-Lindenlaub, and Horst A. Wessel. Klartext, 1997, pp. 82–93.

Read, Paul. "A Short History of Cinema Film Post-Production." *Zur Geschichte des Filmkopierwerks*, edited by Joachim Polzer. Polzer Media Group, 2006, pp. 41–132.

Roepke, Martina. *Privat-Vorstellung. Heimkino in Deutschland vor 1945*. Olms 2006.

Schaller, Hans. *Der Industriefilm schrieb Geschichte. 1895–1995, 100 Jahre Industrie- und Wirtschaftsfilm*. PR Trend Agentur, 1997.

Schmidt, Heinrich. Letters to Hauptbüro des Evangelischen Hilfswerkes Westfalen. June 4, 1947, and June 18, 1947. Landeskirchliches Archiv der Evangelischen Kirche von Westfalen, Bielefeld, Stock 13.110, no. 95/2.

Schneider, Alexandra. *"Die Stars sind wir." Heimkino als filmische Praxis*. Schüren, 2004.

Stadt Dortmund. Letter to Werner Smacka. August 20, 1969. Stadtarchiv Dortmund, Stock 113, no. 124.

Stark, Alexander. *Die „filmende Bäckersfrau" Elisabeth Wilms – Amateurfilmpraktiken und Gebrauchsfilmkultur*. Schüren, 2023.

Tepperman, Charles. *Amateur Cinema: The Rise of North American Moviemaking, 1923–1960*. University of California Press, 2015.

Wilms, Elisabeth. Letter to Aberbach GmbH. August 1, 1973. Stadtarchiv Dortmund, Stock 624, no. 10.

Wilms, Elisabeth. Letter to Dortmunder Stadtwerke AG. January 23, 1953. Stadtarchiv Dortmund, Stock 624, no. 13.

Wilms, Elisabeth. Letter to Thyssen Industrie AG Klönne. September 9, 1977. Stadtarchiv Dortmund, Stock 624, no. 42.

Zimmermann, Yvonne. "Amateur Film in the Factory: Forms and Functions of Amateur Cinema in Corporate Media Culture." *Global Perspectives on Amateur Film Histories and Cultures*, edited by Masha Salazkina and Enrique Fibla. Indiana University Press, 2021.

Zimmermann, Yvonne. "Industriefilme." *Schaufenster Schweiz: Dokumentarische Gebrauchsfilme 1896–1964*, edited by Yvonne Zimmermann. Limmat, 2011, pp. 242–381.

About the Author

Alexander Stark is postdoctoral coordinator in the project "NFDI4Culture" (https://nfdi4culture.de/) at Philipps University Marburg. He studied media and history at the University of Trier and was a doctoral student in the project "Advertising and the Transformation of Screen Cultures" (http://adscreen.org/) and in the graduate research training programme "Configurations of Film" (www.konfigrationen-des-films.de). In his doctoral thesis, he investigated the historic intersections between amateur film and useful cinema in West Germany using the example of the filmmaker Elisabeth Wilms. It was published in 2023 as an Open Access and print publication at Schüren Verlag Marburg, entitled *Die „filmende Bäckersfrau" Elisabeth Wilms – Amateurfilmpraktiken und Gebrauchsfilmkultur.*

24 Movie and Industry in Italy

The "Golden Age" of Italian Industrial Documentary (1950–1970)

Anna Maria Falchero

Abstract
This chapter outlines a brief history of Italian industrial documentary film, from the beginning of the twentieth century to the late 1960s, underlining the role of the main firms in "modern" industry (i.e. chemicals, energy and mobility) and their attitude toward cinema as a communication tool. The aim is to provide some sort of "guide" through this vast and varied source of imagines from the past, partly stored now as files in Archivio del Cinema Industriale e della Comunicazione d'Impresa at the Università Cattaneo in Castellanza.

Keywords: Italy; archives of industrial films; documentary film

Cinema, in all its many, and often controversial, aspects, is the child of what is now, with certain reservations, almost universally defined by historians as the "Second Industrial Revolution": the use of celluloid,[1] also known as nitrocellulose, first cousin from the standpoint of the chemical industry,

1 Celluloid is the trade name for a number of plastic substances invented in the 1860s by the American John Wesley Hyatt, derived from nitrocellulose at 10–11% nitrogen, plasticized with camphor. While it is a flexible material and resistant to moisture, celluloid is highly flammable, which severely limited its use. In 1887 the Episcopal pastor Hannibal Williston Goodwin patented its use as a material for photographic films, opening the way for the birth of cinematography. Two years later, in fact, one of the pioneers in the production of photographic and film equipment, George Eastman, manufactured 35 mm transparent nitrocellulose film, the foundation, until the advent of digital, of the film industry. From 1954, however, celluloid was no longer used in the manufacture of film, precisely because of the aforementioned flammability. It was replaced by cellulose triacetate (no longer in use) and, later, by polyester (polyethylene terephthalate), still in use today for the production of film stock.

Hediger, V., F. Hoof, Y. Zimmermann, with S. Anthony (eds), *Films That Work Harder: The Circulation of Industrial Film*. Taylor & Francis Group 2024
DOI: 10.5117/9789462986534_CH24

of artificial silk, the media of photographic film and therefore of the first 35 mm movies, dates back to the end of the 1890s, and therefore ranks within the category of technological innovation that David Landes defines as "a technological realm whose scientific and intellectual requirements far transcend the possibilities of empiricism and enlightened amateurism, and not only in terms of chemistry, but also mechanics."[2]

It was therefore not entirely by chance that the first "enthusiasts" of the cinematographic document as a medium that conserves the historic memory would number one of the leading exponents of another new sector, the automobile industry, Henry Ford, whose private archive of footage, the Ford Film Collection, was later donated to the National Archives.

As a means of industrial communication, on the other hand, cinema struggled to take off for several decades, primarily due to prohibitive production costs and difficulty of use, but the early years of the twentieth century saw, even in peripheral Italy, "pioneering" attempts at truly industrial filmmaking, in order to document the production processes of individual companies.

Indeed, between 1905, when Ambrosio Film of Turin produced the first industrial documentary, *La fabbrica di salami* (*The Salami Factory*), and in 1914, the beginning of the Great War, more than fifty industrial films were made, few of which survive today, by private film companies, including Ambrosio, Itala, Savoia, Aquila, all based in Turin, Milano Film, Comerio, Croce and Walter, from Milano, and, above all Cines.

The latter, founded in 1906, with financial support from the Bank of Rome, anonymously transformed the Alberini and Santoni cinematographic company of Rome, which had already established a special advertising section in 1907, offering various Italian industrialists the opportunity to cinematically reproduce their factories and processes.[3]

During the First World War the production of short films was largely aimed, if not completely monopolized, by the themes of military propaganda. Despite this there were some projects, or true screenplays, as in the case of *Prometeo* (*Prometheus*) by Ansaldo, involving industrial films.

The post-war period would mark a period of deep crisis in the Italian film sector, which led, after an expensive and unsuccessful attempt to concentrate many of the existing companies, including Cines, in the Unione

2 Landes, David S. *Prometeo liberato*. Einaudi, 1978, pp. 597–603.

3 Cf. Nobili Vitelleschi, Giovanni. "Per un censimento filmografico dei documenti industriali italiani: 1905–1918." *Il ferro e la sua archeologia*, edited by Angelo Nesti and Ivan Tognarini, *Ricerche Storiche*, vol. 31, no. 1–3, January–December 2001, pp. 151–56. The full list of films produced in Italy up to 1913 has been published by ANICA and is available on the website: www.anica.it/archivio.htm. Accessed February 14, 2018.

Cinematografica Italiana (Italian Society of Cinematographers), promoted by the Banca Italiana di Sconto and the Comit,[4] in the constitution of the National Institute LUCE (Union for Educational Cinematography),[5] which the fascist regime would trust, in the context of pervasive totalitarianism, not only with the task of celebrating the glories of the Italy of the Blackshirts, but also with documenting the industrial situation.

LUCE, which was responsible for the majority of the industrial documentaries produced between 1924 and 1943, as conserved in the Historical Archive of the Institute, and available online (research carried out by the Archive of Industrial Cinema has identified about five hundred, but many others have not been accurately dated), was flanked by Cines during the first half of the 1930s. It was reorganized in 1930 by Stefano Pittaluga,[6] a leading player in the production and distribution of movies in Italy, who also filmed, alongside the production of the first commercial film with sound, industrial

4 At the time when Italy entered the war, as many as twenty-two companies in Turin and twenty-four in Rome were in full production. Prominent among these were Caesar and Tiber, which acquired among others shares in Itala of Turin. Following the signing of the armistice, most producers suffered a period of crisis, announcing heavy losses. The two banks most involved in financing the sector, Banca Italiana di Sconto and the Banca Commerciale Italiana, reacted by starting a conglomeration process in 1919, with the establishment of the UCI (Italian Society of Cinematographers) with 30 million lire of capital, which absorbed all the major Italian companies, including Tiber and Itala. The same year also saw the foundation of FERT in Turin, which was funded by SASP (Società Anonima Stefano Pittaluga) and remained independent, as was Lombardo Film in Naples. The UCI did not survive the collapse of the Banca Italiana di Sconto, and it was forced into liquidation in 1923, emerging four years later with a merger with the Pittaluga.

5 The limited company LUCE (a small private film business promoted by the journalist Luciano De Feo with the aim of educating the illiterate Italian population through images) was transformed by Benito Mussolini into a charitable public foundation with Royal Decree Law no. 1985 of November 5, 1925, under the name Istituto LUCE. It remained under the direct control of the Duce until 1936, when passed to the Ministry of Press and Propaganda (Minculpop), LUCE immediately became, thanks to a special circular from the prime minister, the sole supplier of the Ministries of the Interior, Education, Economy and Colonies, which were required to rely exclusively on its technical organization for educational and propaganda materials. The year 1927 saw the launch of the newsreel *Giornale Luce*, for obligatory screening in all Italian cinemas before the feature films, confirming the central role of the institute in the propaganda strategy of the regime. In 1935, with the foundation of ENIC (National Cinematographic Industry Body), LUCE directly entered into film production, on the eve of the construction of Cinecittà.

6 The producer-operator-hirer Stefano Pittaluga, owner of the eponymous company (which had also "inherited," through the Banca Commerciale Italiana, what remained of the liquidation of the UCI), reinstated the Cines brand in 1930, equipping the facilities in Via Vejo for sound. On the death of Pittaluga in 1931 the management of Cines was taken over by Ludovico Toeplitz, who ran the company until 1935, when the company completed its second production cycle. The premises on Via Vejo were destroyed by fire on the night between September 25 and 26 of that year, and later demolished.

documentaries, perhaps "fictionalized," as evidenced by at least two important films: the full-length feature *Acciaio* (*Steel*), shot in the Terni steelworks by a well-known documentary filmmaker, Walter Ruttmann, and the short *Sotto i tuoi occhi* (*Before Your Eyes*), shot in the Fiat factory at Lingotto.

To quantify and qualify the contribution of Cines (whose archive literally went up in smoke, together with the company, in the fire that destroyed its facilities on the night between September 25 and 26, 1935) and other smaller studios requires a great deal of research, made possible by the extensive project promoted by the Department of Performing Arts of the Ministry of Heritage and Culture and carried out by the Film Library in Bologna in collaboration with the National Film Archive and ANICA, which integrates the catalogue already published by the latter.[7]

Between late 1943 and 1948 there were many other problems that Italian industry had to face, but companies had already begun once again to consider cinema to be an important means of communication, both with the "inner public," the workforce, and a broader and undifferentiated "external" public.

The short *Sette canne, un vestito* (*Seven Reeds, One Suit*), produced in 1949 and entrusted by the patron of SNIA Viscose, Franco Marinotti, to a young director, Michelangelo Antonioni, introduced the "golden age" of the Italian industrial film, which lasted until the early 1970s.

Not entirely by chance, the establishment of company film archives and their emergence as an "organic" element in the communication strategy of big businesses coincided with both the resurgence of Italian narrative cinema and the "economic miracle."

While the use of film appeared common and widespread, the choices regarding the type of film and means of production were extremely diverse, largely reflecting the specific context of the company and its sector, as well as confirming the functional organic nature of the corporate image, often previously established, of the various firms.

In this phase, in short, the "medium" did not change the "message." While a comparative analysis of industrial film production would be of interest, it is beyond the remit of this chapter. What can be outlined, in broad terms, is the difference in behaviour between some of the major Italian companies, beginning with a basic distinction between footage from "external" clients and direct production performed by an "internal" facility.[8]

7 The Italia-Taglia project is much broader, and a first complete list, which also includes medium- and short-length films, is available online.

8 See Mosconi, Elena. "Il film industriale." *Comunicazioni sociali*, no. 1–2, January–June 1991, pp. 61–90.

The latter choice, which indicates among other things the intention to use the medium of film in a continuous, and not casual manner, led to the creation of the film libraries of some large companies, such as Fiat, Montecatini, Edison, Carlo Erba, Lepetit and Enel.

Fiat, no stranger to the use of film, dating back to 1909, created a film department at its Press and Advertising Headquarters in 1952, Cinefiat,[9] while in the same year Montecatini founded its Cinema Group, entrusted to Giovanni Cecchinato, who would later manage, after the merger of Edison and Montecatini, the Montedison Cinema Department.[10]

Three years later the Edisonvolta Cinema Department was founded under the direction of Ermanno Olmi, whose experience was well known, thanks in part to a number of interviews he gave on the subject, in which he promoted the overlapping activities of the company, the protagonists of the first rudimentary short films (a type of newsreel for internal use), and the crew of employees who flanked him in the filming of *Sabbioni: una diga a quota 2500* (*Sabbioni: A Dam at 2,500 Metres above Sea Level*) and *La pattuglia del passo San Giacomo* (*The Patrol of St. James's Pass*), both made in 1954. It constituted, in the service of the general secretariat, a true cinematographic service within the group, transformed the following year into the Cinema Section of Edisonvolta SpA,[11] which then passed, at the

9 The Cinefiat film archive, some of which had been transferred to the obsolete laserdisc format, is conserved at the National Archives of Industrial Cinema (a division of the National Film Archive), established in 2005 in Ivrea in agreement between the Experimental Cinema Centre, the Region of Piedmont, the City of Ivrea, and Telecom Italian SpA, for the preservation and dissemination of visual documents made within the company. The archive has the aim of the conservation and restoration of films and has specially equipped facilities.

10 Personnel employed at the group Cinema Montecatini included, besides Giovanni Cecchinato, who was entrusted with the preparation and processing of subjects, screenplays, directing, reviewing, editing, and commentary, Alessandro Bassi and Armando Ambrose, agrarian technicians and operators, Camillo De Paoli, operator and editor, Vincenzo Paradiso, operator and photographer. and Luciano Milesi, operator and sound engineer.
The Montedison Film Archive, which included both Montecatini and Edisonvolta productions, is deposited with the National Archives of Industrial Cinema. Some of the most important films, however, had already been converted to an electronic format, in the mid-1980s, as part of Montedison Cultural Project.

11 On the genesis of the Cinema Section of Edisonvolta, see Sandrini, Oliviero. "La sezione cinema della Edisonvolta S.p.A." *III festival del documentario industrale e artigiano*, edited by Comune di Monza, Associazione industriali di Monza e Brianza, and Unione artigiani di Monza e Brianza. Scuola Tipografica Figli della Provvidenza, 1959, pp. 17–20; Alberto Farassino, „Un metro di pellicola è lungo cinque: Ermanno Olmi alla Edisonvolta", in *Ermanno Olmi, dal cinema industriale al cinema d'autore,* Milano, Obraz Cinestudio e Montedison, 1984.; Corneo, Enrico. "La sezione cinema Edisonvolta. Appunti di memoria." *Ermanno Olmi, dal cinema industriale al cinema d'autore*. Obraz Cinestudio e Montedison, 1984, pp. 22–35. For Olmi's

time of nationalization, to Enel, which would inherit "in toto" the facilities and staff, but not the archive footage.

A similar structure, its history largely unknown, had already been established by Carlo Erba, which had set up its own Film Centre in 1953. This would become one of the most famous in Italy, and certainly the most important in the medical-scientific field, with the direction entrusted to a young doctor who was passionate about film, Mario Scolari, who quickly created, with decidedly pioneering methods, the first documentaries on surgery, laboratory techniques, innovative medicines and treatment methods.

This attention to scientific detail (based on the advice of qualified consultants), combined with objective interest, saw the documentaries awarded numerous awards at international events and festivals, to the extent that within a few years specialized production concentrated on certain defined areas, according to the various methods the company used to promote itself: the main field remained the medical-scientific sector, with materials for training courses for doctors and health personnel, and didactic use in universities, as well as health education films for the general public, and veterinary information shorts for the agricultural sector. These were true industrial documentaries (the only ones shot in 35 mm, as they were intended to be screened in cinemas), produced alongside short films on innovations in the organization of work for "internal use" within the company. These were joined by a number of tourism promotion films that, after illustrating the artistic and natural beauty of a nation, went on to illustrate the activity of the local Carlo Erba facility.

In addition to achieving global fame through representatives and affiliated companies that took care of distribution abroad (particularly in South America), the Carlo Erba Film Centre, managed by Scolari with prerequisites of efficiency and cost containment, and in harmony with the company's strategies, was to prove one of the longest-running facilities, remaining largely unscathed through the crisis in the film industry, to the extent that

version of the story, see Aprà, Adriano. "Ermanno Olmi: un certo giorno." *Cinema e film*, no. 7–8, 1969, pp. 45–63; Mosconi, Elena. "Intervista a Ermanno Olmi." *Comunicazioni sociali*, no. 1–2, January–June 1991, pp. 281–86. For a more in-depth analysis, see Boledi, Luigi. *La cineteca Montedison: immagine aziendale, strategie comunicative*. Thesis, Università Cattolica del Sacro Cuore di Milano, 1993–1994. In particular the first chapter, which outlines that the crew was made up of a production manager (Ugo Franchini), an engineer-electrician (Attilio Torricelli), an operator (Lamberto Caimi), and an assistant director (Walter Locatelli). In 1959 the latter would gather together the legacy of Olmi, which moved to Enel at the time of the nationalization of the electricity industry, while the group would add Roberto Seveso (operator and actor), Oliviero Sandrini (assistant to Olmi), Alberto Soffientini (production coordinator), Gian Piero Viola (editor).

it even increased production in the second half of the 1960s. Indeed, the crisis, and the closure of the Film Centre, only occurred after Carlo Erba was acquired by the Montedison Group in 1971.[12]

As for Lepetit, the first dated films go back to 1957, a time of wide expansion in the genre. They were commissioned to external production facilities, mostly at the Institute for Educational and Scientific Cinematography chaired by Piero Lamperti, while the role of director was entrusted to an employee, Giorgio Gondoni, who would later run the company film centre. This was set up in 1960 with a substantial investment in structures, including a section for shooting animated drawings, a recording studio and a film studio, with a large number of employees.[13]

This substantial investment corresponded with quite limited production, albeit of a very high quality: on average one or two films a year were produced, principally aimed, as was the case at most pharmaceutical companies, at doctors, college students and the general public, for whom informative health education films were produced.

The Lepetit Film Centre was short-lived, probably because of high production and distribution costs, leaving no records after 1966.[14]

While these structures (many of which, such as Cinefiat, made use of prestigious external collaborators) created hundreds of films and tens of thousands of metres of footage, the film libraries of other large public and private companies, which preferred to entrust the creation of their films to external companies or directors, were no less extensive. This was the case of Olivetti, which from the mid-1930s, thanks to Adriano Olivetti, appeared have an organic, coherent and rigorous corporate image policy, entrusted to dedicated structures that answered to top management, and, in particular, Rigo Innocenti.[15] These structures were responsible for the

12 Carlo Erba became part of the Montedison group in 1971. Shortly afterwards, when the final films in the production cycle were completed, the film business was centralized. The archive, however, remained at Carlo Erba, and was recently presented to the Center for the History of Enterprise and Innovation in Milan.

13 Giorgio Gondoni, chosen for his expertise in both medicine and film, was the centre manager and the director of all the films; Marcello Bernardi collaborated with him on the texts (assisted from time to time, by experts in their fields), Piero Pompili for the overall organization, Remo Grisanti for photography, Giovanni Pegoraro as a sound engineer, and Mario Negri, responsible for special effects and animation. See Mosconi. "Il film industriale," pp. 78–79.

14 The fact that 16 and 35 mm editions were made of each film, in various languages, is evidence of the aspirations of the company to distribute the material as widely as possible, despite the associated costs. That said, no footage exists in the archives after 1966.

15 The Publicity Office, headed in the 1950s by Ignazio Weiss, included the Olivetti Technical Advertising Office, of which Giovanni Pintori (who joined as a collaborator in 1937) was the

choice of subjects and texts, while in many cases the direction and creation of the films were commissioned externally.[16]

ENI under the management of Enrico Mattei also made use of the medium of film, from its inception – and, one could argue, even from its conception, if one considers that *Ricerche del metano e del petrolio* (*Finding Methane and Petroleum*), directed by Virgilio Sabel, was produced in 1951, and *Le vie del metano* (*The Ways of Methane*), directed by Ubaldo Magnaghi, was filmed in 1952 – and with uninterrupted continuity. Their documentaries were entrusted to a number of Italian and foreign directors, who were famous to differing degrees, preferring this system to the creation of its own internal cinematographic structure.[17]

Use was made of external companies, such as Icet, the Cinestabilimento Donato,[18] Film Service and Short Film of Milano, by not only companies of various sizes that wanted to produce, in an apparently "casual" manner, a single company documentary,[19] but also large and medium-sized enterprises which produced a number of short films, even without initiating a dedicated film policy: most of these limited their activities to a few short films,[20] whilst only a small number built a repertoire that approached double figures.[21]

In at least one case, that of IBM, Italian subsidiaries of large foreign companies,[22] which possessed a (sometimes vast, and often prestigious)

artistic director from 1950 to 1967. The Literary Office was directed by Franco Fortini and the Zincography Office and the Photographers Office, which led to the company film sector, were entrusted to Aristide Bosio.

16 Bellotto, Adriano. *La memoria del futuro: film d'arte, film e video industriali Olivetti 1949–1992*. Fondazione Adriano Olivetti, 1994. The footage is preserved in the Olivetti Historical Archives.

17 The footage is preserved in the Archivio Storico ENI.

18 The Donato film is preserved at the Fondazione Micheletti in Brescia.

19 Among these, in Milano, AEM, Alfa Romeo, Bassetti, BCS Motofalciatrici, Bemberg, Chatillon Filati, Face Standard, FDM Farmaceutici Balsamici, Luigi Fontana & C., Italstrade, La Prealpina, Metalli. Preziosi, Motta, Virginio Rimoldi & C.; SAE Società Anonima Elettrificazioni, Salcim Società Anonima Lavorazioni Chimiche & Imballaggi Metallici, Società del Linoleum, Techint; Tagliaferri & C. SpA, Van Der Berg SpA, Impregilo, and Rheem Safim.

20 Breda, Campari, Ferrania, Recordati, Marelli, Ceramica Pozzi, Salci, Braibanti, Vetreria Italiana Balzaretti-Modigliani, Innocenti, Singer, Snia Viscosa, and Falk.

21 Carle and Montanari, Farmitalia, Loro and Parisini, Impresit, and Pirelli.

22 In the BP catalogue there are examples of news documentaries (*Sonda n. 20*) and fantasy (*Giuseppina*), accounts of geographical exploration and aeronautical experiments, as well as short educational films on the transportation and refining of petroleum: the company philosophy emphasized projects that "aim to promote a better understanding of the problems involved in the search for and exploitation of petroleum, and to illustrate scientific advances," ibid., p. 76. The extensive production of Philips (about a hundred short films) was mainly based in the informational-educational sector, in particular electronics: *Pan-Tele-Tron* (1958) took

library of documentaries made abroad, would accompany the dubbing of "imported" films into Italian with their own modest "local" production.[23]

Overall, then, between the 1950s and the mid-1960s, film production in the industrial field was both respectable and of a remarkable consistency. It was showcased at a prestigious event, at the initiative of Walter Alberti (the director, along with Gianni Comencini, of the Italian Film Archive in Milan), the Festival of Industrial and Artisanal Documentary, organized by the Association of Industrialists and the Municipality of Monza. Inaugurated in 1957, the event became the European Industrial and Artisan Film Festival in 1959, and then, in 1962, the International Industrial and Artisan Film Festival, before coming to an end two years later.

This initiative also helped the documentary industry find its own space in the specialized press, attracting the attention of major national newspapers and, above all, some notable specialist magazines, primarily the Milan periodical *Close-ups* (which systematically featured news, debates, and interviews) for this "minor" genre of cinematography,[24] while dedicated publications edited by Walter Alberti at the end of each edition of the festival offered visibility for the documentary activity of companies and prolonged the resonance of the event.[25]

The festival organizers also proposed retrospective exhibitions organized in collaboration with the Cineteca of Milan. A number of famous Italian

a humorous approach to the history of telecommunications, while *Os mundi* (1960) covered the history and industrial applications of magnetism. *Interludio elettronico* (1964) by contrast presents the advantages of electronic computers in processing data. The company also has a film archive, Philips Film Library, which distributes material dealing with topical cultural and scientific issues, including to schools and universities.

23 In the late 1960s, IBM Italy, which also offered a catalogue of films about its products (*Amata*, 1946; *IBM Teleprocessing ... Right Now!*, 1964), as well as applicative monographs (*Sistema 357*, 1964) and didactic subjects (*Puppet Show*, 1964), created in the American informative tradition, decided to begin their own production of documentaries that, while modest, continued for a decade, until the creation of a television production centre for the production of programmes for internal use in the mid-1970s.

24 Among these were Bianco & Nero, Cinematografie Specializzate, Ferrania, L'Altro Cinema, and Intermezzo.

25 There were two pamphlets and a book. Respectively: Comune di Monza, Associazione Industriali di Monza and della Brianza, Unione Artigiani di Monza. *Terzo Festival del Documentario Industriale e Artigiano*. Scuola Tipografica Figli della Provvidenza, 1959; idem. *Quarto Festival del Documentario Industriale e Artigiano*. Scuola Tipografica Figli della Provvidenza, 1960; and, finally, Alberti, Walter. *Il film industriale*. Edited by idem and Ezio Croci. Scuola Tipografica Figli della Provvidenza, 1962. The entire collection of films submitted to the festival, during which, on average, fifty to sixty short films a year were presented, selected from the two to three hundred that were submitted, is preserved (unfortunately without an inventory) at the Film Archive of Milan.

documentaries were presented in 1959, and in 1960 the "Masters of European Documentary" event was followed by a historic exhibition of industrial film.[26]

From 1958 onwards the promoters also organized a series of conferences during the festival, on the problems of industrial film, giving rise to an interesting debate on its definition and its intrinsic nature, which involved "insider experts," well beyond the end of the "golden age" of this genre in the 1950s and 1960s.

Industrial cinema is, by its very nature, an extremely composite genre. The Confindustria industrial federation was well aware of this as, from 1959, it organized the longest-running (and most recognized) National Festival of Industrial Film, which was held annually in a series of Italian cities (Turin, Rome, Venice, Florence, Salerno, Bologna, Genoa etc.) with the aim of choosing, among the films in competition, the Italian representative at the International Industrial Film Festival, sponsored by the European Council of Industrial Federations: evidence of this is the breakdown of the films in competition into seven distinct categories, defined not only by their specific contents, but also (if not principally) according to the type of "public" for whom the films were intended, each of which had its own inherent logic that substantially affected the "technical," and indeed, "artistic" characteristics of the films.[27]

26 See Mosconi. "Il film industriale," pp. 85–86.

27 The rules of the festival, which also demonstrate the firm intention to establish a clear dividing line between industrial film and advertising, included:
1) Films about industrial issues (on economic, social, technical, and scientific themes) of general interest and intended to be shown to an indiscriminate audience.
2) Films of particular interest on a material, a project, or an industrial product intended to be shown to an indiscriminate audience.
3) Films that despite not having a specific informative purpose, such as those of the previous categories, add to the prestige of an industrial sector or company, and are intended to be shown to an indiscriminate audience.
4) Films on particular materials, projects and industrial products, intended to be shown to a specialized audience.
5) Films on principles or scientific research with an industrial application, intended to be shown to a specialized audience.
6) Training films for management or the workforce (for example, managerial methods, measures to increase production, rationalization and automation, human relations in the company, orientation and professional training) intended to be shown to an industrial rather than an indiscriminate audience.
7) Films on accident prevention, occupational illnesses, occupational hygiene, rehabilitation, and corporate initiatives of a social character, intended to be shown to an industrial rather than an indiscriminate audience. See Pellizzi, Carlo. "Proposte operative per una nuova impostazione del film industriale." *Cinema industriale e società italiana*. Franco Angeli, 1972, pp. 21–45; a largely

On the other hand, directors and established writers were not the only contributors to the creation of *tecnofilm* (a diminutive definition introduced by some film magazines in the mid-1960s). They were assisted by many of the best Italian cartoonists of the era (from the Gavioli brothers to Gamma Film and Bruno Bozzetto, and the Pagot brothers to Manara), whose characters enlivened the popular Italian sketch show *Carosello*.

As for distribution channels, at least some of the films would find an audience outside corporate circles, arriving (albeit with difficulty) in schools.[28] This was accompanied by a wise policy of "pairing" industrial films with fiction popular with the general public in cinemas, while multiplying "specialist" projections, for specific categories of the "external" public.

An interesting example concerns an extensive series of short educational films produced by Montecatini and screened, with the decisive support of agricultural associations, and the use of a specially equipped vehicle, in the squares of the towns with the highest agricultural concentration, with the intention to not only publicize the various chemicals used in agriculture, but also promote their correct use. Another example involved the dozens of technical videos produced by Carlo Erba for medical and nursing personnel, to illustrate the characteristics and possible applications of pharmaceutical products.

On the other hand, a large number of films, shot in the less expensive 16 mm format,[29] were designed strictly for the "internal" public within the company, and used in the training of skilled workers, technicians and managers, or (another rich vein) to illustrate safety regulations, occupational

similar subdivision is proposed in Verdone, Mario. "Del documentario tecnico-industriale." *Bianco e nero*, vol. 23, no. 3, 1962, pp. 43–46.

28 For example, the superintendent of Milan, while supporting the use of audiovisual materials as a supplementary teaching aid, recognized the Milan Centre for Audio-visual Aids as the sole body responsible for their distribution, which was also affected by the availability of documentaries failing to satisfy demand, thus excluding movies produced in-house by companies, which were considered to be mainly promotional materials.

29 This was the most widely used format for documentaries and industrial cinema as it was more cost effective than 35 mm, with regard to both film stock and equipment (cameras and projectors) and the greater practicality of use, thanks to projectors which were smaller and more portable, which made it possible to project movies in a variety of situations. It was introduced by Eastman Kodak in 1923 in response to 9.5 mm (Neuf-cinq or Pathé Baby), which had been introduced in 1922 by Pathé (a format that was achieving some success in the semi-professional field, especially in Europe). And 16 mm, thanks to new a film stock on the market, reversal film (by which the film used in shooting was automatically processed to become a positive that was ready for projection, thus bypassing the process of developing the negative), which made it possible to reduce the expense of the process of development to 1/6 of the normal cost. The format became increasingly popular and was widely used by "minor" film companies.

illnesses etc., as well as to encourage employees by illustrating and celebrating the "providential" initiatives of the company, (holiday camps, sports teams, business clubs, excursions etc.).

The end of the "economic miracle" was accompanied by the enactment of a cinema law, which was as long awaited as it was disappointing (Law no. 1213 of November 4, 1965). This law did not recognize the validity and cultural function of the industrial documentary and, moreover, obliged major industrial organizations with state participation to entrust the creation and distribution of films to Istituto Luce, which helped to accelerate the decline of an industry that was already penalized by the growing encroachment of television.

Production of training films, while not ceasing altogether, decreased by forty per cent on the previous year in 1965, clearly indicating the beginning of a crisis that lasted twenty years, during which "internal" film structures would disappear, close or be "outsourced," and most resources were directed to the advertising industry, save for some rare (and significant) exceptions.

The introduction, in the course of the second half of the 1970s, of far less expensive electronic formats, not least among them the semi-professional U-matic and Betacam formats, did not initially seem capable of reversing the trend, yet it did considerably simplify the issues related to the screening of such films: judging from the collections consulted in the research for this chapter, U-matic was only used extensively in industrial film from the mid-1980s, when industrial films were featured at the event organized by Confindustria, together with the production of a few large "diehard" companies, commissioned by smaller companies in "light" sectors, such as food, textiles, wood, and ceramics.

The more than fifteen hundred films collected during the various editions of the National Festival of Industrial Film (which became the "Film Selection" in 1984, and was later included, in the 1990s, as part of the Week of Business Communication) represent the major part of the Film Archive of Confindustria, which was presented, at the time of its establishment, to the Archive of Industrial Cinema and Business Communication, which has the task of providing for their digitization, thereby enabling their consultation, as well as their non-commercial exploitation. The archive is also publishing online a "general" catalogue of Italian industrial films.[30]

30 The Archive of Industrial Cinema and Corporate Communications is a non-profit organization launched in 1998 by Confindustria and the Cattaneo University in Castellanza – Liuc, which was joined as a partner in 2000 by the Business Economics and Historical Documentation Study Centre in Rome, headed by Valerio Castronovo. The birth of the new association was

The idea of an archive that would gather together the filmed testimony from the world of Italian industry was suggested many years ago, in the mid-1980s. At the time the proposal attracted little support and was dropped.

In reality, the first laborious steps on the journey involving historical company archives were already being taken a quarter of a century ago, and the amount of work that historians, archivists and willing industry executives were about to undertake initially appeared terrifying, so much so as to deter any attempt to expand the scope to include such unattractive items, at least on first sight, as industrial films.

Fifteen years would pass before this idea would come to be received in a favourable light: thanks to the attention of a university institute, the Cattaneo University in Castellanza, with close links to the business world, taking particular care to preserve its memory, and Confindustria, which had launched a series of projects for the conservation and reorganization of its documentary collections during the 1990s, and proved particularly sensitive to stimuli from the world of research for the advancement of the audiovisual document in the study of business history.

Among archival sources, footage, especially on film, is the most vulnerable: the intrinsic fragility and the objective complexity of appropriate maintenance conspire in causing it to deteriorate, often irremediably, while the complexity of access to content is in many cases fatal. Furthermore, if even the most hardened and indifferent of business "reorganizers," responsible for the loss of a great deal of valuable archival material, hesitates before destroying photographic material, drawings, catalogues, or copies of letters, where the "historical interest" of the contents is immediately evident, they would hardly think twice about throwing out a nondescript grey metal box, perhaps entirely lacking any information about its contents, or a videocassette for which the playback equipment is no longer available. All the more so considering that, except in rare and previously mentioned exceptions, such cases generally involve items that are "isolated" from the archival context of the enterprise, the result of a "casual" decision, the reasons for which are difficult to trace in printed documentation, making it impossible to establish responsibility.

In short, this heritage is an "ugly duckling," a harbinger of countless (and expensive) complications. And, moreover, implicitly devoid, by the very nature of its reproducible source, of any claim of "uniqueness."

made possible thanks to an important financial contribution by the Cariplo Foundation, which provided for the sophisticated computer equipment used to accomplish the digitalization and storage on magnetic media of exhibits which arrived at the archive in traditional formats.

On the other hand, the industrial film is in reality a historical source (and, as such, deserves to be properly submitted to critical analysis) of considerable interest, for a wide variety of disciplines, as well as valuable teaching material.

Provided, of course, that not only is care taken in its conservation, but that accessibility is guaranteed and, above all, will be valued, thus contextualizing it: the golden rule is that the best assurance of preservation is, in the broadest sense, utilization. Therefore, a rule that applies to all corporate archival material has, in the case of industrial films, even greater significance, as they are not easily made use of.

Works Cited

Alberti, Walter. *Il film industriale*. Edited by Walter Alberti and Ezio Croci. Scuola Tipografica Figli della Provvidenza, 1962.

ANICA, www.anica.it/archivio.htm. Accessed February 14, 2018.

Aprà, Adriano. "Ermanno Olmi: un certo giorno." *Cinema e film*, no. 7–8, 1969, pp. 45–63.

Bellotto, Adriano. *La memoria del futuro: film d'arte, film e video industriali Olivetti 1949–1992*. Fondazione Adriano Olivetti, 1994.

Boledi, Luigi. *La cineteca Montedison: immagine aziendale, strategie comunicative*. Thesis, Università Cattolica del Sacro Cuore di Milano, 1993–1994.

Comune di Monza, Associazione Industriali di Monza and della Brianza, Unione Artigiani di Monza. *Quarto Festival del Documentario Industriale e Artigiano*, Scuola Tipografica Figli della Provvidenza, 1960.

Comune di Monza, Associazione Industriali di Monza and della Brianza, Unione Artigiani di Monza. *Terzo Festival del Documentario Industriale e Artigiano*. Scuola Tipografica Figli della Provvidenza, 1959.

Corneo, Enrico. "La sezione cinema Edisonvolta. Appunti di memoria." *Ermanno Olmi, dal cinema industriale al cinema d'autore*. Obraz Cinestudio e Montedison, 1984, pp. 22–35.

Farassino, Alberto. "Un metro di pellicola è lungo cinque: Ermanno Olmi alla Edisonvolta," in *Ermanno Olmi, dal cinema industriale al cinema d'autore*. Milano, Obraz Cinestudio e Montedison, 1984.

Landes, David S. *Prometeo liberato*. Einaudi, 1978.

Mosconi, Elena. "Il film industriale." *Comunicazioni sociali*, no. 1–2, January–June 1991, pp. 61–90.

Mosconi, Elena. "Intervista a Ermanno Olmi." *Comunicazioni sociali*, no. 1–2, January–June 1991, pp. 281–86.

Nobili Vitelleschi, Giovanni. "Per un censimento filmografico dei documenti industriali italiani: 1905–1918." *Ricerche Storiche*, vol. 31, no. 1–3, January–December 2001, pp. 151–56.

Pellizzi, Carlo. "Proposte operative per una nuova impostazione del film industriale." *Cinema industriale e società italiana*. Franco Angeli, 1972, pp. 21–45.

Sandrini, Oliviero. "La sezione cinema della Edisonvolta S.p.A." *III festival del documentario industrale e artigiano*, edited by Comune di Monza, Associazione industriali di Monza e Brianza, and Unione artigiani di Monza e Brianza. Scuola Tipografica Figli della Provvidenza, 1959, pp. 17–20.

Verdone, Mario. "Del documentario tecnico-industriale." *Bianco e nero*, vol. 23, no. 3, 1962, pp. 43–46.

About the Author

Anna Maria Falchero was born in 1954 in Settimo Torinese (Turin) in 1954. She earned a degree in economics at the University of Florence and a PhD in history and civilization at European University Institute. From 1994 until 2003 she was a researcher and teaching professor of economic history at Università Cattaneo (Castellanza) and then at University of Perugia. She has worked as a free archivist for the Centro Studi per la documentazione storica ed economica dell'impresa in the classification and organization of the Archivio Storico Alfa Romeo (1984–1986); the Archivio Storico Breda (1987–1988); the Archivio Storico ENEL (1989–1991); the Archivio Storico Società Metallurgica Italiana e Italtel (1990), and the Archivio Storico Ilva in Terni (1990–1992). From 1997 to 2018 she was the director of the Archivio del Cinema Industriale e della Comunicazione d'Impresa at the Università Cattaneo in Castellanza. She has published six books and more than thirty articles on Italian industrial and banking history.

25 A Film That Doesn't Seem to Work

A Shot of Renault's Early Assembly Line (1920 to 1929) – A Case Study, Methodology and 3D Restitution for Film Analysis

Alain P. Michel

Abstract

For the silent documentary *At the Renault Factory*, produced by the Gaumont Film Company to glorify Renault's effort during the First World War, a short take was shot in a chassis workshop showing a working device that I identify as an early form of assembly line (even though this cannot be certified through usual historic written evidence). To be preserved, the film was transferred in the 1990s onto a new video base, but without cardboards or written subtitles, resulting in it being a totally mute "orphan." Thus, the main argument of this chapter is to suggest a methodology to study these types of archival, or truncated, films, as well as any other types of historic images through systematic cross-analysis and 3D technology.

Keywords: industry; corporate film; utilitarian film (*Gebrauchsfilm*); film archive; Marc Ferro; analytic grid

Introduction

Why and how should historians of manufacturing technology analyse industrial films as first-hand sources? Through the case study of a 1920 "orphan" documentary film and the historical investigation of an assembly line scene shot in a Renault workshop in France, the purpose of this chapter is to highlight the value of such motion pictures, and to suggest a historic way to read archival films.[1]

1 I would like to thank Florian Hoof (Goethe-Universität Frankfurt) for his keen editorial suggestions. Also Eric Alvarez (University of Evry) for his English-language editing and proof reading.

Hediger, V., F. Hoof, Y. Zimmermann, with S. Anthony (eds), *Films That Work Harder: The Circulation of Industrial Film*. Taylor & Francis Group 2024
DOI: 10.5117/9789462986534_CH25

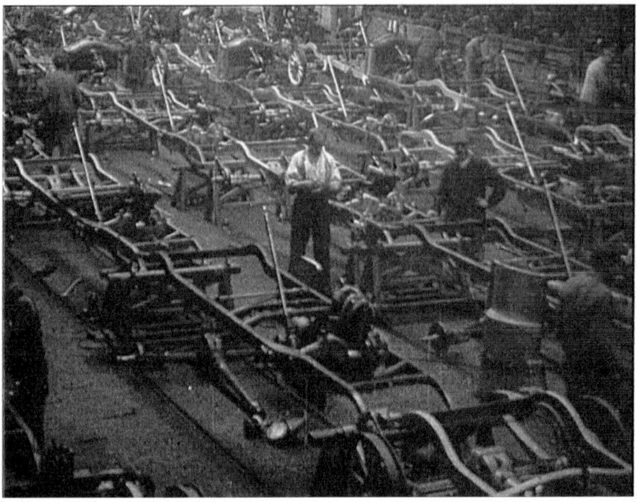

Fig. 25.1. Still frame from the assembly line sequence in the documentary film *Aux Usines Renault* (1920). © Gaumont Pathé Archives, 1920.

Moreover, I will use this documentary to defend an analytic grid that can be used for other films, as well as other types of visual documents whether synthetic (i.e. drawings, paintings etc.), analogue (i.e. photography, film, video, television), and finally digital images. What do these images reveal that is not accessible in other types of documents? Furthermore, by what means can we distinguish the "visible" from the "real," as well as differentiate the "story" that is shown from the "history" that is disclosed? In the 1970s, Marc Ferro and Pierre Sorlin were pioneers in considering films as first-hand historical documents. Studying Soviet archive films and newsreels, Ferro noted that "a film is better analysed when apprehended in its series."[2] He stood out from the aesthetic method of analysing single films as works of art. Sorlin had a comparable approach, which he developed from a collection of neorealist films of the Italian post-World War II period.[3] Through the confrontation of different films, he identified points of "vulnerability" from which an historical analysis could start. It can be a slip, i.e. aspects displayed though not intentionally, or a hint which document history beyond the explicit meaning of images and comments. I draw a comparison of this type of investigation to the work undertaken in an archaeological dig: the film should not be extracted from its original archives without some precaution.

2 Ferro, Marc. "Le film, une contre-analyse de la société." *Annales: Économies, Sociétés, Civilisations*, vol. 28, no. 1, 1973, p. 118. Re-edited in Ferro. *Cinéma et histoire*. Folio, 1977, pp. 31–62.

3 Sorlin, Pierre. *Sociologie du cinéma*. Aubier-Montaigne, 1977, p. 73; with a different approach: Sorlin. *Introduction à une sociologie du cinéma*. Klincksieck, 2015.

It is an investigation into the origins of the image in order to establish its historical roots.[4] It relies on a micro history way of apprehending some material historical evidence left, correlated with other subsidiary traces, in order to deduce what really happened.[5]

The investigation of a short assembly line sequence will be the point of departure into the vulnerability of our mute and exclusively visual documentary, as well as its deconstruction and historic interpretation. Thus, I will start by focusing on the early development of assembly lines and of its cinematographic representation. A methodological film analysis will then be applied to my case study by using an analytic grid which distinguishes the following two points: first, the different scopes of the archival film and, second, the successive scales in its interpretation. I will conclude with the perspectives offered by 3D restitution as a test platform in my understanding of visual documents.

The Cinematographic Fame of the Ford Assembly Lines (through a Corporate Film Collection)

Assembly lines have interested manufacturers, social scientists and cinematographers for their stunning mechanical performances, their industrial efficiency as well as their social impact. Along a conveyor, objects are progressively being manufactured by workers who are stationed at specific places with the proper tools in hand in order to accomplish a defined task. Whether the conveyor is manually or mechanically operated, the range of operations accomplished in each station has to follow the same timing which may seem simple, but which is a technological and collective challenge. Thus, an assembly line is not just a clever concept and a convenient working device. It is also a global and complex system for manufacturing.[6]

First adapted for the automobile production on April 1, 1913, by the Ford Motor Company (FMC), the assembly line was not an intended outcome and

4 Elsaesser, Thomas. "The New Film History as Media Archeology." *CiNéMAS: revue d'études cinématographiques*, vol. 2–3, 2004; Michel, Alain P. "Le travail à la chaîne en images: Les documents visuels face à l'histoire industrielle de Renault." *Images du travail/Travail des images*, no. 3, 2017, https://doi.org/10.4000/itti.1105.

5 Ginzburg, Carlo. *Clues, Myths and the Historical Method*. Johns Hopkins University Press, 1989.

6 Michel, Alain P. "Assembly Line." *The Palgrave Dictionary of Transnational History*, edited by Akira Iriye and Pierre-Yves Saunier. Macmillan, 2009, pp. 67–70; Michel, Alain P. *Travail à la chaîne. Renault 1898/1947*. ETAI, 2007.

emerged gradually.[7] But it was swiftly extended to different manufacturing operations in the Highland Park plant in Detroit and was soon considered as the instrument of Henry Ford's success and fortune.

Though rapidly mechanized, Ford's first assembly lines were manual, i.e. the workers moved the artefacts they assembled at a constant pace. These first manual lines became the symbol of mass production as they were both a technological and a social defy. By 1919, the assembly line concept had wiped out former notions of how things ought to be moved and assembled.[8] It was the nucleus of work organization of Ford's new River Rouge factory in Dearborn in the suburbs of "Motor Town." Indeed, the FMC plants were the subject of a substantial literature, and became the "must-see" destinations of all American business tours.[9] Despite its aforementioned celebrity status, David Hounshell shows that the introduction of the assembly lines was one of the most confusing periods in American history.[10] Indeed, no records remain and posterior reminiscences of the witnesses are contradictory.

Three kinds of assembly lines were identified by the industrial experts. The most famous was the final body assembly line, at the end of which the "auto-mobile" drove out with all its features. The engine assembly was another subject of wonder for the ability to combine very tightly adjusted mechanical parts. Last, the machining lines fascinated for their automatic abduction of metallic material. Assembly lines changed rapidly and concerned very different technical operations.[11] The combination of these multiple tasks was presented by Henry Ford and business historians as a natural flow of sub-assembly lines.[12] His metaphor of rivulets flowing into streams flowing into a great river, was more an image than reality.

7 Nye, David. *America's Assembly Line*. MIT Press, 2013, pp. 21–26.

8 Hounshell, David. *From the American System to Mass Production, 1800–1932: The Development of Manufacturing Technology in the United States*. Johns Hopkins University Press, 1984.

9 Arnold, Horace, and Fay Faurote. *Ford Methods and the Ford Shop* [1915]. The Engineering Magazine Company. 1919; Faurote, Fay. "What Is Going on Behind the Scenes at the Ford Plants." *Factory and Industrial Management*, vol. 74, 1927, pp. 257–62.

10 Hounshell. *From the American System*, p. 247.

11 I believe that what David Nye identifies as the "five features (which) define the assembly line" tend to mix different periods and techniques. Nye. *America's Assembly Line*, pp. 21–26. Subdivision of labour and interchangeability of parts are essential. The importance of electric drive, sequential ordering of machines and automatic movement of work and workers differ from one period to another and whether it concerns machining lines or assembly of mechanical or body parts.

12 In a classical business history approach, see Nevins, Allan, and Frank Ernest Hill. *Ford: The Time, the Man, the Company*. Charles Scribner's Sons, 1954; idem. *Ford: Expansion and Challenge: 1915–1933*. Charles Scribner's Sons, 1957.

The cinematographic flow of images has the ability to represent the pretended flow of assemblies thanks to three features of moving pictures. Film editing can put together operations taking place in separated workshops; it can accomplish a contraction of time; it can have workers play their own part.[13] Ubiquity, metamorphosis and staging thus realistically show the completeness of a work process which is concretely impossible to look at.

Yet, no moving picture depicts Ford's first and famous assembly lines between 1913 and 1916. This eclipse is surprising, because the automobile manufacturer was also an important movie maker.[14] The film department was created at Ford in April 1914,[15] a year after the assembly line was first introduced into the flywheel magneto department, which assembled the electric device to start the motor. There are no reels in the Ford Film Archives that show the initial development of the assembly line experiments.[16] During this first year of existence, the assembly line experiences could have been shot by outside film companies, had they been invited to do so, as was the case for many photographers.[17] The film representation of the 1913 invention was recreated in later documentaries such as the 1955 film called *Ford People*.[18] The reconstitution of this manual assembly line can still be seen in the "Assembly Exhibit" on the third floor the Ford Piquette Plant Museum in Detroit, celebrated as the "Birthplace of the Model T."[19]

It is only in the late 1910s and 1920s that the assembly lines became the stars of a few industrial documentaries, such as *Where and How Fords Are Made* (1919) and *A Tour thru the Rouge Plants* (1926).[20] Even then, the internal

13 Goffman, Erving. *The Presentation of Self in Everyday Life*. Anchor Books, 1959. The first edition (University of Edinburgh, Social Sciences Research Centre, 1956) was a report supported by a Ford Foundation grant. In the "Preface," Goffman indicated that he meant "this report to serve as a sort of handbook detailing one sociological perspective from which social life can be studied, especially the kind of social life that is organised within the physical confines of a building or plant."

14 Lewis, David L. *The Public Image of Henry Ford: An American Folk Hero and His Company*. Wayne State University Press, 1976.

15 "Moving Picture Department." *Ford Times*, vol. 9, no. 12, 1916, pp. 535–40.

16 The Ford Film Collection (1.5 million ft.) is accessible at the NARA, well indexed (*Film Description Blinders*) and inventoried by Mayfield Bray (*Guide to the Ford Film Collection in the National Archives*. The National Archives, Washington, DC, 1970). It is widely consulted and cannibalized for illustrative purposes but not methodologically analysed as a historical document.

17 Arnold and Faurote. *Ford Methods*.

18 *Ford People*. 1955. Composite, colour, 16 mm, 800 ft. Library of Congress, FCA 0481.

19 Ford Piquette Plant Museum, 461 Piquette Ave, Detroit, MI 48202. www.fordpiquetteplant.org

20 Ford Motion Picture Laboratories. *A Visit to the Ford Motor Company*. Produced by Ford Motor Co., Dearborn, 1916. Highland Park, silent, black & white, 35 mm, 214 ft., 2 min 37. NARA,

presentations of industrial scenes were relatively sporadic compared to the film department's massive production of moving images. The FMC only rarely shot industrial scenes inside its own workshops and a minor part of its film collection contains workshop images. Indeed, in its early years, the film department was interested in general news (*Ford Animated Weekly*), and didactics (*Ford Educational Weekly*), and was hardly concerned with what was going on in the workshops.[21] Thus, the activities that took place in the factories were not a major subject of interest to the company's internal film department.[22]

Furthermore, a second sign of this historical perplexity concerns the concrete diffusion of the assembly line outside of the FMC. Ford's self-celebrated achievements were rapidly known by all, however, no other American company copied him, and no other industrial countries simply transferred the Ford recipe into their own markets. Management researchers question whether the FMC ever really had assembly lines working the way that they were promoted.[23] Thus, "exporting line assembly had proved more difficult than Henry Ford has imagined,"[24] and many manufacturers pretended to have adopted the device, but in their own way. Mass production as Henry Ford "had made it and defined it was, for all intents and purpose, dead by 1926."[25]

Of course, there are differences between what is said and what is written, on the one hand, and what is done and what is industrially adapted, on the other hand. Nevertheless, this does not mean that words do not have operational efficiency.[26] Indeed, even though Fordism was a literary myth; assembly lines were actually spread throughout the industrial world. They have been applied, adapted or transformed in many ways, as well as in

Ford Film Collection, 200 FC-2573 (b). A text on a cardboard just indicates that "five immense craneways are a part of the Ford system"; *Where and How Fords Are Made*. Parts 1–5, 1919, silent, black & white, 35 mm. NARA, Ford Film Collection, 200 FC-2608; *Model T Assembly and Launching of the* Henry Ford II [ship]. 1924, silent, black & white, 35 mm, 348 ft., 3 min 53. NARA, Ford Film Collection, 200 FC.424; Ford Motion Picture Laboratories. *A Tour thru the Rouge Plants: Part 1–3*. Produced by Ford Motor Co., Dearborn, 1926. Silent, black & white, 35 mm, 911/1061/1027 ft., 3 x 10 min. NARA, Ford Film Collection, 200 FC-4042/4043/4044.

21 Michel, Alain P. *Les images du travail à la chaîne dans les usines Renault de Boulogne-Billancourt (1917–1939). Une analyse des sources visuelles: cinéma, photographies, plans d'implantation.* Thesis, EHESS, 2001, p. 213.

22 Ibid., pp. 204–45.

23 Williams, Karel, et al. "The Myth of the Line: Ford's Production of the Model T at Highland Park, 1909–16." *Business History*, vol. 35, no. 3, 1993, pp. 66–87.

24 Nye. *America's Assembly Line*, pp. 93–94.

25 Hounshell. *From American System*, p. 12.

26 Austin, John L. *How to Do Things with Words*. Harvard University Press, 1962.

different places for a wide range of production systems. Thus, they have constantly evolved according to the ever-changing industrial contexts. In this perspective we can ask: What was actually happening in 1920 in factories like Renault's C5 workshop?

The Cinematographic Oblivion of the Renault "Chain Work" (Through an "Orphan" Documentary Confronted to a Corpus of Industrial Films)

In France, the American assembly line was celebrated before the end of World War I. It was often referred to as "chain work," and it was associated with the American ally.[27] Nevertheless, most material aspects of this history are not documented by written documents. It is for this reason that I make use of all archival images as first-hand sources.[28]

Here, I will focus on one specific sequence of a rather uncommon, silent filmic document called *Aux usines Renault* (*At the Renault Factory*).[29] The description of this film in the Gaumont catalogue doesn't match, neither in length nor in structure, its material conservation in the producer's archives. It is a specific example of an "orphan film," as discussed by Dan Streible.[30] Paolo Cherchi Usai's family metaphor is nice and friendly, but not historically relevant. For example, it is hyperbolic to say that all "films made before 1908 are technically orphans" just because they were sold to the cinema exhibitors. A film archive is not, or should not be, an orphanage.[31] Gregory Lukow's institutional approach is more precise, as his concern is

27 "Le Fabrication automobile en Amérique." *Bulletin des Usines Renault*, no. 6, October 15, 1918, pp. 1–8; for a general overview of the French industry, see Moutet, Aimée. *Les logiques de l'entreprise: La rationalisation dans l'industrie française de l'entre-deux-guerres*. Éditions de l'École des hautes études en sciences sociales, 1997, p. 12.

28 Michel, Alain P. "On the Cover: An Image Mise en Abyme." *Technology & Culture*, vol. 49, no. 4, 2008, pp. 967–73.

29 "Aux Usines Renault." Série enseignement, no. 5212/Master 3037, April 18, 1920. Silent, 22 min. Gaumont Actualité; the context of the production of this film has been presented in Michel, Alain P. "Corporate Films of Industrial Work: Renault (1916–1939)." *Films That Work: Industrial Film and the Productivity of Media*, edited by Vinzenz Hediger and Patrick Vonderau. Amsterdam University Press, 2009, pp. 165–82.

30 Streible, Dan. "The Role of Orphan Films in the 21st Century Archive." *Cinema Journal*, vol. 46, no. 3, 2007, pp. 124–28.

31 Cherchi Usai, Paolo. "What Is an Orphan Film? Definition, Rationale, and Controversy." Paper presented at "Orphans of the Storm I: Saving Orphan Films in the Digital Age" symposium, University of South Carolina, 1999, www.sc.edu/filmsymposium/archive/orphans2001/usai.html.

to expand the notion to "neglected films" apprehended as collections, not as individual films.[32]

Unlike other "orphan films," *Aux usines Renault* comes directly from inside the official archives of a commercial film producer. Edited in 1920 by the Gaumont Film Company, this sequence presents the French automobile plant of Billancourt that was located in the suburbs of Paris. In 1929, or nine years later, the Gaumont catalogue still presented the same documentary as a twenty-two minute "educational film."[33] Indeed, what is "strange" about this film is that the version that was kept by the Gaumont Archives is not twenty-two minutes long, but rather seventeen minutes long. Thus, this could lead us to believe that there existed short and long versions of the same documentary. However, the twenty-two-minute-long film no longer exists in the Gaumont Archives. The only remaining film is silent and what is most astonishing is that it does not have intertitles, or written cardboards. Now, this archival film adds a succession of austere industrial scenes that are rather hard to identify and to understand even for an audience of specialists. Moreover, this mute educational documentary could not be understood by students it was destined to inform. Only intertitles and texts could convey to the viewers the importance of the industrial scenes that they were looking at. The images were only a selection of what the car producer and the filmmaker considered scenes worth showing.

This raises two questions. First, how could such complex and totally silent industrial scenes be used in a pedagogical documentary nine years after the film was first edited? Second, where have the five minutes of lost footage gone? Indeed, it turns out that in the 1990s, when the document was transferred from its original celluloid film to a new video base, all the intertitles and texts were cut, and one fourth of the film was set aside. Thus, an educational documentary became a simple bank of industrial images. This destruction is an archival catastrophe.[34] Therefore, I provocatively argue that this is an excellent opportunity to

32 Lukow, Gregory. "The Politics of 'Orphanage': The Rise and Impact of the 'Orphan Film' Metaphor on Contemporary Preservation Practice." Paper presented at "Orphans of the Storm I: Saving Orphan Films in the Digital Age" symposium, University of South Carolina, 1999, www.sc.edu/filmsymposium/archive/orphans2001/lukow.html.

33 *Répertoire des films de l'encyclopédie Gaumont*. Société des Etablissements Gaumont, Service enseignement, Imprimerie Gaumont, 1929, p. 72. List of films categorized by topic, with their footage. The film *Aux Usines Renault* is presented as a 631-metre piece of footage.

34 The cost of film transfer was considered more important than the integrity of the historical copy.

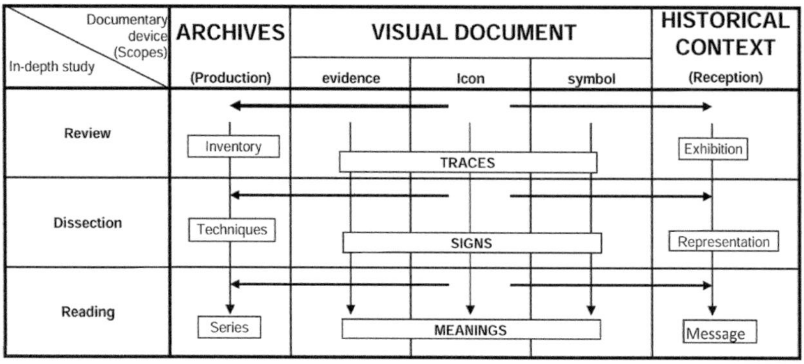

Fig. 25.2. Analytical grid of visual sources. © A. Michel, 2001/2017.

develop a new methodology to film analysis that directly takes images into account, as opposed to what words may say about them (since there are no texts left).

The remaining seventeen minutes of the archived film is now a raw succession of industrial sequences. One of them is an eleven-second high angle sequence shot of a working scene that I identify as the chassis assembly workshop (fig. 25.1 on page 652). It shows five rows of automobile frames being assembled on rolling trestles guided by two rails on the floor. The chassis which are in the process of being assembled do not move and furthermore, no mechanical device seems to exist. Indeed, it is a manual assembly line similar to the first conveyors set up by Ford in 1913 in its Highland Park plant. However, this film sequence only captures a fraction of the activity carried out in this workshop. Moreover, what is most surprising in this extract is the large number of men moving within the range of the screen.

Thus, we can ask: How can we interpret such a short industrial scene the meaning of which is not explicit anymore? My procedure to analyse visual documents from a historical perspective can be summarized in a double entry table with three columns and three lines (fig. 25.2). It combines a vertical and a horizontal approach. Horizontally, the "scope" of the film considers the documentary devices which surround it. The vertical approach is an in-depth progression in the "scale" of each device. Both of these entries, scope and scale, help interpret the meaning of a film.[35]

35 There is, off course, an implicit reference to Alfred Chandler's seminal work on business history and firm organization: Chandler, Alfred. *Scale and Scope*. MIT Press, 1989.

A Visual Document in the Scope of Three Documentary Devices

The "scope" of a historical film refers to the fact that a visual document does not come from nowhere. Even an "archival film" (sic) found on the Internet has been made digitally accessible through a transcription process. The problem is that often, the result is not tied to the original source according to historic rules. Therefore, I discern three different documentary devices which should be taken into consideration. First, of course, the film itself and its contents, or what the motion picture actually shows. Since Charles S. Peirce, semiology indicates that the meaning of a visual document is not only hidden in the images themselves.[36] Thus, part of the information is held within the records that it comes from. Indeed, the archives tell us something about its conservation, and sometimes it reveals information about the conditions of its production. Last, just like any historical document; a film has to be understood in relation to the audience it was meant to be shown to. Indeed, this is the historical context. This socio-historical approach of considering a visual document is connected to other perspectives through which we can study images, such as art history, semiotics, philosophy etc. Nevertheless, its specificity lies in the attention given to the historical conditions of its production as well as of its use.

Original Archives: Source and Preservation?

The "archives" are the first apparatus that we should be careful with, mostly because they are fragile. Thus, we can ask the following questions: Where does the film come from? And, how and why has it been preserved?

In our case, the 1920 film was found in the Gaumont archives because at the time, Renault possessed neither an internal cinema department, nor film records. During the period between the two wars, the automobile manufacturer ordered films when it seemed appropriate, from external cinema companies like Gaumont. For this reason, Gaumont has kept a series of Renault documentaries. *Aux usines Renault*, for example, is just one of the thirty corporate films of this collection.[37] One archival problem, however, is that the Gaumont film archives do not offer written records concerning the demand, the production or the utilization of those motion pictures. Furthermore, we have little information about the authors and their intentions. However, we can deduce that at the beginning of the 1920s,

36 Peirce, Charles Sanders. *Collected Papers*. Harvard University Press, 1931–35.

37 Michel. "Corporate Films," pp. 165–82.

both the automobile company and the film editor were interested in the edition of this documentary.

It is a "*Gebrauchsfilm*," both sponsored and utilitarian.[38] It was produced at this very moment, the end of the war, to serve the glory of the Renault firm and met the demand of Gaumont's cinema audiences in the 1920s. Different concerns were satisfied by the production and distribution of such industrial motion picture scenes. Nevertheless, the use of the film differs from one period to another, as well as according to the partners. Moreover, the nature of this film changed from its edition in 1920 to its educational use in 1929. It also serves to note that the Renault automobile maker did not bother to keep a copy of the film, probably because the 1920 scenes were rapidly outdated. Indeed, a utilitarian film is "ephemeral" with regard to industrial business.[39] The schedule is different for a film producer: outdated technical scenes can still be cinematographically relevant. Although economically obsolete, the 1920 film was still available for students in the 1929 catalogue. It was kept because it still met its educational goal and because Gaumont had no other recent films to show. Indeed, industrial motion pictures were not shot in the Renault plants during the 1920s.[40] Later, in the 1990s, old images could be recycled onto the archival film market. Not very many people cared what was being shown, as long as it looked authentic. Thus, in different contexts, the same moving picture was used and received differently.

The Archival Film Still Extant: The Observation of Situations as a Constructed Flow of Scenes

The second scheme to be considered relates to the images themselves. Thus, we can ask: What is being shown? In this type of documentary, something that really existed is presented in the scene. So, what information can we gather from this representation?

In figure 25.1, for example, were the workers really working? If we carefully view the assembly sequence, we see twelve men coming and going in this part of the workshop during eleven seconds of the sequence shot. On the left part of the scene, a man with a brush and a bucket turns around, and then bustles into the first visible line. Next, someone passes slowly

38 Zimmermann, Yvonne, editor. *Schaufenster Schweiz: Dokumentarische Gebrauchsfilme 1896–1964*. Limmat Verlag, 2011; Prelinger, Rick. "Introduction." *The Field Guide to Sponsored Films*. National Film Preservation Foundation, 2016, pp. iv–xi.

39 Shell, Wanna Rose. "Films in the Archive: Hollywood in Detroit." *Technology and Culture*, vol. 55, no. 3, 2014, pp. 711–15.

40 Michel. "Corporate Film," p. 170.

through the workshop, seems to discuss something with the first worker, and then disappears. On the right side of the picture, a man in a dark blouse does not even move and just stands there staring at the camera. From the right, three other workers arrive in the picture area and join the motionless man in order to be in the picture. The workers know that they are being filmed. Some act as though they are working while others just pose as for a photograph. In this specific situation, the configuration of the activity was transformed by the workers themselves into a collective-fake replay of their genuine working situation.

No real work is actually going on during the shot. Everybody knows that the shooting is taking place. Furthermore, the chassis are not moving and nothing is being assembled. So, if the working scene is simulated, does this mean that the sequence is meaningless?

Indeed, it is not possible to understand the logic of the activity that took place just by looking at it. Nevertheless, it is my belief that this film sequence is precious not only for the information it gives about the activity taking place, but also for the organization of the assembly line installation. Also, the background of the scene certifies that the successive operations accomplished by the teams of workers are different from the operations of the first "gangs" at Peugeot in France, for example, or Ford in America before 1913.[41]

So, is what we see really an assembly line? I say it is for two reasons. My first argument is that this working layout is similar to the installation that was first identified as an "assembly line," according to what the witnesses told and wrote later. The chassis in the background are less assembled than those in the front of the scene. The assembly is progressive. Thus, the film not only confirms the absence of a mechanical power system to convey the chassis, and the flexibility of the tasks between each workstation, but also the obligation for the different tasks to be of equivalent duration. Here, we obviously have a primitive system of assembly which means that it is not easy to operate. Moreover, other photographic evidence shows that this type of assembly line was first adopted by Renault for the production of military vehicles in 1917, even though there are no written records that explicitly refer to them as "chain work."

41 Hounshell. *From the American System*, p. 236. David Hounshell analyses the work of "dynamic assembly teams or gangs [who] moved from station to station down the row"; for the Peugeot case, see Cohen, Yves. "The Modernization of Production in the French Automobile Industry between the Wars: A Photographic Essay." *Business History Review*, vol. 65, no. 4, 1991, p. 756. On the first photograph Yves Cohen described "the teams of specialized workers [who] passed from one fixed car to another."

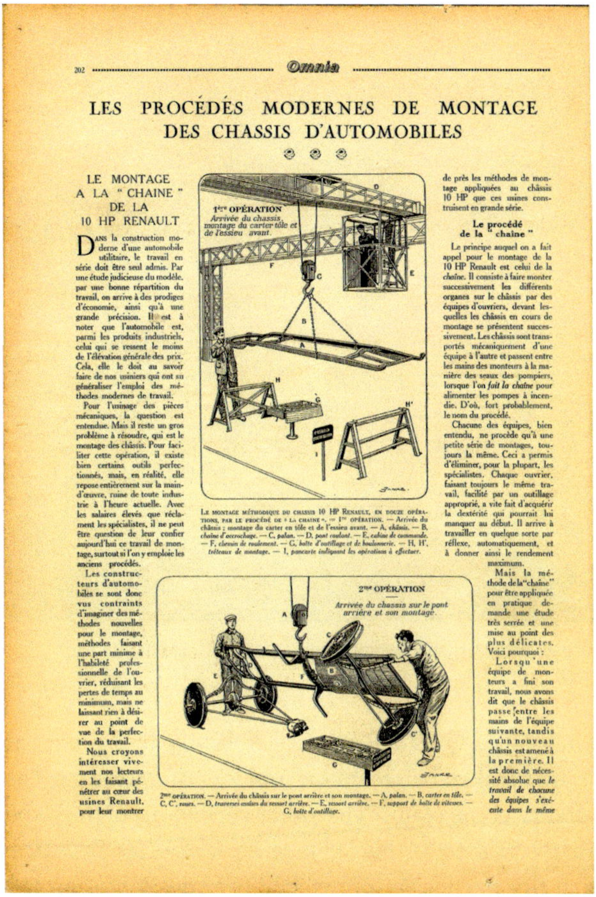

202 Omnia

LES PROCÉDÉS MODERNES DE MONTAGE DES CHASSIS D'AUTOMOBILES

LE MONTAGE A LA "CHAINE" DE LA 10 HP RENAULT

DANS la construction moderne d'une automobile utilitaire, le travail en série doit être seul admis. Par une étude judicieuse du modèle, par une bonne répartition du travail, on arrive à des prodiges d'économie, ainsi qu'à une grande précision. Il est à noter que l'automobile est, parmi les produits industriels, celui qui se ressent le moins de l'élévation générale des prix. Cela, elle le doit au savoir faire de nos usiniers qui ont su généraliser l'emploi des méthodes modernes de travail.

Pour l'usinage des pièces mécaniques, la question est entendue. Mais il reste un gros problème à résoudre, qui est le montage des châssis. Pour faciliter cette opération, il existe bien certains outils perfectionnés, mais, en réalité, elle repose entièrement sur la main-d'œuvre, ruine de toute industrie à l'heure actuelle. Avec les salaires élevés que réclament les spécialistes, il ne peut être question de leur confier aujourd'hui ce travail de montage, surtout si l'on y emploie les anciens procédés.

Les constructeurs d'automobiles se sont donc vus contraints d'imaginer des méthodes nouvelles pour le montage, méthodes faisant une part minime à l'habileté professionnelle de l'ouvrier, réduisant les pertes de temps au minimum, mais ne laissant rien à désirer au point de vue de la perfection du travail.

Nous croyons intéresser vivement nos lecteurs en les faisant pénétrer au cœur des usines Renault, pour leur montrer de près les méthodes de montage appliquées au châssis 10 HP que ces usines construisent en grande série.

1re OPÉRATION
Arrivée du chassis, montage du carter tôle et de l'essieu avant.

LE MONTAGE MÉTHODIQUE DU CHASSIS 10 HP RENAULT, EN DOUZE OPÉRATIONS, PAR LE PROCÉDÉ DE « LA CHAINE ». — 1re OPÉRATION. — Arrivée du châssis ; montage du carter en tôle et de l'essieu avant. — A, *châssis*. — B, *chaîne d'accrochage*. — C, *palan*. — D, *pont roulant*. — E, *cabine de commande*. — F, *chemin de roulement*. — G, *boîte d'outillage et de boulonnerie*. — H, H', *tréteaux de montage*. — I, *pancarte indiquant les opérations à effectuer*.

Le procédé de la "chaine"

Le principe auquel on a fait appel pour le montage de la 10 HP Renault est celui de la *chaîne*. Il consiste à faire monter successivement les différents organes sur le châssis par des équipes d'ouvriers, devant lesquelles les châssis en cours de montage se présentent successivement. Les châssis sont transportés mécaniquement d'une équipe à l'autre et passent entre les mains des monteurs à la manière des seaux des pompiers, lorsque l'on *fait la chaîne* pour alimenter les pompes à incendie. D'où, fort probablement, le nom du procédé.

Chacune des équipes, bien entendu, ne procède qu'à une petite série de montages, toujours la même. Ceci a permis d'éliminer, pour la plupart, les spécialistes. Chaque ouvrier, faisant toujours le même travail, facilité par un outillage approprié, a vite fait d'acquérir la dextérité qui pourrait lui manquer au début. Il arrive à travailler en quelque sorte par réflexe, automatiquement, et à donner ainsi le rendement maximum.

Mais la méthode de la "chaîne" pour être appliquée en pratique demande une étude très serrée et une mise au point des plus délicates. Voici pourquoi :

Lorsqu'une équipe de monteurs a fini son travail, nous avons dit que le châssis passe entre les mains de l'équipe suivante, tandis qu'un nouveau châssis est amené à la première. Il est donc de nécessité absolue que *le travail de chacune des équipes s'exécute dans le même*

2me OPÉRATION
Arrivée du chassis sur le pont arrière et son montage.

2me OPÉRATION. — Arrivée du châssis sur le pont arrière et son montage. — A, *palan*. — B, *carter en tôle*. — C, C', *roues*. — D, *traverses essées du ressort arrière*. — E, *ressort arrière*. — F, *support de boîte de vitesses*. — G, *boîte d'outillage*.

Fig. 25.3. First written attestation of a Renault assembly line, 1922. © *Omnia* (1922), Archives de Renault Histoire.

My second argument is that this working installation is not fundamentally different from the one that is explicitly identified as "chain work" in Pierre Maillard's 1922 article.[42]

Figure 25.3 shows the first written document published on this subject, and this article, or attestation, is often used as evidence of Louis Renault's delay in introducing assembly lines in his factory, whereas their existence is attested for André Citroën as early as 1919. Nevertheless, this assertion is contradicted

42 Maillard, Pierre. "Le montage à la chaîne de la 10 HP Renault." *Omnia*, vol. 28, September 1922. Reproduced in *De Renault Frères, constructeurs d'automobiles, à Renault Régie Nationale*, vol. 3, 1973, pp. 95–100; "Le montage à la chaîne." *L'illustration*, October 1924. Reproduced in *Les grands dossiers de l'Illustration: l'automobile.* Le Livre de Paris, 1993, pp. 90–91. These articles are based on three photographic reports made by the Renault photographic service in the workshops 1-Montage 10HP (2/27/22) SR(42) no. 12237/12252; 2-Montage 6HP (9/1/24) SR(53) no. 15854-15863; and 3-Montage 10HP (7/1/24) SR(52) no. 15639/51 & 15703.

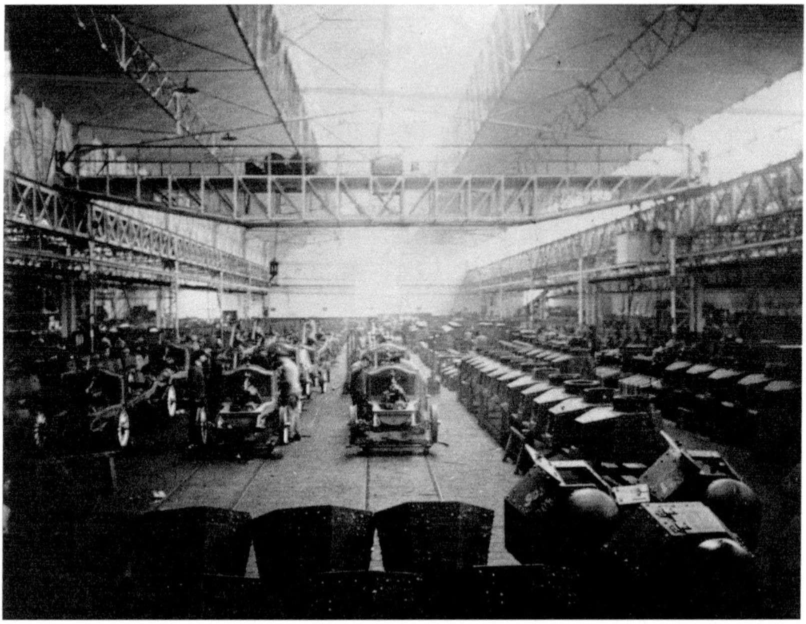

Fig. 25.4. Picture showing a first form of the final assembly line, 1917. © Archives de Renault Histoire, Rights reserved, 1917.

by photographs and motion pictures of Renault as indicated by figure 25.4. Indeed, a series of pictures taken between 1917 and 1921 that were used as proof of notarial inventory of the plant[43] give other viewpoints of the same workshop, and confirm the presence of this rudimentary form of assembly line.

It is my opinion that this purposive approach to the question regarding the issue of delays in introducing the assembly line is pointless. Even at Ford, the word "assembly line" designated a multitude of different devices which evolved all the time, and with no internal records that explicitly referred to those differences. However, what is important to notice is that the assembly system in Renault was referred to as "chain work" in 1922, whereas a similar working device was not named as such earlier on. Indeed, the expression disappeared from Renault's written records after 1924, and then reappeared at the end of the 1920s. Clearly, language changes at a different pace than industrial systems. This fluctuation is meaningful and images help the viewer follow this disconnected chronology.

43 "Chaîne de montage châssis." September 1919, SR(27) no. 7170. Photograph published in the *Bulletin des Usines Renault*, vol. 36, January 15, 1920; SR(30) no. 8404/8408 from 1920. "Inventaire Plousey," Archives of the Renault Histoire Association. In 1921, the Renault firm was to change from a personal affair (Automobiles Louis Renault) to a limited company (Société anonyme des usines Renault).

The Historical Context of a Document Put into Perspective

Ultimately, to be able to interpret images, or to measure their meanings as well as their impact, they must be considered within a broader network of information. The industrial scenes have to be *recontextualized*. Indeed, this common-sense idea was advocated by Marc Ferro in his 1973 paper.[44]

In figure 25.5 we see Ferro's diagram. He calls this context the "invisible reality zone" (*zone de réalité non visible*), located outside the scope of the images. Beyond the "apparent matter" (*contenu apparent*) of the film, stands the "latent content" (*contenu latent*), which can only be revealed through the confrontation of the film to the surrounding situation. Thus, clear knowledge of the historical context is needed to better grasp the meaning of what is shown. In other words, it should be defined in terms of these questions: Who made the images? and Why and to whom were they destined?

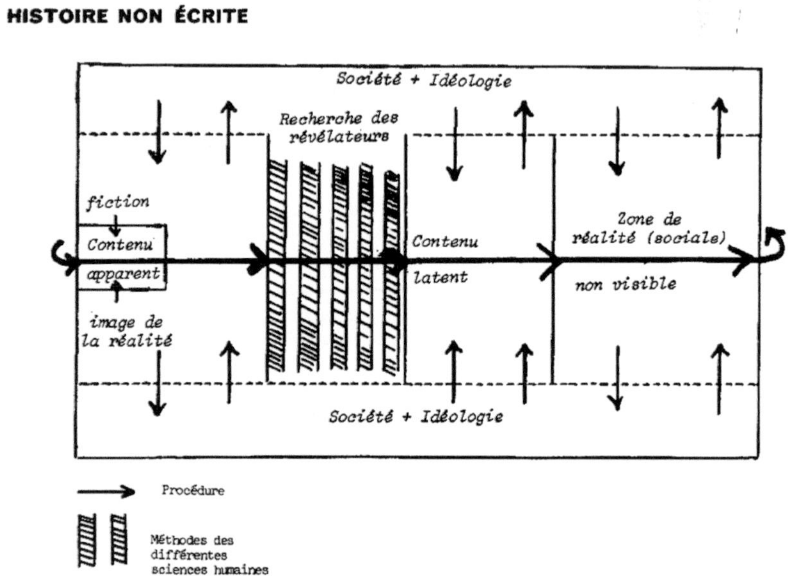

Fig. 25.5. Ferro's analytical grid of films, 1973. © Marc Ferro, Annales ESC, 1973.

In the Renault case, the short crisis of 1921 that was due to the switch from war production to the peace economy postponed Renault's need to promote the assembly line production of its cheaper cars. Moreover, since it was not mechanized, this new assembly line was immediately outdated, when

44 Ferro. "Le film, une contre-analyse de la société."

compared to the Ford standard of the time. Indeed, it had nothing of interest to show in a movie, and photographs and drawings were preferred in order to give the proper vision of this industrial modernity. Thus, it is not through films, but through still images that the Renault firm chose to represent its moving assembly lines in the 1920s.[45]

The different scopes of the visual document help define the role given to the assembly line within a specific documentary. Nevertheless, the interpretation of these images needs an in-depth analysis according to each of the three devices they are a part of.

The Three Scales in the Reading of a Visual Document

My second point is that this "meaning" cannot be directly inferred. The vertical entry method to film studies was inspired by Pierre Sorlin's 1977 *Sociology of Film*.[46] Indeed, Sorlin identified three successive levels of analysis (see Fig. 25.5 p. 665). First, a case-by-case "review" (which Sorlin calls a *mise à plat*) or investigation on what the film, its archives, and its context reveal at first sight. The second step is to investigate the way "films work," which Sorlin calls the *fonctionnement* of each device: its "dissection". Last, it is possible to go even deeper into the interpretation of a film through its "reading" (*lecture*) and interpretation.

The Case-by-Case Review of the Film Sequences

Generally speaking, this first scale of analysis is what we have already done in presenting the different scopes of the film. The review of the film's archives consists in understanding both the way a specific documentary has been preserved, as well as to identify the document inside the series (or corpus of connected films) it comes from. This serialization, as suggested by Marc Ferro, helps with the recontextualization of the document.[47] To do so, I believe that films must be confronted to other visual documents. For example, the same assembly workshop (*atelier* C5 in figure 25.6) was

45 Michel. *Travail à la chaîne*.

46 Sorlin. *Sociologie du cinéma*, pp. 148–287. "Film Analysis and Social History" is the title of the third part of the book. Sorlin gives "the outline of a method that could be applied to any identifiable film set, corresponding to a period, centered on a theme, or defined in any other way." Ibid., p. 158.

47 Ferro, Marc. "Critique des actualités, Histoire parallèle." *Cinéma et histoire*, Folio, 1993, p. 119.

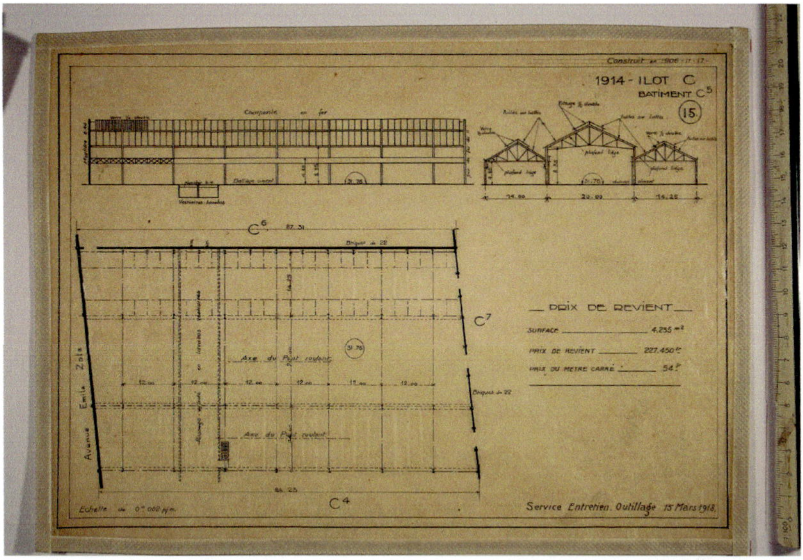

Fig. 25.6. Drawing of the C5 workshop, 1918. © Service Outillage Entretien (SOE, 1918), Archives of Renault Histoire Association.

captured in by many different types of still images. We have seen that photographs shot in 1920 show other viewpoints of the same place since blueprints give the structure of the workshop. Moreover, this specific workshop is only a small part of Renault's large industrial complex. Thus, the historical study reveals its exact location in the factory, and validates the hypothetical date that the moving picture was shot. From a historical point of view, the cinematographic presentation of a "real" industrial scene has to be precisely situated in its specific place and at a precise moment in time.

Thus, the case-by-case review of the film itself has to take into consideration its exact mounting, the origins of the scenes, and the places that they were shot in. We have neither the scenario of our documentary, nor do we have its technical cutting, i.e. the division of this scenario in filmic texts, which would, of course, be of great help.[48] Therefore, in order to compensate for this deficiency, the film review could be established by the manner in which it was put together after mounting, or in other

48 This is the type of document that Vincent Guigueno has remarkably used in his study of the representation of the line work in René Clair (Fonds René Clair, RC 11 of the Bibliothèque de l'Arsenal) and, less directly, in Charlie Chaplin. Guigueno, Vincent. "Le travail à la chaîne à l'épreuve du burlesque." *De l'histoire au cinéma*, edited by Antoine de Baecque and Christian Delage. Complexe, 1998, pp. 127–44.

numéro séquence	Thème de la séquence	Localisation	Durée
	A- Présentation générale		
1	Vue de toits / parc aux aciers	Bât.D 16 / bât. D-A	25 s.
	B- Matières premières et demi-produits		
2	Parc des bois	Ilot K	25 s.
3	Fonderie : vues de l'atelier	Ilot E	40 s.
4	Estacade : déchargement sur embarcadaire	Est. 1 : Quai de Bill.	10 s.
5	Centrale électrique	Bâtiment E8	10 s.
6	Passage par un souterrain	entre C3 et B3	27 s.
7	Parcs à matériaux	Ilot F	30 s.
8	Fonderie (convertisseur Bessemer/coulée)	Aciérie L3 - L8	70 s.
	C- Usinages de la Première Guerre mondiale		
9	Banc d'essai moteurs d'avion	Bât. S1	40 s.
10	Montage des fuselages d'avions	Usine O	5 s.
11	Opérateur sur un tour revolver	Bât. D	45 s.
12	Atelier de soudage / soudeur à lunettes	Bâtiment F2	48 s.
13	Atelier des tours (femmes)	Bâtiment B1	30 s.
14	Usinage de la coiffe d'un obus	Ilot E ou P	120 s.
	D- Usinages de base		
15	Petite presse / cylindre	Ilot L où B3	30 s.
16	Emboutissage /tambour de roue (fonte)	Bâtiment E3	20 s.
17	Martelage d'un vilebrequin		150 s.
	E-Montage châssis		
18	Chaînes de montage châssis	Bât.C5 (At. Richet)	15 s.
	F- Segment mélangé		
19	Sortie de l'usine	Place Nationale	10 s.
20	Atelier de montage des moteurs (d'avions)	Bâtiment B3	12 s.
21	Découpe lingot / presse à filer	Ilot E	40 s.
22	Atelier des tours (hommes et femmes)	Bâtiment B1	30 s.
Source : Gaumont, Série enseignement, n° 5212			

Fig. 25.7. Table of the sequences of the documentary *Aux usines Renault* (1920). © A. Michel, EHESS, 2001.

words, a shot-by-shot survey of the entire documentary film. The fact is that each shot is the "basic unit" of a moving picture (substance of the image, recorded sound, shooting). While the result would be accurate, it would also be limited because it does not reveal by itself the intention behind the film editing. This meticulous decomposition would also be *rédibitoire* for the individual study of a full-length feature film, or for an important series of moving pictures.[49] My argument is that the film review should identify the different "sequences" of the film where each sequence is the assembly of shots that concern a same subject (in the same place).

49 Sorlin. *Sociologie du cinéma*, pp. 153, 181, 224, and 218.

The Coherence of the Editing: The Scenario

Nevertheless, Pierre Sorlin suggests going deeper into the understanding of a historic film in order to "dissect" the images beyond what they represent in the picture, to clarify the way records have been brought together, as well as to identify their ties with the historical situation.

From what we know of the original moving picture archives, we can assume that the truncated film preserved by Gaumont is not just random footage, but rather a succession of sequences that follow the order of the original way in which it was edited. Thus, I have reconstructed its relative structure with the measure of the length of each element, much like rebuilding a monument. Thus, a diagram re-establishes the shape of the film roll (Fig. 25.8).

The film is composed of twenty-two sequences over six chapters (from A to F). Behind its apparent disorder, the documentary offers a fairly comprehensive overview of the conditions under which automobiles and other Renault productions were manufactured. The scenario roughly follows the sequence of steps necessary for the production of an automobile. After a short circular panorama of the plant, the film begins by showing the delivery of the raw materials needed to build cars (sequences 1–7). Next, it shows the metallurgic activities (3, 8, 15, 16, 17, 20), a set of machining operations (11, 12, 13, 21), and finally it ends with a scene where workers are leaving the factory (19). The assembly line sequence (18) is not only very short, but it is also the only one of its kind. The film presents a short view of the chassis assembly line as a recent innovation. Thus, we can ask: Why was this important industrial innovation merely indicated, but not developed in this film?

It is obvious that in this 1920 documentary, emphasis was placed on the machining operations which occupy the greater part of the film. These precision tasks were performed by professional workers on stationary workplaces. Renault's assembly line was only presented discretely at a time when Ford's industrial films emphasized the extreme mechanization of its own lines. In 1919, a series of five successive documentaries detailed the way Ford produced its vehicles (mostly the famous Model T but also a wide range of trucks).[50] *Where and How Fords Are Made* promoted the "modern" organization of its car production and highlighted the logical workflow that the assembly lines should follow. Not only were they, by then, moving mechanically, but also a crowd of employees worked, with a few of them impressively keeping up with the pace of the conveyer, while seated on a stool on wheels (called a "creeper"). The Highland Park plant assembly line

50 *Where and How Fords Are Made.*

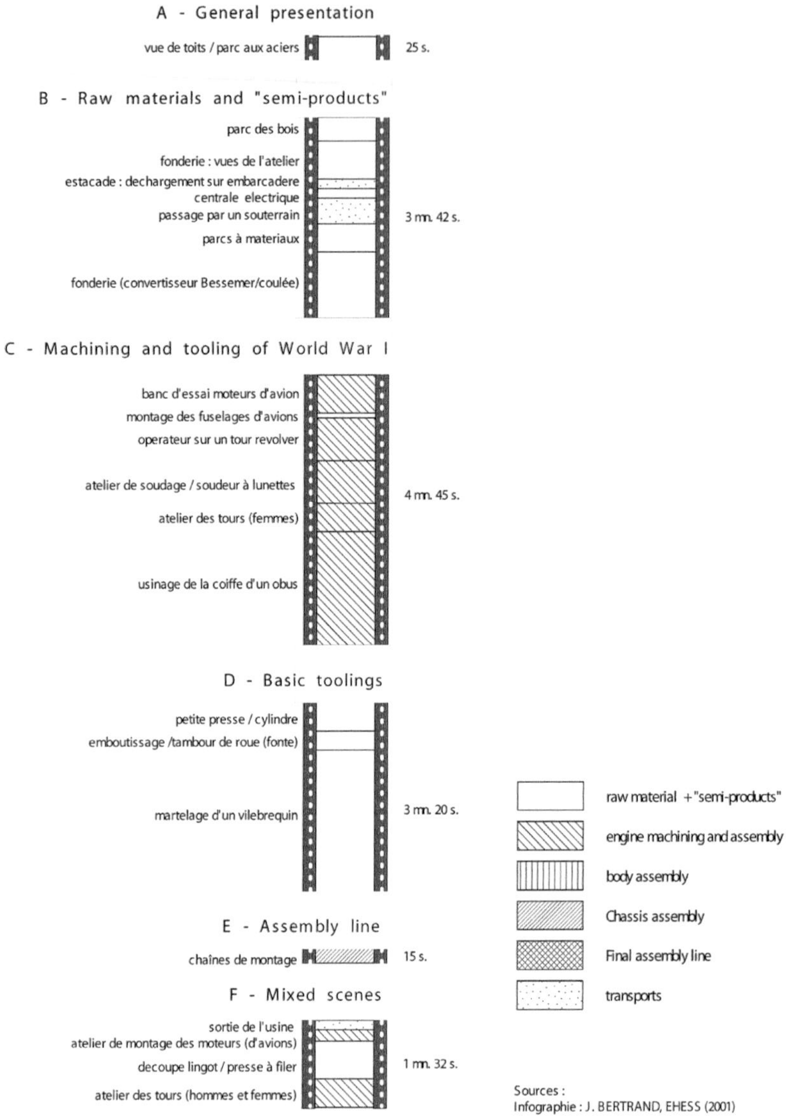

Fig. 25.8. Structure of the scenario of *Aux usines Renault* (1920). © A. Michel & J. Bernard, EHESS, 2001.

was the star of this documentary. In this context, the eleven-second assembly scene frame was too short for the audience to notice that there were actually no mechanic conveyors like in Detroit, making it difficult for anyone to perceive its industrial logic. During that time, the French manufacturer was still transitioning to a peace economy, and was not yet an important automobile manufacturer. Indeed, the total production at Billancourt was

2,500 cars in 1919 and two years later it was still only at 5,800 vehicles. In the meantime, in the US, FMC produced nearly one million Model Ts.

The 1920 film mostly shows a "retrospective" of the wartime activity rather than presenting a factory in the process of its adaptation to peace fabrication. On this front, Renault and Ford had nothing in common as Renault's "assembly line" could not match Detroit's. As suggested by David Nye, "mass production made sense only if there was mass demand."[51] As for many car assemblers around the world, the 1920 Renault case suggests that an assembly line can be adapted to small series.[52] This fact contradicts the widespread Taylorist idea of a "one best way."[53]

The post-war documentary had another goal. It mostly wanted to remind its audience of Renault's participation in the military effort of World War I. Indeed, Renault made huge benefits from 1914 to 1918. Moreover, its factory expanded from 130 to 365 acres in its effort to meet the demands of the war. In 1920, French authorities began to investigate the taxes Renault owed based on the profits it made from war work and the restitution of the parcels appropriated by the firm.[54] A film celebrating Renault's patriotic contributions to victory, as well as its technological innovations, was intended to help people look away from this controversial issue. However, what can be said about the techniques of this representation?

The "Reading" of the Film's Message.

To go even deeper into the interpretation of a film, I suggest that map making, or cartography can be useful. For example, if we locate the successive film sequences on a map, it appears that the documentary is constructed like a cinematographic visit of the plant (Fig. 25.9).

All but one of the scenes were shot in the central part of the factory (*La grande usine*) where the machining workshops were located. At the time, Renault's activities had spread beyond its original site in Billancourt, with a second location in Plant O (*Usine O*) two kilometres north-east. Other

51 Nye. *America's Assembly*, p. 29.

52 Shiomi, Haruhito, and Kazuo Wada, editors. *Fordism Transformed: The Development of Production Methods in the Automobile Industry*. Oxford University Press, 1995.

53 Taylor, Frederick W. *The Principles of Scientific Management*. Harper & Brothers, 1911.

54 In 1919, Renault's 249 million franc business turnover was more than five times that of 1914. Touchelay, Béatrice. "D'une sortie de guerre à l'autre: de l'impôt sur les bénéfices de guerre (1916) à la confiscation des profits illicites (1944–1945), l'État a-t-il appris à compter?" *L'épuration économique en France à la Libération*, edited by Marc Bergère. Presses universitaires de Rennes, 2008, pp. 33–50.

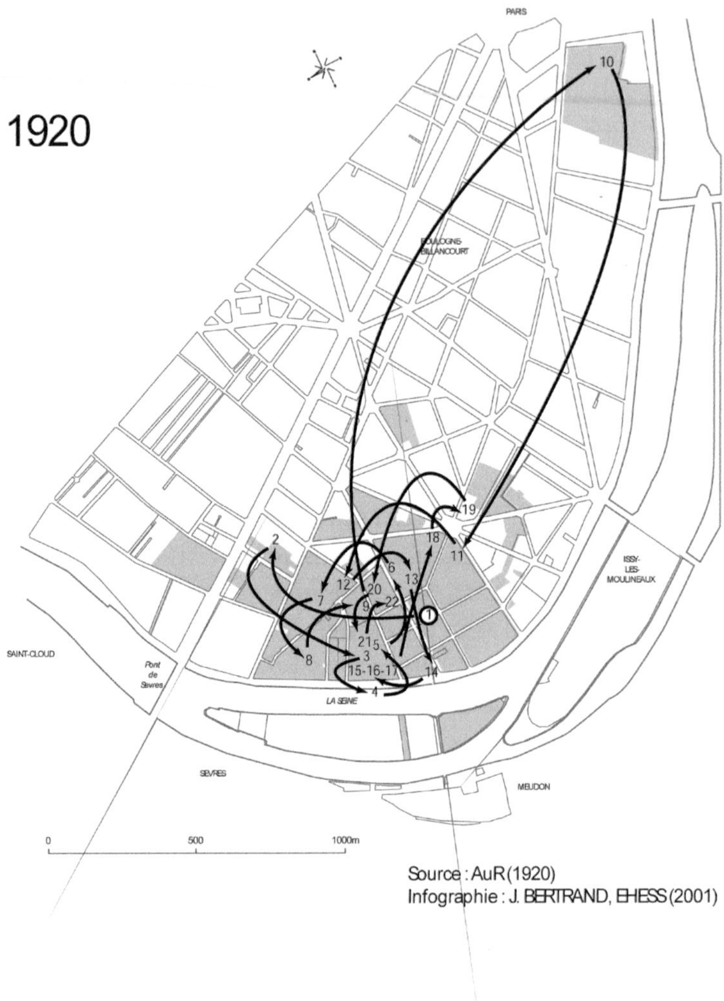

Fig. 25.9. Geography of the scenes in *Aux usines Renault* (1920). © A. Michel & J. Bernard, EHESS, 2001.

factories had been established in French provinces and abroad. The 1920 film mostly celebrated the skills of Renault's qualified workers operating in their spectacular "arena of ability."[55] Indeed, while this type of work was impressive, it remained traditional, as opposed to "modern." On the contrary, Renault's manual assembly line was new but cinematographically disappointing because it was nowhere as spectacular as the Ford film reference.

55 Dodier, Nicolas. *Les hommes et les machines: La conscience collective dans les sociétés technicisées*. Éditions Métailié, 1995, p. 220.

It was surely not appropriate for Renault to promote its new 1920 assembly line since it would look ridiculous when compared to Ford's in Detroit. Therefore, the French car maker could not yet show it in a film because it did not look modern, nor did it move correctly, according to the Ford standard. Indeed, Renault only introduced mechanization later in the mid-1920s.

Furthermore, it serves to mention that during that time, the communist workers' organizations were worried and critical about the American inspired "chain work."

> Comrades will be able to take a look [...] at the application of the new method designated as an American chain (Ford system). They will not have the vision of a workshop working in order, but on the contrary one of a cluster of machines piled one upon the other, pell-mell, not leaving enough space for the workers to move. There they will have an idea of what all the workshops will look like in a little while.[56]

Thus, at this point, it is essential as well as instructive to look at the "position" of the 1920 documentary, compared to the collection of Renault films.[57] The earliest films showing Renault workshops were shot during World War I. In 1916, before the United States entered the war, an American military commission visited the Renault factory and the event was shot by operators from the military film service.[58] Many other scenes show women working in military production in order to promote the civilian effort to support the front lines of the war. Moreover, a visit from Minister of War Albert Thomas on September 1, 1917, is also a subject of interest at a time when the industry and its workforce, in the context of the first Russian revolution, were being re-mobilized for the "last battle before victory." Thus, these first industrial films worked as patriotic propaganda.

The film *Aux usines Renault* is the only one of its kind. No other films were shot in Renault during the 1920s. It reveals that Renault did not want to use films to present its factory. Its assembly line was explicitly promoted in 1922 through an article illustrated by drawings. Moreover, it is the first written account of "chain work," but not the first piece of visual evidence of its existence. I have shown that traces of this set-up can be found as early as 1917 for the assembly of war tanks, and the mass machining of ammunitions.

56 *Bolchevick de chez Renault*. No. 5, October 1924; see Michel. *Travail à la chaîne*, p. 71.

57 See Michel. "Corporate Films," p. 70.

58 About the French Army's Photography and Cinema Service (SPCA), see Veray, Laurent. *La Grande Guerre au cinéma: De la gloire à la mémoire*. Éditions Ramsay, 2008.

In the beginning of the 1920s, still pictures and captions were better at depicting the division of work at each station.

This cinematographic "eclipse" also explains why Gaumont kept an outdated film for students. It was only in 1930 that the Renault firm commissioned a new film so as to show its mechanized assembly lines in the new Seguin Island plant.[59] Indeed, this 1930 assembly line finally fit the Ford standard during America's economic crisis. While Renault's assembly lines were still very different, they looked similar when viewed on a motion picture, and the French manufacturer could use them to celebrate, and as a sign of his prosperity. However, the cinematographic illusion would not last long. By the mid-1930s the economic crisis struck France and no European car maker could pretend to match the Americans.

3D Restitution as an Investigation Device to Interpret Orphan Images

Contradictory to a widespread idea, analogic images are not the simple figurative two-dimensional projection of a past reality. So much so, they are not just an illusion. On the contrary, an image reveals an absence, allowing it to become more real than what it represents, to such a point that it retains the physical trace of a reality that is no longer perceived. Even an orphan documentary film can reveal data through a methodological cross-analysis of what is being presented and why. But how can these different facts be interpreted and put together to document the industrial activity they present? I have used 3D restitution as a tool to conduct the examination of Renault's early images of its final assembly line.[60] Modelization is a useful tool to compensate the mis-restoration of a filmic roll in its original archives and make silent motion pictures talk again.

In this perspective, my 3D mock-up is a research tool elaborated to produce historical knowledge and deepen the history of industrial work and organization. It is an experimental platform to test hypothesis, to cross-check information and to help interpret visual data. The point is to highlight the concrete processes of production through the study of a

59 *Fabrication d'une automobile aux usines Renault*. 1930. Silent, intertitles in French, 41 min 14. Cinémathèque de la ville de Paris, 7, rue Robert Estienne, Paris XIIIe.

60 Michel, Alain P., and Shadia Kilouchi. "Renault-Billancourt's C5 Workshop in the Digital Age: A New Story of the 1922 Assembly Line." *L'histoire contemporaine à l'ère digitale*, edited by Fédéric Clavert and Serge Noiret. PIE-Peter Lang, 2013, pp. 235–49.

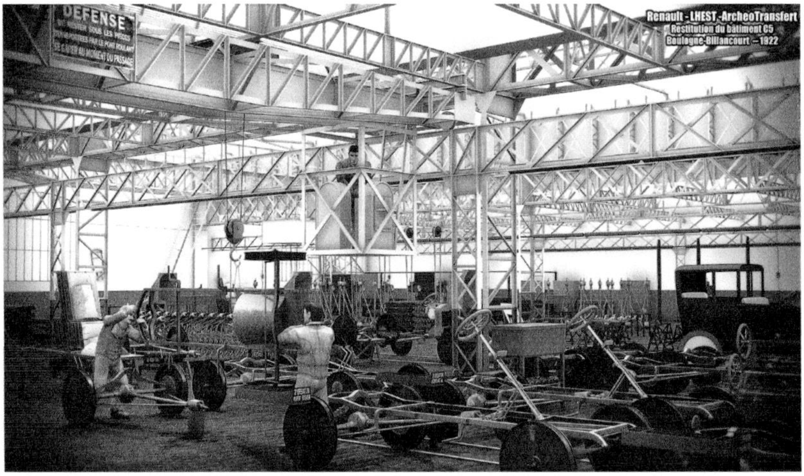

Fig. 25.10. A 3D restitution of an assembly line operation, 1922. © IDHES-Evry et ArchéoTransfert (2010).

relevant techno-industrial site, the vestiges left, the images produced and all other documents concerning the same spot at the same period. The specific contribution of a 3D study is to account for the different scales of the industrial activity in its environment. On this subject, explicit texts are scarce so that most of the production process is hidden in a "black box." But other available sources (visual, oral, archaeological, statistical etc.) offer chances to apprehend part of it. The meaning of these vestiges has to be rediscovered through the creation of multimedia collections (corpus) of documents often coming from scattered archives. The numerical models we extract from the data of these collections of images make it possible to set a workplace in its territory at different levels of its documentary comprehension: from the landscape to the workshop, from the product to the machine, from gestures to processes and through the images to imagination.

The validation of such type of multimedia assembly, data processing and interpretation are part of a documentary heuristic which is in the process of elaboration.[61] It also aims to develop a new methodology for historical modelling. It joins together pluridisciplinary forces to federate partners in the fields of human sciences, digital humanities and engineering techniques. It associates cultural and territorial institutions, business firms and disseminates results to a broader audience. This original coordination between social and computer sciences requires semantic interoperability between

61 Michel, Alain P. "La reconstitution virtuelle d'un atelier de Renault-Billancourt: sources, méthodologie et perspectives." *Documents pour l'histoire des techniques*, vol. 18, 2009, pp. 23–36.

these disciplines and allows the acquisition of expertise richer than the mere sum of skills brought by the partners. The crossing of different sources is used to test hypotheses, to enrich the study procedures, establish a method, develop a modelling ontology and propose standards. This multidisciplinary federation of complementary capabilities leads to mutual enrichment and produces a new expertise. Finally, the results and syntheses of the various historical studies are accessible, both through classical scientific publications but also through public access on a website to documentary corpus and models of similar industrial sites. The goal is to build together a charter of good practices and be a reference label in modelling historical heritage of industrial and technical spots.[62]

Conclusion

In his pioneering film essay *Letter from Siberia* (1958), Chris Marker shows how the voice-off can influence the understanding of a film sequence.[63] Thus, the interpretation of a totally silent film is another challenge. Just like any document, images must be confronted to one another. They can also be used to re-examine traditional written sources. Nevertheless, these various documents are not always complementary; indeed, some are even contradictory.

In a "socio-material" approach of film history, the shift from "old" to "new" media changes, but doesn't revolutionize, the historical method of analysing visual documents from the past.[64] The ties between film representation, filmic technics and filmed technology are blurred. In this perspective, the concept of the "media boundary object" is helpful.[65] The question is: Which is the object and what are the adjacent territories? As Florian Hoof suggests, both the filmic image and the film technology have to be taken into account. I argue that a methodological analysis of an industrial documentary film

62 Michel, Alain P. *"Usines 3D": Reconstitution virtuelle d'ateliers et d'usines du patrimoine industriel.* ANR Corpus 2007, no. 034, 2012, p. 42. Final report of the French National Research Agency Program, www.usines3D.fr.

63 Paggi, Silvia. "Voix-off et commentaire dans le cinéma documentaire et ethnographique." *Cahiers de Narratologie*, vol. 20, 2001, doi:narratologie.revues.org/6321.

64 Hoof, Florian. "'Have We Seen It All Before?' A 'Sociomaterial' Approach to Film History." *Proceedings XXI International Film Studies Conference Udine.* 2015, pp. 347–57; in this perspective, see Delage, Christian, Vincent Guigueno, and André Gunthert, editors. *La fabrique des images contemporaines.* Éditions Cercle d'Art, 2007.

65 Hoof, Florian. "The Media Boundary Objects Concept: Theorizing Film and Media." *Media Matter: The Materiality of Media, Matter as Medium*, edited by Bernd Herzogenrath. Bloomsbury, 2015, pp. 180–200.

also heavily relies on a history of technology attentiveness to the past scenes it presents. Between vision and representation, documentary films must be compared to the practical situation they pretend to show. Images have to be located in space and in time according to what they are supposed to reveal, and often hide. To do so, I argue that 3D restitution is a relevant research tool. Beyond the specific case of *Aux usines Renault*, it can be integrated to the methodological analysis of archival forms of contemporary circulation of orphan films, archival data or source materials.

Works Cited

Arnold, Horace, and Fay Faurote. *Ford Methods and the Ford Shop* [1915]. The Engineering Magazine Company, 1919.

Austin, John L. *How to Do Things with Words*. Harvard University Press, 1962.

Bolchevick de chez Renault. No. 5, October 1924. Archives of the Renault Histoire Association, Billancourt.

Bray, Mayfield. *Guide to the Ford Film Collection in the National Archives*. National Archives, Washington, DC, 1970.

Chandler, Alfred. *Scale and Scope*. MIT Press, 1989.

Cherchi Usai, Paolo. "What Is an Orphan Film? Definition, Rationale, and Controversy." Paper presented at "Orphans of the Storm I: Saving Orphan Films in the Digital Age" symposium, University of South Carolina, 1999, www.sc.edu/filmsymposium/archive/orphans2001/usai.html.

Cohen, Yves. "The Modernization of Production in the French Automobile Industry between the Wars: A Photographic Essay." *Business History Review*, vol. 65, no. 4, 1991, pp. 754–79.

Delage, Christian, Vincent Guigueno, and André Gunthert, editors. *La fabrique des images contemporaines*. Éditions Cercle d'Art, 2007.

Dodier, Nicolas. *Les hommes et les machines: La conscience collective dans les sociétés technicisées*. Éditions Métailié, 1995.

Elsaesser, Thomas. "The New Film History as Media Archeology." *CiNéMAS: revue d'études cinématographiques*, vol. 2–3, 2004.

Faurote, Fay. "What Is Going on behind the Scenes at the Ford Plants." *Factory and Industrial Management*, vol. 74, 1927, pp. 257–62.

Ferro, Marc. "Critique des actualités, Histoire parallèle." *Cinéma et histoire*, Folio, 1993, pp. 109–134.

Ferro, Marc. "Le film, une contre-analyse de la société." *Annales: Économies, Sociétés, Civilisations*, vol. 28, no. 1, 1973. Reedited in *Cinéma et histoire*. Folio, 1977, pp. 31–62.

Ginzburg, Carlo. *Clues, Myths and the Historical Method.* Johns Hopkins University Press, 1989.
Goffman, Erving. *The Presentation of Self in Everyday Life.* Anchor Books, 1959.
Guigueno, Vincent. "Le travail à la chaîne à l'épreuve du burlesque." *De l'histoire au cinéma*, edited by Antoine de Baecque and Christian Delage. Complexe, 1998, pp. 127–44.
Hoof, Florian. "'Have We Seen It All Before?' A 'Sociomaterial' Approach to Film History." *Proceedings XXI International Film Studies Conference Udine.* 2015, pp. 347–57.
Hoof, Florian. "The Media Boundary Objects Concept: Theorizing Film and Media." *Media Matter: The Materiality of Media, Matter as Medium*, edited by Bernd Herzogenrath. Bloomsbury, 2015, pp. 180–200.
Hounshell, David. *From the American System to Mass Production, 1800–1932: The Development of Manufacturing Technology in the United States.* Johns Hopkins University Press, 1984.
"La fabrication automobile en Amérique." *Bulletin des Usines Renault*, no. 6, October 15, 1918, pp. 1–8.
"Le montage à la chaîne." *L'illustration,* October 1924. Reproduced in *Les grands dossiers de l'Illustration: l'automobile.* Le Livre de Paris, 1993, pp. 90–91.
Lewis, David L. *The Public Image of Henry Ford: An American Folk Hero and His Company.* Wayne State University Press, 1976.
Lukow, Gregory. "The Politics of 'Orphanage': The Rise and Impact of the 'Orphan Film' Metaphor on Contemporary Preservation Practice." Paper presented at "Orphans of the Storm I: Saving Orphan Films in the Digital Age" symposium, University of South Carolina, 1999, www.sc.edu/filmsymposium/archive/orphans2001/lukow.html.
Maillard, Pierre. "Le montage à la chaîne de la 10 HP Renault." *Omnia*, vol. 28, September 1922. Reproduced in *De Renault Frères, constructeurs d'automobiles, à Renault Régie Nationale*, vol. 3, 1973, pp. 95–100.
Michel, Alain P. "Assembly Line." *The Palgrave Dictionary of Transnational History*, edited by Akira Iriye and Pierre-Yves Saunier. Macmillan, 2009, pp. 67–70.
Michel, Alain P. "Corporate Films of Industrial Work: Renault (1916–1939)." *Films That Work: Industrial Film and the Productivity of Media*, edited by Vinzenz Hediger and Patrick Vonderau. Amsterdam University Press, 2009, pp. 165–82.
Michel, Alain P. "La reconstitution virtuelle d'un atelier de Renault-Billancourt: sources, méthodologie et perspectives." *Documents pour l'histoire des techniques*, vol. 18, 2009, pp. 23–36.
Michel, Alain P. *Les images du travail à la chaîne dans les usines Renault de Boulogne-Billancourt (1917–1939). Une analyse des sources visuelles: cinéma, photographies, plans d'implantation.* Thesis, EHESS, 2001.

Michel, Alain P. "Le travail à la chaîne en images: Les documents visuels face à l'histoire industrielle de Renault." *Images du travail/Travail des images*, no. 3, 2017, https://doi.org/10.4000/itti.1105.

Michel, Alain P. "On the Cover: An Image Mise en Abyme." *Technology & Culture*, vol. 49, no. 4, 2008, pp. 967–73.

Michel, Alain P. *Travail à la chaîne. Renault 1898/1947*. ETAI, 2007.

Michel, Alain P. *"Usines 3D": Reconstitution virtuelle d'ateliers et d'usines du patrimoine industriel*. ANR Corpus 2007, no. 034, 2012, p. 42. Final report of the French National Research Agency Program, www.usines3D.fr.

Michel, Alain P., and Shadia Kilouchi. "Renault-Billancourt's C5 Workshop in the Digital Age: A New Story of the 1922 Assembly Line." *L'histoire contemporaine à l'ère digitale*, edited by Fédéric Clavert and Serge Noiret. PIE-Peter Lang, 2013, pp. 235–49.

Moutet, Aimée. *Les logiques de l'entreprise: La rationalisation dans l'industrie française de l'entre-deux-guerres*. Éditions de l'École des hautes études en sciences sociales, 1997.

"Moving Picture Department." *Ford Times*, vol. 9, no. 12, 1916, pp. 535–40.

Nevins, Allan, and Frank Ernest Hill. *Ford: Expansion and Challenge: 1915–1933*. Charles Scribner's Sons, 1957.

Nevins, Allan, and Frank Ernest Hill. *Ford: The Time, the Man, the Company*. Charles Scribner's Sons, 1954.

Nye, David. *America's Assembly Line*. MIT Press, 2013.

Paggi, Silvia. "Voix-off et commentaire dans le cinéma documentaire et ethnographique." *Cahiers de Narratologie*, vol. 20, 2001, doi:narratologie.revues.org/6321.

Peirce, Charles Sanders. *Collected Papers*. Harvard University Press, 1931–1935.

Prelinger, Rick. "Introduction." *The Field Guide to Sponsored Films*. National Film Preservation Foundation, 2016, pp. iv–xi.

Répertoire des films de l'encyclopédie Gaumont. Société des Etablissements Gaumont, Service enseignement, Imprimerie Gaumont, 1929.

Shell, Wanna Rose. "Films in the Archive: Hollywood in Detroit." *Technology and Culture*, vol. 55, no. 3, 2014, pp. 711–15.

Shiomi, Haruhito, and Kazuo Wada, editors. *Fordism Transformed: The Development of Production Methods in the Automobile Industry*. Oxford University Press, 1995.

Sorlin, Pierre. *Introduction à une sociologie du cinéma*. Klincksieck, 2015.

Sorlin, Pierre. *Sociologie du cinéma*. Aubier-Montaigne, 1977.

Streible, Dan. "The Role of Orphan Films in the 21st Century Archive." *Cinema Journal*, vol. 46, no. 3, 2007, pp. 124–28.

Taylor, Frederick W. *The Principles of Scientific Management*. Harper & Brothers, 1911.

Touchelay, Béatrice. "D'une sortie de guerre à l'autre: de l'impôt sur les bénéfices de guerre (1916) à la confiscation des profits illicites (1944–1945), l'État a-t-il appris à compter?" *L'épuration économique en France à la Libération*, edited by Marc Bergère. Presses universitaires de Rennes, 2008, pp. 33–50.

Veray, Laurent. *La Grande Guerre au cinéma: De la gloire à la mémoire*. Éditions Ramsay, 2008.

Williams, Karel, Colin Haslam, and John Williams. "The Myth of the Line: Ford's Production of the Model T at Highland Park, 1909–16." *Business History*, vol. 35, no. 3, 1993, pp. 66–87.

Zimmermann, Yvonne, editor. *Schaufenster Schweiz: Dokumentarische Gebrauchsfilme 1896–1964*. Limmat Verlag, 2011.

About the Author

Alain P. Michel is professor of contemporary history of technology at Paris-Saclay University/Evry University. He directs in Evry the History Laboratory "Institutions and Historical Dynamics of the Economy and Society," a joint research unit with the CNRS (IDHES-UMR 8533, https://www.idhes.cnrs.fr). His publications include *Travail à la chaîne: Renault 1898–1947* (Editions ETAI, 2007) and "Out of Control: Telephone Network and Management of Business Talks at Renault (1911–1939)," *Information & Culture*, vol. 51, no. 4, 2016. More recently, he co-edited with Florent Le Bot a special issue: "Ingénieurs et Entreprises (XIXe–XXIe siècle)" in *Artefact. Techniques, histoire et sciences humaines*, no. 13, 2020, http://journals.openedition.org/artefact/6186.

Section 7

Ephemeral Artistry: Ecologies of Authorship in Industrial Cinema

26 Business and Art

Pharmaceutical Industries, Film Production and Circulation, and the French Film Production Company ScienceFilm, 1960–1980

Christian Bonah

Abstract

This chapter addresses commissioned industrial film by enquiring how Éric Duvivier organized the production and distribution network of ScienceFilm from the 1950s to the 1980s. Following the indirect structure of the medical sector, PR is often directed at professionals and ScienceFilm productions navigate vague boundaries between promotional and educational film, with the first intentionally mimicking the second, which is considered as a first meaning of "cinema as a parasite form." A second viral-parasitic form and level of investing and hijacking intentions is opened by filmmakers creating commissioned industrial films that were also circulated in experimental avant-garde cinemas in Paris in the 1970s. This parasite film form – promotion as education, parasitized itself as industrial commissioned experimental art film – is here portrayed.

Keywords: medical film; ScienceFilm; Duvivier; pharmaceutical industry; experimental film; postgraduate medical training

Industrial film may be defined as a production category meaning here films produced by or for use by industrial corporations, ranging in their product dependency from materials including coal, oil or steel, to products like cars, food products or luxury goods or to services including roads, railroads or postal and telephone communication services. One early, longstanding and little studied sector of corporate film production concerns modern medicines producing chemical-pharmaceutical industries in Europe and

Hediger, V., F. Hoof, Y. Zimmermann, with S. Anthony (eds), *Films That Work Harder: The Circulation of Industrial Film.* Taylor & Francis Group 2024
DOI: 10.5117/9789462986534_CH26

North America.[1] Born with the Second Industrial Revolution in the late nineteenth century and kin to business merging and new techniques of public relations (PR) pharmaceutical companies engaged quickly, since the early 1920s, in industrial and medical filmmaking.[2] If their history has attracted limited interest until now, the advent of the so-called therapeutic revolution after 1945 drastically multiplying drug research, development and consumption[3] has increased investments in media communication practices by pharmaceutical giants even more so. This contribution addresses a series of these pharmaceutical films in the golden age of industrial filmmaking between 1960 and 1980 in France.[4]

I will enter the subject of commissioned film not from the point of view of a specific industrial corporation, but from the side of a film company that produced exclusively industrial films for a large variety of different industrial medicines producing sponsors. More precisely my contribution enquires how the French professional industrial filmmaker Éric Duvivier organized his production and distribution company ScienceFilm and how within his network this determined the circulation of pharmaceutical industry sponsored films during the second half of the twentieth century.[5]

1 Hediger, Vinzenz, and Patrick Vonderau, editors. *Films That Work: Industrial Film and the Productivity of Media*. Amsterdam University Press, 2009; Leblanc, Gérard. *Quand l'entreprise fait son cinéma: la médiathèque de Rhône-Poulenc (1972–1981)*. Cinéthique Presses Universitaires de Vincennes, 1983; Zimmermann, Yvonne, editor. *Schaufenster Schweiz: Dokumentarische Gebrauchsfilme 1896–1964*. Limmat, 2011; idem. "Target Group Oriented Corporate Communication: Geigy Films." *Corporate Diversity: Swiss Graphic Design and Advertising by Geigy 1940–1970*, edited by Museum für Gestaltung Zürich. Lars Müller Publishers, 2009, pp. 48–57. In contrast, these excellent volumes have no entry for pharmaceutical industry-sponsored film: Kreimeier, Klaus, Antje Ehmann, and Jeanpaul Goergen, editors. *Geschichte des dokumentarischen Films in Deutschland, Band 2: Weimarer Republik 1918–1933*. Reclam, 2005, and Zimmermann, Peter, and Kay Hoffmann, editors. *Geschichte des dokumentarischen Films in Deutschland, Band 3: Drittes Reich 1933–1945*. Reclam, 2005.

2 Ewen, Stuart. *PR! A Social History of Spin*. Basic Books, 1996; Marchand, Roland. *Creating the Corporate Soul: The Rise of Public Relations and Corporate Imagery in American Big Business*. University of California Press, 1998; Bonah, Christian. "In the Service of Industry and Human Health: The Bayer Corporation, Industrial Film and Promotional Propaganda, 1934–1942." *Communicating Good Health*, edited by Christian Bonah, David Cantor, and Anja Laukötter. Rochester University Press, 2018, pp. 85–142.

3 Greene, Jeremy, et al., editors. *The End of the Therapeutic Revolution?* University of Chicago Press, 2016.

4 Hediger and Vonderau. *Films That Work*; Leblanc. *Quand l'entreprise fait son cinema*; Pessis, Georges. *Entreprise et Cinéma: Cent ans d'images*. La documentation française, 1997; see as well Jacobsen, Brian. "On the Red Carpet in Rouen: Industrial Film Festivals and a World Community of Filmmakers." This volume.

5 For simplification, we design Éric Duvivier's six successive production companies here as ScienceFilm. In fact, they were Films Art et Science (~1950–1960), ScienceFilm (1960–1978), Art et

In the realm of industrial filmmaking pharmaceutical motion pictures stand apart given their audience specificities. Pharmaceutical markets are characterized by indirect consumption mediated by prescribing physicians. Accordingly, much of PR in this sector is directed at professionals of medicine. In this context pharmaceutical industry sponsored films become boundary objects to the point that their identity becomes almost indistinguishable from educational medical films, as suggested in the editorial of the French medical film periodical *Médecine/Cinéma* in 1972:

> The general idea of this [medical cinematographic] production – which is essentially linked to the pharmaceutical industry – consists in reaching practicing physicians: by participating in their information and their training, in return they offer a better opportunity of commercial profit through their prescriptions.[6]

These longstanding connections and manifold promotional practices including film, but as well specific journals as the cited *Médecine/Cinéma* and medical film libraries, have been recently analysed as forms and segments of far-reaching scientific marketing.[7] For film circulation this audience specificity implies that screening and venues were first and foremost medical conferences, postgraduate training sessions and professional gatherings catering to forms of professional sociability including physician's dinners or receptions. Beyond these official and utility-oriented screening events intended and organized by film sponsors my perspective from the film production company side will highlight nevertheless a second, less expected and sometimes symbiotic, sometimes subversive circulation of the industrial

Science (1978–1980), IntersciencеFilm (1980–1988), Telescience (1988–1990), and, finally, VideoScience (1990–). ScienceFilm was the longest lasting and produced the most films. For a historical account of the companies, see Bonah, Christian. "'Réservé strictement au corps médical': les sociétés de production audiovisuelles d'Éric Duvivier, l'industrie pharmaceutique et leurs stratégies de co-production de films médicaux industriels, 1950–1980." *Cahiers d'histoire du CNAM* [special issue: "Le cinématographe pour l'industrie et dans les entreprises (1890–1970): histoire, acteurs, usages et configurations"], 2020, pp. 135–59. Bonah, Christian, and Joël Danet. "Éric Duvivier." *MedFilm*, https://medfilm.unistra.fr/wiki/Pers:%C3%89ric_Duvivier. Accessed July 11, 2023.

6 Chantelou, Philippe. "Comment des cinémathèques scientifiques favorisent un cinéma de création et comment un réalisateur-créateur leur apporte une conception réaliste du film medical." *Médecine/Cinéma*, no. 14, 1972, p. 9.

7 Bonah. Christian. "Marketing Film: Audiovisuals for Scientific Marketing and Medical Training in Psychiatry: The Example of Sandoz in the 1960s." *The Development of Scientific Marketing in the Twentieth Century: Research for Sales in the Pharmaceutical Industry*, edited by Jean-Paul Gaudillière and Ulrike Thoms. Pickering & Chato, 2015, pp. 87–103.

films presented: independent, experimental avant-garde cinema in Paris.[8] It has been acknowledged recently in useful cinema studies that utility films are "multi-sited" and "multi-purpose."[9] Éric Duvivier and his company ScienceFilm may be considered as a poignant case in point for such an understanding.

In the following I will first present ScienceFilm as an industrial film production company. Then I will move to the expected utility film circulation as defined by film sponsors. Fluent and vague boundaries between promotional and educational film,[10] with the first intentionally mimicking the second, are considered here as a first meaning of "cinema as a parasite form" in the sense that industrial utility invested in professional education yet pursued its own interest and objectives within "medical cinematography." Finally, the unexpected circulation and parallel lives of some of the industrial films described in experimental avant-garde cinema circuits in Paris in the 1970s will highlight a second viral-parasitic form and level of investing and hijacking intentions. The artist-filmmaker assuring himself industrial financing to produce commissioned films that he can present in an independent avant-garde film programme and movie theatre portrays in a sense the hijacker being hijacked or the parasite film form being parasitized itself. Nevertheless, in a wider perspective the distinction between the parasitic or symbiotic character of the forms of collaboration between business and film art pursued in this contribution is not essentialist in nature. I will argue that it is rather a subtle equilibrium depending on venues of the screening (sites), specific forms of audiences and the specific content and aesthetic trade-offs of every individual film that established whether respective interests were recognized and mutualized in symbiotic form or rather conflicting, exploiting and harming, alas parasitic, partners of production.

ScienceFilm: Embracing an Industrial Film Production and Distribution Company Point of View

The medical-pharmaceutical film production company Films Art et Science/ ScienceFilm (hereafter ScienceFilm) was set up in the late 1940s by the

8 Kubelka, Peter, and Alain Sayag. *Une histoire du cinéma: exposition.* Musée national d'art moderne, 1976; *Fihman,* Guy, and *Claudine Eizykman. "Xprmntl melba: Images d'un mouvement." Cinédoc Paris Films Coop, www.cinedoc.org/htm/xprmntl/IDMvt.asp. Accessed November 25, 2016.*

9 Waller, Greg. "International Harvester, *Business Screen* and the History of Advertising Film." *Films That Sell: Moving Pictures and Advertising,* edited by Bo Florin et al. Palgrave, 2016, pp. 40–53.

10 Orgeron, Devin, et al., editors. *Learning with the Lights Off: Educational Film in the United States.* Oxford University Press, 2012.

French film director Éric Duvivier. It was active in France in the 1950s and 1960s. The production catalogue of Duvivier's film production company contains a list of approximately seven hundred industrial films produced between 1950 and the 1990s.[11] The man behind ScienceFilm, Éric Duvivier (1928–2018) was the nephew of the better-known film director Julien Duvivier (1896–1967).[12] Growing up in a filmmaking family environment – his father worked with his uncle – Éric Duvivier began to study medicine immediately after World War II and, bridging interests, organized a film club for medical students. He eventually dropped out of medical school and devoted himself entirely to filmmaking.[13] In 1947, he created the Centre international du film medical with the support of medical school deans and the French Mutualité organizing film screenings. Duvivier then set up his first production company, Films Art et Science in the late 1940s. Initially the company was financially supported by a second company, FilmLabo, that thrived by printing and dubbing 16 mm and 35 mm films for major film studios in Paris, a rather lucrative activity immediately after 1945. At the same time, Duvivier mobilized his uncompleted medical education to start producing medical and scientific films for professional audiences. The first medical films he directed involved mainly anatomical subjects, such as, *L'os temporal* (*The Temporal Bone*, 1958), *Le péritoine* (*The Peritoneum*, 1958), or *Anatomie de l'épaule/hanche/genou* (*Anatomy of the Shoulder/Hip/Knee*, 1958). By the late 1950s well established in the field of medical filmmaking and working with physicians, he separated distribution (via his Centre international du film medical) from production by creating a second film company, ScienceFilm, in 1960. He became a natural partner for film library projects that mushroomed in pharmaceutical industry in the early 1960s.

Indeed, the late 1950s and early 1960s witnessed the development of so-called medical film libraries – pharmaceutical corporation-based film distribution units renting out industry-commissioned medical films to

11 Art et Science. *596 films médicaux 16m/m couleurs sonores: Liste N° 17 référence des réalisations d'Éric Duvivier.* ScienceFilm, 1988.

12 Lefebvre, Thierry. "Les films d'Éric Duvivier: un cinema d'auteur." *MEDFILM*, Journée d'étude à l'Université Paris Diderot Grands Moulins, no. 2, September 21, 2012. https://medfilm.unistra.fr/wiki/Accueil:21_septembre_2012: *https://medfilm-stream.u-strasbg.fr/36/duvivier_thierry_lefebvre/public.mp4*. Accessed July 15, 2023; "Entretien avec le réalisateur Éric Duviver, présenté et animé par Thierry Lefebvre et Christian Bonah." *MEDFILM*, Journée d'étude à l'Université Paris Diderot Grands Moulins, no. 2, September 21, 2012. https://medfilm.unistra.fr/wiki/Accueil:21_septembre_2012: https://medfilm-stream.u-strasbg.fr/20/entretien_avec_ric_duvivier/public.mp4. Accessed July 12, 2023.

13 This and the following account are based on an interview with Éric Duvivier by Christian Bonah and Emmanuelle Simon, January 29, 2012.

professional audiences. According to the film scholar Gérard Leblanc, the "extraordinary period of the 1960s" was less about medical films as such than about a golden period for industrial filmmaking in general related to the generalization of corporate films.[14] In the pharmaceutical industry, the movement started when the Swiss pharmaceutical company Sandoz decided to set up the Cinémathèque Sandoz in 1958, as did Ciba Geigy the same year.[15] In the 1970s the Sandoz unit was directed by a pharmacist and employed a film technician for technical affairs and repairs and three administrative agents handling incoming requests, cataloguing and shipping of the film copies to customers as well as returns.[16]

Contrary to what their name suggested, the cinémathèques were not just places for the distribution of films with medical content; they were also the initiators of a new corporate film production strategy by Sandoz and others. They financed, created and distributed free copies of medical films produced by the pharmaceutical industries for physicians. They were therefore as much film production agencies as film distribution libraries. The Cinémathèque Sandoz commissioned films such as *Le horla* (1967) or *L'ordre* (1973), directed by Jean-Daniel Pollet, which today belong to the classics of documentary film.[17] This corporate practice was quickly imitated by other pharmaceutical/chemical companies including, amongst others, Rhône-Poulenc, Lagrange, Beaufour, Delagrange and Boehringer.[18]

14 Leblanc, Gerard. "L'âge d'or du cinéma médical et l'aventure de *Médecine/Cinéma*: Entretien avec Gérard Leblanc." *Sociétés & représentations*, vol. 28, no. 2, 2009, pp. 107–18; idem. *Quand l'entreprise fait son cinéma.*

15 Around the same time Geigy had a film services department that, beginning in 1957, handled technical production work and occasionally produced short medical films. Much of Geigy's film production was outsourced in particular to the Swiss film production company Condor. Even if Geigy claims to have been "the first firm in the industry to produce medical films at regular intervals that, without any product advertising at all, served exclusively as continuing education for practicing physicians," the "Documenta Film Geigy" series of medical films that started in 1963 with the help of French physician Yannic Guéguen seems to follow Sandoz initiatives instead. Zimmermann. "Target Group," p. 49.

16 "Cinémathèque." November 1981, p. 10. Firmenarchiv der Novartis AG, Vorarchiv, Service Audit Groupe Sandoz France.

17 See Leblanc. "L'âge d'or du cinéma médical"; Odin, Roger. *L'âge d'or du documentaire. Europe: Années cinquante. France, Allemagne, Espagne, Italie. Tome 1.* L'Harmattan, 1998, pp. 19–53; Tulard, Jean. *Dictionnaire du Cinéma: Réalisateurs.* Robert Laffont, 2001, p. 726.

18 Evidence for this comes from the ScienceFilm production catalogue. The first entry for a Sandoz-commissioned film dates to 1958; thereafter collaboration became routine; companies like Choay (1962), Delagrange (1963), Squibb (1965), Roche (1965), Hoechst (1967), Glaxo (1968), Servier (1971), Boehringer (1971) or Rhone Poulenc/Specia (1980, 1985, 1986) followed suit. The earliest collaboration between Duvivier and Ciba Geigy appears in 1976: *Pourquoi et comment traiter les hypertendus.* See Zimmermann. "Target Group," p. 52.

Sandoz's 1969 catalogue of medical-scientific films indicates that the library held and offered by that time 116 films that had been produced by Sandoz between 1958 and 1969, averaging ten films produced or sponsored per year.[19] Amongst the authors and production companies in the catalogue, one specific company and its director stand out: ScienceFilm. By 1969, fifty-eight out of the 116 films sponsored by Sandoz were produced by ScienceFilm, i.e. fifty per cent of production. Psychiatric films alone account for roughly one film out of five produced during the first ten years of the cinémathèque's existence.[20] In terms of numbers this is the most outstanding subject in the films produced, a choice understandable since psychoactive substances were of particular interest to the company that had been associated in the 1940s with the invention and development of LSD and in the 1950s with the psychopharmacological revolution of neuroleptics, tranquilizers and antidepressant drugs. Twelve of the twenty-six films listed under the "psychiatry" heading in the catalogue were creations of ScienceFilm and its director.[21]

Gauged from the 1988 Films Art et Science/ScienceFilm production catalogue perspective, Duvivier's film production company listed 596 black-and-white and sound-and-colour films produced since the 1950s and still for rent by his distribution company. Duvivier sustained therefore an average production of fifteen films a year for over forty years. The place of psychiatric films as the ones mentioned above was significant but not exclusive in Duvivier's film corpus. The roughly one hundred films on psychiatric subjects made between 1950 and 1970 account for approximately one-sixth of the film director's activity. Content wise they divided in two sets. One concerned straightforward professional and medical-teaching films for postgraduate physician training on clinical symptoms like *Sémiologie psychiatrique: cinq observations*, *Expérience délirante primaire chez un adolescent*, or *Syndrome hébéphréno-catatonique* (1971), new diagnostic or therapeutic approaches *Le métoclopramide*, or disease descriptions like *Dépression d'automne*, *Psychose alcoolique* or *État démentiel: maladie d'Alzheimer*,[22]

19 *Films médico-scentifiques Sandoz: Edition 1969*. Sandoz, 1969.

20 For the film-corpus approach, see Bonah et al. *Communicating Good Health*.

21 For the corporate history of Sandoz, see Fritz, Hans. *Industrielle Arzneimittelherstellung: Die pharmazeutische Industrie in Basel am Beispiel der Sandoz AG*. Wissenschaftliche Verlagsgesellschaft, 1992; Straumann, Lukas, and Daniel Wildmann. *Schweizer Chemieunternehmen im "Dritten Reich"*. Unabhängige Expertenkommission Schweiz – Zweiter Weltkrieg, vol. 7. Chronos Verlag, 2001; Zeller, Christian. *Globalisierungsstrategien – Der Weg von Novartis*. Springer, 2001; Studer, Tobias. "Die Geschichte der Sandoz im Lichte ihrer Diversifikationen." *Sandoz Bulletin*, vol. 22, 1986, pp. 16–45.

22 Bonah, Christian, and Maike Rotzoll. "Psychopathologie en mouvement. Histoire des films psychiatriques à Strasbourg et à Heidelberg." *Le pré-programme*, edited by Christian Bonah et al.

all in line with commission and production for pharmaceutical industries like the Sandoz film library. The other set involves a highly intriguing series of productions of rather avant-garde experimental films with titles like *La femme 100 têtes*, *Les années folles de Sylvain Fusco*, a French painter who had been committed at the mental institution Asile départemental du Rhône, *Ces maladies qui nous gouvernent*, a critical appraisal of diseases from which leading world politicians suffered, *Images de la folie*, *Images du monde visionnaire* and *Autoportrait d'un schizophrène* (1977), presenting subjective audiovisual and written testimonies of views of patients experiencing drug effects. How can this specific period of film corporation pharmaceutical company collaboration be understood?

To better grasp the cooperation between pharmaceutical companies and ambitious, marginal and experimental documentary filmmakers, let us gauge in detail how ScienceFilm contracted with the pharmaceutical industry for production and distribution of the films. ScienceFilm, through its director and filmmaker Duvivier, enlisted medical professors as opinion leaders to identify medical subjects suitable for filming. Film projects were jointly elaborated and presented to pharmaceutical companies that had established industrial medical film libraries. The pharmaceutical companies commissioned a film to be made by director Duvivier and his film company on the basis of the association between the film's director and a medical professor.

The original and complex dispositive of collaboration introducing a third party in the classical industrial film commission scenario – the opinion-leading physician alongside the film sponsor and director – may be conceived as a first meaning of Michel Serres' conceptualization of the parasite form, "the medium or being through which communication must pass."[23] The third party, its changing presence and role, made the socially inadequate public relation between pharmaceutical industry and physicians possible. It served both sides in the filmmaking process as a go-between and thereby made production itself possible. Let us look at the details of the operation.

In the pharmaceutical advertising and marketing environment corporate messages were addressed at medical professionals who are often self-conscious about their status and self-image as liberal professionals, not businessmen with commercial interests. For many, this difference

Verlag A 25 Rhinfilm, pp. 255–80. Detailed film descriptions are available under film headings at: medfilm.unistra.fr.

23 Brown, Steven D. "In Praise of the Parasite: The Dark Organizational Theory of Michel Serres." *Informática na Educação: teoria e prática, Porto Alegre*, vol. 16, no. 1, 2013, pp. 83–100.

symbolized a deep divide between charlatanism and orthodox medical practice, and throughout the twentieth century physicians have generally been uncomfortable about being directly associated with commercial practices and advertising. Therefore, for Duvivier his two-step strategy consisted in first ensuring the interest and collaboration of a high-profile medical expert whom he addressed as a disinterested artist-filmmaker. Acceptable for the physician from that third party perspective, the proposal to produce a film allowed the physician public representation of himself or public display of his specific working topics and theories. This had been an old incentive of medical filmmaking since the early 1900s, when Eugène Doyen produced his first surgical films, mixing the intention of improving surgical techniques through film observation with public representation and publicity for the medical or surgical author of the film.[24] Then the film director Duvivier, turning producer, used high-profile physicians as incentives for the industry to finance the production of "his" film. Sandoz might not have sponsored a film by Éric Duvivier, but would do so, and did, for a film "by" medical professors Jean Delay, Paul Sivadon or Jacques Duché or artists like Henri Michaux. Leading physicians engaged in film production with Duvivier as a film professional on a medical subject. Leaving financial aspects to the production company ScienceFilm, they apparently stayed clear of pharmaceutical advertising and promotion, a subject always potentially controversial to them. Identified as an opinion leader by the profession and the pharmaceutical industry, the mobilized medical expert's symbolic capital was then transformed by Duvivier into film-production budgets from the pharmaceutical industry. The industry calculated that medical opinion leaders would become message carriers as pharmaceutical opinion leaders in the fields of specific drugs.

The ScienceFilm catalogue corroborates this analysis through a significant detail. Films are listed according to subjects and medical disciplines and, most interestingly, titles are followed by names of prestigious professors of medicine and their city of activity rather than by the identity of the sponsoring pharmaceutical company. Sandoz's film catalogue in return specifies film titles and physician names often dispensing with the film director's name. As disturbing as this might be for film historians, in Duvivier's medical film marketing strategy recruited medical experts became the auteurs of the films. Film credits display revealing ambiguities with films sometimes credited as directed by a medical (Delay, Sivadon,

24 Lefebvre. *La chair et le celluloï: Le cinéma chirurgical du docteur Doyen.* Jean Doyen éditeur, 2004.

Duché, etc) or literary (Michaux) authority and produced by Duvivier and his company. Pharmaceutical companies were described as "presenting" the film. Contrary to the usual practice in the commercial film world – where a film producer is the arranging and financing authority of the production of a film and the film director is the artist making the film and a screenplay writer may produce a script – in the medical film marketing world it was the content-providing physician-professor who was often designated as the film director (or author) yet production contracts reserved film rights to ScienceFilm. Under this arrangement the actual film director was designated as the arranging (but not paying) film producer. The industrial player as a financing authority was the sponsor but often did not retain the film rights – these remained with the "producer," ScienceFilm. The pharmaceutical medical film production rationale of the time required a mediator between the worlds of business and (film)art, the medical expert through which the film creation process had to pass. Authorship acquired a different and differentiated meaning in this specific context. Moral authorship and the symbolic capital of the films were attributed to the medical expert author of the film content. The legal film rights and monetary returns lay with the film's producer, ScienceFilm, and its actual director, Duvivier. Production budget providing corporations were commissioners or sponsors of the films that they acquired distribution rights for and used for image building, corporate identity and prestige[25] or, in Yvonne Zimmermann's words, as "cinematic business cards."[26] The question whether the modern form of patronage portrayed here, understood as the support provided to filmmakers, may be understood in terms of a symbiotic win-win ecology or rather as a concealed parasitic utilization of industrial commission by film directors subverting it for artistic purposes is in part depending on the question whether financial aid came with or without industrial supervision, control or censorship during film production and requests for immediate returns. For another part it depended on the freedom of artistic action granted, which ranged from straightforward commissioning to sponsoring independent film artists, and last but not least on the venues where films were screened, ranging from professional audiences targeted by companies to independent experimental film circuits. Eventually the symbiosis-parasitism divide remains tied to the specific content and aesthetic trade-offs of every individual film. It is on these latter aspects of circulation that I will focus in the following.

25 Zimmermann. *Schaufenster Schweiz*, pp. 42–47.

26 Idem. "Target Group," p. 52.

Official, Intended and Goal-Oriented Forms of Circulation

A second rationale behind the organizing and crediting practice described above was distribution and audience. Where "a film by Éric Duvivier" would not have been identified by practicing physicians, "a film by" Professor Jean Delay, Marcel Bessis, Jean Bernard or Jacques Chretien, for example, worked as a signal in the social world of physicians. With a subject and medical expert at hand, Duvivier then approached a pharmaceutical firm relevant to the professor's medical specialty. In this medical film marketing strategy Sandoz was just one amongst many pharmaceutical companies with which Duvivier and Films Art et Science/ScienceFilm produced films from the late 1950s onwards; others followed suit including Choay (1962), Delagrange (1963), Squibb (1965), Roche (1965), Hoechst (1967), Glaxo (1968), Servier (1971), Boehringer (1971) or Rhone Poulenc/Specia (1980, 1985, 1986).

Attracted by the names of the opinion-leading professors, physicians were conveyed by industry as audience and potential prescribers to medical film events where corporate messages on individual products were additionally presented off screen. The film-sponsoring pharmaceutical companies appeared only in the film credits and product placement was exceptional. Directors like Duvivier were eager to protect their filmmaking freedom from the promotional mission intended by the sponsors of their films. They produced so-called prestige films. This meant that promotional intentions were relegated to how the sponsors then used the films rather than stated through messages embedded directly in the films. In practice, films were used first as corporate films during events, with their screening being part of multimedia alliances in which product promotion was an element of the surrounding organization, i.e. performed by company employees and through stands at the events completing them with catering ranging from refreshments to complete evening dinners. During these social gatherings scientific films were accompanied by leaflets, hand-outs and the like, including specific product-information messages not included in the films, and sales representatives could establish close and relaxed relationships with the physician-prescribers they were to visit in their office at some point or another. An internal audit of Cinémathèque Sandoz was carried out in 1981. This was during a difficult period for the film library; film requests were decreasing and pending technological changes (including VHS replacing 16 mm and informatics revolutionizing library administration) threatened its survival. The audit, however, praised the "excellency of film as tool of promotion," especially in connection with work by pharmaceutical representatives. The audit acknowledged the value of the company's policy of "active placement" of films with chosen spectators, highlighting in particular the rising number of medical students (over eighty thousand) watching Sandoz films annually. But most important for the auditors was that the film library should foster an increased interest in film promotion with the company's pharmaceutical sales representatives themselves serving as transmitters and key motor elements for the targeted distribution of Sandoz's films.[27]

A second form of circulation was organized by Duvivier and his distribution company, Centre international du film medical. This was part and parcel of the film commission and Duvivier organized his own film shows to tour in major towns and cities throughout France and beyond, modelled on itinerant film programmes like "Découverte du Monde" ("Discovering the World"), screened in public cinemas rented for the occasion. Using physician directories (such as the *Guide Rosenwald*), the Centre international invited prescribing physicians in a given location to a film show where art films as good selling attractions were combined with medical information and training films ambiguously mixing entertainment (e.g. an artistic experimental film produced by ScienceFilm) and scientific information (e.g. a film about what was new in psychosis and its treatment – produced by ScienceFilm as well). Sponsored by companies like Sandoz, these events were advertised as an "evening for physicians" and cast in professional terms as meetings about "clinical features and news" (for example, in the field of psychiatry). As Duvivier recounted in 2012, these screenings attracted hundreds of physicians a night in major cities in the 1960s.[28] Medical experts like Delay, Duché or Sivadon served as opinion-leading clinicians and became celebrities in the world of medical film, just like movie stars in mass entertainment.

Finally, because they did not feature any ads for specific products, Duvivier's industrial films were also included in wider distribution circuits, including those of the Ministry of Foreign Affairs, which in the early 1960s established the audiovisual catalogue of Saint-Antoine hospital, a catalogue made available to the cultural sections of French embassies throughout the world.

From Film Circulation to Circulation in between Films

Underlying the forms of circulation described was the fact that these industrial medical films were mimicking educational films, on the one

27 "Cinémathèque." November 1981, p. 10. Firmenarchiv der Novartis AG, Vorarchiv, Service Audit Groupe Sandoz France.

28 Interview with Éric Duvivier by Christian Bonah and Emmanuelle Simon, January 29, 2012.

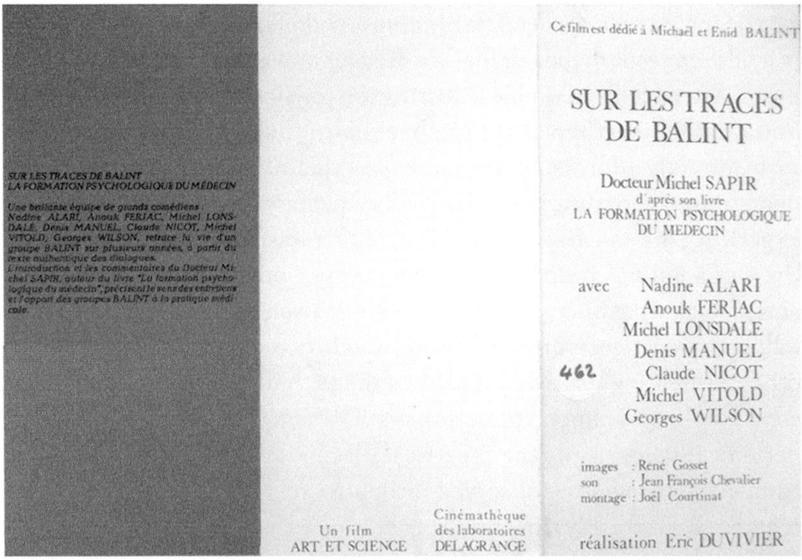

Fig. 26.1. Promotional leaflet for the film *Following Balint* (1976) produced by ScienceFilm and Duvivier for the pharmaceutical company Delagrange. It is an example of authorial ambiguities that could exist in medical films between its medical "author" (Sapir) and the film's "director" (Duvivier).

hand, and that they had aesthetic and artistic qualities that closely connected them to French avant-garde documentary filmmaking, on the other hand. A more detailed analysis of one example shows how films containing artistic qualities could be distributed under the guise of being educational films.

In 1976–1977, Duvivier produced in collaboration with Dr. Michel Sapir the teaching film *Sur les traces de Balint* (*Following Balint*), based on Sapir's 1972 book *The Psychological Education of Physicians*. Sponsored by the firm Delagrange, the film argued for the usefulness of professional peer groups exchanging and self-analysing doctor-patient interactions in a psychoanalytical framework. The group and the method are named after Michael Balint, a psychoanalyst originally from Hungary. He and his wife, Enid Balint, started a series of seminars in London in the 1950s with the aim of helping family physicians reach a better understanding of what they called the "psychological aspect" of general practice. The method consisted of case presentation followed by general discussion with the emphasis on the emotional content of the doctor-patient relationship.

In an intimate atmosphere, a group of six male and female physicians, responding to the prompt "Who has a case history to tell?," relate personal consultation experiences and analyse them under the supervision of a psychoanalyst team leader. The film conveys professional secrets confessed in a mise en scène of the self that are used for professional training amongst peers. After a long preamble featuring the physician-author Michel Sapir and Enid Balint, the wife of Michael Balint (who had passed away in 1970), directly addressing the audience explaining the history of Balint peer groups, the film then conveys the spectator to witness the "Balint experience" by assisting in a series of talking sessions. Original transcriptions of a Balint group were re-enacted by a prestigious set of French actors, including Michel Vitold, Georges Wilson and Michael Lonsdale, telling their patient stories, listening, watching and interacting with each other. Recreating the real intensity of the exchange sessions intended to convince the spectator of the usefulness of this professional exchange. Sapir periodically interrupts the theatrical discussion sessions, commenting from a distance on the observed intimate exchanges. Thereby the film creates two radically different sets of images: one being the acted-out group discussions, the other the theoretical analysis and commentary. Immersing the spectator in the group confessional and distancing him/her to reread the content and present the method, the highly original mise en scène combines a sampling of "reality" and presents the methodological principle.[29]

The Balint meetings take place and are filmed in a single setting: a warmly lit, cosily arranged interior of a bourgeois lounge. The protagonists gather in a circle, sitting in armchairs or on a couch. The most intriguing scene concerns Michael Lonsdale playing a rather hesitant, middle-aged physician relating his case, pausing, gazing, interrupting his narration frequently and accompanying it with empty gestures. It is here that the spectator cannot avoid seeing a parallel.

Indeed, in the same year, the same actor, Michael Lonsdale, appeared in French film director Jean Eustache's *Une sale histoire* (*A Dirty Story*, 1977), a short fiction film based on a true story. The film critic Jean Roy considers this unusual film (it is broken into two halves, one twenty-two and the other twenty-eight minutes long) as Eustache's key work. The film concerns the factual tale of a man who finds a peephole in the female toilets of a café. In the first part of the story it is acted out, and in the

29 Bonah, Christian, and Joël Danet. "Le moment anthropologique du cinéma industriel pharmaceutique. Retour sur le film utilitaire médico-pharmaceutique francophone des années 1970." *Images du travail Travail des images*, no. 4 [special issue: "La relation soignants/soignés à l'épreuve de l'image," edited by Lucia Candelise et al.], 2017, https://journals.openedition.org/itti/993.

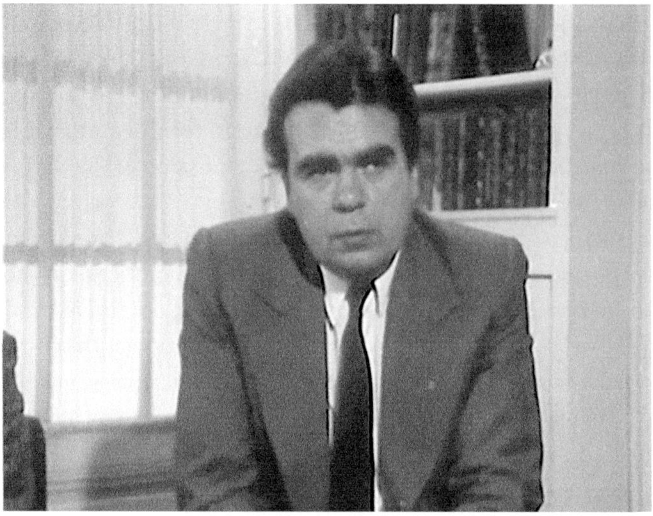

Fig. 26.2. The actor Michael Lonsdale in Éric Duvivier's film *Following Balint* (1976), playing a rather hesitant, middle-aged physician relating his patient case to the peer group.

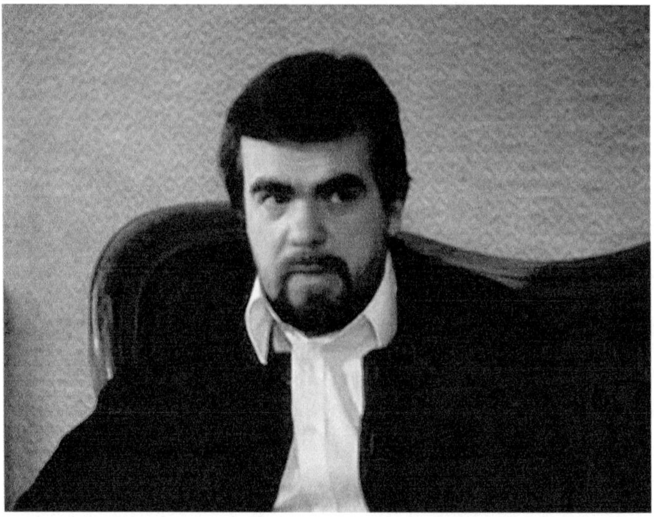

Fig. 26.3. The actor Michael Lonsdale in Jean Eustache's film *A Dirty Story* (1977), playing a man who finds a peephole in the female toilets of a café, conveying in the second part of the film in an off-the-cuff manner.

second part it is conveyed in an off-the-cuff manner by the man it actually happened to.

The mise en scène, artistic devices, motifs and rhetorical patterns of the two films are absolutely similar. Filmed in long static shots, cast in a single interior location, multiplying close-ups and facing audience in an intimate direct address intercut by some short reverse shots providing glimpses of the captivated audience members' faces. Lonsdale plays down his theatrical performance by stressing his nonchalance, hesitation and disturbance, serving characters in both films along similar lines. The dispositive of simply telling a story, dirty or not, in front of a camera is a captivating device that is the fundamental rationale of both Eustache and Duvivier.[30] Their similar mises en scène push this logic to the extreme, systematizing non-action, immobility and the spinning of a verbal web around the spectator. It is this dispositive that Duvivier mobilizes for the transmission of the Balint approach to medical professionals.

What characterized this industrial film production and circulation of and in films thus was the hybrid nature of films and practices, on the one hand, and the ambiguity or multidimensionality of objectives and circulation, on the other hand. This concerned the circulation of the films themselves, ranging from physicians' nights to French embassies around the world, and it involved artistic devices, motifs, rhetorical patterns and even individual actors. Mimicking not only earlier mentioned educational films but as here auteur documentary was achieved to the point that the editor in chief of *Médecine/Cinéma*, Philippe Chantelou, could conclude in 1969: "Medical films basically belong to the pharmaceutical industry, which has given these films a promotional function."[31] These industrial films were not simply imitating educational medical films anymore, they were not just in tune with author documentary either, but by their sheer number, their authors and their quality they had become the mainstream of medical film production. In others words, they were central to medical film far beyond straightforward promotion.

Parallel Lives of Industrial Films

Once the budget was allocated by the pharmaceutical industry, the film director and the production company ScienceFilm requested written content from the physician, establishing for the rest a working space for the film director, Duvivier, almost entirely at his discretion. This strategy ensured significant financing for prestigious film projects and at the same time maintained room for the film director's creativity and his possibility of arguing for limited corporate control by the financing pharmaceutical company over the structure and content of the film. A material detail mentioned by the 1981 Sandoz audit underlines symbolically this point.

30 *Du cinéma à la télévision, propos d'un passeur Serge Daney*. 1995. Documentary by Philippe Roger.

31 Chantelou, Philippe. "Le cinéma médical: essai de situation." *Médecine/Cinéma*, no. 7, 1969, p. 7.

The auditors stressed that the master copies of the films sponsored by Sandoz remained in the hands of the film's producer, ScienceFilm, and further copies had to be purchased through Duvivier's service and at his price from his dubbing company, FilmLabo.[32] In short, the film belonged to ScienceFilm, following the common practice in the sponsored film business in the 1960s and 1970s.[33] Admonishing these "older practices," the audit indicated that by 1981 contracts with film directors had been revised to make Sandoz the owner of the master copies. In a broader perspective, these minor contractual changes were part of a general reconfiguration of the pharmaceutical industry's use of film in the 1980s, which was characterized by refocusing on specific product promotion and less on prestige film, and by enhancing corporate control over communication strategies. The 1981 audit itself was part of a new management and efficiency-based approach to corporate communication.

Between 1960 and 1980, in practice and at the level of the individual films, almost all of the more than seven hundred films produced by ScienceFilm started with a company logo (such as "Sandoz presents") but then made no further mention of the company and were absolutely devoid of product-related promotion or placement. This fact attests to the film director's independence during the golden period of industrial filmmaking of the 1960s and 1970s, a state that overlapped with the sponsors' inclination to minimize their ties to the films they sponsored and even to accept association with prestige films. This instance opened further alternatives for the use of the films.

In November 1968 the Parisian press announced widely the premiere of an experimental surrealist short film programme entitled "Hallucinations" at the Le Ranelagh art gallery and movie theatre in the capital's fashionable sixteenth arrondissement. Accompanied by a surrealist art exhibition (including works by Max Ernst, Henri Michaux and Ferro), the film programme featured *Images du monde visionnaire* (*Visions from a World of Fantasy*, 1963–1964, 28 min/34 min), *La femme 100 têtes* (*The Hundred-Headed Woman*, 1967, BW, 20 min, based on a 1929 graphic novel by Max Ernst) and *Concerto mécanique pour la folie* (*Mechanical Concerto for Insanity*, 1963, 25 min).

32 "Cinémathèque." November 1981, p. 5. Firmenarchiv der Novartis AG, Vorarchiv, Service Audit Groupe Sandoz France.

33 Rather common practice in sponsored film production in Europe, copyrights stayed with the producer while sponsors acquired distribution rights. When films were run down or spoiled the sponsors had to acquire new copies from the producer. Exceptions to the rule existed; for example, there is evidence that Nestlé often acquired all possible rights they could get from the producers. Zimmermann. *Schaufenster Schweiz*, pp. 46–47.

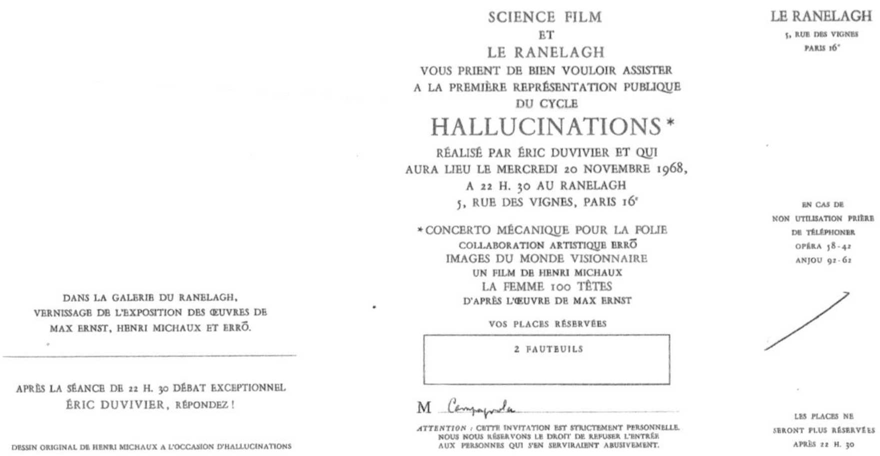

SCIENCE FILM
ET
LE RANELAGH
VOUS PRIENT DE BIEN VOULOIR ASSISTER
A LA PREMIÈRE REPRÉSENTATION PUBLIQUE
DU CYCLE
HALLUCINATIONS *
RÉALISÉ PAR ÉRIC DUVIVIER ET QUI
AURA LIEU LE MERCREDI 20 NOVEMBRE 1968,
A 22 H. 30 AU RANELAGH
5, RUE DES VIGNES, PARIS 16e

* CONCERTO MÉCANIQUE POUR LA FOLIE
COLLABORATION ARTISTIQUE ERRÕ
IMAGES DU MONDE VISIONNAIRE
UN FILM DE HENRI MICHAUX
LA FEMME 100 TÊTES
D'APRÈS L'ŒUVRE DE MAX ERNST

VOS PLACES RÉSERVÉES

2 FAUTEUILS

M Campagnola

ATTENTION : CETTE INVITATION EST STRICTEMENT PERSONNELLE.
NOUS NOUS RÉSERVONS LE DROIT DE REFUSER L'ENTRÉE
AUX PERSONNES QUI S'EN SERVIRAIENT ABUSIVEMENT.

DANS LA GALERIE DU RANELAGH,
VERNISSAGE DE L'EXPOSITION DES ŒUVRES DE
MAX ERNST, HENRI MICHAUX ET ERRÕ.

APRÈS LA SÉANCE DE 22 H. 30 DÉBAT EXCEPTIONNEL
ÉRIC DUVIVIER, RÉPONDEZ !

DESSIN ORIGINAL DE HENRI MICHAUX A L'OCCASION D'HALLUCINATIONS

LE RANELAGH
5, RUE DES VIGNES
PARIS 16e

EN CAS DE
NON UTILISATION PRIÈRE
DE TÉLÉPHONER
OPÉRA 58-42
ANJOU 92-62

LES PLACES NE
SERONT PLUS RÉSERVÉES
APRÈS 22 H. 30

Fig. 26.4. Invitation card announcing the experimental surrealist short film programme "Hallucinations" at the La Ranelagh art gallery and experimental movie theatre Le Ranelagh in the capital's fashionable sixteenth arrondissement in November 1968. The screening was cancelled due to censorship.

The later 1970 version of the programme added *Perception et imaginaire* (*Perception and the Imaginary*, 1964, 25 min). All four films were directed by Éric Duvivier and produced and distributed by ScienceFilm. They had been sponsored, or co-produced as the invitation related, by Sandoz. The short presentation of the programme stated for the audience:

> These four paramedical films united under the title "Hallucination – Hallucinated Worlds" explore the incredible domain of the unconscious in which surrealism after Freud and Lautreamont has found its source of poetic and revolutionary inspiration. It is this descent into hell and into the paradise of the imaginary that these films invite ourselves to. A place where the marvellous, the erotic and the strange compose a fantastic ballet of a surprising facture.[34]

In an interview with Gerard Langlois, Duvivier explained in 1968 his double commitment to medical and experimental filmmaking.[35] The starting point were his medical films. Addressing psychiatric and psycho-pathological themes, Duvivier produced in collaboration with the psychiatrist D.J. Duché

34 Archives Duvivier, Centre Régional de l'Image Nancy, Box N° 21, N° 613–664. Translation by the author.

35 Langlois, Gerard. "Henri Michaux et Éric Duvivier victimes de la censure." *Les lettres françaises*, November 27, 1968.

in an "objective manner the subjective lived experiences of images and sound by a schizophrenic patient."[36] The artistic attempt to explore the unconscious is expanded by Duvivier bridging the artistic and the scientific spheres – as his first production company was named Films Art et Science – to include works by artists recurring to the voluntary use of hallucinogenic, mind-broadening substances. Exploring from the subject's point of view the unconscious expanded thus beyond the medical world of psychiatry to what Duvivier depicts not as a parasite but as "para-medical" in his film programme presentation. He related that the aesthetics of the film made with Michaux in 1963 attracted extra-medical attention and eventually led to the autonomous 1968 experimental film programme.

Not industry, but French state film censorship considered the November 20, 1968, opening night of the film programme as an "uninvited guest or social parasite" – a second meaning of Michel Serres' conceptualization of the parasite form – threatening the 1968 French nation by its subversive infestation (with the use of illegal drugs in the films) and prohibited the extra-medical public screening one hour prior to the event.[37] The motivation of the censorship board was undue "incitation for illegal drug abuse."[38] Representatives of public moral conceived that the "para-medical" circulation of the four medical films that composed the "Hallucinations" programme – the films had been widely screened since 1964 in conferences, medical film festivals and ciné-clubs – was a subversive and dangerous parasite form. From medical to para-medical to parasite, it was the venue and the specific audience that defined the distinguished status of the films.

Duvivier's experimental film *La femme 100 têtes* toured foreign and national film festivals, and received the 1969 prize of the Centre national de la cinématographie in Paris and the Silver Medal at the Venice Film Festival the same year. Eventually with the nomination of Jacques Duhamel to lead the Ministry of Culture in 1971 and the progressive end of French state paternalist film censorship,[39] the art house movie theatre Le Seine (which in its name exchanged the usual female article "La" of the Paris river for the male "Le") screened in early 1970 for an insider audience the four short experimental cinema films of the "Hallucinations" programme.

36 *Films médico-scientifiques Sandoz*, p. 136.

37 *Courrier de notification du refus du visa d'exploitation pour le film "Images du Monde visionnaire"*. Secrétaire d'Etat auprès du Premier Ministre, Chargé de l'Information, no. 5375, Paris, November 22, 1968. Archives personnelles Éric Duvivier, DHVS, Faculté de médecine de Strasbourg.

38 Langlois. "Henri Michaux et Éric Duvivier victimes de la censure."

39 Garreau, Laurent. *Archives secrètes du cinéma français, 1945–1975*. PUF, 2009, p. 245.

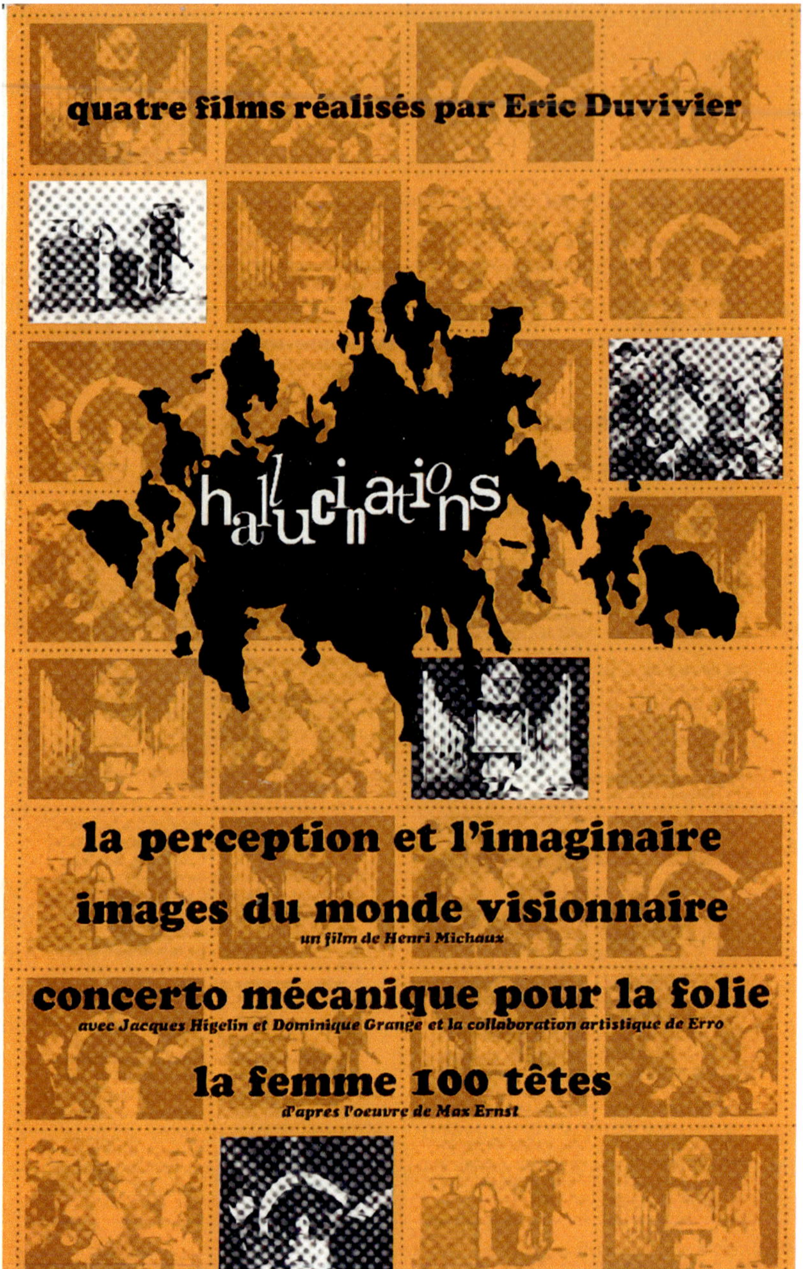

Fig. 26.5. Poster announcing the experimental surrealist short film programme "Hallucinations" at the experimental movie theatre Le Seine in early 1970.

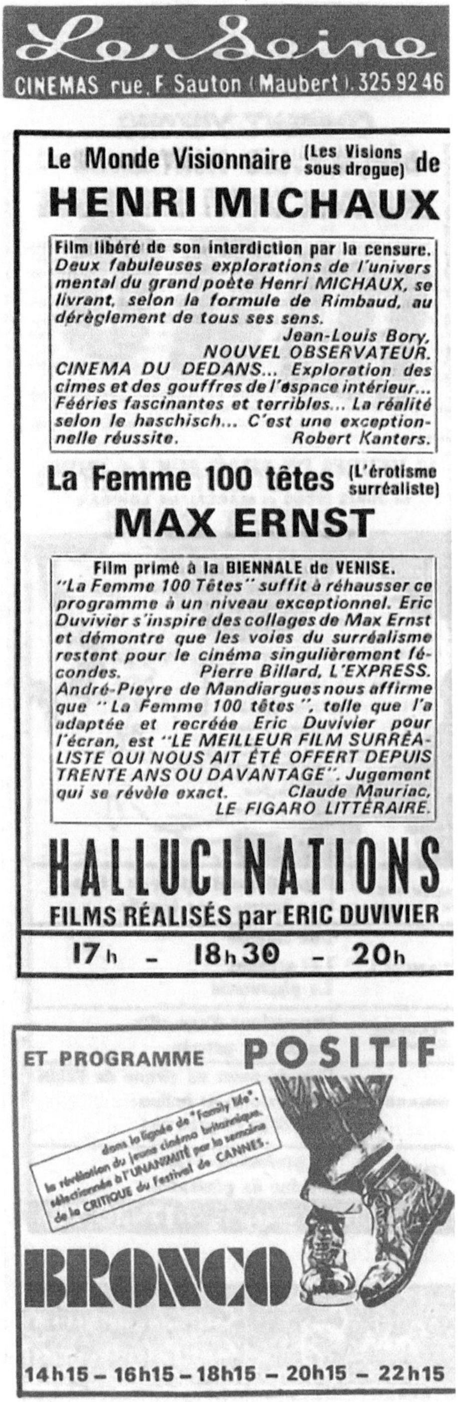

Fig. 26.6. Original entrance ticket to the experimental surrealist short film programme "Hallucinations" at the experimental movie theatre Le Seine in early 1970.

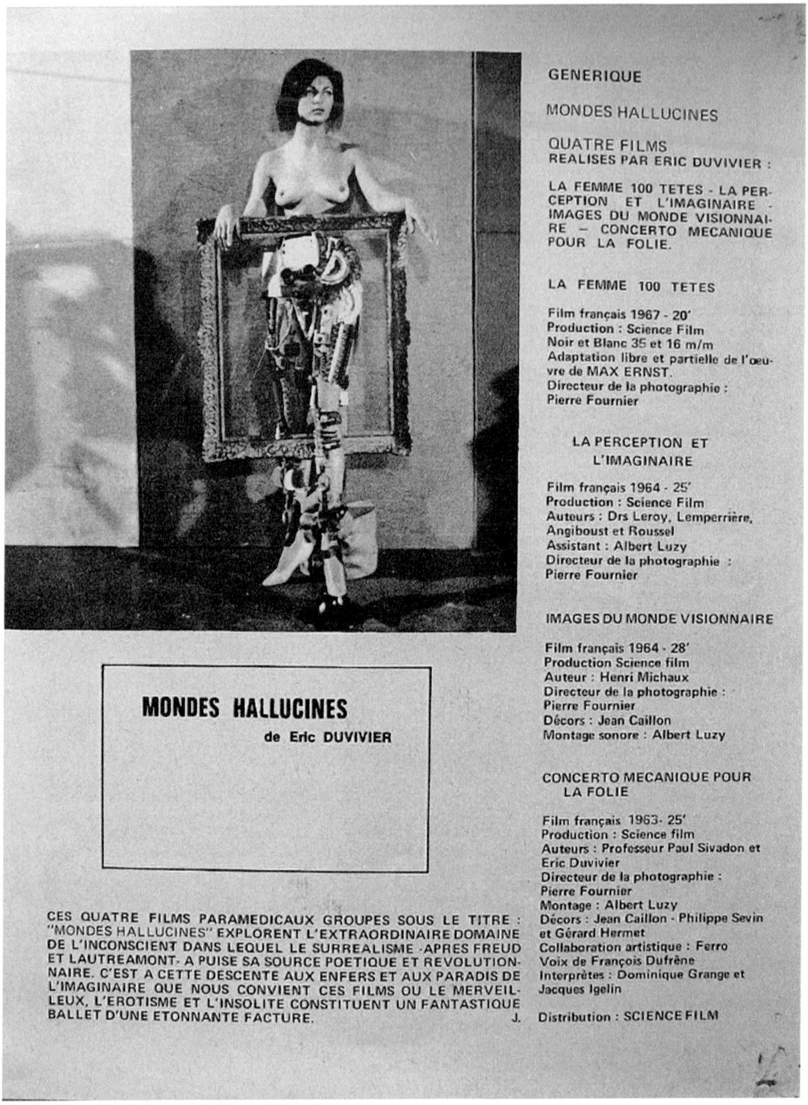

Fig. 26.7. Publicity for the experimental surrealist short film programme "Hallucinations" at the experimental movie theatre Le Seine in early 1970.

Again, exchanges and transfers not only concerned the circulation of films from industrial to avant-garde venues but, as a more detailed presentation of the *Mechanical Concerto for Insanity* reveals, it included as well the involvement of later well-known musicians and actors like Jacques Higelin and Dominique Grange. What is striking about these shorts for today's film historian is not only their surrealist and surprising nature, but the complicities that supported them as well. The functional logic that accommodated a Swiss pharmaceutical industry with post-1968 experimental cinema in Paris and two future French rock stars with an eminent Parisian professor of child psychiatry solidly expands the usual horizons of industrial film history.

Conclusion

My approach from the ScienceFilm company perspective confirms the common appreciation by Rick Prelinger, Vinzenz Hediger and Patrick Vonderau of sponsored "utility" or "ephemeral" films[40] as quantitatively of utmost significance in terms of film production, on the one hand, and as produced and screened in industrial, educational or not-for-profit organization settings, sometimes including commercial film channels, on the other.[41] At the same time, Duvivier's company seems to challenge or at least to complicate to some degree the understanding that these films were simply made by a particular sponsor for a specific purpose other than as a work of art and were designed to serve a specific pragmatic purpose for a limited amount of time. Therein my analysis follows – and Duvivier perfectly illustrates – recent utility cinema studies' acknowledgement that useful films often were multi-sited and multipurpose.

Medical and scientific film directors of the 1950s and 1960s like Éric Duvivier engaged in a strategy to enrol medical opinion leaders and, through them, pharmaceutical companies to finance the production of medical and experimental films otherwise impossible to make. Film directors pitched

40 For "utility," see Hediger, Vinzenz. "'Dann sind Bilder also *nichts*!' Vorüberlegungen zur Konstitution des Forschungsfelds 'Gebrauchsfilm'." *Montage/AV*, vol. 14, no. 2, 2005, pp. 11–22; Zimmermann, Yvonne. "'What Hollywood Is to America, the Corporate Film Is to Switzerland': Remarks on Industrial Film as Utility Film." *Films That Work: Industrial Film and the Productivity of Media*, edited by Vinzenz Hediger and Patrick Vonderau. Amsterdam University Press, 2009, pp. 101–17; for "ephemeral," see Vonderau, Patrick. "Vernacular Archiving: An Interview with Rick Prelinger." *Films That Work: Industrial Film and the Productivity of Media*, edited by Vinzenz Hediger and Patrick Vonderau. Amsterdam University Press, 2009, pp. 51–61; Prelinger, Rick. *The Field Guide to Sponsored Films*. National Film Preservation Foundation, 2006.

41 Hediger and Vonderau, *Films That Work*; idem. *Filmische Mittel, industrielle Zwecke: Das Werk des Industriefilms*. Verlag Vorwerk 8, 2007; Acland, Charles R., and Haidee Wasson, editors. *Useful Cinema*. Duke University Press, 2011; Zimmermann. *Schaufenster Schweiz*; Jung, Uli, and Martin Loiperdinger, editors. *Geschichte des dokumentarischen Films in Deutschland, Band 1: Kaiserreich 1895–1918*. Reclam, 2005; Kreimeier, Klaus, Antje Ehmann, and Jeanpaul Goergen, editors. *Geschichte des dokumentarischen Films in Deutschland, Band 2: Weimarer Republik 1918–1933*. Reclam, 2005.

scientific and medical films that they wished to shoot to the pharmaceutical industry and in the process made their desire to ban specific product promotion from their films match the interests of corporations willing to invest in the arts as institutionalized company policies.[42] It is worthwhile to underline here the wider context of the Swiss (Basel-based) pharmaceutical industry, which was genuinely and extensively involved in business and art collaborations, including Hoffmann-La Roche's longstanding engagement in architecture, art collection and music sponsorship,[43] or Geigy's engagement in film that ran parallel to the Sandoz case described here. The cooperation analysed here suggests that practices described were symbiotic (rather than parasitic) and involved the meeting of mutual and multi-directional interests.

Scientific film marketing strategies consisted in the fact that pharmaceutical companies would buy into the film producer's strategy to use high-profile medical physicians or scientists for the visibility and distribution of their films. Industrial film formats posing as educational used one aspect of parasite strategy: mimicking. Returning to our opening citation this means that "by participating in their [the practicing physicians] information and their training, in return they offer a better opportunity of commercial profit through their prescriptions."[44] At the same time, Sandoz's prestigious production policy granting film directors considerable space for authorship and film expertise, as well as excellent working conditions, led to high-quality film productions that rarely can be equated with the parasite concept of taking without giving. Accommodating aesthetic and useful logics, interactions between Sandoz and ScienceFilm were symbiotic rather than parasitical and, as it turned out, the Sandoz film library itself became an opinion leader amongst pharmaceutical companies when its cinémathèque was quickly imitated by other companies mentioned above as the concept of scientific film marketing became generally accepted in the 1970s.

An interesting point is that in the business and art trade-off, filmmakers and artists did quite well. A number of the films were experimental in nature and would have been impossible to produce in official production circuits that depended on paying audiences. Furthermore, many of the films

42 Chantelou. "Comment des cinémathèques scientifiques favorisent un cinéma de création et comment un réalisateur-créateur leur apporte une conception réaliste du film médical."

43 Bieri, Alexander Lukas. "Roche und die schönen Künste. Eine institutionnalisierte Policy." *CSR und Kultur: Corporate Cultural Responsibility als Erfolgsfaktor in Ihrem Unternehmen*, edited by Vera Steinkellner. Springer Gabler, 2015, pp. 247–62.

44 Chantelou. "Comment des cinémathèques scientifiques favorisent un cinéma de création et comment un réalisateur-créateur leur apporte une conception réaliste du film medical," p. 9.

adopted a resolutely patient-oriented perspective, which was in tune with *direct cinéma* in which directors attempted to get audiences completely immersed in the world that was filmed. Voice-over commentary was limited if not completely absent from these films, and patients often gave their point of view. In this sense, film directors to some extent transformed – and sometime even subverted – the pharmaceutical intentions of their sponsors. To quite an extent they kept control over what in the end remained their film. What was being traded to the pharmaceutical industry was the film and its use more than the content and craftsmanship of the film. As these films were co-produced by a physician for the content and a film director for the images, what mattered most to all the parties involved was that they responded to one of the earliest and fundamental functions of cinema: they had to be attractions. They needed the films to speak to every stakeholder's specific audience, whether corporate, scientific, educational or artistic.

The films mentioned in this chapter are accessible at: https://medfilm.unistra.fr. This research received funding from the European Research Council's "The Healthy Self as Body Capital" (BodyCapital) project under the European Union's Horizon 2020 research and innovation programme (grant agreement no. 694817).

Works Cited

Acland, Charles R., and Haidee Wasson, editors. *Useful Cinema.* Duke University Press, 2011.

Art et Science. *596 films médicaux 16m/m couleurs sonores. Liste no. 17 référence des réalisations d'Éric Duvivier.* ScienceFilm, 1988.

Bieri, Alexander Lukas. "Roche und die schönen Künste. Eine institutionalisierte Policy." *CSR und Kultur: Corporate Cultural Responsibility als Erfolgsfaktor in Ihrem Unternehmen,* edited by Vera Steinkellner. Springer Gabler, 2015, pp. 247–62

Bonah, Christian. "In the Service of Industry and Human Health: The Bayer Corporation, Industrial Film and Promotional Propaganda, 1934–1942." *Communicating Good Health,* edited by Christian Bonah, David Cantor, and Anja Laukötter. Rochester University Press, 2018, pp. 85–142.

Bonah, Christian. "Marketing Film: Audiovisuals for Scientific Marketing and Medical Training in Psychiatry: The Example of Sandoz in the 1960s." *The Development of Scientific Marketing in the Twentieth Century: Research for Sales in the Pharmaceutical Industry,* edited by Jean-Paul Gaudillière and Ulrike Thoms. Pickering & Chato, 2015, pp. 87–103.

Bonah, Christian. "'Réservé strictement au corps médical': les sociétés de production audiovisuelles d'Éric Duvivier, l'industrie pharmaceutique et leurs stratégies de co-production de films médicaux industriels, 1950–1980." *Cahiers d'histoire du CNAM* [special issue: "Le cinématographe pour l'industrie et dans les entreprises (1890–1970): histoire, acteurs, usages et configurations"], 2020, pp. 135–59.

Bonah, Christian, and Joël Danet. "Éric Duvivier." *MedFilm*, https://medfilm.unistra.fr/wiki/%C3%89ric_Duvivier. Accessed April 11, 2021.

Bonah, Christian, and Joël Danet. "Le moment anthropologique du cinéma industriel pharmaceutique. Retour sur le film utilitaire médico-pharmaceutique francophone des années 1970." *Images du travail Travail des images*, no. 4 [special issue: "La relation soignants/soignés à l'épreuve de l'image," edited by Lucia Candelise et al.], 2017, https://journals.openedition.org/itti/993.

Bonah, Christian, and Maike Rotzoll. "Psychopathologie en mouvement. Histoire des films psychiatriques à Strasbourg et à Heidelberg." *Le pré-programme*, edited by Bonah, Alexandre Sumpf, Philipp Osten, Gabriele Moser, Tricica Close-Koenig, and Joël Danet. Verlag A 25 Rhinfilm, pp. 255–80.

Brown, Steven D. "In Praise of the Parasite: The Dark Organizational Theory of Michel Serres." *Informática na Educação: teoria e prática, Porto Alegre*, vol. 16, no. 1, 2013, pp. 83–100.

Chantelou, Philippe. "Comment des cinémathèques scientifiques favorisent un cinéma de création et comment un réalisateur-créateur leur apporte une conception réaliste du film medical." *Médecine/Cinéma*, vol. 14, 1972, pp. 9–11.

Chantelou, Philippe. "Le cinéma médical: essai de situation." *Médecine/Cinéma*, no. 7, 1969, pp. 7–13.

"Entretien avec le réalisateur Éric Duviver, présenté et animé par Thierry Lefebvre et Christian Bonah." *MEDFILM*, Journée d'étude à l'Université Paris Diderot Grands Moulins, no. 2, 21 September 2012. *Canal-U*, www.canal-u.tv/video/cerimes/entretien_avec_eric_duvivier.10645. Accessed 30 November 2016.

Ewen, Stuart. *PR! A Social History of Spin*. Basic Books, 1996.

Fihman, Guy, and Claudine Eizykman. "Xprmntl melba. Images d'un movement." *Cinédoc Paris Films Coop*, www.cinedoc.org/htm/xprmntl/IDMvt.asp. Accessed 25 *November* 2016.

Films médico-scientifiques. Sandoz, 1969.

Fritz, Hans. *Industrielle Arzneimittelherstellung. Die pharmazeutische Industrie in Basel am Beispiel der Sandoz AG*. Wissenschaftliche Verlagsgesellschaft, 1992.

Garreau, Laurent. *Archives secrètes du cinéma français, 1945–1975*. Presses Universitaires de France, 2009.

Greene, Jeremy, Flurin Condrau, and Elizabeth Siegel Watkins, editors. *The End of the Therapeutic Revolution?* University of Chicago Press, 2016.

Hediger, Vinzenz. "'Dann sind Bilder also *nichts*!' Vorüberlegungen zur Konstitution des Forschungsfelds 'Gebrauchsfilm'." *Montage/AV*, vol. 14, no. 2, 2005, pp. 11–22.

Hediger, Vinzenz, and Patrick Vonderau, editors. *Films That Work: Industrial Film and the Productivity of Media*. Amsterdam University Press, 2009.

Hediger, Vinzenz, and Patrick Vonderau, editors. *Filmische Mittel, industrielle Zwecke. Das Werk des Industriefilms*. Vorwerk 8, 2007.

Jung, Uli, and Martin Loiperdinger, editors. *Geschichte des dokumentarischen Films in Deutschland, Band 1: Kaiserreich 1895–1918*. Reclam, 2005.

Kreimeier, Klaus, Antje Ehmann, and Jeanpaul Goergen, editors. *Geschichte des dokumentarischen Films in Deutschland, Band 2: Weimarer Republik 1918–1933*. Reclam, 2005.

Kubelka, Peter, and Alain Sayag. *Une histoire du cinéma: exposition*. Musée national d'art moderne, 1976.

Langlois, Gerard. "Henri Michaux et Éric Duvivier victimes de la censure." *Les lettres françaises*, November 27, 1968.

Leblanc, Gerard. "L'âge d'or du cinéma médical et l'aventure de *Médecine/Cinéma*. Entretien avec Gérard Leblanc." *Sociétés & représentations*, vol. 28, no. 2, 2009, pp. 107–18.

Leblanc, Gerard. *Quand l'entreprise fait son cinéma: la médiathèque de Rhône-Poulenc (1972–1981)*. Cinéthique Presses Universitaires de Vincennes, 1983.

Lefebvre, Thierry. *La chair et le celluloïd: Le cinéma chirurgical du docteur Doyen*. Jean Doyen éditeur, 2004.

Lefebvre, Thierry. "Les films d'Éric Duvivier: un cinema d'auteur." *MEDFILM*, Journée d'étude à l'Université Paris Diderot Grands Moulins, no. 2, September 21, 2012. *Canal-U*, www.canal-u.tv/video/cerimes/les_films_d_eric_duvivier_un_cinema_d_auteur.10646.

Marchand, Roland. *Creating the Corporate Soul: The Rise of Public Relations and Corporate Imagery in American Big Business*. University of California Press, 1998.

Odin, Roger. *L'âge d'or du documentaire. Europe: Années cinquante. France, Allemagne, Espagne, Italie. Tome 1*. L'Harmattan, 1998.

Orgeron, Devin, Orgeron, Marsha, and Dan Streible, editors. *Learning with the Lights Off: Educational Film in the United States*. Oxford University Press, 2012.

Pessis, Georges. *Entreprise et Cinéma. Cent ans d'images*. La documentation française, 1997.

Prelinger, Rick. *The Field Guide to Sponsored Films*. National Film Preservation Foundation, 2006.

Straumann, Lukas, and Daniel Wildmann. *Schweizer Chemieunternehmen im "Dritten Reich"*. Unabhängige Expertenkommission Schweiz – Zweiter Weltkrieg, vol. 7. Chronos Verlag, 2001.

Studer, Tobias. "Die Geschichte der Sandoz im Lichte ihrer Diversifikationen." *Sandoz Bulletin*, vol. 22, 1986, pp. 16–45.
Tulard, Jean. *Dictionnaire du Cinéma: Réalisateurs*. Robert Laffont, 2001.
Vonderau, Patrick. "Vernacular Archiving: An Interview with Rick Prelinger." *Films That Work: Industrial Film and the Productivity of Media*, edited by Vinzenz Hediger and Patrick Vonderau. Amsterdam University Press, 2009, pp. 51–61.
Waller, Greg. "International Harvester, *Business Screen* and the History of Advertising Film." *Films That Sell: Moving Pictures and Advertising*, edited by Bo Florin, Nico de Klerk, and Patrick Vonderau. Palgrave, 2016, pp. 40–53.
Zeller, Christian. *Globalisierungsstrategien – Der Weg von Novartis*. Springer, 2001.
Zimmermann, Peter, and Kay Hoffmann, editors. *Geschichte des dokumentarischen Films in Deutschland, Band 3: Drittes Reich 1933–1945*. Reclam, 2005.
Zimmermann, Yvonne. "Target Group Oriented Corporate Communication: Geigy Films." *Corporate Diversity: Swiss Graphic Design and Advertising by Geigy 1940–1970*, edited by Museum für Gestaltung Zürich. Lars Müller Publishers, 2009, pp. 48–57.
Zimmermann, Yvonne. "'What Hollywood Is to America, the Corporate Film Is to Switzerland': Remarks on Industrial Film as Utility Film." *Films That Work: Industrial Film and the Productivity of Media*, edited by Vinzenz Hediger and Patrick Vonderau. Amsterdam University Press, 2009, pp. 101–17.
Zimmermann, Yvonne, editor. *Schaufenster Schweiz: Dokumentarische Gebrauchsfilme 1896–1964*. Limmat, 2011.

About the Author

Christian Bonah, MD, PhD, is professor of the history of medical and health sciences at the University Strasbourg, a member of the SAGE (Sociétés, Acteurs, Gouvernement en Europe) research lab (https://sage.unistra.fr), director of the Department of Medical Humanities at the University of Strasbourg medical faculty (https://dhvs.unistra.fr), and principal investigator of the ERC Advanced grant BodyCapital. He works on comparative, social, and material history of health, health products and services, and bodies, especially in connection with media and the law. Recent publications include *Body, Capital, and Screens* (Amsterdam University Press, 2020, with Anja Laukötter) and *Health Education Films in the Twentieth Century* (Rochester University Press, 2018, with David Cantor and Anja Laukötter).

27 Transfer of Power

Films Officers in the British Coal Industry

Patrick Russell

Abstract
This chapter argues against reductive understandings of the authorship of industrial filmmaking, and for the importance of sponsors and producers in determining (via complex interplay) its form and content. It further argues for the value of understanding the role of individuals responsible for managing the relationship between major industries and the filmmakers they commission. These themes are explored via the story of the moving image work of Britain's National Coal Board and the men – in turn H.K. Lewenhak, Donald Alexander and Francis Gysin – who managed its relationship with production companies and with its own internal film unit. Their careers are mapped against both the turbulent history of Britain's nationalized coal industry and the rise and fall of the post-war sponsored film.

Keywords: industrial film; authorship in sponsored film; British documentary; coal mining; National Coal Board; Alexander, Donald (1913–1993); Gysin, Francis (1921–1995)

Prospect

> Noel, I am very depressed. (I am writing this thinking-aloud fashion, staring out of the window at plum-trees which bore no plum, and apple-trees without an apple; the black frost which ruined them may have done for me too!) I've looked after the Board's film affairs for more than six years. In that time, I like to think, we've got somewhere – nothing like all the way we *can* go, but still somewhere. It is now, when we have proved our points so far, and when the innovations we require are merely the

Hediger, V., F. Hoof, Y. Zimmermann, with S. Anthony (eds), *Films That Work Harder: The Circulation of Industrial Film*. Taylor & Francis Group 2024
DOI: 10.5117/9789462986534_CH27

> logical consolidation of what we have already established, that I feel like a Channel swimmer who has spent hours in the water and can see the English cliffs, but has already lost the flood. [...] [A] long time ago, you as good as said that I was intent on building an Empire. If you only knew how wrong you were! You could say that to Edgar Anstey, who will grow venerable with a long white beard in his Estate, but not to me. My hero is Cincinnatus, who came out and did what had to be done, and then retired to his vineyards.[1]

These words appear in a 1957 memo from Donald Alexander to Noel Gee. "The Board" was the UK's National Coal Board (NCB), one of Europe's larger national employers, pivotal to Britain's economy, society and self-image. Alexander was its films officer, and Gee, director of public relations, his line manager. Above Gee sat the NCB's executive body, below Alexander a unit of filmmakers as well as contracted external production companies, turning out between them some thirty films a year.

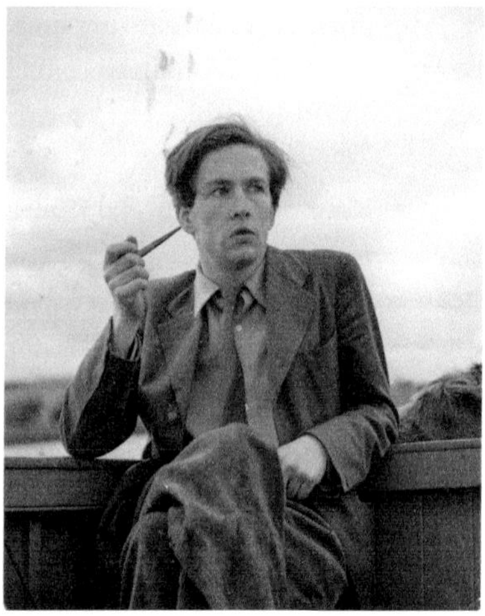

Fig. 27.1. Donald Alexander (1913-93), photographed (by his first wife, the artist and illustrator Isabel Alexander) in 1939 (BFI National Archive; courtesy Robin Alexander).

If, by occurring at the junction of industry and film, industrial films are "*interfaces* between discourses and forms of [...] organization,"[2] then for British coal mining in 1957 that interface was named Donald Alexander.

1 National Library of Scotland, SSA 4:20:41; undated: late 1957 is inferred from contents.

2 Hediger, Vinzenz, and Patrick Vonderau. "Introduction." *Films That Work: Industrial Film and the Productivity of Media*, edited by Vinzenz Hediger and Patrick Vonderau. Amsterdam University Press, 2009, p. 11.

So what might his private remarks to his boss tell us, not just about him but about industrial filmmaking? Erudite, introspective and emotionally, even romantically, invested in his mission, they catch him nearing the edge of a pendulum swinging between hope and despair. They are, in short, the words of someone with a complex relationship to his work. Someone, too, whose position grants him substantial, but much less than total, agency: constrained agency subject to conflicting pressures. To many readers, that will sound like a typical day in the office. A conception of industrial filmmaking bereft of this perspective lacks humanity and common-sense.

It also lacks due complexity. Elsaesser wisely advises that we "look at the industrial film from a pragmatic premise" but his statement of that premise, "the context of the three As: commissioning client, concrete occasion, and target use or target audience,"[3] is too narrow, listing factors that are necessary but insufficient. To communicate with the viewer to fulfil the purpose the client, or contractee, requires a supplier, or contractor, to make the product. Where Hediger and Vonderau correctly assert that industrial films "cannot be divorced from the conditions of their production and the contexts of their use," they go on to present a misleading binary by ascribing to industrial film, following Elsaesser, "an occasion, a purpose, and addressee, or an *Auftrag*, *Anlass*, and *Adressat*, rather than an auteur."[4] Necessarily unpacking contexts of use, it insufficiently unpacks conditions of production, limiting attention to the client side of the contract and conflating consideration of the supplier side with its reduction, which does not follow, to one person: an "authoring" film director.

One need veer nowhere near auteurism to recognize that film offers industry a creative as well as technical service. The twentieth-century film industry's control over technology and expertise obliged industries commissioning those services to delegate some degree of authority to them for the duration of production: a provisional, functional transfer of power. Who pays the piper calls the tune. But it is the piper who plays. Sponsorship may govern *why* filmmaking occurs but within constraints production may determine, technically and creatively, *how*. *What* results is a product of both. Industrial film is, always, the result of a *relationship*.

Enter the films officer, employed to mediate that relationship. Neither the role nor the title was unique to Alexander. He had inherited it from NCB predecessor Kurt Lewenhak and then passed it to successor Francis

3 Elsaesser, Thomas. "Archives and Archaeologies: The Place of Non-fiction Film in Contemporary Media." *Films That Work: Industrial Film and the Productivity of Media*, edited by Vinzenz Hediger and Patrick Vonderau. Amsterdam University Press, 2009, p. 32.

4 Hediger and Vonderau. "Introduction," p. 10.

Gysin, while several contemporaries were similarly contracted elsewhere. Edgar Anstey, "in his Estate," was Alexander's direct counterpart, films officer for another large organization created by post-war nationalization, for which Anstey's British Transport Films (BTF), like Alexander's unit, provided a moving image service reporting to its public relations department. In the private sector, the coal industry's telling counterparts were oil industry giants Shell and BP, for whom Arthur Elton and Ronald Tritton were Alexander's respective equivalents. All had one foot in the corporate world, working with its managers and understanding organizational structures, objectives, budgeting and politics. Their other foot was in the film industry, working with its practitioners and understanding production, exhibition, distribution and aesthetics. Interpreting each to the other, the transfer of power, industry to film and back, was conducted through them.

In post-war Britain, major corporations invested in film over a sustained period through a single budget line. Spending it, the films officer supervised a filmmaking *programme* intended for cumulative effect transcending individual films' immediate purposes. Each company had its particular film strategy, distribution arrangements and combination of production contractors, its predominant on-screen aesthetic, its lineage inherited from earlier filmmaking eras, its house directors and canon of hit films: all yielding screen *personae*. Today we would call them brands – better, sub-brands, discrete from but augmenting their sponsors' master brands. Between sponsors, brand strength varied. So did brand consistency. This should not surprise. Film-commissioning industries, whether transport, oil or coal, are never static. Nor is the film industry receiving their commissions. Dynamic economies, dynamically interacting: the relationship is inherently unstable.

Over some fifty years, the NCB paid for some one thousand moving image productions, their combined running time around 250 hours: one of the world's larger *single-subject, single-source* industrial film collections. This article tells its story, indicating its cinematic lineage then dividing its chronology into six phases, each a momentarily stable model of brand management, derived from a film/industry dynamic differing from preceding and following phases. A phase change is catalysed sometimes from the client, sometimes the supplier side, sometimes both simultaneously, mediated by the incumbent films officer. First Lewenhak's, then Alexander's and then Gysin's careers play out against two histories. In the micro-economic foreground: the industrial documentary industry, waxing then waning. In the macro-economic background: the coal industry, its decline set against, and in part caused by, the oil industry's ascent.

Lineage (1930–1945)

"Let us face the future"

Under this slogan, the Labour government headed by Clement Attlee was elected in 1945. Many were the industries looking ahead, optimism tinged with apprehension. Coal mining, spread across industrial regions, was among the largest. One of the smaller, concentrated in London, was non-fiction filmmaking, most of it (excepting commercial newsreel) institutionally sponsored.

Few industrial films proper were produced during the war but the wartime state's growing communications need had caused its production sector to surge as existing producers' slates expanded and new ones entered the market. Not a homogenous industry, it was woven of several strands from the previous decade. Looming large in film-historical accounts is the Documentary Film Movement indelibly associated with its 1930s leader, producer-strategist-theorist John Grierson. Anstey and Elton were two of Grierson's earliest acolytes, and the Shell Film Unit a product of the Movement's 1930s phase, run via Film Centre, a Griersonian consultancy. The Unit produced government-sponsored films throughout the war.

The NCB's lineage is found not amid this core group but in another strand of the Movement, with arguably more radical leanings, politically and stylistically, connected to Grierson's contemporary, and sometimes critic and rival, Paul Rotha. Donald Alexander entered 1930s filmmaking under Rotha, and like Gysin was in the war employed by him at Paul Rotha Productions, founded to feed the burgeoning state information campaign. But the wartime boom had not only boosted "Movement" production units; it also helped competitors beyond its orbit. As a wartime government information commissioner Ronald Tritton became familiar with both schools and wary of the former, preferring the latter's ostensibly non-ideological, classical approach. Later, at BP, he contracted mainly with film companies of that lineage.

Too little is written about that "commercial" sponsored film sector, but the "Movement" has inspired a large, disputatious literature, often debating the relationship of its motives to its business model.[5] Dominated by men of the educated middle and upper-middle classes, it claimed in 1930s sponsorship a progressive means of putting the working world and class on screen.

5 The literature's contributors include Forsyth Hardy, Rachael Low, Elizabeth Sussex, Eva Orbanz, Brian Winston (the most trenchant critic), Ian Aitken, James Chapman, Timothy Boon, and Scott Anthony.

Now, in the 1940s, it committed wholly to the anti-fascist war effort, and stealthily to the promise of a new political economy to be constructed after victory. Two Alexander-directed Rotha productions, *Five and Under* (1941), a straight "informational" documentary, and *All Those in Favour* (1942), using the scripted "story documentary" mode, exemplify filmmaking implying that promise. Both modes were later frequently echoed in NCB filmmaking.

Coal mining, too, was co-opted to support the war effort under government direction. Once at the heart of the industrial revolution, under growing international competition and chronic corporate underinvestment, the interwar industry had become tired, underproductive and rancorous, its coalfields culturally isolated, its workforce, especially since its central role in 1920s industrial unrest, feared and despised by compatriots. This fragmented, troubled industry sponsored few films. Its imagery informed many: for left-leaning 1930s intelligentsia, filmmakers included, mining (and the figure of The Miner) was the most loaded of industries, iconographically and ideologically. *Coal Face*, by Alberto Cavalcanti, and *The Face of Britain*, by Rotha, both from 1935, are merely two of the most artistically distinguished pieces laden with such symbolism.

Attlee's government implemented coal nationalization on January 1, 1947. Replacing some 750 undertakings, acquiring nearly one thousand collieries employing over 700,000 people, the NCB was overnight one of Britain's largest companies. As a "state corporation" it was (fatefully, as it later panned out) publicly owned, taxpayer-funded, subsidized and price-controlled yet in large measure operating in the marketplace.

First Phase (1947–1952)

World war having awoken institutional awareness of film, NCB Public Relations head Noel Newsome appointed the company's first films officer, Kurt Lewenhak, whom he had known during the war. Their early programme served an immediate objective – via recruitment films (e.g. *Adventure of Coal*, 1947) for mobile cinema distribution – and a broader strategy, both by uncredited investment in mining-themed B feature films and, more propitiously, by creating *Mining Review*. This monthly one-reel newsreel-cum-cinemagazine, distributed to cinemas and non-theatrically, combined national investment and productivity updates, technical items, often shot below ground, and above-ground social and cultural items filmed in Welsh, Scottish and English mining communities. For example, *Mining Review 1st Year No. 6* features underground locomotives, a tour of the Ruhr by UK mining engineers, and the traditional sport of hound trailing in Cumbrian mining districts. It was produced by the government's Crown Film Unit, of directly Griersonian lineage. But, dissatisfied with unpunctual delivery, from the seventh issue Lewenhak contracted with DATA (Documentary and Technicians Alliance), a cooperative founded in 1944 by Rotha staff who were restless under his autocratic leadership – most prominently, Donald Alexander. Their inclinations lay on the left wing of the Movement. *Mining Review*'s attractions were obvious.

Alexander's mining affinities dated to his first steps into filmmaking while studying Classics at Cambridge (he would gain a First). He borrowed a 16 mm camera and took it with him on vacation in 1935. He filmed at locations in mining valleys in South Wales, capturing stark, angry images of the Depression's effects on the people who lived and worked there. Impressed, Rotha screened the film within his then company and hired Alexander, soon giving him an assistant role on an NGO-sponsored film, *Today We Live* (1937), instructing him to recreate on 35 mm a sequence showing desperate men on windswept slagheaps foraging for coal.[6] This reshot footage is infinitely reused by British film and TV as *the* definitive image of 1930s poverty.[7]

For DATA, the contract offered the chance both to support Attlee's flagship nationalization, and to develop closer acquaintance with an industry and class than previous filmmakers. Ironically, both opportunities demanded conservative aesthetics, key to the NCB brand. Its *Adressat* did not share the tastes of those the Movement's more self-conscious "art" classics appealed to. Classical continuity and eye-level shots replaced Rotha's modernist, Soviet-influenced montage and low angles.

That this was a peacetime industry on a war footing had been brought home by 1947's fuel crisis. It was further declared by the government's 1950 *Plan for Coal*, auguring mechanization on top of pit modernization. DATA's 1952 non-*Mining Review* film of the same name is tonally remarkably similar to many of the more stentorian wartime propaganda films. *Mining Review* had taken the theatrical slot previously held by the Rotha-produced government series *Worker and Warfront* and *Britain Can Make It*. Its PR mission represents both progressive continuation, and regressive contraction, of the wartime documentary project: nation-building plus national projection. What was being built was now a nation within a nation: British coalfields under the auspices of a state corporation. And the external world it was

6 See Hogenkamp, Bert. *Deadly Parallels: Film and the Left in Britain 1929–39*. Lawrence & Wishart, 1986; Berry, David. *Wales and Cinema: The First Hundred Years*. University of Wales Press, 1994. Hogenkamp and the late Berry, friends to whom this author owes much, are the only writers to have interviewed Alexander.

7 One relatively recent re-use is in Ken Loach's polemical documentary *The Spirit of '45* (2013).

projected to was not the wider globe but the rest of the UK. NCB films sought to prove to mining communities and the tax-paying public simultaneously that this was an industry and community integrating, organizationally and culturally, and renewing itself through investment, modernization and self-esteem. They have, adapting Elsaesser, two addressees, and addressing one formed part of the rhetoric addressing another. That mineworkers witnessed NCB messages being projected to the stakeholding public sent them a message as important as any projected directly to them, public relations *and as such* industrial relations functions being met concurrently.

> Our films have to fulfil [...] distinct and separate tasks, [...] but films and [...] audiences cannot be divided into watertight compartments. [...] [T]he miner, as a member of the community, obviously sees the film in which you are informing the lay public about the industry, and the way he features in such a film has an obvious effect on his morale.[8]

Far from radical, these films bind the working class to the emergent consensus, from both ends. But if Cavalcanti and Rotha had treated The Worker, The Miner especially, as a cipher, he now, via a formulaic newsreel style imbued with genuine warmth, acquired a modicum of individuality.

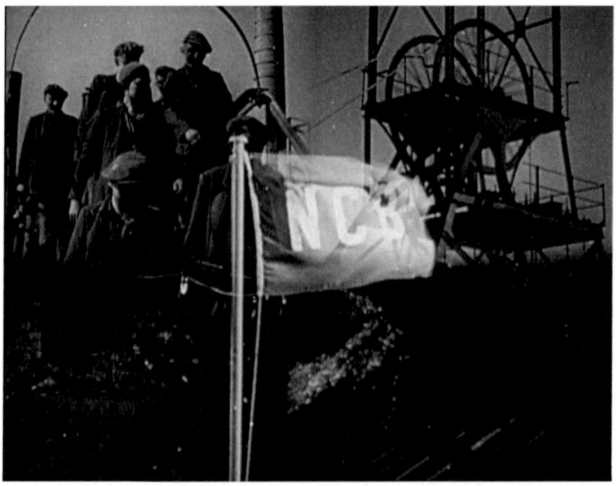

Fig. 27.2. The opening image of *Mining Review 1st Year No. 1* (1947) (BFI National Archive).

The Shell Film Unit contemporaneously reverted to its pre-war status as a corporate production unit; under Elton at Film Centre, it began refining

8 Lewenhak, Kurt. "Films and the National Coal Board." *Documentary Film News*, April 1948, p. 44; "The miner is a highly discriminating filmgoer [...] [and the] producer working for such an audience need not fear that his finer nuances are being wasted on an insensitive audience."

a brand modelled on the austere beauty of pre-war film *Transfer of Power* (1939). Conscious of Shell's head start, Tritton, now in the PR department of the Anglo-Iranian Oil Company (soon to be rebranded BP), dipped his toes into film commissioning. British Transport followed the NCB in forming its own films department, Anstey half-replicating his friend Elton's model: establishing an internal production unit but due to personnel rules running it as employee not Film Centre consultant.

Rotha meanwhile went bankrupt.

Second Phase (1952–1957)

In 1951, Donald Alexander, apparently tiring of the martyrdom of reconciling DATA's idealism with sponsorship, replaced Lewenhak at the NCB. Between his initial years and his predecessor's term there was reasonably seamless client continuity. Maximum output, spurred by massive reinvestment, now embracing mechanization, served a national reliance on coal for energy security and prosperity's return. A changed films programme from 1952 is explained by Alexander's own presence proactively interpreting these continuing concerns.

DATA continued delivering *Mining Review*, aims and editorial balance unchanged (jauntily narrated by actor John Slater, "over a number of issues, it contrives to keep a balance between coalfield and coalfield, between technical stories, cultural items, men's interests and women's interests"[9]). Alexander's innovation, supported by Newsome's successor Guy Nott-Bower, was a massive *parallel programme* of films, begun in 1952, to retrain miners, engineers and managers in the underground implications of the *Plan for Coal*. In Hediger and Vonderau's terminology, Alexander opened a *rationalization* programme, largely faced inwards, augmenting an existing *rhetorical* programme facing both staff and public. Subjects were specific: roof supports, haulage, coal preparation, dirt disposal, first aid. Above all: machinery. "If the period 1840 to 1860 may be regarded as coal mining's first industrial revolution, then the period 1950 to 1970 may be regarded as the second."[10]

Early mechanization films were contracted to DATA, produced by Gysin, Alexander's main liaison there as *Mining Review*'s then-principal producer. But, Alexander persuaded Nott-Bower, in-house production would be more cost-effective and could yield better communication, by developing personnel yet more familiar, from the inside, with organizational culture and

9 Alexander, Donald. "NCB – The First Twelve Years." *Film User*, vol. 13, no. 150, April 1959, p. 174.
10 Griffin, A.R. *The British Coal Mining Industry*. Moorland Publishing, 1976, p. 115.

infrastructure.[11] The NCB Technical Film Unit was created, its achievements impressing Nott-Bower and successor Noel Gee. Its output rapidly assumed the proportions of an encyclopaedic record of an authentic technological revolution, extending from *The Shovel* (1953), then a core tool in widespread use, to *The A-B Meco-Moore Cutter Loader* (1954), a wartime machine steadily advancing through post-war coalfields, to *The Anderton Shearer-Loader* (1955) at the post-war cutting edge, destined within a decade to cut more than half the country's coal.

Alexander insisted he be contracted through Film Centre, meaning the same consultancy was servicing oil and coal.[12] He had, in fact, many sponsors: multiple NCB departments, represented on a film working group assessing annual priorities: "It would have been quite wrong to lay on Film Section the burden of deciding between departments what were the most important subjects," but "once the subjects have been agreed, Film Section can deal with them. [...] In all this, and within the limitations of an annual budget, Film Section has as free a hand as it could possibly wish."[13] Films officers were also responsible for distribution policy. A central library held all productions, prints freely hireable by external non-theatrical users as well as for the larger internal audience. Additionally, each NCB regional division had its own films liaison officer, managing his own prints library and hireable projector. Every large pit had its own screening capacity, with eight hundred projectors altogether said to be in use. For cinemas, *Mining Review* was contracted to a new, enthusiastic distributor who – supposedly – secured distribution to over fifty per cent of UK cinemas.[14]

Associates remember Alexander as a charismatic, driven, somewhat intimidating intellectual, an instinctive teacher with a penetrating stare. His administrator Ken Gay wrote fifty years later of

11 "Film-makers with experience of coal-mining, while not as rare as abominable snowmen, are less common than blackberries." The National Archives (TNA) COAL 32/16: "Working Group on Film Policy: Proposals for 1962."

12 He was later paid through Auxiliary Services Ltd, then Techni-Commercial Services.

13 Alexander, Donald "NCB – The First Twelve Years (2)." *Film User*, vol. 13, no. 151, May 1959, p. 239. National Library of Scotland SSA4:20:75 includes a peculiar, fascinating document, *Some Aspects of the Social Structure, and Other Sociological Aspects of a Documentary Film Unit*, stating: "The Film Section has an independent life from the bureaucracy of this nationalized industry. [...] There are no regular work contacts with other parts of the bureaucracy, except at the highest level. [...] [The] Film Section is very much an independent social unit, and can be thought of as a unity in itself."

14 The distributor was Sam Goodman trading as DATA Film Distributors. As always with industrial film, distribution figures are dubious and unprovable. Figures in trade journals claimed the series reached three hundred cinemas in 1947, over a thousand in 1958, but eight hundred in 1959. In 1983, Gysin stated that "right up to the last issue, Sam Goodman succeeded in having a reel shown in over half the screens in the country." Anecdotally, I have heard that a *Review* issue preceded *Jaws*' UK premiere! Yet many avid filmgoers have no memory of it.

> a strong-willed man of very independent views, despising pretension. [...] Donald was always right. He took this attitude even when discussing films with Board members, one of whom once observed that "we are not making this film according to the gospel of Donald Alexander." Donald was forceful but not arrogant. [...] He could easily lose his temper [...] and could never put up with fools or foolish behaviour, however grand the status of the person concerned. This assessment [...] leaves out his essential kindness and understanding. [...] He changed my life.[15]

Third Phase (1957–1963)

In 1957, affluence taking hold, Britain's first national industrial film festival took place at Harrogate, Yorkshire. This seminal event marked industrial documentary entering its "High Renaissance." Schools of filmmaking had blended, sectarian 1940s divisions between "Movement" and "commercial" sectors ended. Structurally, stability and productivity marked both sides of the contractual relationship, the power balance having somewhat shifted from filmmaker to sponsor, not in absolute terms but relative to the Movement's headier days. Formally, honed classicism characterized their products. Oil, coal and transport had emerged as leading sponsors, among many. Harrogate organizers invented the term "prestige film" to describe the glossiest, least ephemeral sponsored productions, putting Shell, BP and BTF slightly higher up the ladder than the NCB. Personae were now established, proffering differing brands of classical aesthetic. Shell's was magisterial, projecting a scientist and statesman; BP's engaging and middlebrow, a cosmopolitan explorer. Their canvas was a shrinking globe, whereon transnational companies were replacing colonial states as rising powers. Meanwhile BTF painted a lushly coloured, dreamlike Britain reachable by rail. The NCB brand was that of gentle giant: steadfast, practical, proficient but cultured, presenting a firmly national face in keeping with its dependence on wholly domestic manpower. Its films remained black and white, not definitively abandoned until the turn of the 1970s. Safety regulations prevented electric cameras' use underground; even into the 1980s, the NCB persisted (aided by bespoke flameproof lighting) with mute clockwork 35 mm Newman Sinclair cameras, the model Grierson, Rotha and disciples had filmed with in the 1930s and 1940s.

15 Gay, Ken. *Hand in Hand with Time: A Memoir.* Honeysuckle Books, 2009, p. 151.

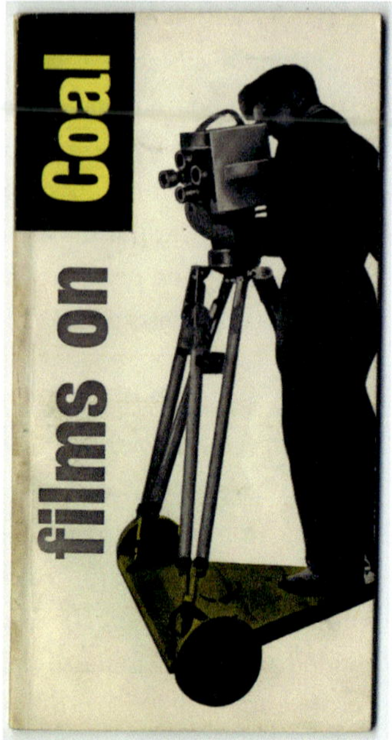

Fig. 27.3. Autumn 1958 edition of NCB films catalogue (BFI National Archive).

Then suddenly: trouble in paradise. If the previous phase resulted from filmmaking intervening into a period of relative continuity in coal mining, dramatic changes in the latter's fortunes now befell a settled films department. In 1957 a sharp fall in oil prices coalescing with anti-pollution legislation triggered mounting piles of unwanted coal, resembling similar scenes across Europe. Ironically, the NCB's investment programme, of which Alexander's unit was part (at one stage absorbing ten per cent of training budgets), was right now unstoppably kicking in. The films programme proceeded apace but in a drastically altered "rationalization" context. DATA's *Mining Reviews* (e.g. *16th Year No. 3*) began registering pit closures and internal migrations. New (for this sponsor) production types emerged. *Arthur Clears the Air* (1961) is a sales film, a hitherto unnecessary category in a sellers' market. Internal films stepped beyond the narrowly "technical" into industrial relations-sensitive topics, like introduction of work study (*Shaft Survey*, 1957[16]).

Two films to which Alexander was closely attentive are worth singling out. First, *Experiment* (1958), directed by Alexander himself, is an internal film applying rhetoric to rationalization, adapting the problem-solving

16 This was directed by Alexander employee Ralph Elton, Arthur's brother. Of an aristocratic family, they shared their ancestral manor house, Clevedon Court. Film industry folklore had it that Arthur's half was heated by an oil-fired burner, Ralph's by a coal furnace.

wartime mode of *All Those in Favour*: a dramatized reconstruction of one colliery's collaborative introduction of method study. The key role is played by manager Philip Weekes, off-screen a close friend of Alexander's, on it personifying benevolent technocracy: phlegmatic, competent and humane in the face of mounting difficulty. *New Power in Their Hands* (1959) is a compilation of footage shot by the Film Unit for its bread-and-butter training films, refashioned as a public-facing prestige essay on the implications of power loading. This was the master purpose of much of the multifaceted technical revolution, replacing discrete processes (undercutting the seam, ripping down its coal and loading it to the surface) with combined operations. The film furthers the PR mission to which Alexander was most committed, positive rebranding of The Miner in the public mind, but at a moment in which an assured role in national fortunes was disrupted. One of the 1950s' most engrossingly ambivalent British industrial films, neither its wintry mood, defensive tone nor quasi-stylized aesthetic resemble the cheery *Mining Review*.

Two important organizational changes occurred in this period. Geoff Kirk replaced Gee as Alexander's boss in 1960, then in 1961 Alexander chose not to renew DATA's contract. Many DATA staff joined the renamed NCB Film Unit, hereon making *Mining Review* as well as most internal productions. Gysin was long gone, since 1958 producing films not for coal but for oil – running Shell's film unit in Venezuela.

Fourth Phase (1963–1972)

In 1963 Alexander received a letter. After five years, Gysin wanted to return to England; could Alexander keep an eye out for film industry opportunities and maybe put a word in for him? Believing Films Section could be entrusted to the combination of Kirk and Gysin, Alexander successfully recommended the latter to the former as his own replacement and abruptly resigned. Gysin was placed on the Board's staff (i.e. not paid through a consultancy). For a few years, before relocating to Scotland, Alexander took several freelance directing jobs from him.

Both were upper-middle-class alumni of elite schools (Shrewsbury and Highgate) and of Cambridge, Rotha and DATA. But the Anglo-Swiss Gysin was less intense than the Anglo-Scottish Alexander. Worldly and sardonic, he applied idiosyncratic nicknames to everyone and everything (Alexander was "The Duck," the NCB's headquarters "The Bordello," two items from a lexicon of hundreds, apparently). Gay on Gysin:

> [L]ife with Francis was less stressful. Possessed of a fast-working brain, [...] like Donald he could not tolerate fools or pretentious people, but under a tough, cynical exterior he was a kind and thoughtful person, solicitous when staff were ill and visiting them when hospitalized. [...] Francis was also notorious for his ability to run things from a bar stool. [...] His usual tipple was Teachers whisky (always ordered as a double), although he preferred Scottish malts when available. [...] Alcohol appeared to have absolutely no effect on him. [...] [D]ecisive and efficient, [he] ran the film unit with brio [...] [and was] greatly respected throughout the Board by staff in all departments, and by Geoff Kirk, his boss. Working in a pub was just his style.[17]

Eight years Alexander's junior, Gysin had entered the Documentary Film Movement after its peak modernist, ideological phase. "He had," his obituarist notes, "the air of a newspaper man – fast in his reactions with the journalist's instinct for a good story. He could make up his mind about a project in a minute, and write an excellent script in half an hour. He also showed immense physical courage in some trying situations."[18]

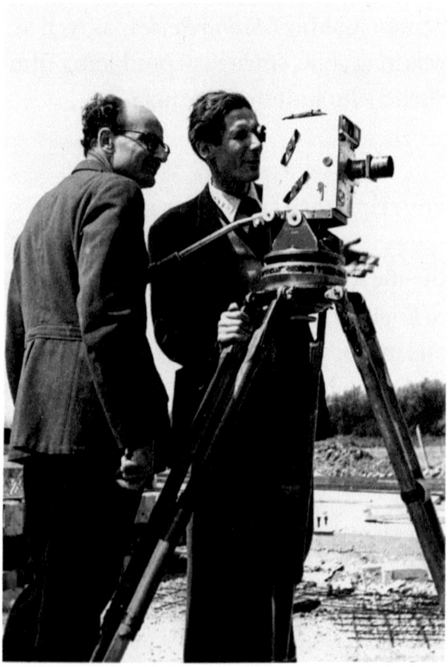

Fig. 27.4. Francis Gysin (1921-95), looking through viewfinder of Newman Sinclair camera, directing an early DATA *Mining Review*. Cinematographer Wolfgang Suschitzky (1912-2016) looks on (BFI National Archive).

17 Gay, *Hand in Hand*, p. 202.
18 "Francis Gysin." *The Times*, December 10, 1995, p. 14.

If Alexander's dominion rested on a theorized worldview, Gysin's ran on the fuel of interpersonal relationships, now embracing several generations of filmmakers, from those with Grierson-Rotha lineage to new hires of the 1960s generation free of Depression-era preoccupations. Gysin was re-entering a booming film industry, but a sponsoring industry in evident trauma. The former's annual productivity hit record levels, estimated at over a thousand industrial films annually, its prestige category most spectacularly vindicated by BTF's Hollywood Oscar (for 1964's *Wild Wings*), following one for BP five years earlier (1959's *Giuseppina*). The latter's predicament intensified. Protectionist government measures (restrictions on imports, coal quotas on the electricity industry) intervened even as productivity continued rising and the workforce shrinking. In 1957, 822 mines had employed 704,000 staff. In 1963 it was 576 and 517,000. By 1972 just 259 collieries employed 252,000 workers. In 1957, 216 million tonnes were consumed by homes and industry, down to 194 million by 1963 and just 121 million tonnes come 1972, the customer base increasingly dominated by electricity-generating stations.[19] Such spectacular downsizing was the more harrowing for how unexpected it would have been several years earlier. Combined with discoveries of natural gas in Britain's North Sea waters and searches for North Sea oil, it definitively overturned the industry's war footing, pushing maximum effort for maximum output, in favour of streamlining processes, plant and staffing. New NCB chairman Alfred Robens, the first from outside the industry and to be media-conscious, embarked on a Janus-faced PR strategy, talking up coal in public while talking tough love to staff. Film was well-positioned to support both. Paradoxically, then, downward pressures, when combined with Gysin's appointment and the cultural changes of the 1960s, widened scope and aesthetics, stretching the brand. *Mining Review* sporadically broke free of newsreel trappings, dabbling with more varied, modish music and camerawork. The unit had a go at prestige films, referencing wider trends: *Two Worlds* (1965) peculiarly mixes BP- and BTF-style historical travelogue with would-be philosophical musings, while *Portrait of a Miner* (1966) reinterprets the PR exercise to rebrand The Miner as coalfield existentialism informed by British New Wave cinema. *A Time to Heal* (1963), outsourced (actually as one of Alexander's final commissions) to youthful production company Derrick Knight & Partners, is Direct Cinema-influenced documentary. A batch of safety shorts contracted to TV Cartoons (headed by noted Canadian animator George Dunning) featured gallows humour animation.

19 Figures from Ashworth, William. *The History of the British Coal Industry, Vol. 5: The Nationalized Industry*. Oxford University Press, 1986.

Unit-produced training films flirted with cynicism (*Nobody's Face*, 1966) and surrealistic gore (*Manfailure*, 1971). Alexander's freelance directorial commission *The 4 M's* (1964), requested by Robens, was centrepiece of a major internal awareness-raising state-of-the-mining-nation campaign. The "M's" are money, markets, men and materials and *Experiment*'s cinematic wartime "story documentary" is replaced by sober, uncomfortable television-style debate between unscripted staff and senior officers – including Alexander's friend Weekes, now director of studies for the NCB's staff college.

For this writer, the NCB's middle period is its most interesting: its third and fourth filmmaking phases, running from 1957 to 1972, marked by its sponsor's traumatic contraction. Divisive memories and political symbolism surrounding the colliery closures and strikes of the 1980s have obscured how crucial the 1960s were in marking, in part instigating, the division between the apparent optimism around this industry in the early post-war years and its descent into strife and despair. The golden eras of BTF and Shell filmmaking came when they and their sponsors navigated relatively calm terrain in apparent harmony. The NCB shows that how film behaves when under sponsorship facing turbulence can be as fascinating, less predictable. This was the NCB's richest, least coherent filmmaking phase: its midlife crisis.

Fifth Phase (1972–1983)

The stresses of the 1960s, managed collaboratively by management and union leaders, ultimately led to abrasion culminating in national mineworkers' strikes in 1972 and 1974. If, pre-1957, official positivity had tended to repress complexity and ambivalence, and its subsequent emergence was a function of flux, now perhaps growing *negativity* fed a different repression.

From 1970, the NCB Film Unit was housed within Gysin's "Bordello," the parent company's corporate headquarters. It was previously located in different premises, often in different central London districts, bolstering its sense of being part of filmmaking as much as coal mining. From 1972, *Mining Review* was renamed *Review*, every other issue now sponsored by non-mining corporations and charities, an ingenious entrepreneurial move by Gysin signalling harder times. Following Slater's death, Gysin himself now both wrote and voiced its scripts; colleagues remember it as the task he took to with greatest gusto. Each issue now consisted of one ten-minute story. *Review 26th Year No. 5*, for instance, is devoted to celebrating the UK's 1973 entry into the European Economic Community (!).

NCB filmmakers still exhibited competence and sporadic flair; film's training use, on undiminished scale, merits continued study. But much of this period's work strikingly stumbles from an elegiac nostalgia in productions looking back at earlier coal industry times (e.g. *40 Years On*, 1978), to a slightly desperate bombast when facing contemporary circumstances. *Review* ran several issues focused on the decade's energy crisis (e.g. *30th Year No. 7*) in which former Shell filmmaker Gysin regards oil, natural gas and nuclear power with a hostile scepticism, and critiques fickle energy policy with a vehemence, surprising in films indirectly funded by governments committed to fuel diversity. In light of the strikes government had fought shy of returning to coal as the basis of security of supply. The oil shock instead redoubled state commitment to indigenous oil and gas supplies.

Nostalgia and bombast were morbid symptoms, as was a flatter, cheaper aesthetic, most obvious in soundtracks (befitting its position slightly below the top end of the production market, the NCB Film Unit always relied a fair bit on library music.[20] Previously orchestral it was now kitsch electric or electronic muzak). *Review*, servicing its diminishing cinema market, was still produced on 35 mm, but internal productions increasingly shot on 16 mm. These were instances of a wider film industry crisis of technique. The industrial documentary industry subsisted for years – not least, delivering a substantial programme of North Sea filming for oil and gas sponsors – but its renaissance was over. Recession, filmmakers and PR officers' deaths (Elton in 1973) and retirements (Anstey in 1974), and most fundamentally non-theatrical distribution's demise caused above all by television's ascendancy, were making film a dubious industrial investment. The prestige film, the most expensive yet least measurably purposive category, was the first to go, hitting the NCB less hard than other big-league sponsors. At the end of the decade, Gysin's team was still turning out one hundred reels a year. Notwithstanding its troubles, the British coal industry enjoyed a high research reputation and NCB films continued lauding technical progress: the historic shift from the largely manual industry of 1947 to wholly mechanized production, now approaching the brink of automation – a climax that never came.

Final Phase (1983–1995)

In 1979, Margaret Thatcher was elected prime minister, heading the most radical government since Attlee's, on the opposite side of politics. In March 1983, she appointed to head the NCB the Scots-American businessman Ian MacGregor, who had recently drastically streamlined British Steel. One year later, the

20 Unlike both oil and transport, it did not deploy talented composers like Edward Williams, Elisabeth Lutyens, or James Stephens.

National Union of Mineworkers began a year-long strike, the most divisive national event since the Depression and the bitterest industrial dispute in British history. Several days after MacGregor's appointment the final edition of the world's longest-running industrial newsreel was released. *Review 36th Year No. 5* is sub-titled "Out of Darkness, Light" and summarizes coal mining's past and prospects. Lucid, loyally (and in hindsight ludicrously) optimistic, but characteristically critical of successive governments' capricious energy policies, it is at once devoid of self-pity and laced with unmistakeable melancholy.

Gysin's Film Unit momentarily continued with internal production, but its imminent demise was hastened by developments it had itself, long before, had a hand in. In the 1960s, Gysin supported a pioneering experiment in industrial use of videotape: locally produced safety programmes shown on closed-circuit television in individual collieries. It seems this led, in the 1970s, to the creation of a Television Unit operating in parallel with the Film Unit. How much dialogue they maintained is unclear, but evidently the former's output stealthily grew. Its own salient production of 1983 was *The Way Ahead*, featuring MacGregor addressing mineworkers on camera about the industry's challenging position with respect to those 4 M's. Compared to the Film Unit's cinema swansong, and even its training films, it palpably lacks craftsmanlike aesthetics. It does not require them and was a more cost-effective outlay.

By the point at which the strike was called on March 12, 1984, plans for the Film Unit's closure were well advanced. It shut on the 30th, its departure unnoticed except in its own specialist corner of the film industry. A victim of technology and cost control it had also, ideologically, come to fall into the crack between increasingly disaffected, militant staff and unsympathetic new management.[21] The 1984–1985 miners' strike is widely considered the event that broke British social democracy, symbolically and permanently. During it, Alexander's old friend Philip Weekes, now South Wales director, was suspected by MacGregor's Board of being too sympathetic to the strikers; while Gysin's boss Kirk was actually dismissed (then used his valedictory press conference to critique the Board's management). Gysin himself remained contracted to retirement age in 1986, to wind up film affairs. Had he recommended that the hundreds of negatives accruing storage costs be destroyed, the Board would doubtless have acquiesced. He insistently oversaw their transfer to the British Film Institute's vaults.

21 One NCB filmmaker wrote a colleague at the time: "FG continues to flounder on gamely through the night of doubt and sorrow, exhorting us to fly the flag and soldier on – but with no proper comprehension of what is happening around him. No one invited him to attend the meetings. All our shell-shocked hero is interested in are despatches from the rear. I don't think he accepts the imminent closure of the unit." David Pitt to Peter Pickering, BFI Special Collections, Pickering Collection.

Post-strike, video production resumed at the British Coal Television Unit, continuing until the industry's privatization in 1995. The early years of "corporate video" are, worldwide, easily the least well preserved, catalogued or commented-upon of all eras of institutionally commissioned filmmaking, despite forming the roots of today's thriving corporate media sector. But it is clear the regular staff video magazine was its most prevalent early genre. In hindsight there is something lugubrious about a dying blue-collar industry embracing it: the centrepiece of the NCB's video programme was monthly in-house video, *National Newscene*. The contrast is with contemporary oil industry staff video-mags (e.g. the BP series *Downstream*) but more so with white-collar sectors on the rise. Financial and retail services, never large-scale industrial film commissioners, were clients for most of the magazines produced by the fledgling video industry. There is an association – less than a cause-and-effect connection but more than coincidence – between that new industry and the financialization of the UK economy, partly underwritten by North Sea oil proceeds turning sterling into petrocurrency.

Where, aesthetically, *Review* borrowed from cinema newsreel, *Newscene* mimicked regional TV news. Where *Review* sought outreach, the "lay public," NCB moving images were now locked behind colliery gates, not distributed publicly. The company had not lost interest in public relations. It merely calculated, correctly, that there was no remaining PR benefit to commissioning visual media, as opposed to seeking to influence media forms overseen by others: the press and TV.

Retrospect

"Art in industrial film [...] is functional art."[22] As such, it is not fully understood if reduced to function or aesthetics alone. Both require the other to operate. Likewise, a credible picture of authorship is a holistic one, recognizing its compound composition of several layers rather than reducing it to one. Our story here could, and should, equally be told from the standpoint of commissioning industry and of on-the-ground (underground) filmmakers.[23] Here it has been told from the vantage point of men in the

22 Zimmermann, Yvonne. "'What Hollywood Is to America, the Corporate Film Is to Switzerland': Remarks on Industrial Film as Utility Film." *Films That Work: Industrial Film and the Productivity of Media*, edited by Vinzenz Hediger and Patrick Vonderau. Amsterdam University Press, 2009, p. 111.

23 Alexander intriguingly commented: "Subject to any higher authority to which he himself is subject, a Producer is ultimately responsible for the film he is producing; but each particular film is its Director's film, and the Director should expect to lean on the Producer's wider experience

middle, an important intermediate level particular to a mid-century, large-corporation manifestation of industrial film.

Whether at commissioning, on-the-ground or intermediary level, the industrial filmmaker is one part artist to two parts working stiff: pursuing the functionality of art, practising the art of functionality, an individual human working with others to do the job as well as brief and budget permit, experiencing as much coherence between work, self and soul as any allows. In 1945, facing the future, Donald Alexander argued:

> In the modern world, dominated by the machine and scientific knowledge, [the] incoherence of the machine itself had inevitable repercussions on the human beings dominated by it. Because work was a mystery, work became separated in their minds from life, [...] a relatively intelligible and warm affair of eating, sleeping, drinking, making love, and alternate laughter and pain. The greatest single need of our time is once again to achieve coherence, to get people to understand and believe that their own lives, scientific knowledge, and the machine are all part and parcel of the same coherent "thing" – humanity.[24]

Alexander was a realist, however, who recognized functional art's contingency and limits as well as its scope. His coal-fuelled journey from student protest to public service contrasts equally with an urbane operator like Tritton, and with a thwarted perfectionist like his own mentor. Paul Rotha's idealism contained cerebral, emotional and somewhat self-centred components. That combination made of his 1930s and 1940s films an artistically compelling argument for a rationalist post-war settlement. That same combination, together with financial ineptitude, rendered him unviable as a professional part of that settlement. Alexander's personal politics lay further to Rotha's left (including time in Britain's Communist party). But he embraced an accommodation with managerialism that Rotha could not,

and knowledge, without fear of the film being taken out of his control. Such a relationship calls for great tact on the part of the Producer, particularly as every Director needs different handling. Some seem to have an eye and no logic, and some are all logic and no eye" (Alexander, Donald. "Paper for Discussion, on the Organisation of Film Section." Undated. Scottish Screen Archives at National Library of Scotland Special Collections, Glasgow, File SSA4:20:75). When credited on screen, Alexander was usually producer or executive producer. A monographic case study of one DATA/NCB director is Russell, Patrick. "Who's Driving? Peter Pickering", in *Shadows of Progress: Documentary Film in Post-war Britain*, edited by Russell and James Piers Taylor (which includes multiple career accounts of sponsored filmmakers). Palgrave Macmillan, 2010, pp. 274–97.

24 Alexander, Donald. "The Documentary Film." Lecture to the British Film Institute Summer School on Film Appreciation, 1945, p. 2.

one informed by his interest in miners' communities and safety, fascination for their technical environment and desire to bolster their public dignity: a project consistent with, but not specifically advancing, his own ideology.

Francis Gysin, in turn, was better suited to mediating mid-life crisis. A director who worked under both reflects:

> Francis never revealed very much about himself personally. [...] [A]t the time I had certain doubts. [...] [I]n any unit people are complaining all the time. [...] [R]etrospectively I think he was very good. [...] [P]eople were always complaining about, you know, the bureaucracy of the Board, but he was in the position of having to deal with it. [...] I think he dealt with it very well. [...] [H]e was having to turn down [...] people's, you know, "why can't we do this," on the one hand, but [...] he was representing these people to the Board, and kept the unit going for a very long time. [...] [I]t was technically a successful unit. [...] [T]echnical films are needed in any society in any situation.[25]

Today, one million Britons work in call centres, more than ever mined the NCB's pits. Their industry has commissioned plenty of corporate video, but it is unlikely a large-canvas programme of production and distribution will be orchestrated, from within over decades, to reveal to employees and public an industry that is economically as important as mining but by comparison atomized and un-cinematic. The tradition that did so was inexorably drawn to richly pictorial masculine heavy "staple" industries handed down by the industrial revolution to the twentieth century. Total war and resulting insolvency had, it is now apparent, artificially altered their situations, temporarily arresting long-term decline. They also distorted the trajectory of sponsored documentary.

Britain's post-war settlement appears today, for good or ill, an aberration from market economics prevailing before and after its thirty-year reign. If we had to date the tipping point for several long-term transfers of power to a single year, it would be 1957. The year the industrial films renaissance crystallized was arguably the one in which Britain passed peak industrialization, certainly the one in which its energy economy took a sharp turn from solid to liquid fuels. Slowly, almost unconsciously, it began its journey to a post-industrial globalized service economy fuelled by oil, gas, nuclear power and imported coal. A journey ending in a consensus forged in reaction to Depression and war unravelling with consequences still playing out.

25 Peter Pickering, interview with author, 2007.

Table 27.1. UK energy industries and film: a tabular digest

SECTOR	COAL	OIL	OIL	OIL	GAS	NUCLEAR
ORGANIZATION	**NATIONAL COAL BOARD**	**ROYAL DUTCH SHELL (Shell)**	**ANGLO-IRANIAN/ BRITISH PETROLEUM (BP)**	**Other oil companies**	**THE GAS COUNCIL; British Gas**	**UK ATOMIC ENERGY AUTHORITY; British Nuclear Fuels**
Films officer	Kurt Lewenhak; Donald Alexander; Francis Gysin	Arthur Elton; Dora Thomas	Ronald Tritton; Dudley Knott; Roly Stafford; Ian Brundle	Various	Stanley Irving; Bob Sullivan	Unknown: Commissioned through Central Office of Information
Est. no. of productions (c. 1945–1985)	1,000	450	350	250	250	150
Key production companies	DATA; NCB (Technical) Film Unit; others	Shell Film Unit via Film Centre	Film Producers Guild (top layer of member companies); World Wide Pictures; Halas & Batchelor; others	Film Producers Guild (mid- and bottom layers); others	Realist Film Unit; Pathé; Public Relationship Films; Anthony Gilkison Associates; others	ACE Film Productions; others
(market position)	*High to middle*	*High*	*High*	*Middle to low*	*Middle; sometimes high*	*Low to middle, occasionally high*
Distribution	Cinema; own library	Cinema; Petroleum Films Bureau	Cinema; Petroleum Films Bureau	Own and generic libraries (rarely cinema)	Own library (rarely cinema)	Own and Central Film Library (rarely cinema)

SECTOR	COAL	OIL	OIL	OIL	GAS	NUCLEAR
ORGANIZATION	**NATIONAL COAL BOARD**	**ROYAL DUTCH SHELL (Shell)**	**ANGLO-IRANIAN/ BRITISH PETROLEUM (BP)**	**Other oil companies**	**THE GAS COUNCIL; British Gas**	**UK ATOMIC ENERGY AUTHORITY; British Nuclear Fuels**
Hit films (examples)	*Mining Review* (series); *Nine Centuries of Coal*	*Frontiers of Friction; The Rival World* (among many others)	*Giusepinna; The Shadow of Progress* (among many others)	*Signorina, It's Important!*	*Family Album, Something Nice to Eat*	*Criticality*
Exemplary directors	Alun Falconer; Peter Pickering (unit members)	Peter de Normanville; Bill Mason (unit members)	James Hill; Derek Williams (freelancers)	David Villiers (freelancers)	Jack Howells; Sarah Erulkar (freelancers)	John Green
Lineage	Documentary Film Movement (Rotha)	Documentary Film Movement (Grierson)	Commercial sponsored filmmaking (Merton Park)	Commercial sponsored filmmaking	Documentary Film Movement; commercial newsreel	Commercial sector (small start-ups)
Aesthetic	Skilled, warm, self-effacing	Polished, austere	Polished, tasteful	Various	Unpretentious, later more artistic	Stolid
Persona (brand)	Stalwart: pragmatic and humane	Scientist and statesman: magisterial	Cosmopolitan explorer: engaging and middlebrow	Various	Avuncular boffin	Modern technician
(brand strength)	*Mid to high*	*High*	*High*	*Mid to low*	*Mid to low*	*Low to mid*
(brand consistency)	*High, with wobbles*	*High*	*High to middling (changes c. 1970)*	*Low*	*Low to middling (changes c. 1967)*	*Middling*

Late that year we encountered Donald Alexander, privately observing a fruitless orchard. Known for direct expression of clear thought, on that occasion he was curiously imprecise about what had triggered his frostbitten dejection.[26] It is tempting to think it was informed by prescience that the familiar world was turning, becoming unfamiliar. More likely, it is explained by budgetary battles and structural frustrations.[27] We can anyway assume his flagging morale was a blip that, with a stiff upper lip, he withheld from staff reporting to him. After as before it, he did what had to be done. So, in different ways under different circumstances, did Lewenhak and Gysin. So, in other ways under other circumstances, did Tritton, Elton, Anstey and others. If, in the long run, it was not enough, maybe it never could have been.

Pragmatists all, incorporating case studies such as theirs into a "look at the industrial film from a pragmatic premise" enriches the account but also, *on pragmatic grounds*, increases its explanatory power. With care, its rigour need not be ruined by the temptation to read them not only as case studies but also as parables, peopled by projections of parts of our conflicted workplace selves. Down the ages, the industrial filmmaker has been most of us. As Sinatra sang: a puppet, a pauper, a pirate, a poet, a pawn and a king. It is not absurd to seek a certain nobility at the junction of industry and film, nor difficult to find a measure of tragedy.

Table 27.1. NCB Film Production: A Tabular Chronology

PHASE	Starts	Pits	Employees	Consumption (tonnes)	Films officer	Production units
1	1947	958	704k	187 million	Kurt Lewenhak	DATA and others
2	1952	880	710k	210 million	Donald Alexander	NCB Technical Film Unit; DATA and others
3	1957	822	704k	216 million	Donald Alexander	NCB Technical Film Unit; DATA and others
4	1963	576	517k	194 million	Francis Gysin	NCB Film Unit; others
5	1972	259	252k	121 million	Francis Gysin	NCB Film Unit; NCB Video Unit
6	1983	191	207k	110 million	N/A?	British Coal Television Unit

26 He notes frustrating below-inflation pay increases but stresses this is a secondary concern.

27 TNA COAL 32/16 contains references to budgetary battles. National Library of Scotland SSA 4:20:75 includes a memo citing rejected structural proposals.

Works Cited

Aitken, Ian. *Film and Reform: John Grierson and the Documentary Film Movement*. Routledge, 1990.
Alexander, Donald. "The Documentary Film." Lecture to the British Film Institute Summer School on Film Appreciation, 1945.
Alexander, Donald. Memo to Noel Gee. Undated [late 1957 is inferred from contents]. Scottish Screen Archives at National Library of Scotland Special Collections, Glasgow, File SSA 4:20:41.
Alexander, Donald. "NCB – the First Twelve Years." *Film User* [British Industrial and Scientific Film Association], vol. 13, no. 150, April 1959, pp. 174–77.
Alexander, Donald. "NCB – the First Twelve Years (2)." *Film User* [British Industrial and Scientific Film Association], vol. 13, no. 151, May 1959, pp. 239–42.
Alexander, Donald. "Paper for Discussion, on the Organisation of Film Section." Undated. Scottish Screen Archives at National Library of Scotland Special Collections, Glasgow, File SSA4:20:75.
Anthony, Scott, and James G. Mansell, editors. *The Projection of Britain: A History of the GPO Film Unit*. Palgrave Macmillan, 2011.
Ashworth, William. *The History of the British Coal Industry, Vol. 5: The Nationalized Industry*. Oxford University Press, 1986.
Berry, David. *Wales and Cinema: The First Hundred Years*. University of Wales Press, 1994.
Boon, Timothy. *Films of Fact: A History of Science Documentary on Film and Television*. Wallflower Press, 2008.
Chapman, James. *A New History of British Documentary*. Palgrave Macmillan, 2015.
Elsaesser, Thomas. "Archives and Archaeologies: The Place of Non-fiction Film in Contemporary Media." *Films That Work: Industrial Film and the Productivity of Media*, edited by Vinzenz Hediger and Patrick Vonderau. Amsterdam University Press, 2009, pp. 19–34.
"Francis Gysin." *The Times*, December 10, 1995, p. 14.
Gay, Ken. *Hand in Hand with Time: A Memoir*. Honeysuckle Books, 2009.
Griffin, A.R. *The British Coal Mining Industry*. Moorland Publishing, 1976.
Hardy, Forsyth. *John Grierson: A Documentary Biography*. Faber & Faber, 1979.
Hediger, Vinzenz, and Patrick Vonderau. "Introduction." *Films That Work: Industrial Film and the Productivity of Media*, edited by Vinzenz Hediger and Patrick Vonderau. Amsterdam University Press, 2009, pp. 9–16.
Hediger, Vinzenz, and Patrick Vonderau, editors. *Films That Work: Industrial Film and the Productivity of Media*. Amsterdam University Press, 2009.
Hogenkamp, Bert. *Deadly Parallels: Film and the Left in Britain 1929–39*. Lawrence & Wishart, 1986.

Lewenhak, Kurt. "Films and the National Coal Board." *Documentary Film News*, April 1948, p. 44.

Low, Rachael. *Documentary and Educational Films of the 1930s*. Allen & Unwin, 1979.

Orbanz, Eva. *Journey to a Legend and Back: The British Realistic Film*. Spiess, 1977.

Pitt, David. Letter to Peter Pickering. 1984. BFI Special Collections, Peter Pickering Papers, London.

Russell, Patrick. "Who's Driving? *Peter Pickering*." *Shadows of Progress: Documentary Film in Post-war Britain*, edited by Patrick Russell and James Piers Taylor. Palgrave Macmillan, 2010, pp. 274–97.

"Some Aspects of the Social Structure, and Other Sociological Aspects of a Documentary Film Unit." Undated. Scottish Screen Archives at National Library of Scotland Special Collections, Glasgow, File SSA4:20:75.

Sussex, Elizabeth. *The Rise and Fall of British Documentary*. University of California Press, 1976.

Winston, Brian. *Claiming the Real: The Documentary Film Revisited*. British Film Institute, 1995.

"Working Group on Film Policy: Proposals for 1962." The National Archives (TNA), London, File series COAL 32/16. Anonymous (presumably Donald Alexander).

Zimmermann, Yvonne. "'What Hollywood Is to America, the Corporate Film Is to Switzerland': Remarks on Industrial Film as Utility Film." *Films That Work: Industrial Film and the Productivity of Media*, edited by Vinzenz Hediger and Patrick Vonderau. Amsterdam University Press, 2009, pp. 101–17.

About the Author

Patrick Russell is senior curator (non-fiction), BFI National Archive. He has led or contributed to many curatorial projects exploring expanding histories of British non-fiction film and TV, and to the UK's major mass digitization initiatives. He is author of *100 British Documentaries* (2007), co-editor of *The Lost World of Mitchell and Kenyon* (2004), co-editor of *Shadows of Progress: Documentary Film in Post-war Britain* (2010), and co-editor of the book series *British Screen Stories*. He regularly blogs about the UK's contemporary corporate films industry and in 2021 was awarded the EVCOM Fellowship by the industry's trade body.

28 Saudi Arabia's John Ford?

Robert Yarnall Richie, *Desert Venture* and Ephemeral Authorship in Industrial Film

Martin Stollery

Abstract

The premiere of *Desert Venture*, a colour film promoting the transformative endeavours of the Arabian-American Oil Company (ARAMCO) in Saudi Arabia, is a prime example of how "ephemeral authorship" can be highly relevant to the study of industrial film. *Desert Venture* was promoted at its premiere and in other contexts as the work of Robert Yarnall Richie, a photographer and filmmaker with a well-established professional reputation within the oil industry. ARAMCO suggested that *Desert Venture* was the creative product of a "frontiersman" who exemplified many qualities outlined in Frederick Jackson Turner's "The Frontier in American History" (1893), and in 1930s "nation-building" Westerns, mirroring the celebration in the film itself of the new American oil frontier in Saudi Arabia.

Keywords: oil culture; authorship; Robert Yarnall Richie; ARAMCO; Westerns; Timothy Mitchell

Thomas Elsaesser recommended, in the first *Films That Work* volume, that industrial films could usefully be considered in terms of their specific occasion, purpose and addressee.[1] Vinzenz Hediger and Patrick Vonderau suggested, in that book's introduction, that this is preferable to searching for industrial film's auteurs.[2] The historiography of non-fiction film has tradi-

1 Elsaesser, Thomas. "Archives and Archaeologies: The Place of Non-fiction Film in Contemporary Media." *Films That Work: Industrial Film and the Productivity of Media*, edited by Vinzenz Hediger and Patrick Vonderau. Amsterdam University Press, 2009.

2 Hediger, Vinzenz, and Patrick Vonderau. "Introduction." *Films That Work: Industrial Film and the Productivity of Media*, edited by Vinzenz Hediger and Patrick Vonderau. Amsterdam University Press, 2009, p. 10.

Hediger, V., F. Hoof, Y. Zimmermann, with S. Anthony (eds), *Films That Work Harder: The Circulation of Industrial Film*. Taylor & Francis Group 2024
DOI: 10.5117/9789462986534_CH28

tionally focused on documentary filmmakers as auteurs, rather than on the varied uses to which some of those same films have been put. Consequently, there are very good reasons for insisting upon a foundational anti-auteurism in the expanded study of industrial film. However, we overlook a significant aspect of the circulation of industrial film if we prohibit any discussion of issues relating to authorship.

A hundred years ago, Boris Tomashevsky made the enduring argument that authors' biographical legends play a role in the function and reception of literature in some periods and contexts, but no role in others. This also applies to industrial film. In many instances biographical legends and conceptions of authorship are of no relevance to the study of industrial film. In other cases, they are. A biographical legend is not a Romantic conception of authorship; it is the image of the author that circulates among readers or audiences and informs their perceptions of the work to which it is attached. A biographical legend may be complex or relatively one-dimensional. The key issue is whether it plays a significant role in the presentation of an industrial film to target audiences. My focus in this chapter is the October 1947 premiere of *Desert Venture*, sponsored by ARAMCO, the Arabian-American Oil Company, produced and directed by the industrial photographer and filmmaker Robert Yarnall Richie, as a prime example of what I call ephemeral authorship. Close attention to the occasion, purpose and addressee of this single screening opens up some broader questions about how industrial film scholars might consider questions of authorship.

1940s US Oil Culture

Opposition to big oil has a long and varied history, as does the oil industry's diverse responses to this opposition. The "muckraking" journalism of the 1890s famously represented John D. Rockefeller, founder of monopolistic Standard Oil, as the epitome of the oil industry robber baron, markedly different to the earlier wildcatters, rugged individualists who drilled speculative oil wells at the dawn of the industry's history. However, more heroic conceptions of big oil emerged during and just after the Second World War. The industry was obliged to work with the federal government's Petroleum Administration for War (PAW). Whatever the realities of an often conflicted relationship, Roger and Diana David Oliens argue that "because emphasis on success was useful to so many groups, it is not surprising that assessments of the war experience were highly upbeat in tone."[3]

Take two examples. On the one hand, *A History of the Petroleum Administration for War* (1946), by former PAW bureaucrats John W. Frey and H. Chandler Ide, asserted that wartime government collaboration with the industry had been a success and advocated continued close federal regulation after the war. On the other hand, a book published by *Look* magazine, *Oil for Victory* (1946), also represented wartime collaboration in a positive manner, but was less committed to post-war federal regulation. *Oil for Victory* was a more populist text, lavishly illustrated with striking photographs, including Richie's. However, both books represented the industry as an integral part of the war effort, with achievements on the production front enabling victories on the battlefront. Alongside this rhetoric of heroic productivity, both books also described the individuals involved in the oil industry, as the Oliens put it, in terms "more descriptive of frontier lawmen than bureaucrats and businessmen."[4] Dramatizing and illustrating the work of industry geologists, for example, in a chapter titled "Searchers in Far Places," *Oil for Victory* described them as "advance scout(s) for the wildcatter,"[5] who sometimes operated in "remote wilderness camps."[6]

Oil companies' direct sponsorship of post-war cultural production capitalized on their contribution to the war effort to consolidate the industry's emergent rebranding as patriotic and public-spirited. Standard Oil (New Jersey), for example, faced public criticism in 1942 for its links with the German company I.G. Farben. Standard commissioned polling from Elmo Roper that indicated this was a widespread perception.[7] Hiring Roy Stryker, the man responsible for developing American New Deal documentary photography at the Farm Security Administration (FSA), associated Standard with American vernacularism rather than international capitalism. Stryker led the Standard Oil photography project from 1943 to 1950. He also commissioned *Louisiana Story* (1948), recruiting Robert J. Flaherty's name recognition among cinephiles, and the prestige of documentary, to Standard's public relations strategy.

3 Olien, Roger M., and Diana David Olien. *Oil and Ideology: The Cultural Creation of the American Petroleum Industry*. University of North Carolina Press, 2000, p. 245.

4 Ibid., p. 246.

5 *Oil for Victory: The Story of Petroleum in War and Peace*. McGraw-Hill, 1946, p. 240.

6 Ibid., p. 235.

7 Plattner, Steven W. *Roy Stryker: USA, 1943–1950: The Standard Oil Photography Project*. University of Texas Press, 1983, pp. 12–13.

ARAMCO and American Exceptionalism

Desert Venture's commentary informs us that oil industry development in Saudi Arabia was relatively new when the film was produced in 1947. The first commercially viable oil well was tapped in 1938. However, as the Saudi industry grew, and became more visible, critics emerged. Some well-informed commentators in reputable, influential publications like the *New York Times* raised the spectre of Yankee imperialism, by comparing ARAMCO's activities to some of the more dubious historic practices of US oil companies in Latin America. From another direction, Secretary of the Interior Harold Ickes, the head of PAW, published articles during the war in the mass circulation magazine *Collier's Weekly* advocating a "government-owned oil company that would claim a share of Near Eastern oil."[8] Federal government recognition of Saudi Arabia's increasing strategic significance was demonstrated by the construction of the Dhahran US military airbase in 1945, and the appointment of the first US ambassador in 1946. Such developments raised the prospect of increased federal regulation of ARAMCO's activities.

In response to these challenges, ARAMCO's publicists developed an alternative narrative, a distinctive variant on the post-war petroleum optimism promoted by publications such as *Oil for Victory*. This narrative, positing ARAMCO as a profoundly American, patriotic venture, a force for good in its own right that did not need close government oversight, was communicated in different ways to diverse audiences across a range of media campaigns. As Robert Vitalis has argued, this narrative had considerable staying power, and persisted for so long partly because it directly emulated wider exceptionalist narratives of American identity.[9]

To take some examples, ARAMCO employed Wallace Stegner to write *Discovery! The Search for Arabian Oil* in 1956, focused on "the story of the pioneering."[10] The book mythologized the role of Max Steineke, ARAMCO's chief geologist during the late 1930s and 1940s. Although *Discovery!*'s publication history was problematic, due to some differences of emphasis between Stegner and ARAMCO executives, Vitalis argues that Stegner did not cross the most fundamental of the "company's red lines," and indeed "breathe[d] new life into the myth of American exceptionalism in the Middle East."[11]

8 Olien and Olien. *Oil and Ideology*, p. 242.

9 Vitalis, Robert, *America's Kingdom: Mythmaking on the Saudi Oil Frontier*. Stanford University Press, 2006.

10 Stegner, Wallace. Qtd. in Vitalis, Robert. "Wallace Stegner's Arabian *Discovery*: Imperial Blind Spots in a Continental Vision." *Pacific Historical Review*, vol. 76, no. 3, 2007, p. 416.

11 Ibid., pp. 432–33.

Hardy, pioneering American geologists, wearing keffiyehs, traversing an unbearably hot desert environment, continued to feature in ARAMCO publicity material even after the company was fully nationalized in 1980. The commemorative film *ARAMCO: Era of Discovery* (1984) is one example. The standard line was that ARAMCO's rugged pioneers had founded a project that again demonstrated America's exceptional world-historical role.

One of the most famous variants of the exceptionalist conception of America is Frederick Jackson's Turner's 1893 "frontier thesis." *Desert Venture*, scripted by two ARAMCO employees, extrapolates aspects of this thesis to the new frontier of Americans working in Saudi Arabia. In this film, rhetoric reminiscent of Turner is aligned with imagery similar to that found in certain types of Western. *Desert Venture*'s commentary praises the first American men who ventured out to work in Saudi Arabia and imbued ARAMCO with its distinctive identity. These men are praised as rugged, ingenious, determined, enterprising, quintessentially American pioneers, qualities Turner sees as having been initially shaped by the historical Western frontier. *Desert Venture* verbalizes and visualizes the references to pioneering found in some earlier 1940s discourse on the American oil industry. The commentary, across a film lasting just under half an hour, frequently invokes "Turneresque" keywords: three references to "pioneering" and two at the end to a "new frontier." Turner argued:

> The wilderness [at the frontier] masters the colonist. [...] It strips off the garments of civilization and arrays him in the hunting shirt and the moccasin. [...] [A]t the frontier the environment is at first too strong for the man. [...] Little by little he transforms the wilderness, but the outcome is not the old Europe, [it] is a new product that is American.[12]

The only named American worker in the film is Paul Strong, who first appears in a business suit and tie, in a small town, declining an offer to work for ARAMCO. Later, he reappears, without a tie, in a khaki work shirt, visually blending with the Saudi Arabian desert sand. The commentary explains Paul had initially forgotten "he had that traditional American characteristic: a hunger for adventure and challenge. He and the others had come to learn of the Middle East, and they found the going was tough."

Desert Venture's emphasis on adventure and challenge in foreign lands shares some common ground with other oil films such as British Petroleum's (BP's) *The New Explorers* (1955), where Abu Dhabi is one of the new frontiers.

12 Turner, Frederick Jackson. *The Frontier in American History*. Henry Holt, 1920, p. 4.

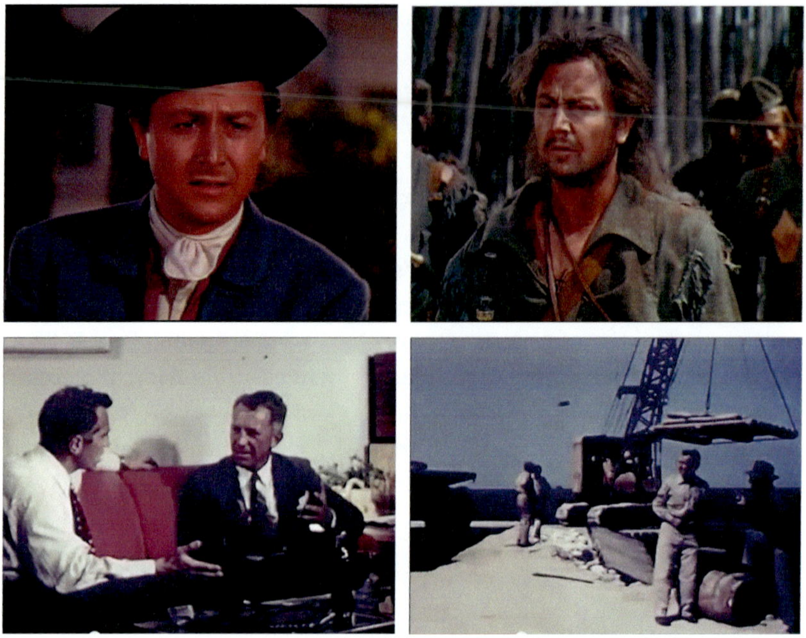

Figs. 28.1–28.4: Sartorial transformations in *Northwest Passage* and *Desert Venture*.

However, certain aspects of *Desert Venture* have a distinctly American inflection. The absolute transformation of Saudi Arabia represented in the film can be seen as an example of what David Nye calls American "second creation" narratives, of which Turner's frontier thesis is one instance. Second creation narratives, consolidated during the nineteenth century, are technological foundation stories in which transformations of supposedly arid, undeveloped landscapes result in personal success and national progress, and the arrival of more settlers. Nye notes how some commentators extrapolated this foundation story beyond the borders of the United States.[13]

Representing Saudi Arabia through American frontier mythology, and in terms of an American second creation narrative, erases the history of previous British imperial involvement in Saudi Arabia; this is never referenced in *Desert Venture*. The apparent novelty of ARAMCO's project disavows possible imperial continuities. *Desert Venture* also shares some elements, relating to exceptionalist narratives and frontier mythology, with what genre historians now call the "nation-building" Westerns of the late 1930s and early 1940s: films such as *Drums along the Mohawk*, *Northwest Passage* and

13 Nye, David. *America as Second Creation: Technology and Narratives of New Beginnings*. MIT Press, 2003, p. 261.

Western Union. Like *Desert Venture*, these films represent historical turning points: the impact of the American Revolutionary War in the Mohawk Valley frontier region; 'Rogers' Rangers' expeditions; the laying of telegraph cables across the continent. Many nation-building Westerns were, like *Desert Venture*, shot in colour. Colour and landscape shots in these contexts offer visual spectacle and emphasize the magnitude of the historical events being represented. Protagonists also frequently undergo sartorial transformation, like Paul Strong in *Desert Venture*, Henry Fonda in *Drums along the Mohawk*, Robert Young in *Northwest Passage*, and Turner's frontiersman.

Ephemeral authorship at the New York Herald Tribune Forum, October 20 to 22, 1947

Desert Venture was produced for a specific occasion in which foregrounding Robert Yarnall Richie's authorship suited ARAMCO's purposes. This differed from Stryker's recruitment of Flaherty for *Louisiana Story*, which attached an established, critically validated reputation to the project. Following Rick Prelinger, we can call the alternative approach adopted for *Desert Venture* ephemeral authorship, because it typically involves a "short-lived" assertion of authorship which falls outside traditional circuits of cultural legitimation, although it may have a parasitical relationship to them.[14] Ephemeral authorship is more closely linked to the moments and specific, restricted contexts in which it is manifested than it is to the impression of permanency and canonicity conferred by academic and certain types of critical commentary, preservation in prestigious archives, and exhibition in culturally legitimated artistic venues. *Desert Venture*'s moment of ephemeral authorship crystallized when it was premiered at a specific event: the sixteenth New York Herald Tribune Forum, October 20 to 22, 1947. At the time, the *Herald Tribune* was the chief rival newspaper to the *New York Times*. The annual Forum was a well-established event that helped to sustain the *Herald Tribune*'s profile and prestige. Speakers at the 1947 event included, among other well-known figures, the poet and journalist Carl Sandburg, George Marshall, of Marshall Plan fame, Chaim Weizmann, soon to become Israel's first president, and the photographer Robert Capa.

14 Vonderau, Patrick. "Vernacular Archiving: An Interview with Rick Prelinger." *Films That Work: Industrial Film and the Productivity of Media*, edited by Vinzenz Hediger and Patrick Vonderau. Amsterdam University Press, 2009, p. 51.

The Forum's venue, the Waldorf Astoria, was associated in the late 1940s with major conferences and meetings relating to the Cold War and international relations. ARAMCO scored a public relations coup in premiering its film at such a high-profile location and event. ARAMCO was the only corporation represented at this gathering, which was otherwise populated by politicians, heads of government agencies, influential journalists and cultural figures. ARAMCO was firmly opposed to anything approaching the level of federal involvement in Middle Eastern oil Ickes advocated during the war. Instead, the company aspired to play a quasi-diplomatic and developmental role in Saudi Arabia. It widely publicized its involvement in modernizing projects such as constructing a railroad, disease prevention and agricultural experiments,[15] all of which are represented in *Desert Venture*. Premiering the film at the Forum therefore amounted to a policy statement; it posited ARAMCO as a company more than capable of acting in America's national interest by driving such projects itself, rather than ceding any significant ground to formal US government regulation of its activities in this increasingly strategic region. *Desert Venture*'s echoes of certain elements of what, at the time, was the most critically legitimated, epic type of Western, the "nation-building" cycle, supported these aspirations.

Desert Venture also promoted specific aspects of the company's current agenda. Animated diagrams represent oil flowing through newly built pipelines across the region, including the Trans-Arabian, described by the commentary as the world's largest. Steel components are shown arriving by sea. Although ARAMCO disdained federal regulation, the company needed US government export licenses to obtain the considerable supplies of steel needed for its pipeline projects. Lobbying, of which *Desert Venture* formed part, resulted in the federal government awarding ARAMCO the necessary license in September 1947. The license was granted in recognition of the strategic significance of ARAMCO's role in the Middle East, despite competing claims by domestic producers. However, the license was not guaranteed in perpetuity; it was suspended in June 1948 and steel shipments did not resume until February 1949.[16] The issue was prominent enough to prompt ARAMCO to commission *Oil across Arabia* (1950), a colour film of more than forty minutes in length about pipeline construction.

15 Parker, Chad. *Making the Desert Modern: Americans, Arabs, and Oil on the Saudi Frontier, 1933–1973*. University of Massachusetts Press, 2015.

16 Stoff, Michael B. *Oil, War, and American Security: The Search for a National Policy on Foreign Oil, 1941–1947*. Yale University Press, 1980, pp. 212–14.

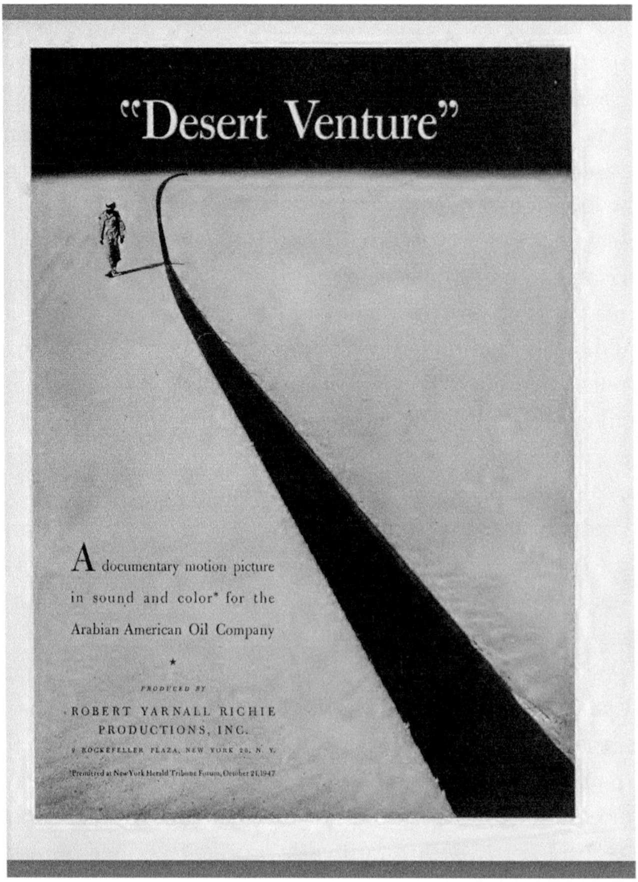

Fig. 28.5. Advertisement for *Desert Venture* in *Business Screen*.

ARAMCO celebrated *Desert Venture*'s release with a full-page advertisement, rarely used for single films within *Business Screen Magazine*, in its December 1947 issue. The advertisement cited the recent premiere at the Forum, and featured an image from the film, of a desert pipeline diagonally bisecting the frame, with a solitary worker walking towards the viewer from the distance, echoing the style of certain images from Westerns. The advertisement also described *Desert Venture* as "a documentary motion picture." By this time, the category "documentary" was associated with a degree of cultural prestige in America. *Desert Venture* points to the historically diverse applications of this term. The audience attending the New York Herald Tribune Forum, broadly interested in current affairs and contemporary cultural developments, is likely to have appreciated the connotations of the term "documentary", but would not necessarily know, or care about the details of which films and filmmakers were considered canonical by cinephile critics and historians.[17] For *Business Screen* readers, "documentary" could also connote, among other things, a quality of film production above the norm for the field covered by the magazine, with reputational benefits for the sponsor. As a 1946 *Business Screen* advertisement for a production house put it: "We are applying twenty-four years of professional production experience to the creation of documentary films, for *unless your motion picture story can compete physically* with the best Hollywood offers, it will suffer by comparison and will simply be branded an 'industrial.'"[18]

Another means through which industrial filmmakers developed a reputation for quality production that was a cut above more anonymous competitors' output was through the shaping of a recognizable professional identity.[19] This could be reflected in a company name, such as Robert Yarnall Richie Productions, which features prominently, as the only fully capitalized text, in the *Desert Venture* advertisement. By the late 1940s, Richie was a well-established industrial photographer, for clients such as *Fortune* and *Life* magazines, as well as a filmmaker, whose professional identity was known to his peers and potential sponsors, and circulated intermittently to wider audiences.

This identity encompassed style and working methods. In Katherine Sullivan's 1944 *Popular Photography* profile, for example, Richie describes himself as "constantly alert to new composition."[20] Richie's professional identity was also partly defined in terms of risk-taking, ruggedness, physical courage, and an orientation towards adventure and challenge similar to the oilmen represented in *Desert Venture*. Sullivan's profile described him as a "man who thinks nothing of climbing a 196-foot derrick in a high wind, carrying all his equipment."[21] Ralph Steiner directly linked Richie's style to his death-defying adventurousness, describing him as a photographer "willing to risk his life to catch just the right angle for a shot."[22] Richie's name was familiar enough to readers of mass market magazines such as *Popular*

17 The Academy Award for Documentary was instituted in 1941. The platform afforded to documentary by prestigious institutions such as MOMA during the 1940s further legitimated the term.

18 Visual Arts Corporation. Advertisement. *Business Screen Magazine*, vol. 5, no. 7, 1946, p. 11.

19 See Cowan, Michael. *Walter Ruttmann and the Cinema of Multiplicity: Avant-garde – Advertising – Modernity*, Amsterdam University Press, 2016, for a discussion of Walter Ruttmann in terms of professional identity that significantly revises established critical perceptions of his career.

20 Sullivan, Katherine. "Assignment: Cotton." *Popular Photography*, vol. 14, no. 5, 1944, p. 28.

21 Ibid.

22 Steiner, Ralph. "His Camera Earns over $500 a Day." *Popular Science*, vol. 161, no. 1, 1952, p. 232.

Photography for him to be quoted in advertisements, endorsing products such as a light meter that could operate in adverse weather conditions.

The 1947 New York Herald Tribune Forum conference report contains the text of all the speeches and lists speakers' names in bold.[23] Richie is listed with the same type face and prominence as everyone else. Temporarily elevated to a position alongside recognized artists such as Sandburg and Capa, elements of Richie's professional identity coalesced into a moment of ephemeral authorship at *Desert Venture*'s premiere. Richie's ephemeral authorship was also intertwined with the representation of ARAMCO's efforts in Saudi Arabia. Helen Rogers Reid, the *Herald Tribune*'s publisher, introduced *Desert Venture* by praising "the company that made the film possible and particularly [...] Mr. Richie, [...] who got the shots in lightning time so as to have the picture here tonight. The difficulties were incredible. All of this is only another proof of why [ARAMCO] has achieved so much in the desert."[24] This was amplified in *Business Screen Magazine*'s December 1947 article on the film. It emphasized how, through ingenuity and determination, Richie adapted to but also mastered a challenging environment in Saudi Arabia: "The desert sand [...] was an even greater location difficulty than the heat, which often hovered at 130 degrees. During the last ten days – when six thousand feet of film covering 1,500 miles over the desert remained to be shot – caution was abandoned for the sake of coverage and industrial masking tape was plastered around all camera openings."[25] The Forum conference report included portrait photographs of all the eminent speakers who appeared at the event, along with the text of their speeches. Richie's photograph, by contrast, emphasizes his status as a man of action, sitting astride a camel in front of a pyramid.[26] This image asserts America's claim to a leading post-war role in the Middle East, while its casual approach to geography, conflating Saudi Arabia with an Egyptian pyramid, supports Edward Said's contention that there was no "deeply invested tradition" of American orientalist knowledge prior to the Second World War, and that American culture displayed a less profound imaginative investment in orientalism than in Britain and France, "perhaps because the American frontier, the one that counted, was the westward one."[27] This photograph

23 *Modern Man: Slave or Sovereign?* New York Herald Tribune, 1947.

24 Reid, Helen Rogers. "Desert Venture." *Modern Man: Slave or Sovereign?* p. 163.

25 "Adventure on Location: Desert Picture Made in Record Time and Temperatures." *Business Screen Magazine*, vol. 8, no. 8, 1947, p. 30.

26 The definitive work on modern visual images of Europeans and Americans and the pyramids remains to be written.

27 Said, Edward W. *Orientalism: Western Conceptions of the Orient*. Vintage Books, 1979, p. 290.

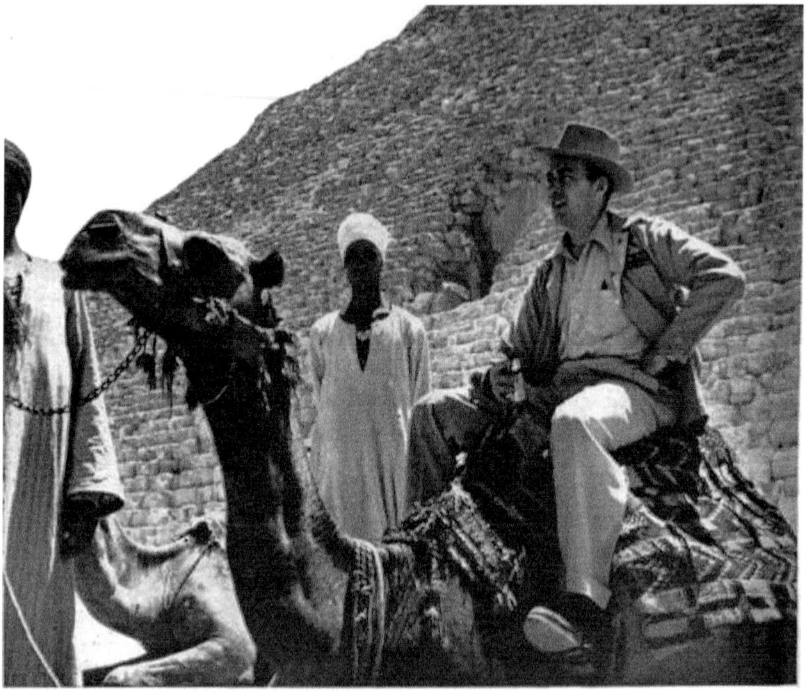

Fig. 28.6. Publicity image of Robert Yarnall Richie for *Desert Venture*.

also accompanied the *Business Screen* article on the film. Featuring an image of a filmmaker was fairly unusual in *Business Screen*; this in itself indicates the emphasis on Richie's ephemeral authorship in relation to *Desert Venture*. Richie's portrait by the pyramid exemplifies adaptation to and mastery of a challenging environment, echoing Turner's frontiersman, and Paul Strong in *Desert Venture*. In this image, Richie has swapped the business suit, homburg and tie he wears in other portrait photographs for more casual attire; an open-necked shirt and cowboy hat. He sits as confidently on the camel as a Western hero on a horse.

Reid's introductory comments about getting shots in "lightning time" drew upon another aspect of Richie's professional identity – speed and efficiency. Profiles of Richie mention the benefits this brought to the corporations for which he worked. Steiner points out, for example, that in a car factory Richie would cause minimal disruption to production, because he always meticulously planned and prepared his shoots, and worked so quickly when on location. Richie is therefore posited as both a rugged frontiersman and the ideal freelancer within a corporate setting. Richie's ephemeral authorship, foregrounded at the premiere of *Desert Venture*, mythically reconciles the maverick individualist and the "organization man" a decade before William

Whyte published his classic book on this topic. Similarly, the film situates expatriate white male American ARAMCO employees, like Paul Strong, as inheritors of the American frontier tradition, while also lauding the company itself for dispensing every possible home comfort to American expats, and for its benign, efficient, bureaucratic regulation of virtually all aspects of its employees' lives.[28]

This latter aspect of the film, emphasized in its second section, occupies a different space to its rhetoric and imagery derived from Turner and the Western. The premiere's audience partly account for this. The Herald Tribune Forum was initially established in 1930 by Reid and journalist and activist Marie Mattingly Meloney, editor of the *Herald Tribune*'s Sunday magazine, for the club women movement. Movement members continued to form a significant part of the audience as it developed into an event with an international scope.[29] These clubs, often loosely associated with mainstream American political parties, held regular discussion meetings which focused on relating national and international issues to the more immediate concerns of American family life.[30]

Desert Venture's second section deals with issues assumed to be of particular interest to club women. It focuses on how ARAMCO now offers American, but not necessarily Arab, employees all the comforts of suburban life, such as air-conditioned homes and offices, good schools, healthcare and jobs for single women – but obviously only as nurses and secretaries. However, the gendered distinctions *Desert Venture* ostensibly proposes do not withstand critical scrutiny. Women only appear in the latter part of the film, firmly associating them with desert suburbia. Nevertheless, the arc of Paul Strong's and ARAMCO's other male American employees' journeys in *Desert Venture* is not only towards a rediscovery of their pioneering qualities, but also from American domestic comfort to a segregated compound dedicated to simulating this in the desert.[31]

28 Nye. *America as Second Creation*, p. 267, highlights an increasing tension within American second creation narratives between assertions of individualism and loss of autonomy. *Desert Venture*'s commentary also blurs temporal boundaries in relation to pioneering. Towards the end of the film, commentator Dan Donaldson refers to ARAMCO's pioneering days being over, yet after this makes a further reference to the ongoing "pioneering [of] a new frontier of progress."

29 Fiske, Jane. "NY Herald Forum Speakers Discuss Racial and Colonial Problems." *Vassar Chronicle*, November 2, 1946, pp. 1, 4–7.

30 Olsen, Margaret Nunnelley. *One Nation, One World: American Clubwomen and the Politics of Internationalism, 1945–1961*. PhD diss., Rice University, 2008, https://scholarship.rice.edu/handle/1911/22244.

31 Vitalis, *America's Kingdom*, makes it abundantly clear this did not apply to ARAMCO's segregated Saudi employees. *Desert Venture* skirts this issues, already the cause of considerable

The Author as Producer/Court Photographer

Another approach to analysing authorship, tracing its lineage to Walter Benjamin's 1934 essay "The Author as Producer," rather than Tomashevsky's "Literature and Biography," affords complementary insights into *Desert Venture*. Situating Richie's authorship within the specific relations of production arising from ARAMCO's involvement in Saudi Arabia requires a shift of focus, away from the film's and his persona's resonance within American culture, and towards what Timothy Mitchell describes as "particular ways of engineering political relations out of flows of energy."[32] The Herald Tribune Forum premiere was the occasion that shaped *Desert Venture*, but part of the film's purpose, to return to Elsaesser's terms, was to cement political relations between ARAMCO, President Truman's administration and the Saudi royal family.

The Arabian-American Oil Company's (ARAMCO's) name, which the company's American executives adopted in 1944, in preference to the initially suggested option of American Arabian Oil Company,[33] indicates the importance of appeasing Saudi elite opinion. *Desert Venture*'s production history also testifies to this; it was preceded by *New Frontier*, also directed by Richie, but produced for a different occasion; Crown Prince Ibn Saud's month-long visit to America in February 1947. ARAMCO paid for the trip, which included a lavish banquet, also at the Waldorf Astoria, a meeting with Truman, and a tour of the set of the Hollywood film *Forever Amber*. The crown prince, however, disapproved of *New Frontier*. According to State department records, he was offended by what he saw as its representation of ARAMCO as Saudi Arabia's salvation from backwardness.[34] *New Frontier* was therefore removed from circulation for the remainder of the crown prince's American tour. *New Frontier* was briefly listed in educational film catalogues thereafter, but superseded by *Desert Venture* as soon as the later film became available on the educational circuit. *New Frontier* does not appear to be extant, but its catalogue description[35] suggests the twenty-eight

industrial unrest by the time the film was produced, with before and after shots accompanied by this commentary: "As rapidly as possible, the company is getting rid of this type of temporary housing, hastily built for the Arabs in the early stages of development, and replacing it with modern, well-built quarters." Needless to say, the modern Arab housing is still segregated from the white American housing.

32 Mitchell, Timothy. *Carbon Democracy: Political Power in the Age of Oil*. Verso, 2011, p. 5.

33 Anderson, Irvine. *Aramco, the United States, and Saudi Arabia: A Study of the Dynamics of Foreign Oil Policy, 1933–1950*. Princeton University Press, 1981, p. 115.

34 Vitalis. *America's Kingdom*, p. 35.

35 "New Frontier." *Educational Film Guide*, edited by Dorothy E. Cook and Katherine M. Holden. H.W. Wilson, 1948, p. 429.

minute long *Desert Venture* was quite probably a re-edited, expanded version of the earlier twenty-minute film, incorporating some newly shot footage.

If this hypothesis is correct, the publicity for *Desert Venture* had to omit any reference to *New Frontier*. The *Business Screen* article, for example, says *Desert Venture* was shot in Saudi Arabia from August to October 1947, and makes no reference to Richie's earlier trip to the kingdom to shoot *New Frontier*. This erasure was necessary to promote Richie's ephemeral authorship as a hyper efficient, lightning fast filmmaker, rather than a recycler of existing footage. *Desert Venture* also addresses the crown prince's criticism of *New Frontier*, partly through its fawning representation of the royal family and their benevolent role in oil industry and Saudi national development. Ibn Saud, the first king of Saudi Arabia, is described in *Desert Venture*'s commentary as a benign visionary, who had the foresight to allow and encourage oil prospecting for the benefit of his faithful people. Richie was invited back to the kingdom in 1948 for a brief stint as court photographer, in which he replicated the kind of images of the royal family included in *Desert Venture*.

Desert Venture represents Saudi Arabia as a profoundly pious society, a model for good Muslims everywhere. The longest shot duration in the entire film, lasting nearly half a minute, compared to an average of six seconds, is of two robed men who pray in the desert, then mount their camels and depart. The commentary says "religion is not worn like a cloak in the Arab land. [...] [I]t is a vital part of every good Muslim's life." We could see this solely as a typically orientalist representation, positing the Middle East as backward-looking, lacking American dynamism. However, viewed from a different standpoint, this shot also conforms to how ARAMCO's business partners, the Saudi ruling elite, wanted their society to be seen. Pascal Ménoret argues: "Bedouinism? Islam? These really are the twin pillars of the image that the Saudi regime wishes to give of the country [...] [paradoxically] the concepts used by the West to criticise Saudi Arabia are the very ones that the Saudi regime has forged for symbolic defence of the country under its sway."[36] The co-presence within *Desert Venture* of American exceptionalism and frontier mythology, representations of simple Bedouin men at prayer amidst the purity of the desert, and virtuous Saudi royalty, can be related to what Mitchell calls "McJihad." This involves a "distinctive understanding [...] of the nature of what we call global capitalism," as far as one of its central industries, oil, is concerned.[37] From this perspective, the dynamic, historically mutable compact between the Saudi royal family, and

36 Ménoret, Pascal. *The Saudi Enigma: A History*. Zed Books, 2005, p. 39.

37 Mitchell. *Carbon Democracy*, p. 213.

the *muwahhidun*, what outsiders refer to as the Wahhabi Islamic movement, does not constitute an atavistic, "pre-capitalist 'cultural' element resisting capitalism from the outside." Instead, the development of these social forces has been integral to the capitalist development of the American/Saudi oil industry, both "essential to and disjunctive with" it.[38]

The relatively autonomous sequence in *Desert Venture* representing Bedouins at prayer can be seen, in part, as a gesture towards containing this essential disjunction. The two men are solitary individuals, rather than part of a wider social movement; they are located in the "natural," empty space of the desert rather than in a mosque, which could signify a broader institutional framework. They appear away from the urban Saudi spaces, more obviously impacted by capitalist development, represented elsewhere in the film. This representation of Islamic practice in *Desert Venture* is a depoliticised one, which replicates Alexis de Tocqueville's classic formulation of American Christianity "clothed with fewer forms, figures, and observances" than elsewhere, occupying a "distinct sphere, in which the priest is sovereign, but out of which he takes care never to go."[39] Other aspects of *Desert Venture*, however, testify more directly to disjunction. As Mitchell puts it, the *muwahhidun* provided legitimacy to the Saudi regime, and a counterweight to Arab nationalist, socialist and communist influences, but "while it was essential to the making of oil profits, political Islam was not itself orientated towards that goal."[40] In broad terms, the *muwahhidun* tacitly accepted the presence of a significant number of non-believing foreigners, in the land of the two holy mosques, in return for well-funded dominance in religious and cultural matters. Hence *Desert Venture*'s commentary refers to ARAMCO personnel being fully aware they are guests in someone else's country. This marks the limit of *Desert Venture*'s pioneer rhetoric, and the limit of any possible comparison between the representation of native Americans in Westerns and Arabs in this film. The pioneers came to America to stay; the Americans in *Desert Venture* are tolerated trespassers.

About John Ford and Robert Yarnall Richie

As Scott Curtis has argued, industrial films, and the publicity material that sometimes surrounds them, can perform "multiple, simultaneous tasks"

38 Ibid., p. 214.

39 Tocqueville, Alexis de. *Democracy in America, Volume Two* [1840]. Alfred Knopf, 1945.

40 Mitchell. *Carbon Democracy*, p. 214.

within certain relations of production.[41] One of these can be to promote a certain image of the filmmaker, as Curtis has demonstrated in relation to Frank B. Gilbreth. *Desert Venture* is similar. The film was designed to serve multiple purposes within the economic, political and social relations between ARAMCO, the US government, US opinion formers, and the Saudi regime during the late 1940s. Promoting Richie's (ephemeral) authorship at its premiere served both his professional and ARAMCO's corporate interests within this context.

John Ford made Westerns, according to legend. Some of the iconography and commentary in *Desert Venture* recall the Western. But there are also more specific parallels between elements in Ford's and Richie's public personae. Ford was sometimes perceived, for example, in Frank Nugent's influential 1949 profile, as a risk-taking, "rough and tumble" director.[42] Later in Ford's career he liked to tell tall stories about having been a cowboy.[43] Yet he also had a reputation for what Tag Gallagher describes as the "disciplined efficiency of his sets," similar to the reputation Richie had for his industrial location work.[44] Hence the title of Nugent's piece on Ford: "Hollywood's Favourite Rebel."

Some recent academic work on Ford's authorship draws on such material to focus on the ways in which, and extent to which, his personae was circulated to cinema audiences.[45] At different historical moments, versions of Ford were more or less visible to non-cinephile publics, through film posters or articles in widely read newspapers and magazines. This approach differs from the older, more traditionally auteurist celebration of Ford pioneered by Lindsay Anderson. At this relatively early point in the development of industrial film studies, we can follow a different route. We can begin by concentrating on the historical utility of ephemeral authorship, across

41 Curtis, Scott. "Images of Efficiency: The Films of Frank B. Gilbreth." *Films That Work: Industrial Film and the Productivity of Media*, edited by Vinzenz Hediger and Patrick Vonderau. Amsterdam University Press, 2009, p. 86.

42 Nugent, Frank S. "Hollywood's Favorite Rebel" [1949]. *John Ford Made Westerns: Filming the Legend in the Sound Era*, edited by Gaylyn Studlar and Matthew Bernstein. Indiana University Press, 2001, p. 266.

43 Haggard, Mark. "Ford in Person" [1971]. *John Ford: Interviews*, edited by Gerald Peary. University Press of Mississippi, 2001, pp. 136.

44 Gallagher, Tag. *John Ford: The Man and His Films*. University of California Press, 1988, p. 38.

45 Maland, Charles J. "From Aesthete to Pappy: The Evolution of John Ford's Public Reputation." *John Ford Made Westerns: Filming the Legend in the Sound Era*, edited by Gaylyn Studlar and Matthew Bernstein. Indiana University Press, 2001, pp. 220–52; Poague, Leland. "That Past, This Present: Historicizing John Ford, 1939." *John Ford's Stagecoach*, edited by Barry Keith Grant. Cambridge University Press, 2003, pp. 82–112.

the diverse circuits within which industrial film circulated, for certain audiences on certain occasions. Constructing a pantheon of great industrial filmmakers can come later, if at all.

Works Cited

"Adventure on Location: Desert Picture Made in Record Time and Temperatures." *Business Screen Magazine*, vol. 8, no. 8, 1947, p. 30.

Anderson, Irvine. *Aramco, the United States, and Saudi Arabia: A Study of the Dynamics of Foreign Oil Policy, 1933–1950*. Princeton University Press, 1981.

Cowan, Michael. *Walter Ruttmann and the Cinema of Multiplicity: Avant-garde – Advertising – Modernity*. Amsterdam University Press, 2016.

Curtis, Scott. "Images of Efficiency: The Films of Frank B. Gilbreth." *Films That Work: Industrial Film and the Productivity of Media*, edited by Vinzenz Hediger and Patrick Vonderau. Amsterdam University Press, 2009, pp. 85–99.

Elsaesser, Thomas. "Archives and Archaeologies: The Place of Non-fiction Film in Contemporary Media." *Films That Work: Industrial Film and the Productivity of Media*, edited by Vinzenz Hediger and Patrick Vonderau. Amsterdam University Press, 2009, pp. 19–34.

Fiske, Jane. "NY Herald Forum Speakers Discuss Racial and Colonial Problems." *Vassar Chronicle*, November 2, 1946, pp. 1, 4–7.

Frey, John W., & Ide, H. Chandler. *A History of the Petroleum Administration for War, 1941–1945*. Government Printing Office, 1946.

Gallagher, Tag. *John Ford: The Man and His Films*. University of California Press, 1988.

Haggard, Mark. "Ford in Person" [1971]. *John Ford: Interviews*, edited by Gerald Peary. University Press of Mississippi, 2001, pp. 127–36.

Hediger, Vinzenz, and Patrick Vonderau. "Introduction." *Films That Work: Industrial Film and the Productivity of Media*, edited by Vinzenz Hediger and Patrick Vonderau. Amsterdam University Press, 2009, pp. 9–16.

Hediger, Vinzenz, and Patrick Vonderau, editors. *Films That Work: Industrial Film and the Productivity of Media*. Amsterdam University Press, 2009.

Maland, Charles J. "From Aesthete to Pappy: The Evolution of John Ford's Public Reputation." *John Ford Made Westerns: Filming the Legend in the Sound Era*, edited by Gaylyn Studlar and Matthew Bernstein. Indiana University Press, 2001, pp. 220–52.

Ménoret, Pascal. *The Saudi Enigma: A History*. Zed Books, 2005.

Mitchell, Timothy. *Carbon Democracy: Political Power in the Age of Oil*. Verso, 2011.

Modern Man: Slave or Sovereign? New York Herald Tribune, 1947.

"New Frontier." *Educational Film Guide*, edited by Dorothy E. Cook and Katherine M. Holden. H.W. Wilson, 1948, p. 429.

Nugent, Frank S. "Hollywood's Favorite Rebel" [1949]. *John Ford Made Westerns: Filming the Legend in the Sound Era*, edited by Gaylyn Studlar and Matthew Bernstein. Indiana University Press, 2001, pp. 261–71.

Nye, David. *America as Second Creation: Technology and Narratives of New Beginnings*. MIT Press, 2003.

Oil for Victory: The Story of Petroleum in War and Peace. McGraw-Hill, 1946.

Olien, Roger M., and Diana David Olien. *Oil and Ideology: The Cultural Creation of the American Petroleum Industry*. University of North Carolina Press, 2000.

Olsen, Margaret Nunnelley. *One Nation, One World: American Clubwomen and the Politics of Internationalism, 1945–1961*. PhD diss., Rice University, 2008, https://scholarship.rice.edu/handle/1911/22244.

Parker, Chad. *Making the Desert Modern: Americans, Arabs, and Oil on the Saudi Frontier, 1933–1973*. University of Massachusetts Press, 2015.

Plattner, Steven W. *Roy Stryker: USA, 1943–1950. The Standard Oil Photography Project*. University of Texas Press, 1983.

Poague, Leland. "That Past, This Present: Historicizing John Ford, 1939." *John Ford's Stagecoach*, edited by Barry Keith Grant. Cambridge University Press, 2003, pp. 82–112.

Reid, Helen Rogers. "Desert Venture." *Modern Man: Slave or Sovereign?* New York Herald Tribune, 1947, pp. 162–63.

Said, Edward W. *Orientalism: Western Conceptions of the Orient*. Vintage Books, 1979.

Steiner, Ralph. "His Camera Earns over $500 a Day." *Popular Science*, vol. 161, no. 1, 1952, pp. 192–95, 228, 230, 232.

Stoff, Michael B. *Oil, War, and American Security: The Search for a National Policy on Foreign Oil, 1941–1947*. Yale University Press, 1980.

Sullivan, Katherine. "Assignment: Cotton." *Popular Photography*, vol. 14, no. 5, 1944, pp. 26–28, 81.

Tocqueville, Alexis de. *Democracy in America, Volume Two* [1840]. Alfred Knopf, 1945.

Turner, Frederick Jackson. *The Frontier in American History*. Henry Holt, 1920.

Visual Arts Corporation. Advertisement. *Business Screen Magazine*, vol. 5, no. 7, 1946, p. 11.

Vitalis, Robert. *America's Kingdom: Mythmaking on the Saudi Oil Frontier*. Stanford University Press, 2006.

Vitalis, Robert. "Wallace Stegner's Arabian *Discovery*: Imperial Blind Spots in a Continental Vision." *Pacific Historical Review*, vol. 76, no. 3, 2007, pp. 405–38.

Vonderau, Patrick. "Vernacular Archiving: An Interview with Rick Prelinger." *Films That Work: Industrial Film and the Productivity of Media*, edited by Vinzenz Hediger and Patrick Vonderau. Amsterdam University Press, 2009, pp. 51–61.

About the Author

Martin Stollery is an independent scholar based in the UK. He is the author of *Alternative Empires: European Modernist Cinemas and Cultures of Imperialism* (2000) and co-author of *British Film Editors* (2004). He has published extensively on documentary and non-fiction film.

29 Sounds Industrial

Understanding the Contribution of Music and Sound in Industrial Films

Annette Davison

Abstract
This chapter seeks to demonstrate the contribution of music, sound, and voice-over commentary to the rhetorical discourse of industrial films. Frank Lewin's article series "The Soundtrack in Industrial Motion Pictures" (1959–60) presents best-practice guidance and thus provides access to the conventions of recording and mixing of music, sound, and voice-over commentary in industrial films of the period. Audiovisual analysis of films made by the Shell Film Unit in the 1950s confirms Lewin's prescriptions: sonic and visual elements are examined separately, in combination, and in the context of production. I argue that such analyses enrich our understanding of the sound worlds of these films and enable the identification and evaluation of techniques of audiovisual rhetoric in films of persuasion.

Keywords: sound; voice-over commentary; audiovisual; sonic analysis; industrial films; Shell Film Unit

In his study of *Promotional Culture* (1991), Andrew Wernick recounts the story of the Wedgwood Portland Vase, a late-eighteenth-century replica of a celebrated Roman glassware pot residing in the British Museum in London. Wernick explains that Josiah Wedgwood undertook to copy the vase with the certain knowledge that its sale, as well as that of the replica vase series made possible through the successful creation of the mould, process, and collection of mixture ingredients, would lose money. Its worth, and that of its replicas, lay in its function as a marketing device.[1] "The direct profits to

1 Wernick, Andrew. *Promotional Culture*. Sage, 1991, p. 6. In 2005 an unnumbered first edition of the "original" series sold at auction for over £14,000. Bonhams. Auction 11904, lot 27,

Hediger, V., F. Hoof, Y. Zimmermann, with S. Anthony (eds), *Films That Work Harder: The Circulation of Industrial Film*. Taylor & Francis Group 2024
DOI: 10.5117/9789462986534_CH29

be made from the sale of a particular product line were inversely related to its promotional value; the production costs of the luxury pieces were subsidized by revenues from the sale of ordinary ware in return for their power to promote."[2] The symbolic power of exclusive items, and particularly the replica Portland Vase(s), was transferred to Josiah Wedgwood and on to his company, thus effectively helping to generate the Wedgwood "brand."[3]

This essay focuses on a small sample of promotional industrial films produced by the Shell Film Unit (SFU) in the 1950s, which contributed to the generation of the Shell "brand." These films do not mention the company's name, except in the opening and closing credits, where the Shell Film Unit is named as the producing company. Sir Arthur Elton, consulting producer at the Unit from the late 1930s through to the 1970s, argued that "[t]he impact of a sponsored film upon its audience will be in inverse ratio to the number of times the sponsor insists on having his name mentioned."[4] Shell also produced sales films, but these were usually handled separately by Shell-Mex and BP Ltd, a joint marketing venture between Royal Dutch Shell (Shell) and British Petroleum established in 1932. The films produced by the SFU, by contrast, were more often technical, educational, or exercises in public relations. In this way the SFU "publicised Shell Petroleum's reputation as a great international commercial company, but one with a lively sense of international responsibility and a leader in the field of science and technology."[5] Many of these films were thus promotional in the sense that they projected the company's image and brand values, or those that it wished to project, whether to employees, the general public, or to governments and other bodies overseas where the company was engaged in business dealings. Indeed, as Vinzenz Hediger and Patrick Vonderau explain, "[c]ompanies use film systematically to project a certain image and create what is usually known as a corporate identity, i.e., a company's internal and external symbolic and social cohesion."[6] Such films may thus be understood as rhetorical, and as examples of soft persuasion.

September 7, 2005, www.bonhams.com/auctions/11904/lot/27/. Accessed January 27, 2017. Until 2017 the Wedgwood Portland Vase series could still be purchased from the company, and thus still functioned as promotional more than two hundred years after it was first produced.

2 Wernick. *Promotional Culture*, p. 6.

3 Ibid., p. 10.

4 This is "Elton's First Law of Industrial Sponsorship." Elton, Arthur. Qtd. in Gordon, Douglas. *The Shell Film Unit: The First Sixty Years*. Shell Film and Video Unit, 1994, p. 1.

5 Elton, Arthur. "How We Use Films in Shell." *Film User*, vol. 10, no. 118, 1956, p. 344.

6 Hediger, Vinzenz, and Patrick Vonderau. "Record, Rhetoric, Rationalization." *Films That Work: Industrial Film and the Productivity of Media*, edited by Vinzenz Hediger and Patrick Vonderau. Amsterdam University Press, 2009, p. 42.

Studies of industrial films, such as those found in the collection *Films That Work*, tend to emphasize the visual over the sonic. This might suggest that the use of sound and music in these films is particularly unremarkable, but this is not the case. While legibility and clarity are primary in most technical and training films, with prestige films, where the goal is to promote a company or corporation as socially and ethically responsible, and/or rich in expertise, examples demonstrate that there is more scope for aesthetic play or experimentation, and exploration of the range of aesthetic choices available. Since the advent of synchronized sound, industrial films have generally been conceived and produced as audiovisual constructs, though as examples of rhetorical discourse, not all are ambitious in terms of their sound worlds. To date, a relatively small number of these films have been subject to detailed audiovisual analysis, and these are most often those judged to be exceptional in some way, such as Harry Watt and Basil Wright's *Night Mail* (1936), or Geoffrey Jones's films for the General Post Office (GPO) and British Transport.[7] With increased accessibility to a significant number of surviving industrial films from the twentieth century, alongside the development of film music and film sound studies as interdisciplinary subfields, the time is ripe to readjust the focus. My aim here is to examine the audiovisuality of these films by interrogating the contribution of each sonic and visual element, and the interrelations between them, in the context of production. This enables an examination and evaluation of audiovisual techniques of persuasion, whether via "argument, appeals, arrangement [...] [or] aesthetics."[8]

I begin with a view from production, summarizing the guidance offered by composer Frank Lewin in a series of three articles published in the British trade journal *Industrial Screen* between 1959 and 1960: "The Soundtrack in Industrial Motion Pictures."[9] The articles provide information concerning the conventions of recording and mixing music, sound, and voice-over commentary in industrial films towards the end of the 1950s. Lewin composed music for film documentaries, the television series *I-Spy* (1955) and later *The Defenders* (1961–1965), as well as writing music for the stage, the concert

7 See, for example, Bulley, James. "Progress Music: Daphne Oram, Geoffrey Jones and Trinidad and Tobago." *Bricks from the Kiln*, no. 2, edited by Andrew Lister and Matthew Stuart. Antenne Books / RAM Publications, 2016, pp. 5–22; Anthony, Scott. *Night Mail*. BFI Publishing, 2007.

8 Herrick, James A. *The History and Theory of Rhetoric: An Introduction*, 3rd ed. Pearson Education, 2005, p. 14.

9 The articles are based on a series of spoken presentations given by Lewin, then of Filmsounds Inc., at the Convention of the Society of Motion Picture and Television Engineers (SMPTE) in Detroit in 1958.

hall, and the opera house. Although his guidance concerning the place of music in industrial films might thus be assumed to be somewhat idealistic, analysis of *Proud Ships* (1954) and other SFU films of the period demonstrate that Lewin's prescriptions reflect practice accurately.[10]

Good Soundtrack Practice in Theory: Frank Lewin, "The Soundtrack in Industrial Motion Pictures"

Voice

The voice "conveys specific information" and the other sonic elements must be organized in such a way that attention is not drawn away from its verbal content. Lewin thus acknowledges the conventional hierarchy that places voice above sound effects and music in these films, but also notes that in the case of all three elements, "effectiveness depends on pacing and moderation."[11] An unremitting flow of commentary can prove "soporific," and a proliferation of adjectives may not prove effective when presented alongside an image. Rather, "[m]usic and [sound] effects may profitably be made to take the place of adjectives, or other rhetorical devices."[12]

Lewin advocates recording the voice prior to the mix separately from the other sonic elements for pragmatic reasons: fluffs can be picked up and fixed easily, and adjustments can be made when running the spoken script to the image. This also enables an easier, more effective, and "harmonious" integration of music, effects, and voice: "When the sound editor knows where music and effects are free to make their points without interfering with the voice he is able to save their punches for these open spaces, rather than sprinkle them throughout the film only to hear an overriding commentary track render them ineffectual."[13] Lewin argues that by giving the mixer control over a pre-recorded voice track at the mix, more time can be given over to "aesthetic considerations."[14]

With films of over eight minutes' duration, attention should be paid to the pacing of the voice track to avoid "holes" that do not give space for reflection.

10 *Proud Ships* can be viewed online at the Internet Archive: archive.org/details/75402ShellFilmProudShips. Accessed January 27, 2017.

11 Lewin, Frank. "The Soundtrack in Industrial Motion Pictures." *Industrial Screen*, vol. 3, no. 5, November–December 1959, p. 257.

12 Ibid.

13 Ibid.

14 Ibid.

One option is to combine some of these to create a section *without* vocal commentary: "Placed at the beginning of a sequence, somewhere in the middle where a breather would be welcome, or at the end, this space affords the music (or sound effects if they are interesting in their own right) a chance to take over."[15] This also means that the voice will be more noticeable on its re-entry.

Lewin also mentions the option of using more than one voice to provide variety, noting that it can be effective if the shift "coincides with changes in the picture."[16] This is more likely the case if the change is "from male to female (or vice versa), or to a child's voice" and the various voices "become identified with objects or actions on the screen – cartoon style – or are used dramatically."[17] If such changes are not motivated by the film's argument, and are used rather to create a momentary effect, this may be achieved more efficiently via the alternation of the various elements of the soundtrack (i.e., music, sound, voice-over), rather than attempting to generate (unmotivated) variety within one element.

Music

Stock music is relatively cheap, and it may be easier to edit a pre-existent recording than to commission and record an original score. Lewin suggests that there are economic and aesthetic benefits to this approach as well, since a recording can be selected that uses a larger ensemble than the budget might otherwise allow were an original score commissioned and recorded. A commissioned score offers uniqueness, however: it avoids potential conflicts caused by the use of the same music for different films; and however carefully it is selected and edited, the "flexibility and power of a truly 'original' score" cannot be equalled by library music.[18] A score's "originality need not thrust itself on the listener's attention by merely trying to sound 'different.' Rather that score may be termed genuinely original which seizes upon the musical possibilities inherent in the subject matter of the film and thus it unfolds in harmony with the voice (and effects)."[19] Indeed, for Lewin, musical "fit" is key. While music may be used to "simulate or imitate" natural sounds, or to "create an illusion of activity," it is also "eloquent" and "often its appeal is emotional." "Fit" should thus occur at

15 Ibid., p. 260.
16 Ibid.
17 Ibid.
18 Ibid., p. 258.
19 Ibid.

the level of matching what the "picture tries to say," rather than simply accompanying cuts, or mickey-mousing screen action.[20]

Music with Voice

When voice-over is also present, music should work with the voice-over rather than with the images, for "[m]usic should not say one thing while the voice tries to say another."[21] If music *must* be heard in combination with the voice, Lewin suggests avoiding powerful, expressive music, the effect of which would be lessened when subdued. Less intense music should be selected instead, which avoids interference with the voice even when played louder. For the same reason, vocal music should be avoided in combination with a voice-over. Vocal music can be used effectively to support the generation of an accurate impression of a scene, even though when distributed overseas, relatively few audioviewers may understand the language in semantic terms.[22]

Where either voice or music can accomplish the same role, Lewin urges the selection of one rather than using both together. For example, if music is used to emphasize a hard cut to a new section, then the voice entry may usefully be delayed so that the change is established by the music first. This can work particularly well when the voice-over was dominant previously: "The new scene will actually rouse the viewer's interest more effectively if he is left guessing for a short while until the voice picks up again."[23] Care must be taken to avoid "splitting" a thought where continuity of narration is important, however. Additionally, "the rhythm of the voice dictates the entrance of the musical cue" so that a rapid-fire delivery should be closely followed by a music entry, with a more languorous approach to scoring when the delivery is more leisurely: "[T]he music carries forward the momentum of the narration and enters in a way which seems natural to the overall pace."[24]

Sound Effects

Lewin highlights the distinction between editing in nontheatrical films such as industrial films by comparison to that enacted for dramatic features;

20 Ibid.
21 Ibid.
22 Ibid.
23 Ibid., p. 259.
24 Ibid., p. 260.

a distinction that Bill Nichols would later describe as evidentiary editing versus continuity editing, respectively.[25] For Lewin this means that sound effects should be used selectively in nontheatrical films.[26] Where the narration mentions particular sounds they should clearly be used, though not necessarily as recorded on location. Indeed, Lewin argues that imagination should be exercised in the production of sound effects, for an effective sound treatment is one in which the sound effects are "interesting as well as plausible."[27] Actions that produce sustained "distinctive" sounds will be more effective if accompanied by appropriate sounds or by music. Where they are "strongly rhythmic or have definite pitch (whistles, bells)," they may usefully be treated as though part of the musical score. They can also make an ideal background for voice-over narration, leaving music to provide other functions, and to make an impact when it enters.

Music and Image and Voice

In addition to the voice-over, onscreen action may cue music entries. While the relationship of the music to the commentary should be prioritized over that with the images, music is good at establishing mood, and if the voice regularly interrupts then the mood can be destroyed. Indeed the "worst thing" is to have the voice crash the excitement established when music and sound effects build to a climax and have to be faded down.[28] Thus Lewin suggests that where sparse enough, consolidating the voice-over into blocks offers more opportunities for "an interesting soundtrack."[29] Music and sound can function as punctuation, "and can deliver a punch much more forcefully, and in a shorter space of time, than the voice."[30] Where loud or active music or sound are used and "belong" to the images, it is better to allow the music or sound to have its moment, with the voice commentating before or after.

When music enters, it should do so – initially (at least) – at strategic points, say something meaningful and "contribute to the pacing of the film."[31] It should be expressive, even when subordinate to other elements. Lewin argues that accents and changes are most "satisfying" when "organic"

25 Nichols, Bill. *Representing Reality: Issues and Concepts in Documentary.* Indiana University Press, 1991.

26 Lewin. "The Soundtrack in Industrial Motion Pictures," p. 279.

27 Ibid.

28 Ibid., p. 260.

29 Ibid.

30 Ibid.

31 Ibid., p. 258.

to the music, and that practices such as entering/exiting at the start or end of a phrase "*will* make a difference in the impression created by the finished project."[32] Indeed, for Lewin, the success of an industrial film's score depends strongly on its integration with the voice and sound effects; he recommends the use of a multiple-head machine for editing tracks where music and voice alternate quickly.

Good Soundtrack Practice in Practice? The Shell Film Unit in the 1950s

The Shell Film Unit produced hundreds of promotional, technical, and educational films, and a significant proportion of those during the 1950s and 1960s. With one exception, the examples discussed here can all be viewed online or have been published on commercially available DVDs. The primary case study is *Proud Ships*.

Proud Ships (1954)

Proud Ships introduces Shell's tanker fleet; how oil is collected from oilfields, transported to refineries, and distributed globally. The film was directed by Philip Armitage and features a score by English composer Edward Williams.[33] In an essay entitled "How to Choose Shell Films" in the 1955 *Shell Film Catalogue*, Elton states that although "almost all the films [...] were made for a purpose, and are designed to inform, train, or teach," *Proud Ships* forms part of a small group which "have been found to command wide general interest and to be what is generally called 'entertaining.'"[34] It is a "lively, popular film of general interest,"[35] which would appeal to "everyone interested in the social logistics of the mid-twentieth century,"[36] yet it could also usefully be programmed as part of an event for "an employee information session on the oil industry."[37]

32 Ibid. Emphasis in original.

33 Film Centre, London, is listed as "production consultant." For more on Film Centre, see Ellis, Jack C. *John Grierson: Life, Contributions, Influence*. Southern Illinois University Press, 2000; and Foxon, Steven. "Film Centre (United Kingdom)." *Encyclopedia of the Documentary Film*, edited by Ian Aitken, 3 vols. Routledge, 2006, vol. 1, pp. 411–13.

34 Shell Petroleum Company. *Shell Film Catalogue*. 1955, p. 84.

35 Ibid., p. 12.

36 Ibid., p. 85.

37 Ibid., p. 87.

The film gives an overview of life on the tanker *Liparus* and the important role tankers play in the transportation of oil prior to and after its refining. It features more than eight different voices (mostly in voice-over) and integrates music, sound, and voice-overs "harmoniously," each taking and ceding primacy in a dynamic manner. As might be expected, music is foregrounded at the beginning during the main title sequence, which features the "christening" of three tankers at their launches (at 00:10; 00:27; 00:40 [mins:secs]). String tremelos and a drum roll build anticipation each time, climaxing with a dramatic flourish synchronized to the moment at which the bottles are smashed against the ships' hulls. This is followed by a repeated bell-like fanfare figure as each ship begins its slide into the sea, pitched higher with each successive launch, and heard atop a more lyrical rising counter-theme on low strings. With the third tanker, the bell-like figure is presented in a more fluid and richly orchestrated version, followed by a more expansive development of the lyrical theme as the ship descends the slipway, filmed first from behind as it slides away, then from the (starboard) side, revealing the vast scale of the tanker as it moves past the camera. The entry of the voice-over commentary (at 01:13) is signalled by the gradual fade of the music lower in the mix. It continues beneath, supporting the imposing framing, but without distracting from the commentary; it assists continuity by drawing attention away from the editing as the sequence shifts from UK waters to Indonesia and beyond. Faded lower, the music is inaudible by c. 01:50. Sound effects replace music as the sequence continues to introduce some of the worldwide locations where tankers are at work, underlining the different conditions the ships face in these locations; from savage to calm seas.

The voice-over closes the first part of the introductory segment by foregrounding how important oil is: "This world runs on oil, and the tanker's business is to keep it running." A second later (at 03:03) a harp glissando signals the start of a new section: a change of image to a simplified 2D world map with the North Pole at its centre. Here the voice-over interacts closely with the score. The commentary offers a chronology of the roles tankers play in the distribution of different types of oil-based materials. The score here functions as a punctuating sound effect, underlining the commentary's explanation of each aspect of the process.

The next section, which explains the role of the control centre in London – "the hub of the whole system" – foregrounds the voice-over commentary, with sound effects in support to indicate the hubbub of the office. This voice sounds subtly different from the previous one, apparently sourced in a body working in the office; "in our case, we work out of London". A similar shift occurs as the action moves to the tanker *Liparus*, with the commentary now spoken by the chief officer. The change of location is signalled by music, as the lyrical theme from the main title sequence returns. On board the ship we also hear the voices of the master, his wife (briefly), the third engineer, Charlie (the pumpman), and other crew members. Sarah Kozloff defines voice-over commentary via both its medium – *voice* – and "the relationship between the source of the sound and images of the screen: the viewer does not see the person who is speaking at the time of hearing his or her voice. [...] [T]he voice comes from another time and space, the time and space of the discourse" – i.e., *over*.[38] None of the commentaries in *Proud Ships* are lip-synchronized; the depicted "characters" whose voices we hear continue to go about their everyday business while explaining various aspects of life aboard a tanker, or the ecology of the tanker fleet. These commentaries remain in the time and space of the discourse, rather than that of the action, and may thus be described accurately as "voice-over" commentaries. There are instances where that distance is shrunk considerably, however.

When the tanker *The Western Planet* runs aground it is decided that the *Liparus* must change its course and deliver its cargo in its place. Its planned voyage from Curaçao to Fall River, Massachusetts, USA (the "dollar run"), is thus rerouted to Dublin and on to the UK. A character in the London office delivers news of the grounding to his superior by means of a typed document. The superior responds with an audible intake of breath. A discussion between this character and an unseen colleague, Mac, ensues, apparently via a telephone call, and decisions are made. The *Liparus* receives the message via Morse code. It is transcribed by the ship's radio operator and presented to the master, who then appears to read it aloud (albeit out of synchronization). This is followed by a shot of the master's wife, knitting, whose expression brightens as we hear her exclaim in an excited voice, "So it's home!" Her lips do not move, however. Are we to assume that these are words spoken in the onscreen character's head (elsewhere described as meta-diegetic or internal focalization)?[39] This sequence not only increases the pace of the film somewhat by its shift into the present (or a proximation of it), but also generates a sense of playfulness with audiovisual conventions of voice-over commentary in documentary films.

A similar situation occurs when we see the chief engineer's wife switch on a radio in the cabin she shares with her husband: music begins. The same piece of music is then heard a little later in the ship's recreation area, where

38 Kozloff, Sarah. *Overhearing Film Dialogue*. University of California Press, 2000, pp. 2–3.

39 See, for example, Gorbman, Claudia. *Unheard Melodies: Narrative Film Music*. Indiana University Press / BFI Publishing, 1987, pp. 22–23; and Branigan, Edward. *Narrative Comprehension and Film*. Routledge, 1992.

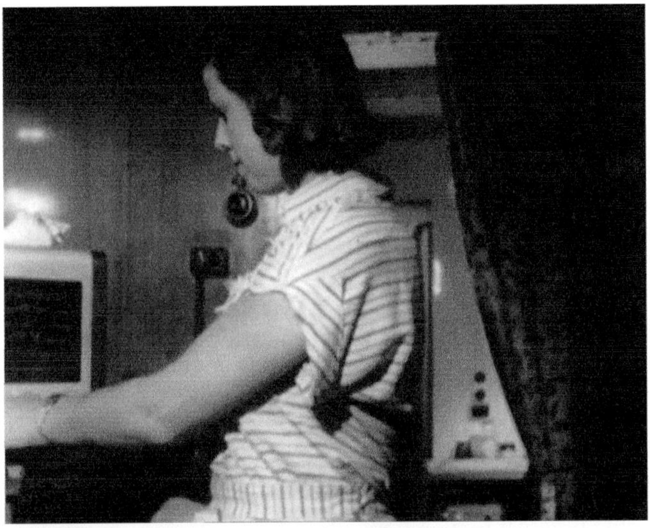

Fig. 29.1. The chief engineer's wife turns on the radio.

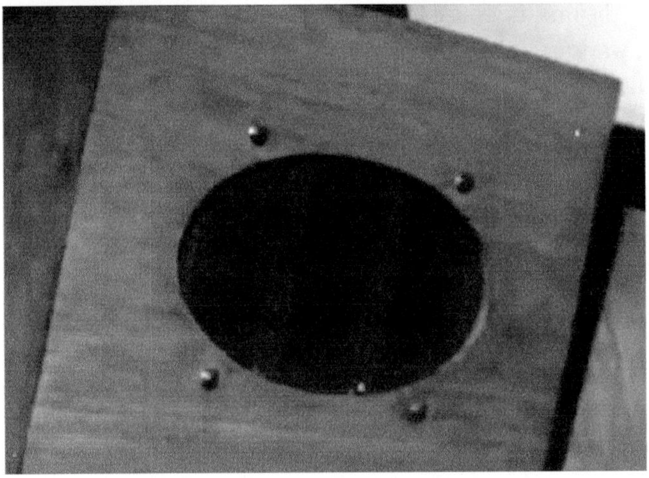

Fig. 29.2. "Do you suffer from corns and bunions?" [close-up of speaker].

radio speakers are also located. Essentially then, music functions as diegetic sound here. The master provides the voice-over commentary at this point, and notes the presence of "culture too, especially on the dollar run," at which point an (amusingly) intrusive advertisement is heard from the radio, the source of the announcer's voice confirmed by its synchronization with a sustained close-up on the radio's speaker. The voice of the chief engineer interrupts the commercial, then situated very high in the mix, decrying it as "personal," implying that it is *too* personal (see Figs. 29.1-2 above).

Less senior crew are distinguished by an array of regional accents quite distinct from the received pronunciation of the master and chief officer, thus imparting class information in a subtle manner. Charlie the pumpman has a Welsh accent, and during an extended segment we hear him explain the routine tasks and drills that are shown onscreen. These shots are interspersed with images of Charlie writing to his family back at home: the words we hear in voice-over are those he appears to write in his letter: "How's Gwen getting on at home? Still painting the town red? A tanker's like a girl, you know. Always demanding attention, especially the tanks."

Music frequently interacts dramatically with the entries and exits of the voice-overs. At the close of one sequence the chief officer notes the "ups and downs" of the job (and the tanker) (08:36). Music returns, synchronized with a hard cut to footage of an up-ended ship sinking, during wartime: a timpani roll sets the tone, as a brass-heavy cue forcefully approximates a barrage of artillery fire. Similarly, when the commentary discusses adjustments made to transform tankers into merchant aircraft carriers (MACs), music is foregrounded in the mix as we see footage of a plane landing dramatically on a MAC's "runway" (c. 09:13), echoing the sounds of the wheels braking as it comes to a halt. Charlie's voice-over concludes with the ship's arrival into the port in Curaçao, here accompanied by a stately brass-led theme that accentuates the port's status, and its spectacular industrial vistas. Supported by a persistent motoric figure, the score is subsequently integrated with noisy and rhythmically regular mechanical sound effects. The sequence culminates with the lowering of the anchor. The music is faded down and the sounds of this noisy activity move to the fore. Silence. Then the voice-over returns: "There we are. Safe and snug."

Another theme depicts the crew's cheer as the rerouted course sends them home earlier than expected. This joyful folk-song-inflected music is organised in four-line strophic song form (AABA), with each line formed of four phrases. The crew are busy making good on promises made to those back home; we see the chief officer creating a jigsaw for his son, for example. With the completion of the theme (and its exit), the commentator's voice (now, unaccompanied) highlights the challenge of a sailor's life, split between the ocean and family, illustrated onscreen by shots of the chief officer's wife and child awaiting his return. Raucous and noisy sound effects of waves crashing over the ship's deck brings the focus back to the tanker fleet. Interior shots during the storm are quieter, avoiding audio conflict with the commentary. These are intercut with exterior shots that foreground the sounds of the raging wind and the sea, without voice-over. Sound effects and voice-over generate continuity through the sequence, taking turns at the helm. The chief officer updates the ship's log, then we hear his expert opinion: "This'll blow itself out in forty-eight hours." A static shot of the tanker in calmer waters follows. There is no voice-over, and no music, just the gentle sound of the waves for fourteen seconds. An anticipatory music cue enters, gently at first, marking time, supporting cuts between shots of the radar mast, its antenna spinning. Officer "Paddy" sips tea from a cup and saucer now the storm has blown over. The master looks through his binoculars as Ireland appears on the horizon. The stately theme heard previously at Curaçao returns, followed by the familiar folk-inflected music soon after, a nod to their destination. "And that's that. Oh well. One trip's much like another." The cue then begins a repeat of the four-line stanza alongside a long shot of the *Liparus* at sea. The voice-over states that the *Liparus* "stands for all the tankers," the delivery of the commentary combined with the (now familiar) four-line musical structure to impressive rhetorical effect, as indicated below:

> (A) And *Liparus* stands for all the tankers. And the men of all the nations who sail them up and down the world.
> (A) Outbound. Homebound. Light or laden. Carrying their vital cargoes, wherever orders take them.
> (B) The ocean tankers. Proud ships all of them. God speed them on their courses and bring them
> (A) safely home.

Having delivered this ship "safely home" Williams' score closes with an emphatic cadence, albeit with some minor discordance en route (much like the tanker's voyage).

Proud Ships illustrates well Lewin's prescriptions for good soundtrack practice. Music, sound effects, and voice-over commentary are interwoven "harmoniously," suggesting that the commentary was probably recorded separately and ahead of the other sonic components, as Lewin recommends. Each sonic component comes to the fore at different moments, thus avoiding the "soporific" effect of an "unremitting flow of commentary." Rather than brief "holes" in the commentary, there are more substantial sections without voice-over in which music or sound effects are foregrounded, as when the latter affirm the calm after the storm.

The film's rejection of a single (unseen) commentator in favour of multiple voices is not unusual or experimental in the context of British documentary films, which had established the practice in the 1930s. Writing about documentary commentary in 1934, John Grierson advocated the use of workmen's own voices "with idiom and accent complete," rather than the "detachment" of a professional: "It makes for intimacy and authenticity."[40] Examples of this approach include *The Cable Ship* (GPO, 1933, dir. Stuart Legg and Alexander Shaw), *Weather Forecast* (GPO, 1934, dir. Evelyn Spice), and *Housing Problems* (British Commercial Gas Association, 1935, dir. Arthur Elton and Edgar Anstey), among others.[41] In *Land of Promise* (1946), Paul Rotha used multiple voice-overs to present dialectical argument, using the voices of a British "everyman," History, Hansard, the Isotype and several others.

In *Proud Ships*, the practice appears to humanize the individuals depicted, creating a sense of a community working together on board the *Liparus*, across the tanker fleet and in the company more broadly, to ensure the delivery of oil and oil-derived materials; a rhetorical effect resulting from an aesthetic decision. *Proud Ships* also makes use of a variety of language types ranging from more rhetorical and poetic content at key structural points, such as the film's beginning and end, through to the more prosaic thoughts of individual crew members during the voyage; different dialects and varieties of pronunciation emphasize these distinctions, while simultaneously emphasizing the unified purpose of the community.

Musical fit occurs at the level of what the "picture tries to say" in *Proud Ships*: it is used to add nuance to the characterization of different locations, activities, and situations. In this way it guides and narrows audience interpretation, much as nondiegetic scoring can and often does in narrative fiction films. Lively folk-inflected music is used to represent the possibility of

40 Grierson, John. "The GPO Gets Sound." *Cinema Quarterly*, vol. 2, no. 4, 1934, p. 216.

41 In *The Cable Ship* multiple non-synchronized voices are heard over the film's opening to indicate the multilingual context of international telephone communication. In addition to the commentary delivered by a white, middle-class male, at one point the cable foreman explains how the cable is hauled onto the ship, and the jointer explains how the joint is repaired: "That's me on the left [of the screen]." In *Weather Forecast*, telephonists and coastguards give notice of gale warnings in discussions with pilots at Croydon Airport in addition to the primary white, middle-class male voice. In both films, some of these voices are dramatized depictions of onscreen characters or roles. A variety of voices are heard in *Housing Problems*. In addition to the individuals who give testimony on screen with synchronized speech, there are two further (unseen) commentators: a white, middle-class male voice that provides the framing commentary, and Councillor Lauder, chairman of Stepney Housing Committee, who explains the context of the slums and slum clearance. For more on voice-overs in British documentaries of this period, see Stollery, Martin. "Housing Problems." *The Encyclopedia of Documentary Film*, edited by Ian Aitken. Routledge, 2006, vol. 2, p pp. 591–93; idem. "Voiceover/Commentary." *The Projection of Britain: A History of the GPO Film Unit*, edited by Scott Anthony and James G. Mansell. Palgrave Macmillan, 2011, pp. 168–78. "The Newsreel Commentator, the Actor, the Intellectual, and the Broadcaster: Celebrity and Personality Voices in Classic British Documentary." *Celebrity Studies*, vol. 4, no. 2, 2013, pp. 202–18.

an earlier-than-expected return home via Dublin; it appeals to the emotions and functions as a cultural reference to Ireland. Stately music is used to present the importance of these ships in the delivery of vital materials and fuel, suggesting a notion of duty, perhaps.

Music is used at strategic points, generating structure and emphasis, and providing rhetorical depth, as in the final sequence wherein the voice-over is matched to the pace, rhythm, and structure of the music (or vice versa). By contrast, during the sequence in which the voice-over explains the role of tankers in the various phases of the transportation of crude oil, its refinement, and the development and delivery of its by-products, the music operates as punctuation: it is integrated thoroughly with the voice-over but in support of the primacy of the voice-over (as well as changes to the map depicted onscreen).

Sound effects are used somewhat sparingly, avoiding sonic clutter, and enhancing their effectiveness when moved to the foreground. The distinctive sound of the anchor and its chains being dropped into the sea offers one such example. The same can be said of music: its absence can help to avoid conflict with, or emphasize the semantic content of, the voice-over, or allow sound effects to move to the fore. There are also occasional moments where "distinctive" sounds are integrated "harmoniously" with music. Such moments not only stand out, but also suggest a high degree of attention to detail in delivering an aesthetically interesting audiovisual whole. The character of the film's audiovisuality impacts its rhetorical delivery directly and thus emphasizes or enhances certain aspects of the company's image and/or brand values via non-verbal means.

The audiovisuality of *Proud Ships* is not unique. Rather, it represents the "norm" in terms of mid- to (relatively) high-budget industrial films produced by the SFU from the late 1940s through into the 1970s; films anticipated to be of interest to a general audience. In this last part of the chapter, I introduce examples from other SFU films of the mid-1950s.[42] These are primarily taken from a four-part film series that charts the life cycle of oil through to extraction, shot by the Dutch director Bert Haanstra for the SFU, mostly on location in Indonesia: *The Changing Earth*, *The Search for Oil*, *The Wildcat*, and *The Oilfield* (1954).[43] Jan Mul composed the music for the first and third

42 For a detailed discussion of the SFU film *The Rival World*, directed by Bert Haanstra, and released in 1955, see also Davison, Annette. "Music That Works: Listening to Prestige British Industrial Films." *Routledge Companion to Screen Media and Sound*, edited by Miguel Mera et al. Routledge, 2017, pp. 213–27.

43 The original Dutch titles are: *Ontstaan en Vergaan*, *De opsporing van Aardolie*, *De verkenningsboring*, and *Het Olieveld*. These films are available as part of a DVD box set celebrating the work of Bert Haanstra: *Bert Haanstra Compleet* (2015). Just Entertainment.

(c. nine and thirty-two minutes duration, respectively) and Max Vredenburg scored the second and fourth (c. thirty and eighteen minutes, respectively). Haanstra and Vredenburg had collaborated on several films previously.[44]

The Changing Earth (SFU, 1954, dir. Bert Haanstra)

The Changing Earth stands out from the group in one conspicuous way: there are no sound effects, diegetic or otherwise. The film explains how oil is formed underground over millennia, and its images illustrate the words of the voice-over. Music is continuous throughout, whereas the voice-over features short breaks. Here images and music are primary, as with the sequence that traces the water running down from mountains to the sea (04:12). This sequence features the second of two main musical themes, the the other presented first during the film's opening, returning at the end of the film, and thus presented during sequences without voice-over.

Lewin states that music should work with voice-over rather than the images, on the basis that "[m]usic should not say one thing while the voice tries to say another."[45] Here, although music supports the onscreen images throughout, music and image should be thought of as a unified whole, working together. The score functions as a musical representation of diegetic sounds that the images imply, but which are not heard. Music and image cohere and combine holistically, with each element the more effective for its relationship with the other, together fitting clearly what the film "tries to say." Examples include an increase in pacing and the introduction of a persistent quaver figure heard on strings at the mention of volcanic eruptions (01:08). At the visual depiction of ice, a plucked line on celli is presented, alongside high register tremolo strings. Timpani strikes are used to indicate a rock fall, pizzicato strings depict rain, and descending figures the rain's erosion of soil. Although the music is continuous, it also assists in the parsing of image sequences into smaller units, indicated by changes in orchestration and texture, pacing, and the scale of orchestral forces. Usually musical decisions fit with conventional (cultural) timbral associations as we understand them in relation to Western concert music, that is, large-scale slow changes are performed on low-register strings; small pebbles moving quickly through water are depicted via lively rhythmic figures performed on high-register woodwind. In one animated sequence, the upward motion

44 *Mirror of Holland* (1950), *Panta Rhei* (1951), and *The Dike Builders* (1952); the latter produced by the Shell Film Unit.

45 Lewin. "The Soundtrack in Industrial Motion Pictures," p. 258.

of oil through the earth's layers is traced by rising musical figures. Chords in parallel motion accompany the visual depiction of the creation of hills and troughs in the strata caused by pressures above and below.

The Search for Oil (SFU, 1954, dir. Bert Haanstra)

The second film of the series, *The Search for Oil* is altogether different audiovisually: the voice-over is supported by both Max Vredenburg's score and sound by Leo Fioole. Although the voice-over dominates overall, the film is in part structured around sonic distinction: sequences are split into those with music and/or with sound.[46] Where *The Changing Earth* provides a geological explanation of how oil is formed, most of this film is shot on location in Indonesia. Exotic sounds from the Indonesian jungle are placed high in the mix, helping to generate a sense of locale while not drawing attention away from the voice-over commentary. Musical entries are more noticeable: they occur at strategic points and contribute to the creation of mood and narrow interpretation of these scenes. One sequence that stands out in audiovisual terms features the momentary musical depiction of animated graphics superimposed over live action, in which the role of the gravimeter in oil exploration is explained (c. 19:00).

The Wildcat (SFU, 1954, dir. Bert Haanstra)

Sound effects play a key role in the third film, *The Wildcat*, which is also parsed into sequences that feature either music or sound effects alongside the voice-over. The film opens with beautiful images and sounds of pristine nature, the voice-over declaring: "Jungle. Green. Mysterious. Unspoilt. [pause] Unspoilt until man planted this pole." A swish pan to a surveying pole is followed by a hard cut to a shot of animals fleeing the monstrous sound of powerful machinery ploughing through vegetation (and habitats) to enable the building of wildcat wells in the search for potentially productive oil fields. Both music and sound feature (separately and loudly) alongside scenes of spectacular camera work and notable framing depicting these machines at work, with the voice-over usually absent during such sequences. In this way both music *and* sound enter at strategic points, drawing attention to key scenes. Music underscores the voice-over a little more in this film, but in such moments is more repetitive and less intense, pushing to the

46 Around twelve and a half minutes of this thirty-minute film are scored with original music, plus approximately one minute and twenty seconds of apparently diegetic local music.

fore only during breaks in the voice-over. The soundtrack is thus highly integrated. In places the relationship between music and voice creates a sense of fluidity, elsewhere it is generated via the integration of music with regular and rhythmic sound effects associated with everyday activities depicted onscreen (e.g., 25:30). This sequence culminates with a rhythm taken up by a (diegetic) drummer just prior to (and appearing to indicate) the Muslim call to prayer.

The Oil Field (SFU, 1954, dir. Bert Haanstra)

This fluid, even graceful integration of sonic elements is also a feature of the final film in the series, *The Oil Field*, which depicts the community needed to construct and service a new oil field and its workers in a remote area. In some sense this film is a return to the sound world of the first, in that more than three-quarters of the film is scored (fourteen of its eighteen minutes). Where there is no music, children play, then recite together in their classroom, a school bell rings, and distinctive sound effects feature dramatically, sometimes providing continuity through assemblages of spatially unconnected shots: hands using adding machines, large plant machinery, and the construction of houses (04:26–05:20). Here, the mechanical rhythms of sound effects organise and structure the film's visual editing. Fluidity in the integration of sonic elements is demonstrated in a scene in which a small child recites rhythmically addition sums on a blackboard (09:23). As the child reaches the last of these – with the answer "10" – there is a cut to a close-up of a till in one of the shops in this new oil town, as the number "10" is rung up (with a diegetic sound effect), its appearance punctuated by a solo flute figure (see Figs. 29.3-5).

The Oilmen (SFU, 1955, dir. Denis Segaller)

Many of the shots from *The Oil Field* appear in *The Oilmen*, directed and edited by Denis Segaller and released a year later (1955). The film features a radically different approach to audiovisuality, however. The original score by James Stevens incorporates two primary themes: one is associated directly with activity, and the other is introduced during the main title sequence. This latter theme is organized in a song structure, which is not coincidental. Towards the end of the film popular American mezzo-soprano Muriel Smith sings lyrics by Paul Dehn to this melody over a montage of shots filmed at Shell Clubs worldwide, which feature employees and their families from a variety of cultures and ethnicities. With a focus on leisure

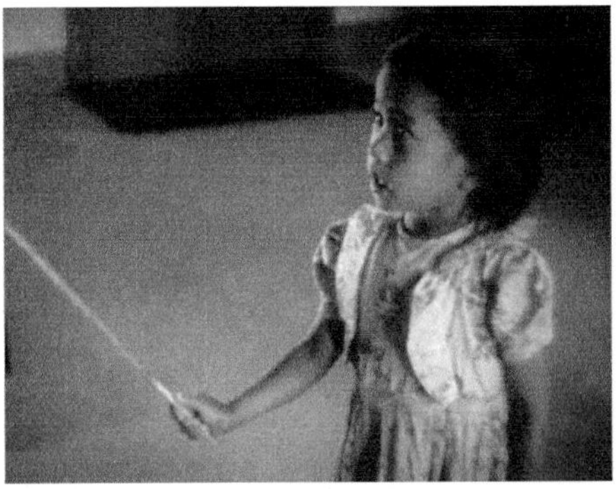

Fig. 29.3. A child recites addition.

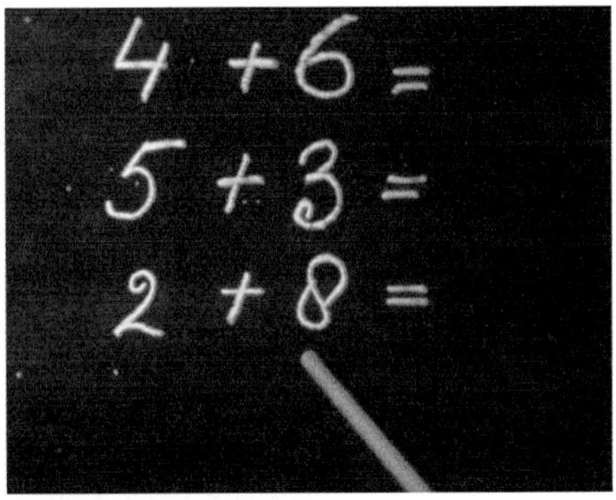

Fig. 29.4. The last sum.

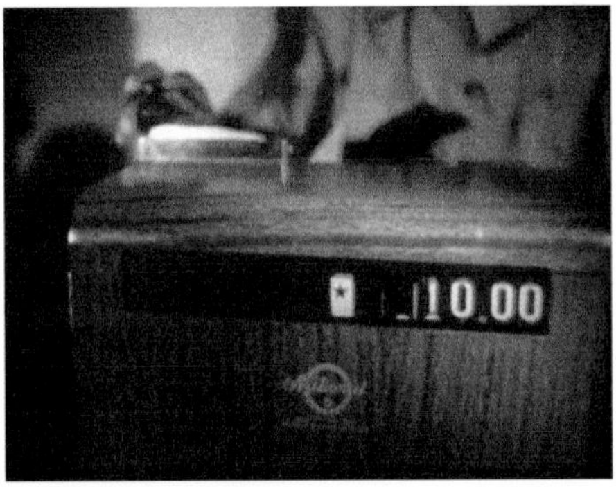

Fig. 29.5. A till presents the answer.

and play, the lyrics relate directly to the images onscreen, and together with Smith's mellow rendering of the song, the sequence implies that a fun and relaxing time can be enjoyed by all.

The inclusion of an original song is unusual in the SFU's output, but there are noticeable sequences of playfulness – even experimentation – in terms of audiovisuality here too. A variety of voice-overs from people of different classes and ethnicities are featured, emphasizing the international aspect of working for the company. Indeed, in a short sequence that takes place at a Shell office, multiple languages are featured in overlapping voice-overs, as though presenting the thoughts of the array of scientists and technicians who are shown undertaking their routine activities onscreen. Somewhat unusually, the film features a female voice-over too, in the form of a middle-class British wife to a doctor (19:16–21:44). Her commentary presents a depiction of family life in an oil town as less unfamiliar than one might expect, while also noting that "in our world, home is where oil demands it." In another sequence an animated male British voice lists (in received pronunciation) the many, many items required when an oil company establishes a sizeable new community in a remote location overseas. A sprightly theme by Stevens provides continuity through a beautifully fluid visual sequence that charts the requisitioning and ordering of the goods and materials to be shipped (15:58–16:30). Here the voice-over is less the provider of information, and more like a sound effect as the recitation of the list increases in speed. Once again, then, an aesthetic decision impacts the rhetorical delivery of the company's self-image and brand values. This is an exciting, modern, and international company to work for! The SFU catalogue describes the film as "a highly stimulating and interesting film with a distinct international flavour, suitable for showing to general audiences."[47] Internal promotional notes for the film suggest that the film also has recruitment potential and could be used productively as part of an induction process for new employees, enabling them to see how their own roles fit into that of the company more broadly.

(Tentative) Conclusion

In his analyses of scores created for "films of persuasion" produced in the USA between 1936 and 1945, musicologist Neil Lerner writes of "one of the expository documentary film score's most subtle functions": music's participation "in the rhetoric of the documentary film text."[48] He suggests that such scores share many of the conventions Claudia Gorman characterizes as principles of composition, mixing and editing of classical Hollywood film scores.[49] There are two exceptions he notes, however: "inaudibility," since music often "assumes a greater role" in expository documentary films, and "narrative cueing/illustrating," because here music often takes "a much more active role."[50] Much the same could be said of the industrial film scores discussed above.

Lerner also notes that the expository documentaries he discusses provide "more opportunities for musical experimentation and the expansion of compositional 'progress'"[51] than Hollywood films of the time, and often allowed "greater participation of the composer and the filmmaker at the pre-production stage."[52] The same cannot be said of the industrial films discussed here: in most cases the musical language is no more modernist

47 Shell Petroleum Company. *Shell Film Catalogue*, p. 142.

48 Lerner, Neil. *The Classical Documentary Score in American Films of Persuasion: Contexts and Case Studies, 1936–45*. PhD thesis, Duke University, 1997, p. 13.

49 Gorbman. *Unheard Melodies*, p. 73.

50 Lerner. *The Classical Documentary Score*, pp. 14–15.

51 Ibid., p. 18.

52 Ibid., p. 19.

or experimental than mainstream film music, and composers appear not to have been involved at the pre-production stage (or rather, I have found little or no evidence of that hitherto). Where I *would* argue that experimentation occurs here is in relation to *audiovisuality*, or, rather, these commercial industrial films continue a tradition of audiovisuality developed in films of the British documentary movement from the 1930s to 1940s. On one hand, this is not surprising, for both the SFU and the Film Centre employed personnel who had been instrumental in that movement and, furthermore, the SFU shared premises in London with the Film Centre from 1951 to 1965.[53] On the other hand, it also demonstrates Shell's belief in this tradition of documentary filmmaking as a tool of public relations and soft persuasion. When considered alongside their visual and verbal language, interrogation of the sound worlds of these films not only enriches our understanding of them, but also enables the identification and evaluation of techniques of audiovisual rhetoric in all manner of films of persuasion.

Works Cited

Anthony, Scott. *Night Mail.* BFI Publishing, 2007.

Bonhams. Auction 11904, lot 27, September 7, 2005, www.bonhams.com/auctions/11904/lot/27/. Accessed January 27, 2017.

Branigan, Edward. *Narrative Comprehension and Film.* Routledge, 1992.

Bulley, James. "Progress Music: Daphne Oram, Geoffrey Jones and Trinidad and Tobago." *Bricks from the Kiln*, no. 2, edited by Andrew Lister and Matthew Stuart. Antenne Books / RAM Publications, 2016, pp. 5–22.

Davison, Annette. "Music That Works: Listening to Prestige British Industrial Films." *Routledge Companion to Screen Media and Sound*, edited by Miguel Mera, Ron Sadoff, and Ben Winters. Routledge, 2017, pp. 213–27.

Ellis, Jack C. *John Grierson: Life, Contributions, Influence.* Southern Illinois University Press, 2000.

Elton, Arthur. "How We Use Films in Shell." *Film User*, vol. 10, no. 118, 1956, p. 344.

Foxon, Steven. "Film Centre (United Kingdom)." *Encyclopedia of the Documentary Film*, edited by Ian Aitken, 3 vols. Routledge, 2006, vol. 1, pp. 411–413.

Gorbman, Claudia. *Unheard Melodies: Narrative Film Music.* Indiana University Press / BFI Publishing, 1987.

Gordon, Douglas. *The Shell Film Unit: The First Sixty Years.* Shell Film and Video Unit, 1994.

53 Foxon. "Film Centre (United Kingdom)."

Grierson, John. "The GPO Gets Sound." *Cinema Quarterly*, vol. 2, no. 4, 1934, p. 216.
Hediger, Vinzenz, and Patrick Vonderau. "Record, Rhetoric, Rationalization." *Films That Work: Industrial Film and the Productivity of Media*, edited by Vinzenz Hediger and Patrick Vonderau. Amsterdam University Press, 2009, pp. 35–49.
Herrick, James A. *The History and Theory of Rhetoric: An Introduction*, 3rd ed. Pearson Education, 2005.
Kozloff, Sarah. *Overhearing Film Dialogue*. University of California Press, 2000.
Lerner, Neil. *The Classical Documentary Score in American Films of Persuasion: Contexts and Case Studies, 1936–45*. PhD thesis, Duke University, 1997.
Lewin, Frank. "The Soundtrack in Industrial Motion Pictures." *Industrial Screen*, vol. 3, no. 5, November–December 1959, p. 257.
Nichols, Bill. *Representing Reality: Issues and Concepts in Documentary*. Indiana University Press, 1991.
Shell Petroleum Company. *Shell Film Catalogue*. 1955.
Stollery, Martin. "Housing Problems." *Encyclopedia of Documentary Film*, edited by Ian Aitken. Routledge, 2006, vol. 2, pp. 591–93.
Stollery, Martin. "The Newsreel Commentator, the Actor, the Intellectual, and the Broadcaster: Celebrity and Personality Voices in Classic British Documentary." *Celebrity Studies*, vol. 4, no. 2, 2013, pp. 202–18.
Stollery, Martin. "Voiceover/Commentary." *The Projection of Britain: A History of the GPO Film Unit*, edited by Scott Anthony and James G. Mansell. Palgrave Macmillan, 2011, pp. 168–78.
Wedgwood. "Prestige: Portland Vase," www.wedgwood.co.uk/prestige-portland-vase. Accessed January 27, 2017.
Wernick, Andrew. *Promotional Culture*. Sage, 1991.

About the Author

Annette Davison is senior lecturer in music at the University of Edinburgh, UK. Her research focuses on the history, theory, and analysis of audiovisual media, particularly cinema and television. She has published monographs on film soundtracks in the 1980s (Ashgate, 2004), and Alex North's score for Kazan's *A Streetcar Named Desire* (Scarecrow, 2009), as well as co-edited collections of essays on the work of David Lynch (2004) and sonic practices in the "silent" era in Britain (2013). After publishing articles and book chapters about the main title and end credit sequences for television serials, her current research centres on energy-focused films and television series.

30 Creative Films for Creative Corporations

Music and Musicians in Experimental Italian Industrial Films[1]

Alessandro Cecchi

Abstract

The chapter proposes to consider the involvement of Italian avant-garde composers as a relevant feature for the definition of an experimental sub-genre of Italian industrial film of the "golden age," ranging from the early 1960s, with the peak around 1965, to the early 1970s, when production was limited to sporadic cases. In that period, prestige films produced or commissioned by such Italian companies as Edison, ENI, Fiat, Innocenti, Italsider, and Olivetti offered experimental composers significant income, which explains their frequent collaboration. This implies that such composers were considered as expressive resources in these films. The presence of their music had a direct influence on the reputation of the companies concerned, and hence on the "corporate image." They contributed with other artistic collaborators to authenticate these films as "avant-garde" productions and provided them with strong musical and sonic impact.

Keywords: industrial film; experimental music; film music; Italian composers; corporate image

In the age of "new media" and "remediation,"[2] during which cinema has reached a "post-medium" and "post-cinema condition,"[3] a number of film

1 After the pandemic, archival research on Italian industrial films made significant progress. The reader is advised that the research status of this chapter is current as of April 2017.

2 See Bolter, Jay David, and Richard Grusin. *Remediation: Understanding New Media*. MIT Press, 1999.

3 See Hediger, Vinzenz, and Miriam De Rosa, editors. "Post-what? Post-when? Thinking Moving Images beyond the Post-medium/Post-cinema Condition." *Cinéma and Cie*, vol. 16, no. 26–27,

Hediger, V., F. Hoof, Y. Zimmermann, with S. Anthony (eds), *Films That Work Harder: The Circulation of Industrial Film*. Taylor & Francis Group 2024
DOI: 10.5117/9789462986534_CH30

forms and genres have resurfaced after decades during which the context of their use and circulation had disappeared, to the point that their audiovisual textuality – apart from intertextual perlustrations of fragments from them in television – had been removed from people's individual memories and from the collective mind of entire nations. Although their presentification occurs through digitalization – i.e. through the filter of remediation and as a "simulation" which is unable to "mirror the technological complexity of the original"[4] – they act as agents enhancing historical knowledge and self-awareness. Even if, or precisely because, they are experienced outside of their historical context and use, these forms and genres emerge in a dense textuality, recasting their meaning through web archives that propose broad selections of films as streaming content.

This is exactly the case of Italian corporate industrial films, that in the last decade have gained increasing attention, particularly after the opening, in 2006, of the Archivio Nazionale Cinema d'Impresa (ANCI) in Ivrea (Piedmont) as a branch of the Centro Sperimentale di Cinematografia, Rome. The ANCI aims at preserving films produced by a significant and still increasing number of Italian corporations, promoting research and disseminating knowledge on them, including making a wide selection of films available on the web – since 2012 on the CinemaimpresaTV YouTube channel.[5] Another symptom of the increasing attention given to these films is the huge set of metadata offered by the Archivio del Cinema Industriale e della Comunicazione d'Impresa in Castellanza, near Varese, a documentation centre created in 1998 by Confindustria and the Università Carlo Cattaneo. This data can be explored through a search engine in the web.[6]

The possibility of exploring metadata about Italian industrial films and experiencing their complete audiovisual configuration – in the recent past also made possible by DVD publications including or being included in books[7] – opened the path to investigations about the artistic collaborators

2016.

4 Negri, Sabrina. "Simulating the Past: Digital Preservation of Moving Images and the 'End of Cinema'." *Cinéma and Cie*, vol. 16, no. 26–27, 2016, pp. 52, 54.

5 See Archivio Nazionale Cinema d'Impresa. "CinemaimpresaTV," www.youtube.com/user/cinemaimpresatv. Accessed April 15, 2017; I thank Elena Testa for mentioning and providing information about Olivetti and Fiat films.

6 See Archivio del Cinema Industriale e della Comunicazione d'Impresa. "Catalogo generale del cinema industriale," www.archiviocinemaindustriale.it/index.php/cataloghi/catalogo-cinema/consultazione. Accessed April 15, 2017.

7 See Olmi, Ermanno. *Gli anni Edison: documentari e cortometraggi 1954–1958*. DVD. Including: Tobagi, Benedetta, editor. *I volti e le mani*. Feltrinelli, 2008; Bertolucci, Bernardo. *La via del petrolio*. DVD. Including: Toffetti, Sergio, editor. *La scoperta dell'altrove*. Feltrinelli, 2010; Canova,

involved, including musicians. The significant role of music in some of these films lead to the hypothesis that the circulation of musicians could be important in identifying and defining specific sub-genres, particularly considering that the most prominent musicians were involved in film productions of prestige. The result of the investigation was the identification of an "experimental" sub-genre of Italian industrial films produced as of the so-called "economic miracle" (1958–1963) and still present in the subsequent decade, even though the latter was characterized by economic stagnation and the rise of protest movements.[8] This sub-genre includes both short ten-minute films whose circulation in cinemas was limited to the paratextual space of newsreels and documentaries, and mostly high-budget, medium- or feature-length film productions that were screened in documentary film festivals including awards, and in some cases broadcast on national television channels. The musicians involved played in many cases an active role in the avant-garde music scene of those years.

In order to apply the category of "experimentation" to an industrial film – and in general to a documentary film – it is not necessary for the whole film to be characterized by experimental aspirations. In terms of "vertical" construction – i.e. the stratification of audio and visual components – experimentation can involve just one or a few layers of the film (shooting technique, synchronization, sound effects, music), while in terms of "horizontal" construction (film editing, formal articulation) experimentation can characterize only specific sequences (a typical example: the title sequences) or take up a limited space in the film. In any case, the presence of experimental music, which is generally accompanied by a high level of audiovisual ingenuity, is a significant marker of the sub-genre.

Striking Music for Striking Film Sequences

The presentation of innovative technologies was generally the occasion to involve avant-garde musicians. This is the case of Olivetti's thirty-minute film production *Elea classe 9000* (dir. Nelo Risi, 1960).[9] In this example,

Gianni, and Giulio Bursi, editors. *Cinema elettrico: i film dell'Archivio AEM (1928–1962)*. DVD + book. Rizzoli, 2011; Latini, Giulio. *L'energia e lo sguardo. Il cinema dell'Eni e i documentari di Gilbert Bovay*. DVD + book. Donzelli, 2011.

8 See Ginsborg, Paul. *A History of Contemporary Italy: Society and Politics 1943–1980*. Palgrave Macmillan, 2003.

9 See "CinemaimpresaTV," www.youtube.com/watch?v=lUhSYR68PqQ. Accessed April 15, 2017.

audiovisual inventiveness reaches an astonishingly high degree. In presenting the Olivetti product Elea 9003, the first Italian transistor computer, the director aimed at integrating documentation, promotion, graphic art, and innovative music in a single film. His collaborators were protagonists of the art scene in those years: a graphic idea by Giovanni Pintori, the industrial designer of Olivetti, became part of the animation by Giulio Gianini and Gianni Polidori, the latter being the set and costume designer of stage director Virginio Puecher. The music was entrusted to Luciano Berio, who composed both electronic and instrumental cues. All these aspects interact in the title sequence, characterized by a strict synchronization between the animation and the music, whose changes correspond to the articulation of the credits: electronic music for the animation of the logo and of the graphic representation of a typewriter, following the renowned design of Pintori; ensemble music for the credit sequence; piano clusters repeated in loop for the acknowledgements. Music is clearly given structural primacy, to which the animation is in strict synchronization. Innovation, experimentation, and the avant-garde are explicitly thematized and become keywords of both the film and the product, just as electronic music evokes the electronic technology of the transistor.

Of course, the same level of experimentation is not maintained over the entire film. Yet, most parts can be characterized as the first example in Italian cinema – fiction films included – of a combination of electronic music created by Berio at the Rai Studio di Fonologia in Milan[10] and music composed for an instrumental ensemble.[11] In any case, *Elea classe 9000* in 1960 was the winner of the first edition of the Rassegna Nazionale del Film Industriale promoted by Confindustria and was presented at the Mostra Internazionale del Documentario e del Cortometraggio in Venice. The film thus received mention in cinema journals, led to a significant promotion in terms of corporate image, and this was one of the reasons for economic investment in those film productions, allowing the involvement of important collaborators. Berio's music is in fact used in another Olivetti industrial film, *La memoria del futuro* (dir. Nelo Risi, 1960),[12] and in one of the first

10 See Novati, Maria Maddalena, and John Dack, editors. *The Studio di Fonologia: A Musical Journey 1954–1983: Update 2008–2012*. Ricordi, 2012.

11 I thank Talia Pecker Berio and the Centro Studi Luciano Berio for kindly allowing me to listen to two tapes that in 2014 were provisionally stored at the Centro Tempo Reale, Florence. These tapes are now conserved in the Luciano Berio collection at the Paul Sacher Foundation, Basel.

12 See "CinemaimpresaTV," www.youtube.com/watch?v=dmoiLXrGXok. Accessed April 15, 2017.

CNRN (Centro Nazionale per la Ricerca sul Nucleare, National Centre for Nuclear Research) films: *Ispra 1* (dir. Gian Luigi Lomazzi, 1959),[13] which separately use tape music (i.e. manipulated recorded music), electronic music (i.e. using synthetic sound) and instrumental music. Electronic music, this time by Bruno Maderna, is also found in the recently rediscovered film *Lavoro a Ferrania* (1962).[14]

Among the early use of electronic music in industrial films, Gino Marinuzzi Jr. must be mentioned as well. The son of conductor and composer Gino Marinuzzi, he was involved in the three-episode feature-length industrial film *L'Italia non è un paese povero* (1960), a PROA (Produttori Associati per la Rai Tv, Associate Producers for Rai Tv) production commissioned by Enrico Mattei, the patron of ENI (Ente Nazionale Idrocarburi, National Hydrocarbons Authority), to Joris Ivens and never released.[15] The film was realized with the help of young assistant directors Tinto Brass and Paolo and Vittorio Taviani – the Tavianis actually directed the third, Sicilian episode of the film. The film was only partially screened at the Venice Mostra Internazionale del Documentario e del Cortometraggio in 1960 and received attention in cinema journals.[16] Gino Marinuzzi Jr., on the one hand, composed the orchestral score and, on the other, created some electronic effects that were later used by Ivens in few sequences of the first episode. Previously, he had been involved as composer and conductor in the Edison industrial film *Toce 28* (dir. Guido Guerrasio, 1954),[17] while subsequently he composed music for the twenty-five-minute prestige film production which won prizes at the Universities of Padua and Rome, *Italsider. Film-relazione 1961* (dir. Valentino Orsini, 1962),[18] whose score has been conserved.[19]

13 See "CinemaimpresaTV," www.youtube.com/watch?v=_IcCSJie-wQ. Accessed April 15, 2017.

14 See "CinemaimpresaTV," www.youtube.com/watch?v=U_2Z3pNkgPI. Accessed April 15, 2017.

15 See Corbella, Maurizio. "Paolo Ketoff e le radici cinematogra che della musica elettronica romana." *AAA–TAC* (*Acoustical Arts and Artifacts – Technology, Aesthetics, Communication*), vol. 6, 2009, pp. 65–75; Cecchi, Alessandro, and Maurizio Corbella. "Experimentation, Documentation, Censorship: A Joris Ivens's Industrial Film and the Italian National Broadcasting Television." Paper presented at the Music and Media 4th Study Group Conference, Turin, June 29, 2012.

16 Verdone, Mario. "Note. *L'Italia non è un paese povero*." *Bianco e nero*, vol. 21, no. 7, 1960, pp. 87–90.

17 See "CinemaimpresaTV," www.youtube.com/watch?v=r7jzelkRnB8. Accessed April 15, 2017.

18 See "CinemaimpresaTV," www.youtube.com/watch?v=3CPPFxASsQY. Accessed April 15, 2017.

19 I thank Anna Maria Marinuzzi for granting permission to study the score and Maurizio Corbella for mentioning and providing it. The score is now preserved at Fondo Marinuzzi, Associazione NoMus, Milan.

In the field of automobile factories, experimental films were called on as competitive factors. The Cinefiat film *F4CB. Acciaio su misura* (dir. Victor De Sanctis, 1965)[20] features music by Angelo Francesco Lavagnino, the film composer chosen by Orson Welles for *The Tragedy of Othello, the Moor of Venice* (1952), and later involved in *Chimes at Midnight* (1965). In this industrial film the composer sometimes uses electronic effects, especially in laboratory scenes, but often he simulates them through a masterful instrumentation of expanded traditional orchestra: long-lasting background sound of string harmonics, an unconventional use of percussion (especially idiophones), pointillist effects, and other special characteristics of Lavagnino's orchestration are similar to compositional devices deploying electronic sound. The end production does not fully correspond to the score conserved in the Lavagnino private collection in Gavi, Alessandria. Particularly, the music for the title sequence – a quite modernist score – was discarded in favour of a more alluring music, more along the lines of Lavagnino's music for other film genres in these same years, that he probably composed at a later stage on request by the film producers.[21] The fact that Lavagnino's involvement in industrial films was limited to film productions of prestige is clearly shown by *Italia 1961 in Circarama* (1961), a film for the Disney Circarama which was offered by Fiat to the participants of the national exposition for the first centenary of Italy in Turin.[22] This was a very expensive film, in collaboration with Walt Disney Productions and Royfilm Rome, by producer Roberto De Leonardis. The director was Elio Piccon, and the script writer Indro Montanelli.

Not many years after Fiat, another relevant Italian automobile factory to produce a corporate film of prestige was Innocenti, with the thirty-minute *Noi continuiamo...* (dir. Mario Damicelli, 1968).[23] Characterized by the presence of many long sequences relying on strict editing and audiovisual synchronization, the music is by Egisto Macchi, using both orchestral and electronic composition. His contribution to the sound given to these sequences serves the director's clear intention of using a higher degree of audiovisual experimentation in progressively longer sections of the

20 See "CinemaimpresaTV," www.youtube.com/watch?v=zfJHlypL5PM. Accessed April 15, 2017.

21 I thank Bianca Lavagnino for kindly allowing me to read the manuscript score, conserved in Gavi.

22 See Vitella, Federico. "Tecnologia, ideologia: il Circarama." *Bianco e nero*, no. 557–58, 2007, pp. 183–95.

23 See "CinemaimpresaTV," www.youtube.com/watch?v=nC8bCC9JwJ4. Accessed April 15, 2017.

film, in spite of its patently promotional purpose. The first part of the film contains a brief overview of the history of Innocenti, beginning with the invention of the famous steel tubes used for scaffolding. This is followed by a representation of the processes involved in the production of steel, shaped in the form of tubes for various applications, introduced by the voice-over, that stresses the importance of these tubes in the construction industry and in industrial society in general. The sequence in question is based on a rapid string of shots that document this phase of production, accompanied by a brusque, even brutal passage of Macchi's orchestral music, replete with brass and percussion instruments, including a prominent presence of anvils. In the second half of the film a sequence lasting roughly three minutes appears, whose audiovisual impact is intense: the video montage is marked by a continuous series of almost instantaneous shots pitted one against the other, and by rapid transitions from one phase to the next of a long and highly mechanized serial production process, which comes to its conclusion with the assemblage of the new Innocenti automobiles. The soundtrack offers an equally rapid succession of electronic and recorded sounds, which partially re-elaborate the noises generated by the production process. There is a clear correspondence here between the principle of film editing and similar techniques used in music, both pervasively present precisely in order to represent of the entire process during which a car is assembled.

In this example the association between the working processes involved in heavy industry and orchestra music rich in percussion and brass instruments is founded on the idea of their timbral uniformity. On the contrary, the close correspondence between electroacoustic experimentation and chemical processes – above all when the footage lingers on electronic measuring instruments – points towards a different kind of uniformity. In some cases, the sound produced by these instruments themselves is what sparks off the relation. More generally, however, what counts is their actual technical and functional similarity with the instruments used in producing electronic sound as of the mid-1950s.

A considerable number of Italian industrial films dating to the 1960s deal with the problem of nuclear energy, all of which were CNEN (National Committee for Nuclear Energy) productions. Among these, *L'atomo in mare* (dir. Virgilio Tosi, 1962),[24] which resorted to the collaboration between the graphic artists Pino Zac and Miro Grisanti, entrusted the sound and the

24 See "CinemaimpresaTV," www.youtube.com/watch?v=TGDGwwMA7z4. Accessed April 15, 2017.

electronic music to composer Franco Potenza. The initial animated sequence, roughly two and a half minutes long, that rolls under the opening credits, has a more specifically narrative function here. That which is presented is a series of prior events: the experimental detonations at Bikini Atoll, in the Pacific Ocean, in June and July 1946, and the ensuing radioactive fallout that spread through the sea and thus made its way into the food chain. Between the initial acknowledgement of the producer and the rest of the opening credits, an animated sequence was inserted that represents the explosion of the H-bomb, preceded by a countdown and followed by a mushroom cloud with radioactive material scattered into the air, which continues for the entire length of the credits. Immediately after this, the animation passes to a representation of the ocean floor, hit by a cascade of radioactive particles that deposit on it and are ingested by fish, one of which winds up on the plate of a man who, eating it, is affected by radioactive contamination. In this and in the other colour-animated sequences of the film – which alternate as usual with more traditionally conceived documentary sequences, here in black and white – one cannot speak of music in the strict sense of the word, but of a calibrated electronic sonorization of the animation.

Potenza's music appears in other CNEN productions, including *Atomi puliti* (dir. Enrico Franceschelli, 1965),[25] in which electronic sound, used in an infinitely more creative way, becomes a sign of danger. If in the opening credits a music mainly played by the piano (but also with faint sustained notes in a sonority that is difficult to identify) is synched with shots of the calm water of a stream and a small waterfall, the voice-over's reflections on environmental damage and smog coincide with the entry of electronic sounds, which only after the piano disappears reveal themselves to have been present since the beginning of the film, while the shots concentrate on the silhouette of a factory, with a long chimney that looms up into the grey sky, laden with industrial fumes. From this moment on, the meditation on environmental disasters is underscored by electronic sounds, whose disturbing presence reaches a climax when the destructive potential of atomic energy is discussed. Naturally, considering who commissioned it, the film then insists on the possibility of prevention, of progress in research and the potential of nuclear energy, underlining its virtues when used in a civil context, benefitting society as a whole; the concerns raised at the beginning are for that matter part of a communicational strategy aimed at providing reassurances and advocating a cautious use of nuclear energy – in

25 See "CinemaimpresaTV," www.youtube.com/watch?v=_eUT1E7rpzw. Accessed April 15, 2017.

the following parts of the film, electronic sound as well becomes a pleasing and harmonious backdrop. Only when deposits of radioactive waste – a long stretch of blue and red bins with conspicuous danger signs, black on a yellow background – are seen at the end of the film do the threatening and disconcerting electronic sounds appear once again, reaching a sort of electronic quiver that concludes the film over a black screen.

Towards a Definition of Experimental Industrial Film

The textual construction that emerges from the examples described above owes many of its connotations to musical and audiovisual experimentation. A structural analysis dealing with this textuality would highlight many aspects that other models tend to pass over: on the one hand, it brings out the single components of the audiovisual text, allowing, for example, the role of sound and music to be detected; on the other, it emphasizes the syntactic and semantic complexity of the audiovisual construction, allowing its textual strategy to be defined. This strategy presupposes an audiovisual linguistic code to the same degree to which it contributes to establishing one; and it is along this double track that textuality proper can significantly contribute to defining the genre.

Let us first have a look at the music. As regards industrial films, we might note that, compared to productions dating to the 1950s, marked by a close adherence to the dominant models of film music of the time, understood as a stylistic definition (orchestral music, with late-romantic or impressionist sonorities, the use of recurring motifs or recognizable thematic constructions which are then subjected to elaboration, and so on), musical and, in particular, electronic experimentation in the 1960s represents a radical change in orientation. Avant-garde music is invariably – and strategically – linked to a segmented form of temporality, favouring a relation with images based on a point-to-point correspondence tending towards a very close synchronization; this creates a continuous interaction between the series of shots and the series of musical events, giving particular attention to sudden interruptions and discontinuity. The experimental sub-genre of industrial film establishes itself thanks to this very kind of syntax, which derives from the conscious choice of giving prominence to musical experimentation, linking it to audiovisual inventiveness.

As to formal strategies, we might note how experimentation favours the appearance of a very wide-reaching typology. These productions' promotional and informative aims anchor them to the principles of documentaries; this is why they invariably alternate strictly documentary and experimental sections. The latter generally include the opening credits, animated sequences and montage sequences based on a rapid pace of the editing (in the latter case, nonetheless, experimentation comes together with the documentation of industrial processing). From the point of view of sound, the characteristics of the two types diverge quite sharply: the documentary sections are dominated by a voice-over, dialogues and direct sound recording of industrial noises, while music is either absent or limited to acting as a backdrop; in the experimental sections, music and sound often move into the foreground. Furthermore, in the latter sections the musical organization is at times given structural priority and dictates the rhythm of the animation shots; in other cases, the two levels of recording (sound and images) are articulated simultaneously, based on a common editing principle.

As regards their visual aspects, industrial films are characterized by a strong discontinuity in the shots and video editing, owing to the need to use framings in a way that serves the documentation of the industrial structures, and the processing and assembly of the finished products. From this point of view, a single industrial film can explore a wide array of technical possibilities, ranging from aerial shots to close-ups of details, advanced photography techniques (such as macro), external shots and internal ones marked by extreme or highly contrasting lighting conditions (which explains the preponderant role given to photographic filters). It is not rare, in the documentary sequences, for an audiovisual syntax inspired by news reports to come into play, above all when historical memories and contexts are to be reconstructed, particularly through repertory images. Hence, the continuous alternation of sequences in black and white and in colour, or again of animated sequences and external shots, with all of these differentiations having their influence on the overall syntax.

As far as the represented content is concerned, industrial films bring into play a wide assortment of peculiar elements, which contribute to establishing the strictly semantic aspects of the text: the facilities, the company, the factory, the workers, scientists, office employees, technicians and managers, the means of production, the machines and the operations, instruments used for measuring and analysis, interactions between man and machine, the representation of mechanized processes, raw materials and their processing, and finally the products, the rhetoric and ideology of industrialization, and the ritual aspect of work; all elements which, while varying considerably, lie at the root of the definition of industrial film.

As regards communicational functionality, it is interesting to note that in experimental productions the distinctions introduced into industrial films' various goals and communication strategies tend to overlap or become irrelevant. Here, contrasting functions and typologies merge: these are "technical-scientific popularization films" with highly pronounced didactic purposes, but at the same time "documentary films" (in particular concerning industrial planning and production); they are "publicity films" but – unlike true advertisements – our attention is not exclusively focused on the finished product, but rather on the processes underlying its production; they are furthermore "public relations films,"[26] not strictly aimed at selling a product but at promoting the company itself, giving particular emphasis to the social and cultural aims of entrepreneurialism, rather than its more specifically economic goals. The Olivetti catalogue explicitly mentions these productions as "corporate image films."[27] Experimental industrial films indeed tend to shun any specific functionality (involving promotion or advertising) in favour of their audiovisual ingenuity, which alternates documentary criteria with those pertaining to artistic communication.

The pragmatic contexts of experimental industrial film corroborate the dominant role given to audiovisual textuality. Those commissioning them were Italy's most significant industrial organizations, both private and public, in the 1950s and the 1960s: companies such as Edison, Enea, ENI, Fiat, and Olivetti were among those that drew heavily on this type of cinematographic communication, while others, such as Alitalia, Ferrovie dello Stato, Ilva, Innocenti and Italsider, turned to it more sporadically. At the origin of some significant projects we find eminent personages from the history of Italian industry, including Adriano Olivetti and Enrico Mattei. Directors, authors, script writers, graphic artists, and composers were, however, the ones who made their intentions concrete. Indeed, audiovisual production consists to a huge degree in a collaboration between several subjects, each of which may be assigned a different degree of responsibility, without necessarily going as far as the idea of multiple or shared authorship: in any case, the result is always the outcome of a convergence of different skills and authors whose background makes a decisive contribution to defining an experimental genre. Not infrequently, effective synergies are created; in so doing, it may happen that a given industry establishes a privileged relationship with one or more collaborators, and this applies to directors no less than composers.

26 Mosconi, Elena. "Il film industriale." *Comunicazioni sociali*, vol. 13, no. 1–2 [special issue: "Il cinema a Milano dal secondo dopoguerra ai primi anni Sessanta," edited by Raffaele De Berti], 1991, pp. 65–66.
27 I thank Elena Testa for this information.

The Circulation of Musicians in Experimental Industrial Films

I would like to further expand upon the role played by some composers, because with their work they decisively contributed to defining the experimental sub-genre. The collaboration of composers belonging to the avant-garde says a great deal about the importance that the works' commissioners, as well as their producers and directors, gave to music and to musical experimentation even during the initial planning phases. Inversely, the remarkable quantity of industrial films in which these composers collaborated bears out the hypothesis that audiovisual works in general and industrial films in particular could be a rather attractive field for experimentation. We should not forget, however, that these collaborations also offered an occasion in which to put one's own skills at disposal, bringing together lesser commitment and an immediate economic return and income; naturally, this is also true for directors and screenplay writers.

The ENI films rely almost exclusively on the collaboration of Egisto Macchi, one of the founding members of the Nuova Consonanza group.[28] The feature-length industrial film in three episodes directed by Bernardo Bertolucci and broadcast by Rai, *La via del petrolio* (1967), is only the most famous chapter in this collaboration. Macchi had indeed already taken part in other ENI productions, from *Oro nero sul Mar Rosso* (dir. Vittorio Gallo, 1962) and *L'isola del petrolio* (dir. Gian Maria Messeri, 1962),[29] to the films directed by Giuseppe Ferrara *CH4 in Lucania* (1963) and *Gela antica e nuova* (1964, text by Leonardo Sciascia),[30] to Gilbert Bovay's *Gli uomini del petrolio* (1965).[31] Other industrial figures also called on his collaboration, Macchi being easily the most prolific composer seen in Italy. In addition to the CNEN production *Re Uranio* (dir. Enrico Franceschelli, 1964)[32] and the Innocenti film *Noi continuiamo...* (dir. Mario Damicelli, 1968), his name appears among the credits of at least one Olivetti industrial film, *Micromondo. Dalle valvole*

28 The Ilva fund in the Cineteca of the Fondazione Ansaldo in Genoa conserves the film *Opsteel* (dir. Emilio Marsili, 1969), whose music is credited to the Gruppo di Improvvisazione Nuova Consonanza, the avant-garde improvisation group including Ennio Morricone and Franco Evangelisti, among many others. I thank Annalaura Burlando for mentioning this film.

29 See "CinemaimpresaTV," www.youtube.com/watch?v=dEhCDRGSSkY; www.youtube.com/watch?v=XoZukXLsQbw. Accessed April 15, 2017.

30 See "CinemaimpresaTV," https://www.youtube.com/watch?v=GyiYSRpkhLE; www.youtube.com/watch?v=UmvsnUAg7r4. Accessed April 15, 2017.

31 See "CinemaimpresaTV," www.youtube.com/watch?v=wAl2xZwsWBQ. Accessed April 15, 2017.

32 See "CinemaimpresaTV," www.youtube.com/watch?v=NPU7ATp9EfA. Accessed: April 15, 2017.

al circuito integrato (dir. Fulvio Tului, 1970),[33] and in the Cinefiat film *Uno stabilimento: grande e subito* (dir. Stefano Calanchi, 1973).[34]

Another very prolific composer in the area of 1960s industrial films was Franco Potenza, who wrote highly interesting instrumental and electronic music and was particularly active in cinema from the late 1950s through to the mid-1970s, at times showing an extraordinary experimentalism. As regards industrial films strictly speaking, Potenza collaborated in numerous CNEN productions, many of which were entrusted to director Virgilio Tosi (*La fusione controllata dell'idrogeno*, 1960; *Sincrotrone*, 1961; *L'atomo in mare*, 1962; *Il quarto stato della materia*, 1963),[35] who was active in science popularization documentaries as well. For the same producer, Potenza also worked with Enrico Franceschelli (*Al di qua dell'Uranio*, 1965; *Atomi puliti*, 1965).[36] As an orchestral music composer, Potenza also appears in Olivetti productions such as *Sud come Nord* (dir. Nelo Risi, 1957)[37] and *Il diavolo della bottiglia* (dir. Sergio Spini, 1968),[38] in the prestigious Italsider film with a script by Dino Buzzati, *Il pianeta acciaio* (dir. Emilio Marsili, 1962),[39] in the Fiat film *Biografia di un aereo* (dir. Ansano Giannarelli and Pietro Nelli, 1964),[40] and in the production by Ferrovie dello Stato, *Uomo macchina uomo* (dir. Marcello Baldi, 1960).[41]

Composer Vittorio Gelmetti's role allows us to reflect on how industrial and auteur films were intertwined. Gelmetti approached cinema for the first time, in fact, when he lent a few of his pieces of electronic music to Michelangelo Antonioni for his first feature-length colour film, *Il deserto rosso* (1964), which also includes music by Giovanni Fusco. Antonioni wished to weave together a complex score of sounds, integrating the noises of

33 See "CinemaimpresaTV," www.youtube.com/watch?v=1cooFXZZe8k. Accessed April 15, 2017.

34 See "CinemaimpresaTV," www.youtube.com/watch?v=HDUBgcQMo8c. Accessed April 15, 2017.

35 See "CinemaimpresaTV," www.youtube.com/watch?v=fmWZixqxZnk; www.youtube.com/watch?v=JjmglddILxc; www.youtube.com/watch?v=TGDGwwMA7z4; www.youtube.com/watch?v=ARo_ugliRNY. Accessed April 15, 2017.

36 See "CinemaimpresaTV," www.youtube.com/watch?v=dSdKBNWnIOw; www.youtube.com/watch?v=_eUT1E7rpzw. Accessed April 15, 2017.

37 See "CinemaimpresaTV," www.youtube.com/watch?v=E4u6Rhq_GFs. Accessed April 15, 2017.

38 See "CinemaimpresaTV," www.youtube.com/watch?v=Rjsx4zMnjao. Accessed April 15, 2017.

39 See "CinemaimpresaTV," www.youtube.com/watch?v=YpD6tLTeRyc. Accessed April 15, 2017.

40 See FCA Group. "Centro Storico Fiat," www.youtube.com/watch?v=7qMvw8er_hw. Accessed April 15, 2017.

41 See Archivio Centrale dello Stato. "Archivio Centrale dello Stato," https://www.youtube.com/watch?v=FGi7jGTQ8Zo. Accessed April 15, 2017.

industrial production, Gelmetti's pre-existing electronic music, and a vocal line composed by Fusco; this is the case in the opening credits, containing extremely blurry images of the industrial landscape. In this work, Antonioni created an artistic sublimation of the industrial film: the landscape, the factories, the complex machinery, and a few specific phases of industrial production, sometimes documented down to their minute technical details, are in some way characters of the film.[42]

In the area of industrial film, Gelmetti's music is only credited in a few titles, concentrated in a couple of years: the Olivetti film *La macchina del tempo* (dir. Antonello Branca, 1968)[43] and the Fiat film *Appunti per l'auto di domani* (dir. Massimo Mida, 1969).[44] Gelmetti's choices in the music (whether it be instrumental or electronic) are never commonplace; even while making an extensive use of electronic sound, the latter never lends itself to stereotyped processes of signification. The same goes for Gelmetti's other collaborations in the area of fiction cinema. In the same years during which he worked on the industrial films mentioned above, in fact, he also collaborated in two feature-length films: *Sotto il segno dello Scorpione* (dir. Paolo and Vittorio Taviani, 1969) and *Sierra Maestra* (dir. Ansano Giannarelli, 1969). His collaboration with Giannarelli was prolonged in later years thanks to two Reiac (Realizzazioni indipendenti autori cinematografici, Independent Realizations Film Authors) short film productions: *Linea di montaggio* and *Analisi del lavoro* (1972).[45] Even though these are not industrial films strictly speaking – considering both their promoters and the highly critical stance taken towards the methods used in industrial production – a link with the topic of factories and the representation of the working environment in general is quite clear. The audiovisual strategy set in place by Giannarelli and Gelmetti is explicitly linked to many points of Marxist critique, abandoning the rhetoric of industrialization and production that characterized industrial film in the 1960s.

42 See Calabretto, Roberto. *Antonioni e la musica*. Marsilio, 2012, pp. 137–62; Corbella, Maurizio. "Sperimentazione elettroacustica e cinema d'autore in Italia negli anni Sessanta. Due casi di studio." *Comunicazioni sociali*, no. 1, 2011, pp. 93–101.

43 See "CinemaimpresaTV," www.youtube.com/watch?v=goFCa449-PM. Accessed April 15, 2017.

44 See "CinemaimpresaTV," www.youtube.com/watch?v=CAZ_R2dC6P8. Accessed April 15, 2017.

45 See Archivio Audiovisivo del Movimento Operaio e Democratico. "Archivio Audiovisivo del Movimento Operaio e Democratico," www.youtube.com/watch?v=ZV_n3jDrVIc; www.youtube.com/watch?v=APwAU-ECOZw. Accessed April 15, 2017.

Historical Context

Some final remarks about the history of corporate industrial films, insofar as historical context can be put in relation with the experimental sub-genre under consideration. From this point of view, it has often been stressed that the years in which the genre flourished chronologically coincide with the period of rapid industrialization seen in Italy in the post-war period, in particular the economic miracle, and are geographically linked to the rich and productive regions in the north, first and foremost the industrial triangle Turin-Milan-Genoa. Indeed, the genre had developed most profoundly by 1964, which saw its quantitative peak in production. Its decline was fostered by both the earliest symptoms of the economic crisis that would extend over the following decades and by a disappointing national law on cinema (Law no. 1213, November 4, 1965). The latter contributed significantly to the demise of the genre, which from a cultural point of view became marginal compared to fiction cinema, and from an economic point of view lost access to public funds. This led, as early as 1965, to a drastic fall in industrial film production compared to the previous year.[46]

These contextual elements and quantitative data must not make us lose sight of the long history of the Italian industrial documentary, which began well before the economic miracle. The first industrial films, obviously silent, date back to at least 1905 (*L'industria di salami*, produced by Ambrosio Film in Turin). Fiat also took up cinematographic communication quite early, as is made clear by a documentary probably produced on occasion of the the Esposizione internazionale dell'industria e del lavoro (International Exhibition of Industry and Work) held in Turin in 1911, restored by the ANCI: *Le officine della "FIAT"* (dir. Luca Comerio, 1911).[47] For the first sound films (by Cines), one has to wait until 1930. The audiovisual construction of a film such as *Sotto i tuoi occhi* (1931),[48] commissioned to promote the Fiat 522, is extremely interesting, and not only for the elements it borrows from fiction film: the sound is rich and complex, giving particular emphasis to the noises of industrial production, without overlooking orchestral music and dialogue, and at times superposing these three levels of sound. In this early phase, moreover, animated films also appear, such as *Non è più un*

46 See Falchero, Anna Maria. "Cinema e industria: i documentari industriali." *Storia Politica Società*, vol. 41, no. 3–4, 2008, pp. 129–42.

47 See "CinemaimpresaTV," www.youtube.com/watch?v=oVHr-d9TdLo. Accessed April 15, 2017.

48 See "CinemaimpresaTV," www.youtube.com/watch?v=OEwOKNpntsw. Accessed April 15, 2017.

sogno (1932),[49] with which the Fiat 508 Sedan was presented. The events surrounding *Acciaio* (dir. Walter Ruttmann, 1933) are rather well-known; one of the first sound films produced by Cines, and partially shot inside the Terni steel mills, this film was based on a short story by Luigi Pirandello ("Giuoca, Pietro!"), adapted by Mario Soldati and given original music by Gian Francesco Malipiero.[50] On account of the standing of those who collaborated in making it, its futuristic slant and some aspects of its synchronization, this production became a touchstone for all Italian industrial cinema in the post-war period, including the experimental production of the 1960s.

As regards the post-war period, it should be mentioned that the 1949 short film *Sette canne, un vestito* (dir. Michelangelo Antonioni, music by Giovanni Fusco) was already a mature product with a significant history lying behind it. This is all the more true for the vast and significant output of Ermanno Olmi, whose collaboration with Edisonvolta took place in the 1950s and came to an end before the years of the economic miracle properly speaking. Obviously, economic growth was closely linked to the appearance of the experimental current in the early 1960s. Italian industry's ability to make projects was then at its peak, and this encouraged the use of advanced forms of cinematic communication. In our context, it may be more interesting to note that industrial film, even more so in its experimental form, abundantly survived the crisis of 1965, which did not lead to a decrease in quality, nor to the use of lesser economic, organizational, or artistic resources for each single production, but only a drop in the number of films created. New productions thrived in the years 1968 and 1969, some of which having a particularly high audiovisual impact, but examples can also be found as late as 1973. This is partially due to the fact that during the 1960s industrial films gained an autonomous textual status, establishing an audiovisual linguistic code that allowed them to resist and overcome the crisis or even to benefit from it, given the creation of a market niche tends to favour avant-garde productions.

Among the reasons that lead to the end of Italian industrial cinema, two appear to be the most striking. First of all, the critical stance towards industrialization and industrial films, starting from 1969, the year of contestation and factory occupations, as documented in *Apollon. Una fabbrica occupata* (dir. Ugo Gregoretti, although half-uncredited), involving free jazz.[51] These

49 See "Centro Storico Fiat," www.youtube.com/watch?v=DkdTkOVLbow. Accessed April 15, 2017.

50 See *Retroscena di "Acciaio": indagine su un'esperienza cinematografica di G. Francesco Malipiero*. Olschki, 1993.

51 See "Archivio Audiovisivo del Movimento Operaio e Democratico," www.youtube.com/watch?v=DsMHZxfoo-k. Accessed April 15, 2017.

productions shed light on the "dark side" of the economic miracle. It is in this context that short documentary films began investigating the consequences of the rapid and wild industrialization in 1960s Italy, such as environmental pollution and the conditions of the working class. Not only the aforementioned Reiac film productions directed by Ansano Giannarelli, dealing with the issue of alienation, but also *La salute in fabbrica* (1972), directed by Giuseppe Ferrara, which specifically deals with health issues in the working class – all elements that find their way in the most significant Italian fiction film on this topic: *La classe operaia va in paradiso* (dir. Elio Petri, 1971), with music by Ennio Morricone. All these aspects speak out against the ideology that for over a decade had sustained the production of industrial films.

Other reasons for the sad epilogue of industrial film productions lie in the economic stagnation in late 1960s and 1970s Italy, particularly after the oil crisis in 1973, which would definitively discourage the production of high-budget industrial films, and the emergence of television communication, in which consumers had to be convinced to buy products in a few minutes. What emerged in this context was the so-called "spot" communication of momentary television commercials, merely focusing attention on the end products, instead of presenting the social enterprise of corporate industry starting from the documentation of production processes.[52] All of these aspects swept away the avant-garde aspirations of experimental industrial film, along with the economic miracle.

Works Cited

Boledi, Luigi, and Elena Mosconi. "Il film industriale." *Un secolo di cinema a Milano*, edited by Raffaele De Berti. Il Castoro, 1996, pp. 295–311.

Bolter, Jay David, and Richard Grusin. *Remediation: Understanding New Media*. MIT Press, 1999.

Calabretto, Roberto. *Antonioni e la musica*. Marsilio, 2012.

Canova, Gianni, and Giulio Bursi, editors. *Cinema elettrico: i film dell'Archivio AEM (1928–1962)*. Rizzoli, 2011.

Cecchi, Alessandro, and Maurizio Corbella. "Experimentation, Documentation, Censorship: A Joris Ivens's Industrial Film and the Italian National Broadcasting Television." Paper presented at the Music and Media 4th Study Group Conference, Turin, June 29, 2012.

52 See Boledi, Luigi, and Elena Mosconi. "Il film industriale." *Un secolo di cinema a Milano*, edited by Raffaele De Berti. Il Castoro, 1996, pp. 295–311.

Corbella, Maurizio. "Paolo Ketoff e le radici cinematografiche della musica elettronica romana." *AAA–TAC (Acoustical Arts and Artifacts – Technology, Aesthetics, Communication)*, vol. 6, 2009, pp. 65–75.
Corbella, Maurizio. "Sperimentazione elettroacustica e cinema d'autore in Italia negli anni Sessanta. Due casi di studio." *Comunicazioni sociali*, no. 1, 2011, pp. 93–101.
Falchero, Anna Maria. "Cinema e industria: i documentari industriali." *Storia Politica Società*, vol. 41, no. 3–4, 2008, pp. 129–42.
Ginsborg, Paul. *A History of Contemporary Italy: Society and Politics 1943–1980*. Palgrave Macmillan, 2003.
Hediger, Vinzenz, and Miriam De Rosa, editors. "Post-what? Post-when? Thinking Moving Images beyond the Post-medium/Post-cinema Condition." *Cinéma and Cie*, vol. 16, no. 26–27, 2016.
Latini, Giulio. *L'energia e lo sguardo. Il cinema dell'Eni e i documentari di Gilbert Bovay*. Donzelli, 2011.
Mosconi, Elena. "Il film industriale." *Comunicazioni sociali*, vol. 13, no. 1–2 [special issue: "Il cinema a Milano dal secondo dopoguerra ai primi anni Sessanta," edited by Raffaele De Berti], 1991, pp. 65–66.
Negri, Sabrina. "Simulating the Past: Digital Preservation of Moving Images and the 'End of Cinema'." *Cinéma and Cie*, vol. 16, no. 26–27, 2016, pp. 45–54.
Novati, Maria Maddalena, and John Dack, editors. *The Studio di Fonologia: A Musical Journey 1954–1983: Update 2008–2012*. Ricordi, 2012.
Retroscena di "Acciaio": indagine su un'esperienza cinematografica di G. Francesco Malipiero. Olschki, 1993.
Tobagi, Benedetta, editor. *I volti e le mani*. Feltrinelli, 2008.
Toffetti, Sergio, editor. *La scoperta dell'altrove*. Feltrinelli, 2010.
Verdone, Mario. "Note. *L'Italia non è un paese povero*." *Bianco e nero*, vol. 21, no. 7, 1960, pp. 87–90.
Vitella, Federico. "Tecnologia, ideologia: il Circarama." *Bianco e nero*, no. 557–58, 2007, pp. 183–95.

About the Author

Alessandro Cecchi is assistant professor of musicology at the University of Pisa, where he teaches music history and film music. His publications on film music include articles in *Music, Sound, and the Moving Image* (vol. 8, no. 2, 2014), *Journal of Film Music* (vol. 8, no. 1–2, 2015), *Biblioteca Teatrale* (no. 129–30, 2019) and *Schermi* (vol. 4, no. 7, 2020) His chapters on music in Italian cinema are in the process of being published in Routledge and Palgrave Macmillan book series. He is series editor of NeoClassica's interdisciplinary book series "musica.performance.media," for which he edited two books, including a collection of Angelo Francesco Lavagnino's writings, conferences, and interviews (*Il difficile mestiere del compositore cinematografico. Scritti, conferenze, interviste (1950–1984)*, NeoClassica, 2022).

Indices

Films and Series

Names

Persons

Legal Entities

Subjects